Robinson Public Library District
606 North Jefferson Street
Robinson, IL 62454-2699

Elementary and Intermediate Algebra

Elementary and Intermediate Algebra

A Unified Approach

Donald Hutchison
Clackamas Community College

▼

Barry Bergman
Clackamas Community College

▼

Louis Hoelzle
Bucks County Community College

Boston   Burr Ridge, IL   Dubuque, IA   Madison, WI   New York   San Francisco   St. Louis
Bangkok   Bogotá   Caracas   Lisbon   London   Madrid
Mexico City   Milan   New Delhi   Seoul   Singapore   Sydney   Taipei   Toronto

McGraw-Hill Higher Education

*A Division of The **McGraw-Hill** Companies*

ELEMENTARY AND INTERMEDIATE ALGEBRA:
A UNIFIED APPROACH, STUDENT VERSION

Copyright © 2000 by the McGraw-Hill Companies, Inc. All rights reserved. Printed in the United States of America. Except as permitted under the United States Copyright Act of 1976, no part of this publication may be reproduced or distributed in any form or by any means, or stored in a data base or retrieval system, without the prior written permission of the publisher.

This book is printed on acid-free paper.

1 2 3 4 5 6 7 8 9 0 VNH/VNH 0 9 8 7 6 5 4 3 2 1 0

ISBN 0–07–366155–4
ISBN 0–07–229654–2 (AIE)

Vice president and editorial director: *Kevin T. Kane*
Publisher: *JP Lenney*
Sponsoring editor: *William K. Barter*
Developmental editor: *Erin Brown*
Marketing manager: *Mary K. Kittell*
Senior project manager: *Kay J. Brimeyer*
Production supervisor: *Laura Fuller*
Design director: *Francis Owens*
Senior photo research coordinator: *Lori Hancock*
Supplement coordinator: *Stacy A. Patch*
Compositor: *York Graphic Services, Inc.*
Typeface: *10/12 Times Roman*
Printer: *Von Hoffmann Press, Inc.*

Text/cover designer: *Vargas/Williams Design*
Cover photograph: *©PhotoDisc*
Photo research: *Shirley Lanners*

Photo Credits
0.00 © R. Lord/The Image Works; 1.00 © Bob Daemmrich/The Image Works; 2.00 © Allan Levenson/Tony Stone Images; 3.00 © Michelle Bridwell/PhotoEdit; 4.00 © Wojnarowiez/The Image Works; 5.00 © Michael Newman/PhotoEdit; 6.00 © Lawrence Migdale/Tony Stone Images; 7.00 © David Young-Wolff/Tony Stone Images; 8.00 © Frank Siteman/Picture Cube/Index Stock; 9.00 © Hank Morgan/SS/Photo Researchers, Inc; 10.00 © Bob Daemmrich/The Image Works; 11.00 © 1982 David Burnett/The Stock Market; 12.00 © Kathy McLaughlin/The Image Works; 13.00 © 1990 Chris Jones/The Stock Market

Library of Congress Cataloging-in-Publication Data

Hutchison, Donald, 1948–
    Elementary and intermediate algebra: a unified approach / Donald Hutchison, Barry Bergman, Louis F. Hoelzle—1st ed.
      p. cm.
    Includes index.
    ISBN 0–07–366155–4 (student version) — ISBN 0–07–229654–2 (AIE version)
      1. Algebra.   I. Bergman, Barry.   II. Hoelzle, Louis F.   III. Title.

QA152.2 .H84 2000
512—dc21
                                                  99–046594
                                                  CIP

www.mhhe.com

About the Authors

Donald Hutchison spent his first ten years of teaching working with disadvantaged students. He taught in an inner-city elementary school and an inner-city high school. Don also worked with physically and mentally challenged children in state agencies in New York and Oregon.

In 1982, Don completed his graduate work in mathematics. He was then hired by Jim Streeter to teach at Clackamas Community College. Through Jim's tutelage, Don developed a fascination with the relationship between teaching and writing mathematics. He has come to believe that his best writing is a result of his classroom experience, and his best teaching is a result of the thinking involved in manuscript preparation.

Don is active in several professional organizations. He was a member of the ACM committee that undertook the writing of computer curriculum for the two-year college. From 1989 to 1994 he was chair of the AMATYC Technology in Mathematics Education Committee. He was President of ORMATYC from 1996 to 1998.

Barry Bergman has enjoyed teaching mathematics to a wide variety of students over the years. He began in the field of adult basic education, and moved into the teaching of high school mathematics in 1977. He taught at that level for 11 years, at which point he served for a year as a K–12 mathematics specialist for his county. This work allowed him the opportunity to help promote the emerging NCTM Standards in his region.

In 1990 Barry began the present portion of his career, having been hired to teach at Clackamas Community College. He maintains a strong interest in the appropriate use of technology in the learning of mathematics.

Throughout the past 22 years, Barry has played an active role in professional organizations. As a member of OCTM, he contributed several articles and activities to the group's journal. Recently, he has served as an officer of ORMATYC for 4 years, and has participated on an AMATYC committee to provide feedback and reactions to NCTM's new revision of the Standards.

Louis Hoelzle has been teaching at Bucks County Community College for 30 years. In 1989, Lou became Chair of the Mathematics Department. He has taught the entire range of courses from arithmetic to calculus, giving him an excellent view of the current and future needs of developmental students.

Over the past 36 years, Lou has also taught physics courses at 4-year colleges, which has enabled him to have the perspective of the practical applications of mathematics. Lou has always focused on the student in his writing.

Lou is active in several professional organizations. He has served on the Placement and Assessment Committee and the Grants Committee for AMATYC. He served as President of PSMATYC from 1997 to 1999.

Dedications

Don Hutchison
This book is dedicated to my life's travel companion, Claudia. She has allowed me to grow as an author, a partner, and as a parent.

Barry Bergman
This book is dedicated to my wife Marcia, who encouraged me to enter into this work, and who, along with our wonderful boys Joel and Adam, gave me support through the duration of the project.

Lou Hoelzle
This book is dedicated to my wife and children who have shared my joys, sorrows, successes, and failures over 34 years. Rosemary, my friend, my inspiration, and my companion on the journey of life, and Beth, Ray, Amy, Oscar, Meg, Johanna, and Patrick, the joys of my life.

> This series is dedicated to the memory of James Arthur Streeter, an artisan with words, a genius with numbers, and a virtuoso with pictures from 1940 until 1989.

Brief Contents

Preface x

Chapter 0
The Arithmetic of Signed Numbers   1

Chapter 1
From Arithmetic to Algebra   55

Chapter 2
Equations and Inequalities   97

Chapter 3
Graphs and Linear Equations   191

Chapter 4
A Beginning Look at Functions   279

Chapter 5
Polynomials   333

Chapter 6
Factoring Polynomials   381

Transition
Moving to the Intermediate Algebra Level   447

Chapter 7
Rational Expressions   479

Chapter 8
Systems of Linear Equations and Inequalities   553

Chapter 9
Graphical Solutions   619

Chapter 10
Radicals and Exponents   665

Chapter 11
Quadratic Functions   713

Chapter 12
Conic Sections   777

Chapter 13
Exponential and Logarithmic Functions   809

Contents

Preface   x

Chapter 0
The Arithmetic of Signed Numbers   1

- **0.1** A Review of Fractions   2
- **0.2** The Integers   10
- **0.3** Adding and Subtracting Signed Numbers   16
- **0.4** Multiplying and Dividing Signed Numbers   28
- **0.5** Exponents and Order of Operations   42

Chapter 1
From Arithmetic to Algebra   55

- **1.1** Transition to Algebra   56
- **1.2** Evaluating Algebraic Expressions   65
- **1.3** Adding and Subtracting Algebraic Expressions   75
- **1.4** Sets   86

Chapter 2
Equations and Inequalities   97

- **2.1** Solving Equations by Adding and Subtracting   99
- **2.2** Solving Equations by Multiplying and Dividing   116
- **2.3** Combining the Rules to Solve Equations   124
- **2.4** Literal Equations and Their Applications   141
- **2.5** Solving Linear Inequalities   156
- **2.6** Absolute Value Equations and Inequalities   173

Chapter 3
Graphs and Linear Equations   191

- **3.1** Solutions of Equations in Two Variables   192
- **3.2** The Cartesian Coordinate System   201
- **3.3** The Graph of a Linear Equation   214
- **3.4** The Slope of a Line   240
- **3.5** Forms of Linear Equations   259

Chapter 4
A Beginning Look at Functions   279

- **4.1** Ordered Pairs and Relations   280
- **4.2** An Introduction to Functions   285
- **4.3** Identifying Functions   296
- **4.4** Reading Values from a Graph   309
- **4.5** Operations on Functions   321

Chapter 5
Polynomials   333

- **5.1** Positive Integer Exponents and Monomials   334
- **5.2** An Introduction to Polynomials   346
- **5.3** Polynomials: Addition and Subtraction   352
- **5.4** Multiplying of Polynomials and Special Products   362

Chapter 6
Factoring Polynomials   381

- **6.1** An Introduction to Factoring   383
- **6.2** Factoring Special Polynomials   393
- **6.3** Factoring Trinomials: The *ac* Method   400
- **6.3*** Factoring Trinomials: Trial and Error   412
- **6.4** Division of Polynomials   422
- **6.5** Solving Quadratic Equations by Factoring   432

Moving to the Intermediate Algebra Level 447

Chapter 7
Rational Expressions 479

- **7.1** Simplifying Rational Expressions *480*
- **7.2** Multiplication and Division of Rational Expressions *493*
- **7.3** Addition and Subtraction of Rational Expressions *503*
- **7.4** Complex Fractions *518*
- **7.5** Rational Equations and Inequalities in One Variable *528*

Chapter 8
Systems of Linear Equations and Inequalities 553

- **8.1** Solving Systems of Linear Equations by Graphing *554*
- **8.2** Systems of Equations in Two Variables with Applications *561*
- **8.3** Systems of Linear Equations in Three Variables *579*
- **8.4** Graphing Linear Inequalities in Two Variables *592*
- **8.5** Systems of Linear Inequalities in Two Variables *602*

Chapter 9
Graphical Solutions 619

- **9.1** Graphical Solutions to Equations in One Variable *621*
- **9.2** Graphing Linear Inequalities in One Variable *630*
- **9.3** Absolute Value Functions *637*
- **9.4** Absolute Value Inequalities *645*

Chapter 10
Radicals and Exponents 665

- **10.1** Zero and Negative Exponents and Scientific Notation *666*
- **10.2** Evaluating Radical Expressions *680*
- **10.3** Rational Exponents *688*
- **10.4** Complex Numbers *699*

Chapter 11
Quadratic Functions 713

- **11.1** Solving Quadratic Equations by Completing the Square *714*
- **11.2** The Quadratic Formula *725*
- **11.3** Solving Quadratic Equations by Graphing *741*
- **11.4** Solving Quadratic Inequalities *759*

Chapter 12
Conic Sections 777

- **12.1** More on the Parabola *778*
- **12.2** The Circle *785*
- **12.3** The Ellipse *791*
- **12.4** The Hyperbola *796*

Chapter 13
Exponential and Logarithmic Functions 809

- **13.1** Inverse Relations and Functions *811*
- **13.2** Exponential Functions *821*
- **13.3** Logarithmic Functions *832*
- **13.4** Properties of Logarithms *844*
- **13.5** Logarithmic and Exponential Equations *861*

Preface

A Unified Text That Serves Your Needs

Most colleges offering elementary and intermediate algebra use two different texts, one for each course. As a result, students may be required to purchase two texts; this can result in a considerable amount of topic overlap. Over the last few years, several publishers have issued "combined" texts that take chapters from two texts and merge them into a single book. This has allowed students to purchase a single text, but it has done little to reduce the overlap. The goal of this author team has been to produce a text that was more than a combined text. We wanted to unify the topics and themes of beginning and intermediate algebra in a fluid, non-repetitive text. We also wanted to produce a text that would prepare a student who comes directly from pre-algebra for college algebra as well as accommodate a student who wants to prepare for college algebra, but who enters at the intermediate algebra level. We believe we have accomplished our goals.

For students entering directly from an arithmetic or pre-algebra course, this is a text that contains all of the material needed to prepare for college algebra. It can be offered in two quarters or in two semesters. The transitional section found between chapters 6 and 7 serves as a mid-book review for students preparing to take a final exam that covers the first seven chapters.

For students who enter with a background in algebra, but need a thorough review, this text can be used in a one-term review course.

Finally, we have produced a text that will accommodate those students placing into the second term of a two-term course sequence. Here is where the transitional material is most valuable. It gives the students an opportunity to check that they have all of the background required to begin in chapter seven. The transitional section is organized so that if students struggle with any of the material from the first seven chapters, they can easily refer back to the appropriate section for review.

Elementary and Intermediate Algebra: A Unified Approach was written with your needs, and the needs of your students, in mind.

Visual Approach

For students of elementary and intermediate algebra, one key to success is the visualization of a problem and its solution. One theme of this text is the teaching of concepts, aided by a visual approach. In some instances, the visualization is encouraged through use of a graphing calculator. At other times, it may be encouraged through a hand-drawn graph. In any event, the text provides students with an opportunity to see the relationship between symbolic manipulation and a visual, concrete model.

Use of Calculators

More and more often, instructors report that students place too much reliance upon calculators. At the same time, most instructors see the graphing calculator as a tool that can enhance students' understanding of algebra by enabling them to visualize concepts more effectively. This text includes exercises and examples that make use of graphing and scientific calculators. These exercises and examples may be considered optional. The graphing calculator exercises and examples are intended to enhance the experience of learning algebra and to actively engage students in mathematical learning. Furthermore, the graphing calculator supports a visual approach to learning algebra.

■ Preface　xi

Important Features

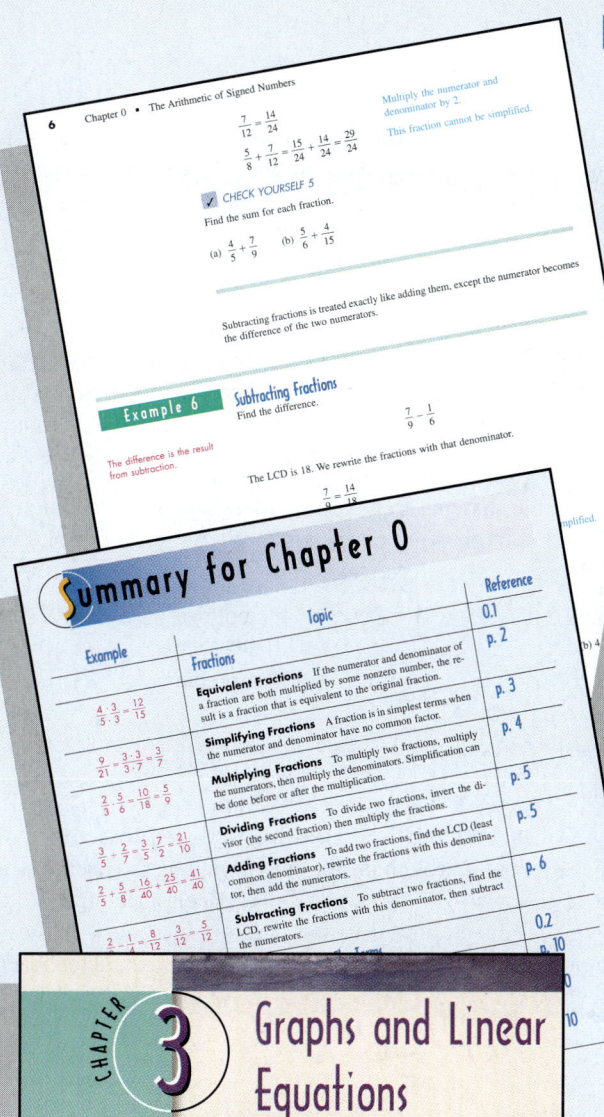

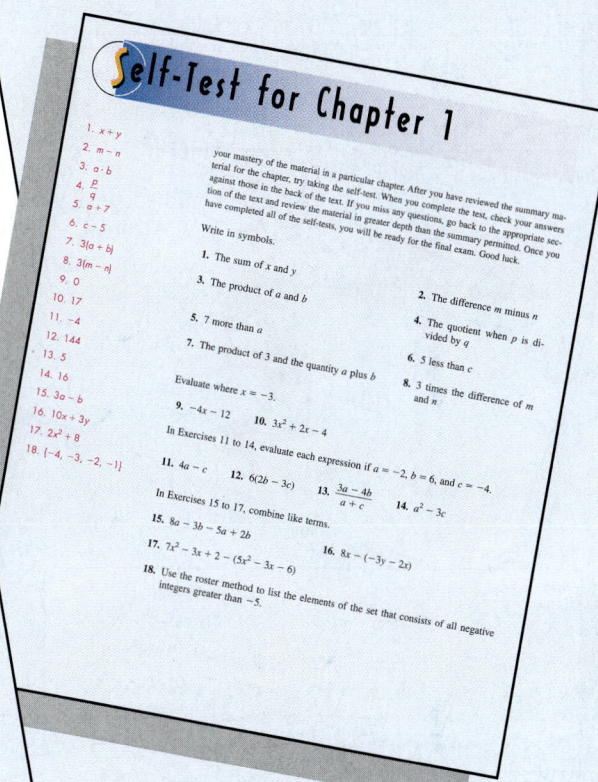

Chapter Zero is provided for students who may elect to brush up on pre-algebra topics.

The **transitional section** is found between Chapters 6 and 7. It contains summaries of each chapter along with self-tests. Students have an opportunity to review their work from chapters 0–6 in this section so that they may be prepared to move on to topics relating to "intermediate algebra."

Each chapter opens with a real-world **vignette** that showcases an example of how mathematics is used in a wide variety of jobs and professions. Problem sets for each section then feature one or more **applications** that relate to the chapter-opening vignette.

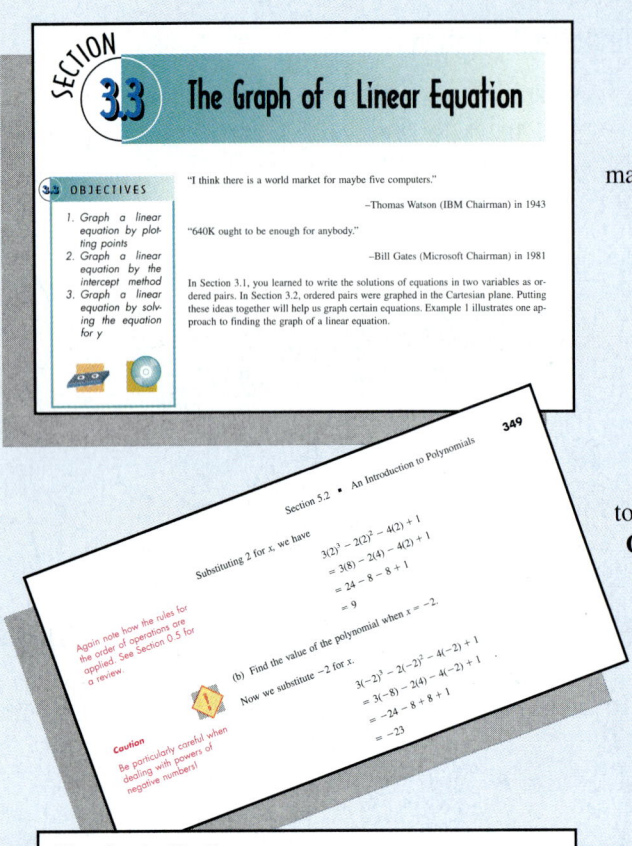

Objectives for each section are clearly identified in the margin. Related media icons are included in these boxes.

Marginal notes are provided throughout and are designed to help the students focus on important topics and techniques. **Caution icons** point out potential trouble spots or common errors.

Check Yourself Exercises are designed to actively involve students throughout the learning process. Each example is followed by an exercise that encourages students to solve a problem similar to the one just presented. Answers are provided at the end of the section for immediate feedback.

Complete exercise sets are at the end of each section. These exercises are designed to reinforce basic skills and develop critical thinking and communication abilities. Exercise sets include **writing exercises and challenge exercises,** all denoted by distinctive icons.

The **calculator** icon points out examples and exercises that illustrate when the calculator can best be used for further understanding of the concept at hand. Optional exercises and hints for graphing calculators are also included.

■ Preface **xiii**

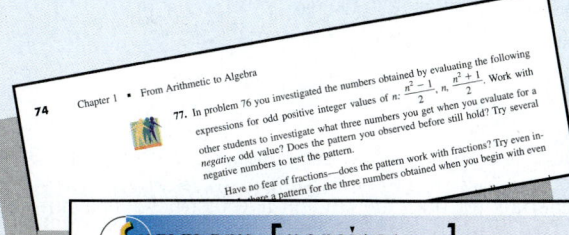

Group Exercises, also denoted by a distinctive icon, provide opportunities for students to develop teamwork, exploration, and conjecture skills.

Summary Exercises are found at the end of each chapter and provide an opportunity for students to review content from each chapter.

Cumulative Tests help students build on what was previously covered and give students additional opportunities to build skills necessary in preparing for midterm and final exams. They can be found at the end of Chapters 2–13.

Self-tests provide students with an opportunity to confirm their understanding of concepts within a particular chapter. For Chapters 0–6, the self-tests are found within the transitional section. For chapters 7–13, the self-tests directly follow the summary exercises.

Supplements

A comprehensive set of ancillary materials for both the student and instructor is available with this text.

Annotated Instructor's Edition

This ancillary includes answers to all exercises and tests. Each answer is printed in a second color for ease of use by the instructor next to each problem on the page where the problem appears.

Instructor's Solutions Manual

This ancillary includes worked-out solutions and answers to all exercises and tests in the text. It also includes worked-out solutions and answers to the summary exercises.

Student Solutions Manual

This ancillary contains worked-out solutions and answers to all odd-numbered problems in the text.

Print and Computerized Testing

The testing materials provide an array of formats that allow the instructor to create tests using both algorithmically generated test questions and those from a standard testbank. This testing system enables the instructor to choose questions either manually or randomly by section, question type, difficulty level, and other criteria. Testing is available for IBM, IBM-compatible, and Macintosh computers. A softcover print version of the testbank provides questions found in the computerized version, along with answer keys. Each chapter of the print version contains three different tests. Additionally, the print testbank contains four final exams.

Hutchison, Bergman, and Hoelzle Video Series

The videotape series contains instructional material and presents opportunities for students to work problems. One of the unique features of the tapes is a question and answer exchange between student participants and instructors. These exchanges help students learn to ask better questions. The tapes are text-specific, covering all chapters of the text.

Hutchison, Bergman, Hoelzle SMART CD-Rom

This interactive CD-ROM is a self-paced tutorial specifically linked to the text and reinforces topics through unlimited opportunities to review concepts and practice problem solving. The CD-ROM contains text-, chapter-, and section-specific tutorials and multiple-choice questions with feedback, as well as algorithmically generated questions. It requires virtually no computer training on the part of students and supports IBM and Macintosh computers.

In addition, a number of other technology and Web-based ancillaries are under development; they will support the ever-changing technology needs in developmental mathematics. For further information about these or any supplements, please contact your local McGraw-Hill sales representative.

Acknowledgments

Those familiar with the publishing process will attest that change is inevitable. The same is true at McGraw-Hill. The difference is that change invariably seems to lead to something positive at McGraw-Hill. We have been most fortunate to work with our editors Bill Barter and Erin Brown. Both manage that fine line by being both demanding managers and supportive coworkers. We have appreciated their talents, energy, and time. As always, we encourage prospective authors to talk with the staff at McGraw-Hill. It will be a valuable use of your time.

In this age of the high-tech mathematics lab, the supplements and the supplement authors have become as important as the text itself. We are proud that our names appear on these supplements together with the following authors:

Lenore Parens, along with Engineering Software Asscociates, Inc., produced the Print Test Bank and Computerized Test Bank.

Laurel Technical Services provided the Student Solutions Manual and Instructor's Solutions Manual, and error-checked the manuscript.

We thank all of the students whom we have taught, talked to, questioned, and tested. This text was created for them. We also thank our community college compatriots. Professionals such as Betsy Farber, Susan Hopkirk, and Jack Scrivener are constantly providing us with both intentional and inadvertent guidance in our writing projects.

Finally, our thanks go to the following people for their important contributions to the development of this edition:

Mary Kay Abbey, Montgomery College
Victor Akatsa, Chicago State University
Dr. Vivian Morgan Alley, Middle Tennessee State University
Dr. John Barker, Oklahoma City Community College
Cynthia Broughton, Arizona Western College
Linda K. Buchanan, Howard College
Walter E. Burlage, Lake City Community College
Marc Campbell, Daytona Beach Community College
Dora S. Cantu, Laredo Community College
Connie Carruthers, Scottsdale Community College
Tim Chappell, Penn Valley Community College
John W. Coburn, St. Louis Community College at Florissant Valley
Patrick S. Cross, University of Oklahoma
Walter Czarnec, Framingham State College
Antonio David, Del Mar College
Yolanda F. Davis, Central Texas College
Andrea DeCosmo, Waubonsee Community College
Dr. Said Fariabi, University of Texas at San Antonio
Joan Finney, Dyersburg State Community College
Barb Gentry, Parkland College
David Graser, Yavapai College
Roberta Grenz, Community College of Southern Nevada

Garry Hart, California State University—Dominquez Hills
Margret M. Hathaway, Kansas City Kansas Community College
Alan T. Hayashi, Oxnard College
Dr. Jeannie Hollar, Lenoir-Rhyne College
Jefferson A. Humphries, San Antonio College
Linda Hunt, Marshall University
Marvin Johnson, College of Lake County
Maryann Justinger, Erie Community College
Eric Paul Kraus, Sinclair Community College
Ann Loving, J. Sargeant Reynolds College
Marva Lucas, Middle Tennessee State University
David P. MacAdam, Cape Cod Community College
Anthony P. Malone, Raymond Walters College
Sarah Martin, Virginia Western Community College
Jeff Mock, Diablo Valley College
Feridoon Moinian, Cameron University
Ann C. Mugavero, College of Staten Island
Carol Murphy, San Diego Miramar College
Linda J. Murphy, Northern Essex Community College
Pinder Naidu, Kennesaw State University
Barbara Napoli, Our Lady of the Lake College
Nancy K. Nickerson, Northern Essex Community College
Sergei Ovchinnikov, San Francisco State University
Dena Perkins, Oklahoma Christian University
Evelyn A. Puaa, Hawaii Pacific University
Dr. Atma Sahu, Coppin State College
Martha W. Scarbrough, Motlow State Community College
Ellen Sawyer, College of DuPage
Wade Sick, Southwestern Community College
Helen Smith, South Mountain Community College
Andrea Spratt, University of Maryland-Baltimore County
Eleanor Storey, Front Range Community College
Nancy Szen, Northern Essex Community College
Lana Taylor, Siena Heights College
Jo Anne Temple, Texas Tech University
Jane H. Theiling, Dyersburg State Community College
John B. Thoo, Yuba College
Victoria C. Wacek, Missouri Western College
Pansy Waycaster, Southwest Virginia Community College
Danny Whited, Virginia Intermont College
Mary Jane Wolfe, University of Rio Grande
Judith B. Wood, Central Florida Community College
Karl Zilm, Lewis and Clark Community College

To the Student

Elementary and Intermediate Algebra: A Unified Approach is the final step in your preparation for college transfer mathematics. In this course, you will cover many new concepts and see the expansion of several familiar ones. All the topics in this text are designed to prepare you for subsequent courses in Pre-Calculus, College Algebra, or Statistics. We have provided situations and applications that will help you understand the relevance of what you are learning. As your mathematical expertise expands in succeeding courses, the applications of your skills will become even more diverse and interesting. Your ability to succeed in those later courses will very much depend on how well you understand the material covered in this text. The following suggestions are designed to help you succeed in this, and future, mathematics classes.

1. If you are in a lecture class, take the time to read the appropriate section before the lecture. Take careful notes of every example your instructor presents in class.

2. When you sit down to study, have your calculator, pencil, and paper ready. Work through the examples in the text. Do every *Check Yourself* exercise, checking your answer against the one at the end of the section. If you have difficulty, go back and reread the previous example. Make certain you understand what you are doing and why. The best test of whether you do understand a concept lies in your ability to explain that concept to a classmate. Try working together.

3. At the end of each section is a set of exercises. Work these carefully in order to check your progress on the section just completed. The answers to odd-numbered exercises are at the end of the section. If you have difficulty with any of the exercises, review the appropriate parts of the section. If your confusion is not completely cleared up, by all means do not become discouraged. Ask your instructor or an available tutor for assistance. A word of caution: Work the exercises on a regular (preferably daily) basis. Learning algebra requires active participation on your part. As is the case with the learning of any skill, the main ingredient is practice.

4. When you have completed a chapter, review by using the *Summary*. Following the summary are *Summary Exercises* for further practice. The exercises are keyed to chapter sections so you will know where to turn if you are still struggling.

5. When finished with the Summary Exercises, try the *Self-Test* that appears at the end of each chapter. Answers, with section references, are in the back of the book.

6. Finally, an important element of success in studying algebra is the process of regular review. We provide a series of *Cumulative Reviews* throughout this book. Use these tests to prepare for any midterms or finals. If it appears that you have forgotten some concepts from earlier chapters, do not worry. Go back and review the section where the idea was first explained, or the appropriate chapter summary; that is the purpose of the *Cumulative Tests*.

We hope you will find these suggestions helpful as you work through the material in this text. Best of luck in this course!

Donald Hutchison
Barry Bergman
Louis Hoelzle

CHAPTER 0
The Arithmetic of Signed Numbers

LIST OF SECTIONS

0.1 A Review of Fractions

0.2 The Integers

0.3 Adding and Subtracting Signed Numbers

0.4 Multiplying and Dividing Signed Numbers

0.5 Exponents and Order of Operations

Anthropologists and archaeologists sometimes investigate cultures that existed so long ago that their characteristics must be inferred from buried objects. When some interesting object is found, often the first question is "How old is this?" With methods such as carbon dating, it has been established that large, organized cultures existed around 3000 BCE in Egypt, 2800 BCE in India, no later than 1500 BCE in China, and around 1000 BCE in the Americas.

How long ago was 1500 BCE? Using the Christian notation for dates, we have to count AD years and BCE years differently. An object from 500 AD is 2000 − 500 years old, or about 1500 years old. But an object from 1500 BCE is 2000 + 1500 years old, or about 3500 years old.

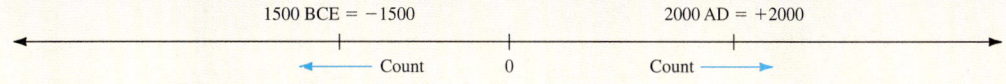

SECTION 0.1 A Review of Fractions

0.1 OBJECTIVES

1. Simplify a fraction
2. Add or subtract two fractions
3. Multiply or divide two fractions

This chapter provides a review of the basic operations, addition, subtraction, division, and multiplication, on signed numbers.

The numbers used for counting are called the **natural numbers.** We write them as 1, 2, 3, 4, ... The three dots indicate that the pattern continues in the same way.

If we include zero in this group of numbers, we then call them the **whole numbers.**

The **numbers of ordinary arithmetic** consist of all the whole numbers and all fractions, whether they are proper fractions such as $\frac{1}{2}$ and $\frac{2}{3}$ or improper fractions such as $\frac{7}{2}$ or $\frac{19}{5}$.

Every number of ordinary arithmetic can be written in fraction form $\frac{a}{b}$.

The number 1 has many different fractional forms. Any fraction in which the numerator and denominator are the same (and not zero) is another name for the number one.

$$1 = \frac{2}{2} \qquad 1 = \frac{12}{12} \qquad 1 = \frac{257}{257}$$

We can multiply any number by one and the result is the same number we started with. It simply has a new name. The following example illustrates this idea.

Example 1

Rewriting Fractions

Write three fractional representations for each number.

(a) $\frac{2}{3}$

Each representation is a numeral, or name for the number. Each number has many names.

Multiplying the numerator and denominator by the same number is the same as multiplying by one.

Numerator and denominator are multiplied by two.

$$\frac{2}{3} = \frac{2}{3} \times \frac{2}{2} = \frac{4}{6}$$

Numerator and denominator are multiplied by three.

$$\frac{2}{3} = \frac{2}{3} \times \frac{3}{3} = \frac{6}{9}$$

Numerator and denominator are multiplied by ten.

$$\frac{2}{3} = \frac{2}{3} \times \frac{10}{10} = \frac{20}{30}$$

(b) 5

$$5 = \frac{5}{1} \times \frac{2}{2} = \frac{10}{2}$$

$$5 = \frac{5}{1} \times \frac{3}{3} = \frac{15}{3}$$

$$5 = \frac{5}{1} \times \frac{100}{100} = \frac{500}{100}$$

✓ **CHECK YOURSELF 1**

Write three fractional representations for each number.

(a) $\dfrac{5}{8}$ (b) $\dfrac{4}{3}$ (c) 3

The simplest fractional representation for a number has the smallest numerator and denominator. Fractions written in this form are said to be **simplified.**

Example 2 Simplifying Fractions

Simplify each fraction.

(a) $\dfrac{22}{55}$ (b) $\dfrac{35}{45}$ (c) $\dfrac{24}{36}$

In each case, we first find the prime factors for the numerator and for the denominator.

(a) $\dfrac{22}{55} = \dfrac{2 \times 11}{5 \times 11}$

We then "remove" the common factor of 11

$$\frac{22}{55} = \frac{2 \times 11}{5 \times 11} = \frac{2}{5}$$

(b) $\dfrac{35}{45} = \dfrac{5 \times 7}{3 \times 3 \times 5}$

Removing the common factor of 5 yields

$$\frac{35}{45} = \frac{7}{3 \times 3} = \frac{7}{9}$$

4 Chapter 0 ▪ The Arithmetic of Signed Numbers

(c) $\dfrac{24}{36} = \dfrac{2 \times 2 \times 2 \times 3}{2 \times 2 \times 3 \times 3}$

Removing the common factor $2 \times 2 \times 3$ yields

$$\dfrac{2}{3}$$

✓ CHECK YOURSELF 2

Simplify each fraction.

(a) $\dfrac{21}{33}$ (b) $\dfrac{15}{30}$ (c) $\dfrac{12}{54}$

When multiplying two fractions, rewrite them in factored form, and then simplify before multiplying.

Example 3 Multiplying Fractions

Find the product of the two fractions.

$$\dfrac{9}{2} \times \dfrac{4}{3}$$

A product is the result from multiplication.

$$\dfrac{9}{2} \times \dfrac{4}{3} = \dfrac{3 \times 3}{2} \times \dfrac{2 \times 2}{3}$$

$$= \dfrac{3 \times 3 \times 2 \times 2}{2 \times 3}$$

$$= \dfrac{3 \times 2}{1}$$

$$= \dfrac{6}{1} \quad \text{The denominator of one is not necessary.}$$

$$= 6$$

✓ CHECK YOURSELF 3

Multiply and simplify each pair of fractions.

(a) $\dfrac{3}{5} \times \dfrac{10}{7}$ (b) $\dfrac{12}{5} \times \dfrac{10}{6}$

To divide two fractions, the divisor is inverted, then the fractions are multiplied.

Example 4

Dividing Fractions

Find the quotient of the two fractions.

A quotient is the result from division.

$$\frac{7}{3} \div \frac{5}{6}$$

$$\frac{7}{3} \div \frac{5}{6} = \frac{7}{3} \times \frac{6}{5} = \frac{7}{3} \times \frac{2 \times 3}{5}$$

$$= \frac{7 \times 2 \times 3}{3 \times 5} = \frac{7 \times 2}{5}$$

$$= \frac{14}{5}$$

✓ CHECK YOURSELF 4

Find the quotient of the two fractions

$$\frac{9}{2} \div \frac{3}{5}$$

When adding two fractions, find the **least common denominator (LCD)** first. The least common denominator is the smallest number that both denominators evenly divide. If you have forgotten how to find the LCD, you might want to review the process from your arithmetic book. After rewriting the fractions with this denominator, add the numerators, then simplify the result.

Example 5

Adding Fractions

Find the sum of the two fractions.

A sum is the result from addition.

$$\frac{5}{8} + \frac{7}{12}$$

The LCD of 8 and 12 is 24. Each fraction should be rewritten as a fraction with that denominator.

$$\frac{5}{8} = \frac{15}{24}$$ Multiply the numerator and denominator by 3.

$$\frac{7}{12} = \frac{14}{24}$$ Multiply the numerator and denominator by 2.

$$\frac{5}{8} + \frac{7}{12} = \frac{15}{24} + \frac{14}{24} = \frac{29}{24}$$ This fraction cannot be simplified.

Chapter 0 • The Arithmetic of Signed Numbers

✓ CHECK YOURSELF 5

Find the sum for each fraction.

(a) $\dfrac{4}{5} + \dfrac{7}{9}$ (b) $\dfrac{5}{6} + \dfrac{4}{15}$

Subtracting fractions is treated exactly like adding them, except the numerator becomes the difference of the two numerators.

Example 6 Subtracting Fractions

Find the difference.

The difference is the result from subtraction.

$$\dfrac{7}{9} - \dfrac{1}{6}$$

The LCD is 18. We rewrite the fractions with that denominator.

$$\dfrac{7}{9} = \dfrac{14}{18}$$

$$\dfrac{1}{6} = \dfrac{3}{18}$$

$$\dfrac{7}{9} - \dfrac{1}{6} = \dfrac{14}{18} - \dfrac{3}{18} = \dfrac{11}{18} \qquad \text{This fraction cannot be simplified.}$$

✓ CHECK YOURSELF 6

Find the difference $\dfrac{11}{12} - \dfrac{5}{8}$.

✓ CHECK YOURSELF ANSWERS

1. Answers will vary. 2. (a) $\dfrac{7}{11}$; (b) $\dfrac{1}{2}$; (c) $\dfrac{2}{9}$. 3. (a) $\dfrac{6}{7}$; (b) 4.

4. $\dfrac{15}{2}$. 5. (a) $\dfrac{71}{45}$; (b) $\dfrac{11}{10}$. 6. $\dfrac{7}{24}$.

Exercises • 0.1

In Exercises 1 to 12, write three fractional representations for each number.

1. $\dfrac{3}{7}$ 2. $\dfrac{2}{5}$ 3. $\dfrac{4}{9}$

4. $\dfrac{7}{8}$ 5. $\dfrac{5}{6}$ 6. $\dfrac{11}{13}$

7. $\dfrac{10}{17}$ 8. $\dfrac{3}{7}$ 9. $\dfrac{9}{16}$

10. $\dfrac{6}{11}$ 11. $\dfrac{7}{9}$ 12. $\dfrac{15}{16}$

Write each fraction in simplest form.

13. $\dfrac{8}{12}$ 14. $\dfrac{12}{15}$ 15. $\dfrac{10}{14}$

16. $\dfrac{15}{50}$ 17. $\dfrac{12}{18}$ 18. $\dfrac{28}{35}$

19. $\dfrac{35}{40}$ 20. $\dfrac{21}{24}$ 21. $\dfrac{11}{44}$

22. $\dfrac{10}{25}$ 23. $\dfrac{12}{36}$ 24. $\dfrac{18}{48}$

25. $\dfrac{24}{27}$ 26. $\dfrac{30}{50}$ 27. $\dfrac{32}{40}$

Answers (in red):

1. $\dfrac{6}{14}, \dfrac{9}{21}, \dfrac{12}{28}$
2. $\dfrac{4}{10}, \dfrac{6}{15}, \dfrac{8}{20}$
3. $\dfrac{8}{18}, \dfrac{16}{36}, \dfrac{40}{90}$
4. $\dfrac{14}{16}, \dfrac{35}{40}, \dfrac{70}{80}$
5. $\dfrac{10}{12}, \dfrac{15}{18}, \dfrac{50}{60}$
6. $\dfrac{22}{26}, \dfrac{55}{65}, \dfrac{110}{130}$
7. $\dfrac{20}{34}, \dfrac{30}{51}, \dfrac{100}{170}$
8. $\dfrac{12}{28}, \dfrac{18}{42}, \dfrac{30}{70}$
9. $\dfrac{18}{32}, \dfrac{27}{48}, \dfrac{90}{160}$
10. $\dfrac{12}{22}, \dfrac{18}{33}, \dfrac{24}{44}$
11. $\dfrac{14}{18}, \dfrac{35}{45}, \dfrac{140}{180}$
12. $\dfrac{30}{32}, \dfrac{45}{48}, \dfrac{150}{160}$
13. $\dfrac{2}{3}$ 14. $\dfrac{4}{5}$ 15. $\dfrac{5}{7}$
16. $\dfrac{3}{10}$ 17. $\dfrac{2}{3}$ 18. $\dfrac{4}{5}$
19. $\dfrac{7}{8}$ 20. $\dfrac{7}{8}$ 21. $\dfrac{1}{4}$
22. $\dfrac{2}{5}$ 23. $\dfrac{1}{3}$ 24. $\dfrac{3}{8}$
25. $\dfrac{8}{9}$ 26. $\dfrac{3}{5}$ 27. $\dfrac{4}{5}$

Chapter 0 ■ The Arithmetic of Signed Numbers

28. $\dfrac{17}{51}$ **29.** $\dfrac{75}{105}$

30. $\dfrac{62}{93}$ **31.** $\dfrac{48}{60}$

32. $\dfrac{48}{66}$ **33.** $\dfrac{105}{135}$

34. $\dfrac{54}{126}$

Multiply. Be sure to simplify each product.

35. $\dfrac{3}{4} \times \dfrac{5}{11}$ **36.** $\dfrac{2}{7} \times \dfrac{5}{9}$ **37.** $\dfrac{3}{4} \times \dfrac{7}{5}$

38. $\dfrac{2}{3} \times \dfrac{8}{5}$ **39.** $\dfrac{3}{5} \times \dfrac{5}{7}$ **40.** $\dfrac{6}{11} \times \dfrac{8}{6}$

41. $\dfrac{6}{13} \times \dfrac{4}{9}$ **42.** $\dfrac{5}{9} \times \dfrac{6}{11}$ **43.** $\dfrac{3}{11} \times \dfrac{7}{9}$

44. $\dfrac{7}{9} \times \dfrac{3}{5}$ **45.** $\dfrac{3}{10} \times \dfrac{5}{9}$ **46.** $\dfrac{5}{21} \times \dfrac{14}{25}$

Divide. Write each result in simplest form.

47. $\dfrac{1}{5} \div \dfrac{3}{4}$ **48.** $\dfrac{2}{5} \div \dfrac{1}{3}$ **49.** $\dfrac{2}{5} \div \dfrac{3}{4}$

50. $\dfrac{5}{8} \div \dfrac{3}{4}$ **51.** $\dfrac{8}{9} \div \dfrac{4}{3}$ **52.** $\dfrac{5}{9} \div \dfrac{8}{11}$

53. $\dfrac{7}{10} \div \dfrac{5}{9}$ **54.** $\dfrac{8}{9} \div \dfrac{11}{15}$ **55.** $\dfrac{8}{15} \div \dfrac{2}{5}$

56. $\dfrac{5}{27} \div \dfrac{15}{54}$ **57.** $\dfrac{5}{27} \div \dfrac{25}{36}$ **58.** $\dfrac{9}{28} \div \dfrac{27}{35}$

Section 0.1 ■ A Review of Fractions

59. $\dfrac{13}{20}$ 60. $\dfrac{29}{30}$

61. $\dfrac{13}{15}$ 62. $\dfrac{53}{60}$

63. $\dfrac{19}{24}$ 64. $\dfrac{31}{72}$

65. $\dfrac{7}{12}$ 66. $\dfrac{47}{42}$

67. $\dfrac{107}{90}$ 68. $\dfrac{167}{150}$

69. $\dfrac{7}{8}$ 70. $\dfrac{19}{30}$

71. $\dfrac{5}{9}$ 72. $\dfrac{3}{10}$

73. $\dfrac{1}{2}$ 74. $\dfrac{1}{3}$

75. $\dfrac{5}{24}$ 76. $\dfrac{7}{30}$

77. $\dfrac{7}{18}$ 78. $\dfrac{7}{12}$

79. $\dfrac{11}{24}$ 80. $\dfrac{11}{36}$

81. $\dfrac{13}{42}$ 82. $\dfrac{23}{90}$

Add.

59. $\dfrac{2}{5} + \dfrac{1}{4}$ 60. $\dfrac{2}{3} + \dfrac{3}{10}$ 61. $\dfrac{2}{5} + \dfrac{7}{15}$

62. $\dfrac{3}{10} + \dfrac{7}{12}$ 63. $\dfrac{3}{8} + \dfrac{5}{12}$ 64. $\dfrac{5}{36} + \dfrac{7}{24}$

65. $\dfrac{2}{15} + \dfrac{9}{20}$ 66. $\dfrac{9}{14} + \dfrac{10}{21}$ 67. $\dfrac{7}{15} + \dfrac{13}{18}$

68. $\dfrac{12}{25} + \dfrac{19}{30}$ 69. $\dfrac{1}{2} + \dfrac{1}{4} + \dfrac{1}{8}$ 70. $\dfrac{1}{3} + \dfrac{1}{5} + \dfrac{1}{10}$

Subtract.

71. $\dfrac{8}{9} - \dfrac{3}{9}$ 72. $\dfrac{9}{10} - \dfrac{6}{10}$

73. $\dfrac{5}{8} - \dfrac{1}{8}$ 74. $\dfrac{11}{12} - \dfrac{7}{12}$

75. $\dfrac{7}{8} - \dfrac{2}{3}$ 76. $\dfrac{5}{6} - \dfrac{3}{5}$

77. $\dfrac{11}{18} - \dfrac{2}{9}$ 78. $\dfrac{5}{6} - \dfrac{1}{4}$

79. $\dfrac{5}{8} - \dfrac{1}{6}$ 80. $\dfrac{13}{18} - \dfrac{5}{12}$

81. $\dfrac{8}{21} - \dfrac{1}{14}$ 82. $\dfrac{13}{18} - \dfrac{7}{15}$

SECTION 0.2 The Integers

0.2 OBJECTIVES

1. Place integers on a number line
2. Find the opposite of a number
3. Find the absolute value of a signed number
4. Identifying integers

In arithmetic, you learned to solve problems that involved working with numbers. In algebra, you will learn to use tools that will help you solve many new types of problems. The first tool we will provide you with involves the expansion of the numbers with which you do computation. Let us look at some important sets of numbers.

The **natural numbers** are all of the counting numbers 1, 2, 3, ...

The **whole numbers** are the natural numbers together with zero.

We can represent whole numbers on a **number line.** Here is the number line.

And here is the number line with the whole numbers 0, 1, 2, and 3 plotted.

Now suppose you want to represent a temperature of 10 degrees below zero, a debt of $50, or an altitude 100 feet below sea level. These situations require a new set of numbers called *negative numbers*. We will expand the number line to include negative numbers.

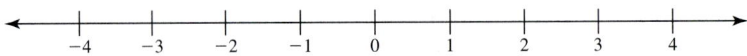

Numbers to the right of (greater than) 0 on the number line are called **positive numbers.** Numbers to the left of (less than) 0 are called **negative numbers.** Zero is neither positive nor negative.

Since -3 is to the left of 0, it is a negative number. Read -3 as "negative three."

To indicate a negative number, we use a minus sign ($-$) in front of the number. Positive numbers may be written with a plus sign ($+$) or with no sign at all, in which case the number is understood to be positive.

Example 1 Identifying Signed Numbers

$+6$ is a positive number.

-9 is a negative number.

If no sign appears, a number (other than 0) is positive.

5 is a positive number.

0 is neither positive nor negative.

Section 0.2 ■ The Integers   11

✓ **CHECK YOURSELF 1**

Label each of the following as positive, negative, or neither.

(a) +3   (b) 7   (c) −5   (d) 0

Together, all positive and negative numbers are called **signed numbers.** An important idea in our work with signed numbers is the *opposite* of a number. Every number has an opposite.

> **Opposite of a Number**
>
> The **opposite** of a number corresponds to a point the same distance from 0 as the given number, but in the opposite direction.

Example 2 — Writing the Opposite of a Signed Number

The opposite of a positive number is negative.

(a) The opposite of 5 is −5.

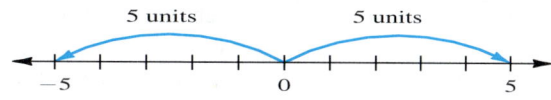

Both numbers are located 5 units from 0.

The opposite of a negative number is positive.

(b) The opposite of −3 is 3.

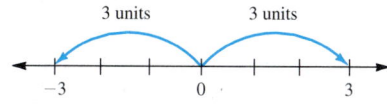

Both numbers correspond to points that are 3 units from 0.

✓ **CHECK YOURSELF 2**

(a) What is the opposite of 8?   (b) What is the opposite of −9?

To represent the opposite of a number, place a minus sign in front of the number.

We write the opposite of 5 as −5. You can now think of −5 in two ways: as negative 5 and as the opposite of 5.

Using the same idea, we can write the opposite of a negative number. The opposite of −3 is −(−3). Since we know from looking at the number line that the opposite of −3 is 3, this means that

Again place a minus sign in front of the number to represent the opposite of that number.

$$-(-3) = 3$$

So the opposite of a negative number must be positive.

Let's summarize our results:

> **The Opposite of a Signed Number**
>
> 1. The opposite of a positive number is negative.
> 2. The opposite of a negative number is positive.
> 3. The opposite of 0 is 0.

We also need to define the *absolute value,* or magnitude, of a signed number.

> **Absolute Value**
>
> The **absolute value** of a signed number is the distance (on the number line) between the number and 0.

Example 3 Finding the Absolute Value

(a) The absolute value of 5 is 5.

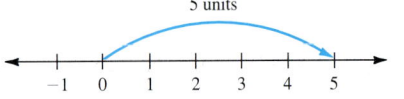

5 is 5 units from 0.

(b) The absolute value of -5 is 5.

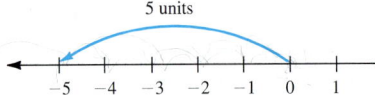

-5 is also 5 units from 0.

We usually write the absolute value of a number by placing vertical bars before and after the number. We can write

$$|5| = 5 \quad \text{and} \quad |-5| = 5$$

|5| is read "the absolute value of 5."

✓ CHECK YOURSELF 3

Complete the following statements.

(a) The absolute value of 9 is _____.

(b) The absolute value of -12 is _____.

(c) $|-6| =$ (d) $|15| =$

All the numbers we looked at in this section were natural numbers, 0, or the negatives of natural numbers. These numbers make up the set of integers.

> **The Set of Integers**
>
> The **set of integers** consists of the natural numbers, their negatives, and 0.

Here we have a graphical representation of the set of integers.

The arrows indicate that the number line extends forever in both directions.

Note that the integers occur at the hash marks on the number line. Any plotted point that would fall on one of the circles on the number line above is an integer. This would be true no matter how far in either direction we extend our number line.

Example 4

Identifying Integers

Which of the following are integers?

$$-3, 5.3, \frac{2}{3}, 4$$

Of these four numbers, only -3 and 4 are integers.

✓ **CHECK YOURSELF 4**

Which of the following are integers?

$$7, 0, \frac{4}{7}, -5, 0.2$$

✓ **CHECK YOURSELF ANSWERS**

1. (a) positive; (b) positive; (c) negative; (d) neither.   2. (a) -8; (b) 9.
3. (a) 9; (b) 12; (c) 6; (d) 15.   4. $7, 0, -5$.

Exercises ▪ 0.2

1. True	2. False
3. True	4. True
5. True	6. True
7. False	8. False
9. False	10. True
11. False	12. False
13. True	14. True
15. True	16. True
17. False	18. False
19. True	20. False
21. True	22. True
23. False	24. True
25. 10	26. 16
27. 20	28. 12
29. 7	30. 9
31. −30	32. 15
33. 6	34. 0
35. 50	36. −18
37. 3	38. −3
39. −7	40. 7

Indicate whether the following statements are true or false.

1. The opposite of 7 is −7.
2. The opposite of −10 is −10.
3. −9 is an integer.
4. 5 is an integer.
5. The opposite of −11 is 11.
6. The absolute value of 8 is 8.
7. $|-6| = -6$
8. $-(-30) = -30$
9. −12 is not an integer.
10. The opposite of −18 is 18.
11. $|7| = -7$
12. The absolute value of −9 is −9.
13. $-(-8) = 8$
14. $\frac{2}{3}$ is not an integer.
15. $|-20| = 20$
16. The absolute value of −3 is 3.
17. $\frac{3}{5}$ is an integer.
18. 0.8 is an integer.
19. 0.15 is not an integer.
20. $|-9| = -9$
21. $\frac{5}{7}$ is not an integer.
22. 0.23 is not an integer.
23. $-(-7) = -7$
24. The opposite of 15 is −15.

Complete each of the following statements.

25. The absolute value of −10 is _____.
26. $-(-16) = $ _____.
27. $|-20| = $ _____.
28. The absolute value of −12 is _____.
29. The absolute value of 7 is _____.
30. The opposite of −9 is _____.
31. The opposite of 30 is _____.
32. $|-15| = $ _____.
33. $-(-6) = $ _____.
34. The absolute value of 0 is _____.
35. $|50| = $ _____.
36. The opposite of 18 is _____.
37. The absolute value of the opposite of 3 is _____.
38. The opposite of the absolute value of 3 is _____.
39. The opposite of the absolute value of −7 is _____.
40. The absolute value of the opposite of −7 is _____.

Section 0.2 ▪ The Integers   15

41. >
42. <
43. <
44. <
45. =
46. >
47. <
48. =
49. −3, 2, 0
50. 2
51. 0, 2
52. −3, −1.5
53. −2, 0, 1
54. 1
55. 0, 1
56. −2, −$\frac{4}{3}$
59. (a) 3; (b) −3; (c) 3; (e) −7

Complete each of the following statements using the symbol < (less than), = (equal to), or > (greater than).

41. −2 _____ −3
42. −10 _____ −5
43. −20 _____ −10
44. −15 _____ −14
45. |3| _____ 3
46. |−5| _____ −5
47. −4 _____ |−4|
48. 7 _____ |7|

For Exercises 49 to 52, use the following numbers: −3, $\frac{2}{3}$, −1.5, 2, and 0.

49. Which of the numbers are integers?
50. Which of the numbers are natural numbers?
51. Which of the numbers are whole numbers?
52. Which of the numbers are negative numbers?

For Exercises 53 to 56, use the following numbers: −2, −$\frac{4}{3}$, 3.5, 0, and 1.

53. Which of the numbers are integers?
54. Which of the numbers are natural numbers?
55. Which of the numbers are whole numbers?
56. Which of the numbers are negative numbers?

57. (a) Every number has an opposite. The opposite of 5 is −5. In English, a similar situation exists for words. For example, the opposite of "regular" is "irregular." Write the opposite of these words:

irredeemable, uncomfortable, uninteresting, uninformed, irrelevant, immoral.

(b) Note that the idea of an opposite is usually expressed by a prefix such as "un" or "ir." What other prefixes can be used to negate or change the meaning of a word to its opposite? List four words using these prefixes, and use the words in a sentence.

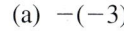

58. (a) What is the difference between positive integers and nonnegative integers?

(b) What is the difference between negative and nonpositive integers?

59. Simplify each of the following:

(a) −(−3)   (b) −(−(−3))   (c) −(−(−(−3)))

(d) Use the results of (a), (b), and (c) to create a rule for simplifying expressions of this type.

(e) Use the rule created in (d) to simplify −(−(−(−(−(−7))))).

SECTION 0.3 Adding and Subtracting Signed Numbers

0.3 OBJECTIVES

1. Add signed numbers
2. Use the commutative property of addition
3. Use the associative property of addition
4. Subtract signed numbers

"Imagination is more important than knowledge."

–Albert Einstein

"Don't play stupid with me. I'm better at it than you are"

–Colonel Flagg

The number line can be used to demonstrate the sum of two signed numbers. To add a positive number, we will move to the right, to add a negative number, we will move to the left.

Example 1 — Finding the Sum of Two Signed Numbers

Find the sum $5 + (-2)$.

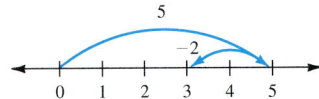

Since the first addend is positive, we start by moving left.

First move 5 units *to the right* of 0. Then, to add 2, move 2 units back *to the left*. We see that

$$5 + (-2) = 3$$

✓ **CHECK YOURSELF 1**

Find the sum.

$$9 + (-7)$$

We can also use the number line to picture addition when two negative numbers are involved. Example 2 illustrates this approach.

Example 2 Finding the Sum of Two Signed Numbers

Find the sum $-2 + (-3)$.

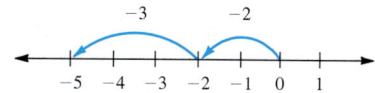

Move 2 units *to the left* of 0. Then move 3 more units *to the left* to add negative 3. We see that

$$-2 + (-3) = -5$$

 CHECK YOURSELF 2

Find the sum.

$$-7 + (-5)$$

You may have noticed some patterns in the previous examples. These patterns will let you do much of the addition mentally. Look at the following rule.

The sum of two positive numbers is positive and the sum of two negative numbers is negative.

> **To Add Signed Numbers**
>
> 1. If two numbers have the same sign, add their absolute values. Give the sum the sign of the original numbers.

Example 3 Finding the Sum of Two Signed Numbers

Find the sums.

(a) $5 + 2 = 7$ The sum of two positive numbers is positive

(b) $-2 + (-6) = -8$ Add the absolute values of the two numbers ($2 + 6 = 8$). Give the sum the sign of the original numbers.

 CHECK YOURSELF 3

Find the sums.

(a) $6 + 7$ (b) $-8 + (-7)$

You have no doubt also noticed that when you add a positive and a negative number, sometimes the answer is positive and sometimes it is negative. This depends on which number has the larger absolute value (the distance from 0). This leads us to the second part of our addition rule.

> **To Add Signed Numbers**
>
> **2.** If two numbers have different signs, subtract the smaller absolute value from the larger. Attach the sign of the number with the larger absolute value to the result.

Example 4

Finding the Sum of Two Signed Numbers

Find the sum.

$6 + (-2) = 4$ The numbers have different signs. Subtract the absolute values $(6 - 2 = 4)$. The result is positive, since 6 has the larger absolute value.

✓ **CHECK YOURSELF 4**

Find the sums.

(a) $8 + (-3)$ (b) $12 + (-10)$

There are three important pieces to the study of algebra. The first is the set of numbers, which we have been expanding in this section. The second is the set of operations, like addition and multiplication. The third is the set of rules, which we call **properties.** The next example enables us to look at an important property of addition.

Example 5

Finding the Sum of Two Signed Numbers

Find the sums.

(a) $2 + (-7) = (-7) + 2 = -5$
(b) $-3 + (-4) = -4 + (-3) = -7$

Note that in both cases the order in which we add the numbers does not affect the sum. This leads us to the following property of signed numbers.

Section 0.3 ■ Adding and Subtracting Signed Numbers

> **The Commutative Property of Addition**
>
> The *order* in which we add two numbers does not change the sum. Addition is **commutative**. In symbols, for any numbers a and b,
>
> $$a + b = b + a$$

✓ CHECK YOURSELF 5

Find the sums $-8 + 2$ and $2 + (-8)$. How do the results compare?

What if we want to add more than two numbers? Another property of addition will be helpful. Look at Example 6.

Example 6 — Finding the Sum of Three Signed Numbers

Find the sum $2 + (-3) + (-4)$. First,

$$[2 + (-3)] + (-4) \quad \text{Add the first two numbers.}$$
$$= \quad -1 \quad + (-4) \quad \text{Then add the third to that sum.}$$
$$= \quad -5$$

Now let's try a second approach.

$$2 + [(-3) + (-4)] \quad \text{This time, add the second and third numbers.}$$
$$= \quad 2 + \quad (-7) \quad \text{Then add the first number to that sum.}$$
$$= -5$$

Do you see that it makes no difference which way we group numbers in addition? The final sum is not changed. We can state the following.

> **The Associative Property of Addition**
>
> The way we *group* numbers does not change the sum. Addition is **associative**. In symbols, for any numbers a, b, and c,
>
> $$(a + b) + c = a + (b + c)$$

✓ CHECK YOURSELF 6

Show that $-2 + (-3 + 5) = [-2 + (-3)] + 5$.

Let's look at a special property of 0.

> **The Additive Identity**
>
> The sum of any number and 0 is just that number. Because of this special property, 0 is called the **additive identity.** In symbols, for any number, a,
>
> $$a + 0 = a$$

Example 7 Finding Sums That Involve Addition of Zero

(a) $-8 + 0 = -8$

(b) $0 + (-20) = -20$

✓ CHECK YOURSELF 7

Find the sum $-6 + 0$.

We saw in Section 0.2 that every number has an opposite. The opposite of a number is also called its **additive inverse.** The additive inverse of a is $-a$.

Example 8 Identifying the Additive Inverse

We can also call -6 the opposite of 6.

(a) The additive inverse of 6 is -6.
(b) The additive inverse of -8 is $-(-8)$, or 8.

✓ CHECK YOURSELF 8

Find the additive inverse for each number.

(a) 8   (b) -9

Section 0.3 ■ Adding and Subtracting Signed Numbers

Use the following rule to add opposite numbers.

> **The Additive Inverse**
>
> The sum of any number and its additive inverse is 0. In symbols, for any number a,
>
> $$a + (-a) = 0$$

Example 9 **Finding the Sum of Two Additive Inverses**

Find the sums.

(a) $6 + (-6) = 0$

(b) $-8 + 8 = 0$

✓ **CHECK YOURSELF 9**

Find the sum.

$$9 + (-9)$$

So far we have looked only at addition of integers. The process is the same if we want to add other types of signed numbers.

Example 10 **Finding the Sum of Two Signed Numbers**

Find the sums.

(a) $\dfrac{15}{4} + \left(-\dfrac{9}{4}\right) = \dfrac{6}{4} = \dfrac{3}{2}$ Subtract the absolute values $\dfrac{15}{4} - \dfrac{9}{4} = \dfrac{6}{4} = \dfrac{3}{2}$.

The sum is positive since $\dfrac{15}{4}$ has the larger absolute value.

(b) $-0.5 + (-0.2) = -0.7$ Add the absolute values $(0.5 + 0.2 = 0.7)$. The sum is negative.

✓ CHECK YOURSELF 10

Find the sums.

(a) $-\dfrac{5}{2} + \left(-\dfrac{7}{2}\right)$ (b) $5.3 + (-4.3)$

Now we turn our attention to the subtraction of signed numbers. Subtraction is called the *inverse* operation to addition. This means that any subtraction problem can be written as a problem in addition. Let's see how it works with the following rule.

> **To Subtract Signed Numbers**
>
> To subtract signed numbers, add the first number and the *opposite* of the number being subtracted. In symbols, by definition
>
> $$a - b = a + (-b)$$

To find the difference $a - b$, we add a and the opposite of b.

Example 11 illustrates this property.

Example 11 Finding the Difference of Two Signed Numbers

(a) Subtract $5 - 3$.

$$5 - 3 = 5 + (-3) = 2$$ To subtract 3, we can add the opposite of 3.

The opposite of 3

(b) Subtract $2 - 5$.

$$2 - 5 = 2 + (-5) = -3$$

The opposite of 5

(c) Subtract $-3 - 4$.

By the definition, add the opposite of 4, -4, to the value -3.

$$-3 - 4 = -3 + (-4) = -7 \qquad -4 \text{ is the opposite of 4.}$$

(d) Subtract $-10 - 15$.

$$-10 - 15 = -10 + (-15) = -25 \qquad -15 \text{ is the opposite of 15.}$$

✓ CHECK YOURSELF 11

Find each difference, using the definition of subtraction.

(a) $8 - 3$ (b) $7 - 9$
(c) $-5 - 9$ (d) $-12 - 6$

Now let's see how the definition is applied in subtracting a negative number.

Example 12 — Finding the Difference of Two Signed Numbers

Find each difference.

By the definition of subtraction, we add the opposite of -3. Remember, the opposite of -3 is 3.

(a) $5 - (-3) = 5 + 3 = 8$ 3 is the opposite of -3.
(b) $7 - (-8) = 7 + 8 = 15$ 8 is the opposite of -8.
(c) $-9 - (-5) = -9 + 5 = -4$ 5 is the opposite of -5.

✓ CHECK YOURSELF 12

Find each difference.

(a) $7 - (-5)$ (b) $5 - (-9)$ (c) $-10 - (-8)$

Signed numbers other than integers are subtracted in exactly the same way, as our final example illustrates.

Example 13 — Finding the Difference of Two Signed Fractions

Subtract.

$$-\frac{11}{4} - \left(-\frac{5}{4}\right) = -\frac{11}{4} + \frac{5}{4} = -\frac{6}{4} = -\frac{3}{2}$$

✓ CHECK YOURSELF 13

Find the difference.

$$\frac{47}{8} - \left(-\frac{19}{8}\right)$$

24 Chapter 0 ■ The Arithmetic of Signed Numbers

Caution

Your graphing calculator can be used to simplify the kinds of problems we've encountered in this section. The negation key is the $\boxed{(-)}$ or the $\boxed{+/-}$ found on the calculator. Do not confuse this with the subtraction key! The subtraction key is used between two numbers. The negation key is used in front of a number.

Example 14 Using a Graphing Calculator to Subtract

Find each difference.

(a) $-4.567 - 3.92$

The key strokes

$\boxed{(-)}$ 4.567 $\boxed{-}$ 3.92 $\boxed{\text{Enter}}$

yields -8.487

(b) $-15.782 - (-19.4)$

The key strokes

$\boxed{(-)}$ 15.782 $\boxed{-}$ $\boxed{(-)}$ 19.4 $\boxed{\text{Enter}}$

yields 3.618

✓ CHECK YOURSELF 14

Find each difference

(a) $-0.936 - 2.75$ (b) $-12.134 - (-1.656)$

✓ CHECK YOURSELF ANSWERS

1. 2.
2. -12.
3. (a) 13; (b) -15.
4. (a) 5; (b) 2.
5. $-6 = -6$.
6. $0 = 0$.
7. -6.
8. (a) -8; (b) 9.
9. 0.
10. (a) -6; (b) 1.
11. (a) 5; (b) -2; (c) -14; (d) -18.
12. (a) 12; (b) 14; (c) -2.
13. $\dfrac{33}{4}$.
14. (a) -3.686; (b) -10.478.

Exercises • 0.3

1. −11
2. 12
3. 4
4. −13
5. −2
6. 7
7. 16
8. 4
9. −6
10. −3
11. −15
12. 1
13. −8
14. 0
15. 1
16. 2
17. 0
18. −15
19. −6
20. −10
21. −3
22. −4
23. 0
24. 0
25. $\frac{9}{4}$
26. $-\frac{9}{8}$
27. $\frac{3}{2}$
28. $-\frac{47}{5}$
29. $\frac{23}{8}$
30. $\frac{17}{8}$
31. −8
32. −5
33. −12
34. 2
35. −7
36. 0
37. −2
38. 2
39. 6
40. −5
41. −11
42. −21
43. −20
44. −6
45. 1
46. −3
47. 0
48. 16
49. 50
50. 75

Add.

1. $-6 + (-5)$
2. $3 + 9$
3. $8 + (-4)$
4. $-6 + (-7)$
5. $4 + (-6)$
6. $9 + (-2)$
7. $7 + 9$
8. $-5 + 9$
9. $(-11) + 5$
10. $5 + (-8)$
11. $-8 + (-7)$
12. $8 + (-7)$
13. $-8 + 0$
14. $7 + (-7)$
15. $-9 + 10$
16. $-6 + 8$
17. $-4 + 4$
18. $5 + (-20)$
19. $7 + (-13)$
20. $0 + (-10)$
21. $-8 + 5$
22. $-7 + 3$
23. $6 + (-6)$
24. $-9 + 9$
25. $\frac{43}{8} + \left(-\frac{25}{8}\right)$
26. $-\frac{35}{16} + \frac{17}{16}$
27. $\frac{29}{8} + \left(-\frac{17}{8}\right)$
28. $-\frac{41}{10} + \left(-\frac{53}{10}\right)$
29. $-\frac{73}{16} + \frac{119}{16}$
30. $-\frac{13}{8} - \left(-\frac{15}{4}\right)$
31. $4 + (-7) + (-5)$
32. $-7 + 8 + (-6)$
33. $-2 + (-6) + (-4)$
34. $12 + (-6) + (-4)$
35. $-3 + (-7) + 5 + (-2)$
36. $7 + (-8) + (-9) + 10$

Subtract.

37. $5 - 7$
38. $7 - 5$
39. $9 - 3$
40. $4 - 9$
41. $-8 - 3$
42. $-15 - 6$
43. $-12 - 8$
44. $9 - 15$
45. $-2 - (-3)$
46. $-9 - (-6)$
47. $-5 - (-5)$
48. $7 - (-9)$
49. $38 - (-12)$
50. $50 - (-25)$

51. $-15 - (-25)$
52. $-20 - (-30)$
53. $-25 - (-15)$
54. $-30 - (-20)$
55. $-(-15) - (-20)$
56. $+18 - (-12)$
57. $48 - (-15)$
58. $-7 - (-12)$
59. $\dfrac{10}{2} - \left(-\dfrac{7}{2}\right)$
60. $-\dfrac{4}{2} - \dfrac{3}{2}$
61. $-\dfrac{9}{4} - \left(-\dfrac{15}{4}\right)$
62. $\dfrac{13}{4} - \left(-\dfrac{7}{4}\right)$
63. $-7 - (-5) - 6$
64. $-5 - (-8) - 10$
65. $-10 - 8 - (-7)$
66. $3 - (-9) - 10$

Find each difference using a graphing calculator.

67. $-11.392 - 13.491$
68. $-8.384 - 17.954$
69. $-7.259 - 4.235$
70. $-6.319 - 2.628$
71. $-18.271 - (-12.569)$
72. $-15.586 - (-9.874)$
73. $-18.681 - (-25.175)$
74. $-11.358 - (-23.145)$

Solve the following applications.

75. Temperature. The temperature in Chicago dropped from 22°F at 4 P.M. to −11°F at midnight. What was the drop in temperature?

76. Banking. Charley's checking account had $225 deposited at the beginning of the month. After he wrote checks for the month, the account was $65 *overdrawn*. What amount of checks did he write during the month?

77. Elevator stops. Micki entered the elevator on the 34th floor. From that point the elevator went up 12 floors, down 27 floors, down 6 floors, and up 15 floors before she got off. On what floor did she get off the elevator?

78. Submarines. A submarine dives to a depth of 500 ft below the ocean's surface. It then dives another 217 ft before climbing 140 ft. What is the depth of the submarine?

79. Military vehicles. A helicopter is 600 ft above sea level and a submarine directly below it is 325 ft below sea level. How far apart are they?

80. Bank balance. Tom has received an overdraft notice from the bank telling him that his account is overdrawn by $142. How much must he deposit in order to have $625 in his account?

81. Change in temperature. At 9:00 A.M., Jose had a temperature of 99.8°. It rose another 2.5° before falling 3.7° by 1:00 P.M. What was his temperature at 1:00 P.M.?

Answers:

51. 10
52. 10
53. −10
54. −10
55. 35
56. 30
57. 63
58. 5
59. $\dfrac{17}{2}$
60. $-\dfrac{7}{2}$
61. $\dfrac{3}{2}$
62. 5
63. −8
64. −7
65. −11
66. 2
67. −24.883
68. −26.338
69. −11.494
70. −8.947
71. −5.702
72. −5.712
73. 6.494
74. 11.787
75. 33°F
76. $290
77. 28th floor
78. 577 ft below sea level
79. 925 ft
80. $767
81. 98.6°

82. $404

82. Bank balance. Aaron had $769 in his bank account on June 1. He deposited $125 and $986 during the month and wrote checks for $235, $529, and $712 during June. What was his balance at the end of the month?

83. Compete the following problems: "$4 - (-9)$ is the same as _____." Write an application problem that might be answered using this subtraction.

84. Explain the difference between these two phrases: "A number less than 7" and "a number subtracted from 7" Use both algebra and English to explain the meaning of these phrases. Write some other ways of expressing subtraction in English.

85. Create an example to show that subtraction of signed numbers is *not* commutative.

86. Create an example to show that subtraction of signed numbers is *not* associative.

87. Do you think that the following statement is true?

$$|a + b| = |a| + |b| \text{ for all numbers } a \text{ and } b$$

When we don't know whether such a statement is true, we refer to the statement as a **conjecture.** We may "test" the conjecture by substituting specific numbers in for the variables.

Test the conjecture using two positive numbers for a and b.

Test again using a positive number for a and 0 for b.

Test again using two negative numbers.

Now try using one positive number and one negative number.

Summarize your results in a rule that you feel is true.

88. (a) positive
    (b) negative
    (c) positive
    (d) positive

88. If a represents a positive number and b represents a negative number, determine whether the given expression is positive or negative.

(a) $|b| + a$ (b) $b + (-a)$ (c) $(-b) + a$ (d) $-b + |-a|$

SECTION 0.4 Multiplying and Dividing Signed Numbers

0.4 OBJECTIVES

1. Multiply signed numbers
2. Use the commutative property of multiplication
3. Use the associative property of multiplication
4. Divide signed numbers
5. Use the distributive property

"Man's mind, once stretched by a new idea, never regains its original dimensions."

–Oliver Wendell Holmes

Multiplication can be seen as repeated addition. We can interpret

$$3 \times 4 = 4 + 4 + 4 = 12$$

We can use this interpretation together with the work of Section 0.3 to find the product of two signed numbers.

Example 1

Finding the Product of Two Signed Numbers

Multiply.

(a) $(3)(-4) = (-4) + (-4) + (-4) = -12$

(b) $(4)(-5) = (-5) + (-5) + (-5) + (-5) = -20$

Note that we use parentheses ( ) to indicate multiplication when negative numbers are involved.

 CHECK YOURSELF 1

Find the product by writing as repeated addition.

$$4(-3)$$

Looking at the products we found by repeated addition in Example 1 should suggest our first rule for multiplying signed numbers.

To Multiply Signed Numbers

1. The product of two numbers with different signs is negative.

28

Example 2 Finding the Product of Two Signed Numbers

Find each product.

$$(5)(-6) = -30$$

$$(10)(-12) = -120$$

$$(-7)(9) = -63$$

$$(1.5)(-0.3) = -0.45$$

The product must have two decimal places.

The product is negative. You can simplify as before in finding the product.

$$\left(-\frac{5}{8}\right)\left(\frac{4}{15}\right) = -\left(\frac{\cancel{5}^1}{\cancel{8}_2} \times \frac{\cancel{4}^1}{\cancel{15}_3}\right)$$

$$= -\frac{1}{6}$$

✓ CHECK YOURSELF 2

Find each product.

(a) $(15)(-5)$ (b) $(-0.8)(0.2)$ (c) $\left(-\frac{2}{3}\right)\left(\frac{6}{7}\right)$

The product of two negative numbers is harder to visualize. The following pattern may help you see how we can determine the sign of the product.

$$(3)(-2) = -6$$
$$(2)(-2) = -4$$
$$(1)(-2) = -2$$
$$(0)(-2) = 0$$
$$(-1)(-2) = 2$$
$$(-2)(-2) = 4$$

Do you see that the product is *increasing* by 2 each time the first number *decreases* by 1?

We already know that the product of two positive numbers is positive.

This suggests that the product of two negative numbers is positive, and this is in fact the case. To extend our multiplication rule, we have the following.

To Multiply Signed Numbers

2. The product of two numbers with the same sign is positive.

Example 3 **Finding the Product of Two Signed Numbers**

Find each product.

$$8 \cdot 7 = 56 \quad \text{Since the numbers have the same sign, the product is positive.}$$

$$(-9)(-6) = 54$$

$$(-0.5)(-2) = 1$$

✓ **CHECK YOURSELF 3**

Find each product.

(a) $(5)(7)$ (b) $(-8)(-6)$ (c) $(9)(-6)$ (d) $(-1.5)(-4)$

Caution

Be Careful! $(-8)(-6)$ tells you to multiply. The parentheses are *next to* one another. The expression $-8 - 6$ tells you to subtract. The numbers are *separated* by the operation sign.

To multiply more than two signed numbers, apply the multiplication rule repeatedly.

Example 4 **Finding the Product of a Set of Signed Numbers**

Multiply.

$$(5)(-7)(-3)(-2)$$
$$= (-35)(-3)(-2) \quad (5)(-7) = -35$$
$$= (105)(-2) \quad (-35)(-3) = 105$$
$$= -210$$

 CHECK YOURSELF 4

Find the product.

$$(-4)(3)(-2)(5)$$

Section 0.4 ■ Multiplying and Dividing Signed Numbers

We saw in Section 0.3 that the commutative and associative properties for addition could be extended to signed numbers. The same is true for multiplication. What about the order in which we multiply? Look at the following examples.

Example 5 — Using the Commutative Property of Multiplication

Find the products.

$$(-5)(7) = (7)(-5) = -35$$
$$(-6)(-7) = (-7)(-6) = 42$$

The order in which we multiply does not affect the product. This gives us the following rule.

> **The Commutative Property of Multiplication**
>
> The order in which we multiply does not change the product. Multiplication is *commutative*. In symbols, for any a and b,
>
> $$a \cdot b = b \cdot a$$

The centered dot represents multiplication. This could have been written as

$$a \times b = b \times a$$

 CHECK YOURSELF 5

Show that $(-8)(-5) = (-5)(-8)$.

What about the way we group numbers in multiplication? Look at Example 6.

Example 6 — Using the Associative Property of Multiplication

Multiply.

$$[(3)(-7)](-2) \quad \text{or} \quad (3)[(-7)(-2)]$$
$$= (-21)(-2) \qquad\qquad = (3)(14)$$
$$= 42 \qquad\qquad\qquad\;\; = 42$$

The symbols [ ] are called brackets and are used to group numbers in the same way as parentheses.

We group the first two numbers on the left and the second two numbers on the right. Note that the product is the same in either case.

The Associative Property of Multiplication

The way we *group* the numbers does not change the product. Multiplication is *associative*. In symbols, for any a, b, and c,

$$(a \cdot b) \cdot c = a \cdot (b \cdot c)$$

✓ CHECK YOURSELF 6

Show that $[(2)(-6)](-3) = (2)[(-6)(-3)]$.

Two numbers, 0 and 1, have special properties in multiplication.

The Multiplicative Identity

The product of 1 and any number is that number. We call 1 the *multiplicative identity*. In symbols, for any a,

$$a \cdot 1 = a$$

Example 7 Multiplying Signed Numbers by 1

Find the products.

$$(-8)(1) = -8$$
$$(1)(-15) = -15$$

✓ CHECK YOURSELF 7

Find the product.

$$(-10)(1)$$

What about multiplication by 0?

Multiplying by Zero

The product of 0 and any number is 0. In symbols, for any a,

$$a \cdot 0 = 0$$

Example 8 Multiplying Signed Numbers by Zero

Find the products.

$$(-9)(0) = 0$$
$$(0)(-23) = 0$$

 CHECK YOURSELF 8

Find the product.

$$(0)(-12)$$

Another important property in mathematics is the **distributive property.** The distributive property involves addition and multiplication together. We can illustrate the property with an application.

Remember: The area of a rectangle is the product of its length and width:

$$A = L \cdot W$$

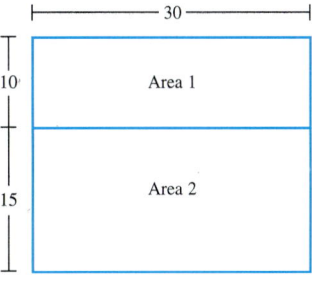

We can find the total area by multiplying the length by the overall width, which is found by adding the two widths.

or

We can find the total area as a sum of the two areas.

Length Overall Width

$$30 \quad \cdot (10 + 15)$$
$$= 30 \cdot 25$$
$$= 750$$

(Area 1)   (Area 2)
Length · Width   Length · Width

$$30 \cdot 10 \quad + \quad 30 \cdot 15$$
$$= 300 \quad + \quad 450$$
$$= 750$$

So

$$30 \cdot (10 + 15) = 30 \cdot 10 + 30 \cdot 15$$

This leads us to the following property.

Note the pattern.
$a(b + c) = a \cdot b + a \cdot c$
We "distributed" the multiplication "over" the addition.

The Distributive Property

If a, b, and c are any numbers,

$$a(b + c) = a \cdot b + a \cdot c \quad \text{and} \quad (b + c)a = b \cdot a + c \cdot a$$

Example 9 — Using the Distributive Property

Use the distributive property to simplify (remove the parentheses in) the following.

(a) $5(3 + 4)$

$5(3 + 4) = 5 \cdot 3 + 5 \cdot 4 = 35$

Note: It is also true that
$5(3 + 4) = 5 \cdot 7 = 35$

(b) $\frac{1}{3}(9 + 12) = \frac{1}{3} \cdot 9 + \frac{1}{3} \cdot 12$

$= 3 + 4 = 7$

Note: It is also true that
$\frac{1}{3}(9 + 12) = \frac{1}{3}(21) = 7$

✓ CHECK YOURSELF 9

Use the distributive property to simplify (remove the parentheses).

(a) $4(6 + 7)$ (b) $\frac{1}{5}(10 + 15)$

The distributive property applies to all signed numbers. First let us look at an example of multiplication distributed over addition.

Example 10 — Distributing Multiplication over Addition

Use the distributive property to remove the parentheses and simplify the following.

(a) $4(-2 + 5) = 4(-2) + 4(5) = -8 + 20 = 12$

(b) $-5(-3 + 2) = (-5)(-3) + (-5)(2) = 15 + (-10) = 5$

✓ CHECK YOURSELF 10

Use the distributive property to remove the parentheses and simplify the following.

(a) $7(-3 + 5)$ (b) $-2(-6 + 3)$

The distributive property can also be used to distribute multiplication over subtraction.

Example 11 — Distributing Multiplication over Subtraction

Use the distributive property to remove the parentheses and simplify the following.

(a) $4(-3 - 6) = 4(-3) - 4(6) = -12 - 24 = -36$

(b) $-7(-3 - 2) = -7(-3) - (-7)(2) = 21 - (-14) = 21 + 14 = 35$

✓ CHECK YOURSELF 11

Use the distributive property to remove the parentheses and simplify the following.

(a) $7(-3 - 4)$ (b) $-2(-4 - 3)$

A detailed explanation of why the product of two negative numbers must be positive concludes our discussion of multiplying signed numbers.

The Product of Two Negative Numbers

The following argument shows why the product of two negative numbers must be positive.

From our earlier work, we know that a number added to its opposite is 0.	$5 + (-5) = 0$
Multiply both sides of the statement by -3.	$(-3)[5 + (-5)] = (-3)(0)$
A number multiplied by 0 is 0, so on the right we have 0.	$(-3)[5 + (-5)] = 0$
We can now use the distributive property on the left.	$(-3)(5) + (-3)(-5) = 0$
Since we know that $(-3)(5) = -15$, the statement becomes	$-15 + (-3)(-5) = 0$

We now have a statement of the form $-15 + \square = 0$. This asks, "What number must we add to -15 to get 0, where $\square$ is the value of $(-3)(-5)$?" The answer is, of course, 15. This means that

$$(-3)(-5) = 15 \qquad \text{The product must be positive.}$$

It doesn't matter what numbers we use in the argument. The product of two negative numbers will always be positive.

Multiplication and division are related operations. So every division problem can be stated as an equivalent multiplication problem.

$$8 \div 4 = 2 \quad \text{Since } 8 = 4 \cdot 2$$

$$\frac{12}{3} = 4 \quad \text{Since } 12 = 3 \cdot 4$$

Since the operations are related, the rules of signs for multiplication are also true for division.

To Divide Signed Numbers

1. If two numbers have the same sign, the quotient is positive.

2. If two numbers have different signs, the quotient is negative.

Example 12 — **Dividing Two Signed Numbers**

Divide 20 by -5.

The numbers 20 and -5 have different signs, and so the quotient is negative.

$$20 \div (-5) = -4 \quad \text{Since } 20 = (-5)(-4)$$

✓ **CHECK YOURSELF 12**

Write the multiplication statement that is equivalent to

$$36 \div (-4) = -9$$

Example 13 — **Dividing Two Signed Numbers**

Divide -20 by -5.

The two numbers have the same sign, and so the quotient is positive.

$$\frac{-20}{-5} = 4 \quad \text{Since } -20 = (-5)(4)$$

✓ **CHECK YOURSELF 13**

Find each quotient.

(a) $\dfrac{48}{-6}$ (b) $(-50) \div (-5)$

As you would expect, division with fractions or decimals uses the same rules for signs. Example 14 illustrates this concept.

Example 14

Dividing Two Signed Numbers

Divide.

First note that the quotient is positive. Then invert the divisor and multiply.

$$\left(-\frac{3}{5}\right) \div \left(-\frac{9}{20}\right) = \frac{\cancel{3}}{\cancel{5}} \cdot \frac{\cancel{20}}{\cancel{9}} = \frac{4}{3}$$

✓ **CHECK YOURSELF 14**

Find each quotient.

(a) $-\dfrac{5}{8} \div \dfrac{3}{4}$ (b) $-4.2 \div (-0.6)$

Be very careful when 0 is involved in a division problem. Remember that 0 divided by any nonzero number is 0. However, division *by 0* is not allowed and will be described as undefined.

Example 15

Dividing Signed Numbers When Zero Is Involved

Divide.

A statement like $-9 \div 0$ has no meaning. There is no answer to the problem. Just write "undefined."

(a) $0 \div 7 = 0$ (b) $\dfrac{0}{-4} = 0$

(c) $-9 \div 0$ is undefined.   (d) $\dfrac{-5}{0}$ is undefined.

✓ **CHECK YOURSELF 15**

Find the quotient if possible.

(a) $\dfrac{0}{-7}$ (b) $\dfrac{-12}{0}$

The result of Example 15 can be confirmed on your calculator. That will be included in the next example.

Example 16　Dividing with a Calculator

Use your calculator to find each quotient.

(a) $\dfrac{-12.567}{0}$

The key stroke sequence on a graphing calculator

[(−)]　12.567　[÷]　0　[Enter]

results in a "Divide by 0" error message. The calculator recognizes that it cannot divide by zero.

On a scientific calculator, 12.567　[+/−]　[÷]　0　[=]　results in an error message.

(b) $-10.992 \div -4.58$

The key stroke sequence

[(−)]　10.992　[÷]　[(−)]　4.58　[Enter]

or 10.992　[+/−]　[÷]　4.58　[+/−]　[=]

yields 2.4

✓ **CHECK YOURSELF 16**

Find each quotient

(a) $\dfrac{-31.44}{6.55}$　　(b) $-23.6 \div 0$

✓ **CHECK YOURSELF ANSWERS**

1. $(-3) + (-3) + (-3) + (-3) = -12$.　　**2.** (a) -75; (b) -0.16; (c) $-\dfrac{4}{7}$.
3. (a) 35; (b) 48; (c) -54; (d) 6.　　**4.** 120.　　**5.** $40 = 40$.　　**6.** $36 = 36$.
7. -10.　　**8.** 0.　　**9.** (a) 52; (b) 5.　　**10.** (a) 14; (b) 6.　　**11.** (a) -49; (b) 14.
12. $36 = (-4)(-9)$.　　**13.** (a) -8; (b) 10.　　**14.** (a) $-\dfrac{5}{6}$; (b) 7.
15. (a) 0; (b) undefined.　　**16.** (a) -4.8; (b) undefined.

Exercises • 0.4

1. 56
2. −72
3. −12
4. 75
5. −72
6. −24
7. 42
8. 24
9. 0
10. −100
11. 64
12. 0
13. −80
14. 200
15. 108
16. 81
17. −20
18. −30
19. −200
20. −125
21. 150
22. 30
23. $\frac{1}{4}$
24. $-\frac{1}{6}$
25. 22
26. 0
27. $\frac{1}{6}$
28. $\frac{2}{3}$
29. 120
30. 60
31. −80
32. −70
33. −150
34. 300
35. 144
36. 240
37. 15
38. 48
39. −48
40. 55
41. 70
42. 36
43. 36
44. 35
45. −5
46. 5
47. 6
48. 10
49. −10
50. −6

Multiply.

1. $7 \cdot 8$
2. $(6)(-12)$
3. $(4)(-3)$
4. $15 \cdot 5$
5. $(-8)(9)$
6. $(-8)(3)$
7. $(-7)(-6)$
8. $(-12)(-2)$
9. $(-10)(0)$
10. $(10)(-10)$
11. $(-8)(-8)$
12. $(0)(-50)$
13. $(20)(-4)$
14. $(-25)(-8)$
15. $(-9)(-12)$
16. $(-9)(-9)$
17. $(-20)(1)$
18. $(1)(-30)$
19. $(-40)(5)$
20. $(-25)(5)$
21. $(-10)(-15)$
22. $(-5)(-6)$
23. $\left(-\frac{7}{10}\right)\left(-\frac{5}{14}\right)$
24. $\left(-\frac{3}{8}\right)\left(\frac{4}{9}\right)$
25. $(-11)(-2)$
26. $\left(-\frac{15}{4}\right)(0)$
27. $\left(-\frac{5}{8}\right)\left(-\frac{4}{15}\right)$
28. $\left(-\frac{8}{21}\right)\left(-\frac{7}{4}\right)$
29. $(-5)(3)(-8)$
30. $(4)(-3)(-5)$
31. $(-2)(-8)(-5)$
32. $(-7)(-5)(-2)$
33. $(2)(-5)(-3)(-5)$
34. $(-2)(-5)(-5)(-6)$
35. $(-4)(-3)(-6)(-2)$
36. $(-8)(3)(-2)(5)$

Use the distributive property to remove parentheses and simplify the following.

37. $5(-6 + 9)$
38. $12(-5 + 9)$
39. $-8(-9 + 15)$
40. $-11(-8 + 3)$
41. $-5(-8 - 6)$
42. $-2(-7 - 11)$
43. $-4(-6 - 3)$
44. $-7(-2 - 3)$

Divide.

45. $15 \div (-3)$
46. $\frac{35}{7}$
47. $\frac{48}{8}$
48. $-20 \div (-2)$
49. $\frac{-50}{5}$
50. $-36 \div 6$

Chapter 0 ■ The Arithmetic of Signed Numbers

51. $\dfrac{-24}{-3}$

52. $\dfrac{42}{-6}$

53. $\dfrac{60}{-15}$

54. $70 \div (-10)$

55. $18 \div (-1)$

56. $\dfrac{-250}{-25}$

57. $\dfrac{0}{-9}$

58. $\dfrac{-12}{0}$

59. $-144 \div (-12)$

60. $\dfrac{0}{-10}$

61. $-7 \div 0$

62. $\dfrac{-25}{1}$

63. $\dfrac{-150}{6}$

64. $\dfrac{-80}{-16}$

65. $-45 \div (-9)$

66. $-\dfrac{2}{3} \div \dfrac{4}{9}$

67. $-\dfrac{7}{9} \div \left(-\dfrac{14}{3}\right)$

68. $(-8) \div (-4)$

69. $\dfrac{7}{10} \div \left(-\dfrac{14}{25}\right)$

70. $\dfrac{75}{-5}$

71. $\dfrac{-75}{15}$

72. $-\dfrac{5}{8} \div \left(-\dfrac{5}{16}\right)$

Divide using a graphing calculator. Round answers to the nearest thousandth.

73. $-5.634 \div 2.398$
74. $-1.897 \div 8.912$
75. $-13.859 \div -4.148$
76. $-39.476 \div -17.629$
77. $32.245 \div -48.298$
78. $43.198 \div -56.249$

79. **Dieting.** A woman lost 42 pounds (lb). If she lost 3 lb each week, how long has she been dieting?

80. **Mowing lawns.** Patrick worked all day mowing lawns and was paid $9 per hour. If he had $125 at the end of a 9-hour day, how much did he have before he started working?

81. **Unit pricing.** A 4.5-lb can of food costs $8.91. What is the cost per pound?

82. **Investment.** Suppose that you and your two brothers bought equal shares of an investment for a total of $20,000 and sold it later for $16,232. How much did each person lose?

83. **Temperature.** Suppose that the temperature outside is dropping at a constant rate. At noon, the temperature is 70°F and it drops to 58°F at 5:00 P.M. How much did the temperature change each hour?

---

51. 8
52. −7
53. −4
54. −7
55. −18
56. 10
57. 0
58. Undefined
59. 12
60. 0
61. Undefined
62. −25
63. −25
64. 5
65. 5
66. $-\dfrac{3}{2}$
67. $\dfrac{1}{6}$
68. 2
69. $-\dfrac{5}{4}$
70. −15
71. −5
72. 2
73. −2.349
74. −0.213
75. 3.341
76. 2.239
77. −0.668
78. −0.768
79. 14 weeks
80. $44
81. $1.98
82. $1256
83. 2.4°F

Section 0.4 ■ Multiplying and Dividing Signed Numbers   **41**

84. 126
85. −2
86. −4
87. −2
88. 5
89. 4
90. −3

84. Test tube count. A chemist has 84 ounces (oz) of a solution. He pours the solution into test tubes. Each test tube holds $\frac{2}{3}$ oz. How many test tubes can he fill?

To evaluate an expression involving a fraction (indicating division), we evaluate the numerator and then the denominator. We then divide the numerator by the denominator as the last step. Using this approach, find the value of each of the following expressions.

85. $\dfrac{5 - 15}{2 + 3}$ 86. $\dfrac{4 - (-8)}{2 - 5}$

87. $\dfrac{-6 + 18}{-2 - 4}$ 88. $\dfrac{-4 - 21}{3 - 8}$

89. $\dfrac{(5)(-12)}{(-3)(5)}$ 90. $\dfrac{(-8)(-3)}{(2)(-4)}$

91. Create an example to show that the division of signed numbers is *not* commutative.

92. Create an example to show that the division of signed numbers is *not* associative.

93. Here is another conjecture to consider:

$$|ab| = |a| \, |b| \text{ for all numbers } a \text{ and } b.$$

(See the discussion in Exercises 0.3, problem 87, concerning testing a conjecture.) Test this conjecture for various values of *a* and *b*. Use positive numbers, negative numbers, and 0. Summarize your results in a rule.

94. Use a calculator (or mental calculations) to compute the following:

$$\frac{5}{0.1}, \frac{5}{0.01}, \frac{5}{0.001}, \frac{5}{0.0001}, \frac{5}{0.00001}$$

In this series of problems, while the numerator is always 5, the denominator is getting smaller (and is getting closer to 0). As this happens, what is happening to the value of the fraction?
Write an argument that explains why $\dfrac{5}{0}$ could not have any finite value.

SECTION 0.5 Exponents and Order of Operations

0.5 OBJECTIVES

1. Plot rational numbers on a number line
2. Evaluate numbers with exponents
3. Use the order of operations
4. Write a product of like factors in exponential form

"The intelligent person is one who has successfully fulfilled many accomplishments, and yet is willing to learn more."

–Ed Parker

To this point, every number that we have encountered in this text could have been written as a ratio of two integers.

For example, -3 could be written as $\dfrac{-3}{1}$ or $3\dfrac{1}{8}$ could have been written as $\dfrac{25}{8}$. Such numbers are called **rational numbers.** On the number line, you can estimate the location of a rational number, as our next example illustrates.

Example 1 Plotting Rational Numbers

Plot each of the following rational numbers on the number line provided.

$$\dfrac{2}{3},\ -3\dfrac{1}{4},\ \dfrac{27}{5},\ -1.445$$

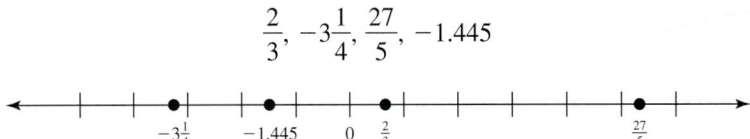

✓ CHECK YOURSELF 1

Plot each of the following rational numbers on the number line provided.

$$-2\dfrac{1}{3},\ \dfrac{37}{11},\ 5.66,\ -\dfrac{1}{4}$$

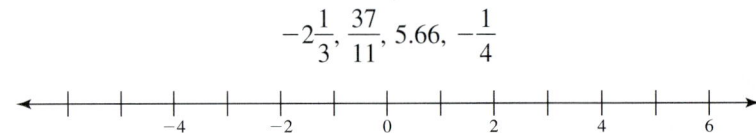

Note that a decimal is really a "decimal fraction." -1.445 is another way of writing

$$\dfrac{-1445}{1000}$$

In Section 0.4, we mentioned that multiplication is a more compact, or "shorthand," form for repeated addition. For example, an expression with repeated addition, such as

Section 0.5 ■ Exponents and Order of Operations   43

$$3 + 3 + 3 + 3 + 3$$

can be rewritten as

$$5 \cdot 3$$

Thus multiplication is shorthand for repeated addition.

In algebra, we frequently have a number or variable that is repeated in an expression several times. For instance, we might have

$$5 \cdot 5 \cdot 5$$

> A factor is a number or a variable that is being multiplied by another number or variable.

To abbreviate this product, we write

$$5 \cdot 5 \cdot 5 = 5^3$$

> Since an exponent represents repeated multiplication, 5^3 is an expression.

This is called **exponential notation** or **exponential form.** The exponent or power, here 3, indicates the number of times that the factor or base, here 5, appears in a product.

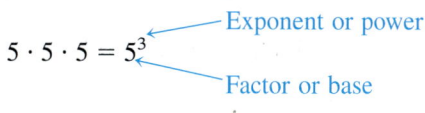

$5 \cdot 5 \cdot 5 = 5^3$ ⟵ Exponent or power
⟵ Factor or base

>
> **Caution**
> Be careful: 5^3 is *not* the same as $5 \cdot 3$. Notice that
> $5^3 = 5 \cdot 5 \cdot 5 = 125$ and
> $5 \cdot 3 = 15$.

Example 2   Writing Expressions in Exponential Form

(a) Write $3 \cdot 3 \cdot 3 \cdot 3$, using exponential form. The number 3 appears 4 times in the product, so

$$3 \cdot 3 \cdot 3 \cdot 3 = 3^4 \quad \text{← Four factors of 3}$$

This is read "3 to the fourth power."

(b) Write $10 \cdot 10 \cdot 10$ using exponential form. Since 10 appears three times in the product, you can write

$$10 \cdot 10 \cdot 10 = 10^3$$

This is read "10 to the third power" or "10 cubed."

✓ CHECK YOURSELF 2

Write in exponential form.

(a) $4 \cdot 4 \cdot 4 \cdot 4 \cdot 4 \cdot 4$ (b) $10 \cdot 10 \cdot 10 \cdot 10$

When evaluating a number raised to a power, it is important to note whether there is a sign attached to the number. Note that

$$(-2)^4 = (-2)(-2)(-2)(-2) = 16$$

whereas,

$$-2^4 = -(2)(2)(2)(2) = -16.$$

Example 3 Evaluating Exponential Terms

Evaluate each expression.

(a) $(-3)^3 = (-3)(-3)(-3) = -27$

(b) $-3^3 = -(3)(3)(3) = -27$

(c) $(-3)^4 = (-3)(-3)(-3)(-3) = 81$

(d) $-3^4 = -(3)(3)(3)(3) = -81$

✓ CHECK YOURSELF 3

Evaluate each expression.

(a) $(-4)^3$

(b) -4^3

(c) $(-4)^4$

(d) -4^4

Your calculator can help you to evaluate expressions containing exponents. If you have a graphing calculator, the appropriate key is the carat, "∧." Enter the base, followed by the carat, followed by the exponent. Other calculators use a key labeled "y^x" in place of the carat.

Example 4 Evaluating Expressions with Exponents

Use your calculator to evaluate each expression.

(a) $3^5 = 243$ Type 3 ∧ 5 Enter

or 3 y^x 5 =

(b) $2^{10} = 1024$ 2 ∧ 10 Enter

or 2 y^x 10 =

✓ CHECK YOURSELF 4

Use your calculator to evaluate each expression.

(a) 3^4

(b) 2^{16}

We have used the term "expression" with numbers taken to powers, like 3^4. But what about something like $4 + 12 - 6$? We call *any* meaningful combination of numbers and operations an **expression.** When we evaluate an expression, we find a number that is equal to the expression. To evaluate an expression, we need to establish a set of rules that tell us the correct order in which to perform the operations. To see why, simplify the expression $5 + 2 \cdot 3$.

Caution

Only one of these results can be correct.

Method 1	or	*Method 2*
$5 + 2 \cdot 3$		$5 + 2 \cdot 3$
Add first.		Multiply first.
$= 7 \cdot 3$		$= 5 + 6$
$= 21$		$= 11$

Since we get different answers depending on how we do the problem, the language of algebra would not be clear if there were no agreement on which method is correct. The following rules tell us the order in which operations should be done.

Parentheses and brackets are both grouping symbols. Fraction bars and radicals are also grouping symbols.

The Order of Operations

Step 1 Evaluate all expressions inside grouping symbols first.

Step 2 Evaluate all expressions involving exponents.

Step 3 Do any multiplication or division in order, working from left to right.

Step 4 Do any addition or subtraction in order, working from left to right.

Example 5 — Evaluating Expressions

Evaluate $5 + 2 \cdot 3$.

There are no parentheses or exponents, so start with step 3: First multiply and then add.

$$5 + 2 \cdot 3$$
$$= 5 + 6 \quad \text{Multiply first.}$$
$$= 11 \quad \text{Then add.}$$

Note: Method 2 shown on the previous page is the correct one.

✓ CHECK YOURSELF 5

Evaluate the following expressions.

(a) $20 - 3 \cdot 4$ (b) $9 + 6 \div 3$

Example 6 — Evaluating Expressions

Evaluate $-5 \cdot 3^2$.

$$-5 \cdot 3^2 = -5 \cdot 9$$

Evaluate the exponent first.

$$= -45$$

✓ CHECK YOURSELF 6

Evaluate $-4 \cdot 2^4$.

Both scientific and graphing calculators correctly interpret the order of operations. This is demonstrated in Example 7.

Example 7 — Using a Calculator to Evaluate Expressions

Use your scientific or graphing calculator to evaluate each expression.

(a) $24.3 + 6.2 \cdot 3.5$

When evaluating expressions by hand, you must consider the order of operations. In this case, the multiplication must be done first, then the addition. With a modern calculator, you need only enter the expression correctly. The calculator is programmed to follow the order of operations.

Entering 24.3 [+] 6.2 [×] 3.5 [Enter]

yields the evaluation 46.

(b) $(2.45)^3 - 49 \div 8000 + 12.2 \cdot 1.3$

As we mentioned earlier, some calculators use the carat ($\wedge$) to designate exponents. Others use the symbol x^y (or y^x).

Entering [(] 2.45 [)] [$\wedge$] 3 [−] 49 [÷] 8000 [+] 12.2 [×] 1.3
or   [(] 2.45 [)] [y^x] 3 [−] 49 [÷] 8000 [+] 12.2 [×] 1.3

yields the evaluation 30.56.

✓ CHECK YOURSELF 7

Use your scientific or graphing calculator to evaluate each expression.

(a) $67.89 - 4.7 \cdot 12.7$ (b) $4.3 \cdot 55.5 - (3.75)^3 + 8007 \div 1600$

Operations inside grouping symbols are done first.

Example 8 Evaluating Expressions

Evaluate $(5 + 2) \cdot 3$.

Do the operation inside the parentheses as the first step.

$$(5 + 2) \cdot 3 = 7 \cdot 3 = 21$$

Add.

✓ CHECK YOURSELF 8

Evaluate $4(9 - 3)$.

The principle is the same when more than two "levels" of operations are involved.

Example 9 — Evaluating Expressions

(a) Evaluate $4(-2 + 7)^3$.

Add inside the parentheses first.

$$4(-2 + 7)^3 = 4(5)^3$$

Evaluate the exponent.

$$= 4 \cdot 125$$

Multiply.

$$= 500$$

(b) Evaluate $5(7 - 3)^2 - 10$.

Evaluate the expression inside the parentheses.

$$5(7 - 3)^2 - 10 = 5(4)^2 - 10$$

Evaluate the exponent.

$$= 5 \cdot 16 - 10$$

Multiply.

$$= 80 - 10 = 70$$

Subtract.

✓ CHECK YOURSELF 9

Evaluate

(a) $4 \cdot 3^3 + 8 \cdot (-11)$.   (b) $12 + 4(2 + 3)^2$.

✓ CHECK YOURSELF ANSWERS

1. [number line with points at $-2\frac{1}{3}$, $-\frac{1}{4}$, 0, $\frac{37}{11}$, 5.66]   2. (a) 4^6; (b) 10^4.
3. (a) -64; (b) -64; (c) 256; (d) -256.
4. (a) 81; (b) 65,536.
5. (a) 8; (b) 11.   6. -64.
7. (a) 8.2; (b) 190.92.   8. 24.   9. (a) 20; (b) 112.

Exercises • 0.5

In Exercises 1 to 4, plot each of the rational numbers on the number line.

1. $\dfrac{3}{4}, -1\dfrac{1}{4}, \dfrac{32}{13}, -4.335$

2. $\dfrac{2}{3}, -2\dfrac{1}{3}, \dfrac{15}{7}, 4.156$

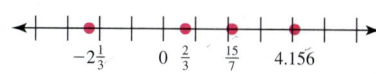

3. $\dfrac{7}{8}, -1\dfrac{5}{6}, \dfrac{35}{14}, -5.156$

4. $\dfrac{3}{5}, 1\dfrac{1}{8}, -\dfrac{17}{8}, 3.165$

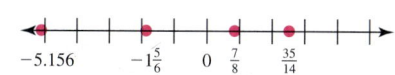
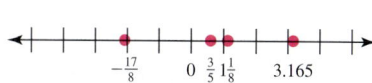

Write each expression using exponential form.

5. $3 \cdot 3 \cdot 3 \cdot 3 \cdot 3$
6. $2 \cdot 2 \cdot 2 \cdot 2 \cdot 2 \cdot 2 \cdot 2$
7. $7 \cdot 7 \cdot 7 \cdot 7 \cdot 7$
8. $10 \cdot 10 \cdot 10 \cdot 10 \cdot 10$
9. $8 \cdot 8 \cdot 8 \cdot 8 \cdot 8 \cdot 8$
10. $5 \cdot 5 \cdot 5 \cdot 5 \cdot 5 \cdot 5$

Evaluate.

11. 3^2
12. 2^3
13. 2^4
14. 2^5
15. 8^3
16. 3^5
17. 1^5
18. 4^4
19. 5^2
20. 6^3
21. 9^4
22. 7^1
23. 10^3
24. 10^2
25. 10^6
26. 10^7
27. 2×4^3
28. $(2 \times 4)^3$
29. 2×3^2
30. $(2 \times 3)^2$
31. $5 + 2^2$
32. $(5 + 2)^2$
33. $(3 \times 2)^4$
34. 3×2^4

5. 3^5
6. 2^7
7. 7^5
8. 10^5
9. 8^6
10. 5^6
11. 9
12. 8
13. 16
14. 32
15. 512
16. 243
17. 1
18. 256
19. 25
20. 216
21. 6561
22. 7
23. 1000
24. 100
25. 1,000,000
26. 10,000,000
27. 128
28. 512
29. 18
30. 36
31. 9
32. 49
33. 1296
34. 48

35. 19	36. 2		
37. 54	38. 12		
39. −14	40. 14		
41. 1	42. 6		
43. 60	44. −1		
45. 144	46. 6		
47. 75	48. 40		
49. 225	50. 1000		
51. 34	52. 40		
53. 21	54. 8		
55. −4	56. −58		
57. 40	58. 195		
59. 256	60. 48		
61. 196	62. 400		
63. 147	64. 40		
65. 21	66. 12		
67. −25	68. 80		
69. 96	70. 89		
71. −15	72. 33		
73. −9	74. −16		
75. 6	76. 17		

Evaluate each of the following expressions.

35. $7 + 2 \cdot 6$ **36.** $10 - 4 \cdot 2$

37. $(7 + 2) \cdot 6$ **38.** $(10 - 4) \cdot 2$

39. $-12 - 8 \div 4$ **40.** $10 + 20 \div 5$

41. $(12 - 8) \div 4$ **42.** $(10 + 20) \div 5$

43. $8 \cdot 7 + 2 \cdot 2$ **44.** $48 \div 8 - 14 \div 2$

45. $8 \cdot (7 + 2) \cdot 2$ **46.** $48 \div (8 - 4) \div 2$

47. $3 \cdot 5^2$ **48.** $5 \cdot 2^3$

49. $(3 \cdot 5)^2$ **50.** $(5 \cdot 2)^3$

51. $4 \cdot 3^2 - 2$ **52.** $3 \cdot 2^4 - 8$

53. $7(2^3 - 5)$ **54.** $4(3^2 - 7)$

55. $3 \cdot 2^4 - 26 \cdot 2$ **56.** $4 \cdot 2^3 - 15 \cdot 6$

57. $(2 \cdot 4)^2 - 8 \cdot 3$ **58.** $(3 \cdot 2)^3 - 7 \cdot 3$

59. $4(2 + 6)^2$ **60.** $3(8 - 4)^2$

61. $(4 \cdot 2 + 6)^2$ **62.** $(3 \cdot 8 - 4)^2$

63. $3(4 + 3)^2$ **64.** $5(4 - 2)^3$

65. $3 \cdot 4 + 3^2$ **66.** $5 \cdot 4 - 2^3$

67. $4(2 + 3)^2 - 125$ **68.** $8 + 2(3 + 3)^2$

69. $(4 \cdot 2 + 3)^2 - 25$ **70.** $8 + (2 \cdot 3 + 3)^2$

71. $-8 - 3^2 + 18 \div 9$ **72.** $14 + 3 \cdot 9 - 28 \div 7 \cdot 2$

73. $4 \cdot 8 \div 2 - 5^2$ **74.** $-12 - 8 \div 4 \cdot 2$

75. $15 + 5 - 3 \cdot 2 + (-2)^3$ **76.** $-8 + 14 \div 2 \cdot 4 - 3$

Section 0.5 ■ Exponents and Order of Operations

Evaluate using your calculator. Round your answer to the nearest tenth.

77. $(1.2)^3 \div 2.0736 \cdot 2.4 + 1.6935 - 2.4896$

78. $(5.21 \cdot 3.14 - 6.2154) \div 5.12 - 0.45625$

79. $1.23 \cdot 3.169 - 2.05194 + (5.128 \cdot 3.15 - 10.1742)$

80. $4.56 + (2.34)^4 \div 4.7896 \cdot 6.93 \div 27.5625 - 3.1269 + (1.56)^2$

81. Population doubling. Over the last 2000 years, the Earth's population has doubled approximately five times. Write the phrase "doubled five times" in exponential form.

82. Volume of a cube. The volume of a cube with each edge of length 9 inches (in.) is given by $9 \cdot 9 \cdot 9$. Write the volume using exponential notation.

Many of the exercise sets in this text have a set of problems marked by the hurdler logo shown here. These are particularly challenging exercises which either introduce ideas that extend the material of the section or require you to generalize from what you have learned.

83. Insert grouping symbols in the proper place so that the value of the expression $36 \div 4 + 2 - 4$ is 2.

84. Work with a small group of students.

Part 1: Write the numbers 1 through 25 on slips of paper and put the slips in a pile, face down. Each of you randomly draws a slip of paper until you have drawn five slips. Turn the papers over and write down the five numbers. Put the five papers back in the pile, shuffle, and then draw one more. This last number is the answer. The first five numbers are the problem. Your task is to arrange the first five into a computation, using all you know about the order of operations, so that the answer is the last number. Each number must be used and may be used only once. If you cannot find a way to do this, pose it as a question to the whole class. Is this guaranteed to work?

Part 2: Use your five numbers in a problem, each number being used and used only once, for which the answer is 1. Try this nine more times with the numbers 2 through 10. You may find more than one way to do each of these. Surprising, isn't it?

Part 3: Be sure that when you successfully find a way to get the desired answer using the five numbers, you can then write your steps using the correct order of operations. Write your 10 problems and exchange them with another group to see if they get these same answers when they do your problems.

77. 1.2
78. 1.5
79. 7.8
80. 5.4
81. 2^5
82. 9^3
83. $36 \div (4 + 2) - 4$

Summary Exercises • 0

This summary exercise set is provided to give you practice with each of the objectives of the chapter. Each exercise is keyed to the appropriate chapter section. Your instructor will give you guidelines on how to best use these exercises in your instructional setting.

[0.1] In Exercises 1 to 3, write three fractional representations for each number.

1. $\dfrac{5}{7}$
2. $\dfrac{3}{11}$
3. $\dfrac{4}{9}$

4. Write the fraction $\dfrac{24}{64}$ in simplest form.

[0.1] In Exercises 5 to 12, perform the indicated operations.

5. $\dfrac{7}{15} \times \dfrac{5}{21}$
6. $\dfrac{10}{27} \times \dfrac{9}{20}$
7. $\dfrac{5}{12} \div \dfrac{5}{8}$

8. $\dfrac{7}{15} \div \dfrac{14}{25}$
9. $\dfrac{5}{6} + \dfrac{11}{18}$
10. $\dfrac{5}{18} + \dfrac{7}{12}$

11. $\dfrac{11}{18} - \dfrac{2}{9}$
12. $\dfrac{11}{27} - \dfrac{5}{18}$

[0.2] In Exercises 13 to 20, complete the statements.

13. The absolute value of 12 is _____
14. The opposite of -8 is _____
15. $|-3| =$ _____
16. $-(-20) =$ _____
17. $-|-4| =$ _____
18. $|-(-5)| =$ _____
19. The absolute value of -16 is _____.
20. The opposite of the absolute value of -9 is _____.

Answers:

1. $\dfrac{10}{14}, \dfrac{20}{28}, \dfrac{50}{70}$

2. $\dfrac{9}{33}, \dfrac{15}{55}, \dfrac{30}{110}$

3. $\dfrac{8}{18}, \dfrac{16}{36}, \dfrac{24}{54}$

4. $\dfrac{3}{8}$
5. $\dfrac{1}{9}$
6. $\dfrac{1}{6}$
7. $\dfrac{2}{3}$
8. $\dfrac{5}{6}$
9. $\dfrac{13}{9}$
10. $\dfrac{31}{36}$
11. $\dfrac{7}{18}$
12. $\dfrac{7}{54}$
13. 12
14. 8
15. 3
16. 20
17. -4
18. 5
19. 16
20. -9

21. <	22. =	
23. >	24. <	
25. 8	26. −5	
27. −11	28. $\frac{1}{2}$	
29. −4	30. −4	
31. 4	32. 7	
33. −5	34. −15	
35. 5	36. −4	
37. $\frac{13}{2}$	38. −5	
39. 1	40. 5	
41. 0	42. 0	
43. 36	44. −80	
45. −15	46. $\frac{3}{10}$	
47. 16	48. 42	
49. −360	50. 54	

[0.2] Complete each of the following statements using the symbol <, >, or =.

21. −3 _____ −1

22. −6 _____ −|−6|

23. |−8| _____ −(−3)

24. −|−5| _____ |−(−5)|

[0.3] In Exercises 25 to 32 add.

25. $15 + (-7)$

26. $4 + (-9)$

27. $-8 + (-3)$

28. $\frac{5}{2} + \left(-\frac{4}{2}\right)$

29. $-3 + (-1)$

30. $5 + (-6) + (-3)$

31. $7 + (-4) + 8 + (-7)$

32. $-6 + 9 + 9 + (-5)$

[0.3] In Exercises 33 to 42, subtract.

33. $15 - 20$

34. $-10 - 5$

35. $2 - (-3)$

36. $-7 - (-3)$

37. $\frac{23}{4} - \left(-\frac{3}{4}\right)$

38. $-3 - 2$

39. $8 - 12 - (-5)$

40. $-6 - 7 - (-18)$

41. $7 - (-4) - 7 - 4$

42. $-9 - (-6) - 8 - (-11)$

[0.4] In Exercises 43 to 50, multiply.

43. $(-12)(-3)$

44. $(-10)(8)$

45. $(-5)(3)$

46. $\left(-\frac{3}{8}\right)\left(-\frac{4}{5}\right)$

47. $(-4)^2$

48. $(-2)(7)(-3)$

49. $(-6)(-5)(4)(-3)$

50. $(-9)(2)(-3)(1)$

51. 4
52. −121
53. −24
54. 36
55. −4
56. 11
57. Undefined
58. 25
59. $\dfrac{7}{6}$
60. 0
61. −2
62. 12
63. 3 · 3 · 3
64. 5 · 5 · 5 · 5
65. 2 · 2 · 2 · 2 · 2 · 2
66. 4 · 4 · 4 · 4 · 4
67. 3
68. 75
69. 80
70. 400
71. 41
72. 25
73. 20
74. 16
75. 324
76. 27
77. 11
78. 169

[0.4] In Exercises 51 to 54 use the distributive property to remove parentheses and simplify.

51. $4(-7 + 8)$ **52.** $11(-15 + 4)$

53. $-8(5 - 2)$ **54.** $-4(-3 - 6)$

[0.4] In Exercises 55 to 62, divide.

55. $48 \div (-12)$ **56.** $\dfrac{-33}{-3}$

57. $-9 \div 0$ **58.** $-75 \div (-3)$

59. $-\dfrac{7}{9} \div \left(-\dfrac{2}{3}\right)$ **60.** $0 \div (-12)$

61. $-8 \div 4$ **62.** $(-12) \div (-1)$

[0.5] Write each of these in expanded form.

63. 3^3 **64.** 5^4

65. 2^6 **66.** 4^5

[0.5] Evaluate each of the following expressions.

67. $18 - 3 \cdot 5$ **68.** $(18 - 3) \cdot 5$ **69.** $5 \cdot 4^2$

70. $(5 \cdot 4)^2$ **71.** $5 \cdot 3^2 - 4$ **72.** $5(3^2 - 4)$

73. $5(4 - 2)^2$ **74.** $5 \cdot 4 - 2^2$ **75.** $(5 \cdot 4 - 2)^2$

76. $3(5 - 2)^2$ **77.** $3 \cdot 5 - 2^2$ **78.** $(3 \cdot 5 - 2)^2$

CHAPTER 1
From Arithmetic to Algebra

LIST OF SECTIONS

1.1 Transition to Algebra

1.2 Evaluating Algebraic Expressions

1.3 Adding and Subtracting Algebraic Expressions

1.4 Sets

Cultures from all over the world have developed number systems and ways to record patterns in their natural surroundings. The Mayans in Central America had one of the most sophisticated number systems in the world in twelfth century AD. The Chinese numbering and recording system dates from around 1200 BCE. The oldest evidence of numerical record is in Africa, where a bone notched in numerical patterns and dating from about 35,000 BCE was found in southern Africa.

The roots of algebra developed among the Babylonians 4000 years ago in an area now part of the country of Iraq. The Babylonians developed ways to record useful numerical relationships so that they were easy to remember, easy to record, and helpful in solving problems. Archaeologists have found instructions for solving problems in engineering, economics, city planning, and agriculture. The writing was on clay tablets. Some of these formulas developed by the Babylonians are still in use today.

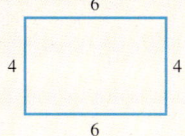 

SECTION 1.1 Transition to Algebra

1.1 OBJECTIVES

1. Introduce the concept of variables
2. Identify algebraic expressions
3. Translate from English to Algebra

In arithmetic, you learned how to do calculations with numbers by using the basic operations of addition, subtraction, multiplication, and division.

In algebra, you will still use numbers and the same four operations. However, you will also use letters to represent numbers. Letters such as *x, y, L,* or *W* are called **variables** when they represent numerical values.

Here we see two rectangles whose lengths and widths are labeled with numbers.

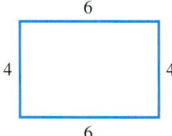 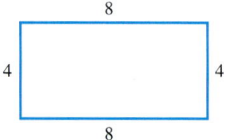

If we need to represent the length and width of *any* rectangle, we can use the variables *L* and *W*.

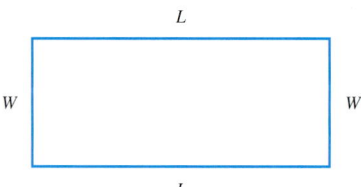

You are familiar with the four symbols ($+$, $-$, $\times$, $\div$) used to indicate the fundamental operations of arithmetic.

Let's look at how these operations are indicated in algebra. We begin by looking at addition.

In arithmetic:
$+$ denotes addition
$-$ denotes subtraction
$\times$ denotes multiplication
$\div$ denotes division.

Addition

$x + y$ means the *sum* of x and y, or x *plus* y.

Section 1.1 ▪ Transition to Algebra 57

Example 1

Writing Expressions That Indicate Addition

Some other words that tell you to add are "more than" and "increased by."

(a) The *sum* of *a* and 3 is written as $a + 3$.

(b) *L plus W* is written as $L + W$.

(c) 5 *more than m* is written as $m + 5$.

(d) *x increased by* 7 is written as $x + 7$.

✓ **CHECK YOURSELF 1**

Write, using symbols.

(a) The sum of *y* and 4 [handwritten: y+4]
(b) *a* plus *b* [handwritten: a+b]
(c) 3 more than *x* [handwritten: 3*x]
(d) *n* increased by 6 [handwritten: n+6]

Let's look at how subtraction is indicated in algebra.

> **Subtraction**
>
> $x - y$ means the *difference* of *x* and *y*, or *x minus y*.

Example 2

Writing Expressions That Indicate Subtraction

Some other words that mean to subtract are "decreased by" and "less than."

(a) *r minus s* is written as $r - s$.

(b) The *difference* of *m* and 5 is written as $m - 5$.

(c) *x decreased by* 8 is written as $x - 8$.

(d) 4 *less than a* is written as $a - 4$.

CHECK YOURSELF 2

Write, using symbols.

(a) *w* minus *z* (b) The difference of *a* and 7

(c) *y* decreased by 3 (d) 5 less than *b*

You have seen that the operations of addition and subtraction are written exactly the same way in algebra as in arithmetic. This is not true in multiplication because the symbol × looks like the letter x. So in algebra we use other symbols to show multiplication to avoid any confusion. Here are some ways to write multiplication.

Multiplication

A raised dot	$x \cdot y$
Parentheses	$(x)(y)$
Writing the letters next to each other	xy

These all indicate the *product* of *x* and *y*, or *x* times *y*.

Note: *x* and *y* are called the **factors** of the product *xy*.

Example 3 — Writing Expressions That Indicate Multiplication

(a) The product of 5 and *a* is written as $5 \cdot a$, $(5)(a)$, or $5a$. The last expression, $5a$, is the shortest and the most common way of writing the product.

(b) 3 times 7 can be written as $3 \cdot 7$ or $(3)(7)$.

(c) Twice *z* is written as $2z$.

(d) The product of 2, *s*, and *t* is written as $2st$.

(e) 4 more than the product of 6 and *x* is written as $6x + 4$.

Note: You can place letters next to each other or numbers and letters next to each other to show multiplication. But you *cannot* place numbers side by side to show multiplication: 37 means the number "thirty-seven," not 3 times 7.

CHECK YOURSELF 3

Write, using symbols.

(a) *m* times *n* *m·n mn (m)(n)*   (b) The product of *h* and *b* *h·b hb*

(c) The product of 8 and 9 *8·9*   (d) The product of 5, *w*, and *y* *5wy*

(e) 3 more than the product of 8 and *a*
 3 + 8a

Before we move on to division, let's look at how we can combine the symbols we have learned so far.

> An **expression** is a meaningful collection of numbers, variables, and symbols of operation.

Not every collection of symbols is an expression.

Example 4

Identifying Expressions

(a) $2m + 3$ is an expression. It means that we multiply 2 and m, then add 3.

(b) $x + \cdot + 3$ is not an expression. The three operations in a row have no meaning.

(c) $y = 2x - 1$ is not an expression. The equals sign is not an operation sign.

(d) $3a + 5b - 4c$ is an expression.

✓ CHECK YOURSELF 4

Identify which are expressions and which are not.

(a) $7 - \cdot x$ (b) $6 + y = 9$

(c) $a + b - c$ (d) $3x - 5yz$

To write more complicated products in algebra, we need some "punctuation marks." Parentheses ( ) mean that an expression is to be thought of as a single quantity. Brackets [ ] are used in exactly the same way as parentheses in algebra. Look at the following example showing the use of these signs of grouping.

Example 5

Expressions with More Than One Operation

(a) 3 times the sum of a and b is written as

$$3(a + b)$$

the sum of a and b is a single quantity, so it is enclosed in parentheses.

This can be read as "3 times the quantity a plus b."

(b) The sum of 3 times a and b is written as $3a + b$.

(c) 2 times the difference of m and n is written as $2(m - n)$.

(d) The product of s plus t and s minus t is written as $(s + t)(s - t)$.

(e) The product of b and 3 less than b is written as $b(b - 3)$.

No parentheses are needed here since the 3 multiplies only the a.

✓ CHECK YOURSELF 5

Write, using symbols.

(a) Twice the sum of p and q

(b) The sum of twice p and q

(c) The product of a and the quantity $b - c$

(d) The product of x plus 2 and x minus 2

(e) The product of x and 4 more than x

In algebra the fraction form is usually used.

Now let's look at the operation of division. In arithmetic, you see the division sign ÷, the long division symbol ⟌, and the fraction notation. For example, to indicate the quotient when 9 is divided by 3, you could write

$$9 \div 3 \quad \text{or} \quad 3\overline{)9} \quad \text{or} \quad \frac{9}{3}$$

> **Division**
>
> $\dfrac{x}{y}$ means x *divided* by y or the *quotient* of x and y.

Example 6 Writing Expressions That Indicate Division

Write, using symbols.

(a) m divided by 3 is written as $\dfrac{m}{3}$.

(b) The quotient of a plus b divided by 5 is written as $\dfrac{a + b}{5}$.

(c) The sum p plus q divided by the difference p minus q is written as $\dfrac{p + q}{p - q}$.

✓ CHECK YOURSELF 6

Write, using symbols.

(a) r divided by s

(b) The quotient when x minus y is divided by 7

(c) The difference a minus 2 divided by the sum a plus 2

Notice that we can use many different letters to represent variables. In Example 6 the letters *m, a, b, p,* and q represented different variables. We often choose a letter that reminds us of what it represents, for example, *L* for *length* or *W* for *width*.

Example 7

Writing Geometric Expressions

Write each geometric expression, using symbols.

(a) *Length* times *width* is written $L \cdot W$.

(b) One-half of *altitude* times *base* is written $\frac{1}{2} a \cdot b$.

(c) *Length* times *width* times *height* is written $L \cdot W \cdot H$.

(d) Pi (π) times *diameter* is written πd.

✓ CHECK YOURSELF 7

Write each geometric expression, using symbols.

(a) Two times *length* plus two times *width*   (b) Two times pi (π) times *radius*

✓ CHECK YOURSELF ANSWERS

1. (a) $y + 4$; (b) $a + b$; (c) $x + 3$; (d) $n + 6$.   **2.** (a) $w - z$; (b) $a - 7$; (c) $y - 3$; (d) $b - 5$.   **3.** (a) mn; (b) hb; (c) $8 \cdot 9$ or $(8)(9)$; (d) $5wy$; (e) $8a + 3$.
4. (a) Not an expression; (b) not an expression; (c) an expression; (d) an expression.
5. (a) $2(p + q)$; (b) $2p + q$; (c) $a(b - c)$; (d) $(x + 2)(x - 2)$; (e) $x(x + 4)$.
6. (a) $\frac{r}{s}$; (b) $\frac{x - y}{7}$; (c) $\frac{a - 2}{a + 2}$.   **7.** (a) $2L + 2W$; (b) $2\pi r$.

Exercises • 1.1

1. $c + d$
2. $a + 7$
3. $w + z$
4. $m + n$
5. $x + 2$
6. $b + 3$
7. $y + 10$
8. $m + 4$
9. $a - b$
10. $s - 5$
11. $b - 7$
12. $r - 3$
13. $r - 6$
14. $x - 3$
15. wz
16. $3c$
17. $5t$
18. $8a$
19. $8mn$
20. $7rs$
21. $3(p + q)$
22. $5(a + b)$
23. $2(x + y)$
24. $3(m + n)$
25. $2x + y$
26. $3m + n$
27. $2(x - y)$
28. $3(c - d)$
29. $(a + b)(a - b)$
30. $(x + y)(x - y)$
31. $m(m - 3)$
32. $a(a + 7)$
33. $\frac{x}{5}$
34. $\frac{b}{8}$
35. $\frac{a + b}{7}$
36. $\frac{x - y}{9}$
37. $\frac{p - q}{4}$
38. $\frac{a + 5}{9}$
39. $\frac{a + 3}{a - 3}$
40. $\frac{m - n}{m + n}$

Write each of the following phrases, using symbols.

1. The sum of c and d
2. a plus 7
3. w plus z
4. the sum of m and n
5. x increased by 2
6. 3 more than b
7. 10 more than y
8. m increased by 4
9. a minus b
10. 5 less than s
11. b decreased by 7
12. r minus 3
13. 6 less than r
14. x decreased by 3
15. w times z
16. The product of 3 and c
17. The product of 5 and t
18. 8 times a
19. The product of 8, m, and n
20. The product of 7, r, and s
21. The product of 3 and the quantity p plus q
22. The product of 5 and the sum of a and b
23. Twice the sum of x and y
24. 3 times the sum of m and n
25. The sum of twice x and y
26. The sum of 3 times m and n
27. Twice the difference of x and y
28. 3 times the difference of c and d
29. The quantity a plus b times the quantity a minus b
30. The product of x plus y and x minus y
31. The product of m and 3 less than m
32. The product of a and 7 more than a
33. x divided by 5
34. The quotient when b is divided by 8
35. The quotient of a plus b, divided by 7
36. The difference x minus y, divided by 9
37. The difference of p and q, divided by 4
38. The sum of a and 5, divided by 9
39. The sum of a and 3, divided by the difference of a and 3
40. The difference of m and n, divided by the sum of m and n

Write each of the following phrases, using symbols. Use the variable x to represent the number in each case.

41. 5 more than a number
42. A number increased by 8
43. 7 less than a number
44. A number decreased by 10
45. 9 times a number
46. Twice a number
47. 6 more than 3 times a number
48. 5 times a number, decreased by 10
49. Twice the sum of a number and 5
50. 3 times the difference of a number and 4
51. The product of 2 more than a number and 2 less than that same number
52. The product of 5 less than a number and 5 more than that same number
53. The quotient of a number and 7
54. A number divided by 3
55. The sum of a number and 5, divided by 8
56. The quotient when 7 less than a number is divided by 3
57. 6 more than a number divided by 6 less than that same number
58. The quotient when 3 less than a number is divided by 3 more than that same number

Write each of the following geometric expressions using symbols.

59. Four times the length of a side (s).
60. $\frac{4}{3}$ times π times the cube of the radius (r)
61. The radius (r) squared times the height (h) times π
62. Twice the length (L) plus twice the width (W)
63. One-half the product of the height (h) and the sum of two unequal sides (b_1 and b_2)
64. Six times the length of a side (s) squared

---

41. $x + 5$
42. $x + 8$
43. $x - 7$
44. $x - 10$
45. $9x$
46. $2x$
47. $3x + 6$
48. $5x - 10$
49. $2(x + 5)$
50. $3(x - 4)$
51. $(x + 2)(x - 2)$
52. $(x - 5)(x + 5)$
53. $\frac{x}{7}$
54. $\frac{x}{3}$
55. $\frac{x + 5}{8}$
56. $\frac{x - 7}{3}$
57. $\frac{x + 6}{x - 6}$
58. $\frac{x - 3}{x + 3}$
59. $4s$
60. $\frac{4}{3}\pi r^3$
61. $\pi r^2 h$
62. $2L + 2W$
63. $\frac{1}{2}h(b_1 + b_2)$
64. $6s^2$

65. Expression
66. Expression
67. Not an expression
68. Not an expression
69. Not an expression
70. Expression
71. Expression
72. Not an expression
73. $2x$
74. $S - 4000$
75. $I = Prt$
76. $KE = \frac{1}{2}mv^2$

Identify which are expressions and which are not.

65. $2(x + 5)$ **66.** $4 + (x - 3)$

67. $4 + \div m$ **68.** $6 + a = 7$

69. $2b = 6$ **70.** $x(y + 3)$

71. $2a + 5b$ **72.** $4x + \cdot 7$

73. Population growth. The Earth's population has doubled in the last 40 years. If we let x represent the Earth's population 40 years ago, what is the population today?

74. Species extinction. It is estimated that the Earth is losing 4000 species of plants and animals every year. If S represents the number of species living last year, how many species are on Earth this year?

75. Interest. The simple interest (I) earned when a principal (P) is invested at a rate (r) for a time (t) is calculated by multiplying the principal times the rate times the time. Write a formula for the interest earned.

76. Kinetic energy. The kinetic energy of a particle of mass m is found by taking one-half of the product of the mass and the square of the velocity (v). Write a formula for the kinetic energy of a particle.

77. Rewrite the following algebraic expressions in English phrases. Exchange papers with another student to edit your writing. Be sure the meaning in English is the same as in algebra. These expressions are not complete sentences, so your English does not have to be in complete sentences. Here is an example.

Algebra: $2(x - 1)$

English: We could write "One less than a number is doubled." Or we might write "A number is diminished by one and then multiplied by two."

(a) $n + 3$ (b) $\dfrac{x + 2}{5}$ (c) $3(5 + a)$

(d) $3 - 4n$ (e) $\dfrac{x + 6}{x - 1}$

SECTION 1.2 Evaluating Algebraic Expressions

1.2 OBJECTIVES

1. Evaluate algebraic expressions given any signed-number value for the variable
2. Use a graphing calculator to evaluate algebraic expressions

"Black holes are where God divided by zero."

–Steven Wright

In applying algebra to problem solving, you will often want to find the value of an algebraic expression when you know certain values for the letters (or variables) in the expression. As we pointed out earlier, finding the value of an expression is called *evaluating the expression* and uses the following steps.

> **To Evaluate an Algebraic Expression**
>
> Step 1 Replace each variable by the given number value.
>
> Step 2 Do the necessary arithmetic operations, following the rules for order of operations.

Example 1   Evaluating Algebraic Expressions

Suppose that $a = 5$ and $b = 7$.

(a) To evaluate $a + b$, we replace a with 5 and b with 7.

$$a + b = 5 + 7 = 12$$

(b) To evaluate $3ab$, we again replace a with 5 and b with 7.

$$3ab = 3 \cdot 5 \cdot 7 = 105$$

✓ **CHECK YOURSELF 1**

If $x = 6$ and $y = 7$, evaluate.

(a) $y - x$ (b) $5xy$

We are now ready to evaluate algebraic expressions that require following the rules for the order of operations.

Example 2 — Evaluating Algebraic Expressions

Evaluate the following expressions if $a = 2$, $b = 3$, $c = 4$, and $d = 5$.

(a) $5a + 7b = 5 \cdot 2 + 7 \cdot 3$ Multiply first.
$ = 10 + 21 = 31$ Then add.

(b) $3c^2 = 3 \cdot 4^2$ Evaluate the power.
$ = 3 \cdot 16 = 48$ Then multiply.

Caution

This is different from
$(3c)^2 = (3 \cdot 4)^2$
$ = 12^2 = 144$

(c) $7(c + d) = 7(4 + 5)$ Add inside the parentheses.
$ = 7 \cdot 9 = 63$

(d) $5a^4 - 2d^2 = 5 \cdot 2^4 - 2 \cdot 5^2$ Evaluate the powers.
$ = 5 \cdot 16 - 2 \cdot 25$ Multiply.
$ = 80 - 50 = 30$ Subtract.

✓ **CHECK YOURSELF 2**

If $x = 3$, $y = 2$, $z = 4$, and $w = 5$, evaluate the following expressions.

(a) $4x^2 + 2$ (b) $5(z + w)$ (c) $7(z^2 - y^2)$

To evaluate algebraic expressions when a fraction bar is used, do the following: Start by doing all the work in the numerator, then do the work in the denominator. Divide the numerator by the denominator as the last step.

Example 3 — Evaluating Algebraic Expressions

If $p = 2$, $q = 3$, and $r = 4$, evaluate:

(a) $\dfrac{8p}{r}$

Replace p with 2 and r with 4.

$$\dfrac{8p}{r} = \dfrac{8 \cdot 2}{4} = \dfrac{16}{4} = 4 \qquad \text{Divide as the last step.}$$

(b) $\dfrac{7q + r}{p + q} = \dfrac{7 \cdot 3 + 4}{2 + 3} \qquad \text{Evaluate the top and bottom separately.}$

$$= \dfrac{21 + 4}{2 + 3} = \dfrac{25}{5} = 5$$

✓ CHECK YOURSELF 3

Evaluate the following if $c = 5$, $d = 8$, and $e = 3$.

(a) $\dfrac{6c}{e}$ (b) $\dfrac{4d + e}{c}$ (c) $\dfrac{10d - e}{d + e}$

Example 4 shows how a calculator can be used to evaluate algebraic expressions.

Example 4 — Using a Calculator to Evaluate Expressions

Use a calculator to evaluate the following expressions.

(a) $\dfrac{4x + y}{z}$ if $x = 2$, $y = 1$, and $z = 3$

Replace x with 2, y with 1, and z with 3:

$$\dfrac{4x + y}{z} = \dfrac{4 \cdot 2 + 1}{3}$$

Remember to use parentheses to group the numerator and denominator.

Now, use the following keystrokes:

$\boxed{(}\ 4\ \boxed{\times}\ 2\ \boxed{+}\ 1\ \boxed{)}\ \boxed{\div}\ 3\ \boxed{\text{ENTER}}$

The display will read 3.

(b) $\dfrac{7x - y}{3z - x}$ if $x = 2$, $y = 6$, and $z = 2$

$$\dfrac{7x - y}{3z - x} = \dfrac{7 \cdot 2 - 6}{3 \cdot 2 - 2}$$

Use the following keystrokes:

(7 × 2 − 6) ÷ (3 × 2 − 2) ENTER

The display will read 2.

✓ CHECK YOURSELF 4

Use a calculator to evaluate the following if $x = 2$, $y = 6$, and $z = 5$.

(a) $\dfrac{2x + y}{z}$ (b) $\dfrac{4y - 2z}{x}$

Example 5 Evaluating Expressions

Evaluate $5a + 4b$ if $a = -2$ and $b = 3$.

Replace a with -2 and b with 3.

$$5a + 4b = 5(-2) + 4(3)$$
$$= -10 + 12$$
$$= 2$$

Remember the rules for the order of operations. Multiply first, then add.

✓ CHECK YOURSELF 5

Evaluate $3x + 5y$ if $x = -2$ and $y = 5$.

We follow the same rules no matter how many variables are in the expression.

Section 1.2 ■ Evaluating Algebraic Expressions 69

Example 6 Evaluating Expressions

Evaluate the following expressions if $a = -4$, $b = 2$, $c = -5$, and $d = 6$.

(a) $7a - 4c = 7(-4) - 4(-5)$ — This becomes $-(-20)$, or $+20$.
$= -28 + 20$
$= -8$

Caution

When a squared variable is replaced by a negative number, square the negative.

$(-5)^2 = (-5)(-5) = 25$

The exponent applies to -5!

$-5^2 = -(5 \cdot 5) = -25$

The exponent applies only to 5!

(b) $7c^2 = 7(-5)^2 = 7 \cdot 25$ — Evaluate the power first, then multiply by 7.
$= 175$

(c) $b^2 - 4ac = 2^2 - 4(-4)(-5)$
$= 4 - 4(-4)(-5)$
$= 4 - 80$
$= -76$

(d) $b(a + d) = 2(-4 + 6)$ — Add inside the parentheses first.
$= 2(2)$
$= 4$

✓ CHECK YOURSELF 6

Evaluate if $p = -4$, $q = 3$, and $r = -2$.

(a) $5p - 3r$ (b) $2p^2 + q$ (c) $p(q + r)$
(d) $-q^2$ (e) $(-q)^2$

If an expression involves a fraction, remember that the fraction bar is a grouping symbol. This means that you should do the required operations first in the numerator and then the denominator. Divide as the last step.

Example 7

Evaluating Expressions

Evaluate the following expressions if $x = 4$, $y = -5$, $z = 2$, and $w = -3$.

(a) $\dfrac{z - 2y}{x} = \dfrac{2 - 2(-5)}{4} = \dfrac{2 + 10}{4}$

$= \dfrac{12}{4} = 3$

(b) $\dfrac{3x - w}{2x + w} = \dfrac{3(4) - (-3)}{2(4) + (-3)} = \dfrac{12 + 3}{8 + (-3)}$

$= \dfrac{15}{5} = 3$

✓ CHECK YOURSELF 7

Evaluate if $m = -6$, $n = 4$, and $p = -3$.

(a) $\dfrac{m + 3n}{p}$ (b) $\dfrac{4m + n}{m + 4n}$

✓ CHECK YOURSELF ANSWERS

1. (a) 1; (b) 210.   **2.** (a) 38; (b) 45; (c) 84.   **3.** (a) 10; (b) 7; (c) 7.
4. (a) 2; (b) 7.   **5.** -31.   **6.** (a) -14; (b) 35; (c) -4; (d) -9; (e) 9.
7. (a) -2; (b) -2.

Exercises · 1.2

1. −22	2. −26
3. 32	4. −6
5. −20	6. 30
7. 12	8. 96
9. 4	10. −4
11. 83	12. 68
13. 6	14. 45
15. −40	16. −108
17. 14	18. 48
19. −9	20. −2
21. 2	22. −7
23. 2	24. −2
25. 11	26. 12
27. 1	28. 4
29. 11	30. 12
31. 91	32. −72
33. 1	34. −216
35. 91	36. −72
37. 29	38. 32
39. 9	40. 64
41. 16	42. 81
43. 2	44. −12
45. −7	46. −144
47. −72	48. −20

Evaluate each of the expressions if $a = -2$, $b = 5$, $c = -4$, and $d = 6$.

1. $3c - 2b$
2. $4c - 2b$
3. $8b + 2c$
4. $7a - 2c$
5. $-b^2 + b$
6. $(-b)^2 + b$
7. $3a^2$
8. $6c^2$
9. $c^2 - 2d$
10. $3a^2 + 4c$
11. $2a^2 + 3b^2$
12. $4b^2 - 2c^2$
13. $2(a + b)$
14. $5(b - c)$
15. $4(2a - d)$
16. $6(3c - d)$
17. $a(b + 3c)$
18. $c(3a - d)$
19. $\dfrac{6d}{c}$
20. $\dfrac{8b}{5c}$
21. $\dfrac{3d + 2c}{b}$
22. $\dfrac{2b + 3d}{2a}$
23. $\dfrac{2b - 3a}{c + 2d}$
24. $\dfrac{3d - 2b}{5a + d}$
25. $d^2 - b^2$
26. $c^2 - a^2$
27. $(d - b)^2$
28. $(c - a)^2$
29. $(d - b)(d + b)$
30. $(c - a)(c + a)$
31. $d^3 - b^3$
32. $c^3 + a^3$
33. $(d - b)^3$
34. $(c + a)^3$
35. $(d - b)(d^2 + db + b^2)$
36. $(c + a)(c^2 - ac + a^2)$
37. $b^2 + a^2$
38. $d^2 - a^2$
39. $(b + a)^2$
40. $(d - a)^2$
41. $a^2 + 2ad + d^2$
42. $b^2 - 2bc + c^2$

Evaluate each of the expression if $x = -2$, $y = -3$, and $z = 4$.

43. $x^2 - 2y^2 + z^2$
44. $4yz + 6xy$
45. $3xy - (y^2 - 2xz)$
46. $3yz - 6xyz + x^2y^2$
47. $2y(z^2 - 2xy) + yz^2$
48. $-z - (-2x - yz)$

72 Chapter 1 ■ From Arithmetic to Algebra

Answers		
49. −15.3	50. −11.4	
51. −11.5	52. 15.3	
53. 1.1	54. −0.8	
55. 14.0	56. −5.6	
57. −90	58. −6732	
59. −77,922		
60. −439,920		
61. True	62. False	
63. False	64. True	
65. 3.75 Ω		
66. 16 cm²		
67. 30 in.		
68. $1440		
69. $1875		
70. 5%		
71. 14°F		
72. 28.26 m²		

Use a calculator to evaluate each expression if $x = -2.34$, $y = -3.14$, and $z = 4.12$. Round your answer to the nearest tenth.

49. $x + yz$ **50.** $y - 2z$

51. $x^2 - z^2$ **52.** $x^2 + y^2$

53. $\dfrac{xy}{z - x}$ **54.** $\dfrac{y^2}{zy}$

55. $\dfrac{2x + y}{2x + z}$ **56.** $\dfrac{x^2 y^2}{xz}$

Use a calculator to evaluate the expression $x^2 - 4x^3 + 3x$ for each given value.

57. $x = 3$ **58.** $x = 12$ **59.** $x = 27$ **60.** $x = 48$

In each of the following problems, decide if the given numbers make the statement true or false.

61. $x - 7 = 2y + 5$; $x = 22$, $y = 5$ **62.** $3(x - y) = 6$; $x = 5$, $y = -3$

63. $2(x + y) = 2x + y$; $x = -4$, $y = -2$ **64.** $x^2 - y^2 = x - y$; $x = 4$, $y = -3$

65. Electrical resistance. The formula for the total resistance in a parallel circuit is given by the formula $R_T = \dfrac{R_1 R_2}{(R_1 + R_2)}$. find the total resistance if $R_1 = 6$ ohms (Ω) and $R_2 = 10$ Ω.

66. Area. The formula for the area of a triangle is given by $A = \dfrac{1}{2}ab$. Find the area of a triangle if $a = 4$ centimeters (cm) and $b = 8$ cm.

67. Perimeter. The perimeter of a rectangle of length L and width W is given by the formula $P = 2L + 2W$. Find the perimeter when $L = 10$ inches (in.) and $W = 5$ in.

68. Simple interest. The simple interest I on a principal of P dollars at interest rate r for time t, in years, is given by $I = Prt$. Find the simple interest on a principal of $6000 at 8% for 3 years. (**Note:** 8% = 0.08)

69. Simple interest. Use the formula $P = \dfrac{I}{r \cdot t}$ to find the principal if the total interest earned was $150 and the rate of interest was 4% for 2 years.

70. Simple interest. Use the formula $r = \dfrac{I}{P \cdot t}$ to find the rate of interest if $10,000 earns $1500 interest in 3 years.

71. Temperature conversion. The formula that relates Celsius and Fahrenheit temperature is $F = \dfrac{9}{5}C + 32$. If the temperature of the day is $-10°C$, what is the Fahrenheit temperature?

72. Geometry. If the area of a circle whose radius is r is given by $A = \pi r^2$, where $\pi = 3.14$, find the area when $r = 3$ meters (m).

Section 1.2 ■ Evaluating Algebraic Expressions

73. Write an English interpretation of each of the following algebraic expressions.

(a) $(2x^2 - y)^3$ (b) $3n = \dfrac{n-1}{2}$ (c) $(2n+3)(n-4)$

74. Is $a^n + b^n = (a+b)^n$? Try a few numbers and decide if you think this is true for all numbers, for some numbers, or never true. Write an explanation of your findings and give examples.

75. (a) 0, 80, 96, 72, 32, 0

(b) 2

(c) 94.5, 95.744, 96.492, 96.768, 96.596, 96, 95.004, 93.632, 91.908, 89.856, 87.5

(d) 1.8

75. (a) Evaluate the expression $4x(5-x)(6-x)$ for $x = 0, 1, 2, 3, 4,$ and 5. Complete the table below.

Value of x	0	1	2	3	4	5
Value of Expression						

(b) For which value of x does the expression value appear to be largest?
(c) Evaluate the expression for $x = 1.5, 1.6, 1.7, 1.8, 1.9, 2.0, 2.1, 2.2, 2.3, 2.4,$ and 2.5. Complete the table.

Value of x	1.5	1.6	1.7	1.8	1.9	2.0	2.1	2.2	2.3	2.4	2.5
Value of Expression											

(d) For which value of x does the expression value appear to be largest?

(e) Continue the search for the value of x that produces the greatest expression value. Determine this value of x to the nearest hundredth.

76. Work with other students on this exercise.

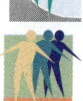

Part 1: Evaluate the three expressions $\dfrac{n^2-1}{2}, n, \dfrac{n^2+1}{2}$ using odd values of n: 1, 3, 5, 7, etc. Make a chart like the one below and complete it.

n	$a = \dfrac{n^2-1}{2}$	$b = n$	$c = \dfrac{n^2+1}{2}$	a^2	b^2	c^2
1						
3						
5						
7						
9						
11						
13						
15						

Part 2: The numbers a, b, and c that you get in each row have a surprising relationship to each other. Complete the last three columns and work together to discover this relationship. You may want to find out more about the history of this famous number pattern.

77. In problem 76 you investigated the numbers obtained by evaluating the following expressions for odd positive integer values of n: $\dfrac{n^2 - 1}{2}$, n, $\dfrac{n^2 + 1}{2}$. Work with other students to investigate what three numbers you get when you evaluate for a *negative* odd value? Does the pattern you observed before still hold? Try several negative numbers to test the pattern.

Have no fear of fractions—does the pattern work with fractions? Try even integers. Is there a pattern for the three numbers obtained when you begin with even integers?

78. Enjoyment of patterns in art, music, and language are common to all cultures, and many cultures also delight in and draw spiritual significance from patterns in numbers. One such set of patterns is that of the "magic" square. One of these squares appears in a famous etching by Albrecht Dürer, who lived from 1471 to 1528 in Europe. He was one of the first artists in Europe to use geometry to give perspective, a feeling of three dimensions, in his work. The magic square in his work is this one:

16	3	2	13
5	10	11	8
9	6	7	12
4	15	14	1

Why is this square "magic?" It is magic because every row, every column, and both diagonals add to the same number. In this square there are 16 spaces for the numbers 1 through 16.

Part 1: What number does each row and column add to?

Write the square that you obtain by adding -17 to each number. Is this still a magic square? If so, what number does each column and row add to? If you add 5 to each number in the original magic square, do you still have a magic square? You have been studying the operations of addition, multiplication, subtraction, and division with integers and with rational numbers. What operations can you perform on this magic square and still have a magic square? Try to find something that will not work. Use algebra to help you decide what will work and what won't. Write a description of your work and explain your conclusions.

Part 2: Here is the oldest published magic square. It is from China, about 250 B.C. Legend has it that it was brought from the River Lo by a turtle to the Emperor Yii, who was a hydraulic engineer.

4	9	2
3	5	7
8	1	6

Check to make sure that this is a magic square. Work together to decide what operation might be done to every number in the magic square to make the sum of each row, column, and diagonal the *opposite* of what it is now. What would you do to every number to cause the sum of each row, column, and diagonal to equal zero?

SECTION 1.3 Adding and Subtracting Algebraic Expressions

1.3 OBJECTIVES

1. Combine like terms
2. Add algebraic expressions
3. Subtract algebraic expressions

To find the perimeter of (or the distance around) a rectangle, we add 2 times the length and 2 times the width. In the language of algebra, this can be written as

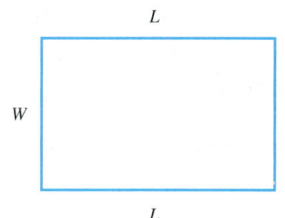

Perimeter = $2L + 2W$

We call $2L + 2W$ an **algebraic expression,** or more simply an **expression.** As we discussed in Section 1.1, an expression allows us to write a mathematical idea in symbols. It can be thought of as a meaningful collection of letters, numbers, and operation symbols.

Some expressions are

1. $5x^2$
2. $3a + 2b$
3. $4x^3 - 2y + 1$
4. $3(x^2 + y^2)$

In algebraic expressions, the addition and subtraction signs break the expressions into smaller parts called *terms.*

If a variable has no exponent, it is raised to the power one.

> A **term** is a number, or the product of a number and one or more variables, raised to a power.

In an expression, each sign (+ or −) is a part of the term that follows the sign.

Example 1 — Identifying Terms

(a) $5x^2$ has one term.

(b) $\underset{\underset{\text{Term}}{\uparrow}}{3a} + \underset{\underset{\text{Term}}{\uparrow}}{2b}$ has two terms: $3a$ and $2b$.

(c) $\underset{\underset{\text{Term}}{\uparrow}}{4x^3} - \underset{\underset{\text{Term}}{\uparrow}}{2y} + \underset{\underset{\text{Term}}{\uparrow}}{1}$ has three terms: $4x^3$, $-2y$, and 1.

Note that each term "owns" the sign that precedes it.

✓ CHECK YOURSELF 1

List the terms of each expression.

(a) $2b^4$ (b) $5m + 3n$ (c) $2s^2 - 3t - 6$

Note that a term in an expression may have any number of factors. For instance, $5xy$ is a term. It has factors of 5, x, and y. The number factor of a term is called the **numerical coefficient.** So for the term $5xy$, the numerical coefficient is 5.

Example 2 — Identifying the Numerical Coefficient

(a) $4a$ has the numerical coefficient 4.

(b) $6a^3b^4c^2$ has the numerical coefficient 6.

(c) $-7m^2n^3$ has the numerical coefficient -7.

(d) Since $1 \cdot x = x$, the numerical coefficient of x is understood to be 1.

✓ CHECK YOURSELF 2

Give the numerical coefficient for each of the following terms.

(a) $8a^2b$ (b) $-5m^3n^4$ (c) y

If terms contain exactly the *same letters* (or variables) raised to the *same powers*, they are called **like terms.**

Section 1.3 ■ Adding and Subtracting Algebraic Expressions

Example 3 Identifying Like Terms

(a) The following are like terms.

$6a$ and $7a$
$5b^2$ and b^2
$10x^2y^3z$ and $-6x^2y^3z$

Each pair of terms has the same letters, with matching letters raised to the same power—the numerical coefficients can be any number.

(b) The following are *not* like terms.

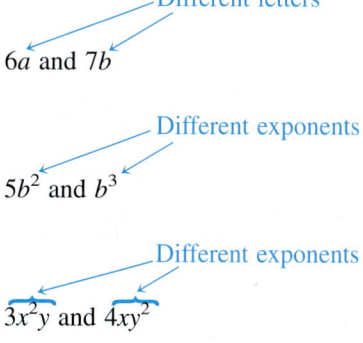

$6a$ and $7b$ — Different letters

$5b^2$ and b^3 — Different exponents

$3x^2y$ and $4xy^2$ — Different exponents

✓ CHECK YOURSELF 3

Circle the like terms.

$$5a^2b \qquad ab^2 \qquad a^2b \qquad -3a^2 \qquad 4ab \qquad 3b^2 \qquad -7a^2b$$

Like terms of an expression can always be combined into a single term. Look at the following:

$$\underbrace{2x}_{x+x} + \underbrace{5x}_{x+x+x+x+x} = \underbrace{7x}_{x+x+x+x+x+x+x}$$

Rather than having to write out all those x's, try

Here we use the distributive property.

$$2x + 5x = (2 + 5)x = 7x$$

In the same way,

You don't have to write all this out—just do it mentally!

$$9b + 6b = (9 + 6)b = 15b$$

and

$$10a - 4a = (10 - 4)a = 6a$$

77

This leads us to the following rule.

> **Combining Like Terms**
>
> To combine like terms, use the following steps.
>
> Step 1 Add or subtract the numerical coefficients.
>
> Step 2 Attach the common variables.

Example 4

Combining Like Terms

Combine like terms.

(a) $8m + 5m = (8 + 5)m = 13m$

(b) $5pq^3 - 4pq^3 = 1pq^3 = pq^3$

Remember that when any factor is multiplied by 0, the product is 0.

(c) $7a^3b^2 - 7a^3b^2 = 0a^3b^2 = 0$

✓ **CHECK YOURSELF 4**

Combine like terms.

(a) $6b + 8b$ (b) $12x^2 - 3x^2$

(c) $8xy^3 - 7xy^3$ (d) $9a^2b^4 - 9a^2b^4$

Let's look at some expressions involving more than two terms. The idea is the same.

Example 5

Combining Like Terms

Combine like terms.

(a) $5ab - 2ab + 3ab$

$= (5 - 2 + 3)ab = 6ab$

The distributive property can be used over any number of like terms.

Section 1.3 ■ Adding and Subtracting Algebraic Expressions

Only like terms can be combined.

(b) $8x - 2x + 5y$

$= 6x + 5y$

With practice you won't be writing out these steps, but doing them mentally.

(c) $5m + 8n + 4m - 3n$ — Like terms, Like terms

$= (5m + 4m) + (8n - 3n)$ Here we have used the associative and commutative properties.

$= 9m + 5n$

Be careful when moving terms. Remember that they own the signs in front of them.

(d) $4x^2 + 2x - 3x^2 + x$

$= (4x^2 - 3x^2) + (2x + x)$

$= x^2 + 3x$

As these examples illustrate, combining like terms often means changing the grouping and the order in which the terms are written. Again all this is possible because of the properties of addition that we introduced in Section 0.3.

✓ CHECK YOURSELF 5

Combine like terms.

(a) $4m^2 - 3m^2 + 8m^2$ (b) $9ab + 3a - 5ab$ (c) $4p + 7q + 5p - 3q$

Addition in always a matter of combining like quantities (two apples plus three apples, four books plus five books, and so on). If you keep that basic idea in mind, adding expressions will be easy. It is just a matter of combining like terms. Suppose that you want to add

$$5x^2 + 3x + 4 \quad \text{and} \quad 4x^2 + 5x - 6$$

Parentheses are sometimes used in adding, so for the sum of these expressions, we can write

$$(5x^2 + 3x + 4) + (4x^2 + 5x - 6)$$

Now what about the parentheses? You can use the following rule.

> **Remove Grouping Symbols Case 1**
>
> If a plus sign (+) or nothing at all appears in front of parentheses, just remove the parentheses. No other changes are necessary.

Now let's return to the addition.

Just remove the parentheses. No other changes are necessary.

$$(5x^2 + 3x + 4) + (4x^2 + 5x - 6)$$
$$= 5x^2 + 3x + 4 + 4x^2 + 5x - 6$$

Like terms — Like terms — Like terms — Like terms

Note the use of the associative and commutative properties in reordering and regrouping.

Collect like terms. (*Remember:* Like terms have the same variables raised to the same power.)

$$= (5x^2 + 4x^2) + (3x + 5x) + (4 - 6)$$

Here we use the distributive property. For example, $5x^2 + 4x^2 = 9x^2$

Combine like terms for the result:

$$= 9x^2 + 8x - 2$$

As should be clear, much of this work can be done mentally. You can then write the sum directly by locating like terms and combining. Example 6 illustrates this property.

Example 6

Combining Like Terms

Add $3x - 5$ and $2x + 3$.

Write the sum.

$$(3x - 5) + (2x + 3)$$
$$= 3x - 5 + 2x + 3 = 5x - 2$$

Like terms   Like terms

✓ **CHECK YOURSELF 6**

Add $6x^2 + 2x$ and $4x^2 - 7x$.

Section 1.3 ■ Adding and Subtracting Algebraic Expressions

Subtracting expressions requires another rule for removing signs of grouping.

> **Removing Grouping Symbols Case 2**
>
> If a minus sign (−) appears in front of a set of parentheses, the parentheses can be removed by changing the sign of each term inside the parentheses.

The use of this rule is illustrated in Example 7.

Example 7 — Removing Parentheses

In each of the following, remove the parentheses.

Note: This uses the distributive property, since

$-(2x + 3y) = (-1)(2x + 3y)$
$\qquad\quad\; = -2x - 3y$

(a) $-(2x + 3y) = -2x - 3y$ Change each sign when removing the parentheses.

(b) $m - (5n - 3p) = m \underbrace{- 5n + 3p}_{\text{Sign changes.}}$

(c) $2x - (-3y + z) = \underbrace{2x + 3y - z}_{\text{Sign changes.}}$

✓ CHECK YOURSELF 7

Remove the parentheses.

(a) $-(3m + 5n)$ (b) $-(5w - 7z)$ (c) $3r - (2s - 5t)$ (d) $5a - (-3b - 2c)$

Subtracting expressions is now a matter of using the previous rule when removing the parentheses and then combining the like terms. Consider Example 8.

Example 8 — Subtracting Expressions

(a) Subtract $5x - 3$ from $8x + 2$.

Write

Note: the expression following "from" is written first in the problem.

$$(8x + 2) - (5x - 3)$$
$$= 8x + 2 \underbrace{- 5x + 3}_{\text{Sign changes.}}$$
$$= 3x + 5$$

(b) Subtract $4x^2 - 8x + 3$ from $8x^2 + 5x - 3$.

Write

$$(8x^2 + 5x - 3) - (4x^2 - 8x + 3)$$
$$= 8x^2 + 5x - 3 \underbrace{- 4x^2 + 8x - 3}_{\text{Sign changes.}}$$
$$= 4x^2 + 13x - 6$$

✓ CHECK YOURSELF 8

(a) Subtract $7x + 3$ from $10x - 7$.

(b) Subtract $5x^2 - 3x + 2$ from $8x^2 - 3x - 6$.

✓ CHECK YOURSELF ANSWERS

1. (a) $2b^4$; (b) $5m$, $3n$; (c) $2s^2$, $-3t$, -6. **2.** (a) 8; (b) -5; (c) 1. **3.** The like terms are $5a^2b$, a^2b, and $-7a^2b$. **4.** (a) $14b$; (b) $9x^2$; (c) xy^3; (d) 0.
5. (a) $9m^2$; (b) $4ab + 3a$; (c) $9p + 4q$.
6. $10x^2 - 5x$. **7.** (a) $-3m - 5n$; (b) $-5w + 7z$; (c) $3r - 2s + 5t$; (d) $5a + 3b + 2c$.
8. (a) $3x - 10$; (b) $3x^2 - 8$.

Exercises ■ 1.3

Answers (left column):

1. 5a, 2   2. 7a, −4b
3. $4x^3$ 4. $3x^2$
5. $3x^2$, 3x, −7
6. $2a^3$, $-a^2$, a
7. 5ab, 4ab   8. $9m^2$, $5m^2$
9. $2x^2y$, $-3x^2y$, $6x^2y$
10. $8a^2b$, $-5a^2b$, $5a^2b$
11. 10m   12. $14a^2$
13. $17b^3$ 14. 20rs
15. 28xyz   16. $19mn^2$
17. $6z^2$ 18. m
19. 0   20. −4xy
21. $-n^2$ 22. 0
23. $-15p^2q$
24. $-9r^3s^2$
25. $6x^2$ 26. 6uv
27. 2a + 4b
28. $11m^2 - 3m$
29. −3x − y
30. $7a^2 - 11a$
31. 4a − 10b − 1
32. $-2p^2 + 3p - 2$
33. −2a − 3b
34. −7x + 4y
35. 5a − 2b + 3c
36. 7x − 4y − 3z
37. 6r − 5s   38. 7m + 2n
39. 8p − 2q   40. 7c + 10d
41. 9a + 4   42. 12x − 1

List the terms of the following expressions.

1. 5a + 2
2. 7a − 4b
3. $4x^3$
4. $3x^2$
5. $3x^2 + 3x - 7$
6. $2a^3 - a^2 + a$

Circle the like terms in the following groups of terms.

7. 5ab, 3b, 3a, 4ab
8. $9m^2$, 8mn, $5m^2$, 7m
9. $4xy^2$, $2x^2y$, $5x^2$, $-3x^2y$, 5y, $6x^2y$
10. $8a^2b$, $4a^2$, $3ab^2$, $-5a^2b$, 3ab, $5a^2b$

Combine the like terms.

11. 3m + 7m
12. $6a^2 + 8a^2$
13. $7b^3 + 10b^3$
14. 7rs + 13rs
15. 21xyz + 7xyz
16. $4mn^2 + 15mn^2$
17. $9z^2 - 3z^2$
18. 7m − 6m
19. $5a^3 - 5a^3$
20. 9xy − 13xy
21. $18n^2 - 19n^2$
22. 7cd − 7cd
23. $6p^2q - 21p^2q$
24. $8r^3s^2 - 17r^3s^2$
25. $10x^2 - 7x^2 + 3x^2$
26. 13uv + 5uv − 12uv
27. 9a − 7a + 4b
28. $5m^2 - 3m + 6m^2$
29. 4x + 4y − 7x − 5y
30. $6a^2 - 11a - 8a^2 + 9a^2$
31. 2a − 7b − 3 + 2a − 3b + 2
32. $5p^2 - 2p - 8 - 7p^2 + 5p + 6$

Remove the parentheses in each of the following expressions, and simplify where possible.

33. −(2a + 3b)
34. −(7x − 4y)
35. 5a − (2b − 3c)
36. 7x − (4y + 3z)
37. 9r − (3r + 5s)
38. 10m − (3m − 2n)
39. 5p − (−3p + 2q)
40. 8d − (−7c − 2d)

Add.

41. 6a − 5 and 3a + 9
42. 9x + 3 and 3x − 4

83

43. $13b^2 - 18b$
44. $8m^2 - 5m$
45. $-2x^2$ 46. $-4p^2$
47. $5x^2 - 2x + 1$
48. $9d^2 - 14d - 2$
49. $2b^2 + 5b + 16$
50. $3x^2 - 5x - 3$
51. $8y^3 - 2y$
52. $9x^4 + 3$
53. $-a^3 + 4a^2$
54. $5m^3 - 8m$
55. $-2x^2 - x + 3$
56. $-2b^3 + 5b^2 - 3b$
57. $x - 7$ 58. $2x + 7$
59. $m^2 - 3m$
60. $2a^2 - 5a$
61. $-2y^2$ 62. $-2n^2$
63. $2x^2 - x + 1$
64. $2x^2 - 6x - 7$
65. $8a^2 - 12a - 7$
66. $x^3 - x^2 - 5x$
67. $-6b^2 + 8b$
68. $6y^2 - 9y$
69. $2x^2 + 12$
70. $-x^2 - 3x$
71. $6b - 1$ 72. $6m - 3$
73. $10x - 9$ 74. $-x^2 + 5$
75. $2x^2 + 5x - 12$
76. $6a + 2$
77. $-6y^2 - 8y$
78. $-2r^3 + 3r^2$
79. $5x^2 - 3x + 9$
80. $5x^2 - 2x + 4$
81. 206.8 82. 67.8

43. $8b^2 - 11b$ and $5b^2 - 7b$
45. $3x^2 - 2x$ and $-5x^2 + 2x$
47. $2x^2 + 5x - 3$ and $3x^2 - 7x + 4$
49. $2b^2 + 8$ and $5b + 8$
51. $8y^3 - 5y^2$ and $5y^2 - 2y$
53. $2a^2 - 4a^3$ and $3a^3 + 2a^2$
55. $4x^2 - 2 + 7x$ and $5 - 8x - 6x^2$

Subtract.

57. $x + 4$ from $2x - 3$
59. $3m^2 - 2m$ from $4m^2 - 5m$
61. $6y^2 + 5y$ from $4y^2 + 5y$
63. $x^2 - 4x - 3$ from $3x^2 - 5x - 2$
65. $3a + 7$ from $8a^2 - 9a$
67. $4b^2 - 3b$ from $5b - 2b^2$
69. $x^2 - 5 - 8x$ from $3x^2 - 8x + 7$

Perform the indicated operations.

71. Subtract $3b + 2$ from the sum of $4b - 2$ and $5b + 3$.
72. Subtract $5m - 7$ from the sum of $2m - 8$ and $9m - 2$.
73. Subtract $3x^2 + 2x - 1$ from the sum of $x^2 + 5x - 2$ and $2x^2 + 7x - 8$.
74. Subtract $4x^2 - 5x - 3$ from the sum of $x^2 - 3x - 7$ and $2x^2 - 2x + 9$.
75. Subtract $2x^2 - 3x$ from the sum of $4x^2 - 5$ and $2x - 7$.
76. Subtract $5a^2 - 3a$ from the sum of $3a - 3$ and $5a^2 + 5$.
77. Subtract the sum of $3y^2 - 3y$ and $5y^2 + 3y$ from $2y^2 - 8y$.
78. Subtract the sum of $7r^3 - 4r^2$ and $-3r^3 + 4r^2$ from $2r^3 + 3r^2$.

Perform the indicated operations.

79. $[(9x^2 - 3x + 5) - (3x^2 + 2x - 1)] - (x^2 - 2x - 3)$
80. $[(5x^2 + 2x - 3) - (-2x^2 + x - 2)] - (2x^2 + 3x - 5)$

Using your calculator, evaluate each of the following for the given values of the variable. Round your answer to the nearest tenth.

81. $7x^2 - 5y^3$ for $x = 7.1695$ and $y = 3.128$
82. $2x^2 + 3y + 5x$ for $x = 3.61$ and $y = 7.91$

44. $2m^2 + 3m$ and $6m^2 - 8m$
46. $3p^2 + 5p$ and $-7p^2 - 5p$
48. $4d^2 - 8d + 7$ and $5d^2 - 6d - 9$
50. $4x - 3$ and $3x^2 - 9x$
52. $9x^4 - 2x^2$ and $2x^2 + 3$
54. $9m^3 - 2m$ and $-6m - 4m^3$
56. $5b^3 - 8b + 2b^2$ and $3b^2 - 7b^3 + 5b$

58. $x - 2$ from $3x + 5$
60. $9a^2 - 5a$ from $11a^2 - 10a$
62. $9n^2 - 4n$ from $7n^2 - 4n$
64. $3x^2 - 2x + 4$ from $5x^2 - 8x - 3$
66. $3x^3 + x^2$ from $4x^3 - 5x$
68. $7y - 3y^2$ from $3y^2 - 2y$
70. $4x - 2x^2 + 4x^3$ from $4x^3 + x - 3x^2$

83. 6.5
84. 212.7
85. $28x + 4$
86. $12x + 4$
87. $-x^2 + 65x - 150$
88. $2y^2 - 3y + 8$
94. 1, 3, 5, 7

83. $4x^2y + 2xy^2 - 5x^3y$ for $x = 1.29$ and $y = 2.56$

84. $3x^3y - 4xy + 2x^2y^2$ for $x = 3.26$ and $y = 1.68$

85. **Geometry.** A rectangle has sides of $8x + 9$ and $6x - 7$. Find the expression that represents its perimeter.

86. **Geometry.** A triangle has sides $3x + 7$, $4x - 9$, and $5x + 6$. Find the expression that represents its perimeter.

87. **Business.** The cost of producing x units of an item is $C = 150 + 25x$. The revenue for selling x units is $R = 90x - x^2$. The profit is given by the revenue minus the cost. Find the expression that represents profit.

88. **Business.** The revenue for selling y units is $R = 3y^2 - 2y + 5$ and the cost of producing y units is $C = y^2 + y - 3$. Find the expression that represents profit.

89. Write a paragraph explaining the difference between n^2 and $2n$.

90. Complete the explanation: "x^3 and $3x$ are not the same because . . . "

91. Complete the statement: "$x + 2$ and $2x$ are different because . . . "

92. Write an English phrase for each algebraic expression:

    (a) $2x^3 + 5x$ (b) $(2x + 5)^3$ (c) $6(n + 4)^2$

93. Work with another student to complete this exercise. Place $>$, $<$, or $=$ in the blank in these statements.

    1^2 _____ 2^1 What happens as the table of numbers is extended? Try more examples.

    2^3 _____ 3^2

    What sign seems to occur the most in your table? $>$, $<$, or $=$?

    3^4 _____ 4^3

    4^5 _____ 5^4 Write an algebraic statement for the pattern of numbers in this table. Do you think this is a pattern that continues? Add more lines to the table and extend the pattern to the general case by writing the pattern in algebraic notation. Write a short paragraph stating your conjecture.

94. Compute and fill in the following blanks.

    Case 1: $1^2 - 0^2 =$ _____

    Case 2: $2^2 - 1^2 =$ _____

    Case 3: $3^2 - 2^2 =$ _____

    Case 4: $4^2 - 3^2 =$ _____

    Based on the pattern you see in these four cases, predict the value of case 5: $5^2 - 4^2$. Compute case 5 to check your prediction. Write an expression for case n. Describe in words the pattern that you see in this exercise.

SECTION 1.4 Sets

1.4 OBJECTIVES

1. Write a set using the roster method
2. Write a set using set-builder notation
3. Plot the elements of a set on a number line

"Whether you think that you can or you can't, you are usually right."

–Henry Ford

For his birthday, Jacob received a jacket, a play ticket, some candy, and a pen. We could call this collection of gifts "Jacob's presents." Such a collection is called a **set**. The things in the set are called **elements** of the set. We can write the set as {jacket, ticket, candy, pen}. The braces tell us where the set begins and ends. Every person could have a set that describes the presents they received on their last birthday.

What if I received no presents on my last birthday? What would my set look like? It would be the set { }, which we call the **empty set**. Sometimes the symbol ∅ is used to indicate the empty set.

Many sets can be written in *roster form*, as was the case with Jacob's presents. The set of prime numbers less than 15 can be written in roster form as {2, 3, 5, 7, 11, 13}. In the first example, you will be asked to list sets in roster form. **Roster form** is a list enclosed in braces.

Example 1

Listing the Elements of a Set

Use the roster form to list the elements of each set described.

(a) The set of all factors of 12.

The factors of 12 are {1, 2, 3, 4, 6, 12}

(b) The set of all integers with an absolute value less than 4.

The integers are {−3, −2, −1, 0, 1, 2, 3}

✓ CHECK YOURSELF 1

Use the roster form to list the elements of each set described.

(a) The set of all factors of 18.
(b) The set of all even prime numbers

Each set that we have examined has had a limited number of elements. If we need to indicate that a set continues in some pattern, we use three dots, called an ellipsis, to indicate that the set continues with the pattern it started.

Example 2

Listing the Elements of a Set

Use the roster form to list the elements of each set described.

(a) The set of all natural numbers less than 100.

The set $\{1, 2, 3, \ldots, 98, 99\}$ indicates that we continue increasing the numbers by one until we get to 99.

(b) The set of all positive multiples of 4.

The set $\{4, 8, 12, 16, \ldots\}$ indicates that we continue counting by fours forever. (There is no indicated stopping point.)

(c) The set of all integers

$\{\ldots, -2, -1, 0, 1, 2, \ldots\}$ indicates that we continue forever in both directions.

✓ CHECK YOURSELF 2

Use the roster form to list the elements of each set described.

(a) The set of all natural numbers between 200 and 300.

(b) The set of all positive multiplies of 3.

(c) The set of all even numbers.

Not all sets can be described using the roster form. What if we want to describe all of the real numbers between 1 and 2? We could not list that set of numbers. Yet another way that we can describe the elements of a set is with **set-builder notation.** To describe the aforementioned set using this notation, we write

$\{x \mid 1 < x < 2\}$. We read this as, "the set of all x, where x is between 1 and 2." Example 3 further illustrates this idea.

Example 3

Using Set-Builder Notation

Use set-builder notation for each set described.

(a) The set of all real numbers less than 100.

We write $\{x \mid x < 100\}$

A statement such as
$1 < x < 2$
is called a "compound inequality." It says that x is greater than 1 and also that x is less than 2.

(b) The set of all real numbers greater than -4 but less than or equal to 9.

$$\{x \mid -4 < x \leq 9\}$$

✓ CHECK YOURSELF 3

Use set-builder notation for each set described.

(a) The set of all real numbers greater than -2.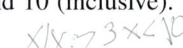

(b) The set of all real numbers between 3 and 10 (inclusive).

Sets of numbers can also be represented graphically. In our final example in this section, we will look at the connection between sets and their graphs.

Example 4 — Plotting the Elements of a Set on a Number Line

Plot the points of each set on the number line.

(a) $\{-2, 1, 5\}$.

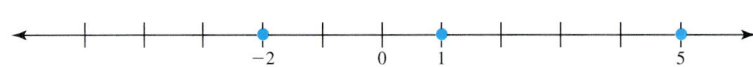

(b) $\{x \mid x < 3\}$

Note that the dark line and dark arrow indicate that we continue forever in the negative direction. The parenthesis at 3 indicates that the 3 is not part of the graph.

(c) $\{x \mid -2 < x < 5\}$

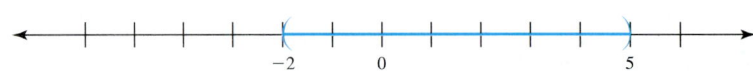

The parentheses indicate that the numbers -2 and 5 are not part of the set.

(d) $\{x \mid x \geq 2\}$

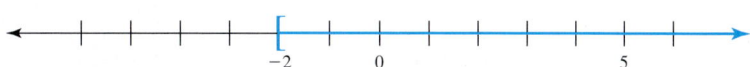

The bracket indicates that 2 is part of the set that is graphed.

✓ CHECK YOURSELF 4

Plot the points of each set on the number line.

(a) $\{-5, -3, 0\}$.

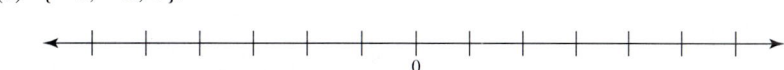

(b) $\{x \mid -3 < x < -1\}$

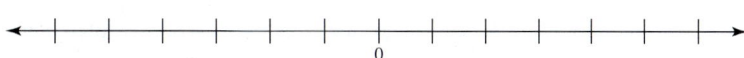

(c) $\{x \mid x \leq 5\}$

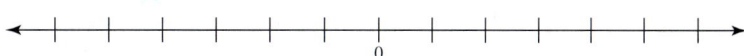

✓ CHECK YOURSELF ANSWERS

1. (a) $\{1, 2, 3, 6, 9, 18\}$; (b) $\{2\}$.
2. (a) $\{201, 202, 203, \ldots, 298, 299\}$; (b) $\{3, 6, 9, 12, \ldots\}$
(c) $\{\ldots, -6, -4, -2, 0, 2, 4, 6, \ldots\}$. 3. (a) $\{x \mid x > -2\}$; (b) $\{x \mid 3 \leq x \leq 10\}$.
4. (a) [number line with points at −5, −3, 0]
(b) [number line with open interval from −3 to −1]
(c) [number line with closed ray ending at 5]

Exercises ▪ 1.4

1. {Monday, Tuesday, Wednesday, Thursday, Friday, Saturday, Sunday}
2. {January, March, May, July, August, October, December}
3. {1, 2, 4, 8, 16}
4. {2, 3, 4, 6, 8, 12, 24}
5. {2, 3, 5, 7, 11, 13, 17, 19, 23, 29}
6. {23, 29, 31, 37}
7. {−5, −4, −3, −2, −1}
8. {1, 2, 3, 4, 5, 6, 7}
9. {0, 2, 4, 6, 8, 10, 12}
10. {1, 3, 5, 7, 9, 11, 13}
11. {3, 4, 5, 6}
12. {6, 7, 8, 9}
13. {−3, −2}
14. {−7, −6, −5, −4}
15. {−5, −4, −3, −2, −1, 0, 1, 2}
16. {−1, 0, 1, 2, 3, 4}
17. {1, 3, 5, 7, . . . }
18. {0, 2, 4, 6, 8, . . . }
19. {0, 2, 4, 6, . . . , 96, 98}
20. {1, 3, 5, . . . , 97, 99}
21. {5, 10, 15, 20, . . . }
22. {6, 12, 18, 24, . . . }

Use the roster method to list the elements of each set.

1. The set of all the days of the week.
2. The set of all months of the year that have 31 days.
3. The set of all factors of 16.
4. The set of all factors of 24.
5. The set of all prime numbers less than 30.
6. The set of all prime numbers between 20 and 40.
7. The set of all negative integers greater than −6.
8. The set of all positive integers less than 8.
9. The set of all even whole numbers less than 13.
10. The set of all odd whole numbers less than 14.
11. The set of integers greater than 2 and less than 7.
12. The set of integers greater than 5 and less than 10.
13. The set of integers greater than −4 and less than −1.
14. The set of integers greater than −8 and less than −3.
15. The set of integers between −5 and 2, inclusive.
16. The set of integers between −1 and 4 inclusive.
17. The set of odd whole numbers.
18. The set of even whole numbers.
19. The set of all even whole numbers less than 100.
20. The set of all odd whole numbers less than 100.
21. The set of all positive multiples of 5.
22. The set of all positive multiples of 6.

23. $\{x \mid x > 8\}$
24. $\{x \mid x < 25\}$
25. $\{x \mid x \geq -5\}$
26. $\{x \mid x \leq -3\}$
27. $\{x \mid 2 < x < 7\}$
28. $\{x \mid -3 < x < -1\}$
29. $\{x \mid -4 \leq x \leq +4\}$
30. $\{x \mid -8 \leq x \leq 3\}$

Use set-builder notation for each set described.

23. The set of all real numbers greater than 8.

24. The set of all real numbers less than 25.

25. The set of all real numbers greater than or equal to -5.

26. The set of all real numbers less than or equal to -3.

27. The set of all real numbers greater than 2 and less than 7.

28. The set of all real numbers greater than -3 and less than -1.

29. The set of all real numbers between -4 and 4, inclusive.

30. The set of all real numbers between -8 and 3, inclusive.

Plot the elements of each set on a number line.

31. $\{-2, -1, 0, 4\}$

32. $\{-5, -1, 2, 3, 5\}$

33. $\{x \mid x > 4\}$

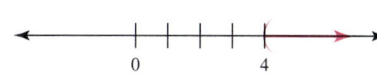

34. $\{x \mid x > -1\}$

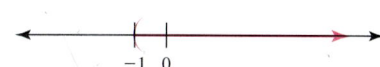

35. $\{x \mid x \geq -3\}$

36. $\{x \mid x \leq 6\}$

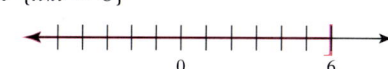

37. $\{x \mid 2 < x < 7\}$

38. $\{x \mid 4 < x < 8\}$

39. $\{x \mid -3 < x \leq 5\}$

40. $\{x \mid -6 < x \leq 1\}$

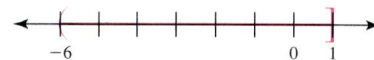

41. $\{x \mid -4 \leq x < 0\}$

42. $\{x \mid -5 \leq x < 2\}$

47. $\{x \mid x \leq 1\}$

48. $\{x \mid x \leq 3\}$

49. $\{x \mid x > -2\}$

50. $\{x \mid x > 2\}$

51. $\{x \mid -2 \leq x \leq 2\}$

52. $\{x \mid -5 \leq x \leq 1\}$

53. $\{x \mid -3 < x < 2\}$

54. $\{x \mid -1 < x < 4\}$

55. $\{x \mid -2 \leq x < 4\}$

56. $\{x \mid -1 \leq x < 4\}$

43. $\{x \mid -7 \leq x \leq -3\}$
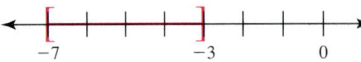

44. $\{x \mid -1 \leq x \leq 4\}$

45. The set of all integers between -7 and -3, inclusive.
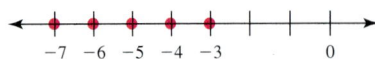

46. The set of all integers between -1 and 4, inclusive.
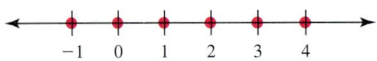

Use set-builder notation to describe each graphed set.

47. 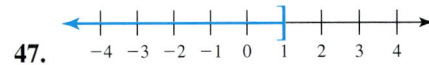   48.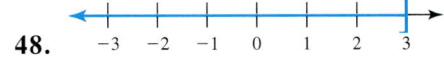

49. 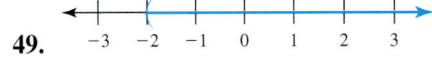   50.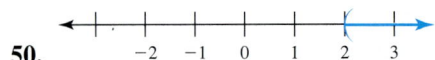

51. 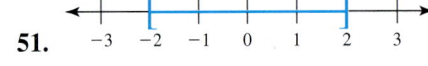   52.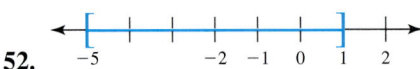

53. 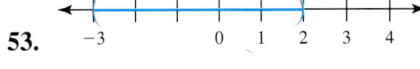   54.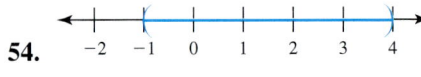

55. 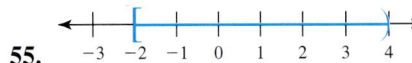   56.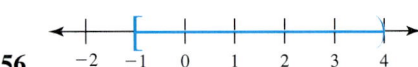

Summary Exercises ▪ 1

This exercise set is provided to give you practice with each of the objectives of the chapter. Each exercise is keyed to the appropriate chapter section. Your instructor will give you guidelines on how to best use these exercises.

Answers:
1. $y + 5$
2. $c - 10$
3. $8a$
4. $\dfrac{y}{3}$
5. $5mn$
6. $a(a - 5)$
7. $17x + 3$
8. $\dfrac{a + 2}{a - 2}$
9. $(x + 6)(x - 6)$
10. $\dfrac{x}{9}$
11. $\dfrac{x - 4}{x + 2}$
12. $x(x - 3)$
13. 3
14. 75
15. 80
16. 400
17. 41
18. 25
19. 20
20. 16
21. 324
22. 27
23. 11
24. 169
25. -7
26. 46
27. 15
28. 80
29. 19
30. -81
31. 25
32. -6
33. 1
34. -3
35. -3
36. 6
37. -48
38. -2313

[1.1] Write, using symbols.

1. 5 more than y
2. c decreased by 10
3. The product of 8 and a
4. The quotient when y is divided by 3
5. 5 times the product of m and n
6. The product of a and 5 less than a
7. 3 more than the product of 17 and x
8. The quotient when a plus 2 is divided by a minus 2
9. The product of 6 more than a number and 6 less than the same number
10. The quotient of a number and 9
11. The quotient when 4 less than a number is divided by 2 more than the same number
12. The product of a number and 3 less than the same number

[1.2] Evaluate each of the following expressions.

13. $18 - 3 \cdot 5$
14. $(18 - 3) \cdot 5$
15. $5 \cdot 4^2$
16. $(5 \cdot 4)^2$
17. $5 \cdot 3^2 - 4$
18. $5(3^2 - 4)$
19. $5(4 - 2)^2$
20. $5 \cdot 4 - 2^2$
21. $(5 \cdot 4 - 2)^2$
22. $3(5 - 2)^2$
23. $3 \cdot 5 - 2^2$
24. $(3 \cdot 5 - 2)^2$

[1.2] Evaluate the expressions if $x = -3$, $y = 6$, $z = -4$, and $w = 2$.

25. $3x + w$
26. $5y - 4z$
27. $x + y - 3z$
28. $5z^2$
29. $3x^2 - 2w^2$
30. $3x^3$
31. $5(x^2 - w^2)$
32. $\dfrac{6z}{2w}$
33. $\dfrac{2x - 4z}{y - z}$
34. $\dfrac{3x - y}{w - x}$
35. $\dfrac{x(y^2 - z^2)}{(y + z)(y - z)}$
36. $\dfrac{y(x - w)^2}{x^2 - 2xw + w^2}$
37. $w^2y^2 - z^3x$
38. $-x^2 - 12xz^3$

93

Answers

39. $4a^3, -3a^2$
40. $5x^2, -7x, 3$
41. $5m^2, -4m^2, m^2$
42. $4ab^2, ab^2, -3ab^2$
43. $12c$
44. $7x$
45. $2a$
46. $3c$
47. $3xy$
48. $7ab^2$
49. $19a + b$
50. $x + 5y$
51. $3x^3 + 9x^2$
52. $a^3 + 2a^2 + 3a$
53. $10a^3$
54. $7x^2$
55. $23 - x$
56. $25 - x$
57. $x + 5$
58. $2x + 5$
59. $x + 4$
60. $6n - 7$
61. $x, 25 - x$
62. $0.10x + 0.25q$

[1.3] List the terms of the expressions.

39. $4a^3 - 3a^2$ **40.** $5x^2 - 7x + 3$

[1.3] Circle like terms.

41. $5m^2, -3m, -4m^2, 5m^3, m^2$

42. $4ab^2, 3b^2, -5a, ab^2, 7a^2, -3ab^2, 4a^2b$

[1.3] Combine like terms.

43. $5c + 7c$ **44.** $2x + 5x$

45. $4a - 2a$ **46.** $6c - 3c$

47. $9xy - 6xy$ **48.** $5ab^2 + 2ab^2$

49. $7a + 3b + 12a - 2b$ **50.** $6x - 2x + 5y - 3x$

51. $5x^3 + 17x^2 - 2x^3 - 8x^2$

52. $3a^3 + 5a^2 + 4a - 2a^3 - 3a^2 - a$

53. Subtract $4a^3$ from the sum of $2a^3$ and $12a^3$.

54. Subtract the sum of $3x^2$ and $5x^2$ from $15x^2$.

[1.1–1.3] Write an expression for each of the following problems.

55. Carpentry. If x feet (ft) are cut off the end of a board that is 23 ft long, how much is left?

56. Money. Sergei has 25 nickels and dimes in his pocket. If x of these are dimes, how many of the coins are nickels?

57. Age. Sam is 5 years older than Angela. If Angela is x years old now, how old is Sam?

58. Money. Margaret has $5 more than twice as much money as Pho. Write an expression for the amount of money that Margaret has.

59. Geometry. The length of a rectangle is 4 meters (m) more than the width. Write an expression for the length of the rectangle.

60. Number problem. A number is 7 less than 6 times the number n. Write an expression for the number.

61. Carpentry. A 25-ft plank is cut into two pieces. Write expressions for the length of each piece.

62. Money. Bernie has x dimes and q quarters in his pocket. Write an expression for the amount of money that Bernie has in his pocket.

- Summary Exercises   **95**

63. {1, 3}
64. {1, 2, 3, 4, 5, 6}
65. {−1, 0, 1, 2, 3}
66. {−4, −3, −2, −1, 0, 1, 2, 3}
67. {1, 3}
68. {−1, 0, 1, 2}
69. $\{x \mid x > 5\}$
70. $\{x \mid -2 < x < 4\}$
71. $\{x \mid x \leq -3\}$
72. $\{x \mid -4 \leq x \leq 3\}$
77. $\{x \mid x \leq 3\}$
78. $\{x \mid x \geq -4\}$
79. $\{x \mid -3 \leq x < 2\}$
80. $\{x \mid -4 < x \leq 2\}$

[1.4] Use the roster method to list the elements of each set.

63. The set of all factors of 3.
64. The set of all positive integers less than 7.
65. The set of integers greater than −2 and less than 4.
66. The set of integers between −4 and 3 inclusive.
67. The set of all odd whole numbers less than 4.
68. The set of all integers greater than −2 and less than 3.

[1.4] Use set-builder notation to represent each set described.

69. The set of all real numbers greater than 5.
70. The set of all real numbers greater than −2 and less than 4.
71. The set of all real numbers less than or equal to −3.
72. The set of all real numbers between −4 and 3 inclusive.

[1.4] Plot the elements of each set on the number line.

73. $\{x \mid x \geq 1\}$ **74.** $\{x \mid x \leq -2\}$

75. $\{x \mid -2 \leq x < 3\}$ **76.** $\{x \mid -7 < x \leq -1\}$

[1.4] Use set-builder notation to describe each set.

77.    **78.**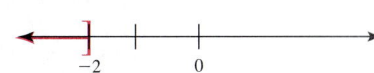

79. 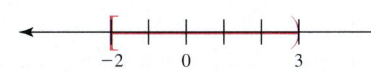   **80.**

CHAPTER 2
Equations and Inequalities

LIST OF SECTIONS

2.1 Solving Equations by Adding and Subtracting

2.2 Solving Equations by Multiplying and Dividing

2.3 Combining the Rules to Solve Equations

2.4 Literal Equations and Their Applications

2.5 Solving Linear Inequalities

2.6 Absolute Value Equations and Inequalities

Many engineers, economists, and environmental scientists are working on the problem of meeting the increasing energy demands of a growing global population. One promising solution to this problem is power generated by wind-driven turbines.

The cost of wind-generated power has fallen from $0.25 per kilowatt hour (kwh) in the early 1980s to about $0.043 per kwh in 2000; thus, using this form of power production is becoming economically feasible. And compared to the cost of pollution from burning coal and oil, wind-generated power may be less expensive.

An economist for a city might use this equation to try to compute the cost for electricity for his city:

$$C = P(0.043)(1000) \quad \text{where } C = \text{cost and } P = \text{power}$$

The equation is an ancient tool for solving problems and writing numerical relationships clearly and accurately. In this chapter you will learn methods to solve linear equations and practice writing equations that accurately describe problem situations.

SECTION 2.1 Solving Equations by Adding and Subtracting

2.1 OBJECTIVES

1. Determine whether a given number is a solution for an equation
2. Use the addition property to solve equations
3. Determine whether a given number is a solution for an application
4. Translate words to equation symbols
5. Solve application problems

An equation such as

$x + 3 = 5$

is called a **conditional equation** because it can be either true or false depending on the value given to the variable.

We could also use set-builder notation. We write $\{x | x = 2\}$, which is read, "Every x such that x equals two." We will use both notations throughout the text.

In this chapter you will begin working with one of the most important tools of mathematics, the equation. The ability to recognize and solve various types of equations is probably the most useful algebraic skill you will learn. We will continue to build upon the methods of this chapter throughout the remainder of the text. To start, let's describe what we mean by an *equation*.

> An **equation** is a mathematical statement that two expressions are equal.

Some examples are $3 + 4 = 7$, $x + 3 = 5$, $P = 2L + 2W$.

As you can see, an equals sign (=) separates the two equal expressions. These expressions are usually called the *left side* and the *right side* of the equation.

An equation may be either true or false. For instance, $3 + 4 = 7$ is true because both sides name the same number. What about an equation such as $x + 3 = 5$ that has a letter or variable on one side? Any number can replace x in the equation. However, only one number will make this equation a true statement.

If $x = \begin{matrix} 1 \\ 2 \\ 3 \end{matrix} \quad \begin{cases} 1 + 3 = 5 \text{ is false} \\ 2 + 3 = 5 \text{ is true} \\ 3 + 3 = 5 \text{ is false} \end{cases}$

The number 2 is called the *solution* (or *root*) of the equation $x + 3 = 5$ because substituting 2 for x gives a true statement. The set $\{2\}$ is called the *solution set*.

> A **solution** for an equation is any value for the variable that makes the equation a true statement.

99

> The **solution set** for an equation is the set of all values for the variables that make the equation a true statement.

Example 1 Verifying a Solution

(a) Is 3 a solution for the equation $2x + 4 = 10$?

To find out, replace x with 3 and evaluate $2x + 4$ on the left.

Note that, until the left side equals the right side, a question mark is placed over the equals sign.

Left side		Right side
$2 \cdot 3 + 4$	$\stackrel{?}{=}$	10
$6 + 4$	$\stackrel{?}{=}$	10
10	$=$	10

Since $10 = 10$ is a true statement, 3 is a solution of the equation. The solution set is $\{3\}$.

(b) Is 5 a solution of the equation $3x - 2 = 2x + 1$?

To find out, replace x with 5 and evaluate each side separately.

Remember the rules for the order of operation. Multiply first; then add or subtract.

Left side		Right side
$3 \cdot 5 - 2$	$\stackrel{?}{=}$	$2 \cdot 5 + 1$
$15 - 2$	$\stackrel{?}{=}$	$10 + 1$
13	$\neq$	11

Since the two sides do not name the same number, we do not have a true statement, and 5 is not a solution.

✓ CHECK YOURSELF 1

For the equation

$$2x - 1 = x + 5$$

(a) Is 4 a solution? (b) Is 6 a solution?

You may be wondering whether an equation can have more than one solution. It certainly can. For instance,

$$x^2 = 9$$

has two solutions. They are 3 and -3 because

$$3^2 = 9 \quad \text{and} \quad (-3)^2 = 9 \qquad \text{The solution set is } \{-3, 3\}.$$

*The equation $x^2 = 9$ is an example of a **quadratic equation.** We will consider methods of solution in Chapter 11.*

In this chapter, however, we will generally work with *linear equations*. These are equations that can be put into the form

$$ax + b = 0$$

where the variable is *x*, where *a* and *b* are any numbers, and *a* is not equal to 0. In a linear equation, the variable can appear only to the first power. No other power (x^2, x^3, etc.) can appear. Linear equations are also called **first-degree equations.** The degree of an equation in one variable is the highest degree to which the variable appears.

> Linear equations in one variable that can be written in the form
>
> $$ax + b = 0 \qquad a \neq 0$$
>
> will have exactly one solution.

Example 2

Identifying Expressions and Equations

Label each of the following as an expression, a linear equation, or an equation that is not linear.

(a) $4x + 5$ is an expression.

(b) $2x + 8 = 0$ is a linear equation.

(c) $3x^2 - 9 = 0$ is not a linear equation.

(d) $5x = 15$ is a linear equation.

✓ CHECK YOURSELF 2

Label each as an expression, a linear equation, or an equation that is not linear.

(a) $2x^2 = 8$ (b) $2x - 3 = 0$ (c) $5x - 10$ (d) $2x + 1 = 7$

One can find the solution for an equation such as $x + 3 = 8$ by guessing the answer to the question "What plus 3 is 8?" Here the answer to the question is 5, and that is also the solution for the equation. But for more complicated equations you are going to need something more than guesswork. A better method is to transform the given equation to an *equivalent equation* whose solution can be found by inspection. Let's make a definition.

> Equations that have the same solution are called **equivalent equations.**

The following are all equivalent equations:

$$2x + 3 = 5 \qquad 2x = 2 \qquad \text{and} \qquad x = 1$$

They all have the same solution, 1. We say that a linear equation is *solved* when it is transformed to an equivalent equation of the form

$$x = \boxed{}$$

The variable is alone on the left side. The right side is some number, the solution.

Note: In some cases we'll write the equation in the form

$$\boxed{} = x$$

The number will be our solution when the equation has the variable isolated on the left or on the right.

The addition property of equality is the first property you will need to transform an equation to an equivalent form.

> **The Addition Property of Equality**
>
> If $\qquad a = b$
>
> then $\qquad a + c = b + c$
>
> In words, adding the same quantity to both sides of an equation gives an equivalent equation.

Remember: An equation is a statement that the two sides are equal. Adding the same quantity to both sides does not change the equality or "balance."

Let's look at an example of applying this property to solve an equation.

Example 3 — Using the Addition Property to Solve an Equation

Solve

$$x - 3 = 9$$

Remember that our goal is to isolate x on one side of the equation. Since 3 is being subtracted from x, we can add 3 to remove it. We must use the addition property to add 3 to both sides of the equation.

Section 2.1 ■ Solving Equations by Adding and Subtracting

To check, replace x with 12 in the original equation:
$$x - 3 = 9$$
$$12 - 3 \stackrel{?}{=} 9$$
$$9 = 9 \quad \text{(True)}$$
Since we have a true statement, 12 is the solution.

$$\begin{array}{r} x - 3 = 9 \\ +3 +3 \\ \hline x = 12 \end{array}$$
{ Adding 3 "undoes" the subtraction and leaves x alone on the left.

Since 12 is the solution for the equivalent equation $x = 12$, it is the solution for our original equation. For either equation, the solution set is {12}.

✓ **CHECK YOURSELF 3**

Solve and check.

$$x - 5 = 4$$

The addition property also allows us to add a negative number to both sides of an equation. This is really the same as subtracting the same quantity from both sides.

Example 4 **Using the Addition Property to Solve an Equation**

Solve

$$x + 5 = 9$$

Recall our comment that we could write an equation in the equivalent forms $x = \Box$ or $\Box = x$, where $\Box$ represents some number. Suppose we have an equation like

$$12 = x + 7$$

Subtracting 7 will isolate x on the right:

$$\begin{array}{r} 12 = x + 7 \\ -7 -7 \\ \hline 5 = x \end{array}$$

and the solution is 5.

In this case, 5 is *added* to x on the left. We can use the addition property to subtract 5 from both sides. This will "undo" the addition and leave the variable x alone on one side of the equation.

$$\begin{array}{r} x + 5 = 9 \\ -5 -5 \\ \hline x = 4 \end{array}$$

The solution set is {4}. To check, replace x with 4:

$$4 + 5 = 9 \quad \text{(True)}$$

✓ **CHECK YOURSELF 4**

Solve and check.

$$x + 6 = 13$$

What if the equation has a variable term on both sides? You will have to use the addition property to add or subtract a term involving the variable to get the desired result.

Example 5 Using the Addition Property to Solve an Equation

Solve

$$5x = 4x + 7$$

We will start by subtracting $4x$ from both sides of the equation. Do you see why? Remember that an equation is solved when we have an equivalent equation of the form $x = \Box$.

$$\begin{array}{rl} 5x = & 4x + 7 \\ -4x & -4x \\ \hline x = & 7 \end{array} \quad \left\{ \begin{array}{l} \text{Subtracting } 4x \text{ from} \\ \text{both sides } removes \\ 4x \text{ from the right.} \end{array} \right.$$

To check: Since 7 is a solution for the equivalent equation $x = 7$, it should be a solution for the original equation. To find out, replace x with 7:

$$5 \cdot 7 \stackrel{?}{=} 4 \cdot 7 + 7$$

$$35 \stackrel{?}{=} 28 + 7$$

$$35 = 35 \quad \text{(True)}$$

 CHECK YOURSELF 5

Solve and check.

$$7x = 6x + 3$$

You may have to apply the addition property more than once to solve an equation. Look at Example 6.

Example 6 Using the Addition Property to Solve an Equation

Solve

$$7x - 8 = 6x$$

We want all variables on *one* side of the equation. If we choose the left, we subtract $6x$ from both sides of the equation. This will remove $6x$ from the right:

$$\begin{array}{rl} 7x - 8 = & 6x \\ -6x & -6x \\ \hline x - 8 = & 0 \end{array}$$

Section 2.1 ■ Solving Equations by Adding and Subtracting 105

We want the variable alone, so we add 8 to both sides. This isolates x on the left.

$$\begin{array}{rr} x - 8 = & 0 \\ +8 & +8 \\ \hline x \quad\; = & 8 \end{array}$$

The solution set is {8}. We'll leave it to you to check this result.

 CHECK YOURSELF 6

Solve and check.

$$9x + 3 = 8x$$

Often an equation will have more than one variable term *and* more than one number. You will have to apply the addition property twice in solving these equations.

Example 7 **Using the Addition Property to Solve an Equation**

Solve

$$5x - 7 = 4x + 3$$

We would like the variable terms on the left, so we start by subtracting $4x$ to remove that term from the right side of the equation:

$$\begin{array}{rr} 5x - 7 = & 4x + 3 \\ -4x & -4x \\ \hline x - 7 = & 3 \end{array}$$

Now, to isolate the variable, we add 7 to both sides to undo the subtraction on the left:

$$\begin{array}{rr} x - 7 = & 3 \\ +7 & +7 \\ \hline x \quad\; = & 10 \end{array}$$

You could just as easily have added 7 to both sides and *then* subtracted $4x$. The result would be the same. In fact, some students prefer to combine the two steps.

The solution set is {10}. To check, replace x with 10 in the original equation:

$$5 \cdot 10 - 7 \stackrel{?}{=} 4 \cdot 10 + 3$$
$$43 = 43 \quad \text{(True)}$$

✓ CHECK YOURSELF 7

Solve and check.

(a) $4x - 5 = 3x + 2$ (b) $6x + 2 = 5x - 4$

Remember, by simplify we mean to combine all like terms.

In solving an equation, you should always simplify each side as much as possible before using the addition property.

Example 8 Combining Like Terms and Solving the Equation

Solve

$$\underset{\text{Like terms}}{5 + 8x - 2} = \underset{\text{Like terms}}{2x - 3 + 5x}$$

Since like terms appear on each side of the equation, we start by combining the numbers on the left (5 and -2). Then we combine the like terms ($2x$ and $5x$) on the right. We have

$$3 + 8x = 7x - 3$$

Now we can apply the addition property, as before:

$$\begin{aligned} 3 + 8x &= 7x - 3 \\ -7x &= -7x & \text{Subtract } 7x. \\ \hline 3 + x &= -3 \\ -3 & -3 & \text{Subtract 3.} \\ \hline x &= -6 & \text{Isolate } x. \end{aligned}$$

The solution set is $\{-6\}$. To check, always return to the original equation. That will catch any possible errors in simplifying. Replacing x with -6 gives

$$5 + 8(-6) - 2 \stackrel{?}{=} 2(-6) - 3 + 5(-6)$$
$$5 - 48 - 2 \stackrel{?}{=} -12 - 3 - 30$$
$$-45 = -45 \quad \text{(True)}$$

✓ CHECK YOURSELF 8

Solve and check.

(a) $3 + 6x + 4 = 8x - 3 - 3x$ (b) $5x + 21 + 3x = 20 + 7x - 2$

Section 2.1 • Solving Equations by Adding and Subtracting   107

We may have to apply some of the properties discussed in Section 0.4 in solving equations. Example 9 illustrates the use of the distributive property to clear an equation of parentheses.

Example 9 — Using the Distributive Property and Solving Equations

Solve

$$2(3x + 4) = 5x - 6$$

Note: $2(3x + 4)$
$= 2(3x) + 2(4)$
$= 6x + 8$

Applying the distributive property on the left, we have

$$6x + 8 = 5x - 6$$

We can then proceed as before:

$$
\begin{aligned}
6x + 8 &= 5x - 6 \\
-5x &= -5x \quad &\text{Subtract } 5x. \\
x + 8 &= -6 \\
-8 & -8 \quad &\text{Subtract } 8. \\
x &= -14
\end{aligned}
$$

Remember that
$x = -14$ and $-14 = x$
are equivalent equations.

The solution set is $\{-14\}$. We will leave the checking of this result to the reader.
Remember: Always return to the original equation to check.

 CHECK YOURSELF 9

Solve and check each of the following equations.

(a) $4(5x - 2) = 19x + 4$ (b) $3(5x + 1) = 2(7x - 3) - 4$

The main reason for learning how to set up and solve algebraic equations is so that we can use them to solve word problems. In fact, algebraic equations were *invented* to make solving word problems much easier. The first word problems that we know about are over 4000 years old. They were literally "written in stone," on Babylonian tablets, about 500 years before the first algebraic equation made its appearance.

Before algebra, people solved word problems primarily by **substitution,** which is a method of finding unknown integers by using trial and error in a logical way. Example 10 shows how to solve a word problem using substitution.

Example 10 Solving a Word Problem by Substitution

The sum of two consecutive integers is 37. Find the two integers.

If the two integers were 20 and 21, their sum would be 41. Since that's more than 37, the integers must be smaller. If the integers were 15 and 16, the sum would be 31. More trials yield that the sum of 18 and 19 is 37.

✓ CHECK YOURSELF 10

The sum of two consecutive integers is 91. Find the two integers.

Most word problems are not so easily solved by substitution. For more complicated word problems, a five-step procedure is used. Using this step-by-step approach will, with practice, allow you to organize your work. Organization is the key to solving word problems. Here are the five steps.

> **To Solve Word Problems**
>
> Step 1 Read the problem carefully. Then reread it to decide what you are asked to find.
>
> Step 2 Choose a letter to represent one of the unknowns in the problem. Then represent all other unknowns of the problem with expressions that use the same letter.
>
> Step 3 Translate the problem to the language of algebra to form an equation.
>
> Step 4 Solve the equation and answer the question of the original problem.
>
> Step 5 Check your solution by returning to the original problem.

We discussed these translations in Section 1.1. You might find it helpful to review that section before going on.

The third step is usually the hardest part. We must translate words to the language of algebra. Before we look at a complete example, the following table may help you review that translation step.

Translating Words to Algebra

Words	Algebra
The sum of x and y	$x + y$
3 plus a	$3 + a$ or $a + 3$
5 more than m	$m + 5$
b increased by 7	$b + 7$
The difference of x and y	$x - y$
4 less than a	$a - 4$
s decreased by 8	$s - 8$
The product of x and y	$x \cdot y$ or xy
5 times a	$5 \cdot a$ or $5a$
Twice m	$2m$
The quotient of x and y	$\dfrac{x}{y}$
a divided by 6	$\dfrac{a}{6}$
One-half of b	$\dfrac{b}{2}$ or $\dfrac{1}{2}b$

Now let's look at some typical examples of translating phrases to algebra.

Example 11 — Translating Statements

Translate each statement to an algebraic expression.

(a) the sum of a and 2 times b $a + 2b$
                                                    Sum   2 times b

(b) 5 times m increased by 1 $5m + 1$
                                                    5 times m Increased by 1

(c) 5 less than 3 times x $3x - 5$
                                              3 times x 5 less than

(d) The product of x and y, divided by 3 $\dfrac{xy}{3}$ — The product of x and y
                                                                            Divided by 3

✓ CHECK YOURSELF 11

Translate to algebra.

(a) 2 more than twice x

(b) 4 less than 5 times n

(c) The product of twice a and b

(d) The sum of s and t, divided by 5

Now let's work through a complete example. Although this problem could be solved by substitution, it is presented here to help you practice the five-step approach.

Example 12   Solving an Application

The sum of a number and 5 is 17. What is the number?

Step 1 *Read carefully.* You must find the unknown number.

Step 2 *Choose letters or variables.* Let x represent the unknown number. There are no other unknowns.

Step 3 *Translate.*

The sum of
$$x + 5 = 17$$
is

Step 4 *Solve.*

$$x + 5 = 17$$
$$x + 5 - 5 = 17 - 5 \qquad \text{Subtract 5.}$$
$$x = 12$$

Always return to the original problem to check your result and not to the equation of step 3. This will prevent possible errors!

So the number is 12.

Step 5 *Check.* Is the sum of 12 and 5 equal to 17? Yes ($12 + 5 = 17$). We have checked our solution.

✓ CHECK YOURSELF 12

The sum of a number and 8 is 35. What is the number?

✓ CHECK YOURSELF ANSWERS

1. (a) 4 is not a solution; (b) 6 is a solution.   **2.** (a) Nonlinear equation; (b) linear equation; (c) expression; (d) linear equation.   **3.** {9}.   **4.** {7}.
5. {3}.   **6.** {−3}.   **7.** (a) {7}; (b) {−6}.   **8.** (a) {−10}; (b) {−3}.
9. (a) {12}; (b) {−13}.   **10.** {45, 46}.
11. (a) $2x + 2$; (b) $5n - 4$; (c) $2ab$; (d) $\dfrac{s + t}{5}$.
12. The equation is $x + 8 = 35$. The number is 27.

Exercises • 2.1

1. Yes
2. No
3. No
4. Yes
5. No
6. Yes
7. Yes
8. No
9. No
10. Yes
11. No
12. Yes
13. Yes
14. No
15. Yes
16. No
17. Yes
18. No
19. No
20. Yes
21. Yes
22. No
23. Linear equation
24. Expression
25. Expression
26. Linear equation
27. Linear equation
28. Not a linear equation
29. {2}
30. {10}
31. {11}
32. {4}
33. {−2}
34. {−3}
35. {6}
36. {−7}
37. {4}
38. {−8}

Is the number shown in parentheses a solution for the given equation?

1. $x + 4 = 9$ (5)
2. $x + 2 = 11$ (8)
3. $x - 15 = 6$ (−21)
4. $x - 11 = 5$ (16)
5. $5 - x = 2$ (4)
6. $10 - x = 7$ (3)
7. $4 - x = 6$ (−2)
8. $5 - x = 6$ (−3)
9. $3x + 4 = 13$ (8)
10. $5x + 6 = 31$ (5)
11. $4x - 5 = 7$ (2)
12. $2x - 5 = 1$ (3)
13. $5 - 2x = 7$ (−1)
14. $4 - 5x = 9$ (−2)
15. $4x - 5 = 2x + 3$ (4)
16. $5x + 4 = 2x + 10$ (4)
17. $x + 3 + 2x = 5 + x + 8$ (5)
18. $5x - 3 + 2x = 3 + x - 12$ (−2)
19. $\frac{3}{4}x = 18$ (20)
20. $\frac{3}{5}x = 24$ (40)
21. $\frac{3}{5}x + 5 = 11$ (10)
22. $\frac{2}{3}x + 8 = -12$ (−6)

Label each of the following as an expression, or a linear equation, or an equation that is not linear.

23. $2x + 1 = 9$
24. $7x + 14$
25. $7x + 2x + 8 - 3$
26. $x + 5 = 13$
27. $2x - 8 = 3$
28. $12x^2 - 5x + 2 = 5$

Solve each equation and check your results. Express each answer in set notation.

29. $x + 9 = 11$
30. $x - 4 = 6$
31. $x - 8 = 3$
32. $x + 11 = 15$
33. $x - 8 = -10$
34. $x + 5 = 2$
35. $11 = x + 5$
36. $x + 7 = 0$
37. $4x = 3x + 4$
38. $7x = 6x - 8$

112

Section 2.1 ■ Solving Equations by Adding and Subtracting

Answers (left column):

39. {−10}
40. {5}
41. {−3}
42. {6}
43. {2}
44. {3}
45. {4}
46. {−7}
47. {6}
48. {−3}
49. {6}
50. {11}
51. {6}
52. {−7}
53. {−18}
54. {13}
55. {16}
56. {−17}
57. {8}
58. {−11}
59. {2}
60. {3}
61. $x + 3 = 7$
62. $x − 5 = 12$
63. $3x − 7 = 2x$
64. $5x + 4 = 6x$
65. $2(x + 5) = x + 18$
66. $3(x + 7) = 4x$
67. c
68. d
69. a
70. d

Exercises:

39. $11x = 10x − 10$
40. $9x = 8x + 5$
41. $6x + 3 = 5x$
42. $12x − 6 = 11x$
43. $2x + 3 = x + 5$
44. $3x − 2 = 2x + 1$
45. $5x − 7 = 4x − 3$
46. $8x + 5 = 7x − 2$
47. $7x − 2 = 6x + 4$
48. $10x − 3 = 9x − 6$
49. $3 + 6x + 2 = 3x + 11 + 2x$
50. $6x − 3 + 2x = 7x + 8$
51. $4x + 7 + 3x = 5x + 13 + x$
52. $5x + 9 + 4x = 9 + 8x − 7$
53. $4(3x + 4) = 11x − 2$
54. $2(5x − 3) = 9x + 7$
55. $3(7x + 2) = 5(4x + 1) + 17$
56. $5(5x + 3) = 3(8x − 2) + 4$
57. $\dfrac{5}{4}x − 1 = \dfrac{1}{4}x + 7$
58. $\dfrac{7}{5}x + 3 = \dfrac{2}{5}x − 8$
59. $\dfrac{9}{2}x − \dfrac{3}{4} = \dfrac{7}{2}x + \dfrac{5}{4}$
60. $\dfrac{11}{3}x + \dfrac{1}{6} = \dfrac{8}{3}x + \dfrac{19}{6}$

In Exercises 61 to 66, translate each statement to an algebraic equation. Let x represent the number in each case.

61. 3 more than a number is 7.
62. 5 less than a number is 12.
63. 7 less than 3 times a number is twice that same number.
64. 4 more than 5 times a number is 6 times that same number.
65. 2 times the sum of a number and 5 is 18 more than that same number.
66. 3 times the sum of a number and 7 is 4 times that same number.
67. Which of the following is equivalent to the equation $8x + 5 = 9x − 4$?
    (a) $17x = −9$ (b) $x = −9$ (c) $8x + 9 = 9x$ (d) $9 = 17x$
68. Which of the following is equivalent to the equation $5x − 7 = 4x − 12$?
    (a) $9x = 19$ (b) $9x − 7 = −12$ (c) $x = −18$ (d) $x − 7 = −12$
69. Which of the following is equivalent to the equation $12x − 6 = 8x + 14$?
    (a) $4x − 6 = 14$ (b) $x = 20$ (c) $20x = 20$ (d) $4x = 8$
70. Which of the following is equivalent to the equation $7x + 5 = 12x − 10$?
    (a) $5x = −15$ (b) $7x − 5 = 12x$ (c) $−5 = 5x$ (d) $7x + 15 = 12x$

114 Chapter 2 ■ Equations and Inequalities

71. True

72. False

73. 26; $x + 7 = 33$

74. 7; $x + 15 = 22$

75. 22; $x - 15 = 7$

76. 25; $x - 8 = 17$

77. 1420; $1840 + x = 3260$

78. $1360; $x + 1440 = 2760$

79. $290; $x + 360 = 650$

80. $1325; $x + 2350 = 3675$

81. $740; $x + 225 = 965$

True or false?

71. Every linear equation with one variable has exactly one solution.

72. Isolating the variable on the right side of the equation will result in a negative solution.

Solve the following word problems. Be sure to label the unknowns and to show the equation you use for the solution.

73. **Number problem.** The sum of a number and 7 is 33. What is the number?

74. **Number problem.** The sum of a number and 15 is 22. What is the number?

75. **Number problem.** The sum of a number and -15 is 7. What is the number?

76. **Number problem.** The sum of a number and -8 is 17. What is the number?

77. **Number of votes cast.** In an election, the winning candidate has 1840 votes. If the total number of votes cast was 3260, how many votes did the losing candidate receive?

78. **Monthly earnings.** Mike and Stefanie work at the same company and make a total of $2760 per month. If Stefanie makes $1400 per month, how much does Mike earn every month?

79. **Appliance costs.** A washer-dryer combination costs $650. If the washer costs $360, what does the dryer cost?

80. **Computer costs.** You have $2350 saved for the purchase of a new computer that costs $3675. How much more must you save?

81. **Price increases.** The price of an item has increased by $225 over last year. If the item is now selling for $965, what was the price last year?

82. An algebraic equation is a complete sentence. It has a subject, a verb, and a predicate. For example, $x + 2 = 5$ can be written in English as "Two more than a number is five." Or, "A number added to two is five." Write an English version of the following equations. Be sure you write complete sentences and that the sentences express the same idea as the equations. Exchange sentences with another student, and see if your interpretation of each other's sentences result in the same equation.

   (a) $2x - 5 = x + 1$ (b) $2(x + 2) = 14$

   (c) $n + 5 = \dfrac{n}{2} - 6$ (d) $7 - 3a = 5 + a$

83. Complete the following explanation in your own words: "The difference between $3(x - 1) + 4 - 2x$ and $3(x - 1) + 4 = 2x$ is . . . "

84. "I make $2.50 an hour more in my new job." If $x =$ the amount I used to make per hour and $y =$ the amount I now make, which equation(s) below say the same thing as the statement above? Explain your choice(s) by translating the equation into English and comparing with the original statement.

   (a) $x + y = 2.50$ (b) $x - y = 2.50$ (c) $x + 2.50 = y$

   (d) $2.50 + y = x$ (e) $y - x = 2.50$ (f) $2.50 - x = y$

85. "The river rose 4 feet above flood stage last night." If a = the river's height at flood stage and b = the river's height now (the morning after), which equations below say the same thing as the statement? Explain your choices by translating the equations into English and comparing the meaning with the original statement.

   (a) $a + b = 4$ (b) $b - 4 = a$ (c) $a - 4 = b$

   (d) $a + 4 = b$ (e) $b + 4 = b$ (f) $b - a = 4$

86. "Surprising Results!" Work with other students to try this experiment. Each person should do the following six steps mentally, not telling anyone else what their calculations are:

   (a) Think of a number.   (b) Add 7.

   (c) Multiply by 3.   (d) Add 3 more than the original number.

   (e) Divide by 4.   (f) Subtract the original number.

   What number do you end up with? Compare your answer with everyone else's. Does everyone have the same answer? Make sure that everyone followed the directions accurately. How do you explain the results? Algebra makes the explanation clear. Work together to do the problem again, using a variable for the number. Make up another series of computations that give "surprising results."

87. (a) Do you think that the following is a linear equation in one variable?

   $$3(2x + 5) = 6(x + 2)$$

   (b) What happens when you use the properties of this section to solve the equation?

   (c) Pick *any* number to substitute for x in this equation. Now try a different number to substitute for x in the equation. Try yet another number to substitute for x in the equation. Summarize your findings.

   (d) Can this equation be called "linear in one variable"? Refer to the definition as you explain your answer.

88. (a) Do you think the following is a linear equation in one variable?

   $$4(3x - 5) = 2(6x - 8) - 3$$

   (b) What happens when you use the properties of this section to solve the equation?

   (c) Do you think it is possible to find a solution for this equation?

   (d) Can this equation be called "linear in one variable"? Refer to the definition as you explain your answer.

SECTION 2.2 Solving Equations by Multiplying and Dividing

2.2 OBJECTIVES

1. Determine whether a given number is a solution for an equation
2. Use the multiplication property to solve equations

Let's look at a different type of equation. For instance, what if we want to solve an equation like the following?

$$6x = 18$$

Using the addition property of the last section won't help. We will need a second property for solving equations.

The Multiplication Property of Equality

If $a = b$ then $ac = bc$ where $c \neq 0$

In words, multiplying both sides of an equation by the same nonzero number gives an equivalent equation.

Again, as long as you do the *same* thing to *both* sides of the equation, the "balance" is maintained. But, multiplying by 0 gives 0 = 0. We have lost the variable!

Let's work through some examples, using this second rule.

Example 1 Solving Equations by Using the Multiplication Property

Solve

$$6x = 18$$

Here the variable x is multiplied by 6. So we apply the multiplication property and multiply both sides by $\frac{1}{6}$. Keep in mind that we want an equation of the form

$$x = \boxed{}$$

$$\frac{1}{6}(6x) = \left(\frac{1}{6}\right)18$$

$\frac{1}{6}(6x) = \left(\frac{1}{6} \cdot 6\right)x$
$\phantom{\frac{1}{6}(6x)} = 1 \cdot x, \text{ or } x$

We can now simplify.

$$1 \cdot x = 3 \quad \text{or} \quad x = 3$$

116

The solution set is {3}. To check, replace x with 3:

$$6 \cdot 3 \stackrel{?}{=} 18$$
$$18 = 18 \quad \text{(True)}$$

 CHECK YOURSELF 1

Solve and check.

$$8x = 32$$

In Example 1 we solved the equation by multiplying both sides by the reciprocal of the coefficient of the variable.

Example 2 illustrates a slightly different approach to solving an equation by using the multiplication property.

Example 2

Solving Equations by Using the Multiplication Property

Solve

$$5x = -35$$

The variable x is multiplied by 5. We *divide* both sides by 5 to "undo" that multiplication:

Since division is defined in terms of multiplication, we can also divide both sides of an equation by the same nonzero number.

$$\frac{5x}{5} = \frac{-35}{5}$$
$$x = -7$$

Note that the right side reduces to -7. Be careful with the rules for signs.

The solution set is $\{-7\}$.

We will leave it to you to check the solution.

 CHECK YOURSELF 2

Solve and check.

$$7x = -42$$

Example 3 Solving Equations by Using the Multiplication Property

Solve

$$-9x = 54$$

In this case, x is multiplied by -9, so we divide both sides by -9 to isolate x on the left:

$$\frac{-9x}{-9} = \frac{54}{-9}$$

$$x = -6$$

The solution set is $\{-6\}$. To check:

$$(-9)(-6) \stackrel{?}{=} 54$$

$$54 = 54 \qquad \text{(True)}$$

CHECK YOURSELF 3

Solve and check.

$$-10x = -60$$

Example 4 illustrates the use of the multiplication property when fractions appear in an equation.

Example 4 Solving Equations by Using the Multiplication Property

(a) Solve

$$\frac{x}{3} = 6$$

Here x is *divided* by 3. We will use multiplication to isolate x.

$$3\left(\frac{x}{3}\right) = 3 \cdot 6$$

$$x = 18$$

This leaves x alone on the left because

$$3\left(\frac{x}{3}\right) = \frac{3}{1} \cdot \frac{x}{3} = \frac{x}{1} = x$$

The solution set is $\{18\}$.

To check:

$$\frac{18}{3} \stackrel{?}{=} 6$$

$$6 = 6 \quad \text{(True)}$$

(b) Solve

$$\frac{x}{5} = -9$$

$$5\left(\frac{x}{5}\right) = 5(-9) \qquad \text{Since } x \text{ is divided by 5,}$$
$$\text{multiply both sides by 5.}$$
$$x = -45$$

The solution set is $\{-45\}$. To check, we replace x with -45:

$$\frac{-45}{5} \stackrel{?}{=} -9$$

$$-9 = -9 \quad \text{(True)}$$

The solution is verified.

✓ **CHECK YOURSELF 4**

Solve and check.

(a) $\frac{x}{7} = 3$ (b) $\frac{x}{4} = -8$

When the variable is multiplied by a fraction that has a numerator other than 1, there are two approaches to finding the solution.

Example 5 **Solving Equations by Using Reciprocals**

Solve

$$\frac{3}{5}x = 9$$

One approach is to multiply by 5 as the first step.

$$5\left(\frac{3}{5}x\right) = 5 \cdot 9$$

$$3x = 45$$

Now we divide by 3.

$$\frac{3x}{3} = \frac{45}{3}$$

$$x = 15$$

To check the solution set {15}, substitute 15 for x.

$$\frac{3}{5} \cdot 15 \stackrel{?}{=} 9$$

$$9 = 9$$

A second approach combines the multiplication and division steps and is generally a bit more efficient. We multiply by $\frac{5}{3}$.

$$\frac{5}{3}\left(\frac{3}{5}x\right) = \frac{5}{3} \cdot 9$$

$$x = \frac{5}{\cancel{3}} \cdot \frac{\cancel{9}^{3}}{1} = 15$$

Recall that $\frac{5}{3}$ is the *reciprocal* of $\frac{3}{5}$, and the product of a number and its reciprocal is 1! So $\left(\frac{5}{3}\right)\left(\frac{3}{5}\right) = 1$

So $x = 15$, as before.

✓ **CHECK YOURSELF 5**

Solve and check.

$$\frac{2}{3}x = 18$$

You may sometimes have to simplify an equation before applying the methods of this section. Example 6 illustrates this property.

Example 6 Combining Like Terms and Solving Equations

Solve and check.

$$3x + 5x = 40$$

Using the distributive property, we can combine the like terms on the left to write

$$8x = 40$$

Section 2.2 ■ Solving Equations by Multiplying and Dividing

We can now proceed as before.

$$\frac{8x}{8} = \frac{40}{8} \quad \text{Divide by 8.}$$

$$x = 5$$

The solution set is {5}. To check, we return to the original equation. Substituting 5 for x yields

$$3 \cdot 5 + 5 \cdot 5 \stackrel{?}{=} 40$$

$$15 + 25 \stackrel{?}{=} 40$$

$$40 = 40 \quad \text{(True)}$$

The solution is verified.

✓ CHECK YOURSELF 6

Solve and check.

$$7x + 4x = -66$$

✓ CHECK YOURSELF ANSWERS

1. {4}. **2.** {−6}. **3.** {6}.

4. (a) {21}, (b) {−32}. **5.** {27}. **6.** {−6}.

Exercises ■ 2.2

Answers (left column):

1. {4} 2. {5}
3. {6} 4. {−7}
5. {7} 6. {11}
7. {−4} 8. {−9}
9. {−8} 10. {−10}
11. {−9} 12. {−7}
13. {3} 14. {−13}
15. {−7} 16. {5}
17. {9} 18. {6}
19. {8} 20. {6}
21. {15} 22. {40}
23. {42} 24. {18}
25. {−20} 26. {−35}
27. {−24} 28. {12}
29. {9} 30. {10}
31. {−20} 32. {−24}
33. {−25} 34. {18}
35. {4} 36. {−10}
37. {−6} 38. {5}
39. {4} 40. {−6}

Solve for x and check your result. Express your answer in set notation.

1. $5x = 20$
2. $6x = 30$
3. $9x = 54$
4. $6x = -42$
5. $63 = 9x$
6. $66 = 6x$
7. $4x = -16$
8. $-3x = 27$
9. $-9x = 72$
10. $10x = -100$
11. $6x = -54$
12. $-7x = 49$
13. $-4x = -12$
14. $52 = -4x$
15. $-42 = 6x$
16. $-7x = -35$
17. $-6x = -54$
18. $-4x = -24$
19. $\dfrac{x}{2} = 4$
20. $\dfrac{x}{3} = 2$
21. $\dfrac{x}{5} = 3$
22. $\dfrac{x}{8} = 5$
23. $6 = \dfrac{x}{7}$
24. $6 = \dfrac{x}{3}$
25. $\dfrac{x}{5} = -4$
26. $\dfrac{x}{7} = -5$
27. $-\dfrac{x}{3} = 8$
28. $-\dfrac{x}{4} = -3$
29. $\dfrac{2}{3}x = 6$
30. $\dfrac{4}{5}x = 8$
31. $\dfrac{3}{4}x = -15$
32. $\dfrac{7}{8}x = -21$
33. $-\dfrac{2}{5}x = 10$
34. $-\dfrac{5}{6}x = -15$
35. $5x + 4x = 36$
36. $8x - 3x = -50$
37. $16x - 9x = -42$
38. $5x + 7x = 60$
39. $4x - 2x + 7x = 36$
40. $6x + 7x - 5x = -48$

Once again, certain equations involving decimal fractions can be solved by the methods of this section. For instance, to solve $2.3x = 6.9$ we simply use our multiplication property to divide both sides of the equation by 2.3. This will isolate x on the left as desired. Use this idea to solve each of the following equations for x.

Section 2.2 ■ Solving Equations by Multiplying and Dividing

41. $3.2x = 12.8$

42. $5.1x = -15.3$

43. $-4.5x = 13.5$

44. $-8.2x = -32.8$

45. $1.3x + 2.8x = 12.3$

46. $2.7x + 5.4x = -16.2$

47. $9.3x - 6.2x = 12.4$

48. $12.5x - 7.2x = -21.2$

Translate each of the following statements to an equation. Let x represent the number in each case.

49. 5 times a number is 40.

50. Twice a number is 36.

51. A number divided by 7 is equal to 6.

52. A number divided by 5 is equal to -4.

53. $\frac{1}{3}$ of a number is 8.

54. $\frac{1}{5}$ of a number is 10.

55. $\frac{3}{4}$ of a number is 18.

56. $\frac{2}{7}$ of a number is 8.

57. Twice a number, divided by 5, is 12.

58. 3 times a number, divided by 4, is 36.

59. Customer satisfaction. Three-fourths of the theater audience left in disgust. If 87 angry patrons walked out, how many were there originally?

60. Auto repair. A mechanic charged $35 an hour to replace the ignition coil on a car plus $125 for parts. If the total bill was $230, how many hours did the repair job take?

61. Telephone charges. A call to Phoenix from Dubuque costs 55 cents for the first minute and 23 cents for each additional minute or portion of a minute. If Barry has $6.30 in change, how long can he talk?

62. Number problem. The sum of 4 times a number and 14 is 34. Find the number.

63. Number problem. If 6 times a number is subtracted from 42, the result is 24. Find the number.

64. Number problem. When a number is divided by -6, the result is 3. Find the number.

65. Geometry. Suppose that the circumference of a tree is measured to be 9 ft 2 in., or 110 in. To find the diameter of the tree at that point, we must solve the equation

$$110 = 3.14d$$

(Note: 3.14 is the approximation for π.) Find the diameter of the tree to the nearest inch.

66. Geometry. Suppose that the circumference of a circular swimming pool is 88 ft. Find the diameter of the pool by solving the equation to the nearest foot).

$$88 = 3.14d$$

---

41. {4}
42. {−3}
43. {−3}
44. {4}
45. {3}
46. {−2}
47. {4}
48. {−4}
49. $5x = 40$
50. $2x = 36$
51. $\frac{x}{7} = 6$
52. $\frac{x}{5} = -4$
53. $\frac{1}{3}x = 8$
54. $\frac{1}{5}x = 10$
55. $\frac{3}{4}x = 18$
56. $\frac{2}{7}x = 8$
57. $\frac{2x}{5} = 12$
58. $\frac{3x}{4} = 36$
59. 116
60. 3
61. 26 min
62. 5
63. 3
64. −18
65. 35 in.
66. 28 ft

SECTION 2.3 Combining the Rules to Solve Equations

2.3 OBJECTIVES

1. Use both addition and multiplication to solve equations
2. Solve equations that involve fractions
3. Solve applications

"Success usually comes to those who are too busy to be looking for it."

–David Henry Thoreau

In Section 2.1 we solved equations using the addition property, which allowed us to solve equations such as $x + 3 = 9$. Then, in Section 2.2, we solved equations using the multiplication property, which allowed us to solve equations such as $5x = 32$. Now, we will solve equations that require using both the addition and multiplication properties.

In our first example, we will check to see that a given value for the variable is a solution to a given equation.

Example 1 Checking a Solution

Test to see if -3 is a solution for the given equation.

$$5x + 6 = 2x - 3$$

The solution set is $\{-3\}$ because replacing x with -3 gives

$$5(-3) + 6 \stackrel{?}{=} 2(-3) - 3$$
$$-15 + 6 \stackrel{?}{=} -6 - 3$$
$$-9 = -9 \quad \text{A true statement}$$

✓ CHECK YOURSELF 1

Verify that 7 is a solution for this equation.

$$5x - 15 = 2x + 6$$

In the next example, we will apply the addition and multiplication properties to find the solution of a linear equation.

Example 2 Applying the Properties of Equality

Solve for x.

$$3x - 5 = 4 \tag{1}$$

We start by using the addition property to add 5 to both sides of the equation.

$$3x - 5 + 5 = 4 + 5$$
$$3x = 9 \tag{2}$$

Why did we add 5? We added 5 because it is the opposite of -5, and the resulting equation will have the variable term on the left and the constant term on the right.

Now we want to get the x term alone on the left with a coefficient of 1 (we call this *isolating* the x). To do this, we use the multiplication property and multiply both sides by $\frac{1}{3}$.

We choose $\frac{1}{3}$ because $\frac{1}{3}$ is the reciprocal of 3 and

$$\frac{1}{3} \cdot 3 = 1$$

$$\frac{1}{3}(3x) = \frac{1}{3}(9)$$

$$\left(\frac{1}{3} \cdot 3\right)(x) = 3$$

So, $x = 3$. The solution set is $\{3\}$. $\tag{3}$

Since any application of the addition or multiplication properties leads to an equivalent equation, equations (1), (2), and (3) all have the same solution, 3.

To check this result, we can replace x with 3 in the original equation:

$$3(3) - 5 \stackrel{?}{=} 4$$
$$9 - 5 \stackrel{?}{=} 4$$
$$4 = 4 \quad \text{A true statement}$$

You may prefer a slightly different approach in the last step of the solution above. From equation (2),

$$3x = 9$$

The multiplication property can be used to *divide* both sides of the equation by 3. Then,

$$\frac{3x}{3} = \frac{9}{3}$$
$$x = 3$$

The result is the same.

✓ CHECK YOURSELF 2

Solve for *x*.

$$4x - 7 = 17$$

The steps involved in using the addition and multiplication properties to solve an equation are the same if more terms are involved in an equation.

Example 3 — Applying the Properties of Equality

Solve for *x*.

$$5x - 11 = 2x - 7$$

Our objective is to use the properties of equality to isolate *x* on one side of an equivalent equation. We begin by adding 11 to both sides.

Again, adding 11 leaves us with the constant term on the right.

$$5x - 11 + 11 = 2x - 7 + 11$$
$$5x = 2x + 4$$

We continue by adding $-2x$ to (or subtracting $2x$ from) both sides.

If you prefer, write
$$5x - 2x = 2x - 2x + 4$$
Again:
$$3x = 4$$

$$5x + (-2x) = 2x + (-2x) + 4$$
$$3x = 4$$

To isolate *x*, we now multiply both sides by $\frac{1}{3}$.

This is the same as dividing both sides by 3. So
$$\frac{3x}{3} = \frac{4}{3}$$
$$x = \frac{4}{3}$$

$$\frac{1}{3}(3x) = \frac{1}{3}(4)$$

$$x = \frac{4}{3}$$

In set notation, we write $\left\{\frac{4}{3}\right\}$. We leave it to you to check this result by substitution.

✓ CHECK YOURSELF 3

Solve for *x*.

$$7x - 12 = 2x - 9$$

Both sides of an equation should be simplified as much as possible *before* the addition and multiplication properties are applied. If like terms are involved on one side (or on both sides) of an equation, they should be combined before an attempt is made to isolate the variable. Example 4 illustrates this approach.

Example 4 — Applying the Properties of Equality with Like Terms

Solve for x.

Note the like terms on the left and right sides of the equation.

$$8x + 2 - 3x = 8 + 3x + 2$$

Here we combine the like terms $8x$ and $-3x$ on the left and the like terms 8 and 2 on the right as our first step. We then have

$$5x + 2 = 3x + 10$$

We can now solve as before.

$$5x + 2 - 2 = 3x + 10 - 2 \quad \text{Subtract 2 from both sides.}$$
$$5x = 3x + 8$$

Then,

$$5x - 3x = 3x - 3x + 8 \quad \text{Subtract } 3x \text{ from both sides.}$$
$$2x = 8$$
$$\frac{2x}{2} = \frac{8}{2} \quad \text{Divide both sides by 2.}$$
$$x = 4$$

The solution set is {4}, which can be checked by returning to the *original equation*.

✓ **CHECK YOURSELF 4**

Solve for x.

$$7x - 3 - 5x = 10 + 4x + 3$$

If parentheses are involved on one or both sides of an equation, the parentheses should be removed by applying the distributive property as the first step. Like terms should then be combined before an attempt is made to isolate the variable. Consider Example 5.

Example 5 Applying the Properties of Equality with Parentheses

Solve for x.

$$x + 3(3x - 1) = 4(x + 2) + 4$$

First, apply the distributive property to remove the parentheses on the left and right sides.

$$x + 9x - 3 = 4x + 8 + 4$$

Combine like terms on each side of the equation.

$$10x - 3 = 4x + 12$$

Now, isolate the variable x on the left side.

Recall that to isolate the x, we must get x alone on the left side with a coefficient of 1.

$10x - 3 + 3 = 4x + 12 + 3$	Add 3 to both sides.
$10x = 4x + 15$	
$10x - 4x = 4x - 4x + 15$	Subtract $4x$ from both sides.
$6x = 15$	
$\dfrac{6x}{6} = \dfrac{15}{6}$	Divide both sides by 6.
$x = \dfrac{5}{2}$	

The solution set is $\left\{\dfrac{5}{2}\right\}$. Again, this can be checked by returning to the original equation.

✓ CHECK YOURSELF 5

Solve for x.

$$x + 5(x + 2) = 3(3x - 2) + 18$$

---

*The LCM of a set of denominators is also called the **lowest common denominator (LCD)**.*

To solve an equation involving fractions, the first step is to multiply both sides of the equation by the **least common multiple (LCM)** of all denominators in the equation. This will clear the equation of fractions, and we can proceed as before.

Example 6 — Applying the Properties of Equality with Fractions

Solve for x.

$$\frac{x}{2} - \frac{2}{3} = \frac{5}{6}$$

First, multiply each side by 6, the least common multiple of 2, 3, and 6.

$$6\left(\frac{x}{2} - \frac{2}{3}\right) = 6\left(\frac{5}{6}\right)$$

$$6\left(\frac{x}{2}\right) - 6\left(\frac{2}{3}\right) = 6\left(\frac{5}{6}\right) \qquad \text{Apply the distributive property.}$$

$$\overset{3}{\cancel{6}}\left(\frac{x}{\cancel{2}}\right) - \overset{2}{\cancel{6}}\left(\frac{2}{\cancel{3}}\right) = \overset{1}{\cancel{6}}\left(\frac{5}{\cancel{6}}\right) \qquad \text{Simplify.}$$

$$3x - 4 = 5$$

The equation is now cleared of fractions.

Next, isolate the variable x on the left side.

$$3x = 9$$

$$x = 3$$

The solution set, $\{3\}$, can be checked as before by returning to the original equation.

CHECK YOURSELF 6

Solve for x.

$$\frac{x}{4} - \frac{4}{5} = \frac{19}{20}$$

Be sure that the distributive property is applied properly so that *every term* of the equation is multiplied by the LCM.

Example 7 Applying the Properties of Equality with Fractions

Solve for x.

$$\frac{2x-1}{5} + 1 = \frac{x}{2}$$

First, multiply each side by 10, the LCM of 5 and 2.

$$10\left(\frac{2x-1}{5} + 1\right) = 10\left(\frac{x}{2}\right) \quad \text{Apply the distributive property on the left. Reduce.}$$

$$\overset{2}{10}\left(\frac{2x-1}{\underset{1}{5}}\right) + 10(1) = \overset{5}{10}\left(\frac{x}{\underset{1}{2}}\right) \quad \text{Next, isolate } x. \text{ Here we isolate } x \text{ on the right side.}$$

$$2(2x-1) + 10 = 5x$$

$$4x - 2 + 10 = 5x$$

$$4x + 8 = 5x$$

$$8 = x \quad \text{The solution set for the original equation is } \{8\}.$$

✓ CHECK YOURSELF 7

Solve for x.

$$\frac{3x+1}{4} - 2 = \frac{x+1}{3}$$

Conditional Equations, Identities, and Contradictions

1. An equation that is true for only particular values of the variable is called a **conditional equation.** Here the equation can be written in the form

    $$ax + b = 0$$

    where $a \neq 0$. This case was illustrated in all our previous examples and exercises.

2. An equation that is true for all possible values of the variable is called an **identity.** In this case, *both a* and *b* are 0, so we get the equation $0 = 0$. This will be the case if both sides of the equation reduce to the same expression (a true statement).

3. An equation that is never true, no matter what the value of the variable, is called a **contradiction.** For example, if *a* is 0 but *b* is 4. This will be the case if the equation simplifies as a false statement.

Example 8 illustrates the second and third cases.

Example 8 — Identities and Contradictions

(a) Solve for x.

$$2(x - 3) - 2x = -6$$

Apply the distributive property to remove the parentheses.

$$2x - 6 - 2x = -6$$
$$-6 = -6 \quad \text{A } \textit{true} \text{ statement}$$

See the definition of an identity, above. By adding 6 to both sides of this equation, we have $0 = 0$.

Because the two sides reduce to the true statement $-6 = -6$, the original equation is an *identity*, and the solution set is the set of all real numbers. This is sometimes written as $\mathbb{R}$, or $\{x | x \in \mathbb{R}\}$, which is read, "the set of all numbers (x) that are elements of the real number set."

(b) Solve for x.

$$3(x + 1) - 2x = x + 4$$

Again, apply the distributive property.

$$3x + 3 - 2x = x + 4$$
$$x + 3 = x + 4$$
$$3 = 4 \quad \text{A } \textit{false} \text{ statement}$$

See the definition of a contradiction. Subtracting 3 from both sides, we have $0 = 1$.

Since the two sides reduce to the false statement $3 = 4$, the original equation is a contradiction. There are no values of the variable that can satisfy the equation. The solution set has nothing in it. We call this the **empty set** and write $\{\ \}$ or $\varnothing$.

✓ CHECK YOURSELF 8

Determine whether each of the following equations is a conditional equation, an identity, or a contradiction.

(a) $2(x + 1) - 3 = x$ (b) $2(x + 1) - 3 = 2x + 1$ (c) $2(x + 1) - 3 = 2x - 1$

An **algorithm** is a step-by-step process for problem solving.

An organized step-by-step procedure is the key to an effective equation-solving strategy. The following algorithm summarizes our work in this section and gives you guidance in approaching the problems that follow.

> **Solving Linear Equations in One Variable**
>
> Step 1 Remove any grouping symbols by applying the distributive property.
>
> Step 2 Multiply both sides of the equation by the LCM of any denominators, to clear the equation of fractions.
>
> Step 3 Combine any like terms that appear on either side of the equation.
>
> Step 4 Apply the addition property of equality to write an equivalent equation with the variable term on *one side* of the equation and the constant term on the *other side*.
>
> Step 5 Apply the multiplication property of equality to write an equivalent equation with the variable isolated on one side of the equation with coefficient one.
>
> Step 6 Check the solution in the *original* equation.
>
> *Note:* If the equation derived in step 5 is always true, the original equation is an *identity*. If the equation is always false, the original equation is a *contradiction*.

When you are solving an equation for which a calculator is recommended, it is often easiest to do all calculations as the last step. For more complex equations, it is usually best to calculate at each step.

Example 9 Evaluating Expressions Using a Calculator

Solve the following equation for x.

$$5(x - 3.25) + \frac{3}{4} = 2110.75$$

Following the steps of the algorithm, we get

$$5x - 16.25 + \frac{3}{4} = 2110.75 \qquad \text{Remove parentheses.}$$

$$20x - 65 + 3 = 8443 \qquad \text{Multiply by the LCM.}$$

$$20x = 8443 + 62 \qquad \text{Isolate the variable.}$$

$$x = \frac{8505}{20}$$

Now, remembering to insert parentheses around the numerator, we use a calculator to simplify the expression on the right.

$$x = 425.25 \quad \text{or} \quad \{425.25\}$$

 CHECK YOURSELF 9

Solve the following equation for x.

$$7(x + 4.3) - \frac{3}{5} = 467$$

Consecutive integers are integers that follow one another, like 10, 11, and 12.

Consecutive Integers

If x is an integer, then $x + 1$ is the next consecutive integer, $x + 2$ is the next, and so on.

If x is an odd integer, the next **consecutive odd integer** is $x + 2$, and the next is $x + 4$.

If x is an even integer, the next **consecutive even integer** is $x + 2$, and the next is $x + 4$.

We'll need this idea in Example 10.

Example 10 — Solving an Application

The sum of two consecutive integers is 41. What are the two integers?

Remember the Steps!

What do you need to find?

Step 1 We want to find the two consecutive integers.

Assign letters to the unknown or unknowns.
Write an equation.

Step 2 Let x be the first integer. Then $x + 1$ must be the next.

Step 3

The first integer → x + $x + 1$ = 41 ← The second integer

The sum ↗   ↑ Is

Solve the equation.

Step 4
$$x + x + 1 = 41$$
$$2x + 1 = 41$$
$$2x = 40$$
$$x = 20$$

The first integer (x) is 20, and the next integer ($x + 1$) is 21.

Check.

Step 5 The sum of the two integers 20 and 21 is 41.

✓ CHECK YOURSELF 10

The sum of three consecutive integers is 51. What are the three integers?

Sometimes algebra is used to reconstruct missing information. Example 11 does just that with some election information.

Example 11 Solving an Application

There were 55 more yes votes than no votes on an election measure. If 735 votes were cast in all, how many yes votes were there? How many no votes?

What do you need to find?

Step 1 We want to find the number of yes votes and the number of no votes.

Assign letters to the unknowns.

Step 2 Let x be the number of no votes. Then

$$\underbrace{x + 55}_{\text{55 more than } x}$$

is the number of yes votes.

Write an equation.

Step 3

$$\underbrace{x}_{\text{No votes}} + \underbrace{x + 55}_{\text{Yes votes}} = \underbrace{735}_{\text{Total votes cast}}$$

Solve the equation.

Step 4

$$x + x + 55 = 735$$
$$2x + 55 = 735$$
$$2x = 680$$
$$x = 340$$
$$\text{No votes } (x) = 340$$
$$\text{Yes votes } (x + 55) = 395$$

Check.

Step 5 Thus 340 no votes plus 395 yes votes equals 735 total votes. The solution checks.

Section 2.3 ■ Combining the Rules to Solve Equations

✓ **CHECK YOURSELF 11**

Francine earns $120 per month more than Rob. If they earn a total of $2680 per month, what are their monthly salaries?

Similar methods will allow you to solve a variety of word problems. Example 12 includes three unknown quantities but uses the same basic solution steps.

Example 12 — Solving an Application

Juan worked twice as many hours as Jerry. Marcia worked 3 more hours than Jerry. If they worked a total of 31 hours, find out how many hours each worked.

Step 1 We want to find the hours each worked, so there are three unknowns.

Step 2 Let x be the hours that Jerry worked.

There are other choices for x, but choosing the smallest quantity will usually give the easiest equation to write and solve.

Juan worked twice Jerry's hours.

Then $2x$ is Juan's hours worked.

Marcia worked 3 more hours than Jerry worked.

and $x + 3$ is Marcia's hours.

Step 3

$$\underbrace{x}_{\text{Jerry}} + \underbrace{2x}_{\text{Juan}} + \underbrace{x + 3}_{\text{Marcia}} = \underset{\uparrow}{31}$$

Sum of their hours

Step 4

$$x + 2x + x + 3 = 31$$
$$4x + 3 = 31$$
$$4x = 28$$
$$x = 7$$

Jerry's hours $(x) = 7$

Juan's hours $(2x) = 14$

Marcia's hours $(x + 3) = 10$

Step 5 The sum of their hours (7 + 14 + 10) is 31, and the solution is verified.

✓ CHECK YOURSELF 12

Lucy jogged twice as many miles (mi) as Paul but 3 less than Isaac. If the three ran a total of 23 mi, how far did each person run?

✓ CHECK YOURSELF ANSWERS

1. $5(7) - 15 \stackrel{?}{=} 2(7) + 6$
   $35 - 15 \stackrel{?}{=} 14 + 6$
   $20 = 20$ ← A true statement.

   2. {6}. 3. $\left\{\dfrac{3}{5}\right\}$. 4. {−8}. 5. $\left\{-\dfrac{2}{3}\right\}$. 6. {7}.

7. {5}. 8. (a) Conditional; (b) contradiction; (c) identity. 9. {62.5}.
10. The equation is $x + x + 1 + x + 2 = 51$. The integers are 16, 17, and 18.
11. The equation is $x + x + 120 = 2680$. Rob's salary is $1280, and Francine's is $1400. 12. Paul: 4 mi; Lucy: 8 mi; Isaac: 11 mi.

Exercises • 2.3

In Exercises 1 to 40, simplify and then solve each equation. Express your answer in set notation.

1. $2x + 1 = 9$
2. $3x - 1 = 17$
3. $3x - 2 = 7$
4. $5x + 3 = 23$
5. $4 - 7x = 18$
6. $8 - 5x = -7$
7. $3 - 4x = -9$
8. $5 - 4x = 25$
9. $\dfrac{x}{2} + 1 = 5$
10. $\dfrac{x}{3} - 2 = 3$
11. $\dfrac{2}{3}x + 5 = 17$
12. $\dfrac{3}{4}x - 5 = 4$
13. $5x = 2x + 9$
14. $7x = 18 - 2x$
15. $9x + 2 = 3x + 38$
16. $8x - 3 = 4x + 17$
17. $4x - 8 = x - 14$
18. $6x - 5 = 3x - 29$
19. $7x - 3 = 9x + 5$
20. $5x - 2 = 8x - 11$
21. $5x + 4 = 7x - 8$
22. $2x + 23 = 6x - 5$
23. $6x + 7 - 4x = 8 + 7x - 26$
24. $7x - 2 - 3x = 5 + 8x + 13$
25. $9x - 2 + 7x + 13 = 10x - 13$
26. $5x + 3 + 6x - 11 = 8x + 25$
27. $5(8 - x) = 3x$
28. $7x = 7(6 - x)$
29. $7(2x - 1) - 5x = x + 25$
30. $9(3x + 2) - 10x = 12x - 7$
31. $2(2x - 1) = 3(x + 1)$
32. $3(3x - 1) = 4(3x + 1)$
33. $8x - 3(2x - 4) = 17$
34. $7x - 4(3x + 4) = 9$
35. $7(3x + 4) = 8(2x + 5) + 13$
36. $-4(2x - 1) + 3(3x + 1) = 9$
37. $9 - 4(3x + 1) = 3(6 - 3x) - 9$
38. $13 - 4(5x + 1) = 3(7 - 5x) - 15$
39. $5.3x - 7 = 2.3x + 5$
40. $9.8x + 2 = 3.8x + 20$

In Exercises 41 to 52, clear fractions and then solve each equation. Express your answer in set notation.

41. $\dfrac{2x}{3} - \dfrac{5}{3} = 3$
42. $\dfrac{3x}{4} + \dfrac{1}{4} = 4$
43. $\dfrac{x}{6} + \dfrac{x}{5} = 11$
44. $\dfrac{x}{6} - \dfrac{x}{8} = 1$

1. {4} 2. {6}
3. {3} 4. {4}
5. {−2} 6. {3}
7. {3} 8. {−5}
9. {8} 10. {15}
11. {18} 12. {12}
13. {3} 14. {2}
15. {6} 16. {5}
17. {−2} 18. {−8}
19. {−4} 20. {3}
21. {6} 22. {7}
23. {5} 24. {−5}
25. {−4} 26. {11}
27. {5} 28. {3}
29. {4} 30. {−5}
31. {5} 32. $\left\{-\dfrac{7}{3}\right\}$
33. $\left\{\dfrac{5}{2}\right\}$ 34. {−5}
35. {5} 36. {2}
37. $\left\{-\dfrac{4}{3}\right\}$ 38. $\left\{\dfrac{3}{5}\right\}$
39. {4} 40. {3}
41. {7} 42. {5}
43. {30} 44. {24}

45. {6}   46. {5/9}
47. {15} 48. {12}
49. {3}   50. {6}
51. {3/2} 52. {−2/3}
53. Conditional
54. Identity
55. Contradiction
56. Conditional
57. Identity
58. Conditional
59. Contradiction
60. Identity
61. Identity
62. Conditional
63. $2x + 3 = 7$
64. $3x − 5 = 25$
65. $4x − 7 = 41$
66. $2x + 10 = 44$
67. $\frac{2}{3}x + 5 = 21$
68. $\frac{3}{4}x − 3 = 24$
69. $3x = x + 12$
70. $5x = x − 8$
71. 13
72. 14
73. 18
74. 16
75. 35, 36
76. 72, 73
77. 20, 21, 22

45. $\dfrac{2x}{3} - \dfrac{x}{4} = \dfrac{5}{2}$

46. $\dfrac{5x}{6} + \dfrac{2x}{3} = \dfrac{5}{6}$

47. $\dfrac{x}{5} - \dfrac{x-7}{3} = \dfrac{1}{3}$

48. $\dfrac{x}{6} + \dfrac{3}{4} = \dfrac{x-1}{4}$

49. $\dfrac{5x-3}{4} - 2 = \dfrac{x}{3}$

50. $\dfrac{6x-1}{5} - \dfrac{2x}{3} = 3$

51. $\dfrac{2x+3}{5} - \dfrac{2x-1}{3} = \dfrac{8}{15}$

52. $\dfrac{3x}{5} - \dfrac{3x-1}{2} = \dfrac{11}{10}$

In Exercises 53 to 62, classify each equation as a conditional equation, an identity, or a contradiction.

53. $3(x − 1) = 2x + 3$
54. $2(x + 3) = 2x + 6$
55. $3(x − 1) = 3x + 3$
56. $2(x + 3) = x + 5$
57. $3(x − 1) = 3x − 3$
58. $2(x + 3) = 3x + 5$
59. $3x − (x − 3) = 2(x + 1) + 2$
60. $5x − (x + 4) = 4(x − 2) + 4$
61. $\dfrac{x}{2} - \dfrac{x}{3} = \dfrac{x}{6}$
62. $\dfrac{3x}{4} - \dfrac{2x}{3} = \dfrac{x}{6}$

Translate each of the following statements to an equation. Let x represent the number in each case.

63. 3 more than twice a number is 7.
64. 5 less than 3 times a number is 25.
65. 7 less than 4 times a number is 41.
66. 10 more than twice a number is 44.
67. 5 more than two-thirds a number is 21.
68. 3 less than three-fourths of a number is 24.
69. 3 times a number is 12 more than that number.
70. 5 times a number is 8 less than that number.
71. **Number addition.** The sum of twice a number and 7 is 33. What is the number?
72. **Number addition.** 3 times a number, increased by 8, is 50. Find the number.
73. **Number subtraction.** 5 times a number, minus 12, is 78. Find the number.
74. **Number subtraction.** 4 times a number, decreased by 20, is 44. What is the number?
75. **Consecutive integers.** The sum of two consecutive integers is 71. Find the two integers.
76. **Consecutive integers.** The sum of two consecutive integers is 145. Find the two integers.
77. **Consecutive integers.** The sum of three consecutive integers is 63. What are the three integers?

Section 2.3 ■ Combining the Rules to Solve Equations 139

78. 30, 31, 32

79. 32, 34

80. 42, 44

81. 25, 27

82. 43, 45

83. 12, 13

84. 20, 22

85. 1550, 1710

86. $1450, $1310

87. $360, $290

88. 9

89. 18, 9

90. 30, 42, 60

92. Solutions are the same.

93. A value for which the original equation is true.

94. An identity is an equation that is true for all possible values of the variable; a contradiction is an equation that is never true, no matter what the variable's value.

78. **Consecutive integers.** If the sum of three consecutive integers is 93, find the three integers.

79. **Even integers.** The sum of two consecutive even integers is 66. What are the two integers? (*Hint:* Consecutive even integers such as 10, 12, and 14 can be represented by $x, x + 2, x + 4$, and so on.)

80. **Even integers.** If the sum of two consecutive even integers is 86, find the two integers.

81. **Odd integers.** If the sum of two consecutive odd integers is 52, what are the two integers? (*Hint:* Consecutive odd integers such as 21, 23, and 25 can be represented by $x, x + 2, x + 4$, and so on.)

82. **Odd integers.** The sum of two consecutive odd integers is 88. Find the two integers.

83. **Consecutive integers.** 4 times an integer is 9 more than 3 times the next consecutive integer. What are the two integers?

84. **Consecutive even integers.** 4 times an even integer is 30 less than 5 times the next consecutive even integer. Find the two integers.

85. **Election votes.** In an election, the winning candidate had 160 more votes than the loser. If the total number of votes cast was 3260, how many votes did each candidate receive?

86. **Monthly salaries.** Jody earns $140 more per month than Frank. If their monthly salaries total $2760, what amount does each earn?

87. **Appliance costs.** A washer-dryer combination costs $650. If the washer costs $70 more than the dryer, what does each appliance cost?

88. **Age.** Yan Ling is 1 year less than twice as old as his sister. If the sum of their ages is 14 years, how old is Yan Ling?

89. **Age.** Diane is twice as old as her brother Dan. If the sum of their ages is 27 years, how old are Diane and her brother?

90. **Gallons of fuel.** The Randolphs used 12 more gallons (gal) of fuel oil in October than in September and twice as much oil in November as in September. If they used 132 gal for the 3 months, how much was used during each month?

91. Complete this statement in your own words: "You can tell that an equation is a linear equation when . . ."

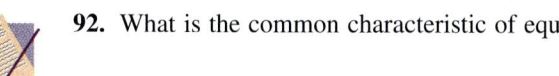

92. What is the common characteristic of equivalent equations?

93. What is meant by a *solution* to a linear equation?

94. Define (a) identity and (b) contradiction.

95. Multiplying by 0 would always give 0 = 0.

96. [6(1500) + 2(1200) + 4(900)](0.0833)

97. (a) $-\dfrac{3}{2}$

(b) $-\dfrac{7}{4}$

(c) $\dfrac{1}{6}$

(d) $\dfrac{2}{5}$

(e) $\dfrac{8}{3}$

(f) $-\dfrac{9}{5}$

(g) $-\dfrac{b}{a}$

98. 2

---

95. Why does the multiplication property of equality not include multiplying both sides of the equation by 0?

96. Maxine lives in Pittsburgh, Pennsylvania, and pays 8.33 cents per kilowatt-hour (kWh) for electricity. During the 6 months of cold winter weather, her household uses about 1500 kWh of electric power per month. During the two hottest summer months, the usage is also high because the family uses electricity to run an air conditioner. During these summer months, the usage is 1200 kWh per month; the rest of the year, usage averages 900 kWh per month.

(a) Write an expression for the total yearly electric bill.

(b) Maxine is considering spending $2000 for more insulation for her home so that it is less expensive to heat and to cool. The insulation company claims that "with proper installation the insulation will reduce your heating and cooling bills by 25%." If Maxine invests the money in insulation, how long will it take her to get her money back in saving on her electric bill? Write to her about what information she needs to answer this question. Give her your opinion about how long it will take to save $2000 on heating bills, and explain your reasoning. What is your advice to Maxine?

97. Solve each of the following equations. Express each solution as a fraction.

(a) $2x + 3 = 0$ (b) $4x + 7 = 0$ (c) $6x - 1 = 0$

(d) $5x - 2 = 0$ (e) $-3x + 8 = 0$ (f) $-5x - 9 = 0$

(g) Based on these problems, express the solution to the equation

$$ax + b = 0$$

where a and b represent real numbers and $a \neq 0$.

98. You are asked to solve the following equation but one number is missing. It reads

$$\dfrac{5x - ?}{4} = \dfrac{9}{2}$$

The solution is 4. What is the missing number?

SECTION 2.4 Literal Equations and Their Applications

2.4 OBJECTIVES

1. Solve a literal equation for any one of its variables
2. Solve applications involving geometric figures
3. Solve mixture problems
4. Solve motion problems

Formulas are extremely useful tools in any field in which mathematics is applied. Formulas are simply equations that express a relationship between more than one letter or variable. You are no doubt familiar with all kinds of formulas, such as

$$A = \frac{1}{2}bh \qquad \text{The area of a triangle}$$

$$I = Prt \qquad \text{Interest}$$

$$V = \pi r^2 h \qquad \text{The volume of a cylinder}$$

Actually a formula is also called a **literal equation** because it involves several letters or variables. For instance, our first formula or literal equation, $A = \frac{1}{2}bh$, involves the three letters A (for area), b (for base), and h (for height).

Unfortunately, formulas are not always given in the form needed to solve a particular problem. Then algebra is needed to change the formula to a more useful equivalent equation, which is solved for a particular letter or variable. The steps used in the process are very similar to those you used in solving linear equations. Let's consider an example.

  **Solving a Literal Equation Involving a Triangle**

Suppose that we know the area A and the base b of a triangle and want to find its height h.

We are given

$$A = \frac{1}{2}bh$$

Our job is to find an equivalent equation with h, the unknown, by itself on one side. We call $\frac{1}{2}b$ the **coefficient** of h. We can remove the two *factors* of that coefficient, $\frac{1}{2}$ and b, separately.

141

Note:

$$2\left(\frac{1}{2}bh\right) = \left(2 \cdot \frac{1}{2}\right)(bh)$$
$$= 1 \cdot bh$$
$$= bh$$

$$2A = 2\left(\frac{1}{2}bh\right)$$ Multiply both sides by 2 to clear the equation of fractions.

or

$$2A = bh$$

$$\frac{2A}{b} = \frac{bh}{b}$$ Divide by b to isolate h.

$$\frac{2A}{b} = h$$

or

$$h = \frac{2A}{b}$$ Reverse the sides to write h on the left.

We now have the height h in terms of the area A and the base b. This is called **solving the equation for h** and means that we are rewriting the formula as an equivalent equation of the form

Here means an expression containing all the numbers or letters *other than* h.

$$h = \square$$

✓ CHECK YOURSELF 1

Solve $V = \frac{1}{3}Bh$ for h.

You have already learned the methods needed to solve most literal equations or formulas for some specified variable. As Example 1 illustrates, the rules of Sections 2.2 and 2.3 are applied in exactly the same way as they were applied to equations with one variable.

You may have to apply both the addition and the multiplication properties when solving a formula for a specified variable. Example 2 illustrates this situation.

Example 2   Solving a Literal Equation

Solve $y = mx + b$ for x.

This is a linear equation in two variables. You will see this again in Chapter 3.

Remember that we want to end up with x alone on one side of the equation. Let's start by subtracting b from both sides to undo the addition on the right.

$$y = mx + b$$
$$y - b = mx + b - b$$
$$y - b = mx$$

If we now divide both sides by m, then x with be alone on the right side.

$$\frac{y - b}{m} = \frac{mx}{m}$$
$$\frac{y - b}{m} = x$$

or

$$x = \frac{y - b}{m}$$

 CHECK YOURSELF 2

Solve $v = a + gt$ for t.

Let's summarize the steps illustrated by our examples.

Solving a Formula or Literal Equation

Step 1 If necessary, multiply both sides of the equation by the same term to clear the equation of fractions.

Step 2 Add or subtract the same term on both sides of the equation so that all terms involving the variable that you are solving for are on one side of the equation and all other terms are on the other side.

Step 3 Divide both sides of the equation by the coefficient of the variable that you are solving for.

Let's look at one more example, using the above steps.

Example 3 Solving a Literal Equation Involving Money

This is a formula for the amount of money in an account after interest has been earned.

Solve $A = P + Prt$ for r.

$$A = P + Prt$$
$$A - P = P - P + Prt \qquad \text{Subtracting } P \text{ from both sides will leave the term involving } r \text{ alone on the right.}$$
$$A - P = Prt$$
$$\frac{A - P}{Pt} = \frac{Prt}{Pt} \qquad \text{Dividing both sides by } Pt \text{ will isolate } r \text{ on the right.}$$
$$\frac{A - P}{Pt} = r$$

or

$$r = \frac{A - P}{Pt}$$

✓ CHECK YOURSELF 3

Solve $2x + 3y = 6$ for y.

Now let's look at an application of solving a literal equation.

Example 4 Solving a Literal Equation Involving Money

Suppose that the amount in an account, 3 years after a principal of $5000 was invested, is $6050. What was the interest rate?

From our previous example,

$$A = P + Prt \qquad (1)$$

where A is the amount in the account, P is the principal, r is the interest rate, and t is the time in years that the money has been invested. By the result of Example 3 we have

$$r = \frac{A - P}{Pt} \qquad (2)$$

Do you see the advantage of having our equation solved for the desired variable?

and we can substitute the known values in equation (2):

$$r = \frac{6050 - 5000}{(5000)(3)}$$

$$= \frac{1050}{15{,}000} = 0.07 = 7\%$$

The interest rate was 7%.

✓ CHECK YOURSELF 4

Suppose that the amount in an account, 4 years after a principal of $3000 was invested, is $3720. What was the interest rate?

In our subsequent applications, we will use the five-step process first described in Section 2.1. As a reminder, here are those steps.

To Solve Word Problems

Step 1 Read the problem carefully. Then reread it to decide what you are asked to find.

Step 2 Choose a letter to represent one of the unknowns in the problem. Then represent all other unknowns of the problem with expressions that use the same letter.

Step 3 Translate the problem to the language of algebra to form an equation.

Step 4 Solve the equation and answer the question of the original problem.

Step 5 Check your solution by returning to the original problem.

Example 5 — Solving a Geometry Application

Whenever you are working on an application involving geometric figures, you should draw a sketch of the problem, including the labels assigned in step 2.

The length of a rectangle is 1 centimeter (cm) less than 3 times the width. If the perimeter is 54 cm, find the dimensions of the rectangle.

Step 1 You want to find the dimensions (the width and length).

Step 2 Let x be the width.

Then $3x - 1$ is the length.

(3 times the width — 1 less than)

Step 3 To write an equation, we'll use this formula for the perimeter of a rectangle:

$$P = 2W + 2L \quad \text{or} \quad 2W + 2L = P$$

So

$$2x + 2(3x - 1) = 54$$

Step 4 Solve the equation.

$$2x + 2(3x - 1) = 54$$
$$2x + 6x - 2 = 54$$
$$8x = 56$$
$$x = 7$$

The width x is 7 cm, and the length, $3x - 1$, is 20 cm. We leave step 5, the check, to you.

Be sure to return to the original statement of the problem when checking your result.

✓ CHECK YOURSELF 5

The length of a rectangle is 5 inches (in.) more than twice the width. If the perimeter of the rectangle is 76 in., what are the dimensions of the rectangle?

You will also often use parentheses in solving *mixture problems*. Mixture problems involve combining things that have a different value, rate, or strength. Look at Example 6.

Example 6

Solving a Mixture Problem

Four hundred tickets were sold for a school play. General admission tickets were $4, while student tickets were $3. If the total ticket sales were $1350, how many of each type of ticket were sold?

Step 1 You want to find the number of each type of ticket sold.

Step 2 Let x be the number of general admission tickets.

We subtract x, the number of general admission tickets, from 400, the total number of tickets, to find the number of student tickets.

Then $400 - x$ student tickets were sold.

400 tickets were sold in all.

Step 3 The sales value for each kind of ticket is found by multiplying the price of the ticket by the number sold.

General admission tickets: $4x$ $4 for each of the x tickets

Student tickets: $3(400 - x)$ $3 for each of the $400 - x$ tickets

So to form an equation, we have

$$4x + 3(400 - x) = 1350$$

Value of general admission tickets — Value of student tickets — Total value

Step 4 Solve the equation.

$$4x + 3(400 - x) = 1350$$
$$4x + 1200 - 3x = 1350$$
$$x + 1200 = 1350$$
$$x = 150$$

This shows that 150 general admission and 250 student tickets were sold. We leave the check to you.

CHECK YOURSELF 6

Beth bought 35¢ stamps and 15¢ stamps at the post office. If she purchased 60 stamps at a cost of $17, how many of each kind did she buy?

The next group of applications we will look at in this section involves *motion problems*. They involve a distance traveled, a rate or speed, and time. To solve motion problems, we need a relationship among these three quantities.

Suppose you travel at a rate of 50 miles per hour (mi/h) on a highway for 6 hours (h). How far (what distance) will you have gone? To find the distance, you multiply:

$$(50 \text{ mi/h})(6 \text{ h}) = 300 \text{ mi}$$

Speed or rate — Time — Distance

> Be careful to make your units consistent. If a rate is given in *miles per hour,* then the time must be given in *hours* and the distance in *miles*.

In general, if r is the rate, t is the time, and d is the distance traveled, then

$$d = r \cdot t$$

148 Chapter 2 ▪ Equations and Inequalities

This is the key relationship, and it will be used in all motion problems. Let's see how it is applied in Example 7.

Example 7 Solving a Motion Problem

On Friday morning Ricardo drove from his house to the beach in 4 h. In coming back on Sunday afternoon, heavy traffic slowed his speed by 10 mi/h, and the trip took 5 h. What was his average speed (rate) in each direction?

Step 1 We want the speed or rate in each direction.

Step 2 Let x be Ricardo's speed to the beach. Then $x - 10$ is his return speed.

It is always a good idea to sketch the given information in a motion problem. Here we would have

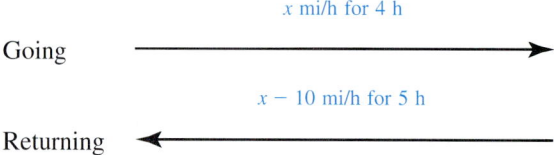

Step 3 Since we know that the distance is the same each way, we can write an equation, using the fact that the product of the rate and the time each way must be the same.

So

$$\text{Distance (going)} = \text{distance (returning)}$$

$$\text{Time} \cdot \text{rate (going)} = \text{time} \cdot \text{rate (returning)}$$

$$\underbrace{4x}_{\substack{\uparrow \\ \text{Time} \cdot \text{rate} \\ \text{(going)}}} = \underbrace{5(x - 10)}_{\substack{\uparrow \\ \text{Time} \cdot \text{rate} \\ \text{(returning)}}}$$

A chart can help summarize the given information. We begin by filling in the information given in the problem.

	Rate	Time	Distance
Going	x	4	
Returning	$x - 10$	5	

Now we fill in the missing information. Here we use the fact that $d = rt$ to complete the chart.

	Rate	Time	Distance
Going	x	4	$4x$
Returning	$x - 10$	5	$5(x - 10)$

From here we set the two distances equal to each other and solve as before.

Step 4 Solve.

$$4x = 5(x - 10)$$
$$4x = 5x - 50$$
$$-x = -50$$
$$x = 50 \text{ mi/h}$$

x was his rate going; x − 10, his rate returning.

So Ricardo's rate going to the beach was 50 mi/h, and his rate returning was 40 mi/h.

Step 5 To check, you should verify that the product of the time and the rate is the same in each direction.

✓ CHECK YOURSELF 7

A plane made a flight (with the wind) between two towns in 2 h. Returning against the wind, the plane's speed was 60 mi/h slower, and the flight took 3 h. What was the plane's speed in each direction?

Example 8 illustrates another way of using the distance relationship.

Example 8 Solving a Motion Problem

Katy leaves Las Vegas for Los Angeles at 10 A.M., driving at 50 mi/h. At 11 A.M. Jensen leaves Los Angeles for Las Vegas, driving at 55 mi/h along the same route. If the cities are 260 mi apart, at what time will they meet?

Step 1 Let's find the time that Katy travels until they meet.

Step 2 Let x be Katy's time.

Then $x - 1$ is Jensen's time.   *Jensen left 1 hr later!*

Again, you should draw a sketch of the given information.

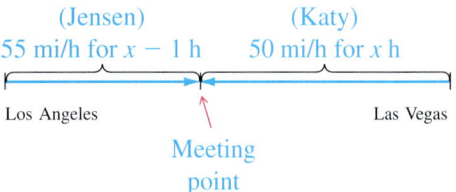

Step 3 To write an equation, we will again need the relationship $d = rt$. From this equation, we can write

$$\text{Katy's distance} = 50x$$

$$\text{Jensen's distance} = 55(x - 1)$$

As before, we can use a table to solve.

	Rate	Time	Distance
Katy	50	x	$50x$
Jensen	55	$x-1$	$55(x-1)$

From the original problem, the sum of those distances is 260 mi, so

$$50x + 55(x - 1) = 260$$

Step 4

$$50x + 55(x - 1) = 260$$
$$50x + 55x - 55 = 260$$
$$105x - 55 = 260$$
$$105x = 315$$
$$x = 3 \text{ h}$$

Be sure to answer the question asked in the problem.

Finally, since Katy left at 10 A.M., the two will meet at 1 P.M. We leave the check of this result to you.

✓ CHECK YOURSELF 8

At noon a jogger leaves one point, running at 8 mi/h. One hour later a bicyclist leaves the same point, traveling at 20 mi/h in the opposite direction. At what time will they be 36 mi apart?

✓ CHECK YOURSELF ANSWERS

1. $h = \dfrac{3V}{B}$.
2. $t = \dfrac{v - a}{g}$.
3. $y = \dfrac{6 - 2x}{3}$ or $y = -\dfrac{2}{3}x + 2$.
4. The interest rate was 6%. 5. The width is 11 in.; the length is 27 in.
6. 40 at 35¢, and 20 at 15¢. 7. 180 mi/h with the wind and 120 mi/h against the wind. 8. At 2 P.M.

Exercises • 2.4

Solve each literal equation for the indicated variable.

1. $P = 4s$ (for s)   Perimeter of a square
2. $V = Bh$ (for B)   Volume of a prism
3. $E = IR$ (for R)   Voltage in an electric circuit
4. $I = Prt$ (for r)   Simple interest
5. $V = LWH$ (for H)   Volume of a rectangular solid
6. $V = \pi r^2 h$ (for h)   Volume of a cylinder
7. $A + B + C = 180$ (for B)   Measure of angles in a triangle
8. $P = I^2 R$ (for R)   Power in an electric circuit
9. $ax + b = 0$ (for x)   Linear equation in one variable
10. $y = mx + b$ (for m)   Slope-intercept form for a line
11. $s = \frac{1}{2}gt^2$ (for g)   Distance
12. $K = \frac{1}{2}mv^2$ (for m)   Energy
13. $x + 5y = 15$ (for y)   Linear equation
14. $2x + 3y = 6$ (for x)   Linear equation
15. $P = 2L + 2W$ (for L)   Perimeter of a rectangle
16. $ax + by = c$ (for y)   Linear equation in two variables
17. $V = \frac{KT}{P}$ (for T)   Volume of a gas
18. $V = \frac{1}{3}\pi r^2 h$ (for h)   Volume of a cone
19. $x = \frac{a+b}{2}$ (for b)   Average of two numbers
20. $D = \frac{C-s}{n}$ (for s)   Depreciation
21. $F = \frac{9}{5}C + 32$ (for C)   Celsius/Fahrenheit
22. $A = P + Prt$ (for t)   Amount at simple interest
23. $S = 2\pi r^2 + 2\pi rh$ (for h)   Total surface area of a cylinder
24. $A = \frac{1}{2}h(B+b)$ (for b)   Area of a trapezoid

---

1. $\dfrac{P}{4}$
2. $\dfrac{V}{h}$
3. $\dfrac{E}{I}$
4. $\dfrac{I}{Pt}$
5. $\dfrac{V}{LW}$
6. $\dfrac{V}{\pi r^2}$
7. $180 - A - C$
8. $\dfrac{P}{I^2}$
9. $\dfrac{b}{a}$
10. $\dfrac{y-b}{x}$
11. $\dfrac{2s}{t^2}$
12. $\dfrac{2K}{v^2}$
13. $\dfrac{15-x}{5}$ or $-\dfrac{1}{5}x + 3$
14. $\dfrac{6-3y}{2}$ or $3 - \dfrac{3}{2}y$
15. $\dfrac{P-2W}{2}$ or $\dfrac{P}{2} - W$
16. $\dfrac{c-ax}{b}$
17. $\dfrac{PV}{K}$
18. $\dfrac{3V}{\pi r^2}$
19. $2x - a$
20. $C - nD$
21. $\dfrac{5}{9}(F-32)$ or $\dfrac{5(F-32)}{9}$
22. $\dfrac{A-P}{Pr}$ or $\dfrac{A}{Pr} - \dfrac{1}{r}$
23. $\dfrac{S - 2\pi r^2}{2\pi r}$ or $\dfrac{S}{2\pi r} - r$
24. $\dfrac{2A - hB}{h}$ or $\dfrac{2A}{h} - B$

Chapter 2 ▪ Equations and Inequalities

25. 3 cm
26. 2.9 in.
27. 5%
28. 18 ft
29. 25°C
30. 8 m
31. $2(x + 4) = 20$
32. $2x + 4 = 20$
33. $3(x - 5) = 21$
34. $3x - 5 = 21$
35. $2x + 3(x + 1) = 48$
36. $4x + 2(x + 2) = 46$
37. 10, 18
38. 13, 16
39. 8, 15

25. **Height of a solid.** A rectangular solid has a base with length 8 centimeters (cm) and width 5 cm. If the volume of the solid is 120 cm^3, find the height of the solid. (See Exercise 5.)

26. **Height of a cylinder.** A cylinder has a radius of 4 inches (in.). If the volume of the cylinder is 144 π in.^{3}, what is the height of the cylinder? (See Exercise 6.)

27. **Interest rate.** A principal of $3000 was invested in a savings account for 3 years. If the interest earned for the period was $450, what was the interest rate? (See Exercise 4.)

28. **Length of a rectangle.** If the perimeter of a rectangle is 60 feet (ft) and the width is 12 ft, find its length.

29. **Temperature conversion.** The high temperature in New York for a particular day was reported at 77°F. How would the same temperature have been given in degrees Celsius? (See Exercise 21.)

30. **Garden length.** Rose's garden is in the shape of a trapezoid. If the height of the trapezoid is 16 meters (m), one base is 20 m, and the area is 224 m^2, find the length of the other base. (See Exercise 24.)

Translate each of the following statements to equations. Let x represent the number in each case.

31. Twice the sum of a number and 4 is 20.

32. The sum of twice a number and 4 is 20.

33. 3 times the difference of a number and 5 is 21.

34. The difference of 3 times a number and 5 is 21.

35. The sum of twice an integer and 3 times the next consecutive integer is 48.

36. The sum of 4 times an odd integer and twice the next consecutive odd integer is 46.

Solve the following word problems.

37. **Number problem.** One number is 8 more than another. If the sum of the smaller number and twice the larger number is 46, find the two numbers.

38. **Number problem.** One number is 3 less than another. If 4 times the smaller number minus 3 times the larger number is 4, find the two numbers.

39. **Number problem.** One number is 7 less than another. If 4 times the smaller number plus 2 times the larger number is 62, find the two numbers.

40. 5, 15

41. 5, 6

42. 9, 11

43. 12 in., 25 in.

44. 7 cm, 16 cm

45. 6 m, 22 m

46. 75 ft, 40 ft

47. Legs, 13 cm; base, 10 cm

48. Legs, 11 in.; base 7 in.

49. 200 $6 tickets, 300 $8 tickets

50. 850 student, 600 nonstudent

51. 30 20¢ stamps, 50 35¢ stamps

52. 85 $10 bills, 40 $20 bills

40. **Number problem.** One number is 10 more than another. If the sum of twice the smaller number and 3 times the larger number is 55, find the two numbers.

41. **Consecutive integers.** Find two consecutive integers such that the sum of twice the first integer and 3 times the second integer is 28. (*Hint:* If x represents the first integer, $x + 1$ represents the next consecutive integer.)

42. **Consecutive integers.** Find two consecutive odd integers such that 3 times the first integer is 5 more than twice the second. (*Hint:* If x represents the first integer, $x + 2$ represents the next consecutive odd integer.)

43. **Dimensions of a rectangle.** The length of a rectangle is 1 inch (in.) more than twice its width. If the perimeter of the rectangle is 74 in., find the dimensions of the rectangle.

44. **Dimensions of a rectangle.** The length of a rectangle is 5 centimeters (cm) less than 3 times its width. If the perimeter of the rectangle is 46 cm, find the dimensions of the rectangle.

45. **Garden size.** The length of a rectangular garden is 4 meters (m) more than 3 times its width. The perimeter of the garden is 56 m. What are the dimensions of the garden?

46. **Size of a playing field.** The length of a rectangular playing field is 5 feet (ft) less than twice its width. If the perimeter of the playing field is 230 ft, find the length and width of the field.

47. **Isosceles triangle.** The base of an isosceles triangle is 3 cm less than the length of the equal sides. If the perimeter of the triangle is 36 cm, find the length of each of the sides.

48. **Isosceles triangle.** The length of one of the equal legs of an isosceles triangle is 3 in. less than twice the length of the base. If the perimeter is 29 in., find the length of each of the sides.

49. **Ticket sales.** Tickets for a play cost $8 for the main floor and $6 in the balcony. If the total receipts from 500 tickets were $3600, how many of each type of ticket were sold?

50. **Ticket sales.** Tickets for a basketball tournament were $6 for students and $9 for nonstudents. Total sales were $10,500, and 250 more student tickets were sold than nonstudent tickets. How many of each type of ticket were sold?

51. **Number of stamps.** Maria bought 80 stamps at the post office in 35¢ and 20¢ denominations. If she paid $23.50 for the stamps, how many of each denomination did she buy?

52. **Money denominations.** A bank teller had a total of 125 $10 bills and $20 bills to start the day. If the value of the bills was $1650, how many of each denomination did he have?

53. 60 coach, 40 berth, 20 sleeping room

54. 500 box seats, 600 grandstand, 1000 bleachers

55. 40 mi/h, 30 mi/h

56. 20 mi/h, 25 mi/h

57. 6 P.M.

58. 4:30 P.M.

59. 2 P.M.

60. 1 P.M.

61. 3 P.M.

62. 4 h

53. **Ticket sales.** Tickets for a train excursion were $120 for a sleeping room, $80 for a berth, and $50 for a coach seat. The total ticket sales were $8600. If there were 20 more berth tickets sold than sleeping room tickets and 3 times as many coach tickets as sleeping room tickets, how many of each type of ticket were sold?

54. **Baseball tickets.** Admission for a college baseball game is $6 for box seats, $5 for the grandstand, and $3 for the bleachers. The total receipts for one evening were $9000. There were 100 more grandstand tickets sold than box seat tickets. Twice as many bleacher tickets were sold as box seat tickets. How many tickets of each type were sold?

55. **Driving speed.** Patrick drove 3 hours (h) to attend a meeting. On the return trip, his speed was 10 miles per hour (mi/h) less and the trip took 4 h. What was his speed each way?

56. **Bicycle speed.** A bicyclist rode into the country for 5 h. In returning, her speed was 5 mi/h faster and the trip took 4 h. What was her speed each way?

57. **Driving speed.** A car leaves a city and goes north at a rate of 50 mi/h at 2 P.M. One hour later a second car leaves, traveling south at a rate of 40 mi/h. At what time will the two cars be 320 mi apart?

58. **Bus distance.** A bus leave a station at 1 P.M., traveling west at an average rate of 44 mi/h. One hour later a second bus leaves the same station, traveling east at a rate of 48 mi/h. At what time will the two buses be 274 mi apart?

59. **Traveling time.** At 8:00 A.M., Catherine leaves on a trip at 45 mi/h. One hour later, Max decides to join her and leaves along the same route, traveling at 54 mi/h. When will Max catch up with Catherine?

60. **Bicycling time.** Martina leaves home at 9 A.M., bicycling at a rate of 24 mi/h. Two hours later, John leaves, driving at the rate of 48 mi/h. At what time will John catch up with Martina?

61. **Traveling time.** Mika leaves Boston for Baltimore at 10:00 A.M., traveling at 45 mi/h. One hour later, Hiroko leaves Baltimore for Boston on the same route, traveling at 50 mi/h. If the two cities are 425 mi apart, when will Mika and Hiroko meet?

62. **Traveling time.** A train leaves town A for town B, traveling at 35 mi/h. At the same time, a second train leaves town B for town A at 45 mi/h. If the two towns are 320 mi apart, how long will it take for the two trains to meet?

63. 360 Douglas fir,
    140 hemlock

64. 350 Douglas fir,
    500 ponderosa pine

63. **Tree inventory.** There are 500 Douglas fir and hemlock trees in a section of forest bought by Hoodoo Logging Co. The company paid an average of $250 for each Douglas fir and $300 for each hemlock. If the company paid $132,000 for the trees, how many of each kind did the company buy?

64. **Tree inventory.** There are 850 Douglas fir and ponderosa pine trees in a section of forest bought by Sawz Logging Co. The company paid an average of $300 for each Douglas fir and $225 for each ponderosa pine. If the company paid $217,500 for the trees, how many of each kind did the company buy?

65. There is a universally agreed on "order of operations" used to simplify expressions. Explain how the order of operations is used in solving equations. Be sure to use complete sentences.

66. A common mistake when solving equations is the following:

    The equation: $\quad\quad 2(x - 2) = x + 3$
    First step in solving: $\quad 2x - 2 = x + 3$

    Write a clear explanation of what error has been made. What could be done to avoid this error?

67. Another very common mistake is in the equation below:

    The equation: $\quad\quad 6x - (x + 3) = 5 + 2x$
    First step in solving: $\quad 6x - x + 3 = 5 + 2x$

    Write a clear explanation of what error has been made and what could be done to avoid the mistake.

68. Write an algebraic equation for the English statement "Subtract 5 from the sum of x and 7 times 3 and the result is 20." Compare you equation with other students. Did you all write the same equation? Are all the equations correct even though they don't look alike? Do all the equations have the same solution? What is wrong? The English statement is *ambiguous*. Write another English statement that leads correctly to more than one algebraic equation. Exchange with another student and see if they think the statement is ambiguous. Notice that the algebra is *not* ambiguous!

SECTION 2.5 Solving Linear Inequalities

2.5 OBJECTIVES

1. Use the notation of inequalities
2. Graph the solution set of an inequality
3. Solve and graph the solution set for an inequality in one variable

"Few things are harder to put up with than a good example."

–Mark Twain

As pointed out earlier in this chapter, an equation is a statement that two expressions are equal. In algebra, an **inequality** is a statement that one expression is less than or greater than another. Four new symbols are used in writing inequalities. The use of two of them is illustrated in Example 1.

Example 1

Reading the Inequality Symbol

To help you remember, the "arrowhead" always points toward the smaller quantity.

$5 < 8$ is an inequality read "5 is less than 8."

$9 > 6$ is an inequality read "9 is greater than 6."

✓ CHECK YOURSELF 1

Fill in the blanks, using the symbols $<$ and $>$.

(a) 12 _____ 8   (b) 20 _____ 25

Just as was the case with equations, inequalities that involve variables may be either true or false depending on the value that we give to the variable. For instance, consider the inequality

$x < 6$

If $x = \begin{cases} 3 & 3 < 6 \text{ is true} \\ 5 & 5 < 6 \text{ is true} \\ -10 & -10 < 6 \text{ is true} \\ 8 & 8 < 6 \text{ is false} \end{cases}$

Therefore 3, 5, and −10 are some *solutions* for the inequality $x < 6$; they make the inequality a true statement. You should see that 8 is *not* a solution. Recall the set of all solutions is the *solution set* for the inequality. Of course, there are many possible solutions.

Since there are so many solutions (an infinite number, in fact), we certainly do not want to try to list them all! A convenient way to show the solution set of an inequality is with the use of a number line.

Example 2

Graphing Inequalities

To graph the solution set for the inequality $x < 6$, we want to include all real numbers that are "less than" 6. This means all numbers *to the left* of 6 on the number line.

We then start at 6 and draw an arrow extending left, as shown:

The colored arrow indicates the direction of the solution set.

Note: The parenthesis at 6 means that we do not include 6 in the solution set (6 is not less than itself). The colored arrow shows all the numbers in the solution set, with the arrowhead indicating that the solution set continues infinitely to the left.

✓ **CHECK YOURSELF 2**

Graph the solution set of $x < -2$.

Two other symbols are used in writing inequalities. They are used with inequalities such as

$$x \geq 5 \quad \text{and} \quad x \leq 2$$

Here $x \geq 5$ is a combination of the two statements $x > 5$ and $x = 5$. It is read "x is greater than or equal to 5." The solution set includes 5 in this case.

The inequality $x \leq 2$ combines the statements $x < 2$ and $x = 2$. It is read "x is less than or equal to 2."

Example 3

Graphing Inequalities

Here the bracket means that we want to include 5 in the solution set.

The solution set for $x \geq 5$ is graphed as follows.

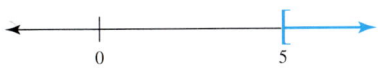

158 Chapter 2 ■ Equations and Inequalities

✓ **CHECK YOURSELF 3**

Graph the solution sets.

(a) $x \leq -4$ (b) $x \geq 3$

We have looked at graphs of the solution sets of some simple inequalities, such as $x < 8$ or $x \geq 10$. Now we will look at more complicated inequalities, such as

$$2x - 3 < x + 4$$

This is called a **linear inequality in one variable.** Only one variable is involved in the inequality, and it appears only to the first power. Fortunately, the methods used to solve this type of inequality are very similar to those we used earlier in this chapter to solve linear equations in one variable. Here is our first property for inequalities.

> **The Addition Property of Inequality**
>
> If $a < b$ then $a + c < b + c$
>
> In words, adding the same quantity to both sides of an inequality gives an **equivalent inequality.**

Equivalent inequalities have exactly the same solution sets.

Example 4 — Solving Inequalities

Solve and graph the solution set for $x - 8 < 7$.

The inequality is solved when an equivalent inequality has the form

$x < \Box$ or $x > \Box$

To solve $x - 8 < 7$, add 8 to both sides of the inequality by the addition property.

$$x - 8 < 7$$
$$x - 8 + 8 < 7 + 8$$
$$x < 15 \qquad \text{The solution}$$

We read

$\{x | x < 15\}$

"every x such that x is less than 15"

The graph of the solution set, $\{x | x < 15\}$, is

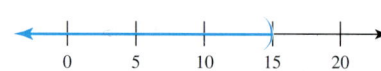

✓ CHECK YOURSELF 4

Solve and graph the solution set for

$$x - 9 > -3$$

Example 5 — Solving Inequalities

Solve and graph the solution set for $4x - 2 \geq 3x + 5$.

As with equations, the addition property allows us to subtract the same quantity from both sides of an inequality.

First, we subtract $3x$ from both sides of the inequality.

$$4x - 2 \geq 3x + 5$$

$$4x - 3x - 2 \geq 3x - 3x + 5$$

$$x - 2 \geq 5 \quad \text{Now we add 2 to both sides.}$$

We subtracted $3x$ and then added 2 to both sides. If these steps are done in the other order, the resulting inequality will be the same.

$$x - 2 + 2 \geq 5 + 2$$

$$x \geq 7$$

The graph of the solution set, $\{x | x \geq 7\}$, is

✓ CHECK YOURSELF 5

Solve and graph the solution set.

$$7x - 8 \leq 6x + 2$$

You will also need a rule for multiplying on both sides of an inequality. Here you'll have to be a bit careful. There is a difference between the multiplication property for inequalities and that for equations. Look at the following:

$$2 < 7 \qquad \text{A true inequality}$$

Let's multiply both sides by 3.

$$2 < 7$$

$$3 \cdot 2 < 3 \cdot 7$$

$$6 < 21 \qquad \text{A true inequality}$$

Now we multiply both sides by -3.

$$2 < 7$$
$$(-3)(2) < (-3)(7)$$
$$-6 < -21 \qquad \textit{Not} \text{ a true inequality}$$

Let's try something different.

$$2 < 7$$
$$(-3)(2) > (-3)(7) \qquad \text{Change the "sense" of the inequality:}$$
$$\qquad\qquad\qquad\qquad < \text{ becomes } >.$$
$$-6 > -21 \qquad \text{This is now a true inequality.}$$

This suggests that multiplying both sides of an inequality by a negative number changes the "sense" of the inequality.

When both sides of an inequality are multiplied by the same negative *number, it is necessary to* reverse the sense *of the inequality to give an equivalent inequality.*

> **The Multiplication Property of Inequality**
>
> If $a < b$ then $ac < bc$ where $c > 0$
>
> and $ac > bc$ where $c < 0$
>
> In words, multiplying both sides of an inequality by the same *positive* number gives an equivalent inequality.

Example 6 — Solving and Graphing Inequalities

(a) Solve and graph the solution set for $5x < 30$.

Multiplying both sides of the inequality by $\frac{1}{5}$ gives

$$\frac{1}{5}(5x) < \frac{1}{5}(30)$$

Simplifying, we have

$$x < 6$$

The graph of the solution set, $\{x | x < 6\}$, is

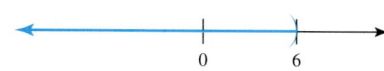

(b) Solve and graph the solution set for $-4x \geq 28$.

In this case we want to multiply both sides of the inequality by $-\dfrac{1}{4}$ to convert the coefficient of x to one on the left.

$$\left(-\dfrac{1}{4}\right)(-4x) \leq \left(-\dfrac{1}{4}\right)(28)$$

Reverse the sense of the inequality because you are multiplying by a negative number!

or $\qquad x \leq -7$

The graph of the solution set, $\{x \mid x \leq -7\}$, is

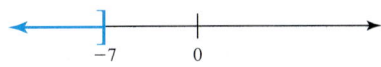

✓ CHECK YOURSELF 6

Solve and graph the solution sets:

(a) $7x > 35$ (b) $-8x \leq 48$

Example 7 illustrates the use of the multiplication property when fractions are involved in an inequality.

Example 7 — Solving and Graphing Inequalities

(a) Solve and graph the solution set for

$$\dfrac{x}{4} > 3$$

Here we multiply both sides of the inequality by 4. This will isolate x on the left.

$$4\left(\dfrac{x}{4}\right) > 4(3)$$

$$x > 12$$

The graph of the solution set, $\{x \mid x > 12\}$, is

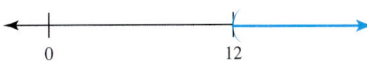

(b) Solve and graph the solution set for

$$-\dfrac{x}{6} \geq -3$$

In this case, we multiply both sides of the inequality by -6:

$$(-6)\left(-\frac{x}{6}\right) \leq (-6)(-3)$$

$$x \leq 18$$

Note that we reverse the sense of the inequality because we are multiplying by a negative number.

The graph of the solution set, $\{x \mid x \leq 18\}$, is

✓ CHECK YOURSELF 7

Solve and graph the solution sets for the following inequalities.

(a) $\dfrac{x}{5} \leq 4$ (b) $-\dfrac{x}{3} < -7$

Example 8   Solving and Graphing Inequalities

(a) Solve and graph the solution set for $5x - 3 < 2x$.

First, add 3 to both sides to undo the subtraction on the left.

$$5x - 3 < 2x$$
$$5x - 3 + 3 < 2x + 3 \qquad \text{Add 3 to both sides to undo the subtraction.}$$
$$5x < 2x + 3$$

Now subtract $2x$, so that only the number remains on the right.

$$5x < 2x + 3$$
$$5x - 2x < 2x - 2x + 3 \qquad \text{Subtract } 2x \text{ to isolate the number on the right.}$$
$$3x < 3$$

Note that the multiplication property also allows us to divide both sides by a nonzero number.

Next *divide* both sides by 3.

$$\frac{3x}{3} < \frac{3}{3}$$
$$x < 1$$

The graph of the solution set, $\{x \mid x < 1\}$, is

(b) Solve and graph the solution set for $2 - 5x < 7$.

$$2 - 5x < 7$$
$$2 - 2 - 5x < 7 - 2 \qquad \text{Subtract 2.}$$
$$-5x < 5$$
$$\frac{-5x}{-5} > \frac{5}{-5} \qquad \text{Divide by } -5. \text{ Be sure to reverse the sense of the inequality.}$$

or $\qquad x > -1$

The graph of the solution set, $\{x | x > -1\}$, is

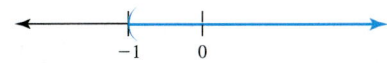

✓ CHECK YOURSELF 8
Solve and graph the solution sets.

(a) $4x + 9 \geq x$ (b) $5 - 6x < 41$

As with equations, when solving an inequality we will collect all variable terms on one side and all constant terms on the other.

Example 9

Solving and Graphing Inequalities

Solve and graph the solution set for $5x - 5 \geq 3x + 4$.

$$5x - 5 \geq 3x + 4$$
$$5x - 5 + 5 \geq 3x + 4 + 5 \qquad \text{Add 5.}$$
$$5x \geq 3x + 9$$
$$5x - 3x \geq 3x - 3x + 9 \qquad \text{Subtract } 3x.$$
$$2x \geq 9$$
$$\frac{2x}{2} \geq \frac{9}{2} \qquad \text{Divide by 2.}$$
$$x \geq \frac{9}{2}$$

The graph of the solution set, $\left\{x | x \geq \dfrac{9}{2}\right\}$, is

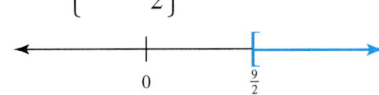

✓ CHECK YOURSELF 9

Solve and graph the solution set for

$$8x + 3 < 4x - 13$$

Be especially careful when negative coefficients occur in the solution process.

Example 10 — Solving and Graphing Inequalities

Solve and graph the solution set for $x + 2 < \frac{5}{2}x - 1$.

$$2(x + 2) < 2(\frac{5}{2}x - 1) \quad \text{Multiply by the LCD.}$$

$$2x + 4 < 5x - 2$$

$$2x + 4 - 4 < 5x - 2 - 4 \quad \text{Subtract 4.}$$

$$2x < 5x - 6$$

$$2x - 5x < 5x - 5x - 6 \quad \text{Subtract } 5x.$$

$$-3x < -6 \quad \text{Divide by } -3, \text{ and reverse the sense of the inequality.}$$

$$\frac{-3x}{-3} > \frac{-6}{-3}$$

$$x > 2$$

The graph of the solution set, $\{x | x > 2\}$, is

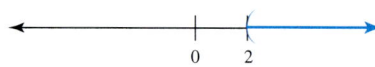

✓ CHECK YOURSELF 10

Solve and graph the solution set.

$$5x + 12 \geq 10x - 8$$

Example 11   Solving and Graphing Inequalities

Solve and graph the solution set for

$$5(x - 2) \geq -8$$

Applying the distributive property on the left yields

$$5x - 10 \geq -8$$

Solving as before yields

$$5x - 10 + 10 \geq -8 + 10 \quad \text{Add 10.}$$
$$5x \geq 2$$

or

$$x \geq \frac{2}{5} \quad \text{Divide by 5.}$$

The graph of the solution set, $\left\{x \mid x \geq \frac{2}{5}\right\}$, is

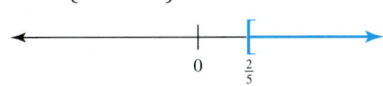

✓ CHECK YOURSELF 11

Solve and graph the solution set.

$$4(x + 3) < 9$$

Some applications are solved by using an inequality instead of an equation. Example 12 illustrates such an application.

Example 12   Solving an Application with Inequalities

Mohammed needs an average score of 92 or higher on four tests to get an A. So far his scores are 94, 89, and 88. What score on the fourth test will get him an A?

What do you need to find?

Step 1 We are looking for the score that will, when combined with the other scores, give Mohammed an A.

Assign a letter to the unknown.

Step 2 Let x represent a fourth-test score that will get him an A.

Write an inequality.

Step 3 The inequality will have the average (mean) on the left side, which must be greater than or equal to the 92 on the right.

$$\frac{94 + 89 + 88 + x}{4} \geq 92$$

Solve the equation for x.

Step 4 First, multiply both sides by 4:

$$94 + 89 + 88 + x \geq 368$$

Then add the test scores:

$$183 + 88 + x \geq 368$$
$$271 + x \geq 368$$

Subtracting 271 from both sides,

$x \geq 97$ Mohammed needs a test result of 97 or higher.

Step 5 To check the solution, we find the mean of the four test scores, 94, 89, 88, and 97.

$$\frac{94 + 89 + 88 + 97}{4} = \frac{368}{4} = 92$$

✓ CHECK YOURSELF 12

Felicia needs an average score of at least 75 on five tests to get a passing grade in her health class. On her first four tests she has scores of 68, 79, 71, and 70. What score on the fifth test will give her a passing grade?

The following outline (or algorithm) summarizes our work in this section.

Solving Linear Inequalities

Step 1 Remove grouping symbols and fraction coefficients and combine any like terms appearing on either side of the inequality.

Step 2 Apply the addition property to write an equivalent inequality with the variable term on one side of the inequality and the number on the other.

Step 3 Apply the multiplication property to write an equivalent inequality with the variable isolated on one side of the inequality. Be sure to reverse the sense of the inequality if you multiply or divide by a negative number. The solution set derived in step 3 can then be graphed on a number line.

Section 2.5 ■ Solving Linear Inequalities

✓ CHECK YOURSELF ANSWERS

1. (a) $>$; (b) $<$.
2. (number line showing $x \leq -2$)
3. (a) (number line showing $x \leq -4$)   (b) (number line showing $x < 3$)
4. $x > 6$
5. $x \leq 10$
6. (a) $x > 5$ (b) $x \geq -6$
7. (a) $x \leq 20$ (b) $x > 21$
8. (a) $x \geq -3$ (b) $x > -6$
9. $x < -4$
10. $x \leq 4$
11. $x < -\dfrac{3}{4}$
12. 87 or greater.

Exercises • 2.5

Complete the statements, using the symbol < or >.

1. 5 _____ 10
2. 9 _____ 8
3. 7 _____ −2
4. 0 _____ −5
5. 0 _____ 4
6. −10 _____ −5
7. −2 _____ −5
8. −4 _____ −11

Write each inequality in words.

9. $x < 3$
10. $x \leq -5$
11. $x \geq -4$
12. $x < -2$
13. $-5 \leq x$
14. $2 < x$

Graph the solution set of each of the following inequalities.

15. $x > 2$
16. $x < -3$

 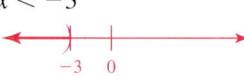

17. $x < 9$
18. $x > 4$

19. $x > 1$
20. $x < -2$

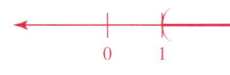 

21. $x < 8$
22. $x > 3$

23. $x > -5$
24. $x < -4$

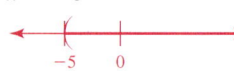 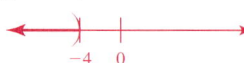

25. $x \geq 9$
26. $x \geq 0$

27. $x < 0$
28. $x \leq -3$

---

1. <
2. >
3. >
4. >
5. <
6. <
7. >
8. >
9. x is less than 3
10. x is less than or equal to −5
11. x is greater than or equal to −4
12. x is less than −2
13. −5 is less than or equal to x
14. 2 is less than x

Section 2.5 • Solving Linear Inequalities **169**

29. $\{x|x<13\}$
30. $\{x|x\leq -1\}$
31. $\{x|x\geq 2\}$
32. $\{x|x>-3\}$
33. $\{x|x<7\}$
34. $\{x|x\geq -4\}$
35. $\{x|x\leq 8\}$
36. $\{x|x>-2\}$
37. $\{x|x\geq 8\}$
38. $\{x|x\leq -8\}$
39. $\{x|x<-9\}$
40. $\{x|x>10\}$
41. $\{x|x\leq 3\}$
42. $\{x|x>4\}$
43. $\{x|x>-7\}$
44. $\{x|x\leq -3\}$
45. $\{x|x\leq -3\}$
46. $\{x|x>-5\}$
47. $\{x|x>6\}$
48. $\{x|x\leq 4\}$
49. $\{x|x>20\}$
50. $\{x|x\leq -9\}$
51. $\{x|x\leq 6\}$
52. $\{x|x>-20\}$

Solve and graph the solution set of each of the following inequalities.

29. $x-7<6$

30. $x+5\leq 4$

31. $x+8\geq 10$
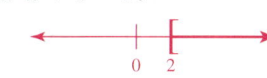

32. $x-11>-14$
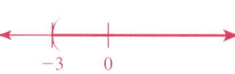

33. $5x<4x+7$
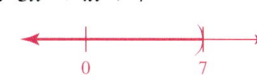

34. $3x\geq 2x-4$

35. $6x-8\leq 5x$

36. $3x+2>2x$
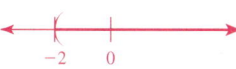

37. $4x-3\geq 3x+5$

38. $5x+2\leq 4x-6$

39. $7x+5<6x-4$

40. $8x-7>7x+3$

41. $3x\leq 9$
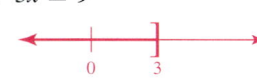

42. $5x>20$

43. $5x>-35$

44. $7x\leq -21$

45. $-6x\geq 18$
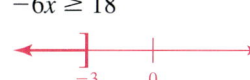

46. $-9x<45$

47. $-10x<-60$
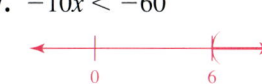

48. $-12x\geq -48$

49. $\dfrac{x}{4}>5$

50. $\dfrac{x}{3}\leq -3$
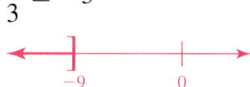

51. $-\dfrac{x}{2}\geq -3$

52. $-\dfrac{x}{5}<4$

Answers (left column)

53. $\{x \mid x < 9\}$
54. $\{x \mid x \geq -12\}$
55. $\{x \mid x > 4\}$
56. $\{x \mid x \leq -3\}$
57. $\{x \mid x < 1\}$
58. $\left\{x \mid x \geq -\dfrac{3}{5}\right\}$
59. $\{x \mid x < -1\}$
60. $\{x \mid x \geq -4\}$
61. $\{x \mid x \leq -6\}$
62. $\{x \mid x > 7\}$
63. $\{x \mid x \leq 9\}$
64. $\{x \mid x > 9\}$
65. $\{x \mid x > -5\}$
66. $\{x \mid x \leq -10\}$
67. $\left\{x \mid x < \dfrac{7}{4}\right\}$
68. $\left\{x \mid x \geq -\dfrac{5}{3}\right\}$
69. $\{x \mid x < 8\}$
70. $\{x \mid x \geq -6\}$
71. $\left\{x \mid x \geq -\dfrac{5}{2}\right\}$
72. $\left\{x \mid x < \dfrac{5}{6}\right\}$
73. $\left\{x \mid x \leq \dfrac{3}{2}\right\}$
74. $\left\{x \mid x > \dfrac{4}{3}\right\}$
75. $\left\{x \mid x < -\dfrac{2}{3}\right\}$
76. $\left\{x \mid x \geq \dfrac{5}{4}\right\}$

Exercises

53. $\dfrac{2x}{3} < 6$
54. $\dfrac{3x}{4} \geq -9$ 
55. $5x > 3x + 8$ 
56. $4x \leq x - 9$ 
57. $5x - 2 < 3x$
58. $7x + 3 \geq 2x$ 
59. $3 - 2x > 5$ 
60. $5 - 3x \leq 17$ 
61. $2x \geq 5x + 18$ 
62. $3x < 7x - 28$
63. $5x - 3 \leq 3x + 15$ 
64. $8x + 7 > 5x + 34$ 
65. $9x + 7 > 2x - 28$ 
66. $10x - 5 \leq 8x - 25$ 
67. $7x - 5 < 3x + 2$
68. $5x - 2 \geq 2x - 7$ 
69. $5x + 7 > 8x - 17$ 
70. $4x - 3 \leq 9x + 27$ 
71. $3x - 2 \leq 5x + 3$ 
72. $2x + 3 > 8x - 2$ 
73. $4(x + 7) \leq 2x + 31$ 
74. $6(x - 5) > 3x - 26$ 
75. $2(x - 7) > 5x - 12$ 
76. $3(x + 4) \leq 7x + 7$

Section 2.5 ■ Solving Linear Inequalities

77. $x + 5 > 3$
78. $x - 3 \leq 5$
79. $2x - 4 \leq 7$
80. $x + 10 > -2$
81. $4x - 15 > x$
82. $2x + 28 \leq 6x$
83. a
84. f
85. c
86. d
87. b
88. e
89. $P < 1000$
90. $C \geq 9M$
91. $x \geq 88$
92. $x \geq 74$
93. Sales $>$ $10,000
94. $0.36 + 0.21(t - 1) \leq 3$

Translate the following statements into inequalities. Let x represent the number in each case.

77. 5 more than a number is greater than 3.

78. 3 less than a number is less than or equal to 5.

79. 4 less than twice a number is less than or equal to 7.

80. 10 more than a number is greater than negative 2.

81. 4 times a number, decreased by 15, is greater than that number.

82. 2 times a number, increased by 28, is less than or equal to 6 times that number.

Match each inequality on the right with a statement on the left.

83. x is nonnegative   (a) $x \geq 0$

84. x is negative   (b) $x \geq 5$

85. x is no more than 5   (c) $x \leq 5$

86. x is positive   (d) $x > 0$

87. x is at least 5   (e) $x < 5$

88. x is less than 5   (f) $x < 0$

89. Panda population. There are fewer than 1000 wild giant pandas left in the bamboo forests of China. Write an inequality expressing this relationship.

90. Forestry. Let C represent the amount of Canadian forest and M represent the amount of Mexican forest. Write an inequality showing the relationship of the forests of Mexico and Canada if Canada contains at least 9 times as much forest as Mexico.

91. Test scores. To pass a course with a grade of B or better, Liza must have an average of 80 or more. Her grades on three tests are 72, 81, and 79. Write an inequality representing the score that Liza must get on the fourth test to obtain a B average or better for the course.

92. Test scores. Sam must have an average of 70 or more in his summer course in order to obtain a grade of C. His first three test grades were 75, 63, and 68. Write an inequality representing the score that Sam must get in the last test in order to get a C grade.

93. Commission. Juanita is a salesperson for a manufacturing company. She may choose to receive $500 or 5 percent commission on her sales as payment for her work. How much does she need to sell to make the 5 percent offer a better deal?

94. Telephone costs. The cost for a long distance telephone call is $0.36 for the first minute and $0.21 for each additional minute or portion thereof. The total cost of the call cannot exceed $3. Write an inequality representing the number of minutes a person could talk without exceeding $3.

95. You are the office manager for a small company. You need to acquire a new copier for the office. You find a suitable one that leases for $250 a month from the copy machine company. It costs 2.5¢ per copy to run the machine. You purchase paper for $3.50 a ream (500 sheets). If your copying budget is no more than $950 per month, is this machine a good choice? Write a brief recommendation to the Purchasing Department. Use equations and inequalities to explain your recommendation.

96. Nutritionists recommend that, for good health, no more than 30% of our daily intake of calories should come from fat. Algebraically, we can write this as $f \leq 0.30(c)$, where f = calories from fat and c = total calories for the day. But this does not mean that everything we eat must meet this requirement. For example, if you eat $\frac{1}{2}$ cup of Ben and Jerry's vanilla ice cream for dessert after lunch, you are eating a total of 250 calories, of which 150 are from fat. This amount is considerably more than 30 percent from fat, but if you are careful about what you eat the rest of the day, you can stay within the guidelines.

Set up an inequality based on your normal caloric intake. Solve the inequality to find how many calories in fat you could eat over the day and still have no more than 30% of your daily calories from fat. The American Heart Association says that to maintain your weight, your daily caloric intake should be 15 calories for every pound. You can compute this number to estimate the number of calories a day you normally eat. Do some research in your grocery store or library to determine what foods satisfy the requirements for your diet for the rest of the day. There are 9 calories in every gram of fat; many food labels give the amount of fat only in grams.

97. Your aunt calls to ask your help in making a decision about buying a new refrigerator. She says that she found two that seem to fit her needs, and both are supposed to last at least 14 years, according to *Consumer Reports*. The initial cost for one refrigerator is $712, but it only uses 88 kilowatt-hours (kWh) per month. The other refrigerator costs $519 and uses an estimated 100 kWh/month. You do not know the price of electricity per kilowatt-hour where your aunt lives, so you will have to decide what in cents per kilowatt-hour will make the first refrigerator cheaper to run for its 14 years of expected usefulness. Write your aunt a letter explaining what you did to calculate this cost, and tell her to make her decision based on how the kilowatt-hour rate she has to pay in her area compares with your estimation.

SECTION 2.6 Absolute Value Equations and Inequalities

2.6 OBJECTIVES

1. Find the absolute value of an expression
2. Solve an absolute value equation
3. Solve an absolute value inequality

Equations may contain absolute value notation in their statements. In this section, we will look at algebraic solutions to statements that include absolute values. First, we will review the concept of absolute value.

The **absolute value** of a signed number is the distance from that signed number to 0. Because absolute value is a distance, it is always positive. Formally, we say

The absolute value of a number x is given by

$$|x| = \begin{cases} -x \text{ if } x < 0 \\ x \text{ if } x \geq 0 \end{cases}$$

Example 1

Finding the Absolute Value of a Number

Find the absolute value for each expression.

(a) $|-3|$ (b) $|7 - 2|$ (c) $|-7 - 2|$

(a) Because $-3 < 0$, $|-3| = -(-3) = 3$.

(b) $|7 - 2| = |5|$ Because $5 \geq 0$, $|5| = 5$.

(c) $|-7 - 2| = |-9|$ Because $-9 < 0$, $|-9| = -(-9) = 9$.

CHECK YOURSELF 1

Find the absolute value for each expression.

(a) $|12|$ (b) $|-9 + 5|$ (c) $|-3 - 4|$

173

174 Chapter 2 ▪ Equations and Inequalities

Given an equation like

$$|x| = 5$$

there are two possible solutions. The value of x could be 5 or -5. In either case, the absolute value would be 5. This can be generalized in the following property of absolute value equations.

> **Absolute Value Equations, Property 1**
>
> If $\quad\quad\quad |x| = p$
>
> then $\quad\quad x = p \text{ or } x = -p$

We'll use this property in the next several examples.

Example 2 — Solving an Absolute Value Equation

Solve the equation

$$|x - 3| = 4$$

From Property 1 above, we know that the expression inside the absolute value signs, $(x - 3)$, must either equal 4 or -4. We set up two equations and solve them both.

Caution

Be Careful! A common mistake is to solve only the equation $(x - 3) = 4$. You must solve *both* equivalent equations to find the **two** required solutions.

$(x - 3) = 4$	$(x - 3) = -4$	
$x - 3 = 4$	$x - 3 = -4$	Add 3 to both sides of the equation.
$x = 7$	$x = -1$	

We arrive at the solution set, $\{-1, 7\}$.

✓ **CHECK YOURSELF 2**

Find the solution set for the equation.

$$|x - 2| = 3$$

We will use Property 1 to solve subsequent examples in this section.

Example 3 Solving an Absolute Value Equation

Solve for x.

$$|3x - 2| = 4$$

From Property 1, we know that $|3x - 2| = 4$ is equivalent to the equations

$$3x - 2 = 4 \quad \text{or} \quad 3x - 2 = -4 \quad \text{Add 2.}$$
$$3x = 6 \quad\quad\quad\quad 3x = -2 \quad\quad \text{Divide by 3.}$$
$$x = 2 \quad\quad\quad\quad x = -\frac{2}{3}$$

The solution set is $\left\{-\frac{2}{3}, 2\right\}$. These solutions are easily checked by replacing x with $-\frac{2}{3}$ and 2 in the original absolute value equation.

✓ CHECK YOURSELF 3

Solve for x.

$$|4x + 1| = 9$$

An equation involving absolute value may have to be rewritten before you can apply Property 1. Consider Example 4.

Example 4 Solving an Absolute Value Equation

Solve for x.

$$|2 - 3x| + 5 = 10$$

To use Property 1, we must first isolate the absolute value on the left side of the equation. This is easily done by subtracting 5 from both sides for the result

$$|2 - 3x| = 5$$

We can now proceed as before by using Property 1.

$$2 - 3x = 5 \quad \text{or} \quad 2 - 3x = -5 \qquad \text{Subtract 2.}$$
$$-3x = 3 \qquad\qquad -3x = -7 \qquad \text{Divide by } -3.$$
$$x = -1 \qquad\qquad x = \frac{7}{3}$$

The solution set is $\left\{-1, \dfrac{7}{3}\right\}$.

✓ CHECK YOURSELF 4

Solve for x.

$$|5 - 2x| - 4 = 7$$

In some applications, there is more than one absolute value in an equation. Consider an equation of the form

$$|x| = |y|$$

Since the absolute values of x and y are equal, x and y are the same distance from 0, which means they are either *equal* or *opposite in sign*. This leads to a second general property of absolute value equations.

Absolute Value Equations, Property 2

If $\quad |x| = |y|$

then $\quad x = y$ or $x = -y$

Let's look at an application of this second property in Example 5.

Example 5 — Solving Equations with Two Absolute Value Expressions

Solve for x.

$$|3x - 4| = |x + 2|$$

By Property 2, we can write

$$3x - 4 = x + 2 \quad \text{or} \quad 3x - 4 = -(x + 2)$$

$$3x - 4 = -x - 2 \quad \text{Add 4 to both sides.}$$

$$3x = x + 6 \qquad\qquad 3x = -x + 2 \quad \text{Isolate } x.$$

$$2x = 6 \qquad\qquad 4x = 2 \quad \text{Divide by 2.}$$

$$x = 3 \qquad\qquad x = \frac{1}{2}$$

The solution set is $\left\{\dfrac{1}{2}, 3\right\}$.

✓ CHECK YOURSELF 5

Solve for x.

$$|4x - 1| = |x + 5|$$

Now, we will look at two types of inequality statements that arise frequently in mathematics. Consider a statement such as

$$-2 < x < 5$$

It is called a **compound** inequality because it combines the two inequalities

$$-2 < x \text{ and } x < 5$$

Because there are two inequality signs in a single statement, these are sometimes called double inequalities.

When we begin with a compound inequality such as

$$-3 \leq 2x + 1 \leq 7$$

we find an equivalent statement in which the variable is isolated in the middle. The next example illustrates.

Example 6 Solving a Compound Inequality

Solve and graph the compound inequality.

$$-3 \leq 2x + 1 \leq 7$$

We are really applying the additive property to each of the two inequalities that make up the double-inequality statement.

First, we subtract 1 from each of the three members of the compound inequality.

$$-3 - 1 \leq 2x + 1 - 1 \leq 7 - 1$$

or

$$-4 \leq 2x \leq 6$$

We now divide by 2 to isolate the variable x.

When we divide by a positive number, the sense of the inequality is preserved.

$$-\frac{4}{2} \leq \frac{2x}{2} \leq \frac{6}{2}$$

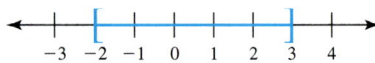

$$-2 \leq x \leq 3$$

The solution set consists of all numbers between -2 and 3, including -2 and 3, and is written

$$\{x | -2 \leq x \leq 3\}$$

That set is graphed below.

Note: Our solution set is equivalent to

$$\{x | x \geq -2 \text{ and } x \leq 3\}$$

Look at the individual graphs.

$\{x | x \geq -2\}$

$\{x | x \leq 3\}$

$\{x | x \geq -2 \text{ and } x \leq 3\}$

Using set notation, we can write

$\{x | x \geq -2\} \cap \{x | x \leq 3\}$

Because the connecting word is "and," we want the *intersection* of the sets, that is, those numbers common to both sets.

✓ CHECK YOURSELF 6

Solve and graph the double inequality.

$$-5 < 3 + 2x < 5$$

A compound inequality may also consist of two inequality statements connected by the word "or." The following example illustrates the solution of that type of compound inequality.

Example 7

Simplifying a Compound Inequality

Solve and graph the inequality

$$2x - 3 < -5 \quad \text{or} \quad 2x - 3 > 5$$

In this case, we must work with each of the inequalities *separately*.

$2x - 3 < -5$	or	$2x - 3 > 5$	Add 3.
$2x < -2$		$2x > 8$	Divide by 2.
$x < -1$		$x > 4$	

The graph of the solution set, $\{x|x < -1 \text{ or } x > 4\}$, is shown.

$$\{x|x < -1 \text{ or } x > 4\}$$

<-+--)--+--+--+--+--(--+--+->
 -3 -2 -1 0 1 2 3 4 5 6

In set notation we can write
$\{x|x < -1\} \cup \{x|x > 4\}$

Note that since the connecting word is "or" in this case, the solution set of the original inequality is the *union* of the two sets, that is, those numbers that belong to either or both of the sets.

✓ CHECK YOURSELF 7

Solve and graph the inequality.

$$3x - 4 \leq -7 \quad \text{or} \quad 3x - 4 \geq 7$$

Imagine that you've received a phone call from a friend who says that his car is stuck on the freeway, within 3 miles of mileage marker 255. Where is he? He must be somewhere between mileage marker 252 and marker 258. What you've actually solved here is an example of an **absolute value inequality.** Absolute value inequalities are in one of two forms.

$$|x - a| < b \quad \text{or} \quad |x - a| > b$$

The mileage marker example is of the first form. We could say

$$|x - 255| < 3$$

To solve an equation of this type, use the following rule.

Solving Absolute Value Inequalities I

If $|x - a| < b$

then $-b < x - a < b$

In our example, because $|x - 255| < 3$,

$$-3 < x - 255 < 3$$

Adding 255 to each member of the inequality, we get

$$-3 + 255 < x - 255 + 255 < 3 + 255$$

or

$$252 < x < 258$$

Example 8 Solving an Absolute Value Inequality

Solve the following inequality, then graph the solution set.

$$|x - 5| < 9$$

According to the rule above, we have

$$|x - 5| < 9$$
$$-9 < x - 5 < 9 \qquad \text{Add 5 to each member.}$$
$$-9 + 5 < x < 9 + 5$$
$$-4 < x < 14$$

Graphing the solution set, $\{x | -4 < x < 14\}$, we get,

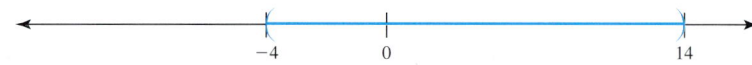

✓ CHECK YOURSELF 8

Solve the following inequality, then graph the solution set.

$$|x - 25| < 136$$

What about inequalities of the form $|x - a| > b$? The following rule applies.

Section 2.6 ■ Absolute Value Equations and Inequalities   181

> **Solving Absolute Value Inequalities II**
>
> If $\quad |x - a| > b$
>
> then $\quad x - a > b \quad \text{or} \quad x - a < -b$

Example 9 Solving Absolute Value Inequalities

Solve the following inequality, then graph the solution set.

$$|x - 7| > 19$$

By the rule above,

$$x - 7 > 19 \quad \text{or} \quad x - 7 < -19$$

Solving each inequality by adding 7 to each side, we get

$$x > 26 \quad \text{or} \quad x < -12$$

Now we can graph the solution set.

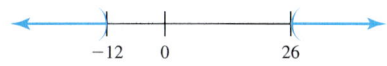

✓ CHECK YOURSELF 9

Solve the following inequality, then graph the solution set.

$$|x - 2| > 7$$

✓ CHECK YOURSELF ANSWERS

1. (a) 12; (b) 4; (c) 7.

2. $|x - 2| = 3$
   $x - 2 = 3 \quad \text{or} \quad x - 2 = -3$
   $\quad x = 5 \qquad\qquad\quad x = -1$
   $\{-1, 5\}$.

3. $\left\{-\dfrac{5}{2}, 2\right\}$.   4. $\{-3, 8\}$.   5. $\left\{-\dfrac{4}{5}, 2\right\}$.

6. $\{x | -4 < x < 1\}$.

7. $\left\{x | x \leq -1 \quad \text{or} \quad x \geq \dfrac{11}{3}\right\}$.

8. $\{x | -111 < x < 161\}$.   9. $\{x | x > 9 \quad \text{or} \quad x < -5\}$.

Exercises ▪ 2.6

1. 15	2. 21
3. 15	4. 18
5. 5	6. 12
7. 4	8. 34
9. 31	10. 46
11. 8	12. 1
13. 25	14. 25
15. {5, −5}	16. {−6, 6}
17. {5, −3}	18. {−4, 8}
19. {3.5, −2.5}	
20. $\left\{-2, \dfrac{10}{3}\right\}$	
21. {1.5, 4.5}	
22. $\left\{-\dfrac{1}{3}, \dfrac{11}{3}\right\}$	
23. No solution	
24. No solution	
25. {0, 4}	26. {−12, 2}
27. {7, 1}	28. {−4, 3}
29. $\left\{-\dfrac{2}{3}, 4\right\}$	30. $\left\{-4, \dfrac{2}{5}\right\}$
31. $\left\{-\dfrac{2}{3}, \dfrac{6}{7}\right\}$	32. $\left\{-\dfrac{5}{4}, \dfrac{1}{6}\right\}$
33. $\left\{2, -\dfrac{4}{9}\right\}$	34. All reals
35. {x|2 ≤ x ≤ 4}	
36. {x|1 < x < 6}	
37. {x|−4 < x < 2}	
38. {x|−2 ≤ x ≤ 3}	
39. $\left\{x \mid 2 \le x \le \dfrac{9}{2}\right\}$	
40. {x|1 < x < 3}	
41. {x|−2 < x < 1}	
42. $-5 \le x \le \dfrac{5}{2}$	
43. x < −2 or x > 4	
44. {x|x < −7 or x > 3}	

In Exercises 1 to 14, find the absolute value for each expression.

1. $|15|$ **2.** $|21|$ **3.** $|-15|$ **4.** $|-18|$

5. $|8 - 3|$ **6.** $|35 - 23|$ **7.** $|-12 + 8|$ **8.** $|-23 - 11|$

9. $|-12 - 19|$ **10.** $|-19 - 27|$ **11.** $-|-19| + |-27|$ **12.** $|-13| - |-12|$

13. $|-13| + |12|$ **14.** $|-13 - 12|$

In Exercises 15 to 34, solve the equations.

15. $|x| = 5$ **16.** $|x| = 6$ **17.** $|x - 1| = 4$

18. $|x - 2| = 6$ **19.** $|2x - 1| = 6$ **20.** $|3x - 2| = 8$

21. $3|2x - 6| = 9$ **22.** $2|3x - 5| = 12$ **23.** $|5x - 2| = -3$

24. $|4x - 9| = -8$ **25.** $|x - 2| + 3 = 5$ **26.** $|x + 5| - 2 = 5$

27. $8 - |x - 4| = 5$ **28.** $10 - |2x + 1| = 3$ **29.** $|2x - 1| = |x + 3|$

30. $|3x + 1| = |2x - 3|$ **31.** $|5x - 2| = |2x - 4|$ **32.** $|5x + 2| = |x - 3|$

33. $|7x - 3| = |2x + 7|$ **34.** $|x - 2| = |2 - x|$

Solve each of the following compound inequalities. Then graph the solution set.

35. $3 \le x + 1 \le 5$ **36.** $-2 < x - 3 < 3$

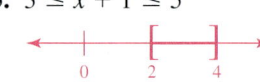   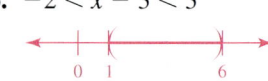

37. $-8 < 2x < 4$ **38.** $-6 \le 3x < 9$

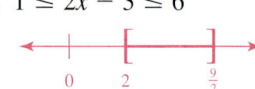   

39. $1 \le 2x - 3 \le 6$ **40.** $-2 < 3x - 5 < 4$

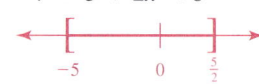

41. $-1 < 5 + 3x < 8$ **42.** $-7 \le 3 + 2x < 8$

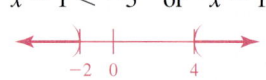   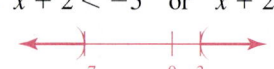

Solve each of the following compound inequalities. Then graph the solution set.

43. $x - 1 < -3$ or $x - 1 > 3$ **44.** $x + 2 < -5$ or $x + 2 > 5$

Section 2.6 ■ Absolute Value Equations and Inequalities 183

45. $\{x | x < -3 \text{ or } x > 4\}$
46. $\{x | x < -3 \text{ or } x > 0\}$
47. $\left\{x | x < -2 \text{ or } x > \dfrac{8}{3}\right\}$
48. $\left\{x | x < -2 \text{ or } x > \dfrac{1}{2}\right\}$
49. $\{x | -5 < x < 5\}$
50. $\{x | x < -3 \text{ or } x > 3\}$
51. $\{x | x \leq -7 \text{ or } x \geq 7\}$
52. $\{x | -4 \leq x \leq 4\}$
53. $\{x | x < 2 \text{ or } x > 6\}$
54. $\{x | -8 < x < -2\}$
55. $\{x | -10 \leq x \leq -2\}$
56. $\{x | x < 2 \text{ or } x > 12\}$
57. $\{x | x < -2 \text{ or } x > 8\}$
58. $\{x | 2 < x < 8\}$
59. No solution
60. All reals
61. $\{x | 1 < x < 4\}$
62. $\left\{x | x < -\dfrac{7}{3} \text{ or } x > 3\right\}$
63. $\left\{x | x \leq -3 \text{ or } x \geq \dfrac{1}{3}\right\}$
64. $\{x | -6 \leq x \leq 3\}$
65. $\left\{x | x < -\dfrac{4}{5} \text{ or } x > 2\right\}$
66. $\left\{x | -\dfrac{4}{3} < x < 3\right\}$
67. $\left\{x | -3 \leq x \leq \dfrac{13}{3}\right\}$
68. $\{x | x < -4 \text{ or } x > 7\}$

45. $2x - 1 < -7$ or $2x - 1 > 7$

46. $2x + 3 < -3$ or $2x + 3 > 3$

47. $3x - 1 < -7$ or $3x - 1 > 7$

48. $4x + 3 < -5$ or $4x + 3 > 5$

In Exercises 49 to 68, solve each inequality. Graph the solution set.

49. $|x| < 5$

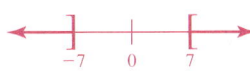

50. $|x| > 3$

51. $|x| \geq 7$

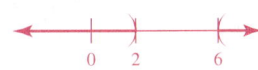

52. $|x| \leq 4$

53. $|x - 4| > 2$

54. $|x + 5| < 3$

55. $|x + 6| \leq 4$

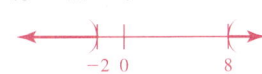

56. $|x - 7| > 5$

57. $|3 - x| > 5$

58. $|5 - x| < 3$

59. $|x - 7| < 0$

60. $|x + 5| \geq 0$

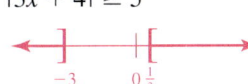
All real numbers

61. $|2x - 5| < 3$

62. $|3x - 1| > 8$

63. $|3x + 4| \geq 5$

64. $|2x + 3| \leq 9$

65. $|5x - 3| > 7$

66. $|6x - 5| < 13$

67. $|2 - 3x| \leq 11$

68. $|3 - 2x| > 11$

69. $|x| < 3$

70. $|x| < 4$

71. $|x| \geq 5$

72. $|x| \geq 2$

73. $|x - (-2)| < 7$

74. $|x - 4| > 6$

75. $|x - (-4)| \geq 3$

76. $|x - 4| \leq 3$

In Exercises 69 to 76, use absolute value notation to write an inequality that represents each sentence.

69. x is within 3 units of 0 on the number line.

70. x is within 4 units of 0 on the number line.

71. x is at least 5 units from 0 on the number line.

72. x is at least 2 units from 0 on the number line.

73. x is less than 7 units from -2 on the number line.

74. x is more than 6 units from 4 on the number line.

75. x is at least 3 units from -4 on the number line.

76. x is at most 3 units from 4 on the number line.

Summary Exercises · 2

This summary exercise set is provided to give you practice with each of the objectives of the chapter. Each exercise is keyed to the appropriate chapter section. Your instructor will give you guidelines on how to best use these exercises in your instructional setting.

[2.1] Tell whether the number shown in parentheses is a solution for the given equation.

1. $7x + 2 = 16$ (2)
2. $5x - 8 = 3x + 2$ (4)
3. $7x - 2 = 2x + 8$ (2)
4. $4x + 3 = 2x - 11$ (−7)
5. $x + 5 + 3x = 2 + x + 23$ (6)
6. $\frac{2}{3}x - 2 = 10$ (21)

[2.1] Solve the following equations and check your results.

7. $x + 5 = 7$
8. $x - 9 = 3$
9. $5x = 4x - 5$
10. $3x - 9 = 2x$
11. $5x - 3 = 4x + 2$
12. $9x + 2 = 8x - 7$
13. $7x - 5 = 6x - 4$
14. $3 + 4x - 1 = x - 7 + 2x$
15. $4(2x + 3) = 7x + 5$
16. $5(5x - 3) = 6(4x + 1)$

[2.2][2.3] Solve the following equations and check your results.

17. $5x = 35$
18. $7x = -28$
19. $-6x = 24$
20. $-9x = -63$
21. $\frac{x}{4} = 8$
22. $-\frac{x}{5} = -3$
23. $\frac{2}{3}x = 18$
24. $\frac{3}{4}x = 24$
25. $5x - 3 = 12$
26. $4x + 3 = -13$
27. $7x + 8 = 3x$
28. $3 - 5x = -17$
29. $3x - 7 = x$
30. $2 - 4x = 5$
31. $\frac{x}{3} - 5 = 1$
32. $\frac{3}{4}x - 2 = 7$
33. $6x - 5 = 3x + 13$
34. $3x + 7 = x - 9$
35. $7x + 4 = 2x + 6$
36. $9x - 8 = 7x - 3$

Answers

1. Yes
2. No
3. Yes
4. Yes
5. No
6. No
7. {2}
8. {12}
9. {−5}
10. {9}
11. {5}
12. {−9}
13. {1}
14. {−9}
15. {−7}
16. {21}
17. {7}
18. {−4}
19. {−4}
20. {7}
21. {32}
22. {15}
23. {27}
24. {32}
25. {3}
26. {−4}
27. {−2}
28. {4}
29. $\left\{\frac{7}{2}\right\}$
30. $\left\{-\frac{3}{4}\right\}$
31. {18}
32. {12}
33. {6}
34. {−8}
35. $\left\{\frac{2}{5}\right\}$
36. $\left\{\frac{5}{2}\right\}$

186 Chapter 2 ▪ Equations and Inequalities

Answers (left column)

37. {6}
38. $\left\{-\dfrac{5}{4}\right\}$
39. {6}
40. {5}
41. {4}
42. {−4}
43. {5}
44. $\left\{-\dfrac{1}{2}\right\}$
45. $\left\{\dfrac{1}{2}\right\}$
46. {−5}
47. $\left\{\dfrac{9}{2}\right\}$
48. {6}
49. $\left\{\dfrac{5}{3}\right\}$
50. {6}
51. {5}
52. {12}
53. {20}
54. {3}
55. {14}
56. $\dfrac{V}{WH}$
57. $\dfrac{P}{2} - W$
58. $\dfrac{c - ax}{b}$
59. $\dfrac{2A}{b}$
60. $\dfrac{A - P}{Pr}$
61. $mq + p$
62. 6
63. 8
64. 42, 43
65. 17, 19, 21
66. $375, $340
67. Susan, 7 yrs; Larry, 9 yrs; Nathan, 14 yrs
68. 54 mi/h, 48 mi/h; 216 mi
69. 3 hr at 28 mi/h; 2 h at 24 mi/h

Problems

37. $2x + 7 = 4x - 5$
38. $3x - 15 = 7x - 10$
39. $\dfrac{10}{3}x - 5 = \dfrac{4}{3}x + 7$
40. $\dfrac{11}{4}x - 15 = 5 - \dfrac{5}{4}x$
41. $3.7x + 8 = 1.7x + 16$
42. $5.4x - 3 = 8.4x + 9$
43. $3x - 2 + 5x = 7 + 2x + 21$
44. $8x + 3 - 2x + 5 = 3 - 4x$
45. $5(3x - 1) - 6x = 3x - 2$
46. $5x + 2(3x - 4) = 14x - 7$
47. $4(2x - 1) = 6x + 5$
48. $7x - 3(x - 2) = 30$
49. $8x - 5(x + 3) = -10$
50. $7(3x + 1) - 13 = 8(2x + 3)$
51. $3(2x - 5) - 2(x - 3) = 11$
52. $\dfrac{2x}{3} - \dfrac{x}{4} = 5$
53. $\dfrac{3x}{4} - \dfrac{2x}{5} = 7$
54. $\dfrac{x}{2} - \dfrac{x + 1}{3} = \dfrac{1}{6}$
55. $\dfrac{x + 1}{5} - \dfrac{x - 6}{3} = \dfrac{1}{3}$

[2.4] Solve for the indicated variable.

56. $V = LWH$ (for L)
57. $P = 2L + 2W$ (for L)
58. $ax + by = c$ (for y)
59. $A = \dfrac{1}{2}bh$ (for h)
60. $A = P + Prt$ (for t)
61. $m = \dfrac{n - p}{q}$ (for n)

[2.2–2.4] Solve the following word problems. Be sure to label the unknowns and to show the equation you used for the solution.

62. The sum of 3 times a number and 7 is 25. What is the number?
63. 5 times a number, decreased by 8, is 32. Find the number.
64. If the sum of two consecutive integers is 85, find the two integers.
65. The sum of three consecutive odd integers is 57. What are the three integers?
66. Rafael earns $35 more per week than Andrew. If their weekly salaries total $715, what amount does each earn?
67. Larry is 2 years older than Susan, while Nathan is twice as old as Susan. If the sum of their ages is 30 years, find each of their ages.
68. Lisa left Friday morning, driving on the freeway to visit friends for the weekend. Her trip took 4 h. When she returned on Sunday, heavier traffic slowed her average speed by 6 mi/h, and the trip took $4\dfrac{1}{2}$ h. What was her average speed in each direction, and how far did she travel each way?
69. A bicyclist started on a 132-mi trip and rode at a steady rate for 3 h. He began to tire at that point and slowed his speed by 4 mi/h for the remaining 2 h of the trip. What was his average speed for each part of the journey?

- Summary Exercises 187

70. 3:20 P.M.

71. 2 P.M.

72. 150

79. {x|x ≤ 11}

80. {x|x > −5}

81. {x|x > −3}

82. {x|x ≥ −3}

83. {x|x > −3}

84. {x|x ≤ −15}

85. $\{x|x \geq \frac{1}{2}\}$

86. {x|x ≥ 3}

87. $\{x|x < -\frac{4}{3}\}$

88. {x|x ≤ 7}

89. $\{x|x \geq \frac{3}{2}\}$

90. {x|x > −6}

91. {x|x < 9}

92. $\{x|x \geq -\frac{5}{3}\}$

93. {−3, 2}

94. $\{-\frac{5}{3}, 3\}$

95. {1, 4}

96. $\{-\frac{5}{2}, 2\}$

97. $\{-\frac{5}{2}, 4\}$

98. {−2, 8}

99. $\{-2, \frac{24}{5}\}$

100. $\{-\frac{4}{3}, \frac{14}{3}\}$

101. No solution

102. No solution

103. {−5, 4}   104. {1, 5}

105. {−1, 3}   106. {1}

107. {x|1 ≤ x ≤ 2}

108. {x|−4 ≤ x ≤ 6}

70. At noon, Jan left her house, jogging at an average rate of 8 mi/h. Two hours later, Stanley left on his bicycle along the same route, averaging 20 mi/h. At what time will Stanley catch up with Jan?

71. At 9 A.M., David left New Orleans for Tallahassee, averaging 47 mi/h. Two hours later, Gloria left Tallahassee for New Orleans along the same route, driving 5 mi/h faster than David. If the two cities are 391 mi apart, at what time will David and Gloria meet?

72. A firm producing running shoes finds that its fixed costs are $3900 per week, and its variable cost is $21 per pair of shoes. If the firm can sell the shoes for $47 per pair, how many pairs of shoes must be produced and sold each week for the company to break even?

[2.5] Graph the solution sets.

73. $x > 5$
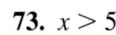

74. $x < -3$

75. $x \leq -4$
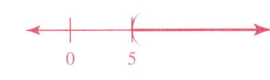

76. $x \geq 9$

77. $x \geq -6$
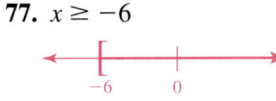

78. $x < 0$

[2.5] Solve the following inequalities.

79. $x - 4 \leq 7$

80. $x + 3 > -2$

81. $5x > 4x - 3$

82. $4x \geq -12$

83. $-12x < 36$

84. $-\frac{x}{5} \geq 3$

85. $2x \leq 8x - 3$

86. $2x + 3 \geq 9$

87. $4 - 3x > 8$

88. $5x - 2 \leq 4x + 5$

89. $7x + 13 \geq 3x + 19$

90. $4x - 2 < 7x + 16$

91. $5(x - 3) < 2x + 12$

92. $4(x + 3) \geq x + 7$

[2.6] Solve the following equations.

93. $|2x + 1| = 5$

94. $|3x - 2| = 7$

95. $|2x - 5| = 3$

96. $|4x + 1| = 9$

97. $|4x - 3| = 13$

98. $|6 - 2x| = 10$

99. $|-5x + 7| = 17$

100. $|-3x + 5| = 9$

101. $|7x - 1| = -3$

102. $|2x + 5| = -1$

103. $|2x + 1| - 3 = 6$

104. $7 - |x - 3| = 5$

105. $|3x - 1| = |x + 5|$

106. $|x - 5| = |x + 3|$

[2.6] Solve each of the following compound inequalities.

107. $4 \leq x + 3 \leq 5$

108. $-8 \leq 2x \leq 12$

109. $\{x|-2 < x \leq 3\}$
110. $\{x|-3 \leq x < 6\}$
111. $\{x|x < -2 \text{ or } x > 6\}$
112. $\{x|x \leq \dfrac{-13}{3} \text{ or } x \geq -3\}$
113. $\{x|-3 \leq x \leq 3\}$
114. $\{x|x < -8 \text{ or } x > 2\}$
115. $\{x|x < 3 \text{ or } x > 11\}$
116. $\{x|-3 < x < 9\}$
117. $\{x|x \leq -6 \text{ or } x \geq -1\}$
118. $\left\{x\middle|-1 \leq x \leq \dfrac{5}{3}\right\}$
119. $\left\{x\middle|-5 < x < \dfrac{7}{3}\right\}$
120. $\{x|x \leq -\dfrac{14}{15} \text{ or } x \geq 2\}$

109. $-3 < 5 + 4x \leq 17$
110. $-5 \leq 4 + 3x < 22$
111. $x - 2 < -4 \text{ or } x - 2 > 4$
112. $3x + 2 \leq -11 \text{ or } 3x + 2 \geq 11$

[2.6] Solve the following inequalities.

113. $|x| \leq 3$
114. $|x + 3| > 5$
115. $|x - 7| > 4$
116. $|3 - x| < 6$
117. $|2x + 7| \geq 5$
118. $|3x - 1| \leq 4$
119. $|3x + 4| < 11$
120. $|5x + 2| \geq 12$

Cumulative Test • 0-2

This test is provided to help you in the process of reviewing the previous chapters. Answers are provided in the back of the book. If you missed any answers, be sure to go back and review the appropriate chapter sections.

Complete each of the following statements.

1. $-|-23| = $ _____   **2.** $-(-11) = $ _____

Evaluate each of the following expressions.

3. $2 \cdot 3^2 - 8 \cdot 2$ **4.** $5(7-3)^2$ **5.** $|12-5|$

6. $|12| - |5|$ **7.** $(-7) + (-9)$ **8.** $\dfrac{17}{3} + \left(-\dfrac{5}{3}\right)$

9. $(-7)(-9)$ **10.** $(-3.2)(5)$ **11.** $\dfrac{0}{-13}$

12. $8 - 12 \div 2 \cdot 3 + 5$ **13.** $5 - 4^2 \div (-8) \cdot 2$

Evaluate each of the following expressions if $x = -2$, $y = 3$, and $z = 5$.

14. $3x - y$ **15.** $4x^2 - y$ **16.** $\dfrac{5z - 4x}{2y + z}$

17. $-y^2 - 8x$

Simplify and combine like terms.

18. $7x - 3y + 2(4x - 3y)$ **19.** $6x^2 - (5x - 4x^2 + 7) - 8x + 9$

Solve the following equations.

20. $12x - 3 = 10x + 5$ **21.** $|x - 3| = 5$

Solve the following inequalities.

22. $7x + 5 \le 4x - 7$ **23.** $-5 \le 2x + 1 < 7$

24. $|x - 1| \le 4$ **25.** $|x + 1| > 8$

Solve the following equations for the indicated variable.

26. $I = Prt$ (for r)   **27.** $A = \dfrac{1}{2}bh$ (for h)   **28.** $ax + by = c$ (for y)

---

1. -23
2. 11
3. 2
4. 80
5. 7
6. 7
7. -16
8. 4
9. 63
10. -16
11. 0
12. -5
13. 9
14. -9
15. 13
16. 3
17. 7
18. $15x - 9y$
19. $10x^2 - 13x + 2$
20. $\{4\}$
21. $\{-2, 8\}$
22. $\{x | x \le -4\}$
23. $\{x | -3 \le x < 3\}$
24. $\{x | -3 \le x \le 5\}$
25. $\{x | x < -9 \text{ or } x > 7\}$
26. $r = \dfrac{I}{Pt}$
27. $h = \dfrac{2A}{b}$
28. $y = \dfrac{c - ax}{b}$

29. 13

30. 42, 43

31. 7

32. $420

33. 5 cm, 17 cm

34. 8 in., 13 in., 16 in.

Solve the following word problems. Be sure to show the equation used for the solution.

29. If 4 times a number decreased by 7 is 45, find that number.

30. The sum of two consecutive integers is 85. What are those two integers?

31. If 3 times an odd integer is 12 more than the next consecutive odd integer, what is that integer?

32. Michelle earns $120 more per week than Dmitri. If their weekly salaries total $720, how much does Michelle earn?

33. The length of a rectangle is 2 centimeters (cm) more than 3 times its width. If the perimeter of the rectangle is 44 cm, what are the dimensions of the rectangle?

34. One side of a triangle is 5 inches (in.) longer than the shortest side. The third side is twice the length of the shortest side. If the triangle perimeter is 37 in., find the length of each leg.

CHAPTER 3
Graphs and Linear Equations

LIST OF SECTIONS

3.1 Solutions of Equations in Two Variables

3.2 The Cartesian Coordinate System

3.3 The Graph of a Linear Equation

3.4 The Slope of a Line

3.5 Forms of Linear Equations

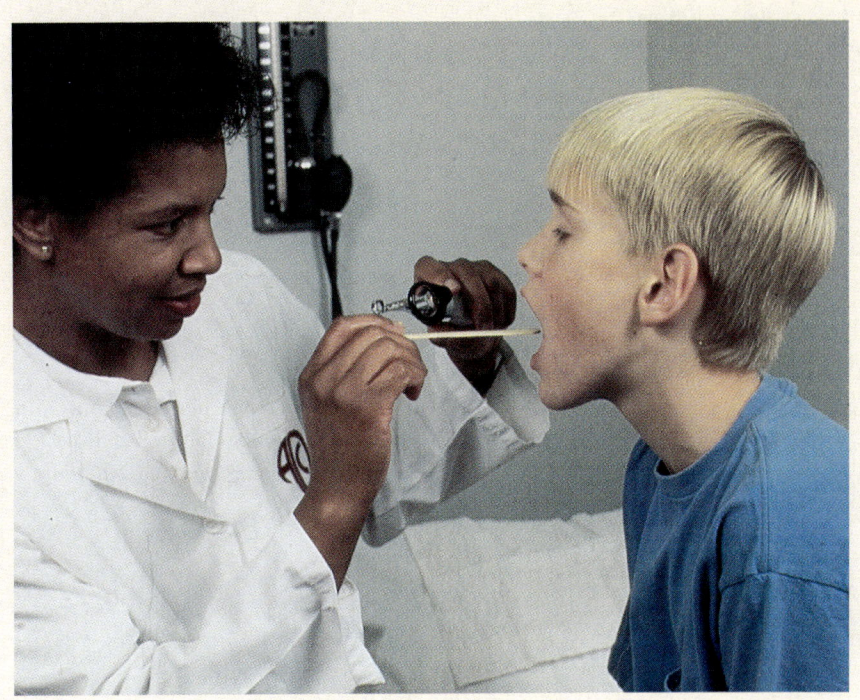

Graphs are used to discern patterns and trends that may be difficult to see when looking at a list of numbers or other kinds of data. The word "graph" comes from Latin and Greek roots and means "to draw a picture." This is what a graph does in mathematics: It draws a picture of a relationship between two or more variables.

In the field of pediatric medicine, there has been controversy about the use of somatotropin (human growth hormone) to help children whose growth has been impeded by various health problems. These children must be distinguished from children who are healthy and simply small of stature and thus should not be subjected to this treatment. Some of the measures used to distinguish between the two groups are blood tests and age and height measurements. The age and height measurements are graphed and monitored over several years of a child's life in order to monitor the rate of growth. If during a certain period the child's rate of growth slows to below 4.5 centimeters per year, this indicates that something may be seriously wrong. The graph can also indicate if the child's size fits within a range considered normal at each age of the child's life.

SECTION 3.1 Solutions of Equations in Two Variables

3.1 OBJECTIVES

1. Find the solution(s) for an equation in two variables
2. Use the ordered pair notation to write solutions for equations in two variables.

We discussed finding solutions for equations in Section 2.1. Recall that a solution is a value for the variable that "satisfies" the equation, or makes the equation a true statement. For example, we know that 4 is a solution of the following equation.

$$2x + 5 = 13$$

We know this is true because, when we replace x with 4, we have

$$2 \cdot 4 + 5 = 13$$
$$8 + 5 = 13$$
$$13 = 13 \quad \text{A true statement}$$

We now want to consider **equations in two variables.** An example is

$$x + y = 5$$

What will the solution look like? It is not going to be a single number, because there are two variables. Here the solution will be a pair of numbers—one value for each of the variables, x and y. Suppose that x has the value 3. In the equation $x + y = 5$, you can substitute 3 for x.

$$3 + y = 5$$

Solving for y gives

$$y = 2$$

So the pair of values $x = 3$ and $y = 2$ satisfies the equation because

$$3 + 2 = 5$$

That pair of numbers is then a *solution* for the equation in two variables.

> An **equation in two variables** is an equation for which *every* solution is a pair of values.

Recall that an equation is two expressions connected by an equal sign.

An equation in two variables "pairs" two numbers, one for x and one for y.

How many such pairs are there? Choose any value for *x* (or for *y*). You can always find the other *paired* or *corresponding* value in an equation of this form. We say that there are an *infinite* number of pairs that will satisfy the equation. Each of these pairs is a solution. We will find some other solutions for the equation $x + y = 5$ in the following example.

Example 1

Solving for Corresponding Values

For the equation $x + y = 5$, find (a) *y* if $x = 5$ and (b) *x* if $y = 4$.

(a) If $x = 5$

$$5 + y = 5 \quad \text{or} \quad y = 0$$

(b) If $y = 4$,

$$x + 4 = 5 \quad \text{or} \quad x = 1$$

So the pairs $x = 5, y = 0$ and $x = 1, y = 4$ are both solutions.

✓ CHECK YOURSELF 1

For the equation $2x + 3y = 26$,

(a) If $x = 4, y = ?$ (b) If $y = 0, x = ?$

To simplify writing the pairs that satisfy an equation, we use the **ordered-pair notation.** The numbers are written in parentheses and are separated by a comma. For example, we know that the values $x = 3$ and $y = 2$ satisfy the equation $x + y = 5$. So we write the pair as

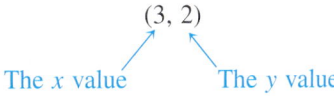

The *x* value   The *y* value

Caution

(3, 2) means $x = 3$ and $y = 2$. (2, 3) means $x = 2$ and $y = 3$. (3, 2) and (2, 3) are entirely different. That's why we call them *ordered pairs*.

The first number of the pair is *always* the value for *x* and is called the **x coordinate.** The second number of the pair is *always* the value for *y* and is the **y coordinate.**

Using this ordered-pair notation, we can say that (3, 2), (5, 0), and (1, 4) are all *solutions* for the equation $x + y = 5$. Each pair gives values for *x* and *y* that will satisfy the equation.

Example 2

Identifying Solutions of Two-Variable Equations

Which of the ordered pairs (2, 5), (5, −1), and (3, 4) are solutions for the equation $2x + y = 9$?

(a) To check whether (2, 5) is a solution, let $x = 2$ and $y = 5$ and see if the equation is satisfied.

$$2x + y = 9 \quad \text{The original equation}$$

$$2 \cdot 2 + 5 = 9 \quad \text{Substitute 2 for } x \text{ and 5 for } y.$$

$$4 + 5 = 9$$

$$9 = 9 \quad \text{A true statement}$$

(2, 5) is a solution because a true statement results.

(2, 5) is a solution for the equation $2x + y = 9$.

(b) For (5, −1), let $x = 5$ and $y = -1$.

$$2 \cdot 5 - 1 = 9$$

$$10 - 1 = 9$$

$$9 = 9 \quad \text{A true statement}$$

So (5, −1) is a solution for $2x + y = 9$.

(c) For (3, 4), let $x = 3$ and $y = 4$. Then

$$2 \cdot 3 + 4 = 9$$

$$6 + 4 = 9$$

$$10 = 9 \quad \text{Not a true statement}$$

So (3, 4) is *not* a solution for the equation.

✓ **CHECK YOURSELF 2**

Which of the ordered pairs (3, 4), (4, 3), (1, −2), and (0, −5) are solutions for the following equation?

$$3x - y = 5$$

If the equation contains only one variable, then the missing variable can take on any value.

Example 3 Identifying Solutions of One-Variable Equations

Which of the ordered pairs, (2, 0), (0, 2), (5, 2), (2, 5), and (2, −1) are solutions for the equation $x = 2$?

A solution is any ordered pair in which the x coordinate is 2. That makes (2, 0), (2, 5), and (2, −1) solutions for the given equation.

✓ CHECK YOURSELF 3

Which of the ordered pairs (3, 0), (0, 3), (3, 3), (−1, 3), and (3, −1) are solutions for the equation $y = 3$?

Remember that, when an ordered pair is presented, the first number is always the x coordinate and the second number is always the y coordinate.

Example 4 Completing Ordered Pair Solutions

Complete the ordered pairs (9,), (, −1), (0,), and (, 0) for the equation $x − 3y = 6$.

*The x coordinate is also called the **abscissa** and the y coordinate is called the **ordinate**.*

(a) The first number, 9, appearing in (9,) represents the x value. To complete the pair (9,), substitute 9 for x and then solve for y.

$$9 − 3y = 6$$
$$−3y = −3$$
$$y = 1$$

The ordered pair (9, 1) is a solution for $x − 3y = 6$.

(b) To complete the pair (, −1), let y be −1 and solve for x.

$$x − 3(−1) = 6$$
$$x + 3 = 6$$
$$x = 3$$

The ordered pair (3, −1) is a solution for the equation $x − 3y = 6$.

(c) To complete the pair (0,), let x be 0.

$$0 - 3y = 6$$
$$-3y = 6$$
$$y = -2$$

(0, −2) is a solution.

(d) To complete the pair (, 0), let y be 0.

$$x - 3 \cdot 0 = 6$$
$$x - 0 = 6$$
$$x = 6$$

(6, 0) is a solution.

✓ CHECK YOURSELF 4

Complete the ordered pairs below so that each is a solution for the equation $2x + 5y = 10$.

(10,), (, 4), (0,), and (, 0)

Example 5 — Finding Some Solutions of a Two-Variable Equation

Find four solutions for the equation

$$2x + y = 8$$

Generally, you'll want to pick values for x (or for y) so that the resulting equation in one variable is easy to solve.

In this case the values used to form the solutions are *up to you*. You can assign any value for x (or for y). We'll demonstrate with some possible choices.

Solution with $x = 2$:

$$2x + y = 8$$
$$2 \cdot 2 + y = 8$$
$$4 + y = 8$$
$$y = 4$$

The ordered pair (2, 4) is a solution for $2x + y = 8$.

Solution with $y = 6$:

$$2x + y = 8$$
$$2x + 6 = 8$$
$$2x = 2$$
$$x = 1$$

$(1, 6)$ is also a solution for $2x + y = 8$.

Solution with $x = 0$:

$$2x + y = 8$$
$$2 \cdot 0 + y = 8$$
$$y = 8$$

The solutions $(0, 8)$ and $(4, 0)$ will have special significance later in graphing. They are also easy to find!

$(0, 8)$ is a solution.

Solution with $y = 0$:

$$2x + y = 8$$
$$2x + 0 = 8$$
$$2x = 8$$
$$x = 4$$

$(4, 0)$ is a solution.

✓ CHECK YOURSELF 5

Find four solutions for $x - 3y = 12$.

✓ CHECK YOURSELF ANSWERS

1. (a) $y = 6$; (b) $x = 13$. **2.** $(3, 4)$, $(1, -2)$, and $(0, -5)$ are solutions.
3. $(0, 3)$, $(3, 3)$, and $(-1, 3)$ are solutions. **4.** $(10, -2)$, $(-5, 4)$, $(0, 2)$, and $(5, 0)$.
5. $(6, -2)$, $(3, -3)$, $(0, -4)$, and $(12, 0)$ are four possibilities.

Exercises • 3.1

Answers (left column):

1. (4, 2), (0, 6), (−3, 9)
2. (13, 1), (12, 0)
3. (5, 2), (4, 0), (6, 4)
4. (10, 2), (20, 0), (25, −1)
5. (2, 0), (1, 3)
6. (8, 0)
7. (3, 0), (6, 2), (0, −2)
8. (2, 0), (0, 4), (6, −8)
9. (4, 0), $\left(\frac{2}{3}, -5\right)$, $\left(5, \frac{3}{2}\right)$
10. $\left(\frac{2}{3}, \frac{5}{2}\right)$, (0, 3)
11. (0, 0), (2, 8)
12. (0, −1), $\left(\frac{1}{2}, 0\right)$
13. (3, 5), (3, 0), (3, 7)
14. (0, 5), (3, 5), (5, 5)
15. 8, 7, 12, 12
16. 11, 8, −7, 7
17. 0, 0, 4, 9
18. 4, 10, 2, 20
19. 3, −5, 5, 2
20. −3, 1, 9, 3
21. 4, 0, −3, 6
22. 4, 2, 10, −5
23. −3, 11, 9, 7
24. 3, 3, 4, 1

Determine which of the ordered pairs are solutions for the given equation.

1. $x + y = 6$ (4, 2), (−2, 4), (0, 6), −3, 9)
2. $x - y = 12$ (13, 1), (13, −1), (12, 0), (6, 6)
3. $2x - y = 8$ (5, 2), (4, 0), (0, 8), (6, 4)
4. $x + 5y = 20$ (10, −2), (10, 2), (20, 0), (25, −1)
5. $3x + y = 6$ (2, 0), (2, 3), (0, 2), (1, 3)
6. $x - 2y = 8$ (8, 0), (0, 4), (5, −1), (10, −1)
7. $2x - 3y = 6$ (0, 2), (3, 0), (6, 2), (0, −2)
8. $8x + 4y = 16$ (2, 0), (6, −8), (0, 4), (6, −6)
9. $3x - 2y = 12$ (4, 0), $\left(\frac{2}{3}, -5\right)$, (0, 6), $\left(5, \frac{3}{2}\right)$
10. $3x + 4y = 12$ (−4, 0), $\left(\frac{2}{3}, \frac{5}{2}\right)$, (0, 3), $\left(\frac{2}{3}, 2\right)$
11. $y = 4x$ (0, 0), (1, 3), (2, 8), (8, 2)
12. $y = 2x - 1$ (0, −2), (0, −1), $\left(\frac{1}{2}, 0\right)$, (3, −5)
13. $x = 3$ (3, 5), (0, 3), (3, 0), (3, 7)
14. $y = 5$ (0, 5), (3, 5), (−2, −5), (5, 5)

Complete the ordered pairs so that each is a solution for the given equation.

15. $x + y = 12$ (4,), (, 5), (0,), (, 0)
16. $x - y = 7$ (, 4), (15,), (0,), (, 0)
17. $3x + y = 9$ (3,), (, 9), (, −3), (0,)
18. $x + 5y = 20$ (0,), (, 2), (10,), (, 0)
19. $5x - y = 15$ (, 0), (2,), (4,), (, −5)
20. $x - 3y = 9$ (0,), (12,), (, 0), (, −2)
21. $3x - 2y = 12$ (, 0), (, −6), (2,), (, 3)
22. $2x + 5y = 20$ (0,), (5,), (, 0), (, 6)
23. $y = 3x + 9$ (, 0), $\left(\frac{2}{3}, \right)$, (0,), $\left(-\frac{2}{3}, \right)$
24. $3x + 4y = 12$ (0,), $\left(, \frac{3}{4}\right)$, (, 0), $\left(\frac{8}{3}, \right)$

Section 3.1 ■ Solutions of Equations in Two Variables   199

25. $-4, 3, \dfrac{4}{3}, 1$

26. 5, 0, 2, 2

27. (0, −7), (2, −5), (4, −3), (6, −1)

28. (0, 18), (6, 12), (12, 6), (18, 0)

29. (0, −6), (3, 0), (6, 6), (9, 12)

30. (0, −12), (3, −3), (6, 6), (9, 15)

31. (8, 0), (−4, 3), (0, 2), (4, 1)

32. (0, 4), (3, 3), (6, 2), (9, 1)

33. (−5, −4), (0, −2), (5, 0), (10, 2)

34. (−7, 4), (0, 2), (7, 0), (14, −2)

35. (0, 3), (1, 5), (2, 7), (3, 9)

36. (0, −5), (1, 3), (2, 11), (3, 19)

37. (−5, 0), (−5, 1), (−5, 2), (−5, 3)

38. (0, 8), (1, 8), (2, 8), (3, 8)

39. 1   40. 0

41. −6   42. 0

43. 5   44. −2

45. $9.50, $11.75, $15.50, $19.25, $23

46. 14°F, 32°F, 59°F, 212°F

47. 25 cm², 100 cm², 144 cm², 225 cm²

48. $22, $47, $57, $67

25. $y = 3x - 4$ (0,), (, 5), (, 0), $\left(\dfrac{5}{3},\ \right)$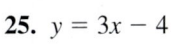

26. $y = -2x + 5$ (0,), (, 5), $\left(\dfrac{3}{2},\ \right)$, (, 1)

Find four solutions for each of the following equations. *Note:* Your answers may vary from those shown in the answer section.

27. $x - y = 7$ 28. $x + y = 18$

29. $2x - y = 6$ 30. $3x - y = 12$

31. $x + 4y = 8$ 32. $x + 3y = 12$

33. $2x - 5y = 10$ 34. $2x + 7y = 14$

35. $y = 2x + 3$ 36. $y = 8x - 5$

37. $x = -5$ 38. $y = 8$

An equation in three variables has an ordered triple as a solution. For example, (1, 2, 2) is a solution to the equation $x + 2y - z = 3$. Complete the ordered-triple solutions for each equation.

39. $x + y + z = 0$ (2, −3,)   40. $2x + y + z = 2$ (, −1, 3)

41. $x + y + z = 0$ (1, , 5)   42. $x + y - z = 1$ (4, , 3)

43. $2x + y + z = 2$ (−2, , 1)   44. $x + y - z = 1$ (−2, 1,)

45. **Hourly wages.** When an employee produces x units per hour, the hourly wage is given by $y = 0.75x + 8$. What are the hourly wages for the following number of units: 2, 5, 10, 15, and 20?

46. **Temperature conversion.** Celsius temperature readings can be converted to Fahrenheit readings using the formula $F = \dfrac{9}{5}C + 32$. What is the Fahrenheit temperature that corresponds to each of the following Celsius temperatures: −10, 0, 15, 100?

47. **Area.** The area of a square is given by $A = s^2$. What is the area of the squares whose sides are 5 centimeters (cm), 10 cm, 12 cm, 15 cm?

48. **Unit pricing.** When x number of units are sold, the price of each unit is given by $p = 5x + 12$. Find the unit price when the following quantities are sold: 2, 7, 9, 11.

49. You now have had practice solving equations with one variable and equations with two variables. Compare equations with one variable to equations with two variables. How are they alike? How are they different?

50. Each of the following sentences describes pairs of numbers that are related. After completing the sentences in parts (a) to (g), write two of your own sentences in (h) and (i).

(a) The *number of hours you work* determines the *amount you are* _____.

(b) The *number of gallons of gasoline* you put in your car determines *the amount you* _____.

(c) The *amount of the* _____ in a restaurant is related to *the amount of the tip.*

(d) The *sales amount of a purchase in a store* determines _____.

(e) The *age of an automobile* is related to _____.

(f) The *amount of electricity you use in a month* determines _____.

(g) The *cost of food for a family of four* and _____.

Think of two more:

(h) _____.

(i) _____.

SECTION 3.2 The Cartesian Coordinate System

3.2 OBJECTIVES

1. Plot ordered pairs
2. Identify plotted points
3. Scale the axes
4. Plot points with a graphing calculator

This system is also called the **Cartesian coordinate system,** named in honor of its inventor, René Descartes (1596–1650), a French mathematician and philosopher.

In Section 3.1, we saw that ordered pairs could be used to write solutions to equations in two variables. The next step is to graph those ordered pairs as points in a plane.

Since there are two numbers (one for x and one for y), we will need two number lines. One line is drawn horizontally, and the other is drawn vertically; their point of intersection (at their respective zero points) is called the *origin*. The horizontal line is called the ***x* axis,** while the vertical line is called the ***y* axis.** Together the lines form the **rectangular coordinate system.**

The axes (pronounced "axees") divide the plane into four regions called **quadrants,** which are numbered (usually by Roman numerals) counterclockwise from the upper right.

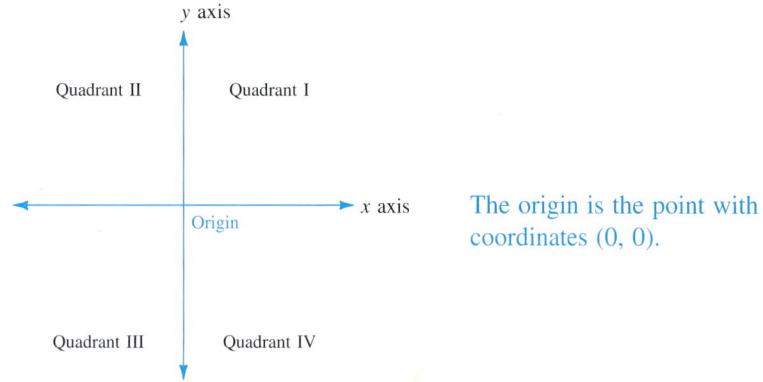

The origin is the point with coordinates (0, 0).

We now want to establish correspondences between ordered pairs of numbers (x, y) and points in the plane.

For any ordered pair,

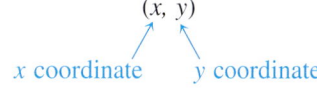

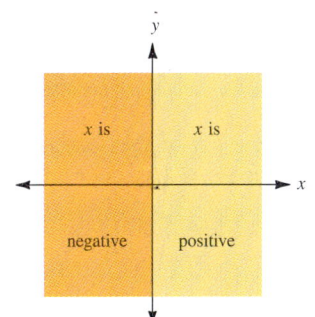

the following are true:

1. If the x coordinate is

   Positive, the point corresponding to that pair is located x units to the *right* of the y axis.

   Negative, the point is x units to the *left* of the y axis.

   Zero, the point is on the y axis.

201

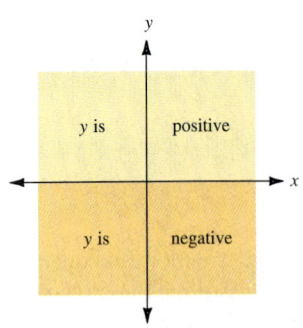

2. If the *y* coordinate is

   Positive, the point is *y* units *above* the *x* axis.

   Negative, the point is *y* units *below* the *x* axis.

   Zero, the point is on the *x* axis.

Example 1 illustrates how to use these guidelines to match coordinates with points in the plane.

Example 1   Identifying the Coordinates for a Given Point

Give the coordinates for the given points.

Remember: The *x* coordinate gives the *horizontal* distance from the *y* axis. The *y* coordinate gives the *vertical* distance from the *x* axis.

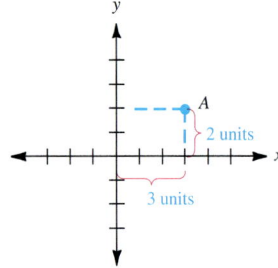

(a) Point *A* is 3 units to the *right* of the *y* axis and 2 units *above* the *x* axis. Point *A* has coordinates (3, 2).

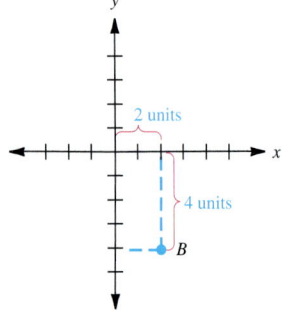

(b) Point *B* is 2 units to the *right* of the *y* axis and 4 units *below* the *x* axis. Point *B* has coordinates (2, −4).

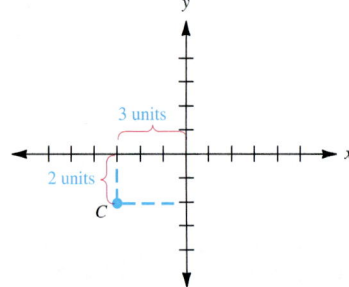

(c) Point *C* is 3 units to the *left* of the *y* axis and 2 units *below* the *x* axis. *C* has coordinates (−3, −2).

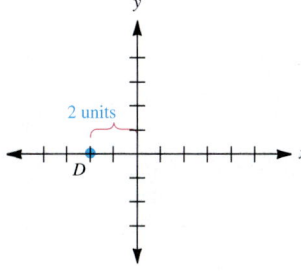

(d) Point *D* is 2 units to the *left* of the *y* axis and *on* the *x* axis. Point *D* has coordinates $(-2, 0)$.

✓ CHECK YOURSELF 1

Give the coordinates of points *P, Q, R,* and *S*.

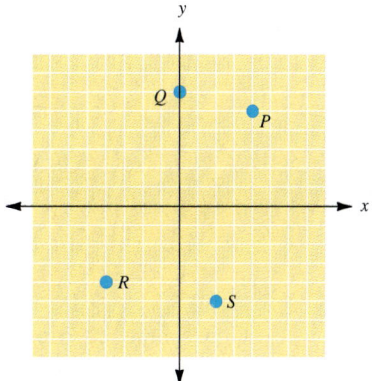

Reversing the process above will allow us to graph (or plot) a point in the plane given the coordinates of the point. You can use the following steps.

To Graph a Point in the Plane

Step 1 Start at the origin.
Step 2 Move right or left according to the value of the *x* coordinate.
Step 3 Move up or down according to the value of the *y* coordinate.

The graphing of individual points is sometimes called **point plotting.**

Example 2 Graphing Points

(a) Graph the point corresponding to the ordered pair (4, 3).

Move 4 units to the right on the *x* axis. Then move 3 units up from the point you stopped at on the *x* axis. This locates the point corresponding to (4, 3).

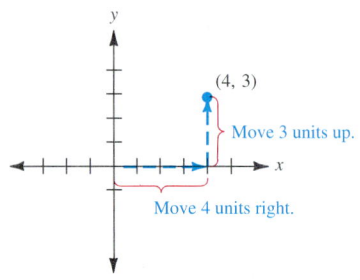

(b) Graph the point corresponding to the ordered pair $(-5, 2)$.

In this case move 5 units *left* (because the x coordinate is negative) and then 2 units *up*.

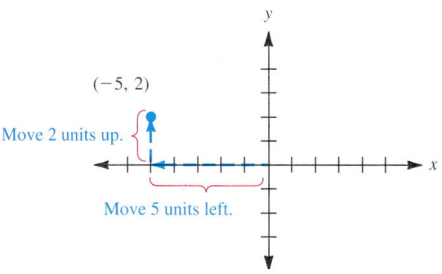

(c) Graph the point corresponding to $(-4, -2)$.

Here move 4 units *left* and then 2 units *down* (the y coordinate is negative).

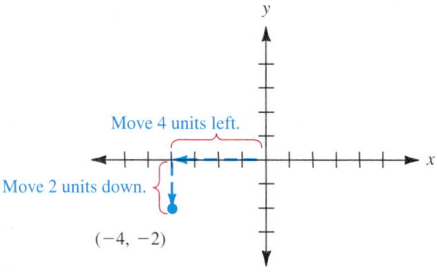

Any point on an axis will have 0 for one of its coordinates.

(d) Graph the point corresponding to $(0, -3)$.

There is *no* horizontal movement because the x coordinate is 0. Move 3 units *down*.

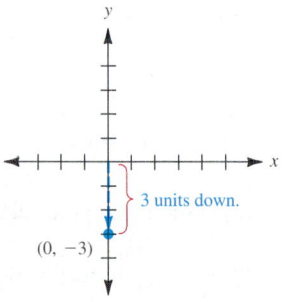

(e) Graph the point corresponding to (5, 0).

Move 5 units *right.* The desired point is on the *x* axis because the *y* coordinate is 0.

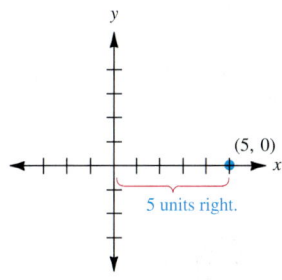

✓ CHECK YOURSELF 2

Graph the points corresponding to $M(4, 3)$, $N(-2, 4)$, $P(-5, -3)$, and $Q(0, -3)$.

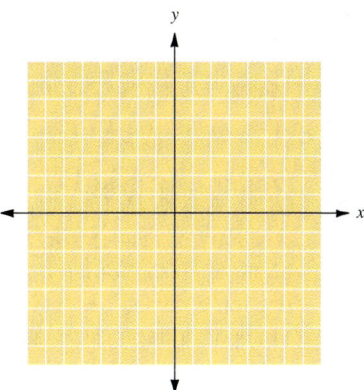

Scaling the Axes

The same decisions must be made when you are using a graphing calculator. When graphing this kind of relation on a calculator, you must decide what the appropriate **viewing window** should be.

It is not necessary, or even desirable, to always use the same scale on both the *x* and *y* axes. For example, if we were plotting ordered pairs in which the first value represented the age of a used car and the second value represented the number of miles driven, it would be necessary to have a different scale on the two axes. If not, the following extreme cases could happen.

Assume that the cars range in age from 1 to 15 years. The cars have mileage from 2000 to 150,000 miles. If we use the same scale on both axes, 0.5 in. between each two counting numbers, how large would the paper have to be on which the points were plotted? The horizontal axis would have to be $15(0.5) = 7.5$ in. The vertical axis would have to be $150,000(0.5) = 75,000$ in. = 6250 feet = almost 1.2 miles long!

So what do we do? We simply use a different, but clearly marked, scale on the axes. In this case, the horizontal axis could be marked in 10s, but the vertical axis would be marked in 50,000s. Additionally, all the numbers would be positive, so we really need only the first quadrant in which *x* and *y* are both always positive. We could draw the graph like this:

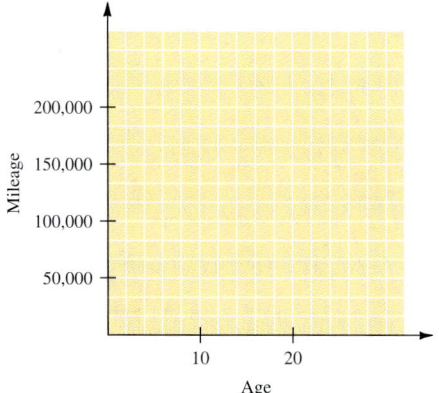

Every ordered pair is either in one of the quadrants or on one of the axes.

Example 3 Scaling the Axes

A survey of residents in a large apartment building was recently taken. The following points represent ordered pairs in which the first number is the number of years of education a person has had, and the second number is their year 2000 income (in thousands of dollars). Estimate, and interpret, each ordered pair represented.

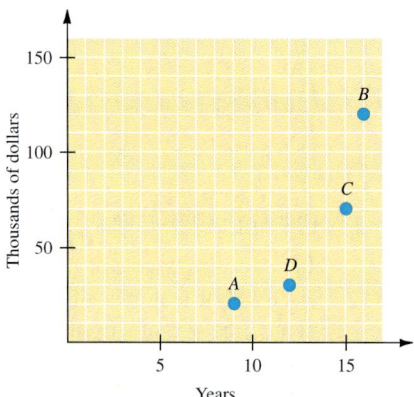

A is (9, 20), *B* is (16, 120), *C* is (15, 70), and *D* is (12, 30). Person *A* completed 9 years of education and made $20,000 in 2000. Person *B* completed 16 years of education and made $120,000 in 2000. Person *C* had 15 years education and made $70,000. Person *D* had 12 years and made $30,000.

Note that there is no obvious "relation" that would allow one to predict income from years of education, but you might suspect that in most cases, more education results in more income.

Section 3.2 ■ The Cartesian Coordinate System   **207**

✓ CHECK YOURSELF 3

Each year on his son's birthday, Armand records his son's weight. The following points represent ordered pairs in which the first number represents his son's age and the second number represents his weight. For example, point A indicates that when his son was 1 year old, the boy weighed 14 pounds. Estimate each ordered pair represented.

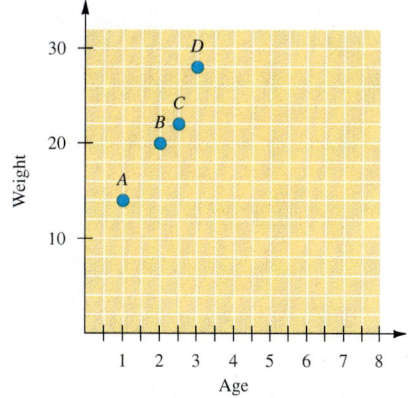

Both the plotting of points and the scaling of the axes can be modeled on a graphing calculator. Our next example describes the process by which one can plot points using a graphing calculator.

✓ CHECK YOURSELF ANSWERS

1. $P(4, 5)$, $Q(0, 6)$, $R(-4, -4)$, and $S(2, -5)$.

2.

3. $A(1, 14)$, $B(2, 20)$, $C\left(\dfrac{5}{2}, 22\right)$, and $D(3, 28)$.

Exercises • 3.2

1. (5, 6)
2. (3, −3)
3. (2, 0)
4. (−5, 1)
5. (−4, −5)
6. (6, 3)
7. (−5, −3)
8. (0, 6)
9. (−3, 5)
10. (2, −6)

Give the coordinates of the points graphed below.

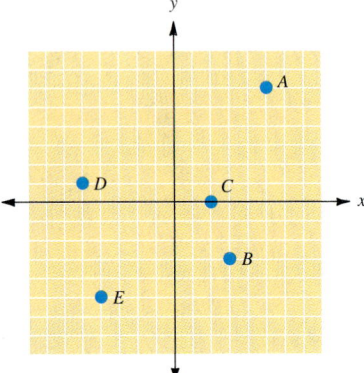

1. A
2. B
3. C
4. D
5. E

Give the coordinates of the points graphed below.

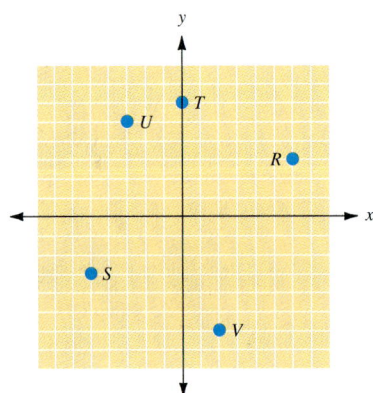

6. R
7. S
8. T
9. U
10. V

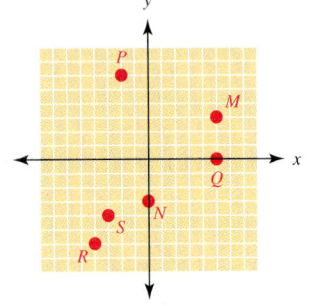

Plot the following points on the rectangular coordinate system.

11. $M(5, 3)$
12. $N(0, -3)$
13. $P(-2, 6)$
14. $Q(5, 0)$
15. $R(-4, -6)$
16. $S(-3, -4)$
17. $F(-3, -1)$
18. $G(4, 3)$
19. $H(5, -2)$
20. $I(-3, 0)$
21. $J(-5, 3)$
22. $K(0, 6)$

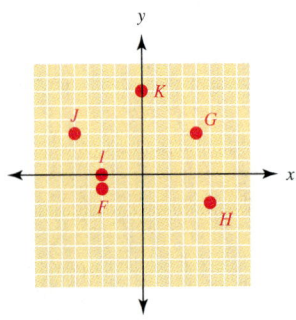

23. I

24. II

25. III

26. IV

27. x axis

28. II

29. II

30. III

31. y axis

32. x axis

33. IV

34. II

35. (1, 30), (2, 45), (3, 60), (4, 60), (5, 75), (6, 90), (7, 95)

In Exercises 23 to 34 give the quadrant in which each of the following points is located or the axis on which the point lies.

23. (4, 5)   **24.** (−3, 2)   **25.** (−4, −3)   **26.** (2, −4)

27. (5, 0)   **28.** (−5, 7)   **29.** (−4, 7)   **30.** (−3, −7)

31. (0, −7)   **32.** (−3, 0)   **33.** $\left(5\frac{3}{4}, -3\right)$ **34.** $\left(-2, 4\frac{5}{6}\right)$

35. A company has kept a record of the number of items produced by an employee as the number of days on the job increases. In the following figure, points correspond to an ordered-pair relationship in which the first number represents days on the job and the second number represents the number of items produced. Estimate each ordered pair produced.

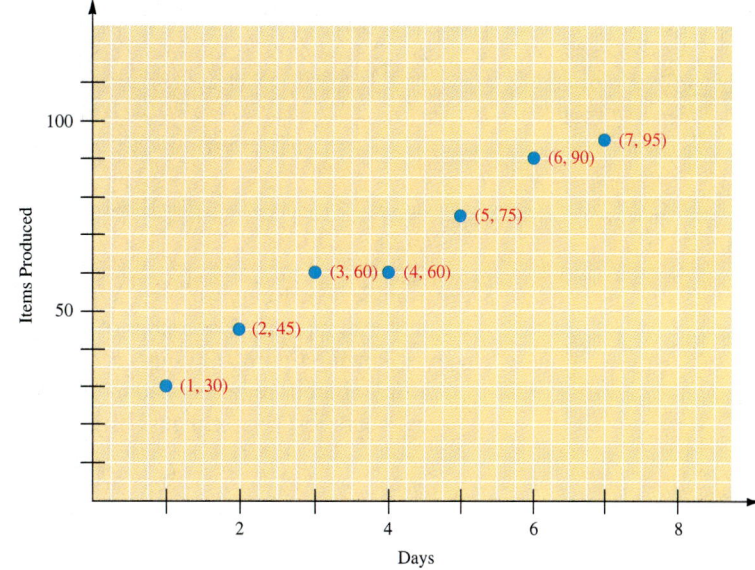

210 Chapter 3 ■ Graphs and Linear Equations

36. (3, 35), (6, 50), (9, 58), (12, 60), (14, 70), (16, 80), (17, 85)

37. (7, 100), (15, 70), (20, 80), (30, 70), (40, 50), (50, 40), (60, 30), (70, 40), (80, 25)

36. In the following figure, points correspond to an ordered-pair relationship between height and age in which the first number represents age and the second number represents height. Estimate each ordered pair represented.

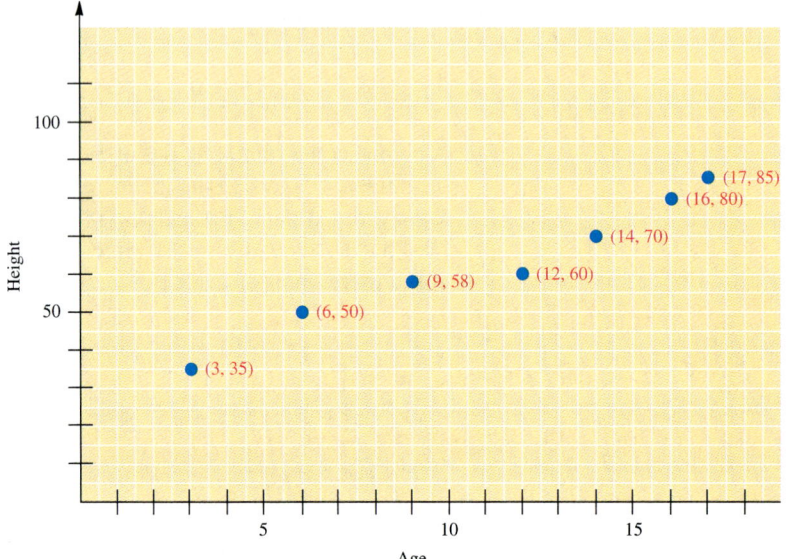

37. An unidentified company has kept a record of the number of hours devoted to safety training and the number of work hours lost due to on-the-job accidents. In the following figure, the points correspond to an ordered pair relationship in which the first number represents hours in safety training and the second number represents hours lost by accidents. Estimate each ordered pair represented.

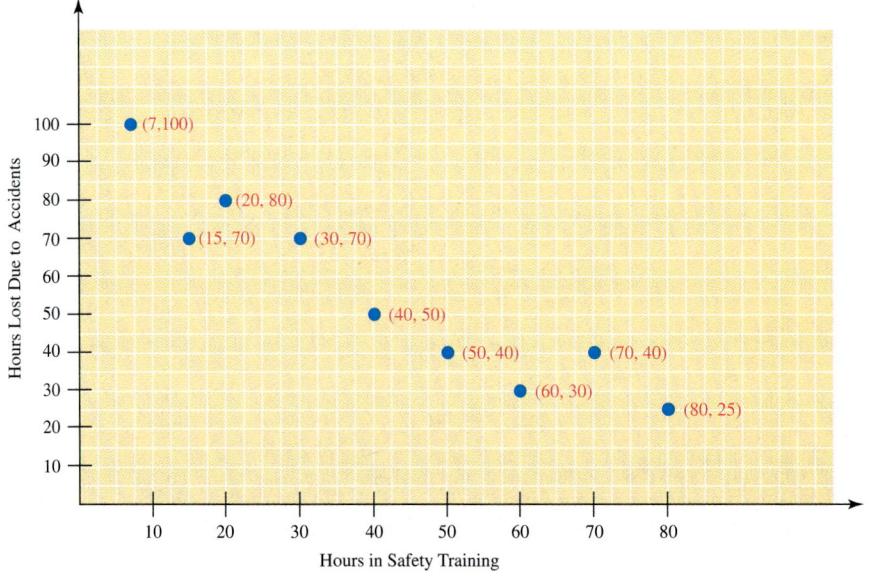

38. (15, 25), (25, 15), (35, 10), (40, 5), (50, 10), (60, 25)

39. The points lie on a line; (1, 2)

40. The points lie on a line; (2, 1)

41. The points lie on a line; (2, −6)

42. The points lie on a line; (3, 7)

38. In the following figure, points correspond to an ordered pair relationship between the age of a person and the annual average number of visits to doctors and dentists for a person that age. The first number represents the age and the second number represents the number of visits. Estimate each ordered pair represented.

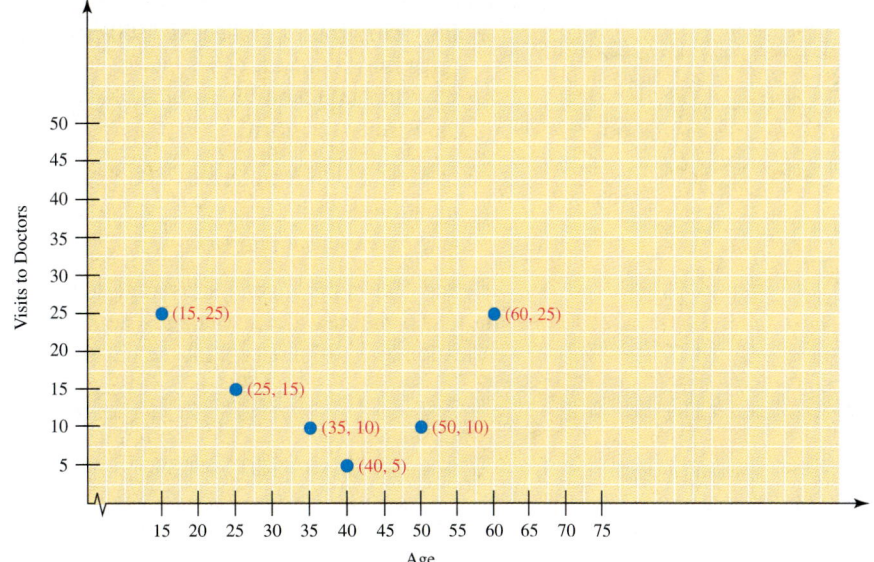

39. Graph points with coordinates (2, 3), (3, 4) and (4, 5). What do you observe? Can you give the coordinates of another point with the same property?

40. Graph points with coordinates (−1, 4), (0, 3), and (1, 2). What do you observe? Can you give the coordinates of another point with the same property?

41. Graph points with coordinates (−1, 3), (0, 0), and (1, −3). What do you observe? Can you give the coordinates of another point with the same property?

42. Graph points with coordinates (1, 5), (−1, 3), and (−3, 1). What do you observe? Can you give the coordinates of another point with the same property?

43.
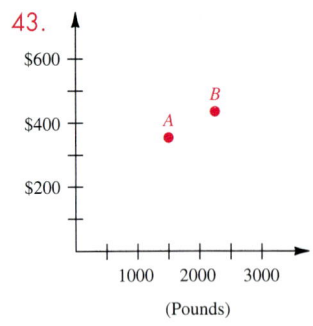

44.
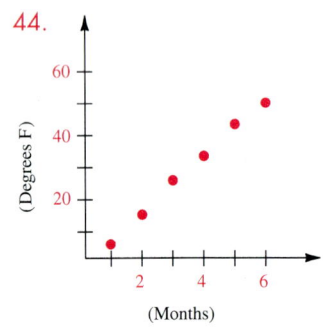

45.
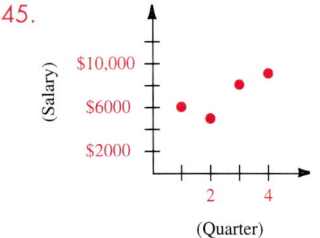

43. Environment. A local plastics company is sponsoring a plastics recycling contest for the local community. The focus of the contest is collecting plastic milk, juice, and water jugs. The company will award $200 plus the current market price of the jugs collected to the group that collects the most jugs in a single month. The number of jugs collected and the amount of money won can be represented as an ordered pair.

(a) In April, group A collected 1500 pounds (lb) of jugs to win first place. The prize for the month was $350. If x represents the pounds of jugs and y represents the amount of money that the group won, graph the point that represents the winner for April.

(b) In May, group B collected 2300 lb of jugs to win first place. The prize for the month was $430. Graph the point that represents the May winner on the same grid you used in part (a).

44. Science. The table gives the average temperature, y (in degrees Fahrenheit), for the first 6 months of the year, x. The months are numbered 1 through 6, with 1 corresponding to January. Plot the data given in the table.

x	1	2	3	4	5	6
y	4	14	26	33	42	51

45. Business. The table gives the total salary of a salesperson, y, for each of the four quarters of the year, x. Plot the data given in the table.

x	1	2	3	4
y	$6000	$5000	$8000	$9000

46. Although high employment is a measure of a country's economic vitality, economists worry that periods of low unemployment will lead to inflation. Look at the following table.

Year	Unemployment Rate (%)	Inflation Rate (%)
1955	4.4	−0.4
1960	5.5	1.7
1965	4.5	1.6
1970	4.9	5.7
1975	8.5	9.1
1980	7.1	13.5
1985	7.2	3.6
1990	5.5	5.4

Plot the figures in the table with unemployment rates on the x axis and inflation rates on the y axis. What do these plots tell you? Do higher inflation rates seem to be associated with lower unemployment rates? Explain.

47. We mentioned that the Cartesian coordinate system was named for the French philosopher and mathematician René Descartes. What philosophy book is Descartes most famous for? Use an encyclopedia as a reference.

48. What characteristic is common to all points on the x axis? On the y axis?

49. How would you describe a rectangular coordinate system? Explain what information is needed to locate a point in a coordinate system.

50. Some newspapers have a special day that they devote to automobile wants ads. Use this special section or the Sunday classified ads from your local newspaper to find all the want ads for a particular automobile model. Make a list of the model year and asking price for 10 ads, being sure to get a variety of ages for this model. After collecting the information, make a scatter plot of the age and the asking price for the car.

Describe your graph, including an explanation of how you decided which variable to put on the vertical axis and which on the horizontal axis. What trends or other information are given by the graph?

SECTION 3.3 The Graph of a Linear Equation

3.3 OBJECTIVES

1. Graph a linear equation by plotting points
2. Graph a linear equation by the intercept method
3. Graph a linear equation by solving the equation for y

"I think there is a world market for maybe five computers."

–Thomas Watson (IBM Chairman) in 1943

"640K ought to be enough for anybody."

–Bill Gates (Microsoft Chairman) in 1981

In Section 3.1, you learned to write the solutions of equations in two variables as ordered pairs. In Section 3.2, ordered pairs were graphed in the Cartesian plane. Putting these ideas together will help us graph certain equations. Example 1 illustrates one approach to finding the graph of a linear equation.

Example 1 Graphing a Linear Equation

Graph $x + 2y = 4$.

Step 1 Find some solutions for $x + 2y = 4$. To find solutions, we choose any convenient values for x, say $x = 0$, $x = 2$, and $x = 4$. Given these values for x, we can substitute and then solve for the corresponding value for y.

We are going to find three solutions for the equation. We'll point out why shortly.

If $x = 0$, then $y = 2$, so (0, 2) is a solution.

If $x = 2$, then $y = 1$, so (2, 1) is a solution.

If $x = 4$, then $y = 0$, so (4, 0) is a solution.

A handy way to show this information is in a table such as this:

The table is a convenient way to display the information. It is the same as writing (0, 2), (2, 1), and (4, 0).

x	y
0	2
2	1
4	0

214

Step 2 We now graph the solutions found in step 1.

$$x + 2y = 4$$

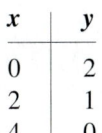 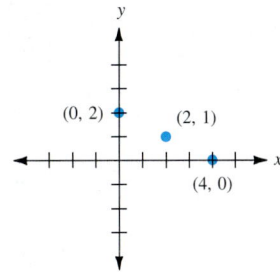

What pattern do you see? It appears that the three points lie on a straight line, and that is in fact the case.

Step 3 Draw a straight line through the three points graphed in step 2.

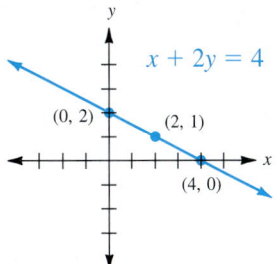

The arrows on the end of the line mean that the line extends infinitely in either direction.

The graph is a "picture" of the solutions for the given equation.

The line shown is the **graph** of the equation $x + 2y = 4$. It represents *all* of the ordered pairs that are solutions (an infinite number) for that equation.

Every ordered pair that is a solution will be plotted as a point on this line. Any point on the line will represent a pair of numbers that is a solution for the equation.

Note: Why did we suggest finding *three* solutions in step 1? Two points determine a line, so technically you need only two. The third point that we find is a check to catch any possible errors.

✓ CHECK YOURSELF 1

Graph $2x - y = 6$, using the steps shown in Example 1.

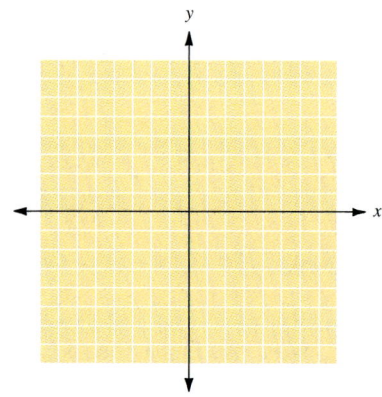

Let's summarize. An equation that can be written in the form

$$Ax + By = C$$

where A, B, and C are real numbers and A and B are not both 0 is called a **linear equation in two variables.** The graph of this equation is a *straight line*.

The steps of graphing follow.

To Graph a Linear Equation

Step 1 Find at least three solutions for the equation, and put your results in tabular form.

Step 2 Graph the solutions found in step 1.

Step 3 Draw a straight line through the points determined in step 2 to form the graph of the equation.

Example 2 Graphing a Linear Equation

Graph $y = 3x$.

Step 1 Some solutions are

x	y
0	0
1	3
2	6

Let $x = 0$, 1, and 2, and substitute to determine the corresponding y values. Again the choices for x are simply convenient. Other values for x would serve the same purpose.

Step 2 Graph the points.

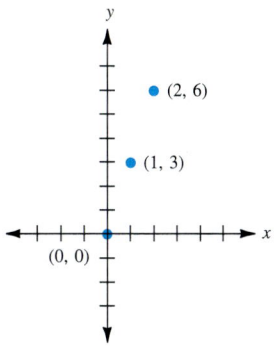

Notice that connecting any two of these points produces the same line.

Step 3 Draw a line through the points.

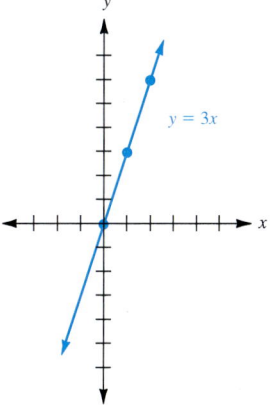

✓ CHECK YOURSELF 2

Graph the equation $y = -2x$ after completing the table of values.

x	y
0	
1	
2	

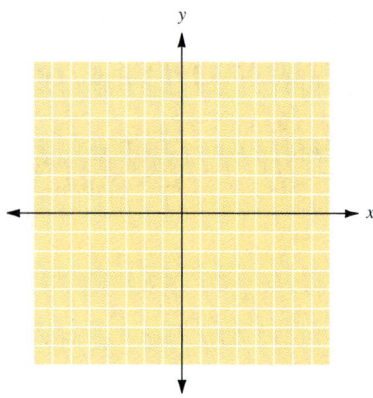

Let's work through another example of graphing a line from its equation.

Example 3 Graphing a Linear Equation

Graph $y = 2x + 3$.

Step 1 Some solutions are

x	y
0	3
1	5
2	7

Step 2 Graph the points corresponding to these values.

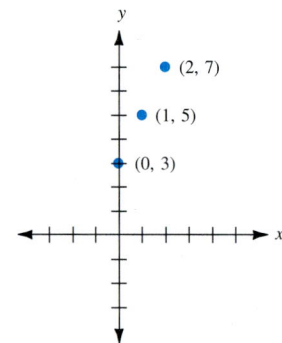

Section 3.3 ■ The Graph of a Linear Equation 219

Step 3 Draw a line through the points.

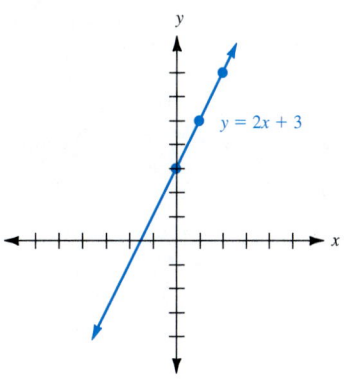

✓ CHECK YOURSELF 3

Graph the equation $y = 3x - 2$ after completing the table of values.

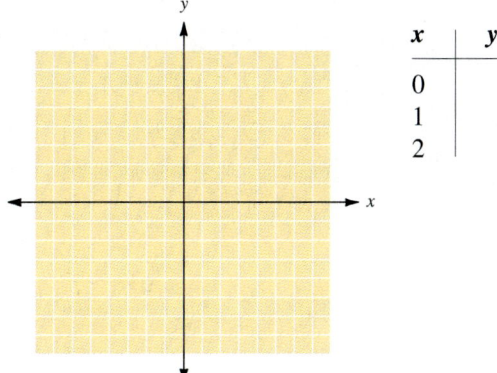

x	y
0	
1	
2	

In graphing equations, particularly when fractions are involved, a careful choice of values for x can simplify the process. Consider Example 4.

Example 4 Graphing a Linear Equation

Graph

$$y = \frac{3}{2}x - 2$$

As before, we want to find solutions for the given equation by picking convenient values for x. Note that in this case, choosing *multiples of 2*, the denominator of the x coefficient, will avoid fractional values for y and make the plotting of those solutions much easier. For instance, here we might choose values of -2, 0, and 2 for x.

Step 1

If $x = -2$:

$$\begin{aligned} y &= \frac{3}{2}x - 2 \\ &= \frac{3}{2}(-2) - 2 \\ &= -3 - 2 = -5 \end{aligned}$$

If $x = 0$:

$$\begin{aligned} y &= \frac{3}{2}x - 2 \\ &= \frac{3}{2}(0) - 2 \\ &= 0 - 2 = -2 \end{aligned}$$

If $x = 2$:

$$\begin{aligned} y &= \frac{3}{2}x - 2 \\ &= \frac{3}{2}(2) - 2 \\ &= 3 - 2 = 1 \end{aligned}$$

Suppose we do *not* choose a multiple of 2, say, $x = 3$. Then

$$\begin{aligned} y &= \frac{3}{2}(3) - 2 \\ &= \frac{9}{2} - 2 \\ &= \frac{5}{2} \end{aligned}$$

$\left(3, \frac{5}{2}\right)$ is still a valid solution, but we must graph a point with fractional coordinates.

In tabular form, the solutions are

x	y
-2	-5
0	-2
2	1

Step 2 Graph the points determined above.

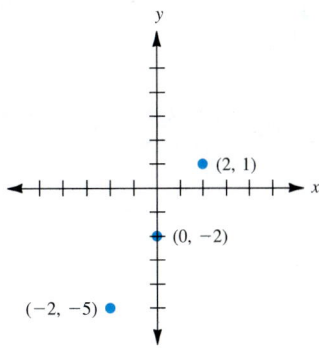

Step 3 Draw a line through the points.

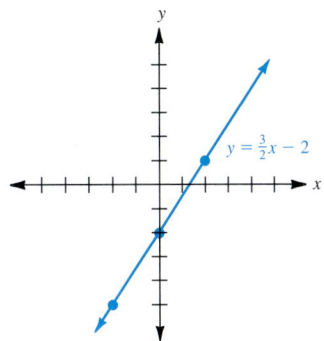

✓ CHECK YOURSELF 4

Graph the equation $y = -\dfrac{1}{3}x + 3$ after completing the table of values.

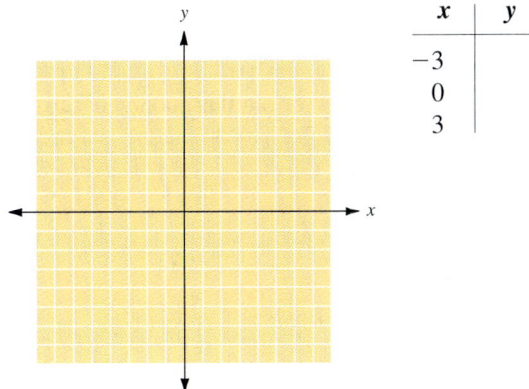

x	y
-3	
0	
3	

Some special cases of linear equations are illustrated in Examples 5 and 6.

Example 5 Graphing an Equation That Results in a Vertical Line

Graph $x = 3$.

The equation $x = 3$ is equivalent to $x + 0 \cdot y = 3$. Let's look at some solutions.

If $y = 1$:   If $y = 4$:   If $y = -2$:

$x + 0 \cdot 1 = 3$ $x + 0 \cdot 4 = 3$ $x + 0(-2) = 3$

$x = 3$ $x = 3$ $x = 3$

In tabular form,

x	y
3	1
3	4
3	-2

What do you observe? The variable x has the value 3, regardless of the value of y. Look at the graph below.

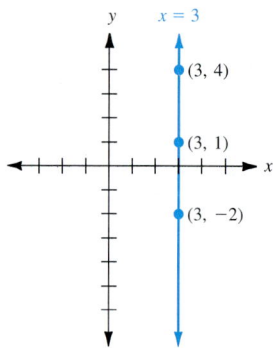

The graph of $x = 3$ is a vertical line crossing the x axis at (3, 0).

Note that graphing (or plotting) points in this case is not really necessary. Simply recognize that the graph of $x = 3$ *must* be a vertical line (parallel to the y axis) which intercepts the x axis at (3, 0).

✓ CHECK YOURSELF 5

Graph the equation $x = -2$.

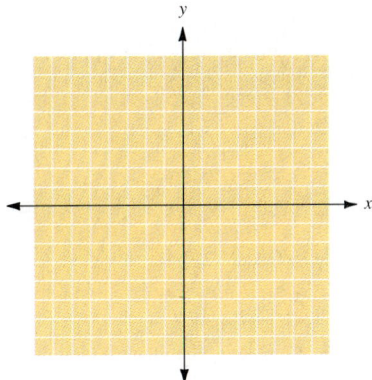

Example 6 is a related example involving a horizontal line.

Example 6 — Graphing an Equation That Results in a Horizontal Line

Graph $y = 4$.

Since $y = 4$ is equivalent to $0 \cdot x + y = 4$, any value for x paired with 4 for y will form a solution. A table of values might be

x	y
-2	4
0	4
2	4

Here is the graph.

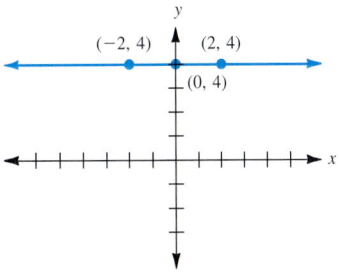

This time the graph is a horizontal line that crosses the y axis at $(0, 4)$. Again the graphing of points is not required. The graph of $y = 4$ *must* be horizontal (parallel to the x axis) and intercepts the y axis at $(0, 4)$.

✓ CHECK YOURSELF 6

Graph the equation $y = -3$.

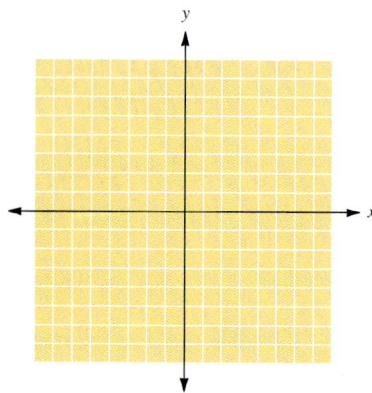

The following box summarizes our work in the previous two examples:

Vertical and Horizontal Lines

1. The graph of $x = a$ is a *vertical line* crossing the x axis at $(a, 0)$.
2. The graph of $y = b$ is a *horizontal line* crossing the y axis at $(0, b)$.

To simplify the graphing of certain linear equations, some students prefer the **intercept method** of graphing. This method makes use of the fact that the solutions that are easiest to find are those with an x coordinate or a y coordinate of 0. For instance, let's graph the equation

$$4x + 3y = 12$$

With practice, all this can be done mentally, which is the big advantage of this method.

First, let $x = 0$ and solve for y.

$$4x + 3y = 12$$
$$4 \cdot 0 + 3y = 12$$
$$3y = 12$$
$$y = 4$$

So (0, 4) is one solution. Now let $y = 0$ and solve for x.

$$4x + 3y = 12$$
$$4x + 3 \cdot 0 = 12$$
$$4x = 12$$
$$x = 3$$

A second solution is (3, 0).

The two points corresponding to these solutions can now be used to graph the equation.

> Remember, only two points are needed to graph a line. A third point is used only as a check.

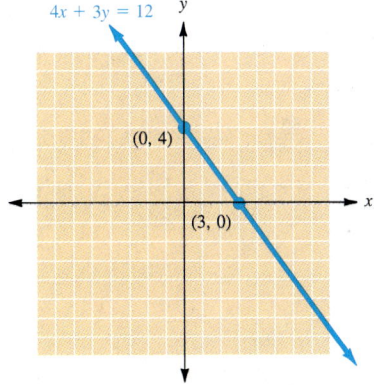

> The intercepts are the points where the line cuts the x and y axes. Here, the x intercept has coordinates (3, 0) and the y intercept has coordinates (0, 4).

The point (3, 0) is called the **x intercept,** and the point (0, 4) is the **y intercept** of the graph. Using these points to draw the graph gives the name to this method. Let's look at a second example of graphing by the intercept method.

Example 7 — Using the Intercept Method to Graph a Line

Graph $3x - 5y = 15$, using the intercept method.

To find the x intercept, let $y = 0$.

$$3x - 5 \cdot 0 = 15$$
$$x = 5$$

The x value of the intercept

To find the y intercept, let $x = 0$.

$$3 \cdot 0 - 5y = 15$$
$$y = -3$$

The y value of the intercept

So (5, 0) and (0, −3) are solutions for the equation, and we can use the corresponding points to graph the equation.

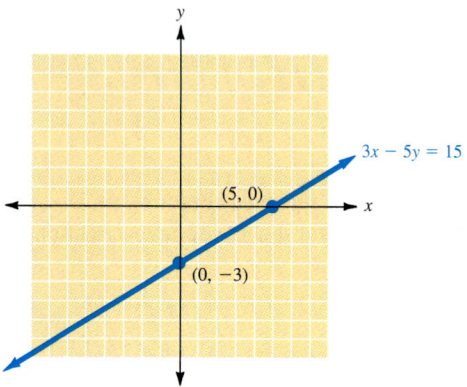

✓ CHECK YOURSELF 7

Graph $4x + 5y = 20$, using the intercept method.

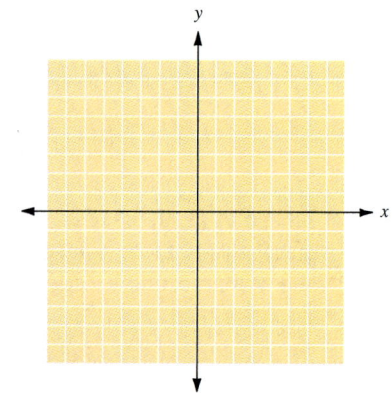

Finding the third "checkpoint" is always a good idea.

This all looks quite easy, and for many equations it is. What are the drawbacks? For one, you don't have a third checkpoint, and it is possible for errors to occur. You can, of course, still find a third point (other than the two intercepts) to be sure your graph is correct. A second difficulty arises when the x and y intercepts are very close to one another (or are actually the same point—the origin). For instance, if we have the equation

$$3x + 2y = 1$$

the intercepts are $\left(\frac{1}{3}, 0\right)$ and $\left(0, \frac{1}{2}\right)$. It is hard to draw a line accurately through these intercepts, so choose other solutions farther away from the origin for your points.

Let's summarize the steps of graphing by the intercept method for appropriate equations.

Graphing a Line by the Intercept Method

Step 1 To find the x intercept: Let $y = 0$, then solve for x.

Step 2 To find the y intercept: Let $x = 0$, then solve for y.

Step 3 Graph the x and y intercepts.

Step 4 Draw a straight line through the intercepts.

A third method of graphing linear equations involves **solving the equation for y.** The reason we use this extra step is that it often will make finding solutions for the equation much easier. Let's look at an example.

Example 8 Graphing a Linear Equation

Graph $2x + 3y = 6$.

Remember that solving for y means that we want to leave y isolated on the left.

Rather than finding solutions for the equation in this form, we solve for y.

$$2x + 3y = 6$$
$$3y = 6 - 2x \qquad \text{Subtract } 2x.$$
$$y = \frac{6 - 2x}{3} \qquad \text{Divide by 3.}$$

or

$$y = 2 - \frac{2}{3}x$$

Now find your solutions by picking convenient values for x.

Again, to pick convenient values for x, we suggest you look at the equation carefully. Here, for instance, picking multiples of 3 for x will make the work much easier.

If $x = -3$:

$$y = 2 - \frac{2}{3}x$$
$$= 2 - \frac{2}{3}(-3)$$
$$= 2 + 2 = 4$$

So $(-3, 4)$ is a solution.

If $x = 0$:

$$y = 2 - \frac{2}{3}x$$

$$y = 2 - \frac{2}{3} \cdot 0$$
$$= 2$$

So (0, 2) is a solution.

If $x = 3$:
$$y = 2 - \frac{2}{3}x$$
$$= 2 - \frac{2}{3} \cdot 3$$
$$= 2 - 2 = 0$$

So (3, 0) is a solution.

We can now plot the points that correspond to these solutions and form the graph of the equation as before.

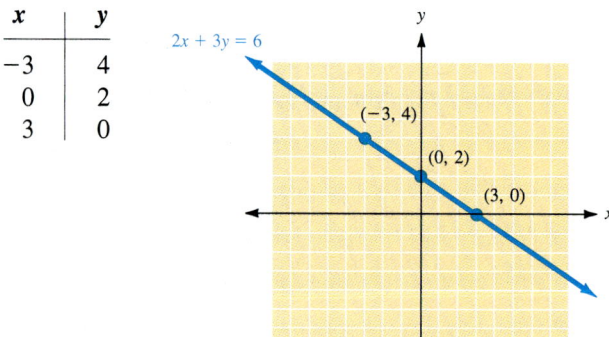

✓ CHECK YOURSELF 8

Graph the equation $5x + 2y = 10$. Solve for y to determine solutions.

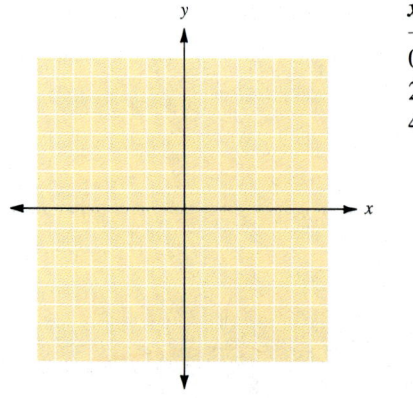

x	y
0	
2	
4	

Section 3.3 ■ The Graph of a Linear Equation   **229**

✓ CHECK YOURSELF ANSWERS

1. | x | y |
    |---|----|
    | 1 | −4 |
    | 2 | −2 |
    | 3 | 0 |

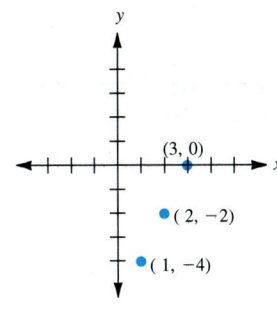 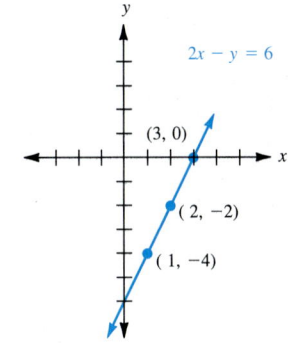

2. | x | y |
    |---|----|
    | 0 | 0 |
    | 1 | −2 |
    | 2 | −4 |

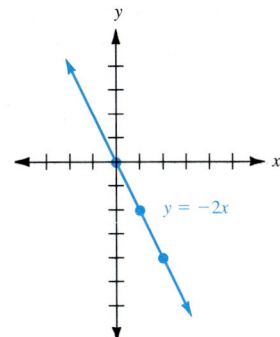
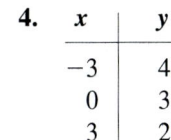

3. | x | y |
    |---|----|
    | 0 | −2 |
    | 1 | 1 |
    | 2 | 4 |

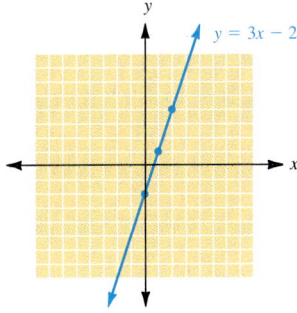

4. | x | y |
    |----|---|
    | −3 | 4 |
    | 0 | 3 |
    | 3 | 2 |

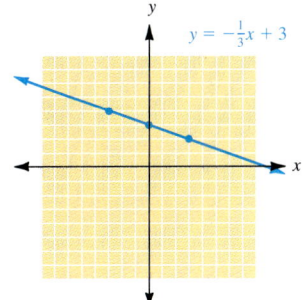

5.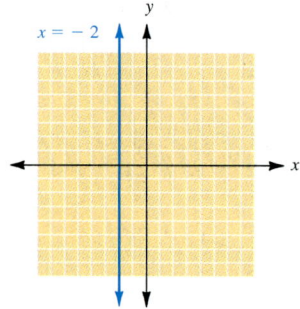

6.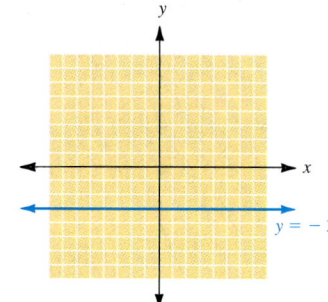

7.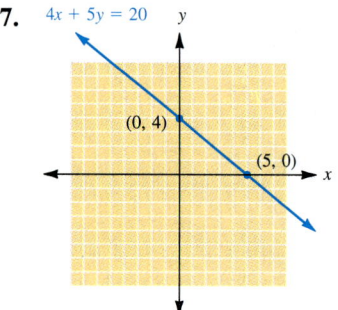

8.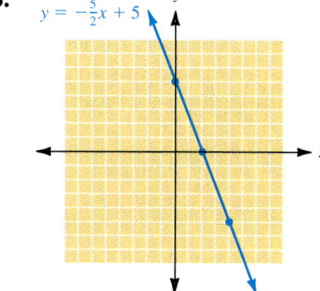

x	y
0	5
2	0
4	−5

Exercises ■ 3.3

Graph each of the following equations.

1. $x + y = 6$

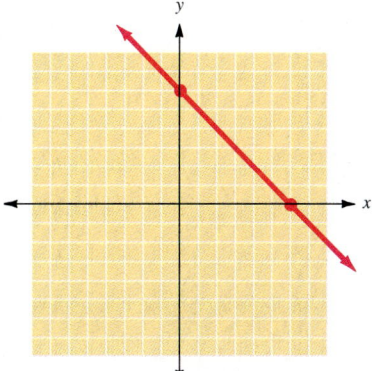

2. $x - y = 5$

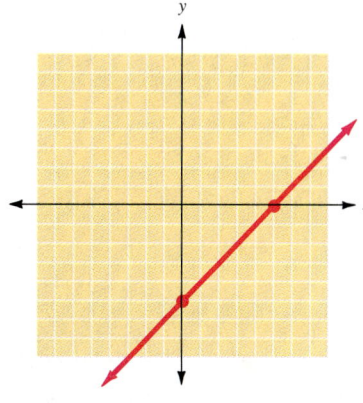

3. $x - y = -3$

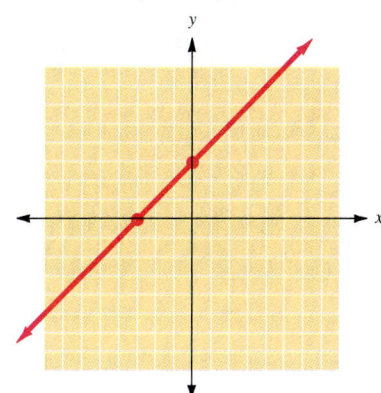

4. $x + y = -3$

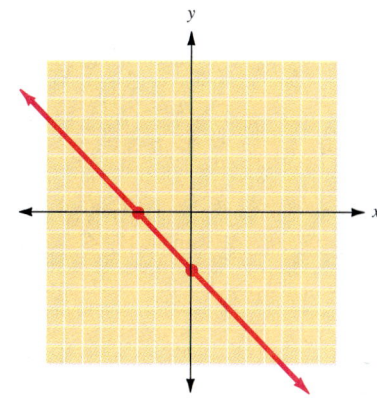

5. $2x + y = 2$

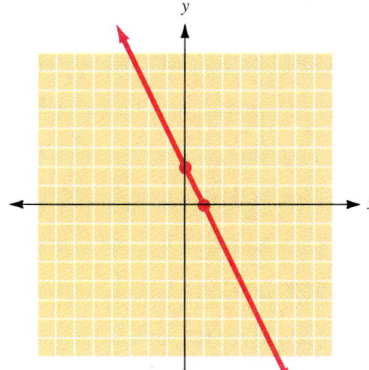

6. $x - 2y = 6$

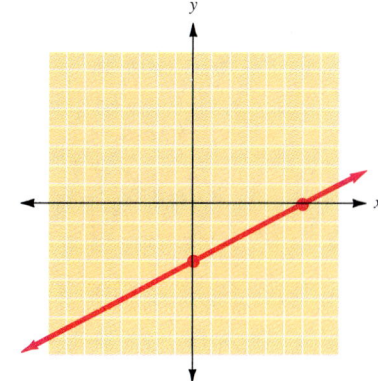

7. $3x + y = 0$

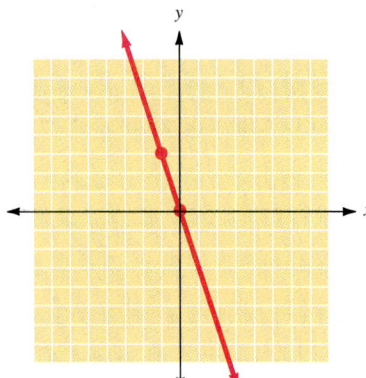

8. $3x - y = 6$

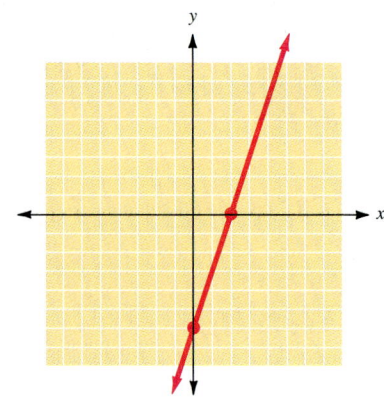

9. $x + 4y = 8$

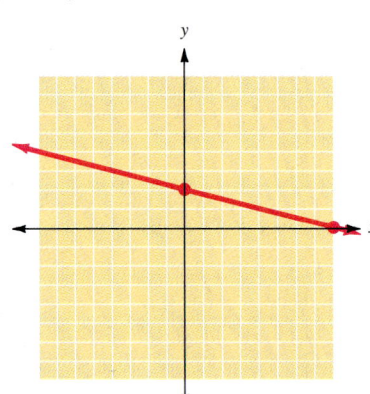

10. $2x - 3y = 6$

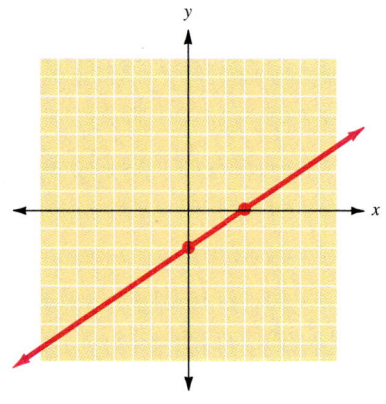

11. $y = 5x$

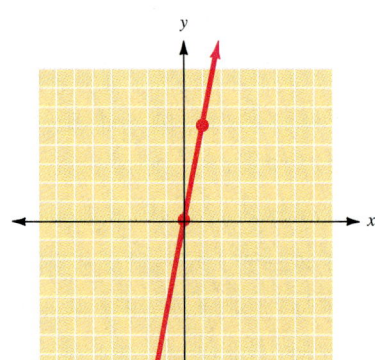

12. $y = -4x$

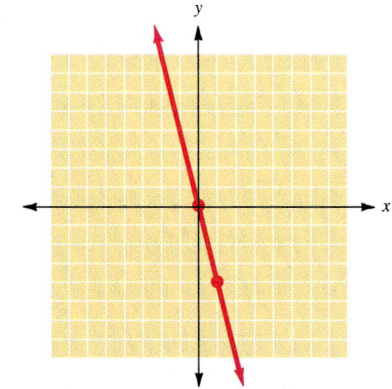

Section 3.3 ▪ The Graph of a Linear Equation   **233**

13. $y = 2x - 1$

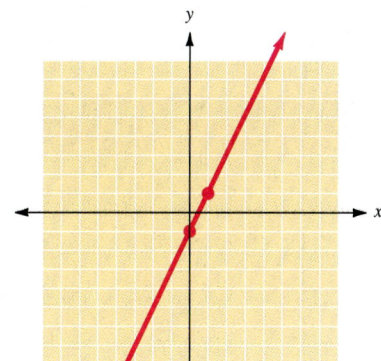

14. $y = 4x + 3$

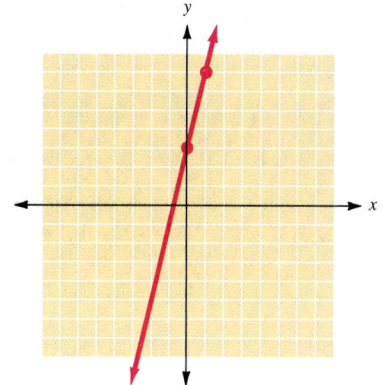

15. $y = -3x + 1$

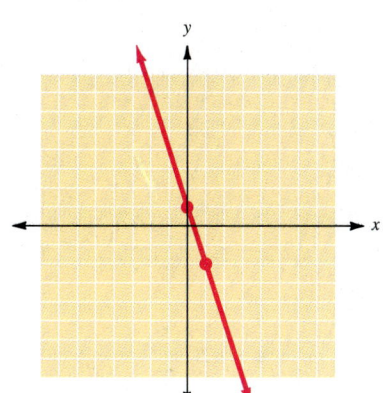

16. $y = -3x - 3$

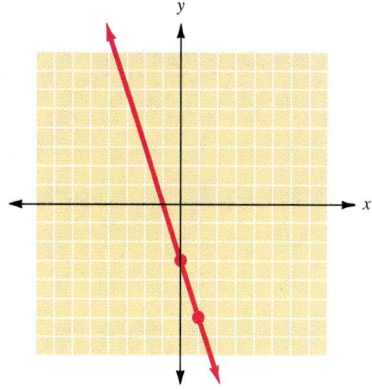

17. $y = \dfrac{1}{3}x$

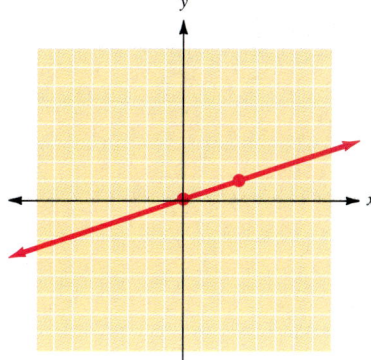

18. $y = -\dfrac{1}{4}x$

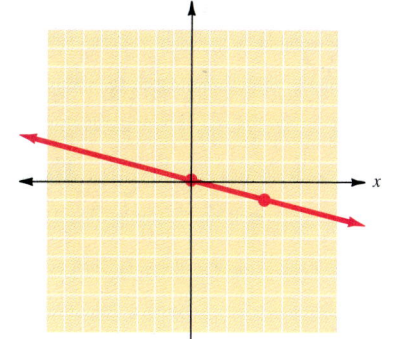

19. $y = \dfrac{2}{3}x - 3$

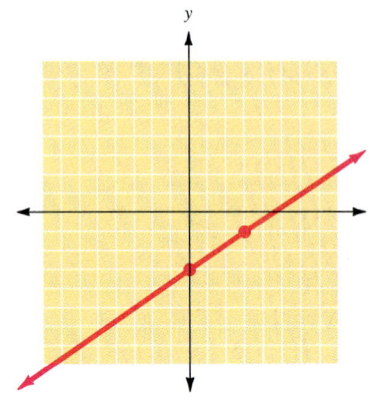

20. $y = \dfrac{3}{4}x + 2$

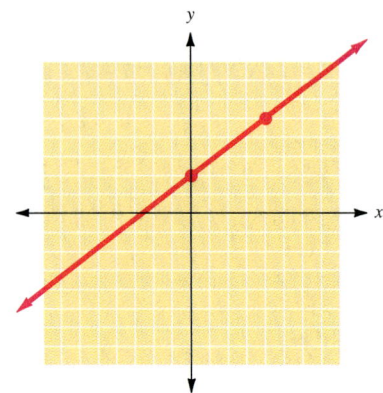

21. $x = 5$

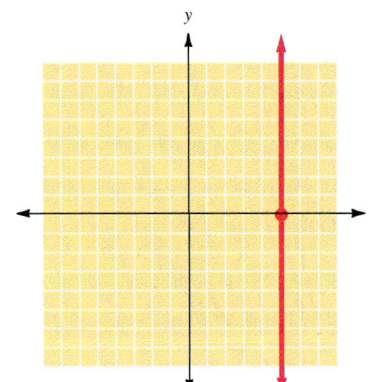

22. $y = -3$

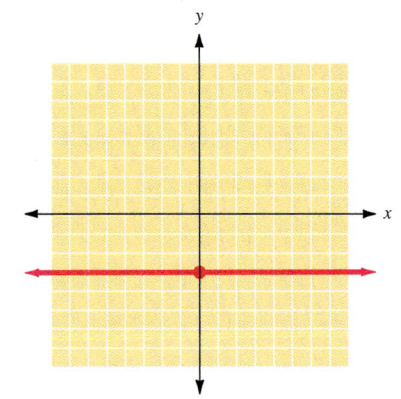

23. $y = 1$

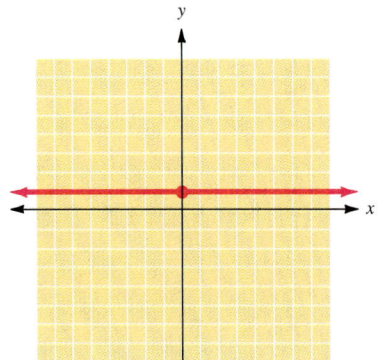

24. $x = -2$

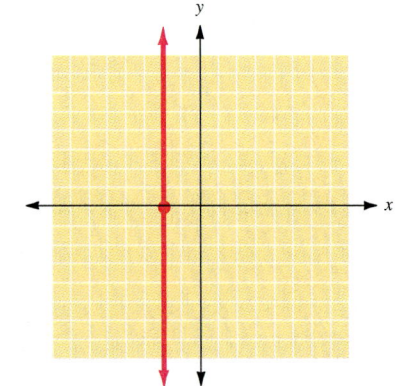

25. $x - 2y = 4$

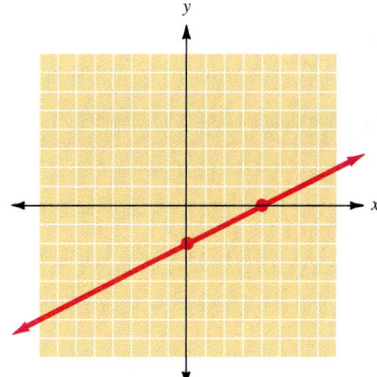

26. $6x + y = 6$

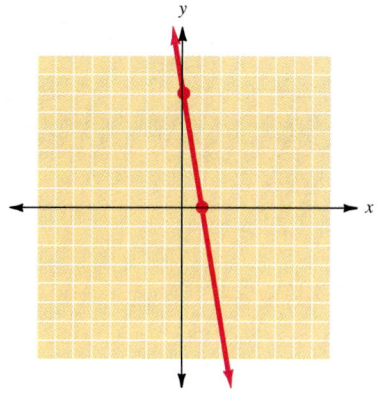

27. $5x + 2y = 10$

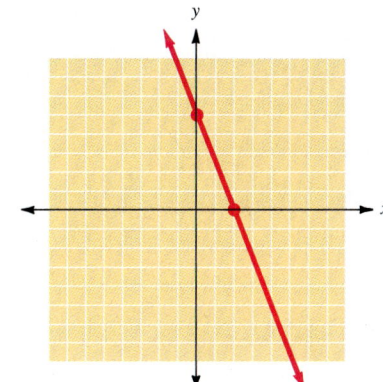

28. $2x + 3y = 6$

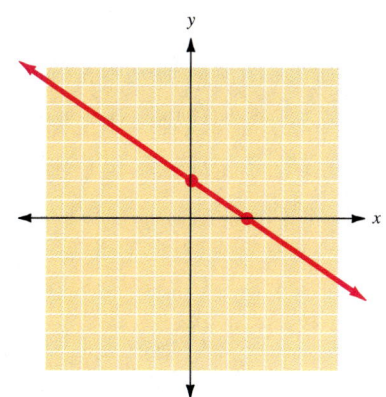

29. $3x + 5y = 15$

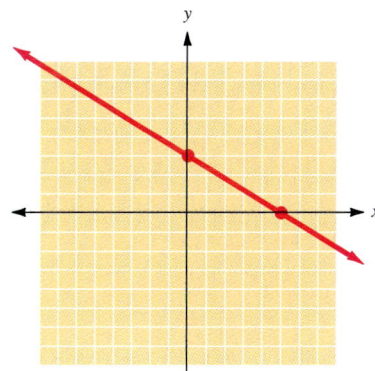

30. $4x + 3y = 12$

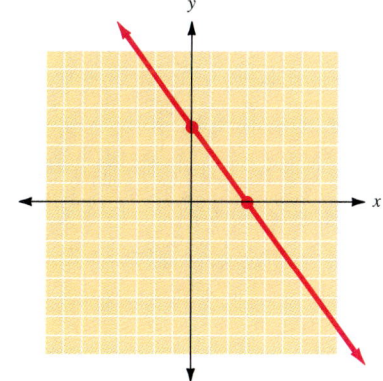

31. $y = 2 - \dfrac{x}{3}$

32. $y = -3 + \dfrac{x}{2}$

33. $y = 3 - \dfrac{3}{4}x$

34. $y = -4 + \dfrac{2}{3}x$

35. $y = -5 + \dfrac{5}{4}x$

36. $y = 7 - \dfrac{7}{3}x$

Graph each of the following equations by first solving for y.

31. $x + 3y = 6$

32. $x - 2y = 6$

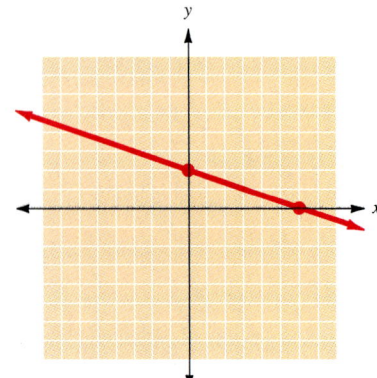
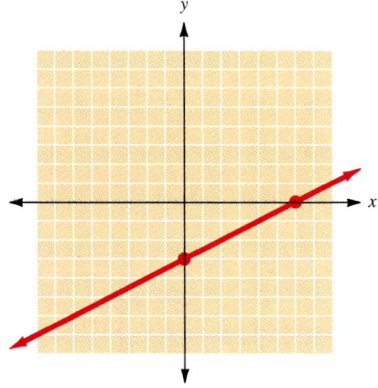

33. $3x + 4y = 12$

34. $2x - 3y = 12$

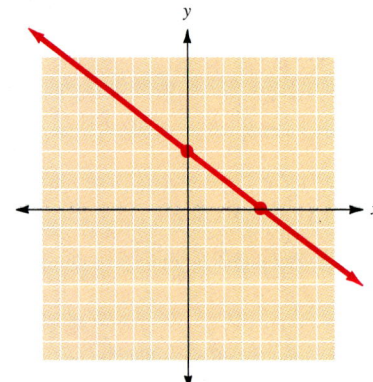
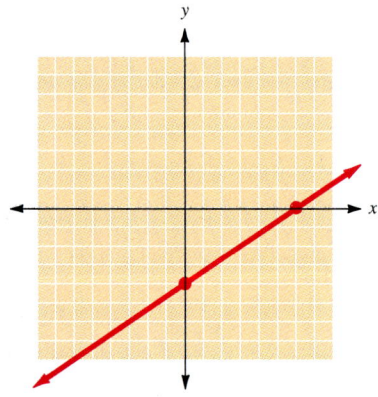

35. $5x - 4y = 20$

36. $7x + 3y = 21$

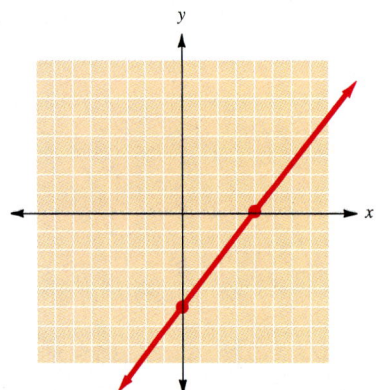
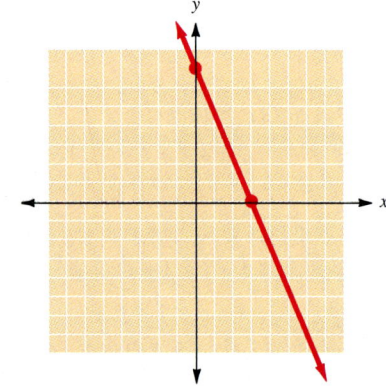

Section 3.3 ■ The Graph of a Linear Equation 237

37. $y = 2x$

38. $y = 3x$

39. $y = x + 3$

40. $y = x - 2$

41. $y = 3x - 3$

42. $y = 2x + 4$

43. $x - 4y = 12$

44. $2x - y = 6$

45. (3, 1)

46. (4, 1)

47. Parallel lines

48. Parallel lines

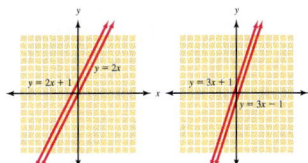

49. Perpendicular lines

50. Perpendicular lines

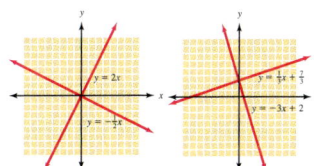

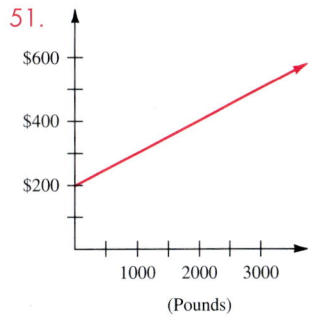

52. $210

Write an equation that describes the following relationships between *x* and *y*. Then graph each relationship.

37. *y* is twice *x*

38. *y* is three times *x*.

39. *y* is 3 more than *x*

40. *y* is 2 less than *x*.

41. *y* is 3 less than 3 times *x*.

42. *y* is 4 more than twice *x*.

43. The difference of *x* and the product of 4 and *y* is 12.

44. The difference of twice *x* and *y* is 6.

Graph each pair of equations on the same grid. Give the coordinates of the point where the lines intersect.

45. $x + y = 4$
$x - y = 2$

46. $x - y = 3$
$x + y = 5$

In each of the following exercises, graph both equations on the same set of axes and report what you observe about the graphs.

47. $y = 2x$ and $y = 2x + 1$

48. $y = 3x + 1$ and $y = 3x - 1$

49. $y = 2x$ and $y = -\dfrac{1}{2}x$

50. $y = \dfrac{1}{3}x + \dfrac{7}{3}$ and $y = -3x + 2$

51. Graph of winnings. The equation $y = 0.10x + 200$ describes the amount of winnings a group earns for collecting plastic jugs in the recycling contest described in Exercise 43 at the end of Section 3.2. Sketch the graph of the line.

52. Minimum values. The contest sponsor will award a prize only if the winning group in the contest collects 100 lb of jugs or more. Use your graph in Exercise 51 to determine the minimum prize possible.

53 (a)

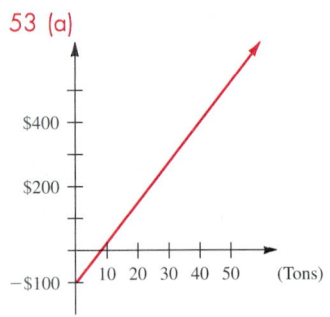

53 (b). $\dfrac{100}{15}$ or ≈ 7 tons

53 (c). $140

53 (d). $y = 17x - 125$

54 (a). $C = 10x + 40$

54 (b) and (c).

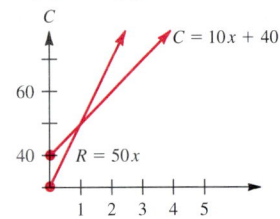

54 (d). 1

55.

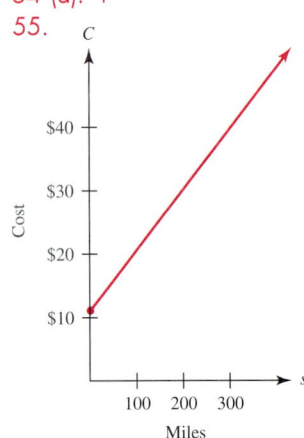

56.

---

53. **Fundraising.** A high school class wants to raise some money by recycling newspapers. They decide to rent a truck for a weekend and to collect the newspapers from homes in the neighborhood. The market price for recycled newsprint is currently $15 per ton. The equation $y = 15x - 100$ describes the amount of money the class will make, where y is the amount of money made in dollars, x is the number of tons of newsprint collected, and 100 is the cost in dollars to rent the truck.

   (a) Draw a graph that represents the relationship between newsprint collected and money earned.

   (b) The truck is costing the class $100. How many tons of newspapers must the class collect to break even on this project?

   (c) If the class members collect 16 tons of newsprint, how much money will they earn?

   (d) Six months later the price of newsprint is $17 dollars a ton, and the cost to rent the truck has risen to $125. Write the equation that describes the amount of money the class might make at that time.

54. **Production costs.** The cost of producing a number of items x is given by $C = mx + b$, where b is the fixed cost and m is the marginal cost (the cost of producing one item).

   (a) If the fixed cost is $40 and the variable cost is $10, write the cost equation.

   (b) Graph the cost equation.

   (c) The revenue generated from the sale of x items is given by $R = 50x$. Graph the revenue equation on the same set of axes as the cost equation.

   (d) How many items must be produced in order for the revenue to equal the cost (the break-even point)?

55. **Consumer affairs.** A car rental agency charges $12 per day and 8¢ per mile for the use of a compact automobile. The cost of the rental C and the number of miles driven per day s are related by the equation

$$C = 0.08s + 12$$

Graph the relationship between C and s. Be sure to select appropriate scaling for the C and s axes.

56. **Checking account charges.** A bank has the following structure for charges on checking accounts. The monthly charges consist of a fixed amount of $8 and an additional charge of 5¢ per check. The monthly cost of an account C and the number of checks written per month n are related by the equation

$$C = 0.05n + 8$$

Graph the relationship between C and n.

Section 3.3 ■ The Graph of a Linear Equation   239

57 (a). $T = 35h + 75$

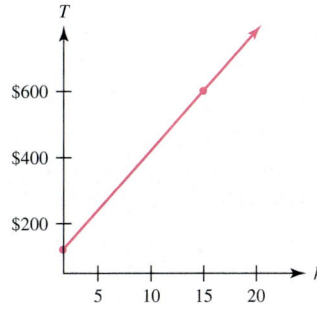

58 (a). $S = 200 + 0.10x$

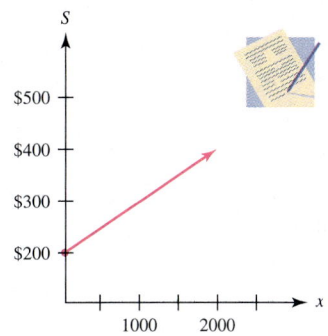

59 (a)

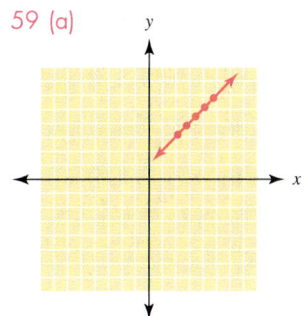

59 (b). Increases by 2

59 (c). Yes

59 (d). Grows by 2 units

60 (b). Increases by 2

60 (c). Yes

60 (d). Grows by 2 units

57. Tuition charges. A college has tuition charges based on the following pattern. Tuition is $35 per credit-hour plus a fixed student fee of $75.

(a) Write a linear equation that shows the relationship between the total tuition charge T and the number of credit-hours taken h.

(b) Graph the relationship between T and h.

58. Weekly salary. A salesperson's weekly salary is based on a fixed amount of $200 plus 10% of the total amount of weekly sales.

(a) Write an equation that shows the relationship between the weekly salary S and the amount of weekly sales x (in dollars).

(b) Graph the relationship between S and x.

59. Consider the equation $y = 2x + 3$.

(a) Complete the following table of values, and plot the resulting points.

Point	x	y
A	5	13
B	6	15
C	7	27
D	8	19
E	9	21

(b) As the x coordinate changes by 1 (for example, as you move from point A to point B), how much do the corresponding y coordinates change?

(c) Is your answer to part b the same if you move from B to C? from C to D? from D to E?

(d) Describe the "growth rate" of the line using these observations. Complete the following statement: When the x value grows by 1 unit, the y value _____.

60. Repeat exercise 59 using $y = 2x + 5$.

61. Repeat exercise 59 using $y = 3x - 2$.

62. Repeat exercise 59 using $y = 3x - 4$.

63. Repeat exercise 59 using $y = -4x + 50$.

64. Repeat exercise 59 using $y = -4x + 40$.

61 (b). Increases by 3   61 (c). Yes   61 (d). Grows by 3 units
62 (b). Increases by 3   62 (c). Yes   62 (d). Grows by 3 units
63 (b). Decreases by 4 units   63 (c). Yes   63 (d). Decreases by 4 units
64 (b). Decreases by 4 units   64 (c). Yes   64 (d). Decreases by 4 units

SECTION 3.4 The Slope of a Line

3.4 OBJECTIVES

1. Find the slope of a line
2. Find the slopes of horizontal and vertical lines
3. Find the slope of a line given an equation
4. Find the slope given a graph
5. Graph linear equations using the slope of a line

Finding the Slope

On the coordinate system below, plot a point, any point.

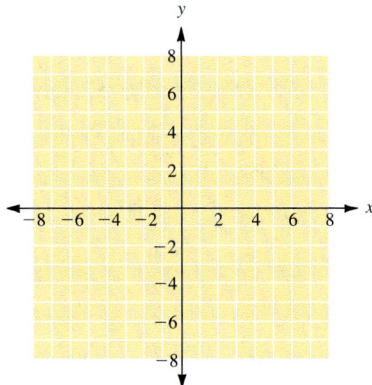

How many different lines can you draw through that point? Hundreds? Thousands? Millions? Actually, there is no limit to the number of different lines that pass through that point.

On the coordinate system below, plot two distinct points.

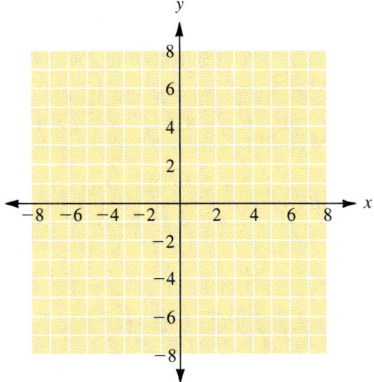

Now, how many different (straight) lines can you draw through those points? Only one! The two points were enough to define the line.

In Section 3.5, we will see how we can find the equation of a line if we are given two of its points. The first part of finding that equation is finding the **slope** of the line, which is a way of describing the steepness of a line.

240

Section 3.4 ■ The Slope of a Line

To define a formula for slope, choose any two distinct points on the line, say, P with coordinates (x_1, y_1) and Q with coordinates (x_2, y_2). As we move along the line from P to Q, the x value, or coordinate, changes from x_1 to x_2. That change in x, also called the **horizontal change**, is $x_2 - x_1$. Similarly, as we move from P to Q, the corresponding change in y, called the **vertical change**, is $y_2 - y_1$. The *slope* is then defined as the ratio of the vertical change to the horizontal change. The letter m is used to represent the slope, which we now define.

The difference, $x_2 - x_1$, is often called the **run.** The difference, $y_2 - y_1$, is the **rise.** So the slope can be thought of as "rise over run."

Note that $x_1 \neq x_2$ or $x_2 - x_1 \neq 0$ ensures that the denominator is nonzero, so that the slope is defined. It also means the line cannot be vertical.

Slope of a Line

The *slope* of a line through two distinct points $P(x_1, y_1)$ and $Q(x_2, y_2)$ is given by

$$m = \frac{\text{change in } y}{\text{change in } x} = \frac{y_2 - y_1}{x_2 - x_1} \tag{1}$$

where $x_1 \neq x_2$.

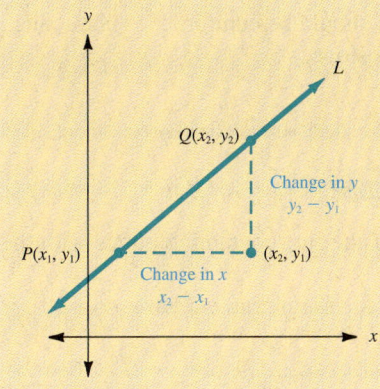

This definition provides exactly the numerical measure of "steepness" that we want. If a line "rises" as we move from left to right, the slope will be positive—the steeper the line, the larger the numerical value of the slope. If the line "falls" from left to right, the slope will be negative.

Let's proceed to some examples.

Example 1 — Finding the Slope

Find the slope of the line containing points with coordinates $(1, 2)$ and $(5, 4)$.

Let $P(x_1, y_1) = (1, 2)$ and $Q(x_2, y_2) = (5, 4)$. By the definition above, we have

$$m = \frac{y_2 - y_1}{x_2 - x_1} = \frac{4 - 2}{5 - 1} = \frac{2}{4} = \frac{1}{2}$$

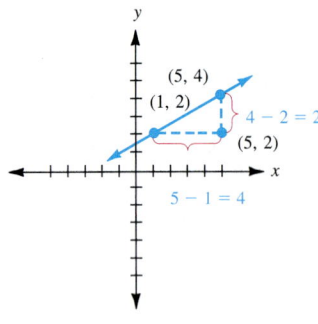

Note: We would have found the same slope if we had reversed P and Q and subtracted in the other order. In that case, $P(x_1, y_1) = (5, 4)$ and $Q(x_2, y_2) = (1, 2)$, so

$$m = \frac{2 - 4}{1 - 5} = \frac{-2}{-4} = \frac{1}{2}$$

It makes no difference which point is labeled (x_1, y_1) and which is (x_2, y_2), the resulting slope will be the same. You must simply stay with your choice once it is made and *not* reverse the order of the subtraction in your calculations.

✓ CHECK YOURSELF 1

Find the slope of the line containing points with coordinates (2, 3) and (5, 5).

By now you should be comfortable subtracting negative numbers. Let's apply that skill to finding a slope.

Example 2 Finding the Slope

Find the slope of the line containing points with the coordinates $(-1, -2)$ and $(3, 6)$.

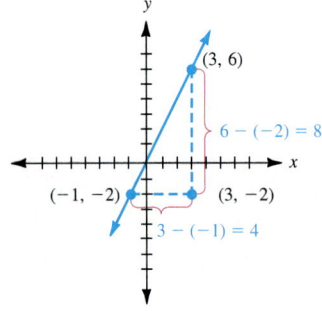

Again, applying the definition, we have

$$m = \frac{6 - (-2)}{3 - (-1)} = \frac{6 + 2}{3 + 1} = \frac{8}{4} = 2$$

The figure compares the slopes found in the two previous examples. Line l_1, from Example 1, had slope $\frac{1}{2}$. Line l_2, from Example 2, had slope 2. Do you see the idea of slope measuring steepness? The greater the slope, the more steeply the line is inclined upward.

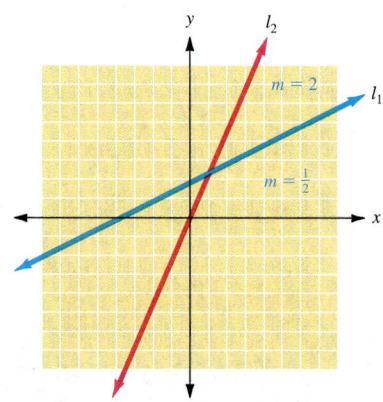

✓ CHECK YOURSELF 2

Find the slope of the line containing points with coordinates $(-1, 2)$ and $(2, 7)$. Draw a sketch of this line and the line of Check Yourself 1. Compare the lines and the two slopes.

Let's look at lines with a negative slope.

Example 3

Finding the Slope

Find the slope of the line containing points with coordinates $(-2, 3)$ and $(1, -3)$.

By the definition,

$$m = \frac{-3 - 3}{1 - (-2)} = \frac{-6}{3} = -2$$

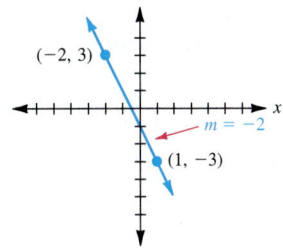

This line has a *negative* slope. The line *falls* as we move from left to right.

✓ CHECK YOURSELF 3

Find the slope of the line containing points with coordinates $(-1, 3)$ and $(1, -3)$.

We have seen that lines with positive slope rise from left to right and lines with negative slope fall from left to right. What about lines with a slope of zero? A line with a slope of 0 is especially important in mathematics.

Example 4

Finding the Slope

Find the slope of the line containing points with coordinates $(-5, 2)$ and $(3, 2)$.

By the definition,

$$m = \frac{2 - 2}{3 - (-5)} = \frac{0}{8} = 0$$

The slope of the line is 0. That will be the case for any horizontal line. Since any two points on the line have the same y coordinate, the vertical change $y_2 - y_1$ must always be 0, and so the resulting slope is 0.

✓ CHECK YOURSELF 4

Find the slope of the line containing points with coordinates $(-2, -4)$ and $(3, -4)$.

Since division by 0 is undefined, it is possible to have a line with an undefined slope.

Example 5

Finding the Slope

Find the slope of the line containing points with coordinates $(2, -5)$ and $(2, 5)$.

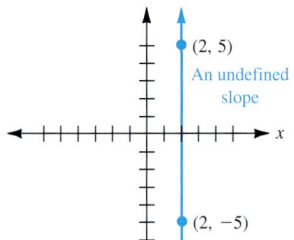

By the definition,

$$m = \frac{5 - (-5)}{2 - 2} = \frac{10}{0}$$

Remember that division by zero is undefined.

We say the vertical line has an undefined slope. On a vertical line, any two points have the same x coordinate. This means that the horizontal change $x_2 - x_1$ must always be 0 and since division by 0 is undefined, the slope of a vertical line will always be undefined.

✓ CHECK YOURSELF 5

Find the slope of the line containing points with the coordinates $(-3, -5)$ and $(-3, 2)$.

The following sketch summarizes the results of our previous examples.

As the slope gets closer to 0, the line gets "flatter."

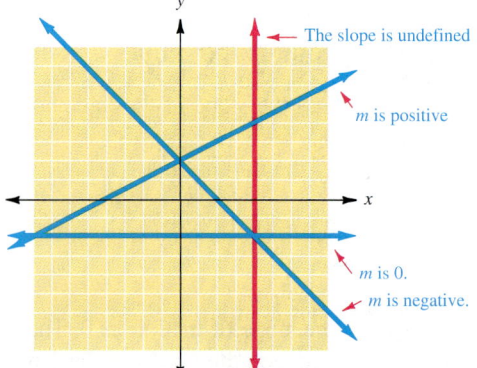

Four lines are illustrated in the figure on the previous page. Note that

1. The slope of a line that rises from left to right is positive.
2. The slope of a line that falls from left to right is negative.
3. The slope of a horizontal line is 0.
4. A vertical line has an undefined slope.

We now want to consider finding the equation of a line when its slope and y intercept are known. Suppose that the y intercept of a line is b. Then the point at which the line crosses the y axis must have coordinates $(0, b)$. Look at the sketch below.

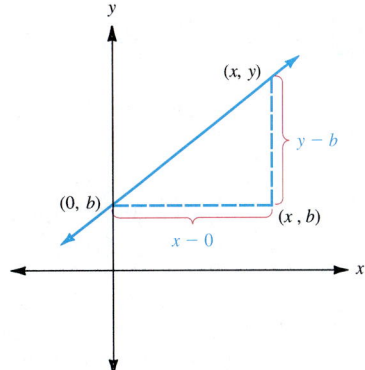

Now, using any other point (x, y) on the line and using our definition of slope, we can write

$$m = \frac{y - b}{x - 0} \quad \text{Change in } y \atop \text{Change in } x \tag{1}$$

or

$$m = \frac{y - b}{x} \tag{2}$$

Multiplying both sides of equation (2) by x, we have

$$mx = y - b \tag{3}$$

Finally, adding b to both sides of equation (3) gives

$$mx + b = y$$

or

$$y = mx + b \tag{4}$$

We can summarize the above discussion as follows:

> **The Slope-Intercept Form for a Line**
>
> An equation of the line with slope m and y intercept b is
>
> $$y = mx + b$$

In this form, the equation is solved for y. The coefficient of x will give you the slope of the line, and the constant term gives the y intercept.

Example 6 Finding the Slope and y Intercept

(a) Find the slope and y intercept for the graph of the equation

$$y = \underset{m}{3}x + \underset{b}{4}$$

The graph has slope 3 and y intercept 4.

(b) Find the slope and y intercept for the graph of the equation

$$y = -\underset{m}{\tfrac{2}{3}}x - \underset{b}{5}$$

The slope of the line is $-\dfrac{2}{3}$; the y intercept is -5.

✓ CHECK YOURSELF 6

Find the slope and y intercept for the graph of each of the following equations.

(a) $y = -3x - 7$ (b) $y = \dfrac{3}{4}x + 5$

As Example 7 illustrates, we may have to solve for y as the first step in determining the slope and the y intercept for the graph of an equation.

Example 7 Finding the Slope and y Intercept

Find the slope and y intercept for the graph of the equation

$$3x + 2y = 6$$

First, we must solve the equation for y.

If we write the equation as

$$y = \frac{-3x + 6}{2}$$

it is more difficult to identify the slope and the intercept.

$$3x + 2y = 6$$
$$2y = -3x + 6 \quad \text{Subtract } 3x \text{ from both sides.}$$
$$y = -\frac{3}{2}x + 3 \quad \text{Divide each term by 2.}$$

The equation is now in slope-intercept form. The slope is $-\frac{3}{2}$, and the y intercept is 3.

 CHECK YOURSELF 7

Find the slope and y intercept for the graph of the equation

$$2x - 5y = 10$$

As we mentioned earlier, knowing certain properties of a line (namely, its slope and y intercept) will also allow us to write the equation of the line by using the slope-intercept form. Example 8 illustrates this approach.

Example 8 Writing the Equation of a Line

(a) Write the equation of a line with slope 3 and y intercept 5.
We know that $m = 3$ and $b = 5$. Using the slope-intercept form, we have

$$y = 3x + 5$$
$$\quad\quad\uparrow\quad\uparrow$$
$$\quad\quad m\quad b$$

which is the desired equation.

(b) Write the equation of a line with slope $-\frac{3}{4}$ and y intercept -3.

We know that $m = -\frac{3}{4}$ and $b = -3$. In this case,

$$y = -\frac{3}{4}x + (-3)$$

where m and b are indicated

or

$$y = -\frac{3}{4}x - 3$$

which is the desired equation.

✓ CHECK YOURSELF 8

Write the equation of a line with the following:

(a) slope -2 and y intercept 7   (b) slope $\frac{2}{3}$ and y intercept -3

We can also use the slope and y intercept of a line in drawing its graph. Consider Example 9.

Example 9 Graphing a Line

Graph the line with slope $\frac{2}{3}$ and y intercept 2.

Since the y intercept is 2, we begin by plotting the point (0, 2). Since the horizontal change (or run) is 3, we move 3 units to the right *from that y intercept.* Then since the vertical change (or rise) is 2, we move 2 units up to locate another point on the desired graph. Note that we will have located that second point at (3, 4). The final step is to draw a line through that point and the y intercept.

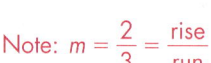

Note: $m = \frac{2}{3} = \frac{\text{rise}}{\text{run}}$

The line rises from left to right because the slope is positive.

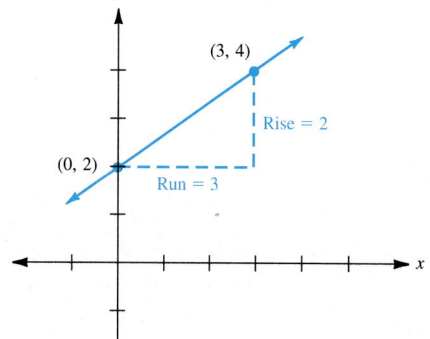

The equation of this line is $y = \frac{2}{3}x + 2$.

Section 3.4 ■ The Slope of a Line   249

✓ **CHECK YOURSELF 9**

Graph the equation of a line with slope $\frac{3}{5}$ and y intercept -2.

The y intercept can pass through the origin, as our next example demonstrates.

Example 10   **Graphing a Line**

Graph the line associated with the equation $y = -3x$, and the line associated with the equation $y = -3x - 3$.

In the first case, the slope is -3 and the y intercept is 0. We begin with the point $(0, 0)$. From there, we move down three units and to the right one unit, arriving at the point $(1, -3)$. Now we draw a line through those two points. On the same axes, we will draw the line with slope -3 through the intercept $(0, -3)$. Note that the two lines are parallel to each other.

Any two lines with the same slope are said to be **parallel lines.**

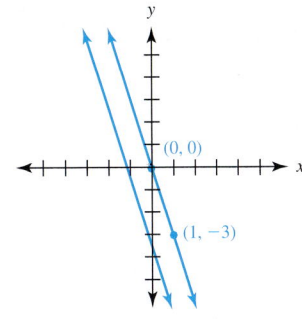

✓ **CHECK YOURSELF 10**

Graph the line associated with the equation $y = \frac{7}{2}x$.

We summarize the use of graphing by the slope-intercept form with the following algorithm.

Graphing by Using the Slope-Intercept Form

Step 1 Write the original equation of the line in slope-interept form, $y = mx + b$.
Step 2 Determine the slope m and the y intercept b.
Step 3 Plot the y intercept at $(0, b)$.
Step 4 Use m (the change in y over the change in x) to determine a second point on the desired line.
Step 5 Draw a line through the two points determined above to complete the graph.

Example 11

Selecting an Appropriate Graphing Method

Decide which of the two methods for graphing lines—the intercept method or the slope-intercept method—is more appropriate for graphing equations (a), (b), and (c).

$$\text{(a) } 2x - 5y = 10$$

Because both intercepts are easy to find, you should choose the intercept method to graph this equation.

$$\text{(b) } 2x + y = 6$$

This equation can be quickly graphed by either method. As it is written, you might choose the intercept method. It can, however, be rewritten as $y = -2x + 6$. In that case the slope-intercept method is more appropriate.

$$\text{(c) } y = \frac{1}{4}x - 4$$

Since the equation is in slope-intercept form, that is the more appropriate method to choose.

✓ CHECK YOURSELF 11

Which would be more appropriate for graphing each equation, the intercept method or the slope-intercept method?

(a) $x + y = -2$ (b) $3x - 2y = 12$ (c) $y = -\frac{1}{2}x - 6$

✓ CHECK YOURSELF ANSWERS

1. $m = \frac{2}{3}$. **2.** $m = \frac{5}{3}$.

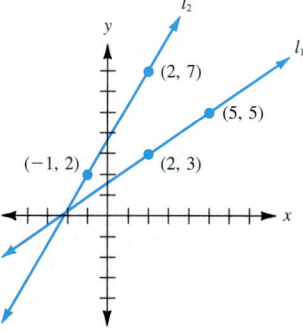

3. $m = -3$.   **4.** $m = 0$.   **5.** m is undefined.   **6.** (a) $m = -3$, $b = -7$;

(b) $m = \dfrac{3}{4}$, $b = 5$.   **7.** $y = \dfrac{2}{5}x - 2$; the slope is $\dfrac{2}{5}$; the y intercept is -2.

8. (a) $y = -2x + 7$; (b) $y = \dfrac{2}{3}x - 3$.

9. 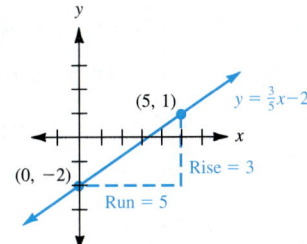   **10.**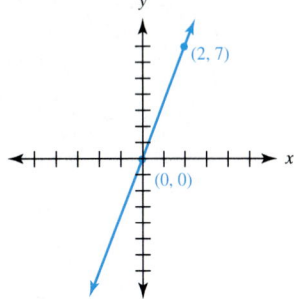

11. (a) Either; (b) intercept; (c) slope-intercept.

Exercises ■ 3.4

1. 1
2. 2
3. 5
4. 5
5. $\dfrac{4}{5}$
6. $\dfrac{1}{3}$
7. −2
8. −1
9. $-\dfrac{3}{2}$
10. $-\dfrac{3}{5}$
11. Undefined
12. 0
13. $\dfrac{5}{7}$
14. Undefined
15. 0
16. $-\dfrac{9}{7}$
17. $-\dfrac{4}{3}$
18. $\dfrac{3}{5}$

19. 3, 5
20. −7, 3
21. −2, −5
22. 5, −2
23. $\dfrac{3}{4}$, 1
24. −4, 0
25. $\dfrac{2}{3}$, 0
26. $-\dfrac{3}{5}$, −2
27. $-\dfrac{4}{3}$, 4
28. $-\dfrac{2}{5}$, 2
29. 0, 9
30. $\dfrac{2}{3}$, −2
31. $\dfrac{3}{2}$, −4
32. Undefined, no y intercept

Find the slope of the line through the following pairs of points.

1. (5, 7) and (9, 11)
2. (4, 9) and (8, 17)
3. (−2, −5) and (2, 15)
4. (−3, 2) and (0, 17)
5. (−2, 3) and (3, 7)
6. (−3, −4) and (3, −2)
7. (−3, 2) and (2, −8)
8. (−6, 1) and (2, −7)
9. (3, 3) and (5, 0)
10. (−2, 4) and (3, 1)
11. (5, −4) and (5, 2)
12. (−5, 4) and (2, 4)
13. (−4, −2) and (3, 3)
14. (−5, −3) and (−5, 2)
15. (−3, −4) and (2, −4)
16. (−5, 7) and (2, −2)
17. (−1, 7) and (2, 3)
18. (−4, −2) and (6, 4)

Find the slope and y intercept of the line represented by each of the following equations.

19. $y = 3x + 5$
20. $y = -7x + 3$
21. $y = -2x - 5$
22. $y = 5x - 2$
23. $y = \dfrac{3}{4}x + 1$
24. $y = -4x$
25. $y = \dfrac{2}{3}x$
26. $y = -\dfrac{3}{5}x - 2$
27. $4x + 3y = 12$
28. $2x + 5y = 10$
29. $y = 9$
30. $2x - 3y = 6$
31. $3x - 2y = 8$
32. $x = 5$

33. $y = 3x + 5$

34. $y = -2x + 4$

35. $y = -3x + 4$

36. $y = 5x - 2$

37. $y = \dfrac{1}{2}x - 2$

38. $y = -\dfrac{3}{4}x + 8$

Write the equation of the line with given slope and y intercept. Then graph each line, using the slope and y intercept.

33. $m = 3, b = 5$ **34.** $m = -2, b = 4$

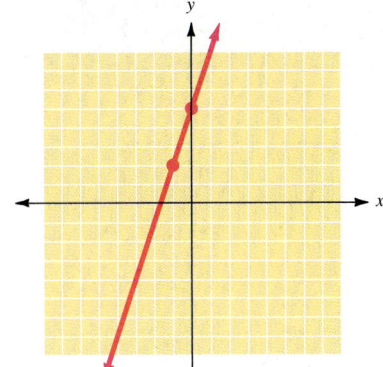 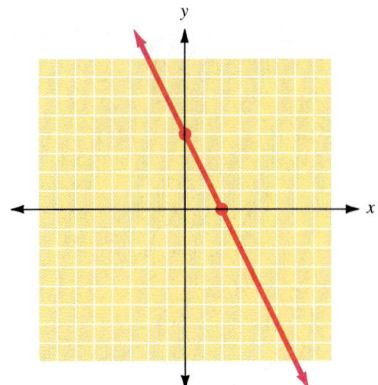

35. $m = -3, b = 4$ **36.** $m = 5, b = -2$

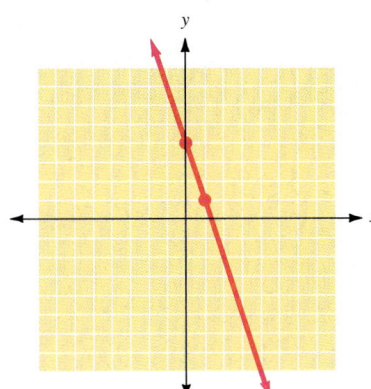 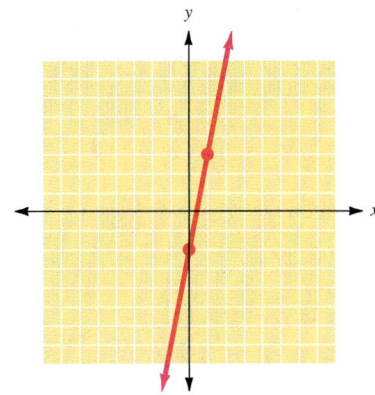

37. $m = \dfrac{1}{2}, b = -2$ **38.** $m = -\dfrac{3}{4}, b = 8$

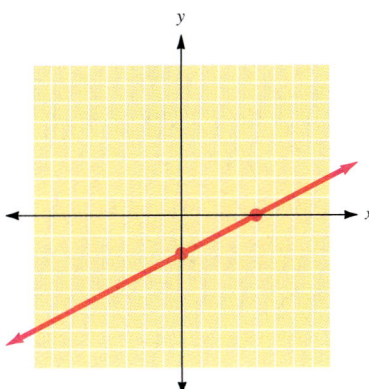 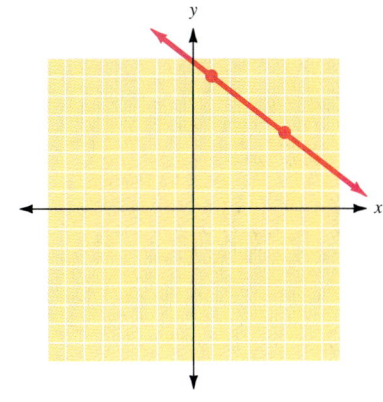

39. $y = \frac{2}{3}x$

40. $y = \frac{2}{3}x - 2$

41. $y = \frac{3}{4}x + 3$

42. $y = -3x$

43. IV

44. IV

45. III

46. III

47. I

48. I

49. III and IV

50. I and IV

39. $m = \frac{2}{3}, b = 0$

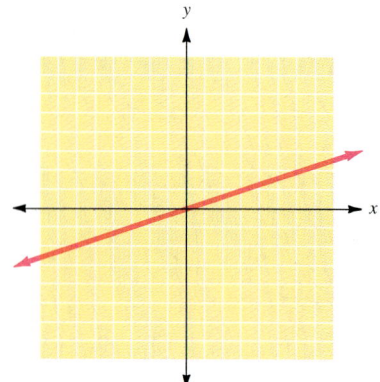

40. $m = \frac{2}{3}, b = -2$

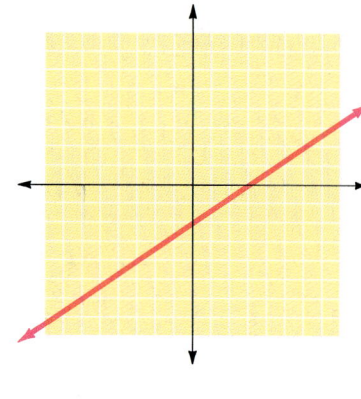

41. $m = \frac{3}{4}, b = 3$

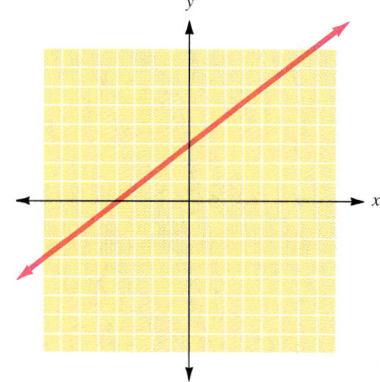

42. $m = -3, b = 0$

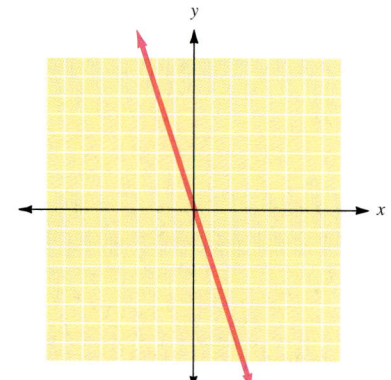

In which quadrant(s) are there no solutions for each equation?

43. $y = 2x + 1$

44. $y = 3x + 2$

45. $y = -x + 1$

46. $y = -2x + 5$

47. $y = -2x - 5$

48. $y = -5x - 7$

49. $y = 3$

50. $x = -2$

Section 3.4 • The Slope of a Line   **255**

51. g

52. d

53. e

54. a

55. h

56. f

57. c

58. b

In Exercises 51 to 58, match the graph with one of the equations below.

(a) $y = 2x$, (b) $y = x + 1$, (c) $y = -x + 3$, (d) $y = 2x + 1$, (e) $y = -3x - 2$,
(f) $y = \dfrac{2}{3}x + 1$, (g) $y = -\dfrac{3}{4}x + 1$, (h) $y = -4x$

51.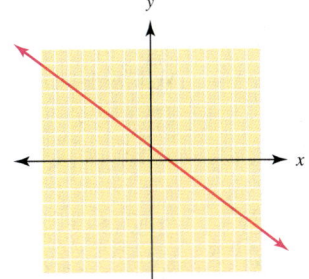

52.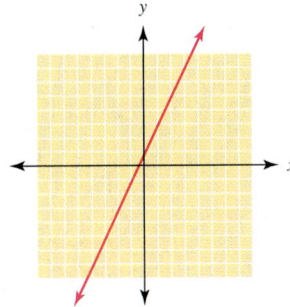

53.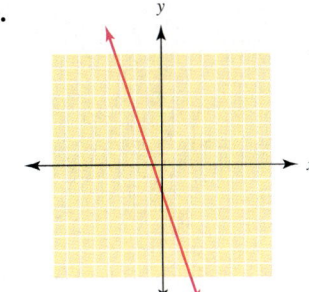

54.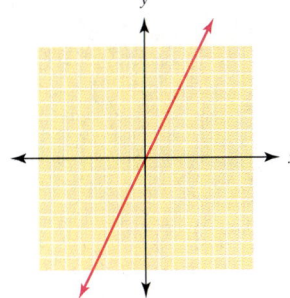

55.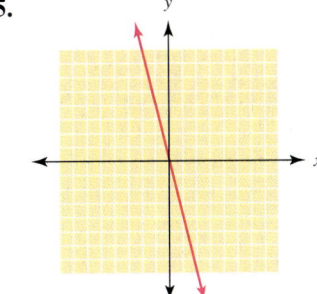

56.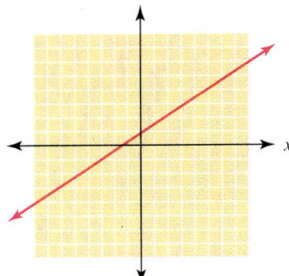

57.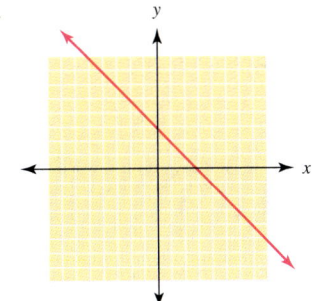

58.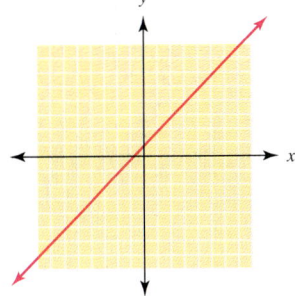

256 Chapter 3 ■ Graphs and Linear Equations

59. $y = -\dfrac{2}{5}x + 2$

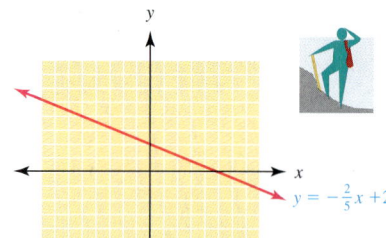

In Exercises 59 to 62, solve each equation for *y*, then use your graphing utility to graph each equation.

59. $2x + 5y = 10$

60. $5x - 3y = 12$

61. $x + 7y = 14$

62. $-2x - 3y = 9$

In Exercises 63 to 70, use the graph to determine the slope of the line.

63.

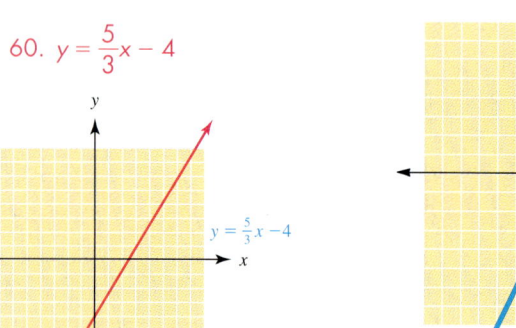

64.

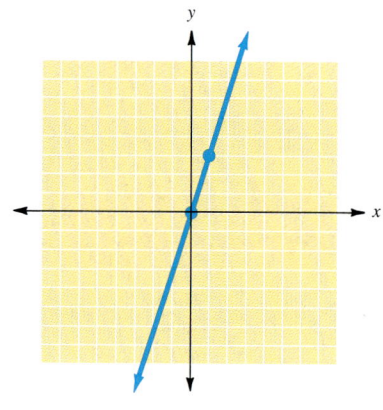

60. $y = \dfrac{5}{3}x - 4$

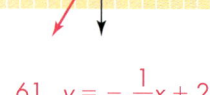

65.

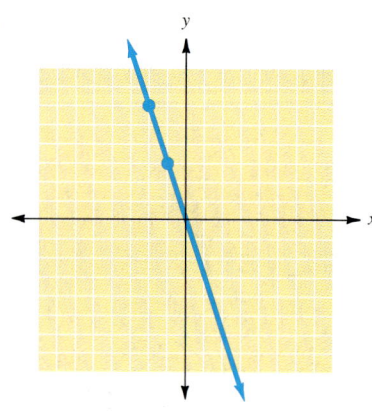

66.

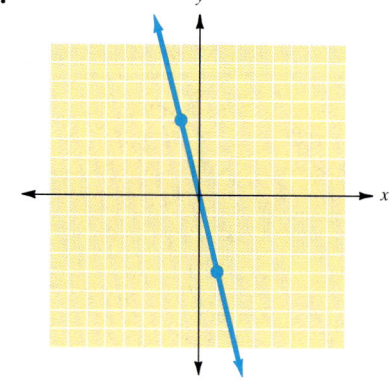

61. $y = -\dfrac{1}{7}x + 2$

62. $y = -\dfrac{2}{3}x - 3$

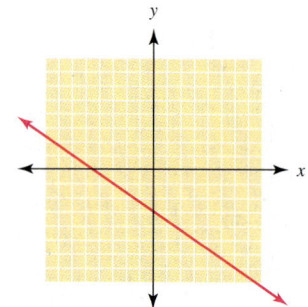

63. 2   64. 3   65. −3   66. −4

67. 3

68. −1

69. $-\dfrac{5}{2}$

70. 2

71. $m = 0.10$, market price per jug; y intercept $= 200$, the $200 award

72. $m = 15$; y intercept $= -100$

73. Slope represents price of newsprint; y intercept represents cost of the truck

74. 0.58

75. −0.30

76. −0.10

77. 2.14¢/yr

67.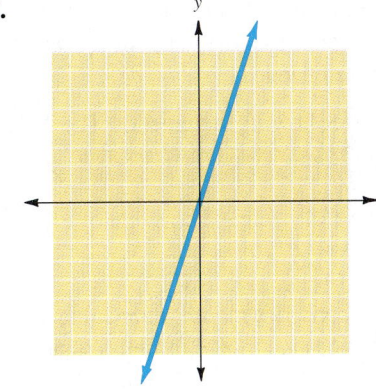

68.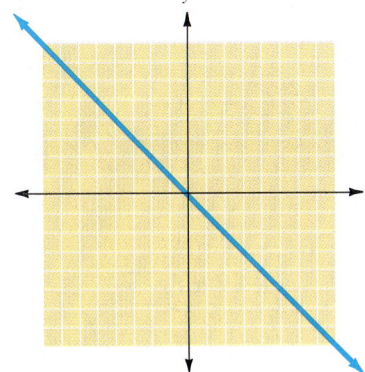

69.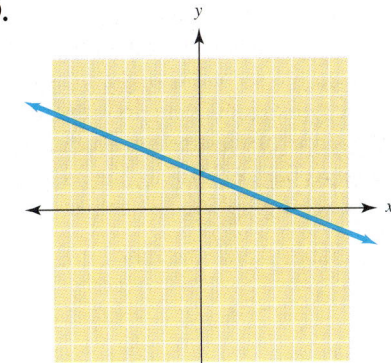

70.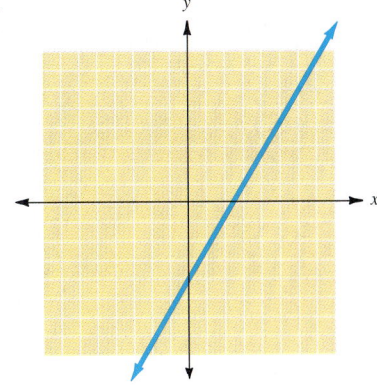

71. **Recycling.** The equation $y = 0.10x + 200$ was used in Section 3.3 to describe the award money in a recycling contest. What are the slope and the y intercept for this equation? What does the slope of the line represent in the equation? What does the y intercept represent?

72. **Fund-raising.** The equation $y = 15x - 100$ was used in Section 3.3 to describe the amount of money a high school class might earn from a paper drive. What are the slope and y intercept for this equation?

73. **Fund-raising.** In the equation in Exercise 72, what does the slope of the line represent? What does the y intercept represent?

74. **Slope of a roof.** A roof rises 8.75 feet (ft) in a horizontal distance of 15.09 ft. Find the slope of the roof to the nearest hundredth.

75. **Slope of airplane descent.** An airplane covered 15 miles (mi) of its route while decreasing its altitude by 24,000 ft. Find the slope of the line of descent that was followed. (1 mi = 5280 ft) Round to the nearest hundredth.

76. **Slope of road descent.** Driving down a mountain, Tom finds that he has descended 1800 ft in elevation by the time he is 3.25 mi horizontally away from the top of the mountain. Find the slope of his descent to the nearest hundredth.

77. **Consumer affairs.** In 1960, the cost of a soft drink was 20¢. By 1995, the cost of the same soft drink had risen to 95¢. During this time period, what was the annual rate of change of the cost of the soft drink?

78. 2°/h

78. Science. On a certain February day in Philadelphia, the temperature at 6:00 A.M. was 10°F. By 2:00 P.M. the temperature was up to 26°F. What was the hourly rate of temperature change?

79. Complete the following statement: "The difference between undefined slope and zero slope is . . . ."

80. Complete the following: "The slope of a line tells you . . . ."

81. On two occasions last month, Sam Johnson rented a car on a business trip. Both times it was the same model from the same company, and both times it was in San Francisco. Sam now has to fill out an expense account form and needs to know how much he was charged per mile and the base rate. On both occasions he dropped the car at the airport booth and just got the total charge, not the details. All Sam knows is that he was charged $210 for 625 miles on the first occasion and $133.50 for 370 miles on the second trip. Sam has called accounting to ask for help. Plot these two points on a graph, and draw the line that goes through them. What question does the slope of the line answer for Sam? How does the y intercept help? Write a memo to Sam explaining the answers to his question and how a knowledge of algebra and graphing has helped you find the answers.

82. Parallel lines; no

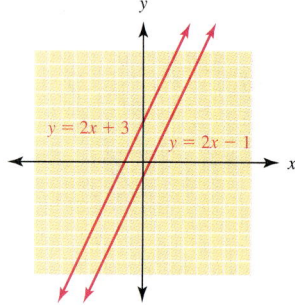

82. On the same graph, sketch the following lines:

$$y = 2x - 1 \text{ and } y = 2x + 3$$

What do you observe about these graphs? Will the lines intersect?

83. Parallel lines; no

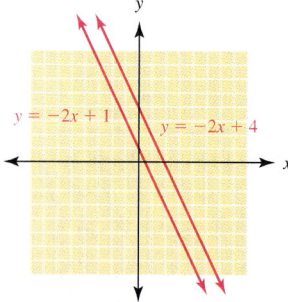

83. Repeat exercise 82 using

$$y = -2x + 4 \text{ and } y = -2x + 1$$

84. Perpendicular lines; -1

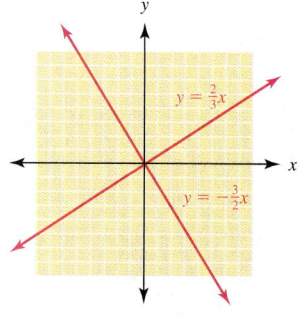

84. On the same graph, sketch the following lines:

$$y = \frac{2}{3}x \text{ and } y = -\frac{3}{2}x$$

What do you observe concerning these graphs? Find the product of the slopes of these two lines.

85. Repeat exercise 84 using

$$y = \frac{4}{3}x \text{ and } y = -\frac{3}{4}x.$$

85. Perpendicular lines; -1

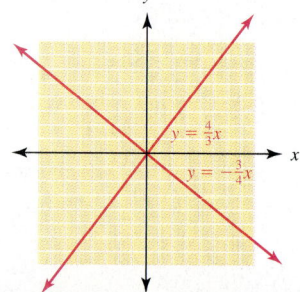

86. Based on exercises 84 and 85, write the equation of a line that is perpendicular to

$$y = \frac{3}{5}x$$

86. $y = -\frac{5}{3}x$

SECTION 3.5 Forms of Linear Equations

3.5 OBJECTIVES

1. Write the equation of a line given its slope and y intercept
2. Write the equation of a line given a slope and any point on the line
3. Graph the equation of a line
4. Write the equation of a line given a point and a slope
5. Write the equation of a line given two points

The special form

$$ax + by = c$$

where a and b cannot both be zero, is called the **standard form for a linear equation.** In Section 3.4 we determined the slope of a line from two ordered pairs. We also used the concept of slope to write the equation of a line.

In this section, we will see that the slope-intercept form clearly indicates whether the graphs of given equations will be parallel, intersecting, or perpendicular lines.

Two lines are **perpendicular** if their slopes are negative reciprocals, so

$$m_1 = -\frac{1}{m_2}$$

Note that $m_1 \cdot m_2 = -1$; the product of their slopes is -1.

Our first example illustrates this concept.

Example 1 Verifying That Two Lines Are Perpendicular

Show that the graphs of $3x + 4y = 4$ and $-4x + 3y = 12$ are perpendicular lines.

First, we solve each equation for y.

$$3x + 4y = 4$$
$$4y = -3x + 4$$
$$y = -\frac{3}{4}x + 1 \tag{1}$$

$$-4x + 3y = 12$$
$$3y = 4x + 12$$
$$y = \frac{4}{3}x + 4 \qquad (2)$$

We now look at the product of the two slopes: $-\frac{3}{4} \cdot \frac{4}{3} = -1$. Any two lines whose slopes have a product of -1 are perpendicular lines. These two lines are perpendicular.

✓ CHECK YOURSELF 1

Show that the graphs of the equations

$$-3x + 2y = 4 \qquad \text{and} \qquad 2x + 3y = 9$$

are perpendicular lines.

The slope-intercept form can also be used in graphing a line, as Example 2 illustrates.

Example 2 — Graphing the Equation of a Line

Graph the line $2x + 3y = 3$.

Solving for y, we find the slope-intercept form for this equation is

$$y = -\frac{2}{3}x + 1$$

To graph the line, plot the y intercept at $(0, 1)$. Because the slope m is equal to $-\frac{2}{3}$, we move from $(0, 1)$ to the right 3 units and then *down* 2 units, to locate a second point on the graph of the line, here $(3, -1)$. We can now draw a line through the two points to complete the graph.

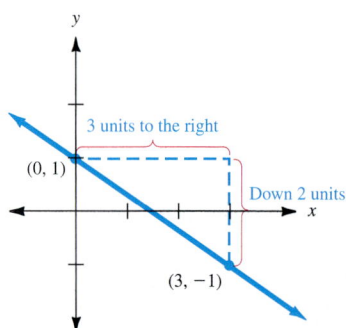

We treat $-\frac{2}{3}$ as $\frac{-2}{+3}$ to move to the right 3 units and down 2 units.

✓ CHECK YOURSELF 2

Graph the line with equation.

$$3x - 4y = 8$$

Hint: First rewrite the equation in slope-intercept form.

Section 3.5 ■ Forms of Linear Equations

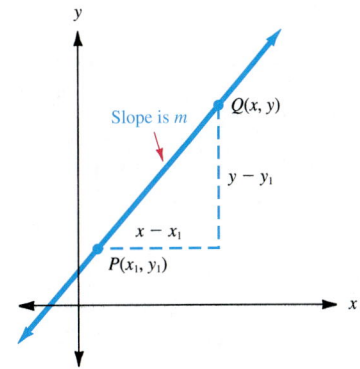

Often in mathematics it is useful to be able to write the equation of a line, given its slope and *any* point on the line. We will now derive a third special form for a line for this purpose.

Suppose a line has slope m and passes through the known point $P(x_1, y_1)$. Let $Q(x, y)$ be any other point on the line. Once again we can use the definition of slope and write

$$m = \frac{y - y_1}{x - x_1} \qquad (1)$$

Multiplying both sides of equation (1) by $x - x_1$, we have

$$m(x - x_1) = y - y_1$$

or

$$y - y_1 = m(x - x_1) \qquad (2)$$

Equation (2) is called the **point-slope form** for the equation of a line, and all points lying on the line [including (x_1, y_1)] will satisfy this equation. We can state the following general result.

The equation of a line with undefined slope passing through the point (x_1, y_1) is given by $x = x_1$.

> **Point-Slope Form for the Equation of a Line**
>
> The equation of a line with slope m that passes through point (x_1, y_1) is given by
>
> $$y - y_1 = m(x - x_1)$$

Example 3 Finding the Equation of a Line

Write the equation for the line that passes through point $(3, -1)$ with a slope of 3.

Letting $(x_1, y_1) = (3, -1)$ and $m = 3$, we will use point-slope form to get

$$y - (-1) = 3(x - 3)$$

or

$$y + 1 = 3x - 9$$

We can write the final result in slope-intercept form as

$$y = 3x - 10$$

✓ CHECK YOURSELF 3

Write the equation of the line that passes through point $(-2, 4)$ with a slope of $\frac{3}{2}$. Write your result in slope-intercept form.

Since we know that two points determine a line, it is natural that we should be able to write the equation of a line passing through two given points. Using the point-slope form together with the slope formula will allow us to write such an equation.

Example 4 Finding the Equation of a Line

Write the equation of the line passing through $(2, 4)$ and $(4, 7)$.

First, we find m, the slope of the line. Here

$$m = \frac{7 - 4}{4 - 2} = \frac{3}{2}$$

Now we apply the point-slope form with $m = \frac{3}{2}$ and $(x_1, y_1) = (2, 4)$:

$$y - 4 = \frac{3}{2}(x - 2)$$

$$y - 4 = \frac{3}{2}x - 3$$

Note: We could just as well have chosen to let

$(x_1, y_1) = (4, 7)$

The resulting equation will be the same in either case. Take time to verify this for yourself.

Write the result in slope-intercept form.

$$y = \frac{3}{2}x + 1$$

✓ CHECK YOURSELF 4

Write the equation of the line passing through $(-2, 5)$ and $(1, 3)$. Write your result in slope-intercept form.

A line with slope zero is a horizontal line. A line with an undefined slope is vertical. Example 5 illustrates the equations of such lines.

Example 5 Finding the Equation of a Line

(a) Find the equation of a line passing through $(7, -2)$ with a slope of zero.

We could find the equation by letting $m = 0$. Substituting into the slope-intercept form, we can solve for the y intercept b.

$$y = mx + b$$
$$-2 = 0(7) + b$$
$$-2 = b$$

So,

$$y = 0x - 2 \quad y = -2$$

It is far easier to remember that any line with a zero slope is a horizontal line and has the form

$$y = b$$

The value for b will always be the y coordinate for the given point.

(b) Find the equation of a line with undefined slope passing through $(4, -5)$.

A line with undefined slope is vertical. It will always be of the form $x = a$, where a is the x coordinate for the given point. The equation is

$$x = 4$$

✓ CHECK YOURSELF 5

(a) Find the equation of a line with zero slope that passes through point $(-3, 5)$.

(b) Find the equation of a line passing through $(-3, -6)$ with undefined slope.

Alternative methods for finding the equation of a line through two points do exist and have particular significance in other fields of mathematics, such as statistics. Example 6 shows such an alternate approach.

Example 6

Finding the Equation of a Line

Write the equation of the line through points $(-2, 3)$ and $(4, 5)$.

First, we find m, as before.

$$m = \frac{5 - 3}{4 - (-2)} = \frac{2}{6} = \frac{1}{3}$$

We now make use of the slope-intercept equation, but in a slightly different form. Since $y = mx + b$, we can write

$$b = y - mx$$

We substitute these values because the line must pass through $(-2, 3)$

Now, letting $x = -2$, $y = 3$, and $m = \frac{1}{3}$, we can calculate b.

$$b = 3 - \left(\frac{1}{3}\right)(-2)$$

$$= 3 + \frac{2}{3} = \frac{11}{3}$$

With $m = \frac{1}{3}$ and $b = \frac{11}{3}$, we can apply the slope-intercept form, to write the equation of the desired line. We have

$$y = \frac{1}{3}x + \frac{11}{3}$$

✓ **CHECK YOURSELF 6**

Repeat the Check Yourself 4 exercise, using the technique illustrated in Example 6.

We now know that we can write the equation of a line once we have been given appropriate geometric conditions, such as a point on the line and the slope of that line. In some applications, the slope may be given not directly but through specified parallel or perpendicular lines.

Example 7

Finding the Equation of a Parallel Line

Find the equation of the line passing through $(-4, -3)$ and parallel to the line determined by $3x + 4y = 12$.

First, we find the slope of the given parallel line, as before.

$$3x + 4y = 12$$

$$4y = -3x + 12$$

$$y = -\frac{3}{4}x + 3$$

The slope of the given line is $-\frac{3}{4}$, the coefficient of the x.

The slopes of two parallel lines will be the same. Because the slope of the desired line must also be $-\frac{3}{4}$, we can use the point-slope form to write the required equation.

The line must pass through $(-4, -3)$, so let
$(x_1, y_1) = (-4, -3)$

$$y - (-3) = -\frac{3}{4}[x - (-4)]$$

This simplifies to

$$y = -\frac{3}{4}x - 6$$

and we have our equation in slope-intercept form.

CHECK YOURSELF 7

Find the equation of the line passing through $(2, -5)$ and parallel to the line determined by $4x - y = 9$.

Example 8   Finding the Equation of a Perpendicular Line

Find the equation of the line passing through $(3, -1)$ and perpendicular to the line $3x - 5y = 2$.

First, find the slope of the perpendicular line.

$$3x - 5y = 2$$

$$-5y = -3x + 2$$

$$y = \frac{3}{5}x - \frac{2}{5}$$

The slope of the perpendicular line is $\frac{3}{5}$. Recall that the slopes of perpendicular lines are negative reciprocals. The slope of our line is the negative reciprocal of $\frac{3}{5}$. It is therefore $-\frac{5}{3}$.

Using the point-slope form, we have the equation

$$y - (-1) = -\frac{5}{3}(x - 3)$$

$$y + 1 = -\frac{5}{3}x + 5$$

$$y = -\frac{5}{3}x + 4$$

✓ CHECK YOURSELF 8

Find the equation of the line passing through (5, 4) and perpendicular to the line with equation $2x - 5y = 10$.

There are many applications of our work with linear equations in various fields. The following is just one of many typical examples.

Example 9 An Application of a Linear Function

In producing a new product, a manufacturer predicts that the number of items produced x and the cost in dollars C of producing those items will be related by a linear equation.

Suppose that the cost of producing 100 items will be $5000 and the cost of producing 500 items will be $15,000. Find the linear equation relating x and C.

To solve this problem, we must find the equation of the line passing through points (100, 5000) and (500, 15,000).

Even though the numbers are considerably larger than we have encountered thus far in this section, the process is exactly the same.

First, we find the slope:

$$m = \frac{15{,}000 - 5000}{500 - 100} = \frac{10{,}000}{400} = 25$$

We can now use the point-slope form as before to find the desired equation.

$$C - 5000 = 25(x - 100)$$

$$C - 5000 = 25x - 2500$$

$$C = 25x + 2500$$

To graph the equation we have just derived, we must choose the scaling on the x and C axes carefully to get a "reasonable" picture. Here we choose increments of 100 on the x axis and 2500 on the C axis since those seem appropriate for the given information.

Note how the change in scaling "distorts" the slope of the line.

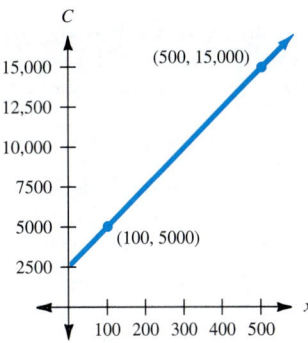

✓ CHECK YOURSELF 9

A company predicts that the value in dollars V and the time that a piece of equipment has been in use t are related by a linear equation. If the equipment is valued at $1500 after 2 years and at $300 after 10 years, find the linear equation relating t and V.

✓ CHECK YOURSELF ANSWERS

1. $m_1 = \dfrac{3}{2}$ and $m_2 = -\dfrac{2}{3}$; $(m_1)(m_2) = -1$. 2.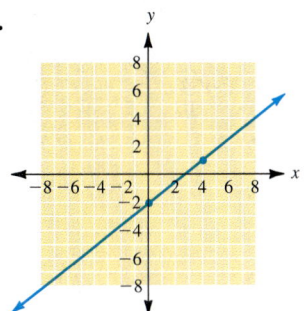

3. $y = \dfrac{3}{2}x + 7$. 4. $y = -\dfrac{2}{3}x + \dfrac{11}{3}$. 5. (a) $y = 5$; (b) $x = -3$.

6. $y = -\dfrac{2}{3}x + \dfrac{11}{3}$. 7. $y = 4x - 13$. 8. $y = -\dfrac{5}{2}x + \dfrac{33}{2}$.

9. $V = -150t + 1800$.

Exercises • 3.5

1. Parallel
2. Perpendicular
3. Neither   4. Parallel
5. Perpendicular
6. Parallel
7. $\dfrac{1}{3}$ 8. $-\dfrac{1}{3}$
9. 12   10. -1
11. $y = \dfrac{5}{4}x - 5$
12. $y = -\dfrac{3}{4}x - 4$
13. $y = 3x - 1$
14. $y = 3x + 5$
15. $y = -3x - 9$
16. $y = -4x$
17. $y = \dfrac{2}{5}x - 5$
18. $y = 3$ 19. $x = 2$
20. $y = \dfrac{1}{4}x - \dfrac{11}{2}$
21. $y = -\dfrac{4}{5}x + 4$
22. $x = -3$ 23. $y = x + 1$
24. $y = 2x - 8$
25. $y = \dfrac{3}{4}x - \dfrac{3}{2}$
26. $y = -x + 2$ 27. $y = 2$
28. $y = -\dfrac{2}{9}x + \dfrac{17}{9}$
29. $y = \dfrac{3}{2}x - 3$
30. $x = 2$ 31. $y = \dfrac{5}{2}x + 4$
32. $y = 1$ 33. $y = 4x - 2$
34. $y = -\dfrac{2}{3}x + 4$

In Exercises 1 to 10, are the pairs of lines parallel, perpendicular, or neither?

1. L_1 through $(-2, -3)$ and $(4, 3)$; L_2 through $(3, 5)$ and $(5, 7)$
2. L_1 through $(-2, 4)$ and $(1, 8)$; L_2 through $(-1, -1)$ and $(-5, 2)$
3. L_1 through $(8, 5)$ and $(3, -2)$; L_2 through $(-2, 4)$ and $(4, -1)$
4. L_1 through $(-2, -3)$ and $(3, -1)$; L_2 through $(-3, 1)$ and $(7, 5)$
5. L_1 with equation $x - 3y = 6$; L_2 with equation $3x + y = 3$
6. L_1 with equation $x + 2y = 4$; L_2 with equation $2x + 4y = 5$
7. Find the slope of any line parallel to the line through points $(-2, 3)$ and $(4, 5)$.
8. Find the slope of any line perpendicular to the line through points $(0, 5)$ and $(-3, -4)$.
9. A line passing through $(-1, 2)$ and $(4, y)$ is parallel to a line with slope 2. What is the value of y?
10. A line passing through $(2, 3)$ and $(5, y)$ is perpendicular to a line with slope $\dfrac{3}{4}$. What is the value of y?

In Exercises 11 to 22, write the equation of the line passing through each of the given points with the indicated slope. Give your results in slope-intercept form, where possible.

11. $(0, -5)$, $m = \dfrac{5}{4}$ 12. $(0, -4)$, $m = -\dfrac{3}{4}$ 13. $(1, 2)$, $m = 3$
14. $(-1, 2)$, $m = 3$ 15. $(-2, -3)$, $m = -3$ 16. $(1, -4)$, $m = -4$
17. $(5, -3)$, $m = \dfrac{2}{5}$ 18. $(4, 3)$, $m = 0$ 19. $(2, -3)$, m is undefined
20. $(2, -5)$, $m = \dfrac{1}{4}$ 21. $(5, 0)$, $m = -\dfrac{4}{5}$ 22. $(-3, 0)$, m is undefined

In Exercises 23 to 32, write the equation of the line passing through each of the given pairs of points. Write your result in slope-intercept form, where possible.

23. $(2, 3)$ and $(5, 6)$ 24. $(3, -2)$ and $(6, 4)$ 25. $(-2, -3)$ and $(2, 0)$
26. $(-1, 3)$ and $(4, -2)$ 27. $(-3, 2)$ and $(4, 2)$ 28. $(-5, 3)$ and $(4, 1)$
29. $(2, 0)$ and $(0, -3)$ 30. $(2, -3)$ and $(2, 4)$ 31. $(0, 4)$ and $(-2, -1)$
32. $(-4, 1)$ and $(3, 1)$

In Exercises 33 to 42, write the equation of the line L satisfying the given geometric conditions.

33. L has slope 4 and y intercept -2.
34. L has slope $-\dfrac{2}{3}$ and y intercept 4.

35. L has x intercept 4 and y intercept 2. **36.** L has x intercept -2 and slope $\dfrac{3}{4}$.

37. L has y intercept 4 and a 0 slope.

38. L has x intercept -2 and an undefined slope.

39. L passes through point $(3, 2)$ with a slope of 5.

40. L passes through point $(-2, -4)$ with a slope of $-\dfrac{3}{2}$.

41. L has y intercept 3 and is parallel to the line with equation $y = 3x - 5$.

42. L has y intercept -3 and is parallel to the line with equation $y = \dfrac{2}{3}x + 1$.

In Exercises 43 to 54, write the equation of each line.

43. L has y intercept 4 and is perpendicular to the line with equation $y = -2x + 1$.

44. L has y intercept 2 and is parallel to the line with equation $y = -1$.

45. L has y intercept 3 and is parallel to the line with equation $y = 2$.

46. L has y intercept 2 and is perpendicular to the line with equation $2x - 3y = 6$.

47. L passes through point $(-3, 2)$ and is parallel to the line with equation $y = 2x - 3$.

48. L passes through point $(-4, 3)$ and is parallel to the line with equation $y = -2x + 1$.

49. L passes through point $(3, 2)$ and is parallel to the line with equation $y = \dfrac{4}{3}x + 4$.

50. L passes through point $(-2, -1)$ and is perpendicular to the line with equation $y = 3x + 1$.

51. L passes through point $(5, -2)$ and is perpendicular to the line with equation $y = -3x - 2$.

52. L passes through point $(3, 4)$ and is perpendicular to the line with equation $y = -\dfrac{3}{5}x + 2$.

53. L passes through $(-2, 1)$ and is parallel to the line with equation $x + 2y = 4$.

54. L passes through $(-3, 5)$ and is parallel to the x axis.

A four-sided figure (quadrilateral) is a parallelogram if the opposite sides have the same slope. If the adjacent sides are perpendicular, the figure is a rectangle. In Exercises 55 to 58, for each quadrilateral *ABCD*, determine whether it is a parallelogram; then determine whether it is a rectangle.

55. $A(0, 0), B(2, 0), C(2, 3), D(0, 3)$

56. $A(-3, 2), B(1, -7), C(3, -4), D(-1, 5)$

57. $A(0, 0), B(4, 0), C(5, 2), D(1, 2)$

58. $A(-3, -5), B(2, 1), C(-4, 6), D(-9, 0)$

---

35. $y = -\dfrac{1}{2}x + 2$

36. $y = \dfrac{3}{4}x + \dfrac{3}{2}$

37. $y = 4$

38. $x = -2$

39. $y = 5x - 13$

40. $y = -\dfrac{3}{2}x - 7$

41. $y = 3x + 3$

42. $y = \dfrac{2}{3}x - 3$

43. $y = \dfrac{1}{2}x + 4$

44. $y = 2$

45. $y = 3$

46. $y = -\dfrac{3}{2}x + 2$

47. $y = 2x + 8$

48. $y = -2x - 5$

49. $y = \dfrac{4}{3}x - 2$

50. $y = -\dfrac{1}{3}x - \dfrac{5}{3}$

51. $y = \dfrac{1}{3}x - \dfrac{11}{3}$

52. $y = \dfrac{5}{3}x - 1$

53. $y = -\dfrac{1}{2}x$

54. $y = 5$

55. Yes, yes

56. Yes, no

57. Yes, no

58. Yes, yes

270 Chapter 3 ■ Graphs and Linear Equations

59. $F = \frac{9}{5} C + 32$

60. $C = 60x + 4000$

61. (a) $C = 35x + 50$
    (b) $172.50
    (c) 3.15 h

62. $S = 15,000t + 12,000$

63. Slope, 1; y intercept, 3

64. Slope, 1; y intercept, −2

65. Slope, 2; y intercept, 1

66. Slope, 3; y intercept, −4

59. Science. A temperature of 10°C corresponds to a temperature of 50°F. Also 40°C corresponds to 104°F. Find the linear equation relating F and C.

60. Business. In planning for a new item, a manufacturer assumes that the number of items produced, x, and the cost in dollars, C, of producing these items are related by a linear equation. Projections are that 100 items will cost $10,000 to produce and that 300 items will cost $22,000 to produce. Find the equation that relates C and x.

61. Business. Mike bills a customer at the rate of $35 per hour plus a fixed service call charge of $50.

(a) Write an equation that will allow you to compute the total bill for any number of hours, x, that it takes to complete a job.

(b) What will the total cost of a job be if it takes 3.5 hours to complete?

(c) How many hours would a job have to take if the total bill were $160.25?

62. Business. Two years after an expansion, a company had sales of $42,000. Four years later the sales were $102,000. Assuming that the sales in dollars, S, and the time, t, in years are related by a linear equation, find the equation relating S and t.

In Exercises 63 to 70, use the graph to determine the slope and y intercept of the line.

63.
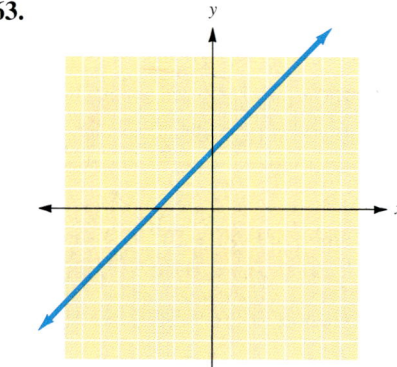

64.
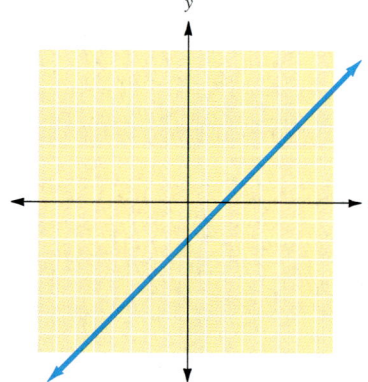

65.
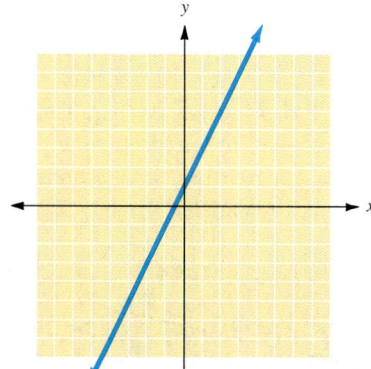

66.
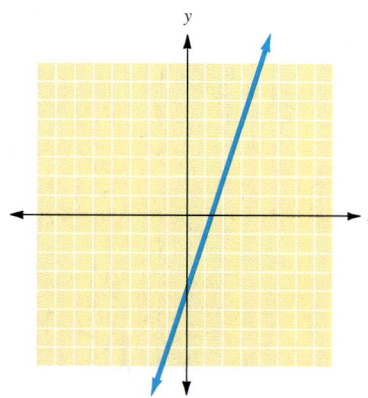

67. Slope, −3; y intercept, 1

68. Slope, −3; y intercept, −2

69. Slope, −2; y intercept, −3

70. Slope, 0; y intercept, 4

67.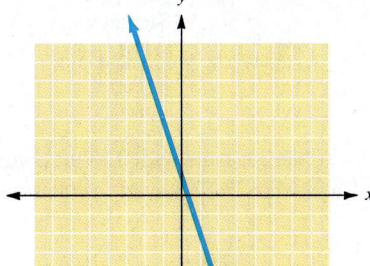

68.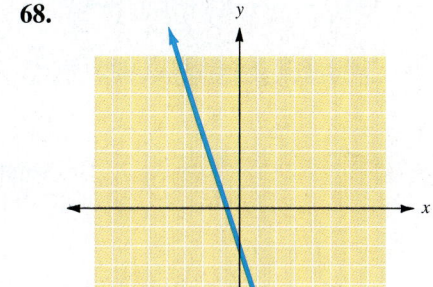

69.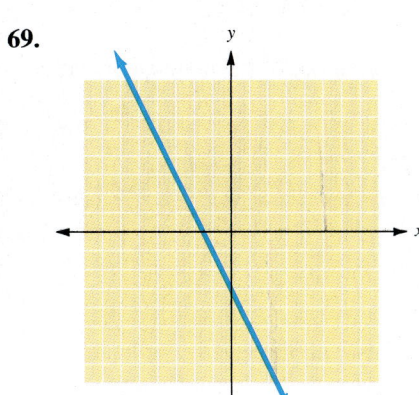

70.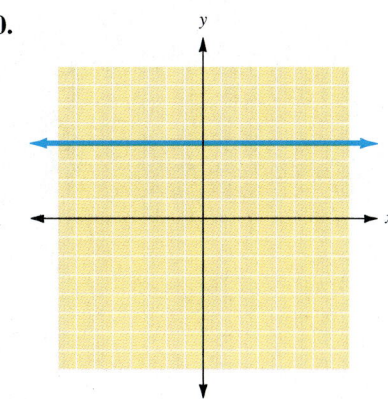

Summary Exercises · 3

This summary exercise set is provided to give you practice with each of the objectives of the chapter. Each exercise is keyed to the appropriate chapter section.

[3.1] Determine which of the ordered pairs are solutions for the given equations.

1. $x - y = 6$ (6, 0), (3, 3), (3, −3), (0, −6)
2. $2x + 3y = 6$ (3, 0), (6, 2), (−3, 4), (0, 2)

[3.2] Give the coordinates of the points graphed below.

3. A
4. B
5. E
6. F

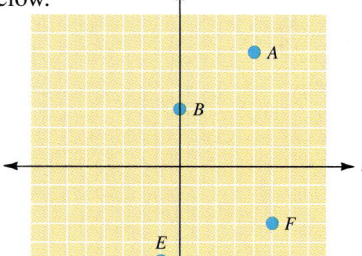

[3.2] Plot the points with the given coordinates.

7. $P(6, 0)$ 8. $Q(5, 4)$ 9. $T(-2, 4)$ 10. $U(4, -2)$

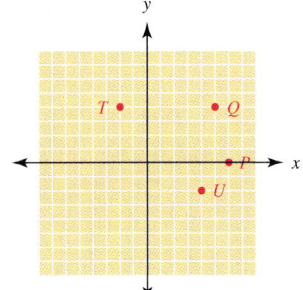

[3.2] In Exercises 9 to 14, give the quadrant in which the following points are located or the axis on which the point lies.

11. (3, 6)   12. (−5, 2)   13. (−1, −6)
14. (−7, 8)   15. (6, 0)   16. (0, −5)

[3.2] Plot the following points.

17. (−2, −3)   18. (−1.25, 3.5)

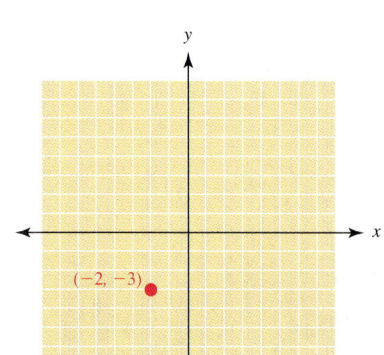 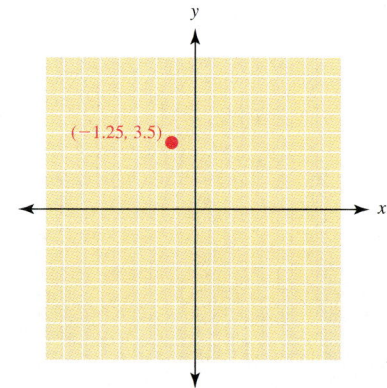

Answers

1. (6, 0) (3, −3) (0, −6)
2. (0, 2) (3, 0) (−3, 4)
3. (4, 6)
4. (0, 3)
5. (−1, −5)
6. (5, −3)
11. I
12. II
13. III
14. II
15. x axis
16. y axis

■ Summary Exercises **273**

19. (6, 3)

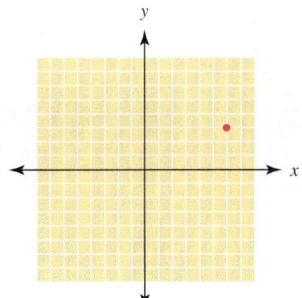

20. (−5, −2)

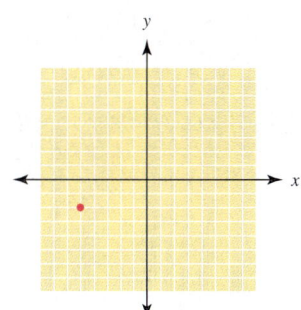

[3.3] Graph each of the following equations.

21. $x + y = 5$

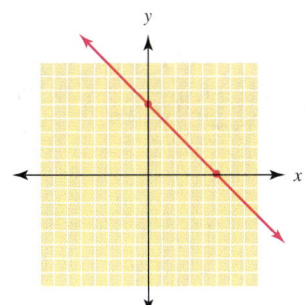

22. $x - y = 6$

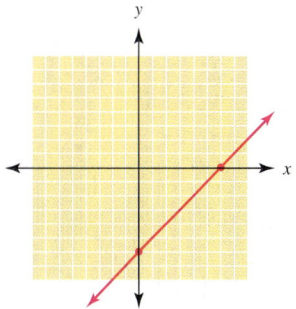

23. $y = 2x$

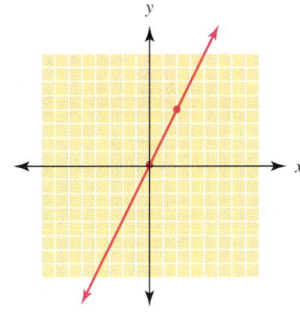

24. $y = -3x$

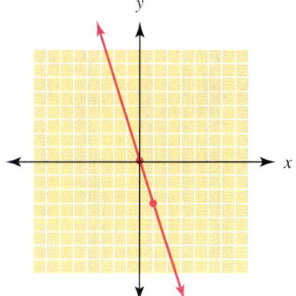

25. $y = \dfrac{3}{2}x$

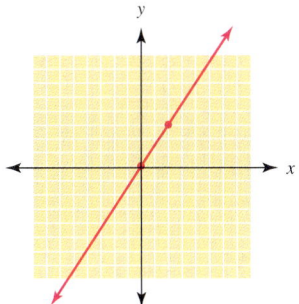

26. $y = 3x + 2$

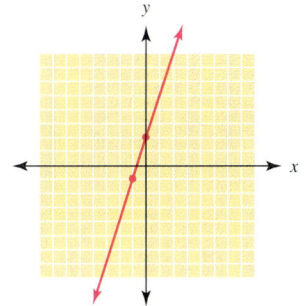

27. $y = 2x - 3$

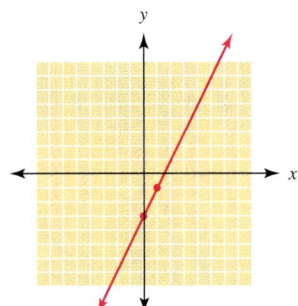

28. $y = -3x + 4$

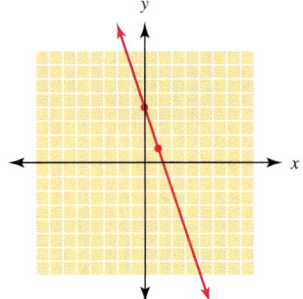

29. $y = \frac{2}{3}x + 2$

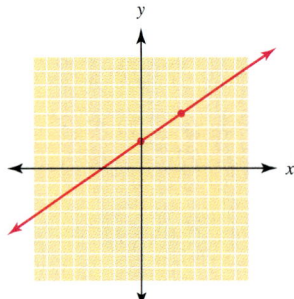

30. $3x - y = 3$

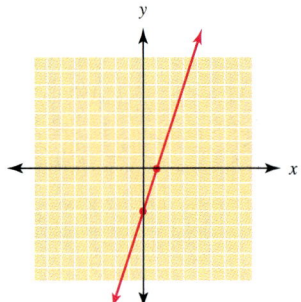

31. $2x + y = 6$

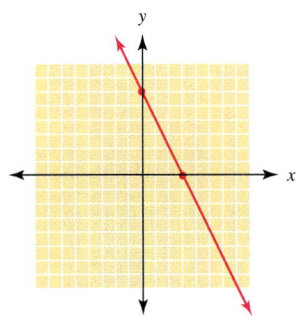

32. $3x + 2y = 12$

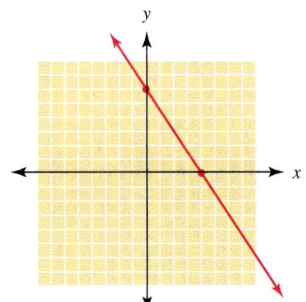

33. $3x - 4y = 12$

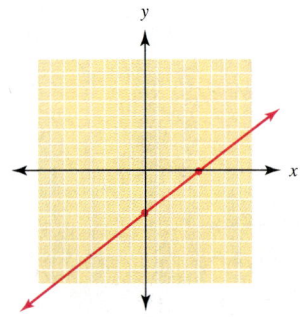

34. $x = 3$

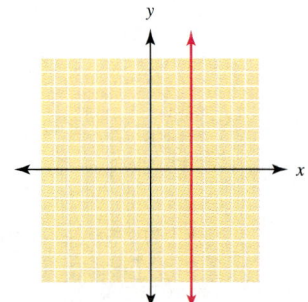

35. $y = -2$

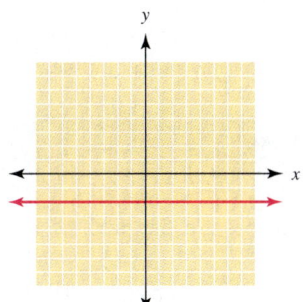

36. $5x - 3y = 15$

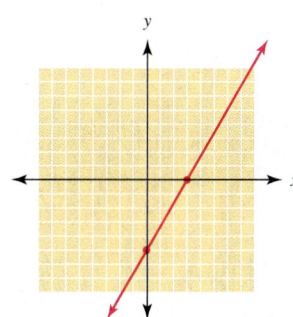

37. $4x + 3y = 12$

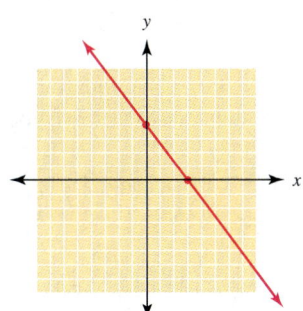

38. $2x + y = 6$

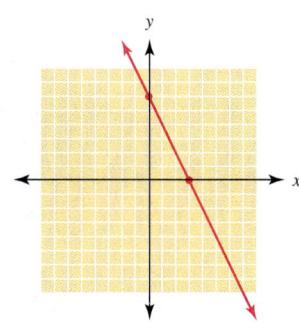

39. $3x + 2y = 6$

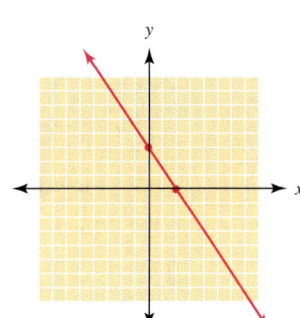

40. $-4x - 5y = 20$

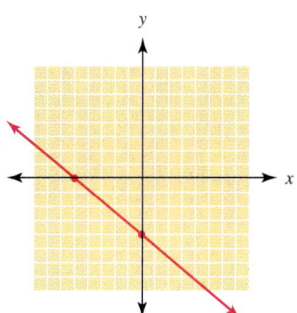

41. 2
42. -3
43. $-\dfrac{1}{2}$
44. $\dfrac{2}{3}$
45. 0
46. $-\dfrac{5}{2}$
47. $\dfrac{1}{2}$
48. Undefined
49. 2, 5
50. $-4, -3$
51. $-\dfrac{3}{4}, 0$
52. $\dfrac{2}{3}, 3$

[3.4] Find the slope of the line through each of the following pairs of points.

41. (3, 4) and (5, 8)   **42.** $(-2, 3)$ and $(1, -6)$ **43.** $(-2, 5)$ and (2, 3)

44. $(-5, -2)$ and (1, 2)   **45.** $(-2, 6)$ and (5, 6)   **46.** $(-3, 2)$ and $(-1, -3)$

47. $(-3, -6)$ and $(5, -2)$ **48.** $(-6, -2)$ and $(-6, 3)$

[3.4] Find the slope and y intercept of the line represented by each of the following equations.

49. $y = 2x + 5$ **50.** $y = -4x - 3$ **51.** $y = -\dfrac{3}{4}x$ **52.** $y = \dfrac{2}{3}x + 3$

53. $-\frac{2}{3}, 2$ 54. $\frac{5}{2}, -5$

55. $0, -3$

56. Undefined slope, no y intercept

57. $y = 2x + 3$

58. $y = \frac{3}{4}x - 2$

59. $y = -\frac{2}{3}x + 2$

60. Parallel

61. Perpendicular

62. Neither 63. Parallel

64. $y = \frac{2}{3}x - 5$

65. $y = -3$ 66. $y = 3x - 3$

67. $x = 4$

68. $y = \frac{5}{3}x - 7$

69. $y = -3$

70. $y = -\frac{5}{2}x - 9$

71. $y = -\frac{4}{3}x - 2$

72. $y = -5$ 73. $x = -\frac{5}{2}$

74. $y = \frac{2}{3}x + 1$

75. $y = -\frac{1}{5}x + 4$

76. $y = \frac{3}{4}x + 3$

77. $y = -\frac{5}{4}x + 2$

78. $y = 3x - 4$

79. $y = -\frac{5}{3}x + 3$

80. $y = -\frac{2}{3}x + \frac{1}{3}$

81. $y = \frac{4}{3}x + \frac{14}{3}$

53. $2x + 3y = 6$ 54. $5x - 2y = 10$ 55. $y = -3$ 56. $x = 2$

[3.4] Write the equation of the line with the given slope and y intercept. Then graph each line, *using* the slope and y intercept.

57. $m = 2, b = 3$ 58. $m = \frac{3}{4}, b = -2$ 59. $m = -\frac{2}{3}, b = 2$

[3.5] In Exercises 60 to 63, are the pairs of lines parallel, perpendicular, or neither?

60. L_1 through $(-3, -2)$ and $(1, 3)$
    L_2 through $(0, 3)$ and $(4, 8)$

61. L_1 through $(-4, 1)$ and $(2, -3)$
    L_2 through $(0, -3)$ and $(2, 0)$

62. L_1 with equation $x + 2y = 6$
    L_2 with equation $x + 3y = 9$

63. L_1 with equation $4x - 6y = 18$
    L_2 with equation $2x - 3y = 6$

[3.5] In Exercises 64 to 73, write the equation of the line passing through the following points with the indicated slope. Give your result in slope-intercept form, where possible.

64. $(0, -5), m = \frac{2}{3}$ 65. $(0, -3), m = 0$ 66. $(2, 3), m = 3$

67. $(4, 3), m$ is undefined 68. $(3, -2), m = \frac{5}{3}$

69. $(-2, -3), m = 0$ 70. $(-2, -4), m = -\frac{5}{2}$

71. $(-3, 2), m = -\frac{4}{3}$ 72. $\left(\frac{2}{3}, -5\right), m = 0$

73. $\left(-\frac{5}{2}, -1\right), m$ is undefined

[3.5] In Exercises 74 to 81, write the equation of the line L satisfying the following sets of geometric conditions.

74. L passes through $(-3, -1)$ and $(3, 3)$.

75. L passes through $(0, 4)$ and $(5, 3)$.

76. L has slope $\frac{3}{4}$ and y intercept 3.

77. L passes through $(4, -3)$ with a slope of $-\frac{5}{4}$.

78. L has y intercept -4 and is parallel to the line with equation $3x - y = 6$.

79. L passes through $(3, -2)$ and is perpendicular to the line with equation $3x - 5y = 15$.

80. L passes through $(2, -1)$ and is perpendicular to the line with equation $3x - 2y = 5$.

81. L passes through the point $(-5, -2)$ and is parallel to the line with equation $4x - 3y = 9$.

Cumulative Test • 0-3

1. 17
2. 8
3. $6x + 3y$
4. $\{-11\}$
5. $\{100\}$
6. $\{3, 4\}$
7. $C = \frac{5}{9}(F - 32)$
8. $\{x | x < 4\}$
9. $\{x | x > -4\}$
10. $\{x | 5 \leq x \leq 9\}$
11. $\{x | x < 2 \text{ or } x > 7\}$
12. $\left\{x | -\frac{13}{2} \leq x \leq \frac{3}{2}\right\}$
13. $\{x | x < 1 \text{ or } x > 11\}$
14. -2
15. $y = 5x - 6$
16. $x + 10y = 86$
17. $y = -\frac{2}{3}x + 6$
18. $4y + 5x = 26$

This test covers selected topics from the first three chapters.

Evaluate each of the following expressions.

1. $2^3 - |-8| \div (-4) \cdot 2 + 5$ 2. $4^2 + (-16 \div 4 \cdot 2)$

3. Combine like terms in the expression $9x - 5y - (3x - 8y)$

Solve each equation. Express your answer in set notation.

4. $5x - 3(2x - 6) + 9 = -2(x - 5) + 6$

5. $\frac{4}{5}x - 2 = 3 + \frac{3}{4}x$

6. $|2x - 7| + 5 = 6$

7. Solve the equation $F = \frac{9}{5}C + 32$ for C.

Solve and graph the solution set for each inequality.

8. $4x - 7 < 9$ 9. $6x + 4 > 3x - 8$ 10. $4 \leq 2x - 6 \leq 12$

 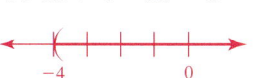 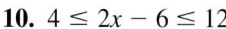

11. $x - 5 < -3$ or $x - 5 > 2$ 12. $|2x + 5| \leq 8$ 13. $|x - 6| > 5$

 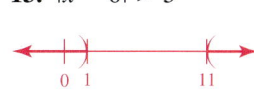

14. Find the slope of the line connecting the points $(6, -4)$ and $(-2, 12)$

Write the equation of the line L that satisfies the given conditions.

15. L has slope 5 and y intercept of -6.

16. L passes through $(-4, 9)$ and $(6, 8)$.

17. L has y intercept 6 and is parallel to the line with the equation $2x + 3y = 6$.

18. L passes through the point $(2, 4)$ and is perpendicular to the line with the equation $4x - 5y = 20$.

19. 9

20. 7

Use the appropriate equation to solve each of the following.

19. If one-third of a number is added to 3 times the number, the result is 30. Find the number.

20. 2 more than 4 times a number is 30. Find the number.

CHAPTER 4
A Beginning Look at Functions

LIST OF SECTIONS

4.1 Ordered Pairs and Relations

4.2 An Introduction to Functions

4.3 Identifying Functions

4.4 Reading Values from a Graph

4.5 Operations on Functions

Economists are among the many professionals who use graphs to show connections between two sets of data. For example, an economist may use a graph to look for a connection between two different measures for the standard of living in various countries.

One way of measuring the standard of living in a country is the per capita gross domestic product (GDP). The GDP is the total value of all goods and services produced by all businesses and individuals over the course of 1 year. To find the per capita GDP, we divide that total value by the population of the country.

Other economists, including some who wrote an article in *Scientific American* in May 1993, question this method. Rather than comparing GDP among countries, they use survival rate (life expectancy) to measure the quality of life.

SECTION 4.1 Ordered Pairs and Relations

4.1 OBJECTIVES

1. Identify the domain of a relation
2. Identify the range of a relation

"The flower of mathematical thought, the notion of a function."

—Thomas McCormack

In Section 3.1, we looked at **ordered pair notation** to designate specific values for x and y in the pair, (x, y). We found that $(2, -3)$ was different from $(-3, 2)$.

In our next example, we will identify ordered pairs.

Example 1

Identifying Ordered Pairs

Which of the following are ordered pairs?

(a) $(2, -\pi)$ (b) $\{2, -4\}$ (c) $(1, 3, -1)$

(d) $\{(1, -5), (9, 0)\}$ (e) $2, 5$

Only (a) is an ordered pair. (b) is a set (it uses braces instead of parentheses), (c) has three numbers instead of two, (d) is a set of ordered pairs, and (e) is simply a list of two numbers.

✓ CHECK YOURSELF 1

Which of the following are ordered pairs?

(a) $\left\{\dfrac{1}{2}, -3\right\}$ (b) $\left(-3, \dfrac{1}{3}\right)$ (c) $\{(5, 0)\}$ (d) $(1, -5)$ (e) $-3, 6$

Relations

A set of ordered pairs is called a **relation**.

280

Section 4.1 ▪ Ordered Pairs and Relations

The ordered pairs can be listed either explicitly or implicitly. Examples of relations include

$$A = \{(-2, 8), (1, 2), (-3, -5)\}$$
$$R = \{(1, 3), (1, 4), (1, 5)\}$$
$$B = \{(x, y) | y = x\}$$

Relation B is composed of every ordered pair in which x and y are equal. This relation includes $(0, 0)$, $\left(\frac{1}{2}, \frac{1}{2}\right)$, $(-3, -3)$, and (π, π).

Domain and Range

In most cases, the domain is the set of all x values, and the range is the set of all y values.

> The **domain** of a relation is the set of all the first elements of the ordered pairs. The **range** of a relation is the set of all the second elements of the ordered pairs.

Example 2 Finding the Domain and Range of a Relation

Find the domain and range for each relation.

(a) $A = \{(-2, 8), (1, 2), (-3, -5)\}$

The domain is the set $\{-2, 1, -3\}$. The range is the set $\{8, 2, -5\}$.

(b) $R = \{(1, 3), (1, 4), (1, 5)\}$

The domain is the set $\{1\}$. The range is the set $\{3, 4, 5\}$.

✓ CHECK YOURSELF 2

Find the domain and range for each relation.

(a) $A = \{(-5, 4), (-4, 7), (-4, 9)\}$ (b) $R = \{(1, 0), (3, 0), (5, 0)\}$

It is also possible to find the domain and the range when the relation is described by a rule rather than by a list. When looking for the domain from a rule, the question to ask is, "What values can I substitute for x?" When looking for the range, we ask, "What values can be substituted for y?"

Example 3 — Identifying the Domain and Range

Find the domain and range for each relation.

(a) $A = \{(x, y) | x = 3\}$

The domain is $\{3\}$. The description of the relation defines x to be 3. The range is all real numbers, $\mathbb{R}$.

The symbol $\mathbb{R}$ represents the real numbers, $\mathbb{N}$ represents the natural numbers, and $\mathbb{Z}$ represents the integers.

(b) $B = \{(x, y) | y = x\}$

The domain is the set of all real numbers, $\mathbb{R}$. The range is also the set of all real numbers, $\mathbb{R}$.

✓ CHECK YOURSELF 3

Find the domain and range for each relation.

(a) $A = \{(x, y) | y = -1\}$ (b) $B = \{(x, y) | y = x - 1\}$

The domain and the range are determined by examining ordered pairs. Recall that a line is the graph of a set of ordered pairs. In Example 4, we will examine the domain and range associated with a line.

Example 4 — Finding the Domain and Range for a Line

This is the same as finding the domain and range of
$A = \{(x, y) | x + y = 5\}$

Find the domain and range for the relation described by the equation

$$x + y = 5$$

We can analyze the domain and range either graphically or algebraically. First, we will look at a graphical analysis. Let's look at the graph of the equation $x + y = 5$.

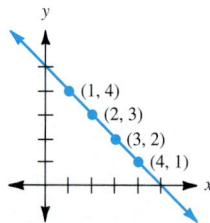

The graph continues forever at both ends. For every value of x, there is an associated point on the line. Therefore, the domain (D) is the set of all real numbers. In set notation, we write

$$D = \{x | x \in \mathbb{R}\}$$

This is read, "The domain is the set of every x that is a real number."

To find the range (R), we look at the graph to see what values are associated with y. Note that every y is associated with some point. The range is written as

$$R = \{y | y \in \mathbb{R}\}$$

This is read, "The range is the set of every y that is a real number."

Let's find the domain and range for the same relation by using an algebraic analysis. Look at the following equation.

$$x + y = 5$$

To determine the domain, we need to find every value of x that allows us to solve for y. That combination will result in an ordered pair (x, y). The set of all those x values is the domain of the relation.

We can find a value for y for *any* real value of x. For example, if $x = -5$,

$$-5 + y = 5$$
$$y = 10$$

The ordered pair $(-5, 10)$ is part of the relation. As in our graphical analysis, the domain is

$$D = \{x | x \in \mathbb{R}\}$$

and the range is

$$R = \{y | y \in \mathbb{R}\}$$

✓ CHECK YOURSELF 4

Find the domain and range for the relation described by the following equation.

$$x - y = 4$$

✓ CHECK YOURSELF ANSWERS

1. (b) and (d) are ordered pairs. **2.** (a) Domain $= \{-5, -4\}$; range $= \{4, 7, 9\}$; (b) domain $= \{1, 3, 5\}$; range $= \{0\}$. **3.** (a) The domain is all real numbers, $\mathbb{R}$. The range is $\{-1\}$. (b) The domain is the set of all real numbers, $\mathbb{R}$. The range is also the set of all real numbers, $\mathbb{R}$. **4.** $D = \{x | x \in \mathbb{R}\}$; $R = \{y | y \in \mathbb{R}\}$.

Exercises • 4.1

In Exercises 1 to 4, identify the ordered pairs.

1. (a) $(3, -5)$ (b) $\{7, 9\}$ (c) $(2, 5)$ (d) 5, 2 (e) $((3, 1), 4)$

2. (a) $\{7, 23\}$ (b) $(1, 0, (5, 6))$ (c) $\left(\dfrac{1}{2}, -1\right)$ (d) $[5, 6]$ (e) $(23, 7)$

3. (a) 18, 67 (b) $(-3, -9)$ (c) $\{3, 9\}$ (d) $(3, 7, -3)$ (e) $[12, 56]$

4. (a) $\{45, 67\}$ (b) $(9, 3)$ (c) 5, 8 (d) $(11, -3, 9)$ (e) $[5, 2]$

In Exercises 5 to 20, identify the domain and range in the sets of ordered pairs.

5. $\{(1, 2), (3, 4), (5, 6), (7, 8), (9, 10)\}$

6. $\{(2, 3), (3, 5), (4, 7), (5, 9), (6, 11)\}$

7. $\{(1, 2), (4, 6), (3, 3), (5, 4), (6, 1)\}$

8. $\{(3, 4), (5, 7), (6, 1), (2, 2), (4, 3)\}$

9. $\{(1, 2), (1, 3), (1, 4), (1, 5), (1, 6)\}$

10. $\{(3, 4), (3, 6), (3, 8), (3, 9), (3, 10)\}$

11. $\{(1, 5), (2, 5), (3, 6), (2, 4), (4, 5)\}$

12. $\{(2, 8), (3, 9), (2, 9), (3, 8), (4, 7)\}$

13. $\{(-1, 3), (-2, 4), (-3, 5), (4, 4), (5, 6)\}$

14. $\{(-2, 4), (1, 4), (-3, 4), (5, 4), (7, 4)\}$

15. $\{(x, y) | x + 2y = 3\}$

16. $\{(x, y) | 3x + 4y = 12\}$

17. $\{(x, y) | y = 5\}$

18. $\{(x, y) | y = -4\}$

19. $\{(x, y) | x = 23\}$

20. $\{(x, y) | x = -9\}$

21. The stock prices for a given stock over a week's time are displayed in a table. List this information as a set of ordered pairs using the day of the week as the domain.

Day	1	2	3	4	5
Price	$9\dfrac{1}{8}$	8	$8\dfrac{7}{8}$	$9\dfrac{1}{4}$	9

22. Food purchases. In the snack department of the local supermarket, candy costs $1.58 per pound. For 1 to 5 pounds, write the cost of candy as ordered pairs.

23. Explain why the ordered pair (2, 3) is different from the ordered pair (3, 2) while the set {2, 3} is the same as {3, 2}.

24. Explain the difference between the relation $\{(x, y) | x + y = 5\}$ and the relation $\{(x, y) | x + y = 5 \text{ and } x \text{ is an integer between 1 and 7}\}$.

---

1. (a) and (c)
2. (c) and (e)
3. (b) 4. (b)
5. D: {1, 3, 5, 7, 9};
   R: {2, 4, 6, 8, 10}
6. D: {2, 3, 4, 5, 6};
   R: {3, 5, 7, 9, 11}
7. D: {1, 3, 4, 5, 6};
   R: {1, 2, 3, 4, 6}
8. D: {2, 3, 4, 5, 6};
   R: {1, 2, 3, 4, 7}
9. D: {1}; R: {2, 3, 4, 5, 6}
10. D: {3};
    R: {4, 6, 8, 9, 10}
11. D: {1, 2, 3, 4};
    R: {4, 5, 6}
12. D: {2, 3, 4}; R: {7, 8, 9}
13. D: {−3, −2, −1, 4, 5};
    R: {3, 4, 5, 6}
14. D: {−3, −2, 1, 5, 7};
    R: {4}
15. D: reals; R: reals
16. D: reals; R: reals
17. D: reals; R: {5}
18. D: reals; R: {−4}
19. D: {23}; R: reals
20. D: {−9};
    R: reals
21. $\left\{\left(1, 9\dfrac{1}{8}\right), (2, 8),\right.$
    $\left(3, 8\dfrac{7}{8}\right), \left(4, 9\dfrac{1}{4}\right),$
    $\left.(5, 9)\right\}$
22. {(1, 1.58), (2, 3.16), (3, 4.74), (4, 6.32), (5, 7.90)}
23. In ordered pairs, the order of the elements is important.
24. The first relation contains an infinite number of ordered pairs. The second contains only 7 ordered pairs.

SECTION 4.2 An Introduction to Functions

4.2 OBJECTIVES

1. Evaluate expressions
2. Evaluate functions
3. Express the equation of a line as a linear function
4. Write an equation as a function
5. Graph a linear function

Variables can be used to represent unknown real numbers. Together with the operations of addition, subtraction, multiplication, division, and exponentiation, these numbers and variables form expressions such as

$$3x + 5 \qquad 7x - 4 \qquad x^2 - 3x - 10 \qquad x^4 - 2x^2 + 3x + 4$$

Four different actions can be taken with expressions. We can

1. Substitute values for the variable(s) and **evaluate the expression.**
2. Rewrite an expression as some simpler equivalent expression. This rewriting is called **simplifying the expression.**
3. Set two expressions equal to each other and **solve for the stated variable.**
4. Set two expressions equal to each other and **graph the equation.**

Throughout this book, everything we do will involve one of these four actions. We now focus on the first item, evaluating expressions. As we saw in Section 1.2, expressions can be evaluated for an indicated value of the variable(s). Example 1 illustrates.

Example 1 Evaluating Expressions

Evaluate the expression $x^4 - 2x^2 + 3x + 4$ for the indicated value of x.

(a) $x = 0$

Substituting 0 for x in the expression yields

$$(0)^4 - 2(0)^2 + 3(0) + 4 = 0 - 0 + 0 + 4$$
$$= 4$$

(b) $x = 2$

Substituting 2 for x in the expression yields

$$(2)^4 - 2(2)^2 + 3(2) + 4 = 16 - 8 + 6 + 4$$
$$= 18$$

(c) $x = -1$

Substituting -1 for x in the expression yields

$$(-1)^4 - 2(-1)^2 + 3(-1) + 4 = 1 - 2 - 3 + 4$$
$$= 0$$

✓ CHECK YOURSELF 1

Evaluate the expression $2x^3 - 3x^2 + 3x + 1$ for the indicated value of x.

(a) $x = 0$ (b) $x = 1$ (c) $x = -2$

Function Notation

We could design a machine whose function would be to crank out the value of an expression for each given value of x. We could call this machine something simple such as f. Our *function* machine might look like this.

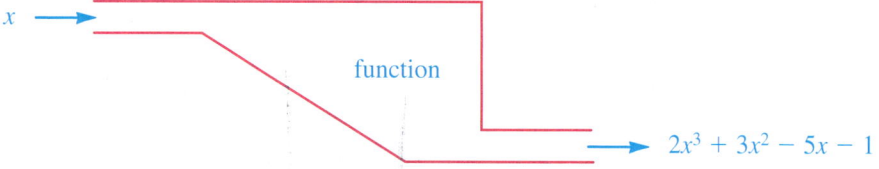

For example, when we put -1 into the machine, the machine would substitute -1 for x in the expression, and 5 would come out the other end because

$$2(-1)^3 + 3(-1)^2 - 5(-1) - 1 = -2 + 3 + 5 - 1 = 5$$

In fact, the idea of the function machine is very useful in mathematics. Your graphing calculator can be used as a function machine. You can enter the expression into the calculator as Y_1 and then evaluate Y_1 for different values of x.

Generally, in mathematics, we do not write $Y_1 = 2x^3 + 3x^2 - 5x - 1$. Instead, we write $f(x) = 2x^3 + 3x^2 - 5x - 1$, which is read as "$f$ of x is equal to . . . ." Instead of calling f a function machine, we say that f is a function of x. The greatest benefit of this notation is that it lets us easily note the input value of x along with the output of the function. Instead of "Evaluate Y_1 for $x = 4$" we say "Find $f(4)$."

Example 2

Evaluating Expressions with Function Notation

Given $f(x) = x^3 - 3x^2 + x + 5$, find the following:

(a) $f(0)$

Substituting 0 for x in the expression on the right, we get

$$(0)^3 - 3(0)^2 + (0) + 5 = 5$$

(b) $f(-3)$

Substituting -3 for x in the expression on the right, we get

$$(-3)^3 - 3(-3)^2 + (-3) + 5 = -27 - 27 - 3 + 5$$
$$= -52$$

(c) $f\left(\dfrac{1}{2}\right)$

Substituting $\dfrac{1}{2}$ for x in the expression on the right, we get

$$\left(\dfrac{1}{2}\right)^3 - 3\left(\dfrac{1}{2}\right)^2 + \left(\dfrac{1}{2}\right) + 5 = \dfrac{1}{8} - 3\left(\dfrac{1}{4}\right) + \dfrac{1}{2} + 5$$
$$= \dfrac{1}{8} - \dfrac{3}{4} + \dfrac{1}{2} + 5$$
$$= \dfrac{1}{8} - \dfrac{6}{8} + \dfrac{4}{8} + 5$$
$$= -\dfrac{1}{8} + 5$$
$$= 4\dfrac{7}{8} \text{ or } \dfrac{39}{8}$$

✓ **CHECK YOURSELF 2**

Given $f(x) = 2x^3 - x^2 + 3x - 2$, find the following.

(a) $f(0)$ (b) $f(3)$ (c) $f\left(-\dfrac{1}{2}\right)$

Given a function f, the pair of numbers $(x, f(x))$ is very significant. We always write them in that order, hence the name *ordered pairs*. In Example 2, part a, we saw that, given $f(x) = x^3 - 3x^2 + x + 5$, $f(0) = 5$, which meant that the ordered pair $(0, 5)$ was associated with the function. The ordered pair consists of the x value first and the function value at that x (the $f(x)$) second.

Example 3

Finding Ordered Pairs

Given the function $f(x) = 2x^2 - 3x + 5$, find the ordered pair $(x, f(x))$ associated with each given value for x.

(a) $x = 0$

$$f(0) = 5$$

so the ordered pair is $(0, 5)$.

(b) $x = -1$

$$f(-1) = 2(-1)^2 - 3(-1) + 5 = 10$$

The ordered pair is $(-1, 10)$.

(c) $x = \dfrac{1}{4}$

$$f\left(\dfrac{1}{4}\right) = 2\left(\dfrac{1}{16}\right) - 3\left(\dfrac{1}{4}\right) + 5 = \dfrac{35}{8}$$

The ordered pair is $\left(\dfrac{1}{4}, \dfrac{35}{8}\right)$.

✓ CHECK YOURSELF 3

Give $f(x) = 2x^3 - x^2 + 3x - 2$, find the ordered pair associated with each given value of x.

(a) $x = 0$ (b) $x = 3$ (c) $x = -\dfrac{1}{2}$

In Chapter 3, we discussed the graph of a linear equation. We saw that the graph for a vertical line had the form $x = a$. The equation for such a line cannot be rewritten as a function, but any nonvertical line can be written as a function.

Example 4 — Writing Equations as Functions

Rewrite each linear equation as a function of *x*.

(a) $y = 3x - 4$

This can be rewritten as

$$f(x) = 3x - 4$$

(b) $2x - 3y = 6$

We must first solve the equation for *y* (recall that this will give us the slope-intercept form).

$$-3y = -2x + 6$$

$$y = \frac{2}{3}x - 2$$

This can be rewritten as

$$f(x) = \frac{2}{3}x - 2$$

✓ CHECK YOURSELF 4

Rewrite each equation as a function of *x*.

(a) $y = -2x + 5$ (b) $3x + 5y = 15$

The process of finding the graph of a linear function is identical to the process of finding the graph of a linear equation.

Example 5 — Graphing a Linear Function

Graph the function

$$f(x) = 3x - 5$$

We could use the slope and *y* intercept to graph the line, or we can find three points (the third is a checkpoint) and draw the line through them. We will do the latter.

$$f(0) = -5 \qquad f(1) = -2 \qquad f(2) = 1$$

We will use the three points (0, −5), (1, −2), and (2, 1) to graph the line.

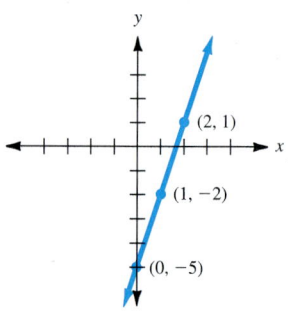

✓ CHECK YOURSELF 5

Graph the function

$$f(x) = 5x - 3$$

One benefit of having a function written in $f(x)$ form is that it makes it fairly easy to substitute values for x. In Example 5, we substituted the values 0, 1, and 2. Sometimes it is useful to substitute nonnumeric values for x.

Example 6 — Substituting Nonnumeric Values for x

Let $f(x) = 2x + 3$. Evaluate f as indicated.

(a) $f(a)$

Substituting a for x in our equation, we see that

$$f(a) = 2a + 3$$

(b) $f(2 + h)$

Substituting $2 + h$ for x in our equation, we get

$$f(2 + h) = 2(2 + h) + 3$$

Distributing the 2, then simplifying, we have

$$f(2 + h) = 4 + 2h + 3$$
$$= 2h + 7$$

Section 4.2 ■ An Introduction to Functions

✓ **CHECK YOURSELF 6**

Let $f(x) = 4x - 2$. Evaluate f as indicated.

(a) $f(b)$ (b) $f(4 + h)$

The TABLE feature on a graphing calculator can also be used to evaluate a function. Our final example illustrates this feature.

Example 7

Using a Graphing Calculator to Evaluate a Function

Evaluate the function $f(x) = 3x^3 + x^2 - 2x - 5$ for each x in the set $\{-6, -5, -4, -3, -2\}$.

1. Enter the function into a Y= screen.
2. Find the table set-up screen.
3. Start the table at -6 with a change of 1.
4. View the table.

The table should look something like this

Although we have assumed that the graphing calculator was a TI, most such calculators have similar capability.

X	Y_1
−6	−605
−5	−345
−4	−173
−3	−71
−2	−21
−1	−5
0	−5

X=−6

The Y_1 column is the function value for each value of x.

✓ **CHECK YOURSELF 7**

Evaluate the function $f(x) = 2x^3 - 3x^2 - x + 2$ for each x in the set $\{-5, -4, -3, -2, -1, 0, 1\}$.

✓ **CHECK YOURSELF ANSWERS**

1. (a) 1; (b) 3; (c) −33.
(b) (3, 52); (c) $(-\frac{1}{2}, -4)$.

2. (a) −2; (b) 52; (c) −4.

3. (a) (0, −2);

4. (a) $f(x) = -2x + 5$; (b) $f(x) = -\frac{3}{5}x + 3$.

5.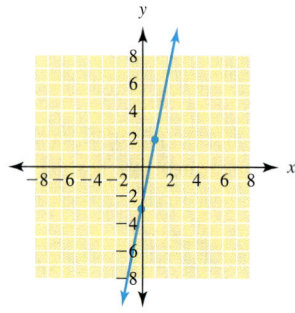

6. (a) $4b - 2$; (b) $4h + 14$.

7.

X	Y₁	
−5	−318	
−4	−170	
−3	−76	
−2	−24	
−1	−2	
0	2	
1	0	
X=−5		

Exercises • 4.2

1. (a) −2; (b) 4; (c) −2
2. (a) 10; (b) 0; (c) 28
3. (a) 9; (b) −1; (c) 3
4. (a) −2; (b) −2; (c) −8
5. (a) −62; (b) −2; (c) 2
6. (a) 7; (b) −1; (c) 1
7. (a) 45; (b) 3; (c) −75
8. (a) 85; (b) −8; (c) −10
9. (a) −1; (b) 2; (c) 13
10. (a) 39; (b) 9; (c) −5
11. $f(x) = -3x + 2$
12. $f(x) = 5x + 7$
13. $f(x) = 4x - 8$
14. $f(x) = -7x - 9$
15. $f(x) = -\frac{3}{2}x + 3$
16. $f(x) = -\frac{4}{3}x + 4$
17. $f(x) = \frac{1}{3}x + \frac{3}{2}$
18. $f(x) = \frac{3}{4}x + \frac{11}{4}$
19. $f(x) = -\frac{5}{8}x + \frac{9}{8}$
20. $f(x) = \frac{4}{7}x + \frac{10}{7}$

In Exercises 1 to 10, evaluate each function for the value specified.

1. $f(x) = x^2 - x - 2$; find (a) $f(0)$, (b) $f(-2)$, and (c) $f(1)$.
2. $f(x) = x^2 - 7x + 10$; find (a) $f(0)$, (b) $f(5)$, and (c) $f(-2)$.
3. $f(x) = 3x^2 + x - 1$; find (a) $f(-2)$, (b) $f(0)$, and (c) $f(1)$.
4. $f(x) = -x^2 - x - 2$; find (a) $f(-1)$, (b) $f(0)$, and (c) $f(2)$.
5. $f(x) = x^3 - 2x^2 + 5x - 2$; find (a) $f(-3)$, (b) $f(0)$, and (c) $f(1)$.
6. $f(x) = -2x^3 + 5x^2 - x - 1$; find (a) $f(-1)$, (b) $f(0)$, and (c) $f(2)$.
7. $f(x) = -3x^3 + 2x^2 - 5x + 3$; find (a) $f(-2)$, (b) $f(0)$, and (c) $f(3)$.
8. $f(x) = -x^3 + 5x^2 - 7x - 8$; find (a) $f(-3)$, (b) $f(0)$, and (c) $f(2)$.
9. $f(x) = 2x^3 + 4x^2 + 5x + 2$; find (a) $f(-1)$, (b) $f(0)$, and (c) $f(1)$.
10. $f(x) = -x^3 + 2x^2 - 7x + 9$; find (a) $f(-2)$, (b) $f(0)$, and (c) $f(2)$.

In Exercises 11 to 20, rewrite each equation as a function of x.

11. $y = -3x + 2$
12. $y = 5x + 7$
13. $y = 4x - 8$
14. $y = -7x - 9$
15. $3x + 2y = 6$
16. $4x + 3y = 12$
17. $-2x + 6y = 9$
18. $-3x + 4y = 11$
19. $-5x - 8y = -9$
20. $4x - 7y = -10$

In Exercises 21 to 26, graph the functions.

21. $f(x) = 3x + 7$

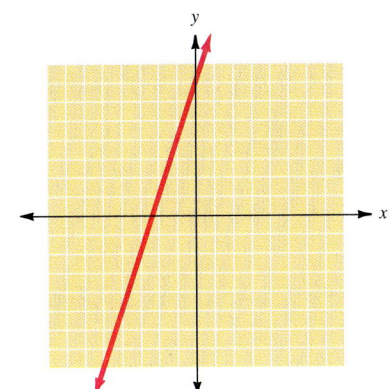

22. $f(x) = -2x - 5$

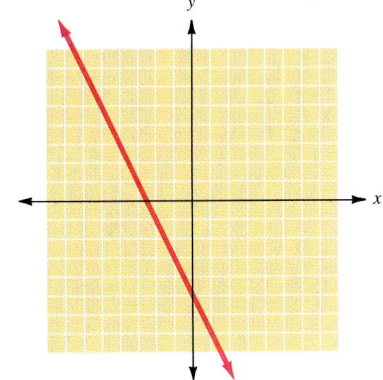

293

27. 17

28. −3

29. 13

30. −7

31. −19

32. −1

33. $5a - 1$

34. $10r - 1$

35. $5x + 4$

36. $5a - 11$

37. $5x + 5h - 1$

38. 5

39. $-3m + 2$

40. $-15n + 2$

41. $-3x - 4$

42. $-3s + 5$

43. 5

44. 9

23. $f(x) = -2x + 7$

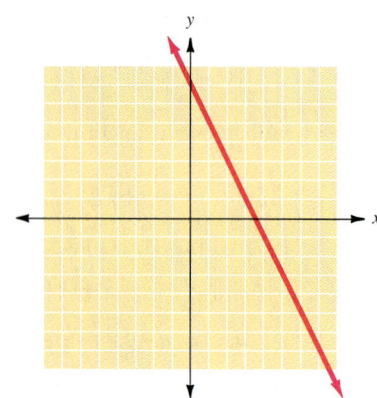

24. $f(x) = -3x + 8$

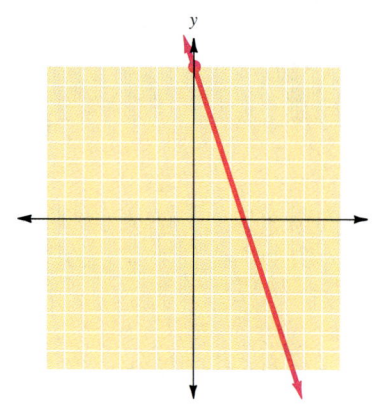

25. $f(x) = -x - 1$

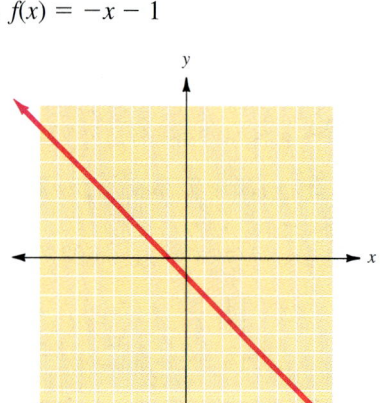

26. $f(x) = -2x - 5$

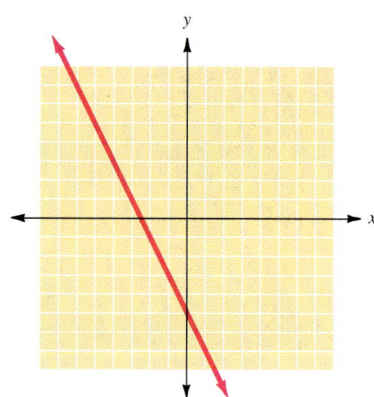

In Exercises 27 to 32, if $f(x) = 4x - 3$, find the following:

27. $f(5)$ 28. $f(0)$ 29. $f(4)$

30. $f(-1)$ 31. $f(-4)$ 32. $f\left(\dfrac{1}{2}\right)$

In Exercises 33 to 38, if $f(x) = 5x - 1$, find the following:

33. $f(a)$ 34. $f(2r)$ 35. $f(x + 1)$

36. $f(a - 2)$ 37. $f(x + h)$ 38. $\dfrac{f(x + h) - f(x)}{h}$

In Exercises 39 to 42, if $g(x) = -3x + 2$, find the following:

39. $g(m)$ 40. $g(5n)$ 41. $g(x + 2)$ 42. $g(s - 1)$

In Exercises 43 to 46, let $f(x) = 2x + 3$.

43. Find $f(1)$.   44. Find $f(3)$.

45. (1, 5), (3, 9)

46. $y = 2x + 3$

47. 107, 75, 49, 29, 15, 7, 5, 9, 19, 35, 57

48. $-162, -51, -2, 9, 6, 13, 54$

49. $-221, -114, -51, -20, -9, -6, 1, 24, 75$

50. $-192, -29, -6, -15, -20, -57, -234$

45. Form the ordered pairs $(1, f(1))$ and $(3, f(3))$.

46. Write the equation of the line passing through the points determined by the ordered pairs in Exercise 45.

In Exercises 47 to 50, use your graphing calculator to evaluate the given function for each value in the given set.

47. $f(x) = 3x^2 - 5x + 7$; $\{-5, -4, -3, -2, -1, 0, 1, 2, 3, 4, 5\}$

48. $f(x) = 4x^3 - 7x^2 + 9$; $\{-3, -2, -1, 0, 1, 2, 3\}$

49. $f(x) = 2x^3 - 4x^2 + 5x - 9$; $\{-4, -3, -2, -1, 0, 1, 2, 3, 4\}$

50. $f(x) = -3x^4 + 5x^2 - 7x - 15$; $\{-3, -2, -1, 0, 1, 2, 3\}$

SECTION 4.3 Identifying Functions

  OBJECTIVES

1. Determine whether a table of values represents a function
2. Use the vertical line test to identify the graph of a function
3. Identify the domain of a function

In Section 4.2, we used a function machine as a model that enabled us to put in a value for x and get out a value that is a function of x. These two values, x and $f(x)$, have a relationship that is usually expressed as an ordered pair.

A similar type of relationship is used in every field in which mathematics is applied.

- The physicist looks for the relationship that uses a planet's mass to predict its gravitational pull.

- The economist looks for the relationship that uses the tax rate to predict the employment rate.

- The business marketer looks for the relationship that uses an item's price to predict the number that will be sold.

- The college board looks for the relationship between tuition costs and the number of students enrolled at the college.

- The biologist looks for the relationship that uses temperature to predict a body of water's nutrient level.

In each of these examples, a researcher matches an item from the given set (the *domain*) with an item from the related set (the *range*). Each pairing becomes an ordered pair.

In Section 4.1, we looked at the concept of a relation, which is a set of ordered pairs. In the preceding list, we mentioned the relationship between a planet's mass and its gravitational pull. This relationship is an example of a function. There cannot be two different gravitational pulls associated with a single planet. If you know a planet's mass, you can find its gravitational pull.

Every set of ordered pairs defines a relation, but not every set of ordered pairs defines a function. A function is a special kind of relation.

> A **function** is a set of ordered pairs (a relation) in which no two first coordinates are equal.

Example 1

Identifying a Function

For each table of values below, decide whether the relation is a function.

(a)
x	y
-2	1
-1	1
1	3
2	3

(b)
x	y
-5	-2
-1	3
-1	6
2	9

(c)
x	y
-3	1
-1	0
0	2
2	4

Part a represents a function. No element of the domain (x) is matched with two different elements of the range (y). Part b is not a function because -1 is matched with two different range elements, 3 and 6. Part c is a function.

✓ CHECK YOURSELF 1

For each table of values below, decide whether the relation is a function.

(a)
x	y
-3	0
-1	1
1	2
3	3

(b)
x	y
-2	-2
-1	-2
1	3
2	3

(c)
x	y
-2	0
-1	1
0	2
0	3

We defined a function in terms of ordered pairs. A set of ordered pairs can be specified in several ways; here are the most common.

1. We can present the ordered pairs in a list or table, as in Example 1.
2. We can give a rule or equation that will generate the ordered pairs.
3. We can use a graph to indicate the ordered pairs. The graph can show distinct ordered pairs, or it can show all the ordered pairs on a line or curve.

Vertical Line Test

Let's look at a graph of the ordered pairs from Example 1 to introduce the **vertical line test,** which is a graphic test for identifying a function.

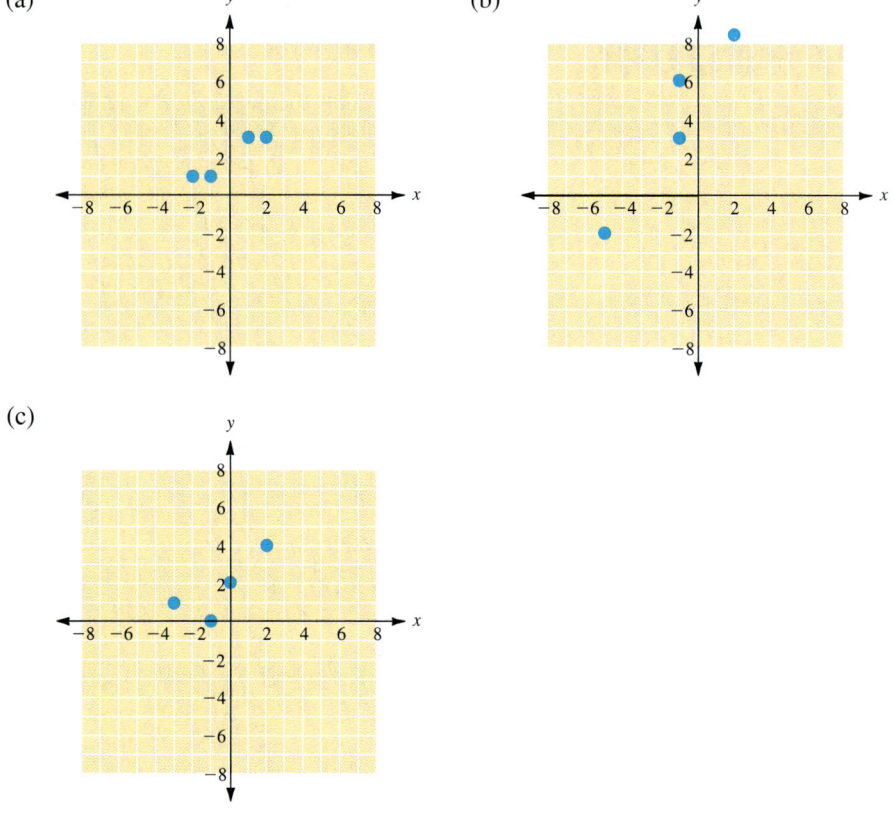

Notice that in the graphs of relations a and c, there is no vertical line that can pass through two different points of the graph. In relation b, a vertical line can pass through the two points that represent the ordered pairs $(-1, 3)$ and $(-1, 6)$. This leads to the following definition.

Vertical Line Test

If no vertical line can pass through two or more points in the graph of a relation, then the relation is a function.

Example 2 — Identifying a Function

For each set of ordered pairs, plot the related points on the provided axes. Then use the vertical line test to determine which of the sets is a function.

(a) {(0, −1), (2, 3), (2, 6), (4, 2), (6, 3)}

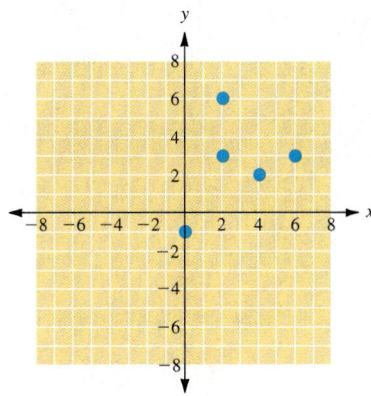

Since a vertical line can be drawn through the points (2, 3) and (2, 6), the relation does not pass the vertical line test. This is not a function.

(b) {(1, 1), (2, 0), (3, 3), (4, 3), (5, 3)}

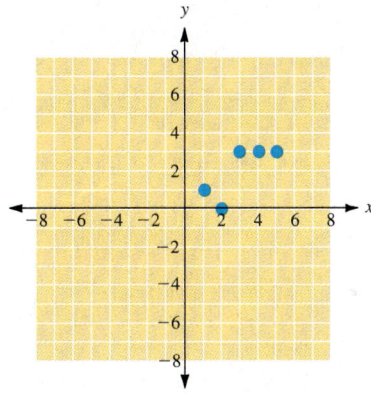

This is a function. Although a horizontal line can be drawn through several points, no vertical line passes through more than one point.

✓ CHECK YOURSELF 2

For each set of ordered pairs, plot the related points. Then use the vertical line test to determine which of the sets is a function.

(a) {(−2, 4), (−1, 4), (0, 4), (1, 3), (5, 5)}

(b) {(−3, −1), (−1, −3), (1, −3), (1, 3)}

The vertical line test can be used to determine whether a graph is the graph of a function.

Example 3

Identifying a Function

Which of the following graphs represents the graph of a function?

(a)
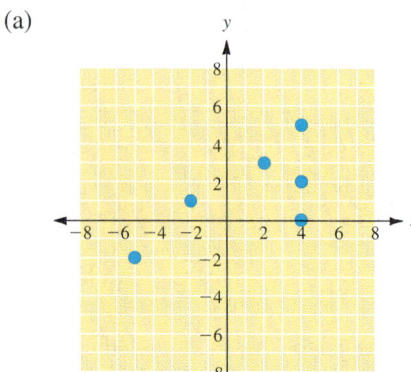

(b)
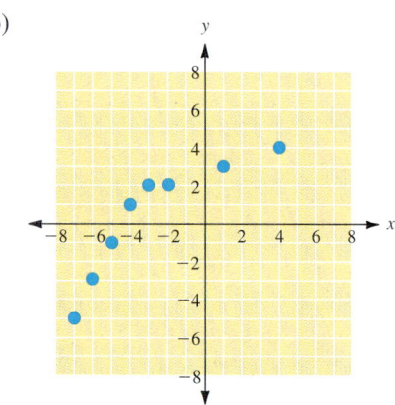

(c)
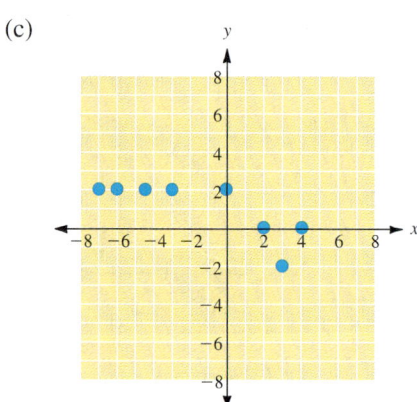

Part a is not a function, part b is a function, and part c is a function.

✓ CHECK YOURSELF 3

Which of the following graphs represents the graph of a function?

(a)
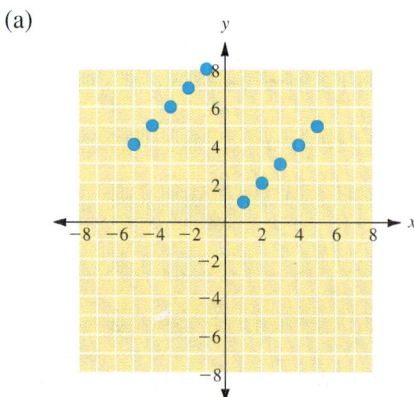

(b)
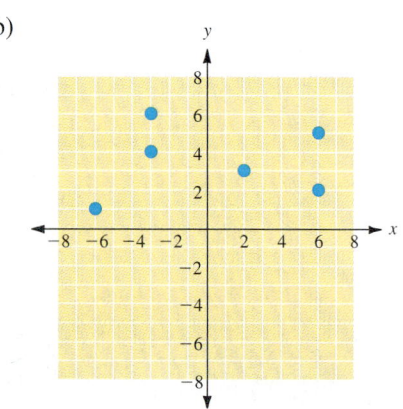

(c)

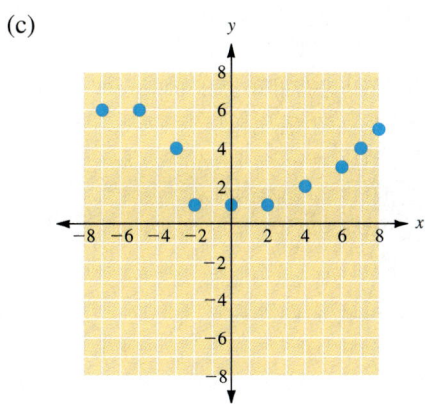

Example 4

Identifying a Function

Which of the following graphs represents the graph of a function?

(a)

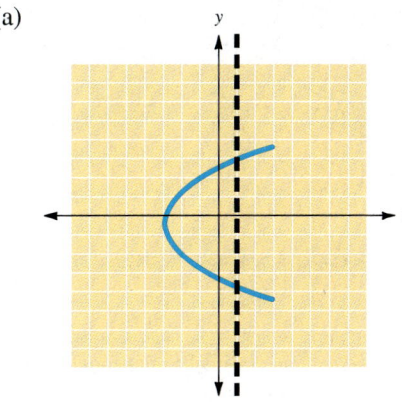

(b)

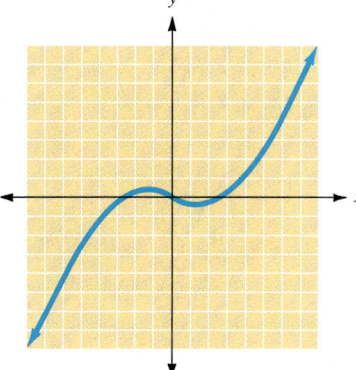

Curves, like the number line, are made up of a continuous set of points.

(c)

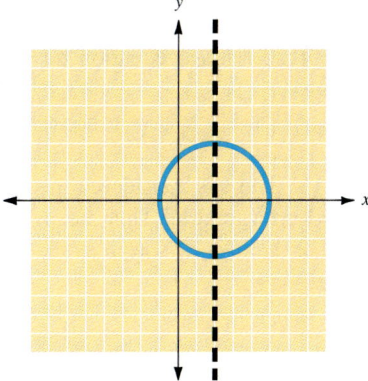

Part a is not a function; it does not pass the vertical line test. Part b is a function because it passes the vertical line test. Part c is not a function.

✓ **CHECK YOURSELF 4**

Which of the following graphs represents the graph of a function?

(a)
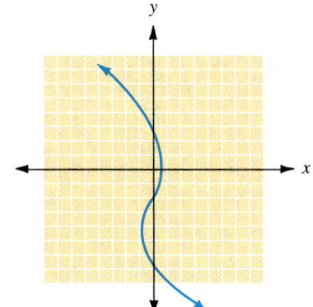

(b)
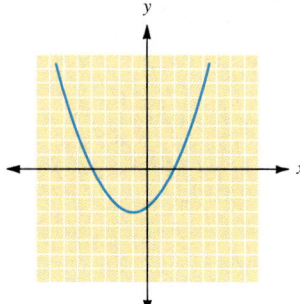

(c)
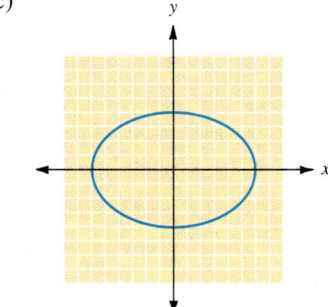

We used the term *function* several times in this chapter. We identified functions, looked at a function machine, used function notation, and found the domain and range for a function. But how does all this relate to the equations in two variables that we studied before this chapter? When is y the same as $f(x)$?

Anytime we solved a linear equation for y, such as

$$y = 3x - 2$$

y was a function of x. The x is considered the **independent variable,** and the y is considered the **dependent variable.** This means that y changes because x has changed. Let's look at some examples of variables that are related and determine which is the dependent variable.

Example 5

Identifying the Dependent Variable

From each pair, identify which variable is dependent on the other.

(a) The age of a car and its resale value.

The value depends on the age, so we would assign the age of the car the independent variable (x) and the value the dependent variable (y).

(b) The amount of interest earned in a bank account and the amount of time the money has been in the bank.

If you think about it, you will see that time will be the independent variable in most ordered pairs. Most everything depends on time rather than the reverse.

The interest depends on the time, so interest is the dependent variable (y) and time is the independent variable (x).

(c) The number of cigarettes one has smoked and the chance of dying from a smoking-related disease.

The number of cigarettes is the independent variable (x), and the chance of dying from a smoking-related disease is the dependent variable (y).

✓ CHECK YOURSELF 5

From each pair, identify which variable is dependent on the other.

(a) The number of credits taken and the amount of tuition paid.

(b) The temperature of a cup of coffee and the length of time since it was poured.

✓ CHECK YOURSELF ANSWERS

1. (a) Is a function; (b) is a function; (c) is not a function.
2. (a) Is a function; (b) is not a function.
3. (a) Is a function; (b) is not a function; (c) is a function.
4. (a) Is not a function; (b) is a function; (c) is not a function.
5. (a) Tuition is dependent on credits taken; (b) the temperature is dependent on the time since the coffee was poured.

Exercises - 4.3

1. Function
2. Function
3. Function
4. Function
5. Not a function
6. Not a function
7. Not a function
8. Not a function
9. Function
10. Function
11. Not a function
12. Not a function
13. Function
14. Function
15. Function
16. Function

In Exercises 1 to 8, determine which of the relations are also functions.

1. {(1, 6), (2, 8), (3, 9)} **2.** {(2, 3), (3, 4), (5, 9)}

3. {(−1, 4), (−2, 5), (−3, 7)} **4.** {(−2, 1), (−3, 4), (−4, 6)}

5. {(1, 3), (1, 2), (1, 1)} **6.** {(2, 4), (2, 5), (3, 6)}

7. {(−1, 1), (2, 1), (2, 3)} **8.** {(2, −1), (3, 4), (3, −1)}

In Exercises 9 to 14, decide whether the relation is a function in each table of values.

9.

x	y
3	1
−2	4
5	3
−7	4

10.

x	y
−2	3
1	4
5	6
2	−1

11.

x	y
2	3
4	2
2	−5
−6	−3

12.

x	y
1	5
3	−6
1	−5
−2	−9

13.

x	y
−1	2
3	6
6	2
−9	4

14.

x	y
4	−6
2	3
−7	1
−3	−6

In Exercises 15 to 20, for each set of ordered pairs, plot the related points. Then use the vertical line test to determine which sets are functions.

15. {(−3, 1), (−1, 2), (−2, 3), (1, 4)}   **16.** {(2, 2), (1, 1), (3, 3), (4, 5)}

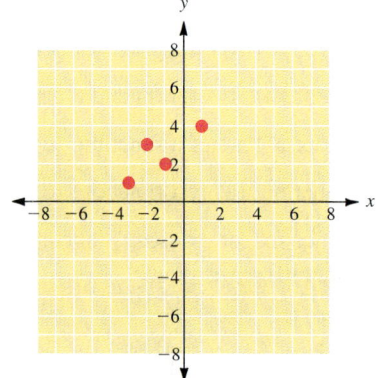
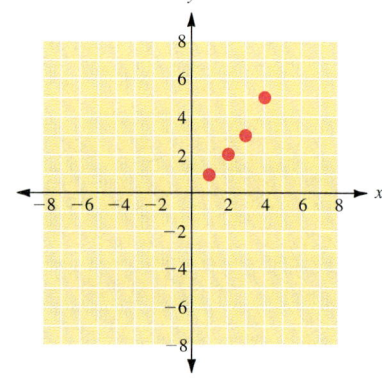

17. Function

18. Function

19. Not a function

20. Not a function

21. Function

22. Not a function

17. $\{(-1, 1), (2, 2), (3, 4), (5, 6)\}$

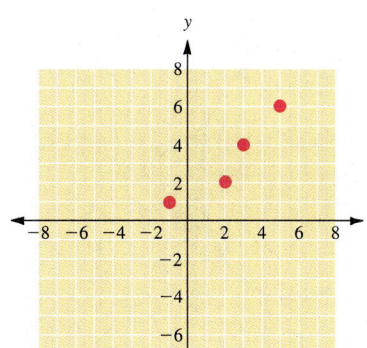

18. $\{(1, 4), (-1, 5), (0, 2), (2, 3)\}$

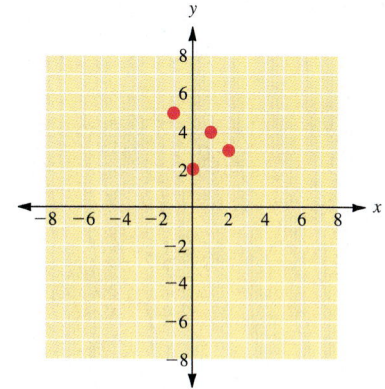

19. $\{(1, 2), (1, 3), (2, 1), (3, 1)\}$

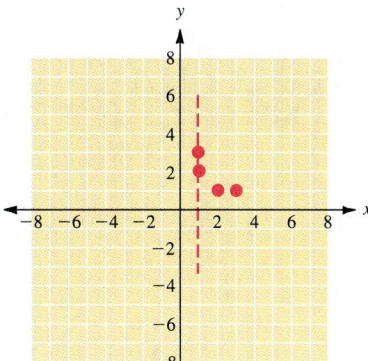

20. $\{(-1, 1), (3, 4), (-1, 2), (5, 3)\}$

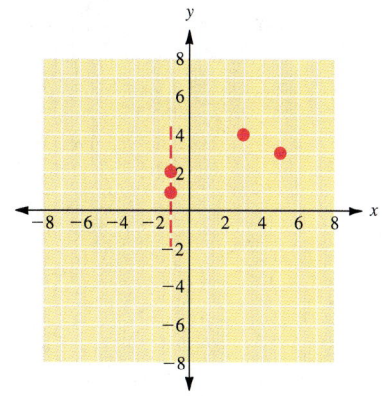

For Exercises 21 to 28, use the vertical line test to determine whether the graphs represent a function.

21.

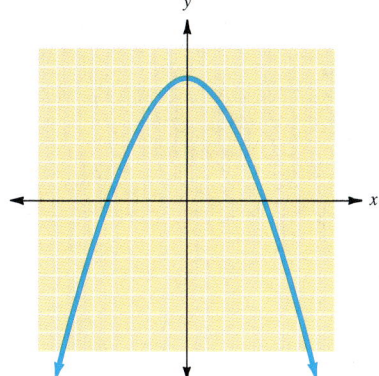

22.

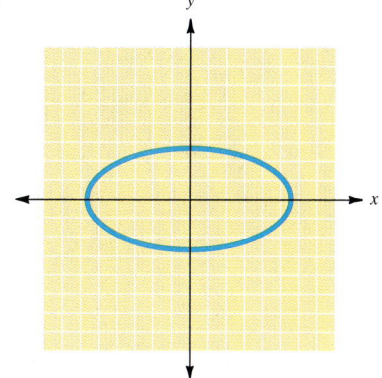

23. Not a function

24. Not a function

25. Function

26. Function

27. Function

28. Function

23.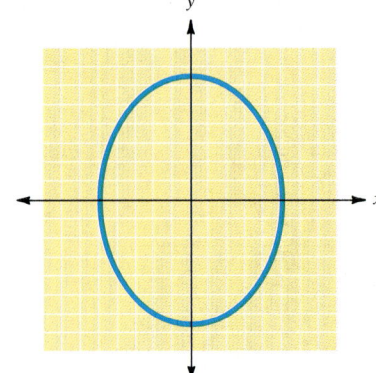

24.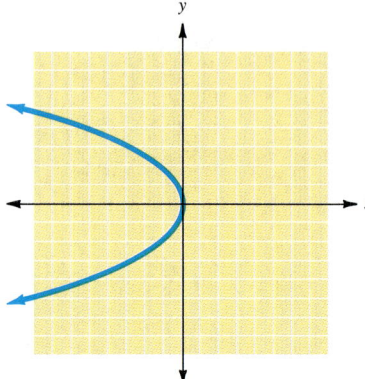

25.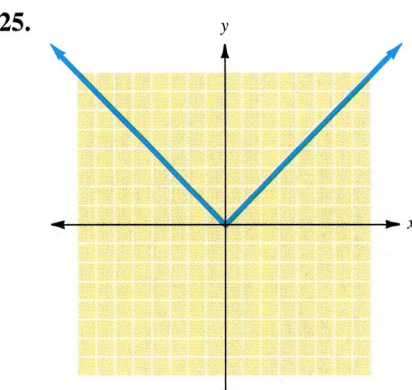

26.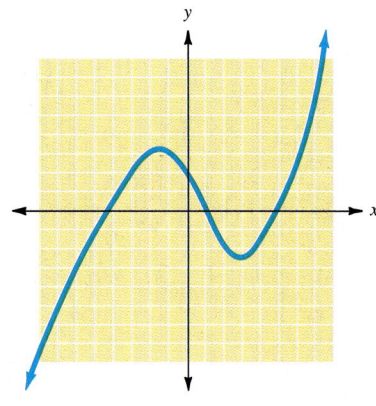

27.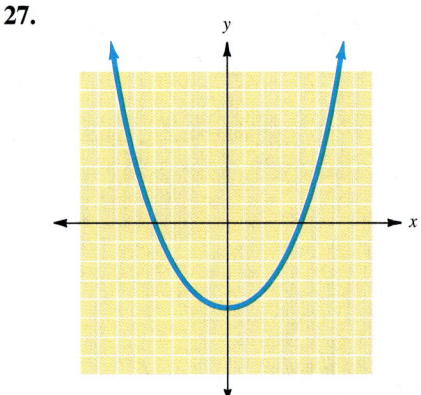

28.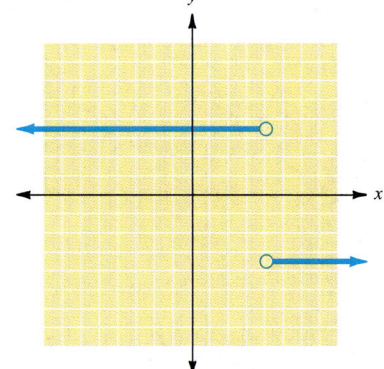

29. (a) D: $-2 < x \leq 2$;
    R: $-1 \leq y \leq 2$
    (b) Yes
    (c) Answers will vary

30. Independent: length of call; dependent: amount of bill

31. Independent: size of tank; dependent: cost

32. Independent: time in air; dependent: height of ball

33. Independent: length of time; dependent: amount of penalty

34. Independent: number of credits; dependent: time to graduate

35. Independent: length of winter; dependent: amount of snowfall

36. Not every relation is a function, but every function is a relation.

29. Consider the following graph.

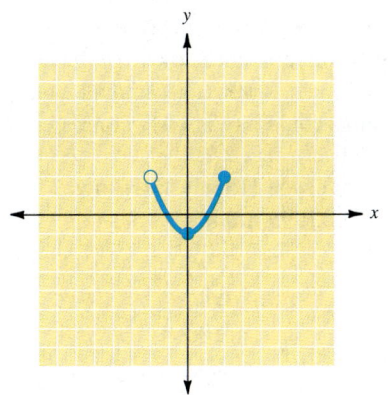

(a) Identify the domain and range of the relation whose graph is given.

(b) Does this graph represent a function? Explain your answer.

(c) How do you use the graph to determine the domain and range of the relation it represents?

In Exercises 30 to 35, from each pair, identify which variable is dependent and which is independent.

30. The amount of a phone bill and the length of the call.

31. The cost of filling a car's gas tank and the size of the tank.

32. The height of a ball thrown in the air and the time in the air.

33. The amount of penalty on an unpaid tax bill and the length of the time unpaid.

34. The length of time needed to graduate from college and the number of credits taken per semester.

35. The amount of snowfall in Boston and the length of the winter.

36. Are all relations functions? Are all functions relations? Explain your answer.

37. The following table shows the average hourly earnings for blue-collar workers from 1947 to 1993. These figures are given in "real" wages, which means that the *purchasing* power of the money is given rather than the actual dollar amount. In other words, the amount earned for 1947 is not the actual amount listed here; in fact, it was much lower. The amount you see here is the amount in dollars that 1947 earnings could buy in 1947 compared to what 1993 wages could buy in 1993.

Year	Average Hourly Earnings (in 1993 dollars)
1947	$ 6.75
1967	10.67
1973	12.06
1979	12.03
1982	11.61
1989	11.26
1991	10.95
1993	10.83

Make a Cartesian coordinate graph of this data, using the year as the domain and the hourly earnings as the range. You will have to decide how to set up the axes so that the data all fit on the graph nicely. (*Hint:* Do not start the year at 0!) In complete sentences, answer the following questions: What are the trends that you notice from reading the table? What additional information does the graph show? Is this relation a function? Why or why not?

SECTION 4.4 Reading Values from a Graph

4.4 OBJECTIVES

1. Given x, find the function value on the graph
2. Given a function value, find the related x value
3. Find the x and y intercepts from a graph

"Reading furnishes the mind only with materials of knowledge; it is thinking that makes what we read ours."

–John Locke

In Section 3.2, we learned to read the coordinates of a point by drawing a vertical line from the point to the *x* axis to find the *x* coordinate and then drawing a horizontal line from the point to the *y* axis to find the *y* coordinate. A graph of a curve (including a graph of a straight line) is actually the graph of an infinite number of connected points. Finding the coordinates of any point on a curve is exactly the same as finding the coordinates of a point.

Keep in mind that although we usually say something like, "Find the coordinates of the point . . . ," every time we read a graph we are able to only *estimate* the coordinates.

Example 1

Reading Values from a Graph

Find the coordinates of the labeled points.

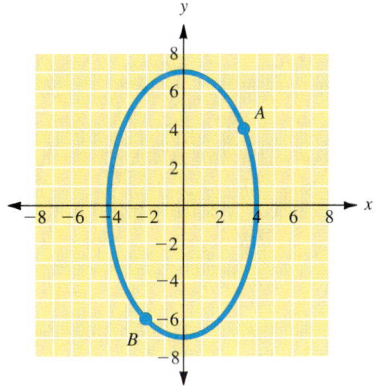

Point *A* has an *x* coordinate of 3 and a *y* coordinate of 4. Point *A* represents the ordered pair (3, 4). Point *B* represents the ordered pair (−2, −6).

✓ **CHECK YOURSELF 1**

Find the coordinates of the labeled points.

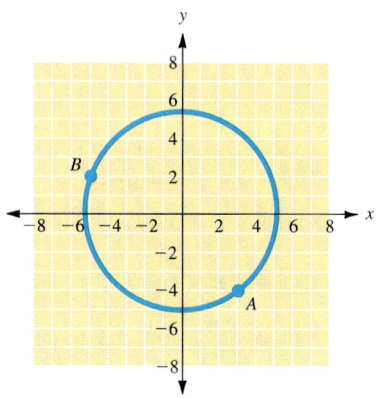

Reading Function Values from Graphs

If a graph is the graph of a function, then every ordered pair (x, y) can be thought of as $(x, f(x))$.

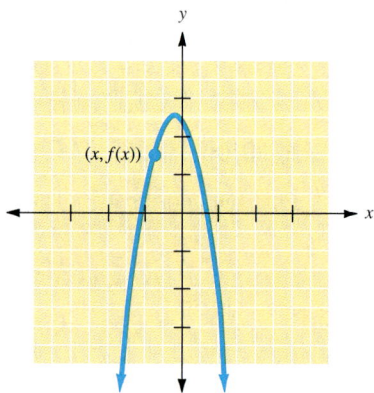

For a specific value of x, let's call it a, we can find $f(a)$ with the following algorithm.

Step 1 Draw a vertical line through a on the x axis.
Step 2 Find the point of intersection of that line with the graph.
Step 3 Draw a horizontal line through the graph at that point.
Step 4 Find the intersection of the horizontal line with the y axis.
Step 5 $f(a)$ is that y value.

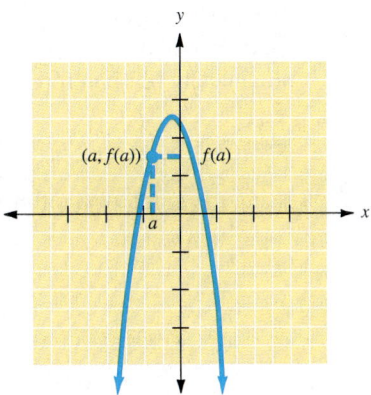

Example 2 illustrates this algorithm.

Example 2 — Finding the Function Value on a Graph Given x

Consider the following graph of the function f. Use the graph to estimate $f(2)$.

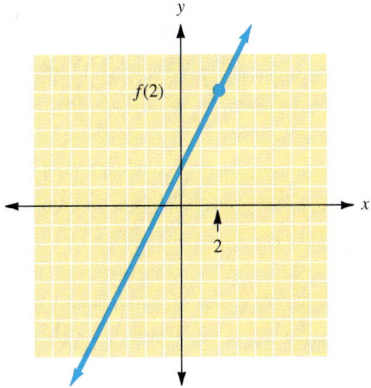

$f(2)$ is a y value. It is the y value that is paired with an x value of 2. Locate the number 2 on the x axis, draw a vertical line to the graph of the function, and then draw a horizontal line to the y axis, as shown below.

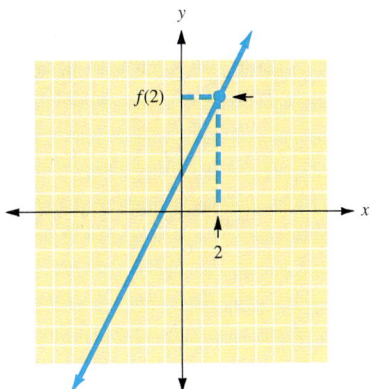

The coordinates of the point are (2, $f(2)$). The y value of the point is $f(2)$. Read the y value of this point on the y axis. It appears that $f(2)$, the y value of the point, is approximately 6. Therefore, $f(2) = 6$.

✓ CHECK YOURSELF 2

Using the graph of the function f in Example 2, estimate each of the following.

(a) $f(1)$ (b) $f(-1)$ (c) $f(-3)$

In the preceding problem, you were given the x value and asked to find the corresponding function value or y value. Now you will do the opposite operation. You will be given the function value and then asked to find the corresponding x value(s). Consider Example 3.

Example 3   Finding the x Value from a Graph Given the Function Value

Use the following graph of the function f to find all values of x such that $f(x) = -5$.

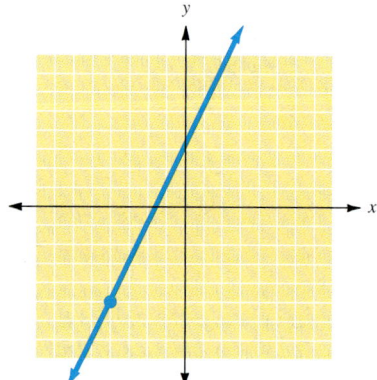

This time -5 is a function value, or y value. Locate -5 on the y axis, and draw a horizontal line to the graph of the function, followed by a vertical line to the x axis, as shown next.

Section 4.4 ■ Reading Values from a Graph   **313**

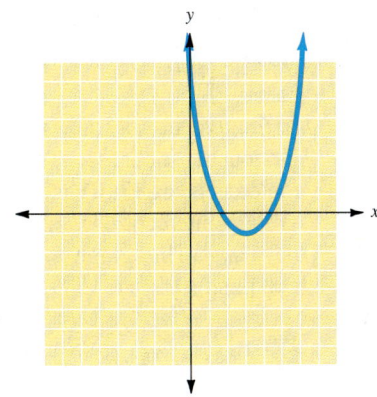

The solution of $f(x) = -5$ is $x = -4$. In particular, $f(-4) = -5$.

✓ CHECK YOURSELF 3

Use the following graph to find all values of x such that

(a) $f(x) = 1$ (b) $f(x) = 7$ (c) $f(x) = -1$

If given the function value, we can find the associated x value by using the following algorithm.

Finding x Values from Function Values

Step 1 Find the given function value on the y axis.
Step 2 Draw a horizontal line through that point.
Step 3 Find every point on the graph that intersects the horizontal line.
Step 4 Draw a vertical line through each of those points of intersection.
Step 5 The x value(s) are each point of intersection of the vertical lines and the x axis.

Reading x and y Intercepts from Graphs

Among the most important values that can be read from graphs are the values of the *x* and *y* intercepts.

Example 4

Finding x and y Intercepts

Find the *x* and *y* intercepts from the graph.

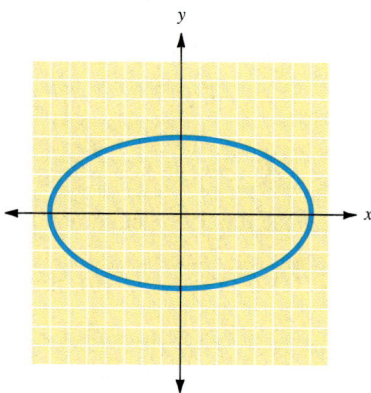

An *x* intercept is a value of *x* for which $y = 0$. It is an *x* value of any point on the graph that touches the *x* axis. This graph touches the *x* axis at (7, 0) and also at (−7, 0). The *x* intercepts are (7, 0) and (−7, 0).

A *y* intercept is a value of *y* when $x = 0$. It is a *y* value of any point that touches the *y* axis. This graph touches the *y* axis at (0, 4) and (0, −4). The *y* intercepts are (0, 4) and (0, −4).

✓ CHECK YOURSELF 4

Find the *x* and *y* intercepts from the graph.

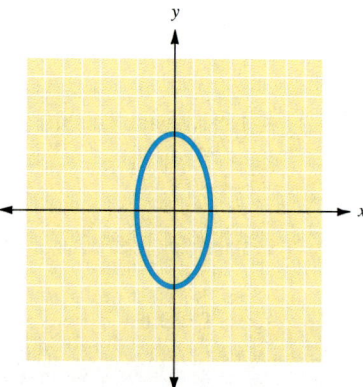

Section 4.4 ■ Reading Values from a Graph

As we did with the lines in Section 4.1, we can find the domain and range from a graph.

Example 5

Finding Domain and Range

Find the domain and range of the graph in Example 4.

The domain is the set of all x values for which there is an ordered pair associated. In this graph, every x value between -7 and 7 is associated with at least one ordered pair, so

$$D = \{x | -7 \leq x \leq 7\}$$

This is read, "The domain consists of every x where x is between -7 and 7, including those values."

The range is the set of y values for which there is an ordered pair associated. In this graph, every y value between -4 and 4 has an ordered pair associated, so

$$R = \{y | -4 \leq y \leq 4\}$$

✓ CHECK YOURSELF 5

Find the domain and range of the graph in Check Yourself 4.

✓ CHECK YOURSELF ANSWERS

1. (a) $(3, -4)$; (b) $(-5, 2)$. **2.** (a) $f(1) = 4$; (b) $f(-1) = 0$; (c) $f(-3) = -4$.
3. (a) $x = 1, 5$; (b) $x = 0, 6$; (c) $x = 3$. **4.** x intercept: $(-2, 0), (2, 0)$; y intercept: $(0, -4), (0, 4)$. **5.** $D = \{x | -2 \leq x \leq 2\}$; $R = \{y | -4 \leq y \leq 4\}$.

Exercises · 4.4

In Exercises 1 to 12, find the coordinates of the labeled points. Assume that each small square is a 1-unit square.

1. A (3, 3); B (2, −4)
2. A (2, 4); B (−6, −2)
3. A (2, 5); B (−2, −4)
4. A (6, 2); B (1, −2)
5. A (0, 5); B (3, 0)
6. A (5, 4); B (0, 5)

1.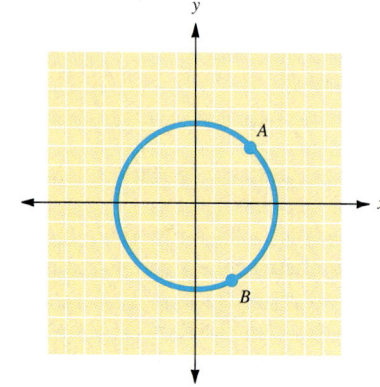

2.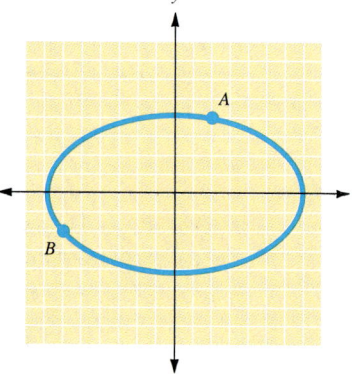

3.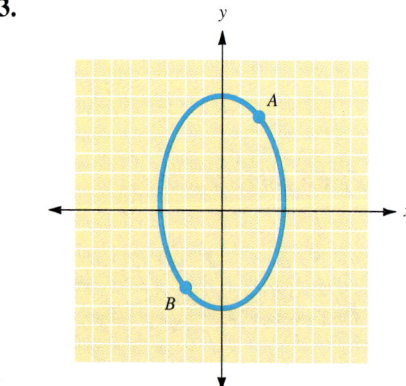

4.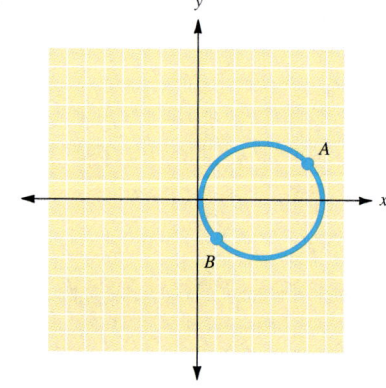

5.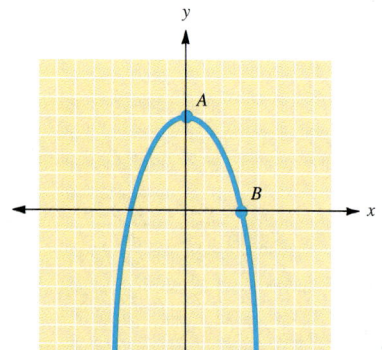

6.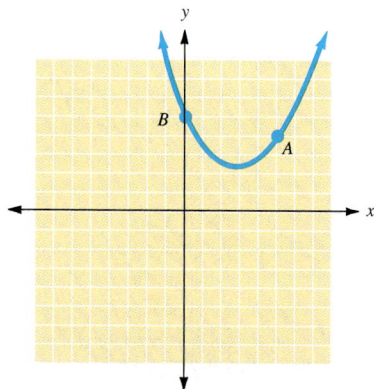

7. A (2, 0); B (6, 4)
8. A (3, 2); B (−3, 2)
9. A (3, 3); B (−3, −3)
10. A (−6, 5); B (−2, 0)
11. A (3, 6); B (3, 0)
12. A (6, 4); B (−5, 2)

7.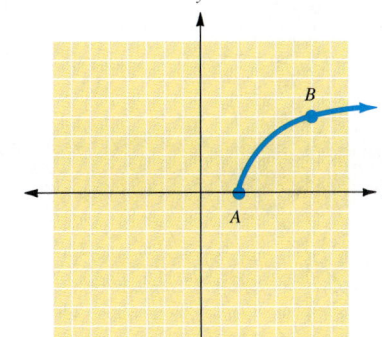

8.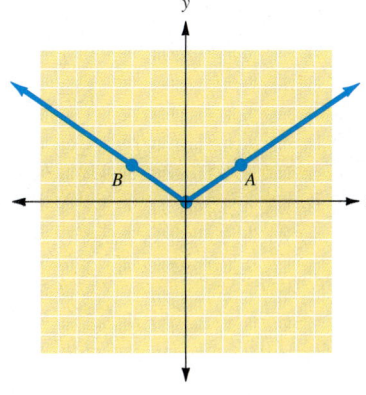

9.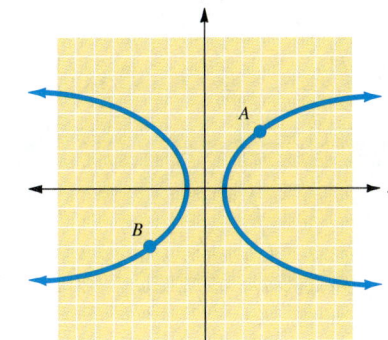

10.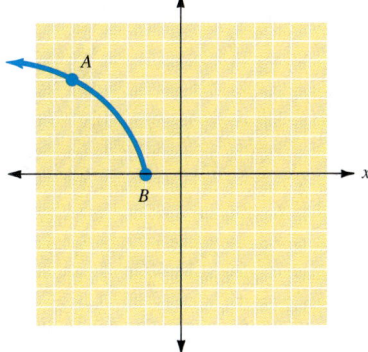

11.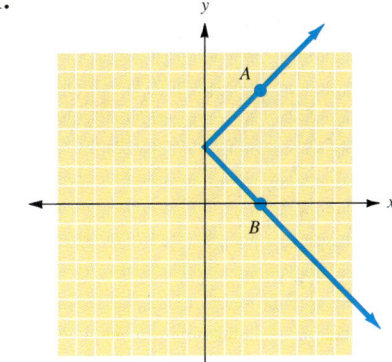

12.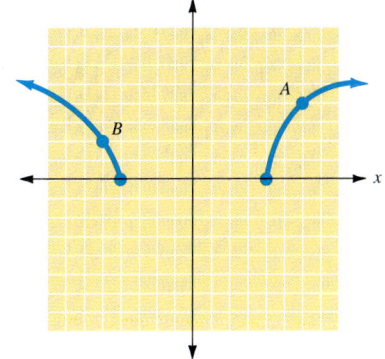

13. (a) 3; (b) 1; (c) 2; (d) 5; (e) 0

14. (a) 0; (b) 2; (c) 1; (d) −2; (e) 3

15. (a) 0.5; (b) 0.5; (c) 0; (d) 4; (e) 1.5

16. (a) 1; (b) 1; (c) 0; (d) 3; (e) 2

17. (a) 1; (b) 3; (c) 2; (d) 1; (e) 4

18. (a) 4.1; (b) 3.9; (c) 4; (d) 4.5; (e) 3.5

In Exercises 13 to 20, use the graph of the function to estimate each of the following values: (a) $f(1)$, (b) $f(-1)$, (c) $f(0)$, (d) $f(3)$, and (e) $f(-2)$

13.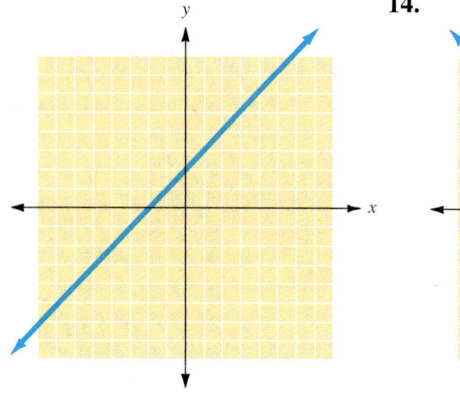

14.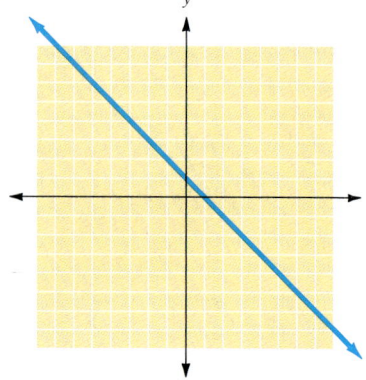

15.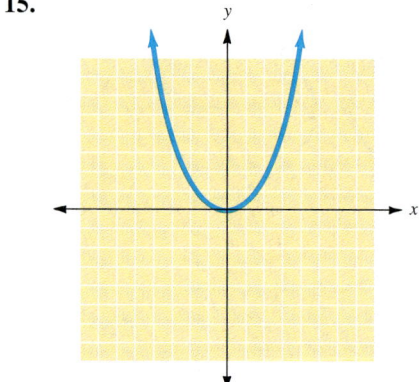

16.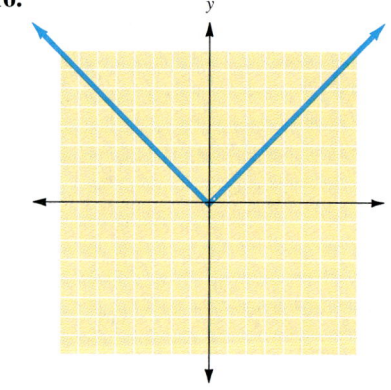

17.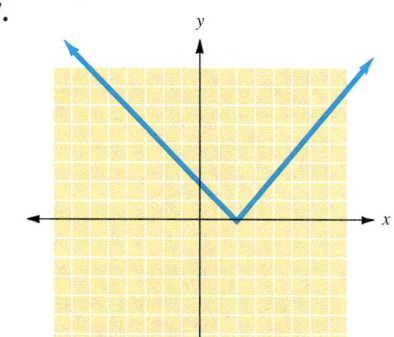

18.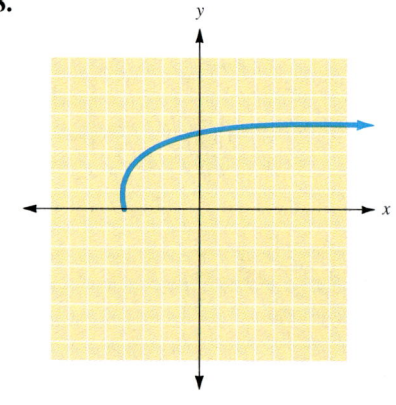

Section 4.4 ■ Reading Values from a Graph   **319**

19. (a) 3; (b) 3; (c) 3; (d) 3; (e) 3

20. (a) −3; (b) −3; (c) −3; (d) −3; (e) −3

21. (a) 1; (b) 2; (c) 4

22. (a) −1; (b) 0; (c) 2

23. (a) 2, −2; (b) 3, −3; (c) 4.5, −4.5

24. (a) −5, 5; (b) −4, 4; (c) −3, 3

19.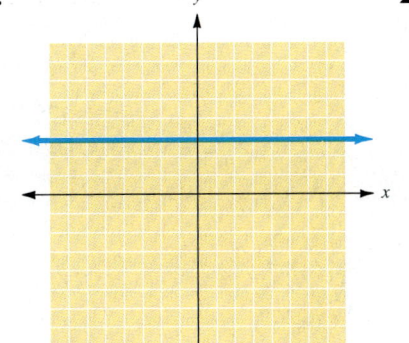

20.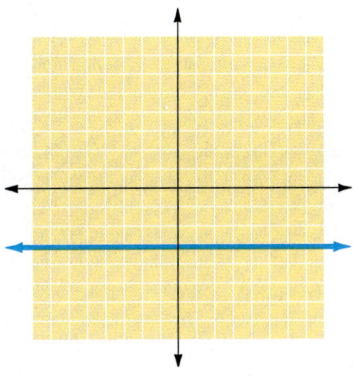

In Exercises 21 to 28, use the graph of $f(x)$ to find all values of x such that (a) $f(x) = -1$, (b) $f(x) = 0$, and (c) $f(x) = 2$.

21. **22.**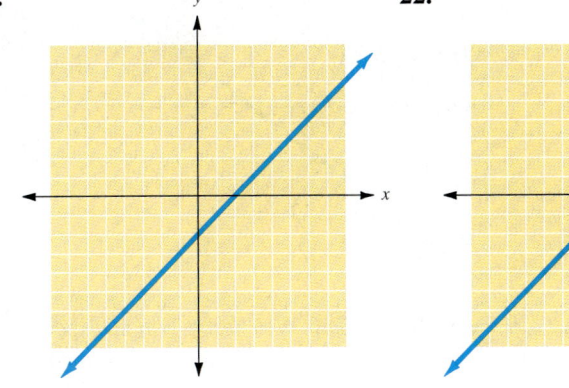

23. 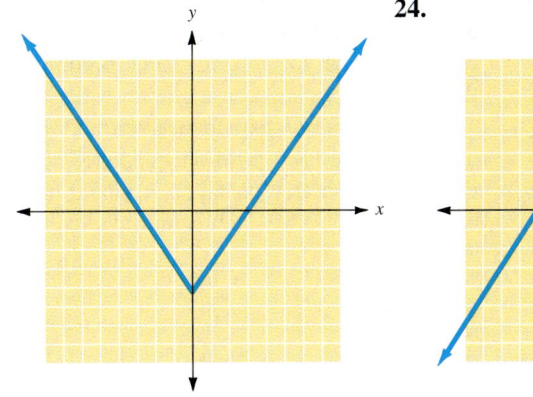 **24.**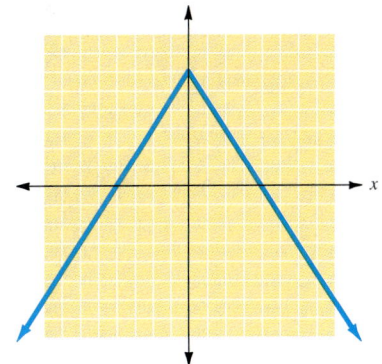

25. (a) −1.5, 1.5;
    (b) 1, −1; (c) 0

26. (a) None; (b) none;
    (c) none

27. (a) 1.5; (b) 2.5; (c) 5.5

28. (a) 0; (b) −2, 2;
    (c) −4, 4

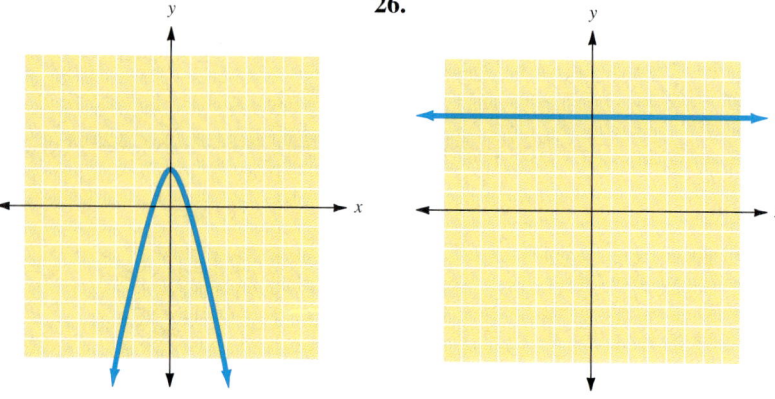

25.

26.

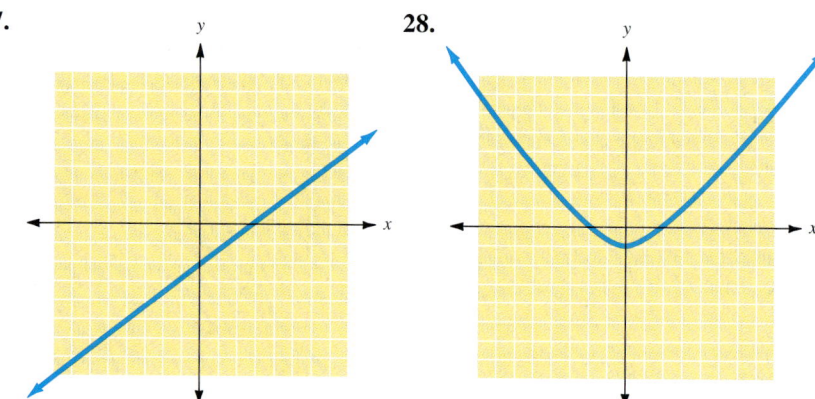

27.

28.

29. Your friend Sam Weatherby is a salesperson and has just been offered two new jobs, one for $600 a month plus 9% of the amount of all his sales over $20,000. The second job offer is $1000 a month plus 3% of the amount of his sales.

Sam and his spouse are about to have a child, and he feels that he has to make $2500 a month just to make ends meet. He has called you to ask for your help in deciding which job to take. To help him picture his options, graph both offers on the same graph, and add a graph of the income of $2500.

Next, write an explanation that answers these questions: How much does he have to sell in each position to earn $2500? When is the first offer better? When is the second better? What sales would he have made to make less than $2000 in each position? Which job should he take?

SECTION 4.5 Operations on Functions

4.5 OBJECTIVES

1. Find the sum or difference of two functions
2. Find the product of two functions
3. Find the quotient of two functions
4. Find the domain of the sum or difference of two functions

In Section 4.2, we first examined the concept of the function. In this section, we will look at operations on functions. For example, the total profit that a company makes on an item is determined by subtracting the cost of making the item from the revenue the company receives in selling the item. This can be written as

$$P(x) = R(x) - C(x)$$

Many applications of functions involve the combining of two or more component functions. In this section, we will look at several properties that allow for the addition, subtraction, multiplication, and division of functions.

The **sum of two functions** f and g is written as $f + g$ and can be defined as

$$f + g: (f + g)(x) = f(x) + g(x)$$

for every value of x that is in the domain of both functions f and g.

The **difference of two functions** f and g is written as $f - g$ and can be defined as

$$f - g: (f - g)(x) = f(x) - g(x)$$

for every value of x that is in the domain of both functions f and g.

Example 1 Finding the Sum or Difference of Two Functions

Given the functions $f(x) = 2x - 1$ and $g(x) = -3x + 4$, (a) find $f + g$, (b) find $f - g$, (c) find $(f + g)(2)$.

(a) $(f + g)(x) = f(x) + g(x)$
$$= (2x - 1) + (-3x + 4) = -x + 3$$

(b) $(f - g)(x) = f(x) - g(x)$
$$= (2x - 1) - (-3x + 4) = 5x - 5$$

(c) If we use the definition of the sum of two functions, we find that

$$(f + g)(2) = f(2) + g(2)$$
$$= 3 + (-2) = 1$$

As an alternative, we could use part a and say

$$(f + g)(x) = -x + 3$$

Therefore,

$$(f + g)(2) = -2 + 3$$
$$= 1$$

✓ CHECK YOURSELF 1

Given the functions $f(x) = -2x - 3$ and $g(x) = 5x - 1$,

(a) find $f + g$.   (b) find $f - g$.   (c) find $(f + g)(2)$.

In defining the sum of two functions, we indicated that the domain was determined by the domain of both functions. We find the domain in Example 2.

Example 2

Finding the Domain of the Sum or Difference of Functions

Given $f(x) = 2x - 4$ and $g(x) = \dfrac{1}{x}$, (a) find $f + g$, (b) find the domain of $f + g$.

(a) $(f + g)(x) = (2x - 4) + \dfrac{1}{x} = 2x - 4 + \dfrac{1}{x}$

(b) The domain of $f + g$ is the set of all numbers in the domain of f and also in the domain of g. The domain of f consists of all real numbers. The domain of g consists of all real numbers except 0 because we cannot divide by 0. The domain of $f + g$ is the set of all real numbers except 0.

✓ CHECK YOURSELF 2

Given $f(x) = -3x + 1$ and $g(x) = \dfrac{1}{x - 2}$,

(a) find $f + g$.   (b) find the domain of $f + g$.

> The **product of two functions** f and g is written as $f \cdot g$ and can be defined as
>
> $$f \cdot g: (f \cdot g)(x) = f(x) \cdot g(x)$$
>
> for every value of x that is in the domain of both functions f and g.

Example 3 Finding the Product of Two Functions

Given $f(x) = x - 1$ and $g(x) = x + 5$, find $f \cdot g$.

$$(f \cdot g)(x) = f(x) \cdot g(x) = (x - 1)(x + 5)$$

In the next chapter, we will learn how to expand the multiplication.

✓ CHECK YOURSELF 3

Given $f(x) = x - 3$ and $g(x) = x + 2$, find $f \cdot g$.

The final operation on functions that we will look at involves the division of two functions.

> The **quotient of two functions** f and g is written as $f \div g$ and can be defined as
>
> $$f \div g: (f \div g)(x) = f(x) \div g(x)$$
>
> for every value of x that is in the domain of both functions f and g, where $g(x) \neq 0$.

Example 4 Finding the Quotient of Two Functions

Given $f(x) = x - 1$ and $g(x) = x + 5$, (a) find $f \div g$, (b) find the domain of $f \div g$.

(a) $(f \div g)(x) = f(x) \div g(x) = (x - 1) \div (x + 5) = \dfrac{x - 1}{x + 5}$

(b) The domain is the set of all real numbers except -5 because $g(-5) = 0$, and division by 0 is undefined.

✓ CHECK YOURSELF 4

Given $f(x) = x - 3$ and $g(x) = x + 2$,

(a) find $f \div g$. (b) find the domain for $f \div g$.

✓ CHECK YOURSELF ANSWERS

1. (a) $3x - 4$; (b) $-7x - 2$; (c) 2. **2.** (a) $-3x + 1 + \dfrac{1}{(x-2)}$;

(b) $D = \{x | x \neq 2\}$. **3.** $(x - 3)(x + 2)$ **4.** (a) $\dfrac{(x-3)}{(x+2)}$;

(b) $D = \{x | x \neq -2\}$.

Exercises • 4.5

In Exercises 1 to 4, find (a) $f + g$, (b) $f - g$, (c) $(f + g)(3)$, and (d) $(f - g)(2)$.

1. $f(x) = -4x + 5 \quad g(x) = 7x - 4$

2. $f(x) = 9x - 3 \quad g(x) = -3x + 5$

3. $f(x) = 8x - 2 \quad g(x) = -5x + 6$

4. $f(x) = -7x + 9 \quad g(x) = 2x - 1$

In Exercises 5 to 8, find (a) $f + g$ and (b) the domain of $f + g$.

5. $f(x) = -9x + 11 \quad g(x) = 15x - 7$

6. $f(x) = -11x + 3 \quad g(x) = 8x - 5$

7. $f(x) = 3x + 2 \quad g(x) = \dfrac{1}{x - 2}$

8. $f(x) = -2x + 5 \quad g(x) = \dfrac{3}{x + 1}$

In Exercises 9 to 12, find (a) $f \cdot g$, (b) $\dfrac{f}{g}$, and (c) the domain of $\dfrac{f}{g}$.

9. $f(x) = 2x - 1 \quad g(x) = x - 3$

10. $f(x) = -x + 3 \quad g(x) = x + 4$

11. $f(x) = 3x + 2 \quad g(x) = 2x - 1$

12. $f(x) = -3x + 5 \quad g(x) = -x + 2$

The velocity, $V(t)$, of a freely falling object is the sum of two functions: the initial velocity, V_0, with which it is thrown, and the acceleration, $a(t)$, which is the change in velocity due to gravity, such that

$$V(t) = V_0 + a(t)$$

In Exercises 13 and 14, find the velocity at any time.

13. $V_0 = 10 \text{ m/s} \quad a(t) = -4.9t^2$

14. $V_0 = 64 \text{ ft/s}^2 \quad a(t) = -16t^2$

---

1. (a) $3x + 1$; (b) $-11x + 9$; (c) 10; (d) -13
2. (a) $6x + 2$; (b) $12x - 8$; (c) 20; (d) 16
3. (a) $3x + 4$; (b) $13x - 8$; (c) 13; (d) 18
4. (a) $-5x + 8$; (b) $-9x + 10$; (c) -7; (d) -8
5. (a) $6x + 4$; (b) all reals
6. (a) $-3x - 2$; (b) all reals
7. (a) $3x + 2 + \dfrac{1}{x - 2}$; (b) $\{x | x \neq 2\}$
8. (a) $-2x + 5 + \dfrac{3}{x + 1}$; (b) $\{x | x \neq -1\}$
9. (a) $(2x - 1)(x - 3)$; (b) $\dfrac{2x - 1}{x - 3}$; (c) $\{x | x \neq 3\}$
10. (a) $(-x + 3)(x + 4)$; (b) $\dfrac{-x + 3}{x + 4}$; (c) $\{x | x \neq -4\}$
11. (a) $(3x + 2)(2x - 1)$; (b) $\dfrac{3x + 2}{2x - 1}$; (c) $\left\{x \big| x \neq \dfrac{1}{2}\right\}$
12. (a) $(-3x + 5)(-x + 2)$; (b) $\dfrac{-3x + 5}{-x + 2}$; (c) $\{x | x \neq 2\}$
13. $V = 10 - 4.9t^2$
14. $V(t) = 64 - 16t^2$

Summary Exercises • 4

This summary exercise set is provided to give you practice with each of the objectives in the chapter. Each exercise is keyed to the appropriate chapter section.

[4.1] In Exercises 1 and 2, identify which are ordered pairs.

1. (a) (2, 1)   (b) {3, 4}   (c) 1, 4   (d) (−4, −3)   (e) ((3, 2), 5)
2. (a) {−1, 4}   (b) 6, 8   (c) (3, 4)   (d) {(3, −1), 4}   (e) (−2, 5)

[4.1] In Exercises 3 to 10, for each set of ordered pairs, identify the domain and range.

3. {(3, 5), (4, 6), (1, 2), (8, 1), (7, 3)}
4. {(−1, 3), (−2, 5), (3, 7), (1, 4), (2, −2)}
5. {(1, 3), (1, 5), (1, 7), (1, 9), (1, 10)}
6. {(2, 4), (−1, 4), (−3, 4), (1, 4), (6, 4)}
7. $\{(x, y) | x + y = 4\}$
8. $\{(x, y) | 3x + 2y = 6\}$
9. $\{(x, y) | y = 5\}$
10. $\{(x, y) | x = 8\}$

[4.2] In Exercises 11 to 16, evaluate $f(x)$ for the value specified.

11. $f(x) = x^2 - 3x + 5$; find (a) $f(0)$, (b) $f(-1)$, and (c) $f(1)$.
12. $f(x) = -2x^2 + x - 7$; find (a) $f(0)$, (b) $f(2)$, and (c) $f(-2)$.
13. $f(x) = x^3 - x^2 - 2x + 5$; find (a) $f(-1)$, (b) $f(0)$, and (c) $f(2)$.
14. $f(x) = -x^2 + 7x - 9$; find (a) $f(-3)$, (b) $f(0)$, and (c) $f(1)$.
15. $f(x) = 3x^2 - 5x + 1$; find (a) $f(-1)$, (b) $f(0)$, and (c) $f(2)$.
16. $f(x) = x^3 + 3x - 5$; find (a) $f(2)$, (b) $f(0)$, and (c) $f(1)$.

[4.2] In Exercises 17 to 20, rewrite each equation as a function of x.

17. $y = 4x + 7$
18. $y = -7x - 3$
19. $4x + 5y = 40$
20. $-3x - 2y = 12$

Answers

1. (a) and (d)
2. (c) and (e)
3. D: {1, 3, 4, 7, 8}; R: {1, 2, 3, 5, 6}
4. D: {−2, −1, 1, 2, 3}; R: {−2, 3, 4, 5, 7}
5. D: {1}; R: {3, 5, 7, 9, 10}
6. D: {−3, −1, 1, 2, 6}; R: {4}
7. D: all real numbers; R: all real numbers
8. D: all real numbers; R: all real numbers
9. D: all real numbers; R: {5}
10. D: {8}; R: all real numbers
11. (a) 5, (b) 9, (c) 3
12. (a) −7; (b) −13; (c) −17
13. (a) 5; (b) 5; (c) 5
14. (a) −39; (b) −9; (c) −3
15. (a) 9; (b) 1; (c) 3
16. (a) −7; (b) −5; (c) −3
17. $f(x) = 4x + 7$
18. $f(x) = -7x - 3$
19. $f(x) = -\dfrac{4}{5}x + 8$
20. $f(x) = -\dfrac{3}{2}x - 6$

■ Summary Exercises 4

[4.2] In Exercises 21 to 26, graph the function.

21. $f(x) = 2x + 5$ **22.** $f(x) = 3x - 6$

23. $f(x) = -5x + 6$ **24.** $f(x) = -x + 3$

25. $f(x) = -3x - 2$ **26.** $f(x) = -2x + 6$

For Ex. 21-26, see the next page.

[4.2] In Exercises 27 to 42, evaluate each function as indicated.

27. $f(x) = 5x + 3$; find $f(2)$ and $f(0)$.

27. 13, 3

28. $f(x) = -3x + 5$; find $f(0)$ and $f(1)$.

28. 5, 2

29. $f(x) = 7x - 5$; find $f\left(\dfrac{5}{4}\right)$ and $f(-1)$.

29. $\dfrac{15}{4}$, -12

30. $f(x) = -2x + 5$; find $f(0)$ and $f(-2)$.

30. 5, 9

31. $f(x) = -5x + 3$; find $f(a)$, $f(2b)$, and $f(x + 2)$.

31. $-5a + 3$, $-10b + 3$, $-5(x + 2) + 3$

32. $f(x) = 7x - 1$; find $f(a)$, $f(3b)$, and $f(x - 1)$.

32. $7a - 1$, $21b - 1$, $7(x - 1) - 1$

[4.3] In Exercises 33 to 38, determine which relations are also functions.

33. Function

33. $\{(1, 3), (2, 4), (5, -1), (-1, 3)\}$

34. Function

34. $\{(-2, 4), (3, 6), (1, 5), (0, 1)\}$

35. Not a function

35. $\{(1, 2), (0, 4), (1, 3), (2, 5)\}$

36. Function

36. $\{(1, 3), (2, 3), (3, 3), (4, 3)\}$

37. Function

37.

x	y
-3	2
-1	1
0	3
1	4
3	5

38. Function

38.

x	y
-1	3
0	2
1	3
2	4
3	5

[4.3] In Exercises 39 to 42, use the vertical line test to determine whether the graph represents a function.

39. Function

39.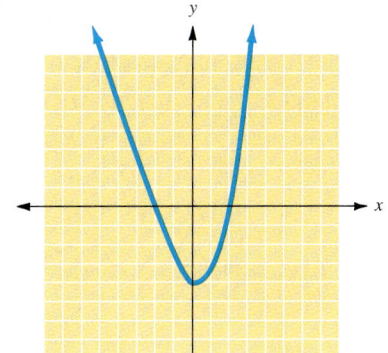

40. Not a function

40.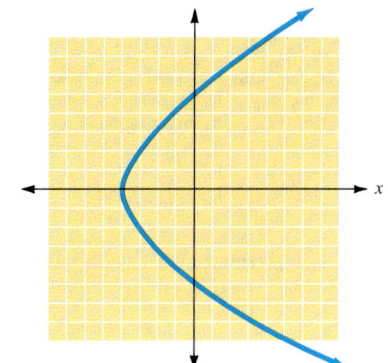

328 Chapter 4 ■ A Beginning Look at Functions

41. Not a function
42. Function
43. A: (3, 4); B: (0, −2)
44. A: (5, 4); B: (−4, −5)
45. A: (3, 6); B: (0, 1)
46. A: (4, 2); B: (−5, −4)

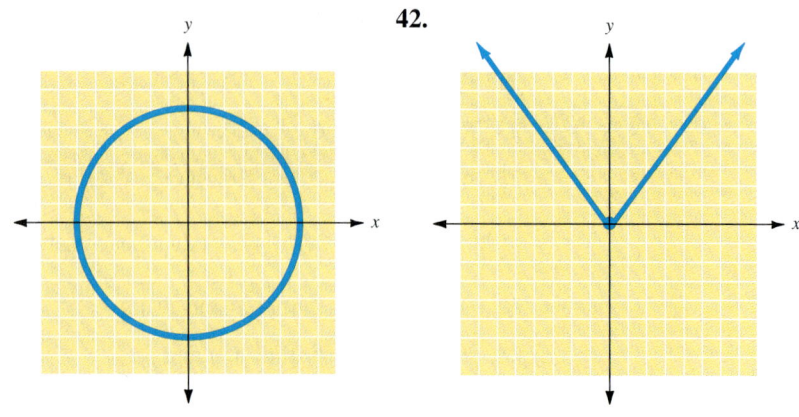

21. $f(x) = 2x + 5$
22. $f(x) = 3x − 6$

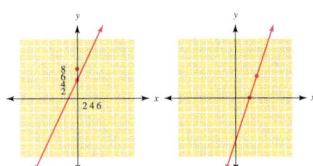

23. $f(x) = −5x + 6$
24. $f(x) = −x + 3$

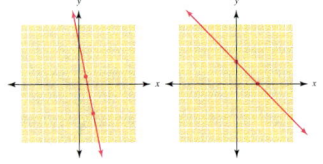

25. $f(x) = −3x − 2$
26. $f(x) = −2x + 6$

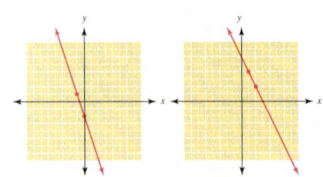

[4.4] In Exercises 43 to 46, find the coordinates of the labeled points.

43.

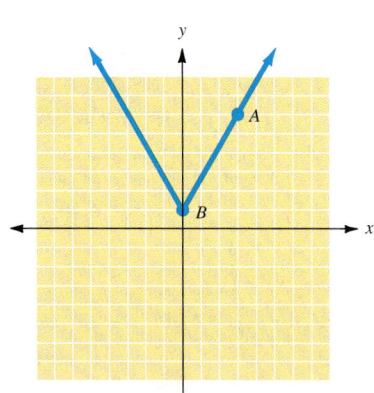

44.

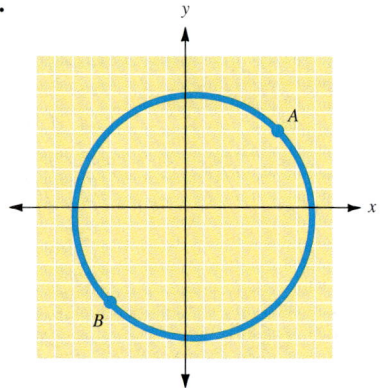

45.

46.

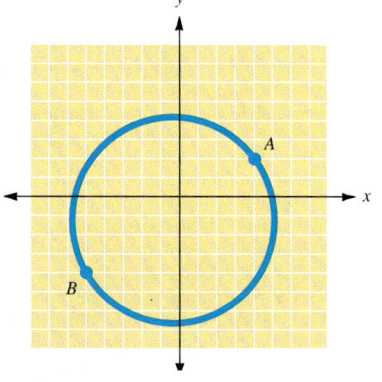

47. (a) −2; (b) 0; (c) 2
48. (a) 5; (b) 0; (c) −5
49. (a) 1; (b) −3; (c) 1
50. (a) 0; (b) 4; (c) 0
51. (a) 3; (b) 0; (c) −3
52. (a) −5, 5; (b) 5; (c) −3.5, 3.5

[4.4] In Exercises 47 to 50, use the graph to estimate (a) $f(-2)$, (b) $f(0)$, and (c) $f(2)$.

47. 48.

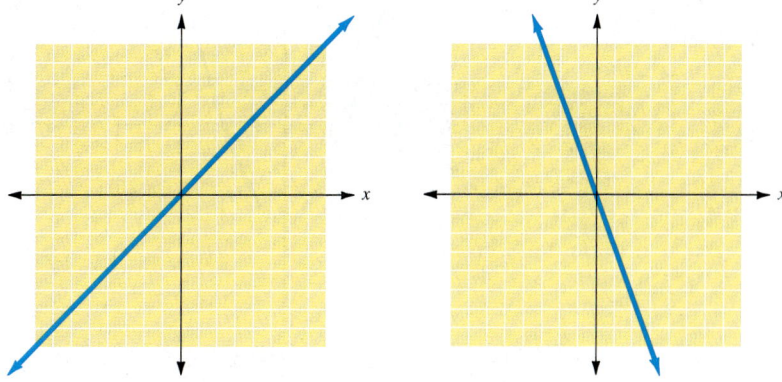

49. 50.

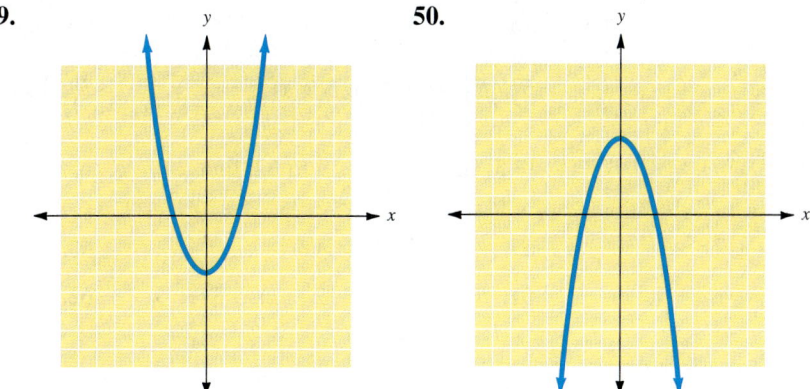

[4.4] In Exercises 51 to 54, use the graph of $f(x)$ to find all values of x such that (a) $f(x) = -1$, (b) $f(x) = 0$, and (c) $f(x) = 1$.

51. 52.

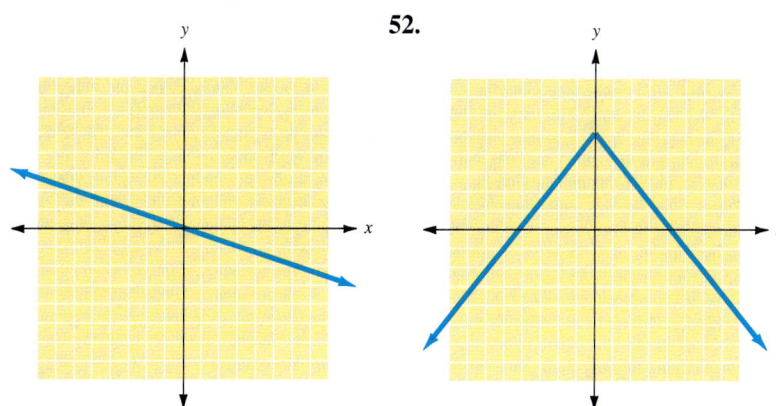

53. (a) −3, 3; (b) −3.5, 3.5; (c) −4, 4

54. (a) No solutions; (b) no solutions; (c) no solutions

55. (a) −6x − 23; (b) 12x + 7; (c) −35; (d) 19; (e) all reals

56. (a) −6x + 17; (b) −10x + 11; (c) 5; (d) 1; (e) all reals

57. (a) $-9x + 5 + \dfrac{3}{x-8}$; (b) $-9x + 5 - \dfrac{3}{x-8}$; (c) −13.5; (d) $\dfrac{-25}{7}$; (e) $\{x \mid x \neq 8\}$

58. (a) $2x + 7 + \dfrac{-4x}{x-9}$; (b) $2x + 7 + \dfrac{4x}{x-9}$; (c) $\dfrac{85}{7}$; (d) $\dfrac{17}{2}$; (e) $\{x \mid x \neq 9\}$

59. (a) $(7x + 3)(x − 2)$; (b) $\dfrac{7x + 3}{x - 2}$; (c) $\{x \mid x \neq 2\}$

60. (a) $(-8x − 3)(x + 11)$; (b) $\dfrac{-8x − 3}{x + 11}$; (c) $\{x \mid x \neq -11\}$

61. (a) $(-9x + 12)(3x − 5)$; (b) $\dfrac{-9x + 12}{3x - 5}$; (c) $\{x \mid x \neq \dfrac{5}{3}\}$

62. (a) $(5x + 3)(4x + 12)$; (b) $\dfrac{5x + 3}{4x + 12}$; (c) $\{x \mid x \neq -3\}$

53.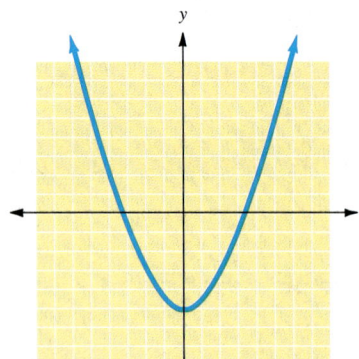

54.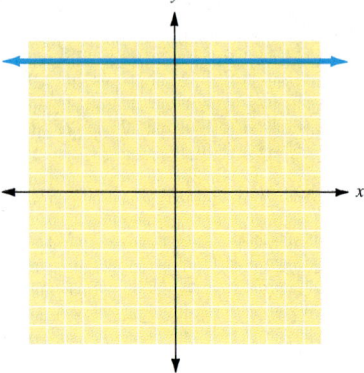

[4.5] In Exercises 55 to 58, find (a) $f + g$, (b) $f − g$, (c) $(f + g)(2)$, (d) $(f − g)(1)$, and (e) the domain of $f + g$.

55. $f(x) = 3x − 8 \quad g(x) = −9x − 15$

56. $f(x) = −8x + 14 \quad g(x) = 2x + 3$

57. $f(x) = −9x + 5 \quad g(x) = \dfrac{3}{x - 8}$

58. $f(x) = 2x + 7 \quad g(x) = \dfrac{-4x}{x - 9}$

[4.5] In Exercises 59 to 62, find (a) $f \cdot g$, (b) $\dfrac{f}{g}$, and (c) the domain of $\dfrac{f}{g}$.

59. $f(x) = 7x + 3 \quad g(x) = x − 2$

60. $f(x) = −8x − 3 \quad g(x) = x + 11$

61. $f(x) = −9x + 12 \quad g(x) = 3x − 5$

62. $f(x) = 5x + 3 \quad g(x) = 4x + 12$

Cumulative Test • 0-4

1. $\{15\}$ 2. $\{-108\}$
3. $\{5, -\frac{5}{3}\}$ 4. $\{-3, 1\}$
5. $R = \dfrac{R_1 R_2}{R_1 + R_2}$
6. $\{x \mid x \leq 1\}$
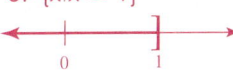
7. $\{x \mid x < -2\}$

8. $\{x \mid -3 < x < 6\}$

9. $\{x \mid x > 13 \text{ or } x < -3\}$

10. $\{x \mid x < \frac{1}{3} \text{ or } x \geq 4\}$

11. $y = 2x - 3$
12. $y = \dfrac{2}{3}x + \dfrac{7}{3}$
13. $y = -\dfrac{5}{4}x - 2$
14. (a) 22; (b) 10
15.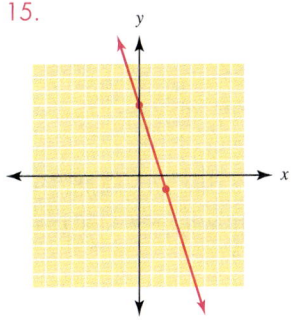

Solve each equation. Express your answer in set notation.

1. $3x - 2(x - 5) + 8 = -3(4 - x)$
2. $\dfrac{2}{3}x - 5 = 4 + \dfrac{3}{4}x$
3. $|3x - 5| = 10$
4. $|x - 5| = |3x + 1|$
5. Solve the equation $\dfrac{1}{R} = \dfrac{1}{R_1} + \dfrac{1}{R_2}$ for R.

In Exercises 6 to 10, solve and graph the solution set for each inequality.

6. $3x + 5 \leq 8$ 7. $2x - 9 > 4x - 5$
8. $|2x - 3| < 9$ 9. $|x - 5| > 8$
10. $3x - 6 < -5$ or $2x - 1 \geq 7$

In Exercises 11 to 13, write the equation of the line L that satisfies the given conditions.

11. L has y intercept -3 and slope of 2.
12. L passes through $(1, 3)$ and $(-2, 1)$.
13. L has y intercept of -2 and is perpendicular to the line with equation $4x - 5y = 20$.
14. If $f(x) = -4x + 18$, evaluate (a) $f(-1)$ and (b) $f(2)$.
15. Graph the function $f(x) = -3x + 5$.

16. Function

17. Not a function

18. (a) Not a function;
    (b) Function

19. (a) −3; (b) 0; (c) 3

20. 5 cm × 17 cm

In each of the following, determine which relations are functions.

16. {(1, 2), (−1, 2), (3, 4), (5, 6)}

17.

x	y
−3	0
−2	1
−1	5
−1	3

18. Use the vertical line test to determine whether each of the following graphs represents a function.

(a)   (b)

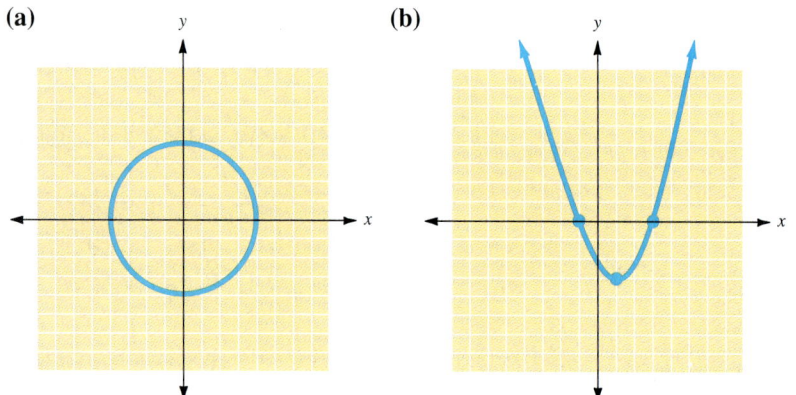

19. Use the given graph to estimate (a) $f(-3)$, (b) $f(0)$, and (c) $f(3)$.

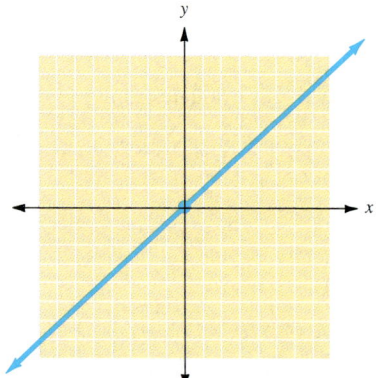

20. The length of a rectangle is 2 centimeters (cm) more than 3 times its width. If the perimeter of the rectangle is 44 cm, what are the dimensions of the rectangle?

CHAPTER 5 Polynomials

LIST OF SECTIONS

5.1 Positive Integer Exponents and Monomials

5.2 An Introduction to Polynomials

5.3 Polynomials: Addition and Subtraction

5.4 Multiplying of Polynomials and Special Products

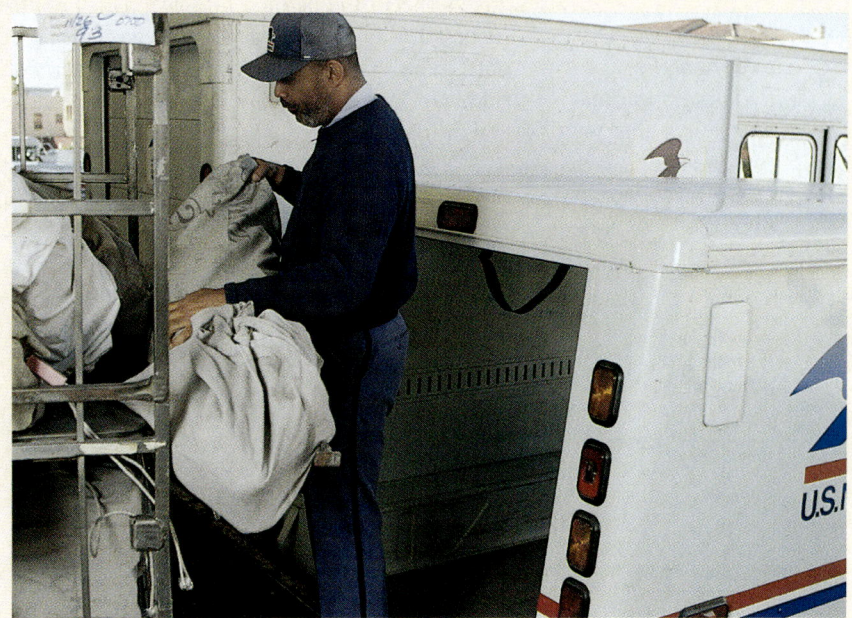

The U.S. Post Office limits the size of rectangular boxes it will accept for mailing. The regulations state the "length plus girth cannot exceed 108 inches." "Girth" means the distance around a cross section; in this case, this measurement is $2h + 2w$. Using the polynomial $l + 2w + 2h$ to describe the measurement required by the Post Office, the regulations say that $l + 2w + 2h \leq 108$ inches.

The volume of a rectangular box is expressed by another polynomial: $V = l \cdot w \cdot h$

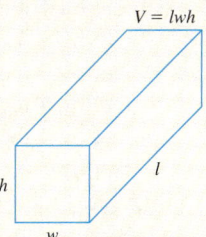

A company that wishes to produce boxes for use by postal patrons must use these formulas as well as do a statistical survey about the shapes that are useful to the most customers. The surface area, expressed by another polynomial expression, $2lw + 2wh + 2lh$, is also used so each box can be manufactured with the least amount of material, in order to help lower costs.

333

SECTION 5.1
Positive Integer Exponents and Monomials

5.1 OBJECTIVES

1. Use properties of exponents
2. Evaluate expressions with integer exponents
3. Use a calculator to evaluate expressions

Doubling

Take a sheet of paper and fold it in half. Fold the resulting paper in half; continue this process for as many folds as you can. Is it possible to get past the seventh fold? Why not? If the sheet of paper were much larger, could you then continue past the seventh fold? No matter how large the paper you start with you will not be able to make the eighth fold. Let's see why this is true.

The first fold doubles the thickness of the paper, so the second fold is like folding two sheets together. Each subsequent fold again doubles the thickness of the paper. The following table will help you see the result of eight folds.

Fold Number	Paper Thickness
1	2 sheets
2	4
3	8
4	16
5	32
6	64
7	128
8	256

The eighth fold is the equivalent of folding 256 sheets of paper. How large would the paper have to be for you to fold 256 sheets? Let's assume that you can fold a 1-foot-square stack of paper 256 sheets thick. How large would the original sheet have to be to equal 1 foot across after seven folds? The next table works the problem backward.

Fold Number	Paper Size
8	1 foot by 1 foot
7	2 feet by 1 foot
6	2 feet by 2 feet
5	4 feet by 2 feet
4	4 feet by 4 feet
3	8 feet by 4 feet
2	8 feet by 8 feet
1	16 feet by 8 feet

The original paper would have had to have been 16 feet long and 8 feet wide to end up with a 1-foot square for the eighth fold.

Exponents

Measuring the paper thickness in the paper folding exercise is an example of repeated multiplication. Exponents are a shorthand form for writing repeated multiplication. Instead of writing

$$2 \cdot 2 \cdot 2 \cdot 2 \cdot 2 \cdot 2 \cdot 2$$

we write

$$2^7$$

Instead of writing

$$a \cdot a \cdot a \cdot a \cdot a$$

we write

$$a^5$$

which we read as "a to the fifth power."

We call a the **base** of the expression and 5 the **exponent,** or **power.**

> In general, for any real number a and any natural number n,
>
> $$a^n = \underbrace{a \cdot a \cdot \ldots \cdot a}_{n \text{ factors}}$$

An expression of this type is said to be in **exponential form.**

Example 1

Using Exponential Notation

Write each of the following, using exponential notation.

(a) $5y \cdot 5y \cdot 5y = (5y)^3$

(b) $w \cdot w \cdot w \cdot w = w^4$

 CHECK YOURSELF 1

Write each of the following, using exponential notation.

(a) $3z \cdot 3z \cdot 3z$ (b) $x \cdot x \cdot x \cdot x$

Let's consider what happens when we multiply two expressions in exponential form with the same base.

We expand the expressions and apply the associative property to regroup.

$$a^4 \cdot a^5 = \underbrace{(a \cdot a \cdot a \cdot a)}_{4 \text{ factors}} \underbrace{(a \cdot a \cdot a \cdot a \cdot a)}_{5 \text{ factors}}$$

$$= \underbrace{a \cdot a \cdot a \cdot a \cdot a \cdot a \cdot a \cdot a \cdot a}_{9 \text{ factors}}$$

$$= a^9$$

Notice that the product is simply the base taken to the power that is the sum of the two original exponents. This leads us to our first property of exponents.

Product Rule for Exponents

For any nonzero real number a and positive integers m and n,

$$a^m \cdot a^n = \underbrace{(a \cdot a \cdot \ldots \cdot a)}_{m \text{ factors}} \underbrace{(a \cdot a \cdot \ldots \cdot a)}_{n \text{ factors}}$$

$$= \underbrace{a \cdot a \cdot \ldots \cdot a}_{m + n \text{ factors}}$$

$$= a^{m+n}$$

This is our first property of exponents

$$a^m \cdot a^n = a^{m+n}$$

Example 2 illustrates the product rule for exponents.

Example 2

Using the Product Rule

Simplify each expression.

(a) $b^4 \cdot b^6 = b^{4+6} = b^{10}$

(b) $(2a)^3 \cdot (2a)^4 = (2a)^{3+4} = (2a)^7$

(c) $(-2)^5(-2)^4 = (-2)^{5+4} = (-2)^9$

(d) $(10^7)(10^{11}) = (10)^{7+11} = (10)^{18}$

✓ **CHECK YOURSELF 2**

Simplify each expression.

(a) $(5b)^6(5b)^5$ (b) $(-3)^4(-3)^3$ (c) $10^8 \cdot 10^{12}$ (d) $(xy)^2(xy)^3$

Applying the commutative and associative properties of multiplication, we know that a product such as

$$2x^3 \cdot 3x^2$$

can be rewritten as

$$(2 \cdot 3)(x^3 \cdot x^2)$$

or as

$$6x^5$$

We expand on these ideas in Example 3.

Example 3 Using Properties of Exponents

Using the product rule for exponents together with the commutative and associative properties, simplify each expression.

(a) $(x^4)(x^2)(x^3)(x) = x^{10}$

(b) $(3x^4)(5x^2) = (3 \cdot 5)(x^4 \cdot x^2) = 15x^6$

(c) $(2x^5y)(9x^3y^4) = (2 \cdot 9)(x^5 \cdot x^3)(y \cdot y^4) = 18x^8y^5$

(d) $(-3x^2y^2)(-2x^4y^3) = (-3)(-2)(x^2 \cdot x^4)(y^2 \cdot y^3) = 6x^6y^5$

Multiply the coefficients and add the exponents by the product rule. With practice you will not need to write the regrouping step.

✓ **CHECK YOURSELF 3**

Simplify each expression.

(a) $(x)(x^5)(x^3)$ (b) $(7x^5)(2x^2)$ (c) $(-2x^3y)(x^2y^2)$ (d) $(-5x^3y^2)(-x^2y^3)$

Now consider the quotient

$$\frac{a^6}{a^4}$$

If we write this in expanded form, we have

$$\frac{\overbrace{a \cdot a \cdot a \cdot a \cdot a \cdot a}^{6 \text{ factors}}}{\underbrace{a \cdot a \cdot a \cdot a}_{4 \text{ factors}}}$$

This can be reduced to

Divide the numerator and denominator by the four common factors of a.

$$\frac{\cancel{a}\cdot\cancel{a}\cdot\cancel{a}\cdot\cancel{a}\cdot a \cdot a}{\cancel{a}\cdot\cancel{a}\cdot\cancel{a}\cdot\cancel{a}} \quad \text{or} \quad a^2$$

This means that

Note that $\frac{a}{a} = 1$, where $a \neq 0$.

$$\frac{a^6}{a^4} = a^2$$

This leads us to our second property of exponents.

> **Quotient Rule for Exponents**
>
> In general, for any real number a ($a \neq 0$) and positive integers m and n,
>
> $$\frac{a^m}{a^n} = a^{m-n} \quad m > n$$

This is our *second property of exponents*. We write $a \neq 0$ to avoid division by 0.

Example 4 illustrates this rule.

Example 4

Using Properties of Exponents

Simplify each expression.

(a) $\dfrac{x^{10}}{x^4} = x^{10-4} = x^6$ *Subtract the exponents, applying the quotient rule.*

Note that $a^1 = a$; there is no need to write the exponent 1 because it is understood.

(b) $\dfrac{a^8}{a^7} = a^{8-7} = a$

(c) $\dfrac{63w^8}{7w^5} = 9w^{8-5} = 9w^3$ *Divide the coefficients and subtract the exponents.*

(d) $\dfrac{-32a^4b^5}{8a^2b} = -4a^{4-2}b^{5-1} = -4a^2b^4$ *Divide the coefficients and subtract the exponents for* each *variable.*

(e) $\dfrac{10^{16}}{10^6} = 10^{16-6} = 10^{10}$

✓ **CHECK YOURSELF 4**

Simplify each expression.

(a) $\dfrac{y^{12}}{y^5}$ (b) $\dfrac{x^9}{x^8}$ (c) $\dfrac{45r^8}{-9r^6}$ (d) $\dfrac{49a^6b^7}{7ab^3}$ (e) $\dfrac{10^{13}}{10^5}$

Section 5.1 ■ Positive Integer Exponents and Monomials   **339**

What happens when a product, such as $(2x)$, is raised to a power? We use the product-power rule.

> **Product-Power Rule for Exponents**
>
> In general, for any real numbers a and b and positive integer n,
>
> $$(ab)^n = a^n b^n$$

This is our third property of exponents.

Example 5 illustrates this rule.

Example 5 Using the Product-Power Rule

Simplify each expression.

(a) $(2x)^3 = 2^3 x^3 = 8x^3$

(b) $(-3x)^4 = (-3)^4 x^4 = 81x^4$

 CHECK YOURSELF 5

Simplify each expression.

(a) $(3x)^3$ (b) $(-2x)^4$

Now, consider the expression

$$(3^2)^3$$

This could be expanded to

$$(3^2)(3^2)(3^2)$$

and then expanded again to

$$(3)(3)(3)(3)(3)(3) = 3^6$$

This leads us to the power rule for exponents.

Power Rule for Exponents

This is our fourth property of exponents.

In general, for any real number a and positive integers m and n,

$$(a^m)^n = a^{mn}$$

Example 6

Using the Power Rule for Exponents

Simplify each expression.

(a) $(2^5)^3 = 2^{15}$

(b) $(x^2)^4 = x^8$

(c) $(2x^3)^3 = 2^3(x^3)^3 = 2^3 x^9 = 8x^9$

✓ **CHECK YOURSELF 6**

Simplify each expression.

(a) $(3^4)^3$ (b) $(x^2)^6$ (c) $(3x^3)^4$

We have one final exponent property to develop. Suppose we have a quotient raised to a power. Consider the following:

$$\left(\frac{x}{3}\right)^3 = \frac{x}{3} \cdot \frac{x}{3} \cdot \frac{x}{3} = \frac{x \cdot x \cdot x}{3 \cdot 3 \cdot 3} = \frac{x^3}{3^3}$$

Note that the power, here 3, has been applied to the numerator x and to the denominator 3. This gives our fifth property of exponents.

Quotient-Power Rule for Exponents

This is our fifth property of exponents.

For any real numbers a and b, where b is not equal to 0, and positive integer m,

$$\left(\frac{a}{b}\right)^m = \frac{a^m}{b^m}$$

In words, to raise a quotient to a power, raise the numerator and denominator to that same power.

Example 7 illustrates the use of this property. Again note that the other properties may also have to be applied in simplifying an expression.

Example 7 — Using the Quotient-Power Rule for Exponents

Simplify each expression.

(a) $\left(\dfrac{3}{4}\right)^3 = \dfrac{3^3}{4^3} = \dfrac{27}{64}$

(b) $\left(\dfrac{x^3}{y^2}\right)^4 = \dfrac{(x^3)^4}{(y^2)^4}$

$= \dfrac{x^{12}}{y^8}$

(c) $\left(\dfrac{r^2 s^3}{t^4}\right)^2 = \dfrac{(r^2 s^3)^2}{(t^4)^2}$

$= \dfrac{(r^2)^2 (s^3)^2}{(t^4)^2}$

$= \dfrac{r^4 s^6}{t^8}$

✓ CHECK YOURSELF 7

Simplify each expression.

(a) $\left(\dfrac{2}{3}\right)^4$ (b) $\left(\dfrac{m^3}{n^4}\right)^5$ (c) $\left(\dfrac{a^2 b^3}{c^5}\right)^2$

The following table summarizes the five properties of exponents that were discussed in this section:

	General Form	Example
1.	$a^m a^n = a^{m+n}$	$x^2 \cdot x^3 = x^5$
2.	$\dfrac{a^m}{a^n} = a^{m-n}$ $(m > n)$	$\dfrac{5^7}{5^3} = 5^4$
3.	$(a^m)^n = a^{mn}$	$(z^5)^4 = z^{20}$
4.	$(ab)^m = a^m b^m$	$(4x)^3 = 4^3 x^3 = 64x^3$
5.	$\left(\dfrac{a}{b}\right)^m = \dfrac{a^m}{b^m}$	$\left(\dfrac{2}{3}\right)^6 = \dfrac{2^6}{3^6} = \dfrac{64}{729}$

In Example 8, we will use the rules of exponents to evaluate some functions.

Example 8

Evaluating Functions

Given the function $f(x) = -2x^3 + 3x^2 - 2x$, find the following:

(a) $f(2) = -2(2)^3 + 3(2)^2 - 2(2) = -2(8) + 3(4) - 2(2) = -16 + 12 - 4 = -8$

(b) $f(2a) = -2(2a)^3 + 3(2a)^2 - 2(2a)$
$= -2(8a^3) + 3(4a^2) - 2(2a)$
$= -16a^3 + 12a^2 - 4a$

✓ **CHECK YOURSELF 8**

Given the function $f(x) = 3x^3 - 3x^2 - 4x$, find

(a) $f(2)$ (b) $f(2a)$

✓ **CHECK YOURSELF ANSWERS**

1. (a) $(3z)^3$; (b) x^4.   **2.** (a) $(5b)^{11}$; (b) $(-3)^7$; (c) 10^{20}; (d) $(xy)^5$.
3. (a) x^9; (b) $14x^7$; (c) $-2x^5y^3$; (d) $5x^5y^5$.   **4.** (a) y^7; (b) x; (c) $-5r^2$;
(d) $7a^5b^4$; (e) 10^8.   **5.** (a) $27x^3$; (b) $16x^4$.   **6.** (a) 3^{12}; (b) x^{12}; (c) $81x^{12}$.
7. (a) $\dfrac{16}{81}$; (b) $\dfrac{m^{15}}{n^{20}}$; (c) $\dfrac{a^4b^6}{c^{10}}$.   **8.** (a) 4; (b) $24a^3 - 12a^2 - 8a$.

Exercises • 5.1

In Exercises 1 to 16, simplify each expression.

1. $x^4 \cdot x^5$
2. $x^7 \cdot x^9$
3. $x^5 \cdot x^3 \cdot x^2$
4. $x^8 \cdot x^4 \cdot x^7$
5. $3^5 \cdot 3^2$
6. $(-3)^4(-3)^6$
7. $(-2)^3(-2)^5$
8. $4^3 \cdot 4^4$
9. $4 \cdot x^2 \cdot x^4 \cdot x^7$
10. $3 \cdot x^3 \cdot x^5 \cdot x^8$
11. $\left(\dfrac{1}{2}\right)^2\left(\dfrac{1}{2}\right)^3\left(\dfrac{1}{2}\right)$
12. $\left(-\dfrac{1}{3}\right)^4\left(-\dfrac{1}{3}\right)\left(-\dfrac{1}{3}\right)^5$
13. $(-2)^2(-2)^3(x^4)(x^5)$
14. $(-3)^4(-3)^2(x)^2(x)^6$
15. $(2x)^2(2x)^3(2x)^4$
16. $(-3x)^3(-3x)^5(-3x)^7$

In Exercises 17 to 28, use the product rule of exponents together with the commutative and associative properties to simplify the products.

17. $(x^2y^3)(x^4y^2)$
18. $(x^4y)(x^2y^3)$
19. $(x^3y^2)(x^4y^2)(x^2y^3)$
20. $(x^2y^3)(x^3y)(x^4y^2)$
21. $(2x^4)(3x^3)(-4x^3)$
22. $(2x^3)(-3x)(-4x^4)$
23. $(5x^2)(3x^3)(x)(-2x^3)$
24. $(4x^2)(2x)(x^2)(2x^3)$
25. $(5xy^3)(2x^2y)(3xy)$
26. $(-3xy)(5x^2y)(-2x^3y^2)$
27. $(x^2yz)(x^3y^5z)(x^4yz)$
28. $(xyz)(x^8y^3z^6)(x^2yz)(xyz^4)$

In Exercises 29 to 36, use the quotient rule of exponents to simplify each expression.

29. $\dfrac{x^{10}}{x^7}$
30. $\dfrac{b^{23}}{b^{18}}$
31. $\dfrac{x^7y^{11}}{x^4y^3}$
32. $\dfrac{x^5y^9}{xy^4}$
33. $\dfrac{x^5y^4z^2}{xy^2z}$
34. $\dfrac{x^8y^6z^4}{x^3yz^3}$
35. $\dfrac{21x^4y^5}{7xy^2}$
36. $\dfrac{48x^6y^6}{12x^3y}$

In Exercises 37 to 48, use your calculator to evaluate each expression.

37. 4^3
38. 5^7
39. $(-3)^4$
40. $(-4)^5$
41. $2^3 \cdot 2^5$
42. $3^4 \cdot 3^6$
43. $(3x^2)(2x^4)$, where $x = 2$
44. $(4x^3)(5x^4)$, where $x = 3$
45. $(2x^4)(4x^2)$, where $x = -2$
46. $(3x^5)(2x^3)$, where $x = -3$
47. $(-2x^3)(-3x^5)$, where $x = 2$
48. $(-3x^2)(-4x^4)$, where $x = 4$

---

1. x^9
2. x^{16}
3. x^{10}
4. x^{19}
5. 3^7
6. $(-3)^{10}$
7. $(-2)^8$
8. 4^7
9. $4x^{13}$
10. $3x^{16}$
11. $\left(\dfrac{1}{2}\right)^6$
12. $\left(-\dfrac{1}{3}\right)^{10}$
13. $(-2)^5 x^9$
14. $(-3)^6 x^8$
15. $(2x)^9$
16. $(-3x)^{15}$
17. x^6y^5
18. x^6y^4
19. x^9y^7
20. x^9y^6
21. $-24x^{10}$
22. $24x^8$
23. $-30x^9$
24. $16x^8$
25. $30x^4y^5$
26. $30x^6y^4$
27. $x^9y^7z^3$
28. $x^{12}y^6z^{12}$
29. x^3
30. b^5
31. x^3y^8
32. x^4y^5
33. x^4y^2z
34. x^5y^5z
35. $3x^3y^3$
36. $4x^3y^5$
37. 64
38. 78, 125
39. 81
40. −1024
41. 256
42. 59, 049
43. 384
44. 43, 740
45. 512
46. 39, 366
47. 1536
48. 49, 152

Chapter 5 ■ Polynomials

49. 51; 7
50. −12; −40
51. 250; 46
52. 39; 12
53. 1; 2; −56
54. −10; −44; 8
55. 3; 372; 15
56. −686; −10; 646
57. −15x^6
58. 10x^4
59. 8x^3
60. −27x^3
61. x^{21}
62. −x^{15}
63. −24x^4
64. −54x^4
65. 32x^{15}
66. −27x^6
67. −216x^{12}
68. 225x^8
69. 50x^9
70. 9x^{14}
71. 144x^{20}
72. $\dfrac{9}{16}$
73. $\dfrac{4}{9}$
74. $\dfrac{x^3}{125}$
75. $\dfrac{a^4}{16}$
76. $\dfrac{m^9}{n^6}$
77. $\dfrac{a^{16}}{b^{12}}$
78. $\dfrac{a^6 b^4}{c^8}$
79. $\dfrac{x^{15} y^6}{z^{12}}$
80. $\dfrac{4x^{10}}{y^6}$
81. $\dfrac{8x^6}{27}$
82. −648$x^{18} y^{21}$
83. −675$x^{19} y^{14}$
84. $\dfrac{3x^8 y^4}{2}$
85. $\dfrac{6 x^{10} y^9}{5}$

In Exercises 49 to 56, use a calculator to evaluate the functions.

49. Given $f(x) = -2x^3 + 7x^2 - 3x + 1$, find $f(-2)$ and $f(2)$.
50. Given $f(x) = -3x^3 - 6x^2 + 5x - 2$, find $f(-2)$ and $f(2)$.
51. Given $f(x) = 2x^4 + 3x^3 - 2x^2 + 7x + 4$, find $f(3)$ and $f(-3)$.
52. Given $f(x) = 4x^4 - 2x^3 + x^2 - 6x - 1$, find $f(2)$ and $f(-1)$.
53. Given $f(x) = -2x^4 + 3x^3 - x^2 - x + 2$, find $f(1)$, $f(0)$, and $f(-2)$.
54. Given $f(x) = 3x^3 - 2x^2 + x - 10$, find $f(0)$, $f(-2)$, and $f(2)$.
55. Given $f(x) = -2x^4 + 6x^3 - 7x^2 + 15$, find $f(2)$, $-f(-3)$, and $f(0)$.
56. Given $f(x) = -3x^4 + 7x^2 - 5x - 10$, find $f(4)$, $f(0)$, and $-f(-4)$.

In Exercises 57 to 85, simplify each expression.

57. $(-3x)(5x^5)$
58. $(-5x^2)(-2x^2)$
59. $(2x)^3$
60. $(-3x)^3$
61. $(x^3)^7$
62. $(-x^3)^5$
63. $(3x)(-2x)^3$
64. $(2x)(-3x)^3$
65. $(2x^3)^5$
66. $(-3x^2)^3$
67. $(-2x^2)^3 (3x^2)^3$
68. $(-3x^2)^2 (5x^2)^2$
69. $(2x^3)(5x^3)^2$
70. $(3x^3)^2 (x^2)^4$
71. $(2x^3)^4 (3x^4)^2$
72. $\left(\dfrac{3}{4}\right)^2$
73. $\left(\dfrac{2}{3}\right)^2$
74. $\left(\dfrac{x}{5}\right)^3$
75. $\left(\dfrac{a}{2}\right)^4$
76. $\left(\dfrac{m^3}{n^2}\right)^3$
77. $\left(\dfrac{a^4}{b^3}\right)^4$
78. $\left(\dfrac{a^3 b^2}{c^4}\right)^2$
79. $\left(\dfrac{x^5 y^2}{z^4}\right)^3$
80. $\left(\dfrac{2x^5}{y^3}\right)^2$
81. $\left(\dfrac{2x^5}{3x^3}\right)^3$
82. $(-8x^2 y)(-3x^4 y^5)^4$
83. $(5x^5 y)^2 (-3x^3 y^4)^3$
84. $\left(\dfrac{3x^4 y^9}{2x^2 y^7}\right)\left(\dfrac{x^6 y^3}{x^3 y^2}\right)^2$
85. $\left(\dfrac{6x^5 y^4}{5xy}\right)\left(\dfrac{x^3 y^5}{xy^3}\right)^3$

Section 5.1 ■ Positive Integer Exponents and Monomials 345

86. $2954.91
87. $5909.82
90. $(x^2)^6$
91. $(y^3)^5$
92. $(a^2)^8$
93. $(m^5)^4$
94. 8^4; 8^6; 8^5; 8^{14}
95. 9^4; 9^7; 9^{20}; 9^{14}
96. $-2x^3y^3z^5$
97. $3x^3y^2z^4$

86. **Business.** The value, P, of a savings account that compounds interest annually is given by the formula

$$P = A(1 + r)^t$$

where

A = original amount

r = interest rate in decimal form

t = time in years

Find the amount of money in the account after 8 years if $2000 was invested initially at 5% compounded annually.

87. **Business.** Using the formula for compound interest in Exercise 86, determine the amount of money in the account if the original investment is doubled.

88. You have learned rules for working with exponents when multiplying, dividing, and raising an expression to a power.

   (a) Explain each rule in your own words. Give numerical examples.

   (b) Is there a rule for raising a *sum* to a power? That is, does $(a + b)^n = a^n + b^n$? Use numerical examples to explain why this is true in general or why it is not. Is it always true or always false?

89. Work with another student to investigate the rate of inflation. The rate of inflation was about 3% from 1990 to 1999. This means that the value of the goods that you could buy for $1 in 1990 would cost 3% more in 1991, 3% more than that in 1992, etc. If a movie ticket cost $5.50 in 1990, what would it cost today if movie tickets just kept up with inflation? Construct a table to solve the problem.

Solve the following problems.

90. Write x^{12} as a power of x^2. 91. Write y^{15} as a power of y^3.

92. Write a^{16} as a power of a^2. 93. Write m^{20} as a power of m^5.

94. Write each of the following as powers of 8 (remember that $8 = 2^3$): 2^{12}, 2^{18}, $(2^5)^3$, $(2^7)^6$.

95. Write each of the following as powers of 9: 3^8, 3^{14}, $(3^5)^8$, $(3^4)^7$.

96. What expression raised to the third power is $-8x^6y^9z^{15}$?

97. What expression raised to the fourth power is $81x^{12}y^8z^{16}$?

SECTION 5.2 An Introduction to Polynomials

5.2 OBJECTIVES

1. Identify types of polynomials
2. Find the degree of a polynomial
3. Write polynomials in descending-exponent form

Our work in this chapter deals with the most common kind of algebraic expression, a *polynomial*. To define a polynomial, let's recall our earlier definition of the word "term."

A **term** is a number or the product of a number and one or more variables.

For example, x^5, $3x$, $-4xy^2$, 8, $\dfrac{5}{x}$, and $-14\sqrt{x}$ are terms. A **polynomial** consists of one or more terms in which the only allowable exponents are the whole numbers, 0, 1, 2, 3, and so on. These terms are connected by addition or subtraction signs. The variable is never used as a divisor in a term of a polynomial.

In a polynomial, terms are separated by + and − signs.

In each term of a polynomial, the number is called the **numerical coefficient,** or more simply the **coefficient,** of that term.

Example 1 Identifying Polynomials

Note: Each sign (+ or −) is attached to the term that *follows* that sign.

(a) $x + 3$ is a polynomial. The terms are x and 3. The coefficients are 1 and 3.

(b) $3x^2 - 2x + 5$ is also a polynomial. Its terms are $3x^2$, $-2x$, and 5. The coefficients are 3, −2, and 5.

(c) $5x^3 + 2 - \dfrac{3}{x}$ is *not* a polynomial because of the division by x in the third term.

 CHECK YOURSELF 1

Which of the following are polynomials?

(a) $5x^2$ (b) $3y^3 - 2y + \dfrac{5}{y}$ (c) $4x^2 - 2x + 3$

Certain polynomials are given special names because of the number of terms that they have.

> The prefix "mono" means 1. The prefix "bi" means 2. The prefix "tri" means 3. There are no special names for polynomials with more than three terms.

A polynomial with exactly one term is called a **monomial.**

A polynomial with exactly two terms is called a **binomial.**

A polynomial with exactly three terms is called a **trinomial.**

Example 2 — Identifying Types of Polynomials

(a) $3x^2y$ is a monomial. It has exactly one term.

(b) $2x^3 + 5x$ is a binomial. It has exactly two terms, $2x^3$ and $5x$.

(c) $5x^2 - 4x + 3$ is a trinomial. Its three terms are $5x^2$, $-4x$, and 3.

✓ CHECK YOURSELF 2

Classify each of these as a monomial, binomial, or trinomial.

(a) $5x^4 - 2x^3$ (b) $4x^7$ (c) $2x^2 + 5x - 3$

> Remember, in a polynomial the allowable exponents are the whole numbers 0, 1, 2, 3, and so on. The degree will be a whole number.

We also classify polynomials by their *degree*. The **degree** of a polynomial that has only one variable is the highest power of that variable appearing in any one term.

Example 3 — Classifying Polynomials by Their Degree

(a) $5x^3 - 3x^2 + 4x$ has degree 3. ← The highest power

(b) $4x - 5x^4 + 3x^3 + 2$ has degree 4. ← The highest power

(c) $8x$ has degree 1 because $8x = 8x^1$.

(d) 7 has degree 0 because $7 = 7 \cdot 1 = 7x^0$.

> In Chapter 10, we will see why $x^0 = 1$.

Note: Polynomials can have more than one variable, such as $4x^2y^3 + 5xy^2$. The degree is then the largest sum of the powers in any single term (here $2 + 3$, or 5). In general, we will be working with polynomials in a single variable, such as x.

✓ CHECK YOURSELF 3

Find the degree of each polynomial.

(a) $6x^5 - 3x^3 - 2$ (b) $5x$ (c) $3x^3 + 2x^6 - 1$ (d) 9

Working with polynomials is much easier if you get used to writing them in **descending-exponent form** (sometimes called *descending-power form*). When a polynomial has only one variable, this means that the term with the highest exponent is written first, then the term with the next highest exponent, and so on.

Example 4   Writing Polynomials in Descending Order

The exponents get smaller from left to right.

(a) $5x^7 - 3x^4 + 2x^2$ is in descending-exponent form.

(b) $4x^4 + 5x^6 - 3x^5$ is *not* in descending-exponent form. The polynomial should be written as

$$5x^6 - 3x^5 + 4x^4$$

Notice that the degree of the polynomial is the power of the *first*, or *leading*, term once the polynomial is arranged in descending-exponent form.

✓ CHECK YOURSELF 4

Write the following polynomials in descending-exponent form.

(a) $5x^4 - 4x^5 + 7$ (b) $4x^3 + 9x^4 + 6x^8$

A polynomial can represent any number. Its value depends on the value given to the variable.

Example 5   Evaluating Polynomials

Given the polynomial

$$3x^3 - 2x^2 - 4x + 1$$

(a) Find the value of the polynomial when $x = 2$.

Substituting 2 for x, we have

$$3(2)^3 - 2(2)^2 - 4(2) + 1$$
$$= 3(8) - 2(4) - 4(2) + 1$$
$$= 24 - 8 - 8 + 1$$
$$= 9$$

Again note how the rules for the order of operations are applied. See Section 0.5 for a review.

(b) Find the value of the polynomial when $x = -2$.

Caution

Be particularly careful when dealing with powers of negative numbers!

Now we substitute -2 for x.

$$3(-2)^3 - 2(-2)^2 - 4(-2) + 1$$
$$= 3(-8) - 2(4) - 4(-2) + 1$$
$$= -24 - 8 + 8 + 1$$
$$= -23$$

 **CHECK YOURSELF 5**

Find the value of the polynomial

$$4x^3 - 3x^2 + 2x - 1$$

when

(a) $x = 3$ (b) $x = -3$

 **CHECK YOURSELF ANSWERS**

1. (a) and (c) are polynomials.   **2.** (a) Binomial; (b) monomial; (c) trinomial.
3. (a) 5; (b) 1; (c) 6; (d) 0.   **4.** (a) $-4x^5 + 5x^4 + 7$; (b) $6x^8 + 9x^4 + 4x^3$.
5. (a) 86; (b) -142.

Exercises • 5.2

Answers (left column):

1. Polynomial
2. Not a polynomial
3.–6. Polynomial
7. Not a polynomial
8. Polynomial
9. $2x^2, -3x$; $2, -3$
10. $5x^3, x$; $5, 1$
11. $4x^3, -3x, 2$; $4, -3, 2$
12. $7x^2$; 7
13. Binomial
14. Monomial
15.–16. Trinomial
17. Not classified
18. Not a polynomial
19. Monomial
20. Not classified
21. Not a polynomial
22. Binomial
23. $4x^5 - 3x^2$; 5
24. $3x^3 + 5x^2 + 4$; 3
25. $-5x^9 + 7x^7 + 4x^3$; 9
26. $x + 2$; 1
27. $4x$; 1
28. $x^{17} - 3x^4$; 17
29. $x^6 - 3x^5 + 5x^2 - 7$; 6
30. 5; 0
31. $7, -5$
32. $5, -15$
33. $4, -4$
34. $34, 34$
35. $62, 30$
36. $-1, 19$
37. $0, 0$
38. $0, 0$

Which of the following expressions are polynomials?

1. $7x^3$
2. $5x^3 - \dfrac{3}{x}$
3. $4x^4y^2 - 3x^3y$
4. 7
5. -7
6. $4x^3 + x$
7. $\dfrac{3 + x}{x^2}$
8. $5a^2 - 2a + 7$

For each of the following polynomials, list the terms and the coefficients.

9. $2x^2 - 3x$
10. $5x^3 + x$
11. $4x^3 - 3x + 2$
12. $7x^2$

Classify each of the following as a monomial, binomial, or trinomial where possible.

13. $7x^3 - 3x^2$
14. $4x^7$
15. $7y^2 + 4y + 5$
16. $2x^2 + 3xy + y^2$
17. $2x^4 - 3x^2 + 5x - 2$
18. $x^4 + \dfrac{5}{x} + 7$
19. $6y^8$
20. $4x^4 - 2x^2 + 5x - 7$
21. $x^5 - \dfrac{3}{x^2}$
22. $4x^2 - 9$

Arrange in descending-exponent form if necessary, and give the degree of each polynomial.

23. $4x^5 - 3x^2$
24. $5x^2 + 3x^3 + 4$
25. $7x^7 - 5x^9 + 4x^3$
26. $2 + x$
27. $4x$
28. $x^{17} - 3x^4$
29. $5x^2 - 3x^5 + x^6 - 7$
30. 5

Find the values of each of the following polynomials for the given values of the variable.

31. $6x + 1$; $x = 1$ and $x = -1$
32. $5x - 5$; $x = 2$ and $x = -2$
33. $x^3 - 2x$; $x = 2$ and $x = -2$
34. $3x^2 + 7$; $x = 3$ and $x = -3$
35. $3x^2 + 4x - 2$; $x = 4$ and $x = -4$
36. $2x^2 - 5x + 1$; $x = 2$ and $x = -2$
37. $-x^2 - 2x + 3$; $x = 1$ and $x = -3$
38. $-x^2 - 5x - 6$; $x = -3$ and $x = -2$

Section 5.2 ■ An Introduction to Polynomials **351**

39. Always
40. Never
41. Sometimes
42. Always
43. Sometimes
44. Always
45. Sometimes
46. Sometimes
47. 4
48. 2
49. 11
50. 11
51. 14
52. 21
53. 5
54. 3
55. 10
56. 10
57. 7
58. 5
59. −2
60. 1
61. $C(x) = 3x + 20$; $170
62. $C(x) = 20x + 150$; $290
63. $337
64. $39.55

Indicate whether each of the following statements is always true, sometimes true, or never true.

39. A monomial is a polynomial.

40. A binomial is a trinomial.

41. The degree of a trinomial is 3.

42. A trinomial has three terms.

43. A polynomial has four or more terms.

44. A binomial must have two coefficients.

45. If x equals 0, the value of a polynomial in x equals 0.

46. The coefficient of the leading term in a polynomial is the largest coefficient of the polynomial.

A polynomial function is simply a function whose defining expression is a polynomial in one variable. P and Q are often used to name such functions.
If $P(x) = x^3 - 2x^2 + 5$ and $Q(x) = 2x^2 + 3$, find

47. $P(1)$

48. $P(-1)$

49. $Q(2)$

50. $Q(-2)$

51. $P(3)$

52. $Q(-3)$

53. $P(0)$

54. $Q(0)$

55. $P(2) + Q(-1)$

56. $P(-2) + Q(3)$

57. $P(3) - Q(-3) \div Q(0)$

58. $Q(-2) \div Q(2) \cdot P(0)$

59. $|Q(4)| - |P(4)|$

60. $\dfrac{P(-1) + Q(0)}{P(0)}$

61. Cost of typing. The cost, in dollars, of typing a term paper is given as 3 times the number of pages plus 20. Use x as the number of pages to be typed and write a polynomial to describe this cost. Find the cost of typing a 50-page paper.

62. Manufacturing. The cost, in dollars, of making suits is described as 20 times the number of suits plus 150. Use x as the number of suits and write a polynomial to describe this cost. Find the cost of making seven suits.

63. Revenue. The revenue, in dollars, when x pairs of shoes are sold is given by $R(x) = 3x^2 - 95$. Find the revenue when 12 pairs of shoes are sold.

64. Manufacturing. The cost in dollars of manufacturing x wing nuts is given by $C(x) = 0.07x + 13.3$. Find the cost when 375 wing nuts are made.

SECTION 5.3 Polynomials: Addition and Subtraction

5.3 OBJECTIVES

1. Add two polynomials
2. Subtract two polynomials

Addition is always a matter of combining like quantities (two apples plus three apples, four books plus five books, and so on). If you keep that basic idea in mind, adding polynomials will be easy. It is just a matter of combining like terms. Suppose that you want to add

$$5x^2 + 3x + 4 \quad \text{and} \quad 4x^2 + 5x - 6$$

Parentheses are sometimes used in adding, so for the sum of these polynomials, we can write

$$(5x^2 + 3x + 4) + (4x^2 + 5x - 6)$$

Now what about the parentheses? You can use the following rule.

The plus sign between the parentheses indicates the addition.

Removing Signs of Grouping: Case 1

If a plus sign (+) or nothing at all appears in front of parentheses, just remove the parentheses. No other changes are necessary.

Now let's return to the addition.

Just remove the parentheses. No other changes are necessary.

$$(5x^2 + 3x + 4) + (4x^2 + 5x - 6)$$
$$= 5x^2 + 3x + 4 + 4x^2 + 5x - 6$$

Like terms — Like terms — Like terms

Note the use of the associative and commutative properties in reordering and regrouping.

Collect like terms. (*Remember:* Like terms have the same variables raised to the same power.)

$$= (5x^2 + 4x^2) + (3x + 5x) + (4 - 6)$$

Combine like terms for the result:

$$= 9x^2 + 8x - 2$$

Here we use the distributive property. For example,

As should be clear, much of this work can be done mentally. You can then write the sum directly by locating like terms and combining.

Section 5.3 ▪ Polynomials: Addition and Subtraction 353

Example 1 Combining Like Terms

Add $3x - 5$ and $2x + 3$.

Write the sum.

$$(3x - 5) + (2x + 3)$$
$$= 3x - 5 + 2x + 3 = 5x - 2$$

Like terms · Like terms

✓ CHECK YOURSELF 1

Add $6x^2 + 2x$ and $4x^2 - 7x$.

The same technique is used to find the sum of two trinomials.

Example 2 Adding Polynomials

Add $4a^2 - 7a + 5$ and $3a^2 + 3a - 4$.

Remember: Only the like terms are combined in the sum.

Write the sum.

$$(4a^2 - 7a + 5) + (3a^2 + 3a - 4)$$
$$= 4a^2 - 7a + 5 + 3a^2 + 3a - 4 = 7a^2 - 4a + 1$$

Like terms
Like terms
Like terms

✓ CHECK YOURSELF 2

Add $5y^2 - 3y + 7$ and $3y^2 - 5y - 7$.

Example 3 Adding Polynomials

Add $2x^2 + 7x$ and $4x - 6$.

Write the sum.

$$(2x^2 + 7x) + (4x - 6)$$
$$= 2x^2 + \underbrace{7x + 4x}_{} - 6$$

These are the only like terms; $2x^2$ and -6 cannot be combined.

$$= 2x^2 + 11x - 6$$

✓ CHECK YOURSELF 3

Add $5m^2 + 8$ and $8m^2 - 3m$.

As we mentioned in Section 5.2, writing polynomials in descending-exponent form usually makes the work easier. Look at Example 4.

Example 4

Adding Polynomials

Add $3x - 2x^2 + 7$ and $5 + 4x^2 - 3x$.

Write the polynomials in descending-exponent form, then add.

$$(-2x^2 + 3x + 7) + (4x^2 - 3x + 5)$$
$$= 2x^2 + 12$$

✓ CHECK YOURSELF 4

Add $8 - 5x^2 + 4x$ and $7x - 8 + 8x^2$.

Subtracting polynomials requires another rule for removing signs of grouping.

Removing Signs of Grouping: Case 2

If a minus sign (−) appears in front of a set of parentheses, the parentheses can be removed by changing the sign of each item inside the parentheses.

The use of this rule is illustrated in Example 5.

Example 5 — Removing Parentheses

In each of the following, remove the parentheses.

(a) $-(2x + 3y) = -2x - 3y$ Change each sign to remove the parentheses.

(b) $m - (5n - 3p) = m - 5n + 3p$
                           $\underbrace{\quad\quad\quad}$
                           Sign changes.

(c) $2x - (-3y + z) = 2x + 3y - z$
                        $\underbrace{\quad\quad\quad}$
                        Sign changes.

Note: This uses the distributive property, since
$-(2x + 3y) = (-1)(2x + 3y)$
$\quad\quad\quad\;\; = -2x - 3y$

✓ CHECK YOURSELF 5

Remove the parentheses.

(a) $-(3m + 5n)$ (b) $-(5w - 7z)$ (c) $3r - (2s - 5t)$ (d) $5a - (-3b - 2c)$

Subtracting polynomials is now a matter of using the previous rule to remove the parentheses and then combining like terms. Consider Example 6.

Example 6 — Subtracting Polynomials

(a) Subtract $5x - 3$ from $8x + 2$.

Write

$$(8x + 2) - (5x - 3)$$
$$= 8x + 2 \underbrace{- 5x + 3}_{\text{Sign changes.}}$$
$$= 3x + 5$$

Note: The expression following "from" is written first in the problem.

(b) Subtract $4x^2 - 8x + 3$ from $8x^2 + 5x - 3$.

Write

$$(8x^2 + 5x - 3) - (4x^2 - 8x + 3)$$
$$= 8x^2 + 5x - 3 \underbrace{- 4x^2 + 8x - 3}_{\text{Sign changes.}}$$
$$= 4x^2 + 13x - 6$$

✓ **CHECK YOURSELF 6**

(a) Subtract $7x + 3$ from $10x - 7$.

(b) Subtract $5x^2 - 3x + 2$ from $8x^2 - 3x - 6$.

Again, writing all polynomials in descending-exponent form will make locating and combining like terms much easier. Look at Example 7.

Example 7 **Subtracting Polynomials**

(a) Subtract $4x^2 - 3x^3 + 5x$ from $8x^3 - 7x + 2x^2$.

Write

$$(8x^3 + 2x^2 - 7x) - (-3x^3 + 4x^2 + 5x)$$
$$= 8x^3 + 2x^2 - 7x \underbrace{+ 3x^3 - 4x^2 - 5x}_{\text{Sign changes.}}$$
$$= 11x^3 - 2x^2 - 12x$$

(b) Subtract $8x - 5$ from $-5x + 3x^2$.

Write

$$(3x^2 - 5x) - (8x - 5)$$
$$= 3x^2 \underbrace{- 5x - 8x}_{} + 5$$

Only the like terms can be combined.

$$= 3x^2 - 13x + 5$$

✓ **CHECK YOURSELF 7**

(a) Subtract $7x - 3x^2 + 5$ from $5 - 3x + 4x^2$.

(b) Subtract $3a - 2$ from $5a + 4a^2$.

If you think back to addition and subtraction in arithmetic, you'll remember that the work was arranged vertically. That is, the numbers being added or subtracted were placed under one another so that each column represented the same place value. This meant that in adding or subtracting columns you were always dealing with "like quantities."

Section 5.3 ■ Polynomials: Addition and Subtraction

It is also possible to use a vertical method for adding or subtracting polynomials. First rewrite the polynomials in descending-exponent form, then arrange them one under another, so that each column contains like terms. Then add or subtract in each column.

Example 8 — Adding Using the Vertical Method

(a) Add $3x - 5$ and $x^2 + 2x + 4$

$$\begin{array}{r} 3x - 5 \\ x^2 + 2x + 4 \\ \hline x^2 + 5x - 1 \end{array}$$

(b) Add $2x^2 - 5x$, $3x^2 + 2$, and $6x - 3$.

$$\begin{array}{r} 2x^2 - 5x \\ 3x^2 + 2 \\ 6x - 3 \\ \hline 5x^2 + x - 1 \end{array}$$

✓ **CHECK YOURSELF 8**

Add $3x^2 + 5$, $x^2 - 4x$, and $6x + 7$.

The following example illustrates subtraction by the vertical method.

Example 9 — Subtracting Using the Vertical Method

(a) Subtract $5x - 3$ from $8x - 7$.

Write

$$\begin{array}{r} 8x - 7 \\ (-)5x - 3 \\ \hline \end{array}$$

To subtract, change each sign of $5x - 3$ to get $-5x + 3$, then add.

$$\begin{array}{r} 8x - 7 \\ -5x + 3 \\ \hline 3x - 4 \end{array}$$

(b) Subtract $5x^2 - 3x + 4$ from $8x^2 + 5x - 3$.

Write

$$\begin{array}{r} 8x^2 + 5x - 3 \\ (-)5x^2 - 3x + 4 \\ \hline \end{array}$$

To subtract, change each sign of $5x^2 - 3x + 4$ to get $-5x^2 + 3x - 4$, then add.

$$\begin{array}{r} 8x^2 + 5x - 3 \\ -5x^2 + 3x - 4 \\ \hline 3x^2 + 8x - 7 \end{array}$$

Subtracting using the vertical method takes some practice. Take time to study the method carefully. You'll be using it in long division in Section 6.4.

✓ CHECK YOURSELF 9

Subtract, using the vertical method.

(a) $4x^2 - 3x$ from $8x^2 + 2x$ (b) $8x^2 + 4x - 3$ from $9x^2 - 5x + 7$

✓ CHECK YOURSELF ANSWERS

1. $10x^2 - 5x$.   **2.** $8y^2 - 8y$.   **3.** $13m^2 - 3m + 8$.   **4.** $3x^2 + 11x$.
5. (a) $-3m - 5n$; (b) $-5w + 7z$; (c) $3r - 2s + 5t$; (d) $5a + 3b + 2c$.
6. (a) $3x - 10$; (b) $3x^2 - 8$.   **7.** (a) $7x^2 - 10x$; (b) $4a^2 + 2a + 2$.
8. $4x^2 + 2x + 12$.   **9.** (a) $4x^2 + 5x$; (b) $x^2 - 9x + 10$.

Exercises • 5.3

Answers (left column):
1. $9a + 4$
2. $12x - 1$
3. $13b^2 - 18b$
4. $8m^2 - 5m$
5. $-2x^2$
6. $-4p^2$
7. $5x^2 - 2x + 1$
8. $9d^2 - 14d - 2$
9. $2b^2 + 5b + 16$
10. $3x^2 - 5x - 3$
11. $8y^3 - 2y$
12. $9x^4 + 3$
13. $-a^3 + 4a^2$
14. $5m^3 - 8m$
15. $-2x^2 - x + 3$
16. $-2b^3 + 5b^2 - 3b$
17. $-2a - 3b$
18. $-7x + 4y$
19. $5a - 2b + 3c$
20. $7x - 4y - 3z$
21. $6r - 5s$
22. $7m + 2n$
23. $8p - 2q$
24. $7c + 10d$
25. $x - 7$
26. $2x + 7$
27. $m^2 - 3m$
28. $2a^2 - 5a$
29. $-2y^2$
30. $-2n^2$
31. $2x^2 - x + 1$
32. $2x^2 - 6x - 7$
33. $8a^2 - 12a - 7$
34. $x^3 - x^2 - 5x$
35. $-6b^2 + 8b$
36. $6y^2 - 9y$
37. $2x^2 + 12$
38. $-x^2 - 3x$

Add.

1. $6a - 5$ and $3a + 9$
2. $9x + 3$ and $3x - 4$
3. $8b^2 - 11b$ and $5b^2 - 7b$
4. $2m^2 + 3m$ and $6m^2 - 8m$
5. $3x^2 - 2x$ and $-5x^2 + 2x$
6. $3p^2 + 5p$ and $-7p^2 - 5p$
7. $2x^2 + 5x - 3$ and $3x^2 - 7x + 4$
8. $4d^2 - 8d + 7$ and $5d^2 - 6d - 9$
9. $2b^2 + 8$ and $5b + 8$
10. $4x - 3$ and $3x^2 - 9x$
11. $8y^3 - 5y^2$ and $5y^2 - 2y$
12. $9x^4 - 2x^2$ and $2x^2 + 3$
13. $2a^2 - 4a^3$ and $3a^3 + 2a^2$
14. $9m^3 - 2m$ and $-6m - 4m^3$
15. $4x^2 - 2 + 7x$ and $5 - 8x - 6x^2$
16. $5b^3 - 8b + 2b^2$ and $3b^2 - 7b^3 + 5b$

Remove the parentheses in each of the following expressions, and simplify where possible.

17. $-(2a + 3b)$
18. $-(7x - 4y)$
19. $5a - (2b - 3c)$
20. $7x - (4y + 3z)$
21. $9r - (3r + 5s)$
22. $10m - (3m - 2n)$
23. $5p - (-3p + 2q)$
24. $8d - (-7c - 2d)$

Subtract.

25. $x + 4$ from $2x - 3$
26. $x - 2$ from $3x + 5$
27. $3m^2 - 2m$ from $4m^2 - 5m$
28. $9a^2 - 5a$ from $11a^2 - 10a$
29. $6y^2 + 5y$ from $4y^2 + 5y$
30. $9n^2 - 4n$ from $7n^2 - 4n$
31. $x^2 - 4x - 3$ from $3x^2 - 5x - 2$
32. $3x^2 - 2x + 4$ from $5x^2 - 8x - 3$
33. $3a + 7$ from $8a^2 - 9a$
34. $3x^3 + x^2$ from $4x^3 - 5x$
35. $4b^2 - 3b$ from $5b - 2b^2$
36. $7y - 3y^2$ from $3y^2 - 2y$
37. $x^2 - 5 - 8x$ from $3x^2 - 8x + 7$
38. $4x - 2x^2 + 4x^3$ from $4x^3 + x - 3x^2$

359

39. $6b - 1$
40. $6m - 3$
41. $10x - 9$
42. $-x^2 + 5$
43. $2x^2 + 5x - 12$
44. $6a + 2$
45. $-6y^2 - 8y$
46. $-2r^3 + 3r^2$
47. $6w^2 - 2w + 2$
48. $5x^2 + 2x + 3$
49. $9x^2 - x$
50. $8x^2 - 4x - 10$
51. $2a^2 + 5a$
52. $-2r^3 - 6r^2$
53. $3x^2 + x$
54. $x^2 - 4x + 4$
55. $3x^2 + 3x - 9$
56. $2x^2 - 6x - 3$
57. $5x^2 - 3x + 9$
58. $5x^2 - 2x + 4$
59. (a) $9x + 4$; (b) 13; (c) 13
60. (a) $12x^2 - 4x$; (b) 8; (c) 8
61. (a) $7x^2 + 7x$; (b) 14; (c) 14
62. (a) $15x^2 - x - 2$; (b) 12; (c) 12

Perform the indicated operations.

39. Subtract $3b + 2$ from the sum of $4b - 2$ and $5b + 3$.
40. Subtract $5m - 7$ from the sum of $2m - 8$ and $9m - 2$.
41. Subtract $3x^2 + 2x - 1$ from the sum of $x^2 + 5x - 2$ and $2x^2 + 7x - 8$.
42. Subtract $4x^2 - 5x - 3$ from the sum of $x^2 - 3x - 7$ and $2x^2 - 2x + 9$.
43. Subtract $2x^2 - 3x$ from the sum of $4x^2 - 5$ and $2x - 7$.
44. Subtract $5a^2 - 3a$ from the sum of $3a - 3$ and $5a^2 + 5$.
45. Subtract the sum of $3y^2 - 3y$ and $5y^2 + 3y$ from $2y^2 - 8y$.
46. Subtract the sum of $7r^3 - 4r^2$ and $-3r^3 + 4r^2$ from $2r^3 + 3r^2$.

Add using the vertical method.

47. $2w^2 + 7$, $3w - 5$, and $4w^2 - 5w$
48. $3x^2 - 4x - 2$, $6x - 3$, and $2x^2 + 8$
49. $3x^2 + 3x - 4$, $4x^2 - 3x - 3$, and $2x^2 - x + 7$
50. $5x^2 + 2x - 4$, $x^2 - 2x - 3$, and $2x^2 - 4x - 3$

Subtract using the vertical method.

51. $3a^2 - 2a$ from $5a^2 + 3a$
52. $6r^3 + 4r^2$ from $4r^3 - 2r^2$
53. $5x^2 - 6x + 7$ from $8x^2 - 5x + 7$
54. $8x^2 - 4x + 2$ from $9x^2 - 8x + 6$
55. $5x^2 - 3x$ from $8x^2 - 9$
56. $7x^2 + 6x$ from $9x^2 - 3$

Perform the indicated operations.

57. $[(9x^2 - 3x + 5) - (3x^2 + 2x - 1)] - (x^2 - 2x - 3)$
58. $[(5x^2 + 2x - 3) - (-2x^2 + x - 2)] - (2x^2 + 3x - 5)$

In Exercises 59 to 66, $f(x)$ and $g(x)$ are given. Let $h(x) = f(x) + g(x)$. Find (a) $h(x)$, (b) $f(1) + g(1)$, and (c) use the results of part (a) to find $h(1)$.

59. $f(x) = 5x - 3$ and $g(x) = 4x + 7$
60. $f(x) = 7x^2 + 3x$ and $g(x) = 5x^2 - 7x$
61. $f(x) = 5x^2 + 3x$ and $g(x) = 4x + 2x^2$
62. $f(x) = 8x^2 - 3x + 10$ and $g(x) = 7x^2 + 2x - 12$

Section 5.3 ■ Polynomials: Addition and Subtraction

63. (a) $-x^2 - 2x - 2$; (b) -5; (c) -5

64. (a) $-3x^3 + 3x^2 + 7x + 8$; (b) 15; (c) 15

65. $5x^3 - 5x^2 - 11x - 25$; (b) -36; (c) -36

66. (a) $3x^3 + 5x^2 + 19x - 14$; (b) 13; (c) 13

67. (a) $2x + 13$; (b) 15; (c) 15

68. (a) $-3x - 5$; (b) -8; (c) -8

69. (a) $12x^2 - x$; (b) 11; (c) 11

70. (a) $-7x^2 + 6x$; (b) -1; (c) -1

71. (a) $3x^2 - 2x - 7$; (b) -6; (c) -6

72. (a) $-2x^3 + x^2 - 8x + 8$; (b) -1; (c) -1

73. (a) $-3x^2 + 7x - 5$; (b) -1; (c) -1

74. (a) $-4x^2 - 7x + 3$; (b) -8; (c) -8

75. $a = 3$; $b = 5$; $c = 0$; $d = -1$

76. $a = 1$; $b = -4$; $c = -2$; $d = 6$

77. $28x + 4$

78. $12x + 4$

79. $-x^2 + 65x - 150$

80. $2y^2 - 3y + 8$

63. $f(x) = -3x^2 - 5x - 7$ and $g(x) = 2x^2 + 3x + 5$

64. $f(x) = 2x^3 + 5x^2 + 8$ and $g(x) = -5x^3 - 2x^2 + 7x$

65. $f(x) = -5x^2 - 3x - 15$ and $g(x) = 5x^3 - 8x - 10$

66. $f(x) = 5x^2 + 12x - 5$ and $g(x) = 3x^3 + 7x - 9$

In Exercises 67 to 74, $f(x)$ and $g(x)$ are given. Let $h(x) = f(x) - g(x)$. Find (a) $h(x)$, (b) $f(1) - g(1)$, and (c) use the results of part (a) to find $h(1)$.

67. $f(x) = 7x + 10$ and $g(x) = 5x - 3$

68. $f(x) = 5x - 12$ and $g(x) = 8x - 7$

69. $f(x) = 7x^2 - 3x$ and $g(x) = -5x^2 - 2x$

70. $f(x) = -10x^2 + 3x$ and $g(x) = -3x^2 - 3x$

71. $f(x) = 8x^2 - 5x - 7$ and $g(x) = 5x^2 - 3x$

72. $f(x) = 5x^3 - 2x^2 - 8x$ and $g(x) = 7x^3 - 3x^2 - 8$

73. $f(x) = 5x^2 - 5$ and $g(x) = 8x^2 - 7x$

74. $f(x) = 5x^2 - 7x$ and $g(x) = 9x^2 - 3$

Find values for *a*, *b*, *c*, and *d* so that the following equations are true.

75. $3ax^4 - 5x^3 + x^2 - cx + 2 = 9x^4 - bx^3 + x^2 - 2d$

76. $(4ax^3 - 3bx^2 - 10) - 3(x^3 + 4x^2 - cx - d) = x^3 - 6x + 8$

77. Geometry. A rectangle has sides of $8x + 9$ and $6x - 7$. Find the polynomial that represents its perimeter.

78. Geometry. A triangle has sides $3x + 7$, $4x - 9$, and $5x + 6$. Find the polynomial that represents its perimeter.

79. Business. The cost of producing x units of an item is $C = 150 + 25x$. The revenue for selling x units is $R = 90x - x^2$. The profit is given by the revenue minus the cost. Find the polynomial that represents profit.

80. Business. The revenue for selling y units is $R = 3y^2 - 2y + 5$ and the cost of producing y units is $y^2 + y - 3$. Find the polynomial that represents profit.

SECTION 5.4 Multiplying of Polynomials and Special Products

5.4 OBJECTIVES

1. Find the product of a monomial and a polynomial
2. Find the product of two polynomials
3. Square a polynomial
4. Find the product of two binomials that differ only in sign

You have already had some experience in multiplying polynomials. In Section 5.1 we stated the product rule for exponents and used that rule to find the product of two monomials. Let's review briefly.

To Find the Product of Monomials

Step 1 Multiply the coefficients.
Step 2 Use the product rule for exponents to combine the variables:

$$ax^m \cdot bx^n = abx^{m+n}$$

Let's look at an example in which we multiply two monomials.

Example 1

Multiplying Monomials

Multiply $3x^2y$ and $2x^3y^5$.

Write

Once again we have used the commutative and associative properties to rewrite the problem.

$$(3x^2y)(2x^3y^5)$$
$$= (3 \cdot 2)(x^2 \cdot x^3)(y \cdot y^5)$$

  Multiply the coefficients.   Add the exponents.

$$= 6x^5y^6$$

 CHECK YOURSELF 1

Multiply.

(a) $(5a^2b)(3a^2b^4)$ (b) $(-3xy)(4x^3y^5)$

Section 5.4 ■ Multiplying of Polynomials and Special Products

Our next task is to find the product of a monomial and a polynomial. Here we use the distributive property, which we introduced in Section 0.5. That property leads us to the following rule for multiplication.

You might want to review Section 0.5 before going on.

To Multiply a Polynomial by a Monomial

Use the distributive property to multiply each term of the polynomial by the monomial and simplify the result.

Distributive property:
$a(b + c) = ab + ac$

Example 2

Multiplying a Monomial and a Binomial

(a) Multiply $2x + 3$ by x.

Write

$$x(2x + 3)$$
$$= x \cdot 2x + x \cdot 3$$
$$= 2x^2 + 3x$$

Multiply x by $2x$ and then by 3, the terms of the polynomial. That is, "distribute" the multiplication over the sum.

Note: With practice you will do this step mentally.

(b) Multiply $2a^3 + 4a$ by $3a^2$.

Write

$$3a^2(2a^3 + 4a)$$
$$= 3a^2 \cdot 2a^3 + 3a^2 \cdot 4a = 6a^5 + 12a^3$$

✓ **CHECK YOURSELF 2**

Multiply.

(a) $2y(y^2 + 3y)$ (b) $3w^2(2w^3 + 5w)$

The patterns of Example 2 extend to *any* number of terms.

Example 3

Multiplying a Monomial and a Polynomial

Multiply the following.

(a) $3x(4x^3 + 5x^2 + 2)$
$= 3x \cdot 4x^3 + 3x \cdot 5x^2 + 3x \cdot 2 = 12x^4 + 15x^3 + 6x$

Again we have shown all the steps of the process. With practice you can write the product directly, and you should try to do so.

(b) $-5c(4c^2 - 8c)$

$= (-5c)(4c^2) - (-5c)(8c) = -20c^3 + 40c^2$

(c) $3c^2d^2(7cd^2 - 5c^2d^3)$

$= 3c^2d^2 \cdot 7cd^2 - (3c^2d^2)(5c^2d^3) = 21c^3d^4 - 15c^4d^5$

✓ **CHECK YOURSELF 3**

Multiply.

(a) $3(5a^2 + 2a + 7)$ (b) $4x^2(8x^3 - 6)$

(c) $-5m(8m^2 - 5m)$ (d) $9a^2b(3a^3b - 6a^2b^4)$

Example 4 **Multiplying Binomials**

(a) Multiply $x + 2$ by $x + 3$.

We can think of $x + 2$ as a single quantity and apply the distributive property.

Note that this ensures that each term, x and 2, of the first binomial is multiplied by each term, x and 3, of the second binomial.

$(x + 2)(x + 3)$ Multiply $x + 2$ by x and then by 3.

$= (x + 2)x + (x + 2)3$

$= x \cdot x + 2 \cdot x + x \cdot 3 + 2 \cdot 3$

$= x^2 + 2x + 3x + 6$

$= x^2 + 5x + 6$

(b) Multiply $a - 3$ by $a - 4$.

$(a - 3)(a - 4)$ (Think of $a - 3$ as a single quantity and distribute.)

$= (a - 3)a - (a - 3)(4)$

$= a \cdot a - 3 \cdot a - [(a \cdot 4) - (3 \cdot 4)]$

$= a^2 - 3a - (4a - 12)$ Note that the parentheses are needed here because a *minus sign* precedes the binomial.

✓ **CHECK YOURSELF 4**

Multiply.

(a) $(x + 4)(x + 5)$ (b) $(y + 5)(y - 6)$

Section 5.4 ■ Multiplying of Polynomials and Special Products

Fortunately, there is a pattern to this kind of multiplication that allows you to write the product of the two binomials directly without going through all these steps. We call it the **FOIL method** of multiplying. The reason for this name will be clear as we look at the process in more detail.

To multiply $(x + 2)(x + 3)$:

Remember this by F!

1. $(x + 2)(x + 3)$
   $x \cdot x$ Find the product of the *first* terms of the factors.

Remember this by O!

2. $(x + 2)(x + 3)$
   $x \cdot 3$ Find the product of the *outer* terms.

Remember this by I!

3. $(x + 2)(x + 3)$
   $2 \cdot x$ Find the product of the *inner* terms.

Remember this by L!

4. $(x + 2)(x + 3)$
   $2 \cdot 3$ Find the product of the *last* terms.

Combining the four steps, we have

$$(x + 2)(x + 3)$$
$$= x^2 + 3x + 2x + 6$$
$$= x^2 + 5x + 6$$

It's called FOIL to give you an easy way of remembering the steps: *First, Outer, Inner,* and *Last*.

With practice, the FOIL method will let you write the products quickly and easily. Consider Example 5, which illustrates this approach.

Example 5 — Using the FOIL Method

Find the following products, using the FOIL method.

(a) $(x + 4)(x + 5)$

F: $x \cdot x$
L: $4 \cdot 5$
I: $4x$
O: $5x$

When possible, you should combine the outer and inner products mentally and write just the final product.

$= x^2 + 5x + 4x + 20$
     F   O   I   L
$= x^2 + 9x + 20$

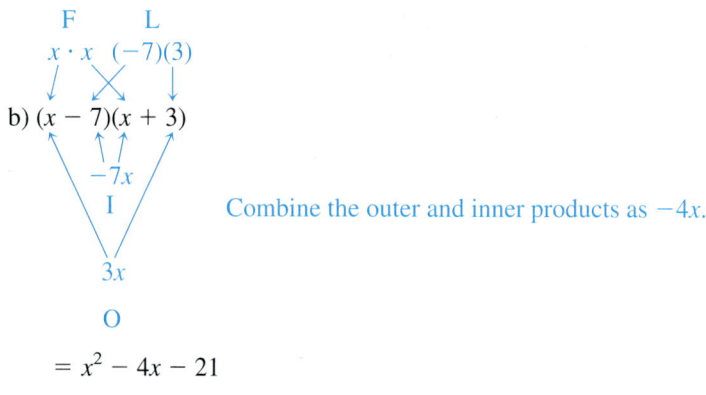

b) $(x - 7)(x + 3)$

Combine the outer and inner products as $-4x$.

$= x^2 - 4x - 21$

✓ CHECK YOURSELF 5

Multiply.

(a) $(x + 6)(x + 7)$ (b) $(x + 3)(x - 5)$ (c) $(x - 2)(x - 8)$

Using the FOIL method, you can also find the product of binomials with leading coefficients other than 1 or with more than one variable.

Example 6

Using the FOIL Method

Find the following products, using the FOIL method.

(a) $(4x - 3)(3x + 2)$

Combine: $-9x + 9x = -x$

$= 12x^2 - x - 6$

(b) $(3x - 5y)(2x - 7y)$

Combine: $-10xy + -21xy = -31xy$

$= 6x^2 - 31xy + 35y^2$

Section 5.4 ■ Multiplying of Polynomials and Special Products

The following rule summarizes our work in multiplying binomials.

> **To Multiply Two Binomials**
>
> Step 1 Find the first term of the product of the binomials by multiplying the first terms of the binomials (F).
> Step 2 Find the middle term of the product as the sum of the outer and inner products (O + I).
> Step 3 Find the last term of the product by multiplying the last terms of the binomials (L).

✓ **CHECK YOURSELF 6**

Multiply.

(a) $(5x + 2)(3x - 7)$ (b) $(4a - 3b)(5a - 4b)$ (c) $(3m + 5n)(2m + 3n)$

The FOIL method works well when multiplying any two binomials. But what if one of the factors has three or more terms? The vertical format, shown in Example 7, works for factors with any number of terms.

Example 7

Using the Vertical Method

Multiply $x^2 - 5x + 8$ by $x + 3$.

Step 1
$$\begin{array}{r} x^2 - 5x + 8 \\ x + 3 \\ \hline 3x^2 - 15x + 24 \end{array}$$

Multiply each term of $x^2 - 5x + 8$ by 3.

Step 2
$$\begin{array}{r} x^2 - 5x + 8 \\ x + 3 \\ \hline 3x^2 - 15x + 24 \\ x^3 - 5x^2 + 8x \end{array}$$

Now multiply each term by x.

Note that this line is shifted over so that like terms are in the same columns.

Note: Using this vertical method ensures that each term of one factor multiples each term of the other. That's why it works!

Step 3
$$\begin{array}{r} x^2 - 5x + 8 \\ x + 3 \\ \hline 3x^2 - 15x + 24 \\ x^3 - 5x^2 + 8x \\ \hline x^3 - 2x^2 - 7x + 24 \end{array}$$

Now add to combine like terms to write the product.

✓ **CHECK YOURSELF 7**

Multiply $2x^2 - 5x + 3$ by $3x + 4$.

Certain products occur frequently enough in algebra that it is worth learning special formulas for dealing with them. First, let's look at the **square of a binomial,** which is the product of two equal binomial factors.

$$(x + y)^2 = (x + y)(x + y)$$
$$= x^2 + 2xy + y^2$$
$$(x - y)^2 = (x - y)(x - y)$$
$$= x^2 - 2xy + y^2$$

The patterns above lead us to the following rule.

> **To Square a Binomial**
>
> Step 1   Find the first term of the square by squaring the first term of the binomial.
> Step 2   Find the middle term of the square as twice the product of the two terms of the binomial.
> Step 3   Find the last term of the square by squaring the last term of the binomial.

Example 8 **Squaring a Binomial**

(a) $(x + 3)^2 = x^2 + 2 \cdot x \cdot 3 + 3^2$

    Square of first term | Twice the product of the two terms | Square of the last term

$$= x^2 + 6x + 9$$

Caution

A very common mistake in squaring binomials is to forget the middle term.

(b) $(3a + 4b)^2 = (3a)^2 + 2(3a)(4b) + (4b)^2$
$$= 9a^2 + 24ab + 16b^2$$

(c) $(y - 5)^2 = y^2 + 2 \cdot y \cdot (-5) + (-5)^2$
$$= y^2 - 10y + 25$$

Section 5.4 ■ Multiplying of Polynomials and Special Products

Again we have shown all the steps. With practice you can write just the square.

(d) $(5c - 3d)^2 = (5c)^2 + 2(5c)(-3d) + (-3d)^2$
$= 25c^2 - 30cd + 9d^2$

✓ CHECK YOURSELF 8

Multiply.

(a) $(2x + 1)^2$ (b) $(4x - 3y)^2$

Example 9

Squaring a Binomial

Find $(y + 4)^2$.

$(y + 4)^2$ is *not* equal to $y^2 + 4^2$ or $y^2 + 16$

The correct square is

$(y + 4)^2 = y^2 + 8y + 16$

The middle term is twice the product of y and 4.

You should see that $(2 + 3)^2 \neq 2^2 + 3^2$ because $5^2 \neq 4 + 9$

✓ CHECK YOURSELF 9

Multiply.

(a) $(x + 5)^2$ (b) $(3a + 2)^2$ (c) $(y - 7)^2$ (d) $(5x - 2y)^2$

A second special product will be very important in the next chapter, which deals with factoring. Suppose the form of a product is

$(x + y)(x - y)$

The two factors differ only in sign.

Let's see what happens when we multiply.

$(x + y)(x - y)$
$= x^2 - xy + xy - y^2$
$\quad\quad\quad\quad\;\; = 0$
$= x^2 - y^2$

Since the middle term becomes 0, we have the following rule.

> **Special Product**
>
> The product of two binomials that differ only in the sign between the terms is the square of the first term minus the square of the second term.

Let's look at the application of this rule in Example 10.

Example 10 Multiplying Polynomials

Multiply each pair of factors.

(a) $(x + 5)(x - 5) = x^2 - 5^2$

   ↑ Square of the first term   ↑ Square of the second term

   $= x^2 - 25$

Note: $(2y)^2 = (2y)(2y) = 4y^2$

(b) $(x + 2y)(x - 2y) = x^2 - (2y)^2$

   ↑ Square of the first term   ↑ Square of the second term

   $= x^2 - 4y^2$

(c) $(3m + n)(3m - n) = 9m^2 - n^2$

(d) $(4a - 3b)(4a + 3b) = 16a^2 - 9b^2$

✓ CHECK YOURSELF 10

Find the products.

(a) $(a - 6)(a + 6)$ (b) $(x - 3y)(x + 3y)$

(c) $(5n + 2p)(5n - 2p)$ (d) $(7b - 3c)(7b + 3c)$

When finding the product of three or more factors, it is useful to first look for the pattern in which two binomials differ only in their sign. Finding this product first will make it easier to find the product of all the factors.

Example 11

Multiplying Polynomials

(a) $x(x - 3)(x + 3)$ — These binomials differ only in the sign.

$= x(x^2 - 9)$

$= x^3 - 9x$

(b) $(x + 1)(x - 5)(x + 5)$ — These binomials differ only in the sign.

$= (x + 1)(x^2 - 25)$ — With two binomials, use the FOIL method.

$= x^3 + x^2 - 25x - 25$

(c) $(2x - 1)(x + 3)(2x + 1)$

$(2x - 1)\,(x + 3)\,(2x + 1)$ — These two binomials differ only in the sign of the second term. We can use the commutative property to rearrange the terms.

$= (x + 3)(2x - 1)(2x + 1)$

$= (x + 3)(4x^2 - 1)$

$= 4x^3 + 12x^2 - x - 3$

✓ CHECK YOURSELF 11

Multiply.

(a) $3x(x - 5)(x + 5)$

(b) $(x - 4)(2x + 3)(2x - 3)$

(c) $(x - 7)(3x - 1)(x + 7)$

✓ CHECK YOURSELF ANSWERS

1. (a) $15a^4b^5$; (b) $-12x^4y^6$. **2.** (a) $2y^3 + 6y^2$; (b) $6w^5 + 15w^3$.
3. (a) $15a^2 + 6a + 21$; (b) $32x^5 - 24x^2$; (c) $-40m^3 + 25m^2$; (d) $27a^5b^2 - 54a^4b^5$.
4. (a) $x^2 + 9x + 20$; (b) $y^2 - y - 30$.   **5.** (a) $x^2 + 13x + 42$; (b) $x^2 - 2x - 15$;
(c) $x^2 - 10x + 16$.   **6.** (a) $15x^2 - 29x - 14$; (b) $20a^2 - 31ab + 12b^2$;
(c) $6m^2 + 19mn + 15n^2$.   **7.** $6x^3 - 7x^2 - 11x + 12$.   **8.** (a) $4x^2 + 4x + 1$;
(b) $16x^2 - 24xy + 9y^2$.   **9.** (a) $x^2 + 10x + 25$; (b) $9a^2 + 12a + 4$;
(c) $y^2 - 14y + 49$; (d) $25x^2 - 20xy + 4y^2$.   **10.** (a) $a^2 - 36$; (b) $x^2 - 9y^2$;
(c) $25n^2 - 4p^2$; (d) $49b^2 - 9c^2$.   **11.** (a) $3x^3 - 75x$; (b) $4x^3 - 16x^2 - 9x + 36$;
(c) $3x^3 - x^2 - 147x + 49$.

Exercises - 5.4

1. $15x^5$
2. $28a^{11}$
3. $-28b^{10}$
4. $-56y^{10}$
5. $40p^{13}$
6. $-54m^{15}$
7. $-12m^6$
8. $15r^8$
9. $32x^5y^3$
10. $21r^6s^7$
11. $-6m^9n^3$
12. $-42a^7b^6$
13. $10x + 30$
14. $28b - 20$
15. $12a^2 + 15a$
16. $10x^2 - 35x$
17. $12s^4 - 21s^3$
18. $27a^5 + 45a^3$
19. $8x^3 - 4x^2 + 2x$
20. $20m^4 - 15m^3 + 10m$
21. $6x^3y^2 + 3x^2y^3 + 15x^2y^2$
22. $5a^2b^3 - 15a^2b^2 + 25ab^3$
23. $18m^4n^2 - 12m^3n^2 + 6m^3n^3$
24. $16p^2q^3 - 24p^2q^2 + 40pq^3$
25. $x^2 + 5x + 6$
26. $a^2 - 10a + 21$
27. $m^2 - 14m + 45$
28. $b^2 + 12b + 35$
29. $p^2 - p - 56$
30. $x^2 - x - 90$
31. $w^2 + 30w + 200$
32. $s^2 - 20s + 96$
33. $3x^2 - 29x + 40$
34. $4w^2 + 13w - 35$
35. $6x^2 - x - 12$
36. $15a^2 + 38a + 7$
37. $12a^2 - 31ab + 9b^2$
38. $21s^2 + 47st - 24t^2$
39. $21p^2 - 13pq - 20q^2$
40. $10x^2 - 13xy + 4y^2$
41. $6x^2 + 23xy + 20y^2$
42. $16x^2 - 8xy - 15y^2$

Multiply.

1. $(5x^2)(3x^3)$
2. $(7a^5)(4a^6)$
3. $(-2b^2)(14b^8)$
4. $(14y^4)(-4y^6)$
5. $(-10p^6)(-4p^7)$
6. $(-6m^8)(9m^7)$
7. $(4m^5)(-3m)$
8. $(-5r^7)(-3r)$
9. $(4x^3y^2)(8x^2y)$
10. $(-3r^4s^2)(-7r^2s^5)$
11. $(-3m^5n^2)(2m^4n)$
12. $(7a^3b^5)(-6a^4b)$
13. $5(2x + 6)$
14. $4(7b - 5)$
15. $3a(4a + 5)$
16. $5x(2x - 7)$
17. $3s^2(4s^2 - 7s)$
18. $9a^2(3a^3 + 5a)$
19. $2x(4x^2 - 2x + 1)$
20. $5m(4m^3 - 3m^2 + 2)$
21. $3xy(2x^2y + xy^2 + 5xy)$
22. $5ab^2(ab - 3a + 5b)$
23. $6m^2n(3m^2n - 2mn + mn^2)$
24. $8pq^2(2pq - 3p + 5q)$

Multiply.

25. $(x + 3)(x + 2)$
26. $(a - 3)(a - 7)$
27. $(m - 5)(m - 9)$
28. $(b + 7)(b + 5)$
29. $(p - 8)(p + 7)$
30. $(x - 10)(x + 9)$
31. $(w + 10)(w + 20)$
32. $(s - 12)(s - 8)$
33. $(3x - 5)(x - 8)$
34. $(w + 5)(4w - 7)$
35. $(2x - 3)(3x + 4)$
36. $(5a + 1)(3a + 7)$
37. $(3a - b)(4a - 9b)$
38. $(7s - 3t)(3s + 8t)$
39. $(3p - 4q)(7p + 5q)$
40. $(5x - 4y)(2x - y)$
41. $(2x + 5y)(3x + 4y)$
42. $(4x - 5y)(4x + 3y)$

Section 5.4 ■ Multiplying of Polynomials and Special Products

43. $x^2 + 10x + 25$

44. $y^2 + 18y + 81$

45. $w^2 - 12w + 36$

46. $a^2 - 16a + 64$

47. $z^2 + 24z + 144$

48. $p^2 - 40p + 400$

49. $4a^2 - 4a + 1$

50. $9x^2 - 12x + 4$

51. $36m^2 + 12m + 1$

52. $49b^2 - 28b + 4$

53. $9x^2 - 6xy + y^2$

54. $25m^2 + 10mn + n^2$

55. $4r^2 + 20rs + 25s^2$

56. $9a^2 - 24ab + 16b^2$

57. $64a^2 - 144ab + 81b^2$

58. $49p^2 + 84pq + 36q^2$

59. $x^2 + x + \dfrac{1}{4}$

60. $w^2 - \dfrac{1}{2}w + \dfrac{1}{16}$

61. $x^2 - 36$ 62. $y^2 - 64$

63. $m^2 - 144$

64. $w^2 - 100$

65. $x^2 - \dfrac{1}{4}$ 66. $x^2 - \dfrac{4}{9}$

67. $p^2 - 0.16$

68. $m^2 - 0.36$

69. $a^2 - 9b^2$

70. $p^2 - 16q^2$

71. $16r^2 - s^2$

72. $49x^2 - y^2$

73. $64w^2 - 25z^2$

74. $49c^2 - 4d^2$

75. $25x^2 - 81y^2$

76. $36s^2 - 25t^2$

Find each of the following squares.

43. $(x + 5)^2$

44. $(y + 9)^2$

45. $(w - 6)^2$

46. $(a - 8)^2$

47. $(z + 12)^2$

48. $(p - 20)^2$

49. $(2a - 1)^2$

50. $(3x - 2)^2$

51. $(6m + 1)^2$

52. $(7b - 2)^2$

53. $(3x - y)^2$

54. $(5m + n)^2$

55. $(2r + 5s)^2$

56. $(3a - 4b)^2$

57. $(8a - 9b)^2$

58. $(7p + 6q)^2$

59. $\left(x + \dfrac{1}{2}\right)^2$

60. $\left(w - \dfrac{1}{4}\right)^2$

Find each of the following products.

61. $(x - 6)(x + 6)$

62. $(y + 8)(y - 8)$

63. $(m + 12)(m - 12)$

64. $(w - 10)(w + 10)$

65. $\left(x - \dfrac{1}{2}\right)\left(x + \dfrac{1}{2}\right)$

66. $\left(x + \dfrac{2}{3}\right)\left(x - \dfrac{2}{3}\right)$

67. $(p - 0.4)(p + 0.4)$

68. $(m - 0.6)(m + 0.6)$

69. $(a - 3b)(a + 3b)$

70. $(p + 4q)(p - 4q)$

71. $(4r - s)(4r + s)$

72. $(7x - y)(7x + y)$

73. $(8w + 5z)(8w - 5z)$

74. $(7c + 2d)(7c - 2d)$

75. $(5x - 9y)(5x + 9y)$

76. $(6s - 5t)(6s + 5t)$

Chapter 5 • Polynomials

77. $24x^3 - 10x^2 - 4x$

78. $12x^3 - 3x$

79. $80a^3 - 45a$

80. $54m^3 - 162m^2 + 84m$

81. $60s^3 - 39s^2 + 6s$

82. $28w^3 - 63w$

83. $x^3 - 4x^2 + x + 6$

84. $y^3 - 3y^2 - 10y + 24$

85. $a^3 - 3a^2 + 3a - 1$

86. $x^3 + 3x^2 + 3x + 1$

87. $\dfrac{x^2}{3} + \dfrac{11x}{45} - \dfrac{4}{15}$

88. $\dfrac{x^2}{4} + \dfrac{29}{80}x - \dfrac{9}{20}$

89. $x^2 - y^2 + 4y - 4$

90. $x^2 - y^2 + 6y - 9$

91. False

92. False

93. True

94. True

95. $6x^2 - 11x - 35$ cm²

96. $3y^2 + \dfrac{5}{2}y - \dfrac{21}{2}$ in²

97. $10x - 3x^2$

98. $100x - 2x^3$

99. $25x^2 - 40x + 16$

Multiply.

77. $2x(3x - 2)(4x + 1)$

78. $3x(2x + 1)(2x - 1)$

79. $5a(4a - 3)(4a + 3)$

80. $6m(3m - 2)(3m - 7)$

81. $3s(5s - 2)(4s - 1)$

82. $7w(2w - 3)(2w + 3)$

83. $(x - 2)(x + 1)(x - 3)$

84. $(y + 3)(y - 2)(y - 4)$

85. $(a - 1)^3$

86. $(x + 1)^3$

Multiply the following.

87. $\left(\dfrac{x}{2} + \dfrac{2}{3}\right)\left(\dfrac{2x}{3} - \dfrac{2}{5}\right)$

88. $\left(\dfrac{x}{3} + \dfrac{3}{4}\right)\left(\dfrac{3x}{4} - \dfrac{3}{5}\right)$

89. $[x + (y - 2)][x - (y - 2)]$

90. $[x + (3 - y)][x - (3 - y)]$

Label the following as true or false.

91. $(x + y)^2 = x^2 + y^2$

92. $(x - y)^2 = x^2 - y^2$

93. $(x + y)^2 = x^2 + 2xy + y^2$

94. $(x - y)^2 = x^2 - 2xy + y^2$

95. **Length.** The length of a rectangle is given by $3x + 5$ centimeters (cm) and the width is given by $2x - 7$ cm. Express the area of the rectangle in terms of x.

96. **Area.** The base of a triangle measures $3y + 7$ inches (in.) and the height is $2y - 3$ in. Express the area of the triangle in terms of y.

97. **Revenue.** The price of an item is given by $p = 10 - 3x$. If the revenue generated is found by multiplying the number of items (x) sold by the price of an item, find the polynomial which represents the revenue.

98. **Revenue.** The price of an item is given by $p = 100 - 2x^2$. Find the polynomial that represents the revenue generated from the sale of x items.

99. **Tree planting.** Suppose an orchard is planted with trees in straight rows. If there are $5x - 4$ rows with $5x - 4$ trees in each row, how many trees are there in the orchard?

Section 5.4 ■ Multiplying of Polynomials and Special Products   **375**

100. $9x^2 - 12x + 4$ cm^2

101. $x(x + 2)$ or $x^2 + 2x$

102. $x(3x - 6)$ or $3x^2 - 6x$

100. Area of a square. A square has sides of length $3x - 2$ centimeters (cm). Express the area of the square as a polynomial.

101. Area of a rectangle. The length and width of a rectangle are given by two consecutive odd integers. Write an expression for the area of the rectangle.

102. Area of a rectangle. The length of a rectangle is 6 less than three times the width. Write an expression for the area of the rectangle.

103. Work with another student to complete this table and write the polynomial. A paper box is to be made from a piece of cardboard 20 inches (in.) wide and 30 in. long. The box will be formed by cutting squares out of each of the four corners and folding up the sides to make a box.

If x is the dimension of the side of the square cut out of the corner, when the sides are folded up, the box will be x inches tall. You should use a piece of paper to try this to see how the box will be made. Complete the following chart.

Length of Side of Corner Square	Length of Box	Width of Box	Depth of Box	Volume of Box
1 in.				
2 in.				
3 in.				
n in.				

Write general formulas for the width, length, and height of the box and a general formula for the *volume* of the box, and simplify it by multiplying. The variable will be the height, the side of the square cut out of the corners. What is the highest power of the variable in the polynomial you have written for the volume? Extend the table to decide what the dimensions are for a box with maximum volume. Draw a sketch of this box and write in the dimensions.

104. Complete the following statement: $(a + b)^2$ is not equal to $a^2 + b^2$ because. . . . But, wait! Isn't $(a + b)^2$ *sometimes* equal to $a^2 + b^2$? What do you think?

105. Is $(a + b)^3$ ever equal to $a^3 + b^3$? Explain.

106. In the following figures, identify the length and the width of the square:

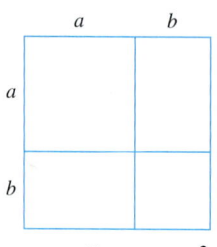

Length = _____

Width = _____

Area = _____

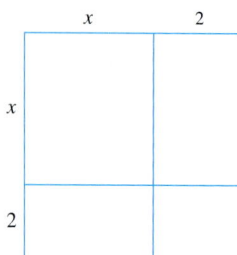

Length = _____

Width = _____

Area = _____

107. The square shown is x units on a side. The area is _____.

Draw a picture of what happens when the sides are doubled. The area is _____.

Continue the picture to show what happens when the sides are tripled. The area is _____.

If the sides are quadrupled, the area is _____.

In general, if the sides are multiplied by n, the area is _____.

If each side is increased by 3, the area is increased by _____.

If each side is decreased by 2, the area is decreased by _____.

In general, if each side is increased by n, the area is increased by _____, and if each side is decreased by n, the area is decreased by _____.

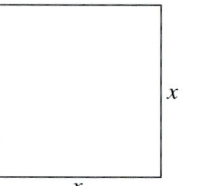

For each of the following problems, let x represent the number, then write an expression for the product.

108. The product of 6 more than a number and 6 less than that number

108. $x^2 - 36$

109. The square of 5 more than a number

109. $x^2 + 10x + 25$

110. The square of 4 less than a number

110. $x^2 - 8x + 16$

111. The product of 5 less than a number and 5 more than that number

111. $x^2 - 25$

Note that $(28)(32) = (30 - 2)(30 + 2) = 900 - 4 = 896$. Use this pattern to find each of the following products.

112. 2499
113. 891
114. 884
115. 9996
116. 3575
117. 3584

112. $(49)(51)$ **113.** $(27)(33)$

114. $(34)(26)$ **115.** $(98)(102)$

116. $(55)(65)$ **117.** $(64)(56)$

Summary Exercises · 5

This summary exercise set is provided to give you practice with each of the objectives in the chapter. Each exercise is keyed to the appropriate chapter section.

[5.1] Simplify each expression, using the properties of exponents.

1. $r^4 \cdot r^9$
2. $(-3)^2(-3)^3$
3. $(a^3b^2)(a^8b^3)$
4. $(6c^0d^4)(-3c^2d^2)$
5. $\dfrac{x^{12}}{x^5}$
6. $\dfrac{(2x-1)^8}{(2x-1)^5}$
7. $(x^2y)^3$
8. $(2c^3d^4)^3$
9. $(2a^3)^0(-3a^4)^2$
10. $\left(\dfrac{x}{y^2}\right)^2$
11. $\left(\dfrac{3m^2n^3}{p^4}\right)^3$
12. $\left(\dfrac{a^8}{b^4}\right)\left(\dfrac{b^2}{2a^2}\right)^3$

13. Given $f(x) = 4x^2 + 5x - 6$, find $f(-2)$.
14. Given $f(x) = 5x^2 - 4x + 8$, find $f(3)$.
15. Given $f(x) = -2x^3 + 4x^2 - 5x + 8$, find $f(3)$.
16. Given $f(x) = 3x^3 + 5x^2 - 2x + 6$, find $f(-1)$.
17. Given $f(x) = -2x^4 - 5x^2 + 3x - 6$, find $f(2)$.
18. Given $f(x) = -x^5 - 3x^4 + 5x^3 - x - 4$, find $f(-2)$.

[5.2] Classify each of the following polynomials as a monomial, binomial, or trinomial.

19. $5x^3 - 2x^2$
20. $7x^5$
21. $4x^5 - 8x^3 + 5$
22. $x^3 + 2x^2 - 5x + 3$
23. $9a^2 - 18a$

[5.2] Arrange in descending-exponent form, and give the degree of each polynomial.

24. $5x^5 + 3x^2$
25. $9x$
26. $6x^2 + 4x^4 + 6$
27. $5 + x$
28. -8
29. $9x^4 - 3x + 7x^6$

[5.3] Add or subtract as indicated.

30. $9a^2 - 5a$ and $12a^2 + 3a$
31. $5x^2 + 3x - 5$ and $4x^2 - 6x - 2$
32. $5y^3 - 3y^2$ and $4y + 3y^2$
33. $4x^2 - 3x$ from $8x^2 + 5x$
34. $2x^2 - 5x - 7$ from $7x^2 - 2x + 3$
35. $5x^2 + 3$ from $9x^2 - 4x$

[5.3] Perform the indicated operations.

36. Subtract $5x - 3$ from the sum of $9x + 2$ and $-3x - 7$.
37. Subtract $5a^2 - 3a$ from the sum of $5a^2 + 2$ and $7a - 7$.
38. Subtract the sum of $16w^2 - 3w$ and $8w + 2$ from $7w^2 - 5w + 2$.

---

1. r^{13}
2. $(-3)^5$
3. $a^{11}b^5$
4. $-18c^2d^6$
5. x^7
6. $(2x-1)^3$
7. x^6y^3
8. $8c^9d^{12}$
9. $9a^8$
10. $\dfrac{x^2}{y^4}$
11. $\dfrac{27m^6n^9}{p^{12}}$
12. $\dfrac{a^2b^2}{8}$
13. 0
14. 41
15. −25
16. 10
17. −52
18. −58
19. Binomial
20. Monomial
21. Trinomial
22. Not classified
23. Binomial
24. $5x^5 + 3x^2$, 5
25. $9x$, 1
26. $4x^4 + 6x^2 + 6$, 4
27. $x + 5$, 1
28. −8, 0
29. $7x^6 + 9x^4 - 3x$, 6
30. $21a^2 - 2a$
31. $9x^2 - 3x - 7$
32. $5y^3 + 4y$
33. $4x^2 + 8x$
34. $5x^2 + 3x + 10$
35. $4x^2 - 4x - 3$
36. $x - 2$
37. $10a - 5$
38. $-9w^2 - 10w$

39. $3x^2 + 9x - 6$
40. $9b^2 + 8b - 2$
41. $5x^2 - 5x + 5$
42. $2x^2 - 2x - 9$
43. $9m^2 - 8m$
44. $5a^5$
45. $6x^7$
46. $54p^5$
47. $-21a^5b^7$
48. $15x - 40$
49. $12a^2 + 28a$
50. $-10r^3s^2 + 25r^2s^2$
51. $21m^3n^2 - 14m^2n^3 + 35m^2n^2$
52. $x^2 + 9x + 20$
53. $w^2 - 19w + 90$
54. $a^2 - 49b^2$
55. $p^2 - 6pq + 9q^2$
56. $a^2 + 7ab + 12b^2$
57. $2b^2 - 13b - 24$
58. $6x^2 - 19xy + 15y^2$
59. $15r^2 - 24rs - 63s^2$
60. $y^3 - y + 6$
61. $b^3 - 2b^2 - 22b - 21$
62. $x^3 - 8$
63. $m^4 + 4m^2 - 21$
64. $2x^3 - 2x^2 - 60x$
65. $4a^3 - 24a^2b + 35ab^2$
66. $x^2 + 14x + 49$
67. $a^2 - 16a + 64$
68. $4w^2 - 20w + 25$
69. $9p^2 + 24p + 16$
70. $a^2 + 14ab + 49b^2$
71. $64x^2 - 48xy + 9y^2$

[5.3] Add using the vertical method.

39. $x^2 + 5x - 3$ and $2x^2 + 4x - 3$
40. $9b^2 - 7$ and $8b + 5$
41. $x^2 + 7$, $3x - 2$, and $4x^2 - 8x$

[5.3] Subtract using the vertical method.

42. $5x^2 - 3x + 2$ from $7x^2 - 5x - 7$
43. $8m - 7$ from $9m^2 - 7$

[5.4] Multiply.

44. $(5a^3)(a^2)$
45. $(2x^2)(3x^5)$
46. $(-9p^3)(-6p^2)$
47. $(3a^2b^3)(-7a^3b^4)$
48. $5(3x - 8)$
49. $4a(3a + 7)$
50. $(-5rs)(2r^2s - 5rs)$
51. $7mn(3m^2n - 2mn^2 + 5mn)$
52. $(x + 5)(x + 4)$
53. $(w - 9)(w - 10)$
54. $(a - 7b)(a + 7b)$
55. $(p - 3q)^2$
56. $(a + 4b)(a + 3b)$
57. $(b - 8)(2b + 3)$
58. $(3x - 5y)(2x - 3y)$
59. $(5r + 7s)(3r - 9s)$
60. $(y + 2)(y^2 - 2y + 3)$
61. $(b + 3)(b^2 - 5b - 7)$
62. $(x - 2)(x^2 + 2x + 4)$
63. $(m^2 - 3)(m^2 + 7)$
64. $2x(x + 5)(x - 6)$
65. $a(2a - 5b)(2a - 7b)$

[5.4] Find the following products.

66. $(x + 7)^2$
67. $(a - 8)^2$
68. $(2w - 5)^2$
69. $(3p + 4)^2$
70. $(a + 7b)^2$
71. $(8x - 3y)^2$
72. $(x - 5)(x + 5)$
73. $(y + 9)(y - 9)$
74. $(2m + 3)(2m - 3)$
75. $(3r - 7)(3r + 7)$
76. $(5r - 2s)(5r + 2s)$
77. $(7a + 3b)(7a - 3b)$
78. $2x(x - 5)^2$
79. $3c(c + 5d)(c - 5d)$

This test covers selected topics from the first five chapters.

Solve the following equations and check your results.

72. $x^2 - 25$
73. $y^2 - 81$
74. $4m^2 - 9$
75. $9r^2 - 49$
76. $25r^2 - 4s^2$
77. $49a^2 - 9b^2$
78. $2x^3 - 20x^2 + 50x$
79. $3c^3 - 75cd^2$

Cumulative Test • 0-5

1. $11x - 7 = 10x$
2. $-\frac{2}{3}x = 24$
3. $7x - 5 = 3x + 11$
4. $\frac{3}{5}x - 8 = 15 - \frac{2}{5}x$
5. $|x - 4| = 5$
6. $|3x + 5| - 5 = 9$

Solve the following inequalities.

7. $7x - 5 > 8x + 10$
8. $6x - 9 < 3x + 6$
9. $|x - 8| < 5$
10. $|3x - 10| \geq 8$

11. Write the equation of the line that passes through the point (3, 2) and is parallel to the line $4x - 5y = 20$.

12. Graph the function $f(x) = 4x - 5$.

Perform the indicated operations.

13. $(x^2 - 3x + 5) + (2x^2 + 5x - 9)$
14. $(3x^2 - 8x - 7) - (2x^2 - 5x + 11)$
15. $4x(3x - 5)$
16. $(2x - 5)(3x + 8)$
17. $(x + 2)(x^2 - 3x + 5)$
18. $(2x + 7)(2x - 7)$
19. $(3x - 5)^2$
20. $5x(2x - 5)^2$

21. Use the given graph to estimate (a) $f(-2)$, (b) $f(0)$, and (c) $f(3)$.

---

1. $\{7\}$
2. $\{-36\}$
3. $\{4\}$
4. $\{23\}$
5. $\{9, -1\}$
6. $\left\{3, -\frac{19}{3}\right\}$
7. $\{x | x < -15\}$
8. $\{x | x < 5\}$
9. $\{x | 3 < x < 13\}$
10. $\left\{x | x \geq 6 \text{ or } x \leq \frac{2}{3}\right\}$
11. $y = \frac{4}{5}x - \frac{2}{5}$
12. 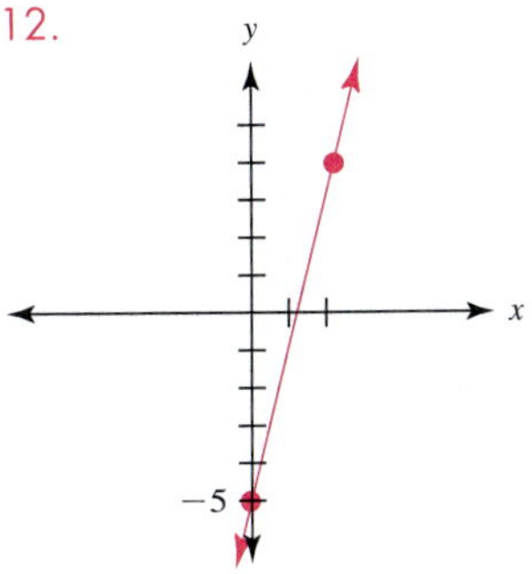
13. $3x^2 + 2x - 4$
14. $x^2 - 3x - 18$
15. $12x^2 - 20x$
16. $6x^2 + x - 40$
17. $x^3 - x^2 - x + 10$
18. $4x^2 - 49$
19. $9x^2 - 30x + 25$
20. $20x^3 - 100x^2 + 125x$

380   Chapter 5 ▪ Polynomials

21. 4; 0; −6

22. 9

23. 8

24. 65, 67

25. 4 cm by 24 cm

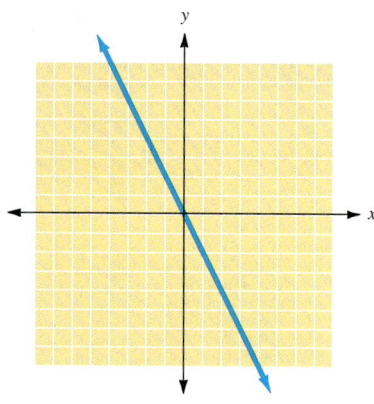

22. If $f(x) = -2x + 5$, evaluate $f(-2)$.

Solve the following problems.

23. If 7 times a number decreased by 9 is 47, find the number.

24. The sum of two consecutive odd integers is 132. What are the two integers?

25. The length of a rectangle is 4 centimeters (cm) more than 5 times the width. If the perimeter is 56 cm, what are the dimensions of the rectangle?

CHAPTER 6
Factoring Polynomials

LIST OF SECTIONS

6.1 An Introduction to Factoring

6.2 Factoring Special Polynomials

6.3 Factoring Trinomials: The ac Method

6.3* Factoring Trinomials: Trial and Error

6.4 Division of Polynomials

6.5 Solving Quadratic Equations by Factoring

Civil engineers use "polynomial splines" when designing tunnels and highways. These splines help in the design of a roadway, ensuring that changes in direction and altitude occur smoothly and gradually. For example: a road passes through a valley at 75-meters altitude and then climbs through some hills, reaching an altitude of 350 meters before descending again. The graph on the next page shows the change in altitude for 22 km of the roadway.

The road seems very steep in this graph, but remember that the y axis is the altitude measured in meters, and the x axis is horizontal distance measured in kilometers. To get a true feeling for the vertical change, the horizontal axis would have to be stretched by a factor of 1000.

Based on measurements of the distance, altitude, and change in the slope taken at intervals along the planned path of the road,

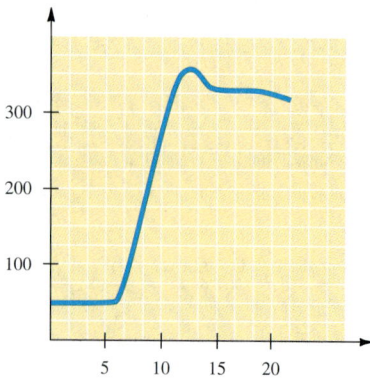

formulas are developed that model the roadway for short sections. The formulas are then pieced together to form a model of the road over several kilometers. Such formulas are found by using algebraic methods to solve systems of equations. Here are some of the splines that could be fit together to create the roadway needed in our graph; y is the altitude measured in meters and x is the horizontal distance, measured in kilometers.

First 5 km: $\quad y = 75 + 0.15x$

Next 7.5 km: $\quad y = 21.5975x^3 + 40.726x^2 - 287.296x + 693.521$

Final 5.5 km: $\quad y = 20.290101x^3 + 14.9427x^2 - 255.559x + 1771.94$

By entering these equations in a graphing calculator, you can see part of the design for the road. Be certain that you adjust your graphing window appropriately.

Polynomials are useful in many areas. In this chapter, you will learn how to solve problems involving polynomials.

SECTION 6.1 An Introduction to Factoring

6.1 OBJECTIVES

1. Remove the greatest common factor (GCF)
2. Factor by grouping

In Chapter 5 you were given factors and asked to find a product. We are now going to reverse the process. You will be given a polynomial and asked to find its factors. This is called **factoring**.

Let's start with an example from arithmetic. To *multiply* $5 \cdot 7$, you write

$$5 \cdot 7 = 35$$

To *factor* 35, you would write

$$35 = 5 \cdot 7$$

Factoring is the *reverse* of multiplication.

Now let's look at factoring in algebra. You have used the distributive property as

$$a(b + c) = ab + ac$$

For instance,

$$3(x + 5) = 3x + 15$$

3 and $x + 5$ are the factors of $3x + 15$.

To use the distributive property in factoring, we apply that property in the opposite fashion, as

$$ab + ac = a(b + c)$$

The property lets us remove the common monomial factor a from the terms of $ab + ac$. To use this in factoring, the first step is to see whether each term of the polynomial has a common monomial factor. In our earlier example,

$$3x + 15 = 3 \cdot x + 3 \cdot 5$$

Common factor

383

So, by the distributive property,

$$3x + 15 = 3(x + 5)$$

The original terms are each divided by the greatest common factor to determine the terms in parentheses.

Again, factoring is the reverse of multiplication.

To check this, multiply $3(x + 5)$.

Here is a diagram that will relate the idea of multiplication and factoring.

Multiplying →
$$3(x + 5) = 3x + 15$$
← Factoring

The first step in factoring is to identify the *greatest common factor* (GCF) of a set of terms. This is the monomial with the largest common numerical coefficient and the largest power common to both variables.

In fact, we will see that factoring out the GCF is the first method to try in any of the factoring problems we will discuss.

> The **greatest common factor (GCF)** of a polynomial is the monomial with the highest degree and the largest numerical coefficient that is a factor of each term of the polynomial.

Example 1

Finding the GCF

Find the GCF for each set of terms.

(a) 9 and 12   The largest number that is a factor of both is 3.

(b) 10, 25, 150   The GCF is 5.

(c) x^4 and x^7 The largest power common to the two variables is x^4.

(d) $12a^3$ and $18a^2$ The GCF is $6a^2$.

✓ **CHECK YOURSELF 1**

Find the GCF for each set of terms.

(a) 14, 24   (b) 9, 27, 81   (c) a^9, a^5 (d) $10x^5, 35x^4$

Checking your answer is always important and perhaps is never easier than after you have factored.

> **To Factor a Monomial from a Polynomial**
>
> Step 1 Find the *greatest common factor* (GCF) for all the terms.
> Step 2 Factor the GCF from each term, then apply the distributive property.
> Step 3 Mentally check your factoring by multiplication.

Section 6.1 • An Introduction to Factoring   **385**

Example 2 Finding the GCF of a Binomial

(a) Factor $8x^2 + 12x$.

The largest common numerical factor of 8 and 12 is 4, and x is the variable factor with the largest common power. So $4x$ is the GCF. Write

$$8x^2 + 12x = 4x \cdot 2x + 4x \cdot 3$$

  GCF

It is always a good idea to check your answer by multiplying to make sure that you get the original polynomial. Try it here. Multiply $4x$ by $2x + 3$.

Now, by the distributive property, we have

$$8x^2 + 12x = 4x(2x + 3)$$

(b) Factor $6a^4 - 18a^2$.

The GCF in this case is $6a^2$. Write

$$6a^4 - 18a^2 = 6a^2 \cdot a^2 - 6a^2 \cdot 3$$

  GCF

It is also true that
$$6a^4 - 18a^2 = 3a(2a^3 - 6a)$$
However, this is not completely factored. Do you see why? You want to find the common monomial factor with the largest possible coefficient and the largest exponent, in this case $6a^2$.

Again, using the distributive property yields

$$6a^4 - 18a^2 = 6a^2(a^2 - 3)$$

You should check this by multiplying.

✓ **CHECK YOURSELF 2**

Factor each of the following polynomials.

(a) $5x + 20$ (b) $6x^2 - 24x$ (c) $10a^3 - 15a^2$

The process is exactly the same for polynomials with more than two terms. Consider Example 3.

Example 3 Finding the GCF of a Polynomial

(a) Factor $5x^2 - 10x + 15$.

The GCF is 5.

$$5x^2 - 10x + 15 = 5 \cdot x^2 - 5 \cdot 2x + 5 \cdot 3$$

(arrows pointing to the GCF)

$$= 5(x^2 - 2x + 3)$$

(b) Factor $6ab + 9ab^2 - 15a^2$.

The GCF is $3a$.

$$6ab + 9ab^2 - 15a^2 = 3a \cdot 2b + 3a \cdot 3b^2 - 3a \cdot 5a$$

(arrows pointing to the GCF)

$$= 3a(2b + 3b^2 - 5a)$$

(c) Factor $4a^4 + 12a^3 - 20a^2$.

The GCF is $4a^2$.

$$4a^4 + 12a^3 - 20a^2 = 4a^2 \cdot a^2 + 4a^2 \cdot 3a - 4a^2 \cdot 5$$

(arrows pointing to the GCF)

$$= 4a^2(a^2 + 3a - 5)$$

(d) Factor $6a^2b + 9ab^2 + 3ab$.

Mentally note that 3, a, and b are factors of each term, so

$$6a^2b + 9ab^2 + 3ab = 3ab(2a + 3b + 1)$$

In each of these examples, you will want to check the result by multiplying the factors.

✓ CHECK YOURSELF 3

Factor each of the following polynomials.

(a) $8b^2 + 16b - 32$ (b) $4xy - 8x^2y + 12x^3$
(c) $7x^4 - 14x^3 + 21x^2$ (d) $5x^2y^2 - 10xy^2 + 15x^2y$

Factoring by Grouping

A related factoring method is called **factoring by grouping**. We introduce this method in Example 4.

Example 4 — Finding a Common Factor

(a) Factor $3x(x + y) + 2(x + y)$.

We see that *the binomial $x + y$ is a common factor* and can be removed.

Because of the commutative property, the factors can be written in either order.

$$3x(x + y) + 2(x + y)$$
$$= (x + y) \cdot 3x + (x + y) \cdot 2$$
$$= (x + y)(3x + 2)$$

(b) Factor $3x^2(x - y) + 6x(x - y) + 9(x - y)$.

We note that here the GCF is $3(x - y)$. Factoring as before, we have

$$3(x - y)(x^2 + 2x + 3)$$

✓ CHECK YOURSELF 4

Completely factor each of the polynomials.

(a) $7a(a - 2b) + 3(a - 2b)$ (b) $4x^2(x + y) - 8x(x + y) - 16(x + y)$

If the terms of a polynomial have no common factor (other than 1), factoring by grouping is the preferred method, as illustrated in Example 5.

Example 5 — Factoring by Grouping Terms

Suppose we want to factor the polynomial

$$ax - ay + bx - by$$

As you can see, the polynomial has no common factors. However, look at what happens if we separate the polynomial into *two groups* of *two terms*.

Note that our example has four terms. That is the clue for trying the factoring by grouping method.

$$ax - ay + bx - by$$
$$= \underline{ax - ay} + \underline{bx - by}$$
$$\quad\quad (1) \quad\quad\quad (2)$$

Now *each* group has a common factor, and we can write the polynomial as

$$a(x - y) + b(x - y)$$

In this form, we can see that $x - y$ is the GCF. Factoring out $x - y$, we get

$$a(x - y) + b(x - y) = (x - y)(a + b)$$

✓ CHECK YOURSELF 5

Use the factoring by grouping method.

$$x^2 - 2xy + 3x - 6y$$

Be particularly careful of your treatment of algebraic signs when applying the factoring by grouping method. Consider Example 6.

Example 6

Factoring by Grouping Terms

Factor $2x^3 - 3x^2 - 6x + 9$.
We group the polynomial as follows.

$$\underbrace{2x^3 - 3x^2}_{(1)} \underbrace{- 6x + 9}_{(2)}$$ Remove the common factor of -3 from the second two terms.

Note that $9 = (-3)(-3)$.

$$= x^2(2x - 3) - 3(2x - 3)$$
$$= (2x - 3)(x^2 - 3)$$

✓ CHECK YOURSELF 6

Factor by grouping.

$$3y^3 + 2y^2 - 6y - 4$$

It may also be necessary to change the order of the terms as they are grouped. Look at Example 7.

Example 7

Factoring by Grouping Terms

Factor $x^2 - 6yz + 2xy - 3xz$.
Grouping the terms as before, we have

$$\underbrace{x^2 - 6yz}_{(1)} + \underbrace{2xy - 3xz}_{(2)}$$

Do you see that we have accomplished nothing because there are no common factors in the first group?

We can, however, rearrange the terms to write the original polynomial as

$$\underbrace{x^2 + 2xy}_{(1)} - \underbrace{3xz - 6yz}_{(2)}$$

$= x(x + 2y) - 3z(x + 2y)$ We can now remove the common factor of $x + 2y$ in group (1) and group (2).
$= (x + 2y)(x - 3z)$

Note: It is often true that the grouping can be done in more than one way. The factored form will be the same.

✓ CHECK YOURSELF 7

We can write the polynomial of Example 7 as

$$x^2 - 3xz + 2xy - 6yz$$

Factor, and verify that the factored form is the same in either case.

✓ CHECK YOURSELF ANSWERS

1. (a) 2; (b) 9; (c) a^5; (d) $5x^4$.   **2.** (a) $5(x + 4)$; (b) $6x(x - 4)$; (c) $5a^2(2a - 3)$.
3. (a) $8(b^2 + 2b - 4)$; (b) $4x(y - 2xy + 3x^2)$; (c) $7x^2(x^2 - 2x + 3)$;
(d) $5xy(xy - 2y + 3x)$.   **4.** (a) $(a - 2b)(7a + 3)$; (b) $4(x + y)(x^2 - 2x + 4)$.
5. $(x - 2y)(x + 3)$.   **6.** $(3y + 2)(y^2 - 2)$ **7.** $(x - 3z)(x + 2y)$.

Exercises • 6.1

Answers

1. 2
2. 5
3. 8
4. 11
5. x^2
6. y^7
7. a^3
8. b^4
9. $5x^4$
10. $8y^3$
11. $2a^4$
12. $3b^3$
13. $3xy$
14. $6a^2b^2$
15. $5b$
16. 3
17. $3abc^2$
18. $9xy^2z$
19. $(x+y)^2$
20. $4(a+b)^3$
21. $4(2a+1)$
22. $5(x-3)$
23. $8(3m-4n)$
24. $7(p-3q)$
25. $4m(3m+2)$
26. $8n(3n-4)$
27. $5s(2s+1)$
28. $6y(2y-1)$
29. $12x(x+2)$
30. $14b(b-2)$
31. $5a^2(3a-5)$
32. $12b^2(3b^2+2)$
33. $6pq(1+3p)$
34. $8ab(1-3b)$
35. $7mn(m^2-3n^2)$
36. $9pq(4pq-1)$
37. $6(x^2-3x+5)$
38. $7(a^2+3a-6)$
39. $3a(a^2+2a-4)$
40. $5x(x^2-3x+5)$

Find the greatest common factor for each of the following sets of terms.

1. 10, 12
2. 15, 35
3. 16, 32, 88
4. 55, 33, 132
5. x^2, x^5
6. y^7, y^9
7. a^3, a^6, a^9
8. b^4, b^6, b^8
9. $5x^4, 10x^5$
10. $8y^9, 24y^3$
11. $8a^4, 6a^6, 10a^{10}$
12. $9b^3, 6b^5, 12b^4$
13. $9x^2y, 12xy^2, 15x^2y^2$
14. $12a^3b^2, 18a^2b^3, 6a^4b^4$
15. $15ab^3, 10a^2bc, 25b^2c^3$
16. $9x^2, 3xy^3, 6y^3$
17. $15a^2bc^2, 9ab^2c^2, 6a^2b^2c^2$
18. $18x^3y^2z^3, 27x^4y^2z^3, 81xy^2z$
19. $(x+y)^2, (x+y)^3$
20. $12(a+b)^4, 4(a+b)^3$

Factor each of the following polynomials.

21. $8a+4$
22. $5x-15$
23. $24m-32n$
24. $7p-21q$
25. $12m^2+8m$
26. $24n^2-32n$
27. $10s^2+5s$
28. $12y^2-6y$
29. $12x^2+24x$
30. $14b^2-28b$
31. $15a^3-25a^2$
32. $36b^4+24b^2$
33. $6pq+18p^2q$
34. $8ab-24ab^2$
35. $7m^3n-21mn^3$
36. $36p^2q^2-9pq$
37. $6x^2-18x+30$
38. $7a^2+21a-42$
39. $3a^3+6a^2-12a$
40. $5x^3-15x^2+25x$

Section 6.1 ■ An Introduction to Factoring

41. $3m(2 + 3n - 5n^2)$

42. $2s(2 + 3t - 7t^2)$

43. $5xy(2x + 3 - y)$

44. $3ab(b + 2 - 5a)$

45. $5r^2s^2(2r + 5 - 3s)$

46. $7x^2y(4y^2 - 5y + 6x)$

47. $3a(3a^4 - 5a^3 + 7a^2 - 9)$

48. $8p^2(p^4 - 5p^2 + 3p - 2)$

49. $5mn(3m^2n - 4m + 7n^2 - 2)$

50. $7ab^2(2b^2 + 3ab - 5a^2 + 4)$

51. $(x - 2)(x + 3)$

52. $(y + 5)(y - 3)$

53. $(p - 2q)(p - q)$

54. $(c + d)(2c + 3d)$

55. $(y - z)(x + 3)$

56. $(c - d)(2a - b)$

57. $(b - c)(a + b)$

58. $(x + 2)(a + b)$

59. $(r + 2s)(6r - 1)$

60. $(n - 2m)(2m + 3)$

61. $(a - 2)(b^2 + 3)$

62. $(r^2 - 3)(s^2 - 2)$

63. $(x + 2)(x - 5y)$

64. $(a + 3b)(a - 4)$

65. $(m - 3n)(m + 2n^2)$

66. $(r - 3s^2)(r + 4s)$

67–68. Correct

69–70. Incorrect

71–72. Correct

73. 10

74. 12

---

41. $6m + 9mn - 15mn^2$

42. $4s + 6st - 14st^2$

43. $10x^2y + 15xy - 5xy^2$

44. $3ab^2 + 6ab - 15a^2b$

45. $10r^3s^2 + 25r^2s^2 - 15r^2s^3$

46. $28x^2y^3 - 35x^2y^2 + 42x^3y$

47. $9a^5 - 15a^4 + 21a^3 - 27a$

48. $8p^6 - 40p^4 + 24p^3 - 16p^2$

49. $15m^3n^2 - 20m^2n + 35mn^3 - 10mn$

50. $14ab^4 + 21a^2b^3 - 35a^3b^2 + 28ab^2$

51. $x(x - 2) + 3(x - 2)$

52. $y(y + 5) - 3(y + 5)$

53. $p(p - 2q) - q(p - 2q)$

54. $2c(c + d) + 3d(c + d)$

55. $x(y - z) + 3(y - z)$

56. $2a(c - d) - b(c - d)$

In Exercises 57 to 62, factor each polynomial by grouping.

57. $ab - ac + b^2 - bc$

58. $ax + 2a + bx + 2b$

59. $6r^2 + 12rs - r - 2s$

60. $2mn - 4m^2 + 3n - 6m$

61. $ab^2 - 2b^2 + 3a - 6$

62. $r^2s^2 - 3s^2 - 2r^2 + 6$

In Exercises 63 to 66, factor each polynomial by grouping. (*Hint:* Consider a rearrangement of terms.)

63. $x^2 - 10y - 5xy + 2x$

64. $a^2 - 12b + 3ab - 4a$

65. $m^2 - 6n^3 + 2mn^2 - 3mn$

66. $r^2 - 3rs^2 - 12s^3 + 4rs$

Determine if the factoring in each of the following is correct.

67. $x^2 - x - 6 = (x - 3)(x + 2)$

68. $x^2 + x - 12 = (x + 4)(x - 3)$

69. $x^2 + x - 12 = (x + 6)(x - 2)$

70. $x^2 + 2x - 8 = (x + 8)(x - 1)$

71. $2x^2 - 5x - 3 = (2x + 1)(x - 3)$

72. $6x^2 - 13x + 6 = (3x - 2)(2x - 3)$

73. The GCF of $2x - 6$ is 2. The GCF of $5x + 10$ is 5. Find the greatest common factor of the product $(2x - 6)(5x + 10)$.

74. The GCF of $3z + 12$ is 3. The GCF of $4z + 8$ is 4. Find the GCF of the product $(3z + 12)(4z + 8)$.

75. 6x

76. The GCF for the product of two factors is the product of the GCFs.

79. 33 − t

80. 3x + 5

75. The GCF of $2x^3 - 4x$ is $2x$. The GCF of $3x + 6$ is 3. Find the GCF of the product $(2x^3 - 4x)(3x + 6)$.

76. State, in a sentence, the rule that the previous three exercises illustrated.

77. For the monomials x^4y^2, x^8y^6, and x^9y^4, explain how you can determine the GCF by inspecting exponents.

78. It is not possible to use the grouping method to factor $2x^3 + 6x^2 + 8x + 4$. Is it correct to conclude that the polynomial is prime? Justify your answer.

79. Area of a rectangle. The area of a rectangle with width t is given by $33t - t^2$. Factor the expression and determine the length of the rectangle in terms of t.

80. Area of a rectangle. The area of a rectangle of length x is given by $3x^2 + 5x$. Find the width of the rectangle.

81. For centuries, mathematicians have found factoring numbers into prime factors a fascinating subject. A prime number is a number that cannot be written as a product of any numbers but 1 and itself. The list of positive primes begins with 2 because 1 is not considered a prime number and then goes on: 3, 5, 7, 11, ...What are the first 10 primes? What are the primes less than 100? If you list the numbers from 1 to 100 and then cross out all numbers that are factors of 2, 3, 5, and 7, what is left? Are all the numbers not crossed out prime? Write a paragraph to explain why this might be so. You might want to investigate the Sieve of Eratosthenes, a system from 230 B.C. for finding prime numbers.

82. If we made a list of all the prime numbers, what number would be at the end of the list? Because there are an infinite number of prime numbers, there is no "largest prime number" at the end of the list. But is there some formula that will give us all the primes? Here are some formulas proposed over the centuries:

$$n^2 + n + 17 \qquad 2n^2 + 29 \qquad n^2 - n + 11$$

In all these expressions, $n = 1, 2, 3, 4, \ldots$, that is, a positive integer beginning with 1. Investigate these expressions with a partner. Do the expressions give prime numbers when they are evaluated for these values of n? Do the expressions give *every* prime in the range of resulting numbers? Can you put in *any* positive number for n?

83. How are primes used in coding messages and for security? Work together to decode the messages. The messages are coded using this code: After the numbers are factored into prime factors, the power of 2 gives the number of the letter in the alphabet. This code would be easy for a code breaker to figure out, but you might make up a code that would be more difficult to break.

(a) 1310720, 229376, 1572864, 1760, 460, 2097152, 336

(b) 786432, 143, 4608, 278528, 1344, 98304, 1835008, 352, 4718592, 5242880

(c) Code a message using this rule. Exchange your message with a partner to decode it.

SECTION 6.2 Factoring Special Polynomials

6.2 OBJECTIVES

1. Factor the difference of two squares
2. Factor the sum and difference of two cubes

In Section 5.4, we introduced some special products. Recall the following formula for the product of a sum and difference of two terms:

$$(a + b)(a - b) = a^2 - b^2$$

This also means that a binomial of the form $a^2 - b^2$, called a **difference of two squares,** has as its factors $a + b$ and $a - b$.

To use this idea for factoring, we can write

$$a^2 - b^2 = (a + b)(a - b)$$

A **perfect-square** term has a coefficient that is a square (1, 4, 9, 16, 25, 36, etc.), and any variables will have exponents that are multiples of 2 (x^2, y^4, z^6, etc.)

Example 1

Factoring the Difference of Two Squares

Factor $x^2 - 16$.

Think $x^2 - 4^2$

Since $x^2 - 16$ is a difference of squares, we have

$$x^2 - 16 = (x + 4)(x - 4)$$

You could also write $(x - 4)(x + 4)$. The order doesn't matter because multiplication is commutative.

 CHECK YOURSELF 1

Factor $m^2 - 49$.

Any time an expression is a difference of two squares, it can be factored.

393

Example 2 — Factoring the Difference of Two Squares

Factor $4a^2 - 9$.

Think $(2a)^2 - 3^2$

So
$$4a^2 - 9 = (2a)^2 - (3)^2$$
$$= (2a + 3)(2a - 3)$$

✓ CHECK YOURSELF 2

Factor $9b^2 - 25$.

The process for factoring a difference of squares does not change when more than one variable is involved.

Example 3 — Factoring the Difference of Two Squares

Think $(5a)^2 - (4b^2)^2$

Factor $25a^2 - 16b^4$.

$$25a^2 - 16b^4 = (5a + 4b^2)(5a - 4b^2)$$

✓ CHECK YOURSELF 3

Factor $49c^4 - 9d^2$.

We will now consider an example that combines common-term factoring with difference-of-squares factoring. Note that the common factor is always removed as the *first step*.

Example 4 — Removing the GCF First

Factor $32x^2y - 18y^3$.

Note that $2y$ is a common factor, so

Step 1
Remove the GCF.
Step 2
Factor the remaining binomial.

$$32x^2y - 18y^3 = 2y(\underbrace{16x^2 - 9y^2}_{\text{Difference of squares}})$$
$$= 2y(4x + 3y)(4x - 3y)$$

CHECK YOURSELF 4

Factor $50a^3 - 8ab^2$.

You may also have to apply the difference of two squares method *more than once* to completely factor a polynomial.

Example 5

Factoring the Difference of Two Squares

Factor $m^4 - 81n^4$

$$m^4 - 81n^4 = (m^2 + 9n^2)(m^2 - 9n^2)$$

Do you see that we are not done in this case? Since $m^2 - 9n^2$ is still factorable, we can continue to factor as follows.

$$m^4 - 81n^4 = (m^2 + 9n^2)(m + 3n)(m - 3n)$$

Note: The other binomial factor, $m^2 + 9n^2$, is a *sum of two squares*, which cannot be factored further.

CHECK YOURSELF 5

Factor $x^4 - 16y^4$.

Factoring the Sum or Difference of Two Cubes

Two additional factoring patterns include the sum or difference of two cubes.

The Sum or Difference of Two Cubes

$$a^3 + b^3 = (a + b)(a^2 - ab + b^2) \qquad (1)$$

$$a^3 - b^3 = (a - b)(a^2 + ab + b^2) \qquad (2)$$

Be sure you take the time to expand the product on the right-hand side to confirm the identity.

Example 6 — Factoring the Sum or Difference of Two Cubes

(a) Factor $x^3 + 27$.

The first term is the cube of x, and the second is the cube of 3, so we can apply equation (1). Letting $a = x$ and $b = 3$, we have

We are now looking for perfect cubes—the exponents must be multiples of 3 and the coefficients perfect cubes—1, 8, 27, 64, and so on.

$$x^3 + 27 = (x)^3 + (3)^3 = (x + 3)(x^2 - 3x + 9)$$

(b) Factor $8w^3 - 27z^3$.

This is a difference of cubes, so use equation (2).

Again, looking for a common factor should be your first step.

$$8w^3 - 27z^3 = (2w)^3 - (3z)^3 = (2w - 3z)[(2w)^2 + (2w)(3z) + (3z)^2]$$
$$= (2w - 3z)(4w^2 + 6wz + 9z^2)$$

(c) Factor $5a^3b - 40b^4$.

Remember to write the GCF as a part of the final factored form.

First note the common factor of $5b$. The binomial is the difference of cubes, so use equation (2).

$$5a^3b - 40b^4 = 5b(a^3 - 8b^3)$$
$$= 5b\,[(a)^3 - (2b)^3]$$
$$= 5b(a - 2b)(a^2 + 2ab + 4b^2)$$

✓ CHECK YOURSELF 6

Factor completely.

(a) $27x^3 + 8y^3$ (b) $3a^4 - 24ab^3$

In each example in this section, we factored a polynomial expression. If we are given a polynomial function to factor, there is no change in the ordered pairs represented by the function after it is factored.

Example 7

Factoring a Polynomial Function

Given the function $f(x) = 9x^2 + 15x$, complete the following.

(a) Find $f(1)$.

$$f(1) = 9(1)^2 + 15(1)$$
$$= 9 + 15$$
$$= 24$$

(b) Factor $f(x)$.

$$f(x) = 9x^2 + 15x$$
$$= 3x(3x + 5)$$

(c) Find $f(1)$ from the factored form of $f(x)$.

$$f(1) = 3(1)(3(1) + 5)$$
$$= 3(8)$$
$$= 24$$

✓ CHECK YOURSELF 7

Given the function $f(x) = 16x^5 + 10x^2$, complete the following:

(a) Find $f(1)$.   (b) Factor $f(x)$.   (c) Find $f(1)$ from the factored form of $f(x)$.

✓ CHECK YOURSELF ANSWERS

1. $(m + 7)(m - 7)$.   **2.** $(3b + 5)(3b - 5)$.   **3.** $(7c^2 + 3d)(7c^2 - 3d)$.
4. $2a(5a + 2b)(5a - 2b)$.   **5.** $(x^2 + 4y^2)(x + 2y)(x - 2y)$.
6. (a) $(3x + 2y)(9x^2 - 6xy + 4y^2)$; (b) $3a(a - 2b)(a^2 + 2ab + 4b^2)$.   **7.** (a) 26;
(b) $2x^2(8x^3 + 5)$; (c) 26.

Exercises ■ 6.2

Answers (left column):

1. No
2. No
3. Yes
4. Yes
5. No
6. No
7. No
8. Yes
9. Yes
10. No
11. $(m+n)(m-n)$
12. $(r+3)(r-3)$
13. $(x+7)(x-7)$
14. $(c+d)(c-d)$
15. $(7+y)(7-y)$
16. $(9+b)(9-b)$
17. $(3b+4)(3b-4)$
18. $(6+x)(6-x)$
19. $(4w+7)(4w-7)$
20. $(2x+5)(2x-5)$
21. $(2s+3r)(2s-3r)$
22. $(8y+x)(8y-x)$
23. $(3w+7z)(3w-7z)$
24. $(5x+9y)(5x-9y)$
25. $(4a+7b)(4a-7b)$
26. $(8m+3n)(8m-3n)$
27. $(x^2+6)(x^2-6)$
28. $(y^3-7)(y^3+7)$
29. $(xy+4)(xy-4)$
30. $(mn+8)(mn-8)$
31. $(5+ab)(5-ab)$
32. $(7+wz)(7-wz)$
33. $(r^2+2s)(r^2-2s)$
34. $(p+3q^2)(p-3q^2)$
35. $(9a+10b^3)(9a-10b^3)$
36. $(8x^2+5y^2)(8x^2-5y^2)$
37. $2x(3x+y)(3x-y)$
38. $2b(5a+b)(5a-b)$
39. $3mn(2m+5n)(2m-5n)$
40. $7p^2(3p+q)(3p-q)$
41. $(a^2+4b^2)(a+2b)(a-2b)$
42. $(9x^2+y^2)(3x+y)(3x-y)$

For each of the following binomials, state whether the binomial is a difference of squares.

1. $3x^2 + 2y^2$
2. $5x^2 - 7y^2$
3. $16a^2 - 25b^2$
4. $9n^2 - 16m^2$
5. $16r^2 + 4$
6. $p^2 - 45$
7. $16a^2 - 12b^3$
8. $9a^2b^2 - 16c^2d^2$
9. $a^2b^2 - 25$
10. $4a^3 - b^3$

Factor the following binomials.

11. $m^2 - n^2$
12. $r^2 - 9$
13. $x^2 - 49$
14. $c^2 - d^2$
15. $49 - y^2$
16. $81 - b^2$
17. $9b^2 - 16$
18. $36 - x^2$
19. $16w^2 - 49$
20. $4x^2 - 25$
21. $4s^2 - 9r^2$
22. $64y^2 - x^2$
23. $9w^2 - 49z^2$
24. $25x^2 - 81y^2$
25. $16a^2 - 49b^2$
26. $64m^2 - 9n^2$
27. $x^4 - 36$
28. $y^6 - 49$
29. $x^2y^2 - 16$
30. $m^2n^2 - 64$
31. $25 - a^2b^2$
32. $49 - w^2z^2$
33. $r^4 - 4s^2$
34. $p^2 - 9q^4$
35. $81a^2 - 100b^6$
36. $64x^4 - 25y^4$
37. $18x^3 - 2xy^2$
38. $50a^2b - 2b^3$
39. $12m^3n - 75mn^3$
40. $63p^4 - 7p^2q^2$
41. $a^4 - 16b^4$
42. $81x^4 - y^4$

Section 6.2 ▪ Factoring Special Polynomials **399**

43. $(x + 4)(x^2 - 4x + 16)$

44. $(y - 2)(y^2 + 2y + 4)$

45. $(m - 5)(m^2 + 5m + 25)$

46. $(b + 3)(b^2 - 3b + 9)$

47. $(ab - 3)(a^2b^2 + 3ab + 9)$

48. $(pq - 4)(p^2q^2 + 4pq + 16)$

49. $(2w + z)(4w^2 - 2wz + z^2)$

50. $(c - 3d)(c^2 + 3cd + 9d^2)$

51. $(r - 4s)(r^2 + 4rs + 16s^2)$

52. $(5x + y)(25x^2 - 5xy + y^2)$

53. $(2x - 3y)(4x^2 + 6xy + 9y^2)$

54. $(4m + 3n)(16m^2 - 12mn - 9n^2)$

55. $3(a + 3b)(a^2 - 3ab + 9b^2)$

56. $(m^2 - 3n)(m^4 + 3m^2n + 9n^2)$

57. $4(x - 2y)(x^2 + 2xy + 4y^2)$

58. (a) 33;
   (b) $3x^2(4x^3 + 7)$;
   (c) 33

59. (a) -16;
   (b) $-2x(3x^2 + 5)$;
   (c) -16

60. (a) 12;
   (b) $-4x(2x^4 - 5)$;
   (c) 12

61. (a) -30;
   (b) $5x^3(x^2 - 7)$;
   (c) -30

62. (a) 4; (b) $x^2(x^3 + 3)$;
   (c) 4

43. $x^3 + 64$

46. $b^3 + 27$

49. $8w^3 + z^3$

52. $125x^3 + y^3$

55. $3a^3 + 81b^3$

44. $y^3 - 8$

47. $a^3b^3 - 27$

50. $c^3 - 27d^3$

53. $8x^3 - 27y^3$

56. $m^6 - 27n^3$

45. $m^3 - 125$

48. $p^3q^3 - 64$

51. $r^3 - 64s^3$

54. $64m^3 + 27n^3$

57. $4x^3 - 32y^3$

In each of the following, (a) find $f(1)$, (b) factor $f(x)$, and (c) find $f(1)$ from the factored form.

58. $f(x) = 12x^5 + 21x^2$

59. $f(x) = -6x^3 - 10x$

60. $f(x) = -8x^5 + 20x$

61. $f(x) = 5x^5 - 35x^3$

62. $f(x) = x^5 + 3x^2$

63. $f(x) = 6x^6 - 16x^5$

Factor each expression.

64. $x^2(x + y) - y^2(x + y)$

65. $a^2(b - c) - 16b^2(b - c)$

66. $2m^2(m - 2n) - 18n^2(m - 2n)$

67. $3a^3(2a + b) - 27ab^2(2a + b)$

68. Find a value for k so that $kx^2 - 25$ will have the factors $2x + 5$ and $2x - 5$.

69. Find a value for k so that $9m^2 - kn^2$ will have the factors $3m + 7n$ and $3m - 7n$.

70. Find a value for k so that $2x^3 - kxy^2$ will have the factors $2x$, $x - 3y$, and $x + 3y$.

71. Find a value for k so that $20a^3b - kab^3$ will have the factors $5ab$, $2a - 3b$, and $2a + 3b$.

72. Complete the following statement in complete sentences: "To factor a number you . . . ."

73. Complete this statement: "To factor an algebraic expression into prime factors means . . . ."

74. Verify the formula for factoring the sum of two cubes by finding the product $(a + b)(a^2 - ab + b^2)$.

75. Verify the formula for factoring the difference of two cubes by finding the product $(a - b)(a^2 + ab + b^2)$.

76. What are the characteristics of a monomial that is a perfect cube?

77. Suppose you factored the polynomial $4x^2 - 16$ as follows:

$$4x^2 - 16 = (2x + 4)(2x - 4)$$

Would this be in completely factored form? If not, what would be the final form?

63. (a) -10; (b) $2x^5(3x - 8)$; (c) -10 64. $(x + y)^2(x - y)$
65. $(b - c)(a + 4b)(a - 4b)$ 66. $2(m - 2n)(m + 3n)(m - 3n)$
67. $3a(2a + b)(a + 3b)(a - 3b)$ 68. 4   69. 49   70. 18   71. 45   74. $a^3 + b^3$
75. $a^3 - b^3$

SECTION 6.3 Factoring Trinomials: The ac Method

 6.3 OBJECTIVES

1. Factor a trinomial of the form $ax^2 + bx + c$
2. Completely factor a trinomial

The product of two binomials of the form

$$(\_\_x + \_\_)(\_\_x + \_\_)$$

will always be a trinomial. In Section 5.4, we used the FOIL method to find the product of two binomials. In this section we will use the factoring-by-grouping method to find the binomial factors for a trinomial.

First let's look at some factored trinomials.

Example 1

Matching Trinomials and Their Factors

Determine which of the following are true statements.

(a) $x^2 - 2x - 8 = (x - 4)(x + 2)$

This is a true statement. Using the FOIL method, we see that

$$(x - 4)(x + 2) = x^2 + 2x - 4x - 8 = x^2 - 2x - 8$$

(b) $x^2 - 6x + 5 = (x - 2)(x - 3)$

This is not a true statement.

$$(x - 2)(x - 3) = x^2 - 3x - 2x + 6 = x^2 - 5x + 6$$

(c) $x^2 + 5x - 14 = (x - 2)(x + 7)$

This is true: $(x - 2)(x + 7) = x^2 + 7x - 2x - 14 = x^2 + 5x - 14$

(d) $x^2 - 8x - 15 = (x - 5)(x - 3)$

This is false: $(x - 5)(x - 3) = x^2 - 3x - 5x + 15 = x^2 - 8x + 15$

✓ **CHECK YOURSELF 1**

Determine which of the following are true statements.

(a) $2x^2 - 2x - 3 = (2x - 3)(x + 1)$

(b) $3x^2 + 11x - 4 = (3x - 1)(x + 4)$

(c) $2x^2 - 7x + 3 = (x - 3)(2x - 1)$

The first step in learning to factor a trinomial is to identify its coefficients. So that we are consistent, we first write the trinomial in standard $ax^2 + bx + c$ form, then label the three coefficients as a, b, and c.

Example 2

Identifying the Coefficients of $ax^2 + bx + c$

First, where necessary, rewrite the trinomial in $ax^2 + bx + c$ form. Then give the values for a, b, and c, where a is the coefficient of the x^2 term, b is the coefficient of the x term, and c is the constant.

(a) $x^2 - 3x - 18$

$$a = 1 \quad b = -3 \quad c = -18$$

(b) $x^2 - 24x + 23$

Note that the negative sign is attached to the coefficients.

$$a = 1 \quad b = -24 \quad c = 23$$

(c) $x^2 + 8 - 11x$

First rewrite the trinomial in descending order:

$$x^2 - 11x + 8$$
$$a = 1 \quad b = -11 \quad c = 8$$

✓ **CHECK YOURSELF 2**

First, where necessary, rewrite the trinomials in $ax^2 + bx + c$ form. Then label a, b, and c, where a is the coefficient of the x^2 term, b is the coefficient of the x term, and c is the constant.

(a) $x^2 + 5x - 14$ (b) $x^2 - 18x + 17$ (c) $x - 6 + 2x^2$

Chapter 6 ■ Factoring Polynomials

Not all trinomials can be factored. To discover if a trinomial is factorable, we try the *ac* **test.**

> **The *ac* Test**
>
> A trinomial of the form $ax^2 + bx + c$ is factorable if (and only if) there are two numbers, *m* and *n*, such that
>
> $$ac = mn \quad \text{and} \quad b = m + n$$

In Example 3 we will look for *m* and *n* to determine whether each trinomial is factorable.

Example 3 — Using the *ac* Test

Use the *ac* test to determine which of the following trinomials can be factored. Find the values of *m* and *n* for each trinomial that can be factored.

(a) $x^2 - 3x - 18$

First, we find the values of *a*, *b*, and *c*, so that we can find *ac*.

$$a = 1 \qquad b = -3 \qquad c = -18$$
$$ac = 1(-18) = -18 \quad \text{and} \quad b = -3$$

Then, we look for two numbers, *m* and *n*, such that $mn = ac$, and $m + n = b$. In this case, that means

$$mn = -18 \quad \text{and} \quad m + n = -3$$

We now look at all pairs of integers with a product of -18. We then look at the sum of each pair of integers.

mn	$m + n$	
$1(-18) = -18$	$1 + (-18) = -17$	
$2(-9) = -18$	$2 + (-9) = -7$	We need look no further than
$3(-6) = -18$	$3 + (-6) = -3$	3 and -6.
$6(-3) = -18$		
$9(-1) = -18$		
$18(-1) = -18$		

3 and -6 are the two integers with a product of *ac* and a sum of *b*. We can say that

$$m = 3 \quad \text{and} \quad n = -6$$

We could have chosen $m = -6$ and $n = 3$ as well.

Because we found values for *m* and *n*, we know that $x^2 - 3x - 18$ is factorable.

(b) $x^2 - 24x + 23$

We find that
$$a = 1 \qquad b = -24 \qquad c = 23$$
$$ac = 1(23) = 23 \qquad \text{and} \qquad b = -24$$

So
$$mn = 23 \qquad \text{and} \qquad m + n = -24$$

We now calculate integer pairs, looking for two numbers with a product of 23 and a sum of -24.

mn	$m + n$
$1(23) = 23$	$1 + 23 = 24$
$-1(-23) = 23$	$-1 + (-23) = -24$

$$m = -1 \qquad \text{and} \qquad n = -23$$

So, $x^2 - 24x + 23$ is factorable.

(c) $x^2 - 11x + 8$

We find that $a = 1$, $b = -11$, and $c = 8$. Therefore, $ac = 8$ and $b = -11$. Thus $mn = 8$ and $m + n = -11$. We calculate integer pairs:

mn	$m + n$
$1(8) = 8$	$1 + 8 = 9$
$2(4) = 8$	$2 + 4 = 6$
$-1(-8) = 8$	$-1 + (-8) = -9$
$-2(-4) = 8$	$-2 + (-4) = -6$

There are no other pairs of integers with a product of 8, and none of these pairs has a sum of -11. The trinomial $x^2 - 11x + 8$ is not factorable.

(d) $2x^2 + 7x - 15$

We find that $a = 2$, $b = 7$, and $c = -15$. Therefore, $ac = 2(-15) = -30$ and $b = 7$. Thus $mn = -30$ and $m + n = 7$. We calculate integer pairs:

mn	$m + n$
$1(-30) = -30$	$1 + (-30) = -29$
$2(-15) = -30$	$2 + (-15) = -13$
$3(-10) = -30$	$3 + (-10) = -7$
$5(-6) = -30$	$5 + (-6) = -1$
$6(-5) = -30$	$6 + (-5) = 1$
$10(-3) = -30$	$10 + (-3) = 7$

There is no need to go any further. We see that 10 and -3 have a product of -30 and a sum of 7, so

$$m = 10 \quad \text{and} \quad n = -3$$

Therefore, $2x^2 + 7x - 15$ is factorable.

It is not always necessary to evaluate all the products and sums to determine whether a trinomial is factorable. You may have noticed patterns and shortcuts that make it easier to find m and n. By all means, use them to help you find m and n. This is essential in mathematical thinking. You are taught a mathematical process that will always work for solving a problem. Such a process is called an **algorithm.** It is very easy to teach a computer to use an algorithm. It is very difficult (some would say impossible) for a computer to have insight. Shortcuts that you discover are *insights*. They may be the most important part of your mathematical education.

✓ CHECK YOURSELF 3

Use the *ac* test to determine which of the following trinomials can be factored. Find the values of m and n for each trinomial that can be factored.

(a) $x^2 - 7x + 12$ (b) $x^2 + 5x - 14$

(c) $3x^2 - 6x + 7$ (d) $2x^2 + x - 6$

So far we have used the results of the *ac* test only to determine whether a trinomial is factorable. The results can also be used to help factor the trinomial.

Example 4 Using the Results of the ac Test to Factor

Rewrite the middle term as the sum of two terms, then factor by grouping.

(a) $x^2 - 3x - 18$

We find that $a = 1$, $b = -3$, and $c = -18$, so $ac = -18$ and $b = -3$. We are looking for two numbers, m and n, where $mn = -18$ and $m + n = -3$. In Example 3, part a, we looked at every pair of integers whose product (mn) was -18, to find a pair that had a sum ($m + n$) of -3. We found the two integers to be 3 and -6, because $3(-6) = -18$ and $3 + (-6) = -3$, so $m = 3$ and $n = -6$. We now use that result to rewrite the middle term as the sum of $3x$ and $-6x$.

$$x^2 + 3x - 6x - 18$$

We then factor by grouping:

$$x^2 + 3x - 6x - 18 = x(x + 3) - 6(x + 3)$$
$$= (x + 3)(x - 6)$$

(b) $x^2 - 24x + 23$

We use the results from Example 3, part b, in which we found $m = -1$ and $n = -23$, to rewrite the middle term of the equation.

$$x^2 - 24x + 23 = x^2 - x - 23x + 23$$

Then we factor by grouping:

$$\begin{aligned} x^2 - x - 23x + 23 &= (x^2 - x) - (23x - 23) \\ &= x(x - 1) - 23(x - 1) \\ &= (x - 1)(x - 23) \end{aligned}$$

(c) $2x^2 + 7x - 15$

From Example 3, part d, we know that this trinomial is factorable, and $m = 10$ and $n = -3$. We use that result to rewrite the middle term of the trinomial.

$$\begin{aligned} 2x^2 + 7x - 15 &= 2x^2 + 10x - 3x - 15 \\ &= (2x^2 + 10x) - (3x + 15) \\ &= 2x(x + 5) - 3(x + 5) \\ &= (x + 5)(2x - 3) \end{aligned}$$

Careful readers will note that we did not ask you to factor Example 3, part c, $x^2 - 11x + 8$. Recall that, by the *ac* method, we determined that this trinomial was not factorable.

✓ CHECK YOURSELF 4

Use the results of Check Yourself 3 to rewrite the middle term as the sum of two terms, then factor by grouping.

(a) $x^2 - 7x + 12$ (b) $x^2 + 5x - 14$ (c) $2x^2 + x - 6$

Let's look at some examples that require us to first find m and n, then factor the trinomial.

Example 5 — Rewriting Middle Terms to Factor

Rewrite the middle term as the sum of two terms, then factor by grouping.

(a) $2x^2 - 13x - 7$

We find that $a = 2$, $b = -13$, and $c = -7$, so $mn = ac = -14$ and $m + n = b = -13$. Therefore,

mn	$m + n$
$1(-14) = -14$	$1 + (-14) = -13$

So, $m = 1$ and $n = -14$. We rewrite the middle term of the trinomial as follows:

$$2x^2 - 13x - 7 = 2x^2 + x - 14x - 7$$
$$= (2x^2 + x) - (14x + 7)$$
$$= x(2x + 1) - 7(2x + 1)$$
$$= (2x + 1)(x - 7)$$

(b) $6x^2 - 5x - 6$

We find that $a = 6$, $b = -5$, and $c = -6$, so $mn = ac = -36$ and $m + n = b = -5$.

mn	$m + n$
$1(-36) = -36$	$1 + (-36) = -35$
$2(-18) = -36$	$2 + (-18) = -16$
$3(-12) = -36$	$3 + (-12) = -9$
$4(-9) = -36$	$4 + (-9) = -5$

So, $m = 4$ and $n = -9$. We rewrite the middle term of the trinomial:

$$6x^2 + 5x - 6 = 6x^2 + 4x - 9x - 6$$
$$= (6x^2 + 4x) - (9x + 6)$$
$$= 2x(3x + 2) - 3(3x + 2)$$
$$= (3x + 2)(2x - 3)$$

✓ CHECK YOURSELF 5

Rewrite the middle term as the sum of two terms, then factor by grouping.

(a) $2x^2 - 7x - 15$ (b) $6x^2 - 5x - 4$

Be certain to check trinomials and binomial factors for any common monomial factor. (There is no common factor in the binomial unless it is also a common factor in the original trinomial.) Example 6 shows the removal of monomial factors.

Example 6

Removing Common Factors

Completely factor the trinomial.

$$3x^2 + 12x - 15$$

We could first remove the common factor of 3:

$$3x^2 + 12x - 15 = 3(x^2 + 4x - 5)$$

Finding m and n for the trinomial $x^2 + 4x - 5$ yields $mn = -5$ and $m + n = 4$.

mn	$m + n$
$1(-5) = -5$	$1 + (-5) = -4$
$5(-1) = -5$	$-1 + (-5) = 4$

So, $m = 5$ and $n = -1$. This gives us

$$\begin{aligned} 3x^2 + 12x - 15 &= 3(x^2 + 4x - 5) \\ &= 3(x^2 + 5x - x - 5) \\ &= 3[(x^2 + 5x) - (x + 5)] \\ &= 3[x(x + 5) - (x + 5)] \\ &= 3[(x + 5)(x - 1)] \\ &= 3(x + 5)(x - 1) \end{aligned}$$

✓ CHECK YOURSELF 6

Completely factor the trinomial.

$$6x^3 + 3x^2 - 18x$$

Not all possible product pairs need to be tried to find m and n. A look at the sign pattern of the trinomial will eliminate many of the possibilities. Assuming the leading coefficient is positive, there are four possible sign patterns.

Pattern	Example	Conclusion
1. b and c and both positive.	$2x^2 + 13x + 15$	m and n must both be positive.
2. b is negative and c is positive.	$x^2 - 7x + 12$	m and n must both be negative.
3. b is positive and c is negative.	$x^2 + 3x - 10$	m and n are of opposite signs. (The value with the larger absolute value is positive.)
4. b is negative and c is negative.	$x^2 - 3x - 10$	m and n are of opposite signs. (The value with the larger absolute value is negative.)

Trial and Error

Sometimes the factors of a trinomial seem obvious. At other times you might be certain that there are only a couple of possible sets of factors for a trinomial. It is perfectly acceptable to check these proposed factors to see if they work. If you find the factors in this manner, we say that you have used the trial-and-error method. The difficulty with using this method exclusively is found in the reason that we call this method trial and *error*.

The next section offers instruction on the trial and error technique. Even if it has not been assigned, you may find it useful to examine that section.

✓ CHECK YOURSELF ANSWERS

1. (a) False; (b) true; (c) true.   2. (a) $a = 1, b = 5, c = -14$; (b) $a = 1, b = -18, c = 17$; (c) $a = 2, b = 1, c = -6$.   3. (a) Factorable, $m = -3, n = -4$; (b) factorable, $m = 7, n = -2$; (c) not factorable; (d) factorable, $m = 4, n = -3$.
4. (a) $x^2 - 3x - 4x + 12 = (x - 3)(x - 4)$; (b) $x^2 + 7x - 2x - 14 = (x + 7)(x - 2)$; (c) $2x^2 + 4x - 3x - 6 = (2x - 3)(x + 2)$.
5. (a) $2x^2 - 10x + 3x - 15 = (2x + 3)(x - 5)$;
(b) $6x^2 - 8x + 3x - 4 = (3x - 4)(2x + 1)$.   6. $3x(2x - 3)(x + 2)$.

Exercises ■ 6.3

State whether each of the following is true or false.

1. $x^2 + 2x - 3 = (x + 3)(x - 1)$
2. $y^2 - 3y - 18 = (y - 6)(y + 3)$
3. $x^2 - 10x - 24 = (x - 6)(x + 4)$
4. $a^2 + 9a - 36 = (a - 12)(a + 4)$
5. $x^2 - 16x + 64 = (x - 8)(x - 8)$
6. $w^2 - 12w - 45 = (w - 9)(w - 5)$
7. $25y^2 - 10y + 1 = (5y - 1)(5y + 1)$
8. $6x^2 + 5xy - 4y^2 = (6x - 2y)(x + 2y)$
9. $10p^2 - pq - 3q^2 = (5p - 3q)(2p + q)$
10. $6a^2 + 13a + 6 = (2a + 3)(3a + 2)$

For each of the following trinomials, label *a*, *b*, and *c*.

11. $x^2 + 4x - 9$
12. $x^2 + 5x + 11$
13. $x^2 - 3x + 8$
14. $x^2 + 7x - 15$
15. $3x^2 + 5x - 8$
16. $2x^2 + 7x - 9$
17. $4x^2 + 8x + 11$
18. $5x^2 + 7x - 9$
19. $-3x^2 + 5x - 10$
20. $-7x^2 + 9x - 18$

Use the *ac* test to determine which of the following trinomials can be factored. Find the values of *m* and *n* for each trinomial that can be factored.

21. $x^2 + x - 6$
22. $x^2 + 2x - 15$
23. $x^2 + x + 2$
24. $x^2 - 3x + 7$
25. $x^2 - 5x + 6$
26. $x^2 - x + 2$
27. $2x^2 + 5x - 3$
28. $3x^2 - 14x - 5$
29. $6x^2 - 19x + 10$
30. $4x^2 + 5x + 6$

Rewrite the middle term as the sum of two terms and then factor by grouping.

31. $x^2 + 6x + 8$
32. $x^2 + 3x - 10$
33. $x^2 - 9x + 20$
34. $x^2 - 8x + 15$
35. $x^2 - 2x - 63$
36. $x^2 + 6x - 55$

---

1. True
2. True
3. False
4. False
5. True
6. False
7. False
8. False
9. True
10. True
11. $a = 1; b = 4; c = -9$
12. $a = 1; b = 5; c = 11$
13. $a = 1; b = -3; c = 8$
14. $a = 1; b = 7; c = -15$
15. $a = 3; b = 5; c = -8$
16. $a = 2; b = 7; c = -9$
17. $a = 4; b = 8; c = 11$
18. $a = 5; b = 7; c = -9$
19. $a = -3; b = 5; c = -10$
20. $a = -7; b = 9; c = -18$
21. Factorable; 3, −2
22. Factorable; 5, −3
23. Not factorable
24. Not factorable
25. Factorable; −3, −2
26. Not factorable
27. Factorable; 6, −1
28. Factorable; −15, 1
29. Factorable; −15, −4
30. Not factorable
31–33. $2x + 4x; (x + 2)(x + 4)$
$5x - 2x; (x - 2)(x + 5)$
$-5x - 4x; (x - 5)(x - 4)$

34–36. $-5x - 3x;$
$(x - 5)(x - 3)$
$-9x + 7x; (x - 9)(x + 7)$
$11x - 5x; (x + 11)(x - 5)$

409

37. $(x + 3)(x + 5)$
38. $(x - 3)(x - 8)$
39. $(x - 4)(x - 7)$
40. $(y - 5)(y + 4)$
41. $(s + 10)(s + 3)$
42. $(b + 3)(b + 11)$
43. $(a - 8)(a + 6)$
44. $(x - 12)(x - 5)$
45. $(x - 1)(x - 7)$
46. $(x + 9)(x - 2)$
47. $(x - 10)(x + 4)$
48. $(x - 1)(x - 10)$
49. $(x - 7)(x - 7)$
50. $(s - 8)(s + 4)$
51. $(p - 12)(p + 2)$
52. $(x - 15)(x + 4)$
53. $(x + 11)(x - 6)$
54. $(a + 10)(a - 8)$
55. $(c + 4)(c + 15)$
56. $(t - 10)(t + 6)$
57. $(n + 10)(n - 5)$
58. $(x - 9)(x - 7)$
59. $(x + 2y)(x + 5y)$
60. $(x - 6y)(x - 2y)$
61. $(a - 7b)(a + 6b)$
62. $(m - 4n)(m - 4n)$
63. $(x - 5y)(x - 8y)$
64. $(r - 12s)(r + 3s)$
65. $(3x + 2)(2x + 5)$
66. $(2x - 3)(3x + 1)$
67. $(5x - 3)(3x + 2)$
68. $(4w + 1)(3w + 4)$
69. $(6m - 5)(m + 5)$

Rewrite the middle term as the sum of two terms and then factor completely.

37. $x^2 + 8x + 15$
38. $x^2 - 11x + 24$
39. $x^2 - 11x + 28$
40. $y^2 - y - 20$
41. $s^2 + 13s + 30$
42. $b^2 + 14b + 33$
43. $a^2 - 2a - 48$
44. $x^2 - 17x + 60$
45. $x^2 - 8x + 7$
46. $x^2 + 7x - 18$
47. $x^2 - 6x - 40$
48. $x^2 - 11x + 10$
49. $x^2 - 14x + 49$
50. $s^2 - 4s - 32$
51. $p^2 - 10p - 24$
52. $x^2 - 11x - 60$
53. $x^2 + 5x - 66$
54. $a^2 + 2a - 80$
55. $c^2 + 19c + 60$
56. $t^2 - 4t - 60$
57. $n^2 + 5n - 50$
58. $x^2 - 16x + 63$
59. $x^2 + 7xy + 10y^2$
60. $x^2 - 8xy + 12y^2$
61. $a^2 - ab - 42b^2$
62. $m^2 - 8mn + 16n^2$
63. $x^2 - 13xy + 40y^2$
64. $r^2 - 9rs - 36s^2$
65. $6x^2 + 19x + 10$
66. $6x^2 - 7x - 3$
67. $15x^2 + x - 6$
68. $12w^2 + 19w + 4$
69. $6m^2 + 25m - 25$
70. $8x^2 - 6x - 9$
71. $9x^2 - 12x + 4$
72. $20x^2 - 23x + 6$
73. $12x^2 - 8x - 15$
74. $16a^2 + 40a + 25$
75. $3y^2 + 7y - 6$
76. $12x^2 + 11x - 15$
77. $8x^2 - 27x - 20$
78. $24v^2 + 5v - 36$
79. $2x^2 + 3xy + y^2$
80. $3x^2 - 5xy + 2y^2$
81. $5a^2 - 8ab - 4b^2$
82. $5x^2 + 7xy - 6y^2$

70. $(4x + 3)(2x - 3)$ 71. $(3x - 2)(3x - 2)$ 72. $(5x - 2)(4x - 3)$
73. $(6x + 5)(2x - 3)$ 74. $(4a + 5)(4a + 5)$ 75. $(3y - 2)(y + 3)$
76. $(3x + 5)(4x - 3)$ 77. $(8x + 5)(x - 4)$ 78. $(8v - 9)(3v + 4)$
79. $(2x + y)(x + y)$ 80. $(3x - 2y)(x - y)$ 81. $(5a + 2b)(a - 2b)$
82. $(5x - 3y)(x + 2y)$

83. $(9x - 5y)(x + y)$
84. $(4x + 3y)(4x + 5y)$
85. $(3m - 4n)(2m - 3n)$
86. $(5x + 3y)(3x - 2y)$
87. $(12a - 5b)(3a + b)$
88. $(3q + r)(q - 6r)$
89. $(x + 2y)^2$
90. $(5b - 8c)^2$
91. $5(2x - 3)(2x + 1)$
92. $6(4x + 1)(x - 1)$
93. $4(2m + 1)(m + 1)$
94. $2(7x - 3)(x - 1)$
95. $3(5r - 2s)(r - s)$
96. $5(2x - 3y)(x + 2y)$
97. $2x(x - 2)(x + 1)$
98. $y(2y + 3)(y - 1)$
99. $y^2(2y + 3)(y + 1)$
100. $2z(2z + 1)(z - 5)$
101. $6a(3a - 1)(2a - 3)$
102. $2n^2(2n - 3)(5n + 2)$
103. $3(p + q)(3p + 7q)$
104. $2(2x + 3y)(3x - 4y)$
105. 6 or 9
106. 6 or 10
107. 8 or 10 or 17
108. 18   109. 4
110. 6   111. 2
112. 4 or 6
113. 3, 8, 15, 24, . . .
114. 2, 6, 12, 20, . . .
115. (a) $(x - 4)(x + 2)$;
   (b) $4, -2$
116. (a) $(x - 5)(x + 2)$;
   (b) $5, -2$

83. $9x^2 + 4xy - 5y^2$
85. $6m^2 - 17mn + 12n^2$
87. $36a^2 - 3ab - 5b^2$
89. $x^2 + 4xy + 4y^2$
91. $20x^2 - 20x - 15$
93. $8m^2 + 12m + 4$
95. $15r^2 - 21rs + 6s^2$
97. $2x^3 - 2x^2 - 4x$
99. $2y^4 + 5y^3 + 3y^2$
101. $36a^3 - 66a^2 + 18a$
103. $9p^2 + 30pq + 21q^2$

84. $16x^2 + 32xy + 15y^2$
86. $15x^2 - xy - 6y^2$
88. $3q^2 - 17qr - 6r^2$
90. $25b^2 - 80bc + 64c^2$
92. $24x^2 - 18x - 6$
94. $14x^2 - 20x + 6$
96. $10x^2 + 5xy - 30y^2$
98. $2y^3 + y^2 - 3y$
100. $4z^3 - 18z^2 - 10z$
102. $20n^4 - 22n^3 - 12n^2$
104. $12x^2 + 2xy - 24y^2$

Find a positive integer value for k for which each of the following can be factored.

105. $x^2 + kx + 8$
106. $x^2 + kx + 9$
107. $x^2 - kx + 16$
108. $x^2 - kx + 17$
109. $x^2 - kx - 5$
110. $x^2 - kx - 7$
111. $x^2 + 3x + k$
112. $x^2 + 5x + k$
113. $x^2 + 2x - k$
114. $x^2 + x - k$

In each of the following, (a) factor the given function, (b) identify the values of x for which $f(x) = 0$, (c) graph $f(x)$ using the graphing calculator and determine where the graph crosses the x axis, and (d) compare the results of (b) and (c).

115. $f(x) = x^2 - 2x - 8$
116. $f(x) = x^2 - 3x - 10$
117. $f(x) = 2x^2 - x - 3$
118. $f(x) = 3x^2 - x - 2$

117. (a) $(2x - 3)(x + 1)$; (b) $\frac{3}{2}, -1$
118. (a) $(3x + 2)(x - 1)$; (b) $-\frac{2}{3}, 1$

SECTION 6.3* Factoring Trinomials: Trial and Error

6.3* OBJECTIVES

1. Factor a trinomial of the form $ax^2 + bx + c$
2. Completely factor a trinomial

Recall that the product of two binomials may be a trinomial of the form

$$ax^2 + bx + c$$

This suggests that some trinomials may be factored as the product of two binomials. And, in fact, factoring trinomials in this way is probably the most common type of factoring that you will encounter in algebra. One process for factoring a trinomial into a product of two binomials is called *trial and error*.

As before, let's introduce the factoring technique with an example from multiplication. Consider

$$(x + 3)(x + 4) = x^2 + 4x + 3x + 12$$
$$= x^2 + 7x + 12$$

- Product of first terms, x and x
- Sum of inner and outer products, $3x$ and $4x$
- Product of last terms, 3 and 4

To reverse the multiplication process to one of factoring, we see that the product of the *first* terms of the binomial factors is the *first* term of the given trinomial, the product of the *last* terms of the binomial factors is the *last* term of the trinomial, and the *middle* term of the trinomial must equal the sum of the *outer* and *inner* products. That leads us to the following sign patterns in factoring a trinomial.

Factoring Trinomials

		Factoring Sign Pattern
$x^2 + bx + c$	Both signs are positive.	$(x +)(x +)$
$x^2 - bx + c$	The constant is positive, and the x coefficient is negative.	$(x -)(x -)$
$x^2 + bx - c$ or $x^2 - bx - c$	The constant is negative	$(x +)(x -)$

Given the above information let's work through an example.

Example 1 Factoring Trinomials, $a = 1$

To factor

$$x^2 + 7x + 10$$

the desired sign pattern is

$$(x + \_\_)(x + \_\_)$$

From the constant, 10, and the x coefficient of our original trinomial, 7, for the second terms of the binomial factors, we want two numbers whose product is 10 and whose sum is 7.

Consider the following:

Factors of 10	Sum
1, 10	11
2, 5	7

With practice, you will do much of this work mentally. We show the factors and their sums here, and in later examples, to emphasize the process.

We can see that the correct factorization is

$$x^2 + 7x + 10 = (x + 2)(x + 5)$$

Note: To check, multiply the factors using the method of Section 5.4.

✓ CHECK YOURSELF 1

Factor $x^2 + 8x + 15$.

Example 2 Factoring Trinomials, $a = 1$

Factor $x^2 - 9x + 14$. Do you see that the sign pattern must be as follows?

$$(x - \_\_)(x - \_\_)$$

We then want two factors of 14 whose sum is -9.

Factors of 14	Sum
$-1, -14$	-15
$-2, -7$	-9

Here we use two negative factors of 14 since the coefficient of the x term is negative while the constant is positive.

Since the desired middle term is $-9x$, the correct factors are

$$x^2 - 9x + 14 = (x - 2)(x - 7)$$

✓ CHECK YOURSELF 2

Factor $x^2 - 12x + 32$.

Let's turn now to applying our factoring technique to a trinomial whose constant term is negative. Consider the following example.

Example 3 — Factoring Trinomials, $a = 1$

Factor

$$x^2 + 4x - 12$$

In this case, the sign pattern is

$$(x - \_\_)(x + \_\_)$$

Since the constant is now negative, the signs in the binomial factors must be opposite.

Here we want two numbers whose product is -12 and whose sum is 4. Again let's look at the possible factors:

Factors of -12	Sum
1, -12	-11
-1, 12	11
3, -4	-1
-3, 4	1
2, -6	-4
-2, 6	4

From the information above, we see that the correct factors are

$$x^2 + 4x - 12 = (x - 2)(x + 6)$$

✓ CHECK YOURSELF 3

Factor $x^2 - 7x - 18$.

Thus far we have considered only trinomials of the form $x^2 + bx + c$. Suppose that the leading coefficient is *not* 1. In general, to factor the trinomial $ax^2 + bx + c$ (with $a \neq 1$), we must consider binomial factors of the form

$$(\_\_ x + \_\_)(\_\_ x + \_\_)$$

where one or both of the coefficients of x in the binomial factors are greater than 1. Again let's look at a multiplication example for some clues to the technique. Consider

$$(2x + 3)(3x + 5) = 6x^2 + 19x + 15$$

Product of $2x$ and $3x$ — Sum of outer and inner products, $10x$ and $9x$ — Product of 3 and 5

Now, to reverse the process to factoring, we can proceed as in the following example.

Example 4

Factoring Trinomials, $a \neq 1$

To factor $5x^2 + 9x + 4$, we must have the pattern

$$(\_x + \_)(\_x + \_) \quad \text{This product must be 4.}$$

This product must be 5.

Factors of 5	Factors of 4
1, 5	1, 4
	4, 1
	2, 2

Now that the lead coefficient is no longer 1, we must be prepared to try both 1, 4 and 4, 1.

Therefore the possible binomial factors are

$$(x + 1)(5x + 4)$$
$$(x + 4)(5x + 1)$$
$$(x + 2)(5x + 2)$$

Checking the middle terms of each product, we see that the proper factorization is

$$5x^2 + 9x + 4 = (x + 1)(5x + 4)$$

 CHECK YOURSELF 4

Factor $6x^2 - 17x + 7$.

The sign patterns discussed before remain the same when the leading coefficient is not 1. Look at the following example involving a trinomial with a negative constant.

Example 5

Factoring Trinomials, $a \neq 1$

(a) Factor $6x^2 + 7x - 3$. The sign patterns are

$$(\_x + \_)(\_x - \_)$$

Factors of 6	Factors of -3
1, 6	1, -3
2, 3	-1, 3

Again, as the number of factors for the first coefficient and the constant increase, the number of possible factors becomes larger. Can we reduce the search? One clue: If the trinomial has no common factors (other than 1), then a binomial factor can have no common factor. This means that $6x - 3$, $6x + 3$, $3x - 3$, and $3x + 3$ need not be considered. They are shown here to completely illustrate the possibilities.

There are eight possible binomial factors:

$$(x + 1)(6x - 3)$$
$$(x - 1)(6x + 3)$$
$$(x + 3)(6x - 1)$$
$$(x - 3)(6x + 1)$$
$$(2x + 1)(3x - 3)$$
$$(2x - 1)(3x + 3)$$
$$(3x + 1)(2x - 3)$$
$$(3x - 1)(2x + 3)$$

Again, checking the middle terms, we have the correct factors:

$$6x^2 + 7x - 3 = (3x - 1)(2x + 3)$$

Factoring certain trinomials in more than one variable involves similar techniques, as is illustrated below.

(b) Factor

$$4x^2 - 16xy + 7y^2$$

From the first term of the trinomial we see that possible first terms for our binomial factors are $4x$ and x or $2x$ and $2x$. The last term of the trinomial tells us that the only choices for the last terms of our binomial factors are y and $7y$. So given the sign of the middle and last terms, the only possible factors are

Find the middle term of each product.

$$(4x - 7y)(x - y)$$
$$(4x - y)(x - 7y)$$
$$(2x - 7y)(2x - y)$$

From the middle term of our original trinomial we see that $2x - 7y$ and $2x - y$ are the proper factors.

✓ CHECK YOURSELF 5

Factor $6a^2 + 11ab - 10b^2$.

Example 6 — Factoring Trinomials, $a \neq 1$

Recall our earlier comment that the *first step* in any factoring problem is to remove any existing common factors. As before, it may be necessary to combine common-term factoring with other methods (such as factoring a trinomial into a product of binomials) to completely factor a polynomial. Look at the following example.

(a) Factor

$$2x^2 - 16x + 30$$

First note the common factor of 2. So we can write

"Remove" the common factor of 2.

$$2x^2 - 16x + 30 = 2(x^2 - 8x + 15)$$

Now, as the second step, examine the trinomial factor. By our earlier methods we know that

$$x^2 - 8x + 15 = (x - 3)(x - 5)$$

and we have

$$2x^2 - 16x + 30 = 2(x - 3)(x - 5)$$

in completely factored form.

(b) Factor

$$6x^3 + 15x^2y - 9xy^2$$

As before, note the common factor of $3x$ in each term of the trinomial. Removing that common factor, we have

$$6x^3 + 15x^2y - 9xy^2 = 3x(2x^2 + 5xy - 3y^2)$$

Again, considering the trinomial factor, we see that $2x^2 + 5xy - 3y^2$ has factors of $2x - y$ and $x + 3y$. And our original trinomial becomes

$$3x(2x - y)(x + 3y)$$

in completely factored form.

✓ CHECK YOURSELF 6

Factor.

$$\text{(a)} \quad 9x^2 - 39x + 36$$

$$\text{(b)} \quad 24a^3 + 4a^2b - 8ab^2$$

One final note. When factoring, we require that all coefficients be integers. Given this restriction, not all polynomials are factorable over the integers. The following illustrates.

To factor $x^2 - 9x + 12$, we know that the only possible binomial factors (using integers as coefficients) are

$$(x - 1)(x - 12)$$
$$(x - 2)(x - 6)$$
$$(x - 3)(x - 4)$$

You can easily verify that *none* of these pairs gives the correct middle term of $-9x$. We then say that the original trinomial is not factorable using integers as coefficients.

✓ CHECK YOURSELF ANSWERS

1. $(x + 3)(x + 5)$. **2.** $(x - 4)(x - 8)$. **3.** $(x - 9)(x + 2)$.
4. $(2x - 1)(3x - 7)$. **5.** $(3a - 2b)(2a + 5b)$. **6.** (a) $3(x - 3)(3x - 4)$;
(b) $4a(3a + 2b)(2a - b)$.

Exercises • 6.3*

Answers (left column):

1. True
2. True
3. False
4. False
5. True
6. False
7. False
8. False
9. True
10. True
11. $a = 1; b = 4; c = -9$
12. $a = 1; b = 5; c = 11$
13. $a = 1; b = -3; c = 8$
14. $a = 1; b = 7; c = -15$
15. $a = 3; b = 5; c = -8$
16. $a = 2; b = 7; c = -9$
17. $a = 4; b = 8; c = 11$
18. $a = 5; b = 7; c = -9$
19. $a = -3; b = 5; c = -10$
20. $a = -7; b = 9; c = -18$
21. $(x + 2)(x + 4)$
22. $(x - 2)(x + 5)$
23. $(x - 5)(x - 4)$
24. $(x - 5)(x - 3)$
25. $(x - 9)(x + 7)$
26. $(x + 11)(x - 5)$
27. $(x + 3)(x + 5)$
28. $(x - 3)(x - 8)$
29. $(x - 4)(x - 7)$
30. $(y - 5)(y + 4)$
31. $(s + 10)(s + 3)$
32. $(b + 3)(b + 11)$
33. $(a - 8)(a + 6)$
34. $(x - 12)(x - 5)$
35. $(x - 1)(x - 7)$
36. $(x + 9)(x - 2)$

State whether each of the following is true or false.

1. $x^2 + 2x - 3 = (x + 3)(x - 1)$
2. $y^2 - 3y - 18 = (y - 6)(y + 3)$
3. $x^2 - 10x - 24 = (x - 6)(x + 4)$
4. $a^2 + 9a - 36 = (a - 12)(a + 4)$
5. $x^2 - 16x + 64 = (x - 8)(x - 8)$
6. $w^2 - 12w - 45 = (w - 9)(w - 5)$
7. $25y^2 - 10y + 1 = (5y - 1)(5y + 1)$
8. $6x^2 + 5xy - 4y^2 = (6x - 2y)(x + 2y)$
9. $10p^2 - pq - 3q^2 = (5p - 3q)(2p + q)$
10. $6a^2 + 13a + 6 = (2a + 3)(3a + 2)$

For each of the following trinomials, label a, b, and c.

11. $x^2 + 4x - 9$
12. $x^2 + 5x + 11$
13. $x^2 - 3x + 8$
14. $x^2 + 7x - 15$
15. $3x^2 + 5x - 8$
16. $2x^2 + 7x - 9$
17. $4x^2 + 8x + 11$
18. $5x^2 + 7x - 9$
19. $-3x^2 + 5x - 10$
20. $-7x^2 + 9x - 18$

Factor completely.

21. $x^2 + 6x + 8$
22. $x^2 + 3x - 10$
23. $x^2 - 9x + 20$
24. $x^2 - 8x + 15$
25. $x^2 - 2x - 63$
26. $x^2 + 6x - 55$

Factor completely.

27. $x^2 + 8x + 15$
28. $x^2 - 11x + 24$
29. $x^2 - 11x + 28$
30. $y^2 - y - 20$
31. $s^2 + 13s + 30$
32. $b^2 + 14b + 33$
33. $a^2 - 2a - 48$
34. $x^2 - 17x + 60$
35. $x^2 - 8x + 7$
36. $x^2 + 7x - 18$

37. $(x - 10)(x + 4)$
38. $(x - 1)(x - 10)$
39. $(x - 7)(x - 7)$
40. $(s - 8)(s + 4)$
41. $(p - 12)(p + 2)$
42. $(x - 15)(x + 4)$
43. $(x + 11)(x - 6)$
44. $(a + 10)(a - 8)$
45. $(c + 4)(c + 15)$
46. $(t - 10)(t + 6)$
47. $(n + 10)(n - 5)$
48. $(x - 9)(x - 7)$
49. $(x + 2y)(x + 5y)$
50. $(x - 6y)(x - 2y)$
51. $(a - 7b)(a + 6b)$
52. $(m - 4n)(m - 4n)$
53. $(x - 5y)(x - 8y)$
54. $(r - 12s)(r + 3s)$
55. $(3x + 2)(2x + 5)$
56. $(2x - 3)(3x + 1)$
57. $(5x - 3)(3x + 2)$
58. $(4w + 1)(3w + 4)$
59. $(6m - 5)(m + 5)$
60. $(4x + 3)(2x - 3)$
61. $(3x - 2)(3x - 2)$
62. $(5x - 2)(4x - 3)$
63. $(6x + 5)(2x - 3)$
64. $(4a + 5)(4a + 5)$
65. $(3y - 2)(y + 3)$
66. $(3x + 5)(4x - 3)$
67. $(8x + 5)(x - 4)$
68. $(8v - 9)(3v + 4)$
69. $(2x + y)(x + y)$

37. $x^2 - 6x - 40$
39. $x^2 - 14x + 49$
41. $p^2 - 10p - 24$
43. $x^2 + 5x - 66$
45. $c^2 + 19c + 60$
47. $n^2 + 5n - 50$
49. $x^2 + 7xy + 10y^2$
51. $a^2 - ab - 42b^2$
53. $x^2 - 13xy + 40y^2$
55. $6x^2 + 19x + 10$
57. $15x^2 + x - 6$
59. $6m^2 + 25m - 25$
61. $9x^2 - 12x + 4$
63. $12x^2 - 8x - 15$
65. $3y^2 + 7y - 6$
67. $8x^2 - 27x - 20$
69. $2x^2 + 3xy + y^2$
71. $5a^2 - 8ab - 4b^2$
73. $9x^2 + 4xy - 5y^2$

38. $x^2 - 11x + 10$
40. $s^2 - 4s - 32$
42. $x^2 - 11x - 60$
44. $a^2 + 2a - 80$
46. $t^2 - 4t - 60$
48. $x^2 - 16x + 63$
50. $x^2 - 8xy + 12y^2$
52. $m^2 - 8mn + 16n^2$
54. $r^2 - 9rs - 36s^2$
56. $6x^2 - 7x - 3$
58. $12w^2 + 19w + 4$
60. $8x^2 - 6x - 9$
62. $20x^2 - 23x + 6$
64. $16a^2 + 40a + 25$
66. $12x^2 + 11x - 15$
68. $24v^2 + 5v - 36$
70. $3x^2 - 5xy + 2y^2$
72. $5x^2 + 7xy - 6y^2$
74. $16x^2 + 32xy + 15y^2$

70. $(3x - 2y)(x - y)$
71. $(5a + 2b)(a - 2b)$
72. $(5x - 3y)(x + 2y)$
73. $(9x - 5y)(x + y)$
74. $(4x + 3y)(4x + 5y)$

75. $(3m - 4n)(2m - 3n)$
76. $(5x + 3y)(3x - 2y)$
77. $(12a - 5b)(3a + b)$
78. $(3q + r)(q - 6r)$
79. $(x + 2y)^2$
80. $(5b - 8c)^2$
81. $5(2x - 3)(2x + 1)$
82. $6(4x + 1)(x - 1)$
83. $4(2m + 1)(m + 1)$
84. $2(7x - 3)(x - 1)$
85. $3(5r - 2s)(r - s)$
86. $5(2x - 3y)(x + 2y)$
87. $2x(x - 2)(x + 1)$
88. $y(2y + 3)(y - 1)$
89. $y^2(2y + 3)(y + 1)$
90. $2z(2z + 1)(z - 5)$
91. $6a(3a - 1)(2a - 3)$
92. $2n^2(2n - 3)(5n + 2)$
93. $3(p + q)(3p + 7q)$
94. $2(2x + 3y)(3x - 4y)$
95. 6 or 9
96. 6 or 10
97. 8 or 10 or 17
98. 18   99. 4
100. 6   101. 2
102. 4 or 6
103. 3, 8, 15, 24, . . .
104. 2, 6, 12, 20, . . .
105. (a) $(x - 4)(x + 2)$; (b) 4, −2
106. (a) $(x - 5)(x + 2)$; (b) 5, −2
107. (a) $(2x - 3)(x + 1)$; (b) $\frac{3}{2}$, −1
108. (a) $(3x + 2)(x - 1)$; (b) $-\frac{2}{3}$, 1

75. $6m^2 - 17mn + 12n^2$
76. $15x^2 - xy - 6y^2$
77. $36a^2 - 3ab - 5b^2$
78. $3q^2 - 17qr - 6r^2$
79. $x^2 + 4xy + 4y^2$
80. $25b^2 - 80bc + 64c^2$
81. $20x^2 - 20x - 15$
82. $24x^2 - 18x - 6$
83. $8m^2 + 12m + 4$
84. $14x^2 - 20x + 6$
85. $15r^2 - 21rs + 6s^2$
86. $10x^2 + 5xy - 30y^2$
87. $2x^3 - 2x^2 - 4x$
88. $2y^3 + y^2 - 3y$
89. $2y^4 + 5y^3 + 3y^2$
90. $4z^3 - 18z^2 - 10z$
91. $36a^3 - 66a^2 + 18a$
92. $20n^4 - 22n^3 - 12n^2$
93. $9p^2 + 30pq + 21q^2$
94. $12x^2 + 2xy - 24y^2$

Find a positive integer value for k for which each of the following can be factored.

95. $x^2 + kx + 8$
96. $x^2 + kx + 9$
97. $x^2 - kx + 16$
98. $x^2 - kx + 17$
99. $x^2 - kx - 5$
100. $x^2 - kx - 7$
101. $x^2 + 3x + k$
102. $x^2 + 5x + k$
103. $x^2 + 2x - k$
104. $x^2 + x - k$

In each of the following, (a) factor the given function, (b) identify the values of x for which $f(x) = 0$, (c) graph $f(x)$ using the graphing calculator and determine where the graph crosses the x axis, and (d) compare the results of (b) and (c).

105. $f(x) = x^2 - 2x - 8$
106. $f(x) = x^2 - 3x - 10$
107. $f(x) = 2x^2 - x - 3$
108. $f(x) = 3x^2 - x - 2$

SECTION 6.4 Division of Polynomials

6.4 OBJECTIVES

1. Find the quotient when a polynomial is divided by a monomial
2. Find the quotient of two polynomials

In Section 5.1, we introduced the quotient rule for exponents, which was used to divide one monomial by another monomial. Let's review that process.

> **To Divide a Monomial by a Monomial**
> Step 1 Divide the coefficients.
> Step 2 Use the quotient rule for exponents to combine the variables.

Example 1

Dividing Monomials

The quotient rule says: If x is not zero and $m > n$,

$$\frac{x^m}{x^n} = x^{m-n}$$

(a) $\dfrac{8x^4}{2x^2} = 4x^{4-2}$ Divide: $\dfrac{8}{2} = 4$ Subtract the exponents.

$= 4x^2$

(b) $\dfrac{45a^5b^3}{9a^2b} = 5a^3b^2$

✓ CHECK YOURSELF 1

Divide.

(a) $\dfrac{16a^5}{8a^3}$ (b) $\dfrac{28m^4n^3}{7m^3n}$

422

Now let's look at how this can be extended to divide any polynomial by a monomial. For example, to divide $12a^3 + 8a^2$ by $4a$, proceed as follows:

$$\frac{12a^3 + 8a^2}{4a} = \frac{12a^3}{4a} + \frac{8a^2}{4a}$$

Divide each term in the numerator by the denominator, $4a$.

Technically, this step depends on the distributive property and the definition of division.

Now do each division.

$$= 3a^2 + 2a$$

The preceding work leads us to the following rule.

To Divide a Polynomial by a Monomial

Divide each term of the polynomial by the monomial. Then combine the results.

Example 2   **Dividing by Monomials**

(a) Divide each term by 2.
$$\frac{4a^2 + 8}{2} = \frac{4a^2}{2} + \frac{8}{2}$$
$$= 2a^2 + 4$$

(b) Divide each term by $6y$.
$$\frac{24y^3 - 18y^2}{6y} = \frac{24y^3}{6y} - \frac{18y^2}{6y}$$
$$= 4y^2 - 3y$$

(c) Remember the rules for signs in division.
$$\frac{15x^2 + 10x}{-5x} = \frac{15x^2}{-5x} + \frac{10x}{-5x}$$
$$= -3x - 2$$

With practice you can write just the quotient.

(d) $\dfrac{14x^4 + 28x^3 - 21x^2}{7x^2} = \dfrac{14x^4}{7x^2} + \dfrac{28x^3}{7x^2} - \dfrac{21x^2}{7x^2}$

Chapter 6 ■ Factoring Polynomials

$$= 2x^2 + 4x - 3$$

(e) $\dfrac{9a^3b^4 - 6a^2b^3 + 12ab^4}{3ab} = \dfrac{9a^3b^4}{3ab} - \dfrac{6a^2b^3}{3ab} + \dfrac{12ab^4}{3ab}$

$$= 3a^2b^3 - 2ab^2 + 4b^3$$

✓ CHECK YOURSELF 2

Divide.

(a) $\dfrac{20y^3 - 15y^2}{5y}$ 　　(b) $\dfrac{8a^3 - 12a^2 + 4a}{-4a}$

(c) $\dfrac{16m^4n^3 - 12m^3n^2 + 8mn}{4mn}$

We are now ready to look at dividing one polynomial by another polynomial (with more than one term). The process is very much like long division in arithmetic, as Example 3 illustrates.

Example 3 — Dividing by Binomials

Divide $x^2 + 7x + 10$ by $x + 2$.

The first term in the dividend, x^2, is divided by the first term in the divisor, x.

Step 1

$$\begin{array}{r} x \\ x+2\overline{\smash{\big)}\,x^2 + 7x + 10} \end{array}$$

Divide x^2 by x to get x.

Step 2

$$\begin{array}{r} x \\ x+2\overline{\smash{\big)}\,x^2 + 7x + 10} \\ \underline{x^2 + 2x} \end{array}$$

Multiply the divisor, $x + 2$, by x.

Remember: To subtract $x^2 + 2x$, mentally change each sign to $-x^2 - 2x$, and add. Take your time and be careful here. It's where most errors are made.

Step 3

$$\begin{array}{r} x \\ x+2\overline{\smash{\big)}\,x^2 + 7x + 10} \\ \underline{x^2 + 2x} \\ 5x + 10 \end{array}$$

Subtract and bring down 10.

Step 4

$$\begin{array}{r} x + 5 \\ x+2\overline{\smash{\big)}\,x^2 + 7x + 10} \\ \underline{x^2 + 2x} \\ 5x + 10 \end{array}$$

Divide $5x$ by x to get 5.

Note that we repeat the process until the degree of the remainder is less than that of the divisor or until there is no remainder.

Step 5

$$\begin{array}{r} x + 5 \\ x + 2 \overline{\smash{)}x^2 + 7x + 10} \\ \underline{x^2 + 2x } \\ 5x + 10 \\ \underline{5x + 10} \\ 0 \end{array}$$

Multiply $x + 2$ by 5 and then subtract.

The quotient is $x + 5$.

✓ CHECK YOURSELF 3

Divide $x^2 + 9x + 20$ by $x + 4$.

In Example 3, we showed all the steps separately to help you see the process. In practice, the work can be shortened.

Example 4

Dividing by Binomials

Divide $x^2 + x - 12$ by $x - 3$.

You might want to write out a problem like $408 \div 17$, to compare the steps.

$$\begin{array}{r} x + 4 \\ x - 3 \overline{\smash{)}x^2 + x - 12} \\ \underline{x^2 - 3x } \\ 4x - 12 \\ \underline{4x - 12} \\ 0 \end{array}$$

The Steps
1. Divide x^2 by x to get x, the first term of the quotient.
2. Multiply $x - 3$ by x.
3. Subtract and bring down -12. Remember to mentally change the signs to $-x^2 + 3x$ and add.
4. Divide $4x$ by x to get 4, the second term of the quotient.
5. Multiply $x - 3$ by 4 and subtract.

The quotient is $x + 4$.

✓ CHECK YOURSELF 4

Divide.

$$(x^2 + 2x - 24) \div (x - 4)$$

You may have a remainder in algebraic long division just as in arithmetic. Consider Example 5.

Example 5 Dividing by Binomials

Divide $4x^2 - 8x + 11$ by $2x - 3$.

$$\begin{array}{r}
2x - 1 \\
2x - 3\overline{)\,4x^2 - 8x + 11} \\
\underline{4x^2 - 6x \phantom{{}+11}} \\
-2x + 11 \\
\underline{-2x + 3} \\
8
\end{array}$$

Quotient: $2x - 1$ Divisor: $2x - 3$ Remainder: 8

This result can be written as

$$\frac{4x^2 - 8x + 11}{2x - 3}$$

$$= 2x - 1 + \frac{8}{2x - 3}$$

Quotient: $2x - 1$ Remainder: 8 Divisor: $2x - 3$

✓ *CHECK YOURSELF 5*

Divide.

$$(6x^2 - 7x + 15) \div (3x - 5)$$

The division process shown in our previous examples can be extended to dividends of a higher degree. The steps involved in the division process are exactly the same, as Example 6 illustrates.

Example 6 Dividing by Binomials

Divide $6x^3 + x^2 - 4x - 5$ by $3x - 1$.

$$
\begin{array}{r}
2x^2 + x - 1 \\
3x - 1 \overline{\smash{)}\, 6x^3 + x^2 - 4x - 5} \\
\underline{6x^3 - 2x^2} \\
3x^2 - 4x \\
\underline{3x^2 - x} \\
-3x - 5 \\
\underline{-3x + 1} \\
-6
\end{array}
$$

This result can be written as

$$\frac{6x^3 + x^2 - 4x - 5}{3x - 1} = 2x^2 + x - 1 + \frac{-6}{3x - 1}$$

✓ CHECK YOURSELF 6

Divide $4x^3 - 2x^2 + 2x + 15$ by $2x + 3$.

Suppose that the dividend is "missing" a term in some power of the variable. You can use 0 as the coefficient for the missing term. Consider Example 7.

Example 7 Dividing by Binomials

Divide $x^3 - 2x^2 + 5$ by $x + 3$.

$$
\begin{array}{r}
x^2 - 5x + 15 \\
x + 3 \overline{\smash{)}\, x^3 - 2x^2 + 0x + 5} \\
\underline{x^3 + 3x^2} \\
-5x^2 + 0x \\
\underline{-5x^2 - 15x} \\
15x + 5 \\
\underline{15x + 45} \\
-40
\end{array}
$$

Write $0x$ for the "missing" term in x.

This result can be written as

$$\frac{x^3 - 2x^2 + 5}{x + 3} = x^2 - 5x + 15 + \frac{-40}{x + 3}$$

✓ CHECK YOURSELF 7

Divide.

$$(4x^3 + x + 10) \div (2x - 1)$$

You should always arrange the terms of the divisor and the dividend in descending-exponent form before starting the long division process, as illustrated in Example 8.

Example 8

Dividing by Binomials

Divide $5x^2 - x + x^3 - 5$ by $-1 + x^2$.

Write the divisor as $x^2 - 1$ and the dividend as $x^3 + 5x^2 - x - 5$.

$$\begin{array}{r} x + 5 \\ x^2 - 1 \overline{) x^3 + 5x^2 - x - 5} \\ \underline{x^3 - x} \\ 5x^2 - 5 \\ \underline{5x^2 - 5} \\ 0 \end{array}$$

Write $x^3 - x$, the product of x and $x^2 - 1$, so that like terms fall in the same columns.

✓ CHECK YOURSELF 8

Divide.

$$(5x^2 + 10 + 2x^3 + 4x) \div (2 + x^2)$$

✓ CHECK YOURSELF ANSWERS

1. (a) $2a^2$; (b) $4mn^2$.   2. (a) $4y^2 - 3y$; (b) $-2a^2 + 3a - 1$; (c) $4m^3n^2 - 3m^2n + 2$.
3. $x + 5$.   4. $x + 6$.   5. $2x + 1 + \dfrac{20}{3x - 5}$.   6. $2x^2 - 4x + 7 + \dfrac{-6}{2x + 3}$.
7. $2x^2 + x + 1 + \dfrac{11}{2x - 1}$.   8. $2x + 5$.

Exercises · 6.4

Answers (left column):

1. $2x^4$
2. $4a^2$
3. $5m^2$
4. $7x^2y$
5. $a+2$
6. $x-2$
7. $3b^2-4$
8. $2m^2+m$
9. $4a^2-6a$
10. $3x^2+4x$
11. $-4m-2$
12. $-4b^2+5b$
13. $3a^3+2a^2-a$
14. $3x^4-4x^3+2x^2$
15. $4x^2y-3y^2+2x$
16. $2m^2n+3m-5n$
17. $x+3$
18. $x+5$
19. $x-5$
20. $x-7$
21. $x+3$
22. $x+8$
23. $2x+3+\dfrac{4}{x-3}$
24. $3x-1+\dfrac{-6}{x+6}$

Divide.

1. $\dfrac{18x^6}{9x^2}$
2. $\dfrac{20a^7}{5a^5}$
3. $\dfrac{35m^3n^2}{7mn^2}$
4. $\dfrac{42x^5y^2}{6x^3y}$
5. $\dfrac{3a+6}{3}$
6. $\dfrac{4x-8}{4}$
7. $\dfrac{9b^2-12}{3}$
8. $\dfrac{10m^2+5m}{5}$
9. $\dfrac{16a^3-24a^2}{4a}$
10. $\dfrac{9x^3+12x^2}{3x}$
11. $\dfrac{12m^2+6m}{-3m}$
12. $\dfrac{20b^3-25b^2}{-5b}$
13. $\dfrac{18a^4+12a^3-6a^2}{6a}$
14. $\dfrac{21x^5-28x^4+14x^3}{7x}$
15. $\dfrac{20x^4y^2-15x^2y^3+10x^3y}{5x^2y}$
16. $\dfrac{16m^3n^3+24m^2n^2-40mn^3}{8mn^2}$

Perform the indicated divisions.

17. $\dfrac{x^2+5x+6}{x+2}$
18. $\dfrac{x^2+8x+15}{x+3}$
19. $\dfrac{x^2-x-20}{x+4}$
20. $\dfrac{x^2-2x-35}{x+5}$
21. $\dfrac{2x^2+5x-3}{2x-1}$
22. $\dfrac{3x^2+20x-32}{3x-4}$
23. $\dfrac{2x^2-3x-5}{x-3}$
24. $\dfrac{3x^2+17x-12}{x+6}$

25. $4x + 2 + \dfrac{-5}{x-5}$

26. $3x + 6 + \dfrac{16}{x-8}$

27. $2x + 3 + \dfrac{5}{3x-5}$

28. $2x - 4 + \dfrac{3}{2x+7}$

29. $x^2 - x - 2$

30. $x^2 + x + 7$

31. $x^2 + 2x + 3 + \dfrac{8}{4x-1}$

32. $x^2 - 2x + 3 + \dfrac{1}{2x+1}$

33. $x^2 + x + 2 + \dfrac{9}{x-2}$

34. $x^2 - 3x + 13 + \dfrac{-42}{x+3}$

35. $5x^2 + 2x + 1 + \dfrac{2}{5x-2}$

36. $2x^2 - 2x + 1 + \dfrac{-1}{4x+1}$

37. $x^2 + 4x + 5 + \dfrac{2}{x-2}$

38. $2x^2 - 7x + 10 + \dfrac{-8}{x+4}$

39. $x^3 + x^2 + x + 1$

40. $x^3 - 2x^2 + 5x - 10 + \dfrac{4}{x+2}$

41. $x - 3$

42. $x + 2$

43. $x^2 - 1 + \dfrac{1}{x^2+3}$

44. $x^2 + 3 + \dfrac{1}{x^2-2}$

45. $y^2 - y + 1$

46. $y^2 + 2y + 4$

47. $x^2 + 1$

48. $x^3 + 1$

---

25. $\dfrac{4x^2 - 18x - 15}{x-5}$

27. $\dfrac{6x^2 - x - 10}{3x-5}$

29. $\dfrac{x^3 + x^2 - 4x - 4}{x+2}$

31. $\dfrac{4x^3 + 7x^2 + 10x + 5}{4x-1}$

33. $\dfrac{x^3 - x^2 + 5}{x-2}$

35. $\dfrac{25x^3 + x}{5x-2}$

37. $\dfrac{2x^2 - 8 - 3x + x^3}{x-2}$

39. $\dfrac{x^4 - 1}{x-1}$

41. $\dfrac{x^3 - 3x^2 - x + 3}{x^2 - 1}$

43. $\dfrac{x^4 + 2x^2 - 2}{x^2 + 3}$

45. $\dfrac{y^3 + 1}{y+1}$

47. $\dfrac{x^4 - 1}{x^2 - 1}$

49. Find the value of c so that $\dfrac{y^2 - y + c}{y+1} = y - 2$

50. Find the value of c so that $\dfrac{x^3 + x^2 + x + c}{x^2 + 1} = x + 1$

26. $\dfrac{3x^2 - 18x - 32}{x-8}$

28. $\dfrac{4x^2 + 6x - 25}{2x+7}$

30. $\dfrac{x^3 - 2x^2 + 4x - 21}{x-3}$

32. $\dfrac{2x^3 - 3x^2 + 4x + 4}{2x+1}$

34. $\dfrac{x^3 + 4x - 3}{x+3}$

36. $\dfrac{8x^3 - 6x^2 + 2x}{4x+1}$

38. $\dfrac{x^2 - 18x + 2x^3 + 32}{x+4}$

40. $\dfrac{x^4 + x^2 - 16}{x+2}$

42. $\dfrac{x^3 + 2x^2 + 3x + 6}{x^2 + 3}$

44. $\dfrac{x^4 + x^2 - 5}{x^2 - 2}$

46. $\dfrac{y^3 - 8}{y - 2}$

48. $\dfrac{x^6 - 1}{x^3 - 1}$

49. $c = -2$

50. $c = 1$

53. (a) $(x^2 + 3x - 4)$;
(b) $(x + 4)(x - 1)$;
(c) $-2, -4, 1$

54. (a) $(x^2 - 5x + 4)$;
(b) $(x - 4)(x - 1)$;
(c) $-3, 4, 1$

55. (a) $x^2 - 4$;
(b) $(x + 2)(x - 2)$;
(c) $-1, 2, -2$

56. (a) $x^2 - 16$;
(b) $(x + 4)(x - 4)$;
(c) $3, 4, -4$

51. Write a summary of your work with polynomials. Explain how a polynomial is recognized, and explain the rules for the arithmetic of polynomials—how to add, subtract, multiply, and divide. What parts of this chapter do you feel you understand very well, and what part(s) do you still have questions about, or feel unsure of? Exchange papers with another student and compare your answers.

52. An interesting (and useful) thing about division of polynomials: To find out about this interesting thing, do this division. Compare your answer with another student.

$$(x - 2) \overline{) 2x^2 + 3x - 5} \qquad \text{Is there a remainder?}$$

Now, evaluate the polynomial $2x^2 + 3x - 5$ when $x = 2$. Is this value the same as the remainder?

Try $(x + 3) \overline{) 5x^2 - 2x + 1}$. Is there a remainder?

Evaluate the polynomial $5x^2 - 2x + 1$ when $x = -3$. Is this value the same as the remainder?

What happens when there is no remainder?

Try $(x - 6) \overline{) 3x^3 - 14x^2 - 23x - 6}$. Is the remainder zero?

Evaluate the polynomial $3x^3 - 14x^2 - 23x + 6$ when $x = 6$. Is this value zero? Write a description of the patterns you see. When does the pattern hold? Make up several more examples, and test your conjecture.

In Exercises 53 to 56, (a) divide the polynomial $f(x)$ by the given linear factor and (b) factor the quotient obtained in part a. Then, (c) using your graphing calculator, graph the polynomial $f(x)$ and determine where the graph passes through the x axis and (d) compare the results of parts b and c.

53. $f(x) = x^3 + 5x^2 + 2x - 8$; $x + 2$

54. $f(x) = x^3 - 2x^2 - 11x + 12$; $x + 3$

55. $f(x) = x^3 + x^2 - 4x - 4$; $x + 1$

56. $f(x) = x^3 - 3x^2 - 16x + 48$; $x - 3$

SECTION 6.5 Solving Quadratic Equations by Factoring

6.5 OBJECTIVES

1. Solve quadratic equations by factoring
2. Find the zeros of a quadratic function

The factoring techniques you have learned provide us with tools for solving equations that can be written in the form

$$ax^2 + bx + c = 0 \qquad a \neq 0$$

This is a quadratic equation in one variable, here x. You can recognize such a quadratic equation by the fact that the highest power of the variable x is the second power.

where a, b, and c are constants.

An equation written in the form $ax^2 + bx + c = 0$ is called a **quadratic equation in standard form.** Using factoring to solve quadratic equations requires the **zero-product principle**, which says that if the product of two factors is 0, then one or both of the factors must be equal to 0. In symbols:

> **Zero-Product Principle**
> If $a \cdot b = 0$, then $a = 0$ or $b = 0$ or $a = b = 0$.

Let's see how the principle is applied to solving quadratic equations.

Example 1

Solving Equations by Factoring

Solve.

$$x^2 - 3x - 18 = 0$$

Factoring on the left, we have

$$(x - 6)(x + 3) = 0$$

To use the zero-product principle, 0 must be on one side of the equation.

By the zero-product principle, we know that one or both of the factors must be zero. We can then write

$$x - 6 = 0 \qquad \text{or} \qquad x + 3 = 0$$

Solving each equation gives

$$x = 6 \quad \text{or} \quad x = -3$$

The two solutions are 6 and -3 and the solution set is written as $\{x | x = -3, 6\}$ or simply $\{-3, 6\}$.

The solutions are sometimes called the **zeros,** or **roots,** of the equation. They represent the x coordinates of the points where the graph of the equation $y = x^2 - 3x - 18$ crosses the x axis. In this case, the graph crosses the x axis at $(-3, 0)$ and $(6, 0)$.

Quadratic equations can be checked in the same way as linear equations were checked: by substitution. For instance, if $x = 6$, we have

$$6^2 - 3 \cdot 6 - 18 \stackrel{?}{=} 0$$
$$36 - 18 - 18 \stackrel{?}{=} 0$$
$$0 = 0$$

which is a true statement. We leave it to you to check the solution of -3.

Graph the function

$$y = x^2 - 3x - 18$$

on your graphing calculator. The solutions to the equation $0 = x^2 - 3x - 18$ will be those values on the curve at which $y = 0$. Those are the points at which the graph intercepts the x axis.

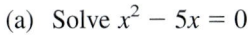 **CHECK YOURSELF 1**

Solve $x^2 - 9x + 20 = 0$.

Other factoring techniques are also used in solving quadratic equations. Example 2 illustrates this concept.

Example 2

Solving Equations by Factoring

(a) Solve $x^2 - 5x = 0$.

Again, factor the left side of the equation and apply the zero-product principle.

$$x(x - 5) = 0$$

Caution

A common mistake is to forget the statement $x = 0$ when you are solving equations of this type. Be sure to include the *two*

Now

$$x = 0 \quad \text{or} \quad x - 5 = 0$$
$$x = 5$$

The two solutions are 0 and 5. The solution set is $\{0, 5\}$.

434 Chapter 6 ■ Factoring Polynomials

The symbol ± is read "plus or minus."

(b) Solve $x^2 - 9 = 0$.

Factoring yields

$$(x + 3)(x - 3) = 0$$

$$x + 3 = 0 \quad \text{or} \quad x - 3 = 0$$

$$x = -3 \qquad\qquad x = 3$$

The solution set is $\{-3, 3\}$, which may be written as $\{\pm 3\}$.

✓ CHECK YOURSELF 2

Solve by factoring.

(a) $x^2 + 8x = 0$ (b) $x^2 - 16 = 0$

Example 3 illustrates a crucial point. Our solution technique depends on the zero-product principle, which means that the product of factors *must be equal to 0*. The importance of this is shown now.

Example 3

Solving Equations by Factoring

Solve $2x^2 - x = 3$.

The first step in the solution is to write the equation in standard form (that is, when one side of the equation is 0). So start by adding -3 to both sides of the equation. Then,

Caution

Consider the equation

$x(2x - 1) = 3$

Students are sometimes tempted to write

$x = 3$ or $2x - 1 = 3$

This is *not correct*. Instead, subtract 3 from both sides of the equation *as the first step* to write

$x^2 - 2x - 3 = 0$

in standard form. Only now can you factor and proceed as before.

$2x^2 - x - 3 = 0$ *Make sure all terms are on one side of the equation. The other side will be 0.*

You can now factor and solve by using the zero-product principle.

$$(2x - 3)(x + 1) = 0$$

$2x - 3 = 0 \quad \text{or} \quad x + 1 = 0$

$2x = 3 \qquad\qquad x = -1$

$x = \dfrac{3}{2}$

The solution set is $\left\{\dfrac{3}{2}, -1\right\}$.

✓ CHECK YOURSELF 3

Solve $3x^2 = 5x + 2$.

In all the previous examples, the quadratic equations had two distinct real number solutions. That may or may not always be the case, as we shall see.

Example 4 Solving Equations by Factoring

Solve $x^2 - 6x + 9 = 0$.

Factoring, we have

$$(x - 3)(x - 3) = 0$$

and

$$x - 3 = 0 \quad \text{or} \quad x - 3 = 0$$
$$x = 3 \qquad\qquad\qquad x = 3$$

The solution set is $\{3\}$.

A quadratic (or second-degree) equation always has *two* solutions. When an equation such as this one has two solutions that are the same number, we call 3 the **repeated** (or **double**) **solution** of the equation.

Even though a quadratic equation will always have two solutions, they may not always be real numbers. More about this in a later section.

✓ CHECK YOURSELF 4

Solve $x^2 + 6x + 9 = 0$.

Always examine the quadratic member of an equation for common factors. It will make your work much easier, as Example 5 illustrates.

Example 5

Solving Equations by Factoring

Solve $3x^2 - 3x - 60 = 0$.

First, note the common factor 3 in the quadratic member of the equation. Factoring out the 3, we have

$$3(x^2 - x - 20) = 0$$

Now divide both sides of the equation by 3.

Note the advantage of dividing both members by 3. The coefficients in the quadratic member become smaller, and that member is much easier to factor.

$$\frac{3(x^2 - x - 20)}{3} = \frac{0}{3}$$

or

$$x^2 - x - 20 = 0$$

We can now factor and solve as before.

$$(x - 5)(x + 4) = 0$$

$x - 5 = 0$ or $x + 4 = 0$

$x = 5$ $x = -4$ or $\{5, -4\}$

✓ CHECK YOURSELF 5

Solve $2x^2 - 10x - 48 = 0$.

In Chapter 4, we introduced the concept of a function and expressed the equation of a line in function form. Another type of function is called a quadratic function.

> A **quadratic function** is a function that can be written in the form
>
> $$f(x) = ax^2 + bx + c$$
>
> where a, b, and c are real numbers and $a \neq 0$.

For example, $f(x) = 3x^2 - 2x - 1$ and $g(x) = x^2 - 2$ are quadratic functions. In working with functions, we often want to find the values of x for which $f(x) = 0$. As in quadratic equations, these values are called the **zeros of the function.** They represent the x coordinates of the points where the graph of the function crosses the x axis. To find the zeros of a quadratic function, a quadratic equation must be solved.

Section 6.5 ■ Solving Quadratic Equations by Factoring **437**

Example 6

Finding the Zeros of a Function

Find the zeros of $f(x) = x^2 - x - 2$.

To find the zeros of $f(x) = x^2 - x - 2$, we must solve the quadratic equation $f(x) = 0$.

$$f(x) = 0$$
$$x^2 - x - 2 = 0$$
$$(x - 2)(x + 1) = 0$$
$$x - 2 = 0 \quad \text{or} \quad x + 1 = 0$$
$$x = 2 \qquad\qquad x = -1 \quad \text{or} \quad \{-1, 2\}$$

The zeros of the function are -1 and 2.

The graph of
$f(x) = x^2 - x - 2$
intercepts the x axis at the points $(2, 0)$ and $(-1, 0)$, so 2 and -1 are the *zeros* of the function.

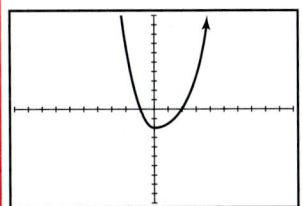

✓ CHECK YOURSELF 6

Find the zeros of $f(x) = 2x^2 - x - 3$.

✓ CHECK YOURSELF ANSWERS

1. $\{4, 5\}$.   **2.** (a) $\{0, -8\}$; (b) $\{4, -4\}$.   **3.** $\left\{-\dfrac{1}{3}, 2\right\}$.   **4.** $\{-3\}$.

5. $\{-3, 8\}$.   **6.** $\left\{-1, \dfrac{3}{2}\right\}$.

Exercises • 6.5

In Exercises 1 to 46, solve the quadratic equations by factoring.

1. $x^2 + 4x + 3 = 0$
2. $x^2 - 5x + 4 = 0$
3. $x^2 - 2x - 15 = 0$
4. $x^2 + 4x - 32 = 0$
5. $x^2 - 11x + 30 = 0$
6. $x^2 + 14x + 48 = 0$
7. $x^2 - 4x - 21 = 0$
8. $x^2 + 5x - 36 = 0$
9. $x^2 - 5x = 50$
10. $x^2 + 14x = -33$
11. $x^2 = 2x + 35$
12. $x^2 = 6x + 27$
13. $x^2 - 8x = 0$
14. $x^2 + 7x = 0$
15. $x^2 + 10x = 0$
16. $x^2 - 9x = 0$
17. $x^2 = 5x$
18. $4x = x^2$
19. $x^2 - 25 = 0$
20. $x^2 - 49 = 0$
21. $x^2 = 64$
22. $x^2 = 36$
23. $4x^2 + 12x + 9 = 0$
24. $9x^2 - 30x + 25 = 0$
25. $2x^2 - 17x + 36 = 0$
26. $5x^2 + 17x - 12 = 0$
27. $3x^2 - x = 4$
28. $6x^2 = 13x - 6$
29. $6x^2 = 7x - 2$
30. $4x^2 - 3 = x$
31. $2m^2 = 12m + 54$
32. $5x^2 - 55x = 60$
33. $4x^2 - 24x = 0$
34. $6x^2 - 9x = 0$
35. $5x^2 = 15x$
36. $7x^2 = -49x$

1. $\{-3, -1\}$
2. $\{4, 1\}$
3. $\{-3, 5\}$
4. $\{-8, 4\}$
5. $\{5, 6\}$
6. $\{-8, -6\}$
7. $\{-3, 7\}$
8. $\{-9, 4\}$
9. $\{-5, 10\}$
10. $\{-11, -3\}$
11. $\{-5, 7\}$
12. $\{-3, 9\}$
13. $\{0, 8\}$
14. $\{-7, 0\}$
15. $\{0, -10\}$
16. $\{0, 9\}$
17. $\{0, 5\}$
18. $\{0, 4\}$
19. $\{-5, 5\}$
20. $\{-7, 7\}$
21. $\{-8, 8\}$
22. $\{-6, 6\}$
23. $\left\{-\dfrac{3}{2}\right\}$
24. $\left\{\dfrac{5}{3}\right\}$
25. $\left\{4, \dfrac{9}{2}\right\}$
26. $\left\{-4, \dfrac{3}{5}\right\}$
27. $\left\{-1, \dfrac{4}{3}\right\}$
28. $\left\{\dfrac{2}{3}, \dfrac{3}{2}\right\}$
29. $\left\{\dfrac{1}{2}, \dfrac{2}{3}\right\}$
30. $\left\{-\dfrac{3}{4}, 1\right\}$
31. $\{-3, 9\}$
32. $\{-1, 12\}$
33. $\{0, 6\}$
34. $\left\{0, \dfrac{3}{2}\right\}$
35. $\{0, 3\}$
36. $\{-7, 0\}$

Answers (left column)

37. $\{-3, 5\}$
38. $\{-7, 4\}$
39. $\left\{-\dfrac{3}{2}, 3\right\}$
40. $\left\{-\dfrac{13}{3}, 4\right\}$
41. $\left\{-\dfrac{7}{3}, 2\right\}$
42. $\left\{-2, \dfrac{5}{2}\right\}$
43. $\{-2, 6\}$
44. $\{-7, 2\}$
45. $\left\{-\dfrac{3}{2}, 5\right\}$
46. $\left\{-3, \dfrac{8}{3}\right\}$
47. $\{2, 6\}$
48. $\{-4, 7\}$
49. $\{-5, 1\}$
50. $\{2, 9\}$
53. $x^2 + x - 6 = 0$
54. $x^2 - 5x = 0$
55. $x^2 - 8x + 12 = 0$
56. $x^2 - 16 = 0$
57. $\{-2, 0, 5\}$
58. $\{-5, -3, 0\}$
59. $\{-3, 0, 3\}$
60. $\{-4, 0, 4\}$
61. $\{-2, -1, 2\}$
62. $\{-1, 1, 5\}$
63. $\{-3, -1, 1, 3\}$
64. $\{-2, -1, 1, 2\}$
65. $\{0$ cm, 100 cm$\}$
66. $\{0$ cm, 125 cm$\}$
67. $\{0$ cm, 132 cm$\}$
68. $\{0$ cm, 146 cm$\}$

Exercises

37. $x(x - 2) = 15$ **38.** $x(x + 3) = 28$ **39.** $x(2x - 3) = 9$

40. $x(3x + 1) = 52$ **41.** $2x(3x + 1) = 28$ **42.** $3x(2x - 1) = 30$

43. $(x - 3)(x - 1) = 15$ **44.** $(x + 4)(x + 1) = 18$ **45.** $(2x + 1)(x - 4) = 11$

46. $(3x - 5)(x + 2) = 14$

In Exercises 47 to 50, find the zeros of the functions.

47. $f(x) = 3x^2 - 24x + 36$ **48.** $f(x) = 2x^2 - 6x - 56$

49. $f(x) = 4x^2 + 16x - 20$ **50.** $f(x) = 3x^2 - 33x + 54$

51. Explain the differences between solving the equations $3(x - 2)(x + 5) = 0$ and $3x(x - 2)(x + 5) = 0$.

52. How can a graphing calculator be used to determine the zeros of a quadratic function?

In Exercises 53 to 56, write an equation that has the following solutions. (*Hint:* Write the binomial factors and then the quadratic member of the equation.)

53. $\{2, -3\}$ **54.** $\{0, 5\}$

55. $\{6, 2\}$ **56.** $\{-4, 4\}$

The zero-product rule can be extended to three or more factors. If $a \cdot b \cdot c = 0$, then at least one of these factors is 0. In Exercises 57 to 60, use this information to solve the equations.

57. $x^3 - 3x^2 - 10x = 0$ **58.** $x^3 + 8x^2 + 15x = 0$

59. $x^3 - 9x = 0$ **60.** $x^3 = 16x$

In Exercises 61 to 64, extend the ideas in the previous exercises to find solutions for the following equations. (*Hint:* Apply factoring by grouping in Exercises 61 and 62.)

61. $x^3 + x^2 - 4x - 4 = 0$ **62.** $x^3 - 5x^2 - x + 5 = 0$

63. $x^4 - 10x^2 + 9 = 0$ **64.** $x^4 - 5x^2 + 4 = 0$

The net productivity of a forested wetland as related to the amount of water moving through the wetland can be expressed by a quadratic equation. In Exercises 65 to 68, if y represents the amount of wood produced, in grams per square meter, and x represents the amount of water present, in centimeters, determine where the productivity is zero in each wetland represented by the equations.

65. $y = -3x^2 + 300x$ **66.** $y = -4x^2 + 500x$

67. $y = -6x^2 + 792x$ **68.** $y = -7x^2 + 1022x$

440 Chapter 6 ■ Factoring Polynomials

69. {30}

70. {10, 15}

71. 2, 6 are the zeros.

72. −4, 7 are the zeros.

69. Break-even analysis. The manager of a bicycle shop knows that the cost of selling x bicycles is $C = 20x + 60$ and the revenue from selling x bicycles is $R = x^2 - 8x$. Find the break-even value of x.

70. Break-even analysis. A company that produces computer games has found that its operating cost in dollars is $C = 40x + 150$ and its revenue in dollars is $R = 65x - x^2$. For what value(s) of x will the company break even?

Graphing calculator. Use your calculator to graph $f(x)$.

71. $f(x) = 3x^2 - 24x + 36$. Note the x values at which the graph crosses the x axis. Compare your answers to the solution for Exercise 47.

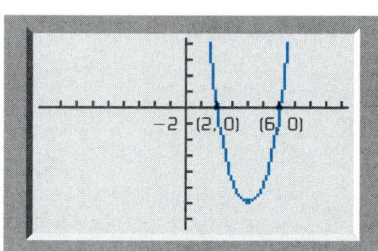

2, 6 are the zeros of Exercise 47.

72. $f(x) = 2x^2 - 6x - 56$. Note the x values at which the graph crosses the x axis. Compare your answer to the solution for Exercise 48.

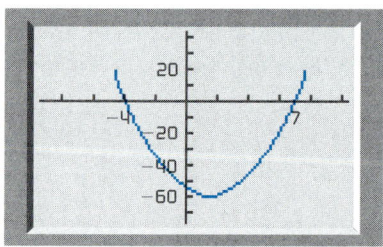

−4, 7 are the zeros of Exercise 48.

73. Work with another student and use your calculator to solve this exercise. Assume that each graph is the graph of a quadratic function. Find an equation for each graph given. There could be more than one equation for some of the graphs. Remember the connection between the x intercepts and the zeros.

(a)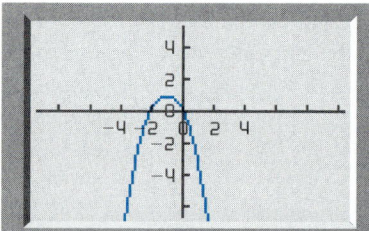

(b)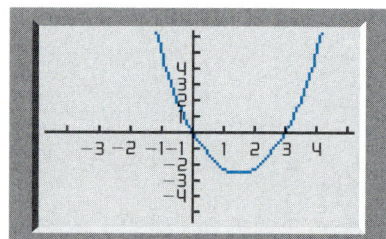

Section 6.5 ■ Solving Quadratic Equations by Factoring **441**

(c)

(d)

(e)

(f)

(g)

(h)

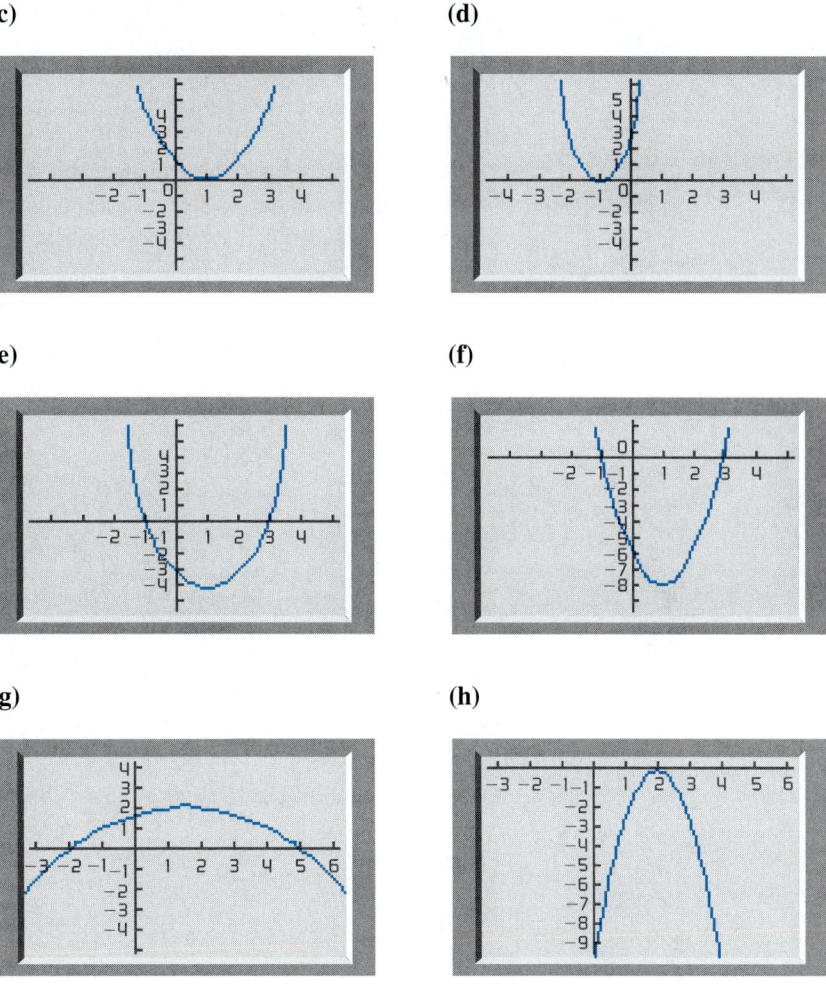

Summary Exercises • 6

This summary exercise set is provided to give you practice with each of the objectives of the chapter. Each exercise is keyed to the appropriate chapter section. Your instructor will give you guidelines on how to best use these exercises in your instructional setting.

[6.1] Factor each of the following polynomials.

1. $18a + 24$
2. $9m^2 - 21m$
3. $24s^2t - 16s^2$
4. $18a^2b + 36ab^2$
5. $35s^3 - 28s^2$
6. $3x^3 - 6x^2 + 15x$
7. $18m^2n^2 - 27m^2n + 45m^2n^3$
8. $121x^8y^3 + 77x^6y^3$
9. $8a^2b + 24ab - 16ab^2$
10. $3x^2y - 6xy^3 + 9x^3y - 12xy^2$
11. $x(2x - y) + y(2x - y)$
12. $5(w - 3z) - w(w - 3z)$

[6.2] Factor each of the following binomials completely.

13. $p^2 - 49$
14. $25a^2 - 16$
15. $m^2 - 9n^2$
16. $16r^2 - 49s^2$
17. $25 - z^2$
18. $a^4 - 16b^2$
19. $25a^2 - 36b^2$
20. $x^6 - 4y^2$
21. $3w^3 - 12wz^2$
22. $16a^4 - 49b^2$
23. $2m^2 - 72n^4$
24. $3w^3z - 12wz^3$

[6.2] Factor the following polynomials completely.

25. $x^2 - 4x + 5x - 20$
26. $x^2 + 7x - 2x - 14$
27. $6x^2 + 4x - 15x - 10$
28. $12x^2 - 9x - 28x + 21$
29. $6x^3 + 9x^2 - 4x^2 - 6x$
30. $3x^4 + 6x^3 + 5x^3 + 10x^2$

---

1. $6(3a + 4)$
2. $3m(3m - 7)$
3. $8s^2(3t - 2)$
4. $18ab(a + 2b)$
5. $7s^2(5s - 4)$
6. $3x(x^2 - 2x + 5)$
7. $9m^2n(2n - 3 + 5n^2)$
8. $11x^6y^3(11x^2 + 7)$
9. $8ab(a + 3 - 2b)$
10. $3xy(x - 2y^2 + 3x^2 - 4y)$
11. $(2x - y)(x + y)$
12. $(w - 3z)(5 - w)$
13. $(p + 7)(p - 7)$
14. $(5a + 4)(5a - 4)$
15. $(m + 3n)(m - 3n)$
16. $(4r + 7s)(4r - 7s)$
17. $(5 + z)(5 - z)$
18. $(a^2 + 4b)(a^2 - 4b)$
19. $(5a + 6b)(5a - 6b)$
20. $(x^3 + 2y)(x^3 - 2y)$
21. $3w(w + 2z)(w - 2z)$
22. $(4a^2 + 7b)(4a^2 - 7b)$
23. $2(m + 6n^2)(m - 6n^2)$
24. $3wz(w + 2z)(w - 2z)$
25. $(x - 4)(x + 5)$
26. $(x + 7)(x - 2)$
27. $(2x - 5)(3x + 2)$
28. $(4x - 3)(3x - 7)$
29. $x(2x + 3)(3x - 2)$
30. $x^2(x + 2)(3x + 5)$

31. (a) 18; (b) $3x^2(x+5)$; (c) 18
32. (a) -8; (b) $-2x^2(3x+1)$; (c) -8
33. (a) 4; (b) $4x^4(3x^2-2)$; (c) 4
34. (a) 0; (b) $2x(x^4-1)$; (c) 0
35. $(x+8)(x-8)$
36. $(5a+4)(5a-4)$
37. $(4m+7n)(4m-7n)$
38. $3w(w+2z)(w-2z)$
39. $(a^2+4b^2)(a+2b)(a-2b)$
40. $(m-4)(m^2+4m+16)$
41. $(2x+1)(4x^2-2x+1)$
42. $(2c-3d)(4c^2+6cd+9d^2)$
43. $(5m+4n)(25m^2-20mn+16n^2)$
44. $2x(x+3)(x^2-3x+9)$
45. Factorable; $m=-6$, $n=5$
46. Factorable; $m=2$, $n=1$
47. Factorable; $m=-3$, $n=-8$
48. Factorable; $m=-3$, $n=-20$
49. $(x+10)(x+2)$
50. $(a-4)(a+3)$
51. $(w-8)(w-5)$
52. $(r-12)(r+3)$
53. $(x-12y)(x+4y)$
54. $(a+15b)(a+2b)$
55. $(5x-2)(x+3)$
56. $(2a-7)(a+5)$

[6.2] For each of the following functions, (a) find $f(1)$, (b) factor $f(x)$, and (c) find $f(1)$ from the factored form.

31. $f(x) = 3x^3 + 15x^2$
32. $f(x) = -6x^3 - 2x^2$
33. $f(x) = 12x^6 - 8x^4$
34. $f(x) = 2x^5 - 2x$

[6.2] Factor each of the following binomials completely.

35. $x^2 - 64$
36. $25a^2 - 16$
37. $16m^2 - 49n^2$
38. $3w^3 - 12wz^2$
39. $a^4 - 16b^4$
40. $m^3 - 64$
41. $8x^3 + 1$
42. $8c^3 - 27d^3$
43. $125m^3 + 64n^3$
44. $2x^4 + 54x$

[6.3] Use the *ac* test to determine which of the following trinomials can be factored. Find the values of *m* and *n* for each trinomial that can be factored.

45. $x^2 - x - 30$
46. $x^2 + 3x + 2$
47. $2x^2 - 11x + 12$
48. $4x^2 - 23x + 15$

[6.3] Completely factor each of the following trinomials.

49. $x^2 + 12x + 20$
50. $a^2 - a - 12$
51. $w^2 - 13w + 40$
52. $r^2 - 9r - 36$
53. $x^2 - 8xy - 48y^2$
54. $a^2 + 17ab + 30b^2$
55. $5x^2 + 13x - 6$
56. $2a^2 + 3a - 35$

57. $(x + 4)(x + 5)$
58. $(x - 4)(x - 6)$
59. $(a - 4)(a - 3)$
60. $(w + 8)(w + 5)$
61. $(x + 6)(x + 6)$
62. $(r - 12)(r - 3)$
63. $(b - 7c)(b + 3c)$
64. $n(m + 8)(m - 4)$
65. $m(m + 7)(m - 5)$
66. $2(x - 5)(x + 4)$
67. $3y(y - 7)(y - 9)$
68. $3b(b - 7)(b + 2)$
69. $(3x + 5)(x + 1)$
70. $(5w - 2)(w + 3)$
71. $(2b - 3)(b - 3)$
72. $(4x + 3)(2x - 1)$
73. $(5x - 3)(2x - 1)$
74. $(4a - 5)(a + 3)$
75. $(3y - 5z)(3y + 4z)$
76. $(2x + 5y)(4x - 3y)$
77. $4x(2x + 1)(x - 5)$
78. $3(3x + 1)(x - 2)$
79. $3x(2x - 3)(x + 1)$
80. $5(w - 2z)(w - 3z)$
81. $3a^3$
82. $4m^2n$
83. $3a - 2$
84. $4a^2 + 3$
85. $-3rs + 6r^2$
86. $5xy - 3y^2 + 2x$
87. $x - 5$
88. $x + 7$

[6.3] Factor each of the following trinomials completely.

57. $x^2 + 9x + 20$
58. $x^2 - 10x + 24$
59. $a^2 - 7a + 12$
60. $w^2 - 13w + 40$
61. $x^2 + 12x + 36$
62. $r^2 - 15r + 36$
63. $b^2 - 4bc - 21c^2$
64. $m^2n + 4mn - 32n$
65. $m^3 + 2m^2 - 35m$
66. $2x^2 - 2x - 40$
67. $3y^3 - 48y^2 + 189y$
68. $3b^3 - 15b^2 - 42b$
69. $3x^2 + 8x + 5$
70. $5w^2 + 13w - 6$
71. $2b^2 - 9b + 9$
72. $8x^2 + 2x - 3$
73. $10x^2 - 11x + 3$
74. $4a^2 + 7a - 15$
75. $9y^2 - 3yz - 20z^2$
76. $8x^2 + 14xy - 15y^2$
77. $8x^3 - 36x^2 - 20x$
78. $9x^2 - 15x - 6$
79. $6x^3 - 3x^2 - 9x$
80. $5w^2 - 25wz + 30z^2$

[6.4] Divide.

81. $\dfrac{9a^5}{3a^2}$
82. $\dfrac{24m^4n^2}{6m^2n}$
83. $\dfrac{15a - 10}{5}$
84. $\dfrac{32a^3 + 24a}{8a}$
85. $\dfrac{9r^2s^3 - 18r^3s^2}{-3rs^2}$
86. $\dfrac{35x^3y^2 - 21x^2y^3 + 14x^3y}{7x^2y}$

[6.4] Perform the indicated long division.

87. $\dfrac{x^2 - 2x - 15}{x + 3}$
88. $\dfrac{2x^2 + 9x - 35}{2x - 5}$

Summary Exercises

89. $\dfrac{x^2 - 8x + 17}{x - 5}$

90. $\dfrac{6x^2 - x - 10}{3x + 4}$

91. $\dfrac{6x^3 + 14x^2 - 2x - 6}{6x + 2}$

92. $\dfrac{4x^3 + x + 3}{2x - 1}$

93. $\dfrac{3x^2 + x^3 + 5 + 4x}{x + 2}$

94. $\dfrac{2x^4 - 2x^2 - 10}{x^2 - 3}$

[6.5] Solve each of the following equations by factoring.

95. $x^2 + 5x - 6 = 0$
96. $x^2 - 2x - 8 = 0$
97. $x^2 + 7x = 30$
98. $x^2 - 6x = 40$
99. $x^2 = 11x - 24$
100. $x^2 = 28 - 3x$
101. $x^2 - 10x = 0$
102. $x^2 = 12x$
103. $x^2 - 25 = 0$
104. $x^2 = 144$
105. $2x^2 - x - 3 = 0$
106. $3x^2 - 4x = 15$
107. $3x^2 + 9x - 30 = 0$
108. $4x^2 + 24x = -32$
109. $x(x - 3) = 18$
110. $(x - 2)(2x + 1) = 33$
111. $x^3 - 2x^2 - 15x = 0$
112. $x^3 + x^2 - 4x - 4 = 0$

113. Suppose that the cost, in dollars, of producing x stereo systems is given by

$$C(x) = 3000 - 60x + 3x^2$$

How many systems can be produced for $7500?

114. The demand equation for a certain type of computer paper is predicted to be

$$D = -3p + 69$$

The supply equation is predicted to be

$$S = -p^2 + 24p - 3$$

Find the equilibrium price.

Answers

89. $x - 3 + \dfrac{2}{x - 5}$

90. $2x - 3 + \dfrac{2}{3x + 4}$

91. $x^2 + 2x - 1 + \dfrac{-4}{6x + 2}$

92. $2x^2 + x + 1 + \dfrac{4}{2x - 1}$

93. $x^2 + x + 2 + \dfrac{1}{x + 2}$

94. $2x^2 + 4 + \dfrac{2}{x^2 - 3}$

95. {−6, 1}
96. {−2, 4}
97. {−10, 3}
98. {−4, 10}
99. {3, 8}
100. {−7, 4}
101. {0, 10}
102. {0, 12}
103. {−5, 5}
104. {−12, 12}
105. $\left\{-1, \dfrac{3}{2}\right\}$
106. $\left\{-\dfrac{5}{3}, 3\right\}$
107. {−5, 2}
108. {−4, −2}
109. {−3, 6}
110. $\left\{-\dfrac{7}{2}, 5\right\}$
111. {0, −3, 5}
112. {−2, −1, 2}
113. 50
114. $3 or $24

Cumulative Test ■ 0-6

1. $2x^2y + 5xy$
2. $3a^4b^6$
3. $4x^2 + x + 3$
4. $x^2 + 9x - 2$
5. 27
6. -1
7. $a^2 - 9b^2$
8. $x^2 - 4xy + 4y^2$
9. $x^2 + 3x - 10$
10. $a^2 + a - 12$
11. $3x + 2$
12. $3x + 3 + \dfrac{1}{x-1}$
13. $\{11\}$
14. $\{-33\}$
15. $\{15, -3\}$
16. $\{1, -10\}$
17. $\{x | x > -6\}$
18. $\left\{x \mid \dfrac{1}{2} < x < \dfrac{13}{2}\right\}$
19. $\{x | x > -2 \text{ or } x < -8\}$
20. $4(3x + 5)$
21. $(5x + 7y)(5x - 7y)$
22. $(4x - 5)(3x + 2)$
23. $(x - 5)(2x - 3)$
24. $3x - 5 = 46; \ x = 17$
25. $x + (x - 5) = 81; \ x = 43$

This test covers selected topics from the first six chapters.

Simplify the expression.

1. $7x^2y + 3xy - 5x^2y + 2xy$
2. $\dfrac{27a^5b^7}{9ab}$
3. $(3x^2 - 2x + 5) + (x^2 + 3x - 2)$
4. $(5x^2 + 4x - 3) - (4x^2 - 5x - 1)$

Evaluate the expression.

5. $3(5 - 2)^2$
6. $|3| - |-4|$

Multiply.

7. $(a - 3b)(a + 3b)$
8. $(x - 2y)^2$
9. $(x - 2)(x + 5)$
10. $(a - 3)(a + 4)$

Divide.

11. $(9x^2 + 12x + 4) \div (3x + 2)$
12. $(3x^2 - 2) \div (x - 1)$

Solve each equation and check your results.

13. $7a - 3 = 6a + 8$
14. $\dfrac{2}{3}x = -22$
15. $|x - 6| = 9$
16. $|2x + 9| - 5 = 6$

Solve each of the following inequalities.

17. $2x - 7 < 5 + 4x$
18. $|2x - 7| < 6$
19. $|x + 5| > 3$

Factor each of the following expressions.

20. $12x + 20$
21. $25x^2 - 49y^2$
22. $12x^2 - 15x + 8x - 10$
23. $2x^2 - 13x + 15$

Solve the following word problems. Show the equation used for the solution.

24. 3 times a number decreased by 5 is 46. Find the number.
25. Juan's biology text cost $5 more than his mathematics text. Together they cost $81. Find the cost of the biology text.

446

Moving to the Intermediate Algebra Level

To Students Beginning the Text at This Point

If you are beginning your studies at this point (through placement or transfer) we want to help you correct any deficiencies you have before those deficiencies interfere with your ability to learn. Your instructor will advise you as to the appropriate course of action before continuing. One option, suggested by the authors, is described below.

You have undoubtedly taken many tests (too many?) as you have progressed through your mathematical studies. In the next few pages, we are going to strongly suggest that you take eight more. The first seven of these tests are chapter tests, one for each of the chapters to this point of the text. Each of these tests is preceded by a summary of the material in that chapter. A perusal of the summary should be sufficient review for you to pass the test. If it is not, you need to go back and work on the material in that chapter. If that seems like a daunting task, you should reconsider the placement. You may be better off starting with the material in the beginning of this text.

The eighth test is a final exam for the material to this point. Successfully completing this exam will indicate that you are ready to begin the material in Chapter 7.

To Students Continuing Study in This Text

The material in this transition section provides you an excellent opportunity to prepare for a final exam on the material presented to this point in the text. There is a summary of each chapter, followed by a test of the material in that chapter. These seven tests are followed by a practice final.

If you successfully complete the seven tests and the final, you are undoubtedly prepared for a final exam. If not, more review is in order. This can be done effectively with the StreeterSmart software. Talk to your instructor to find the appropriate way for you to deal with your deficiencies.

Summary for Chapter 0

Example	Topic	Reference
	Fractions	0.1
$\dfrac{4 \cdot 3}{5 \cdot 3} = \dfrac{12}{15}$	**Equivalent Fractions** If the numerator and denominator of a fraction are both multiplied by some nonzero number, the result is a fraction that is equivalent to the original fraction.	p. 2
$\dfrac{9}{21} = \dfrac{3 \cdot 3}{3 \cdot 7} = \dfrac{3}{7}$	**Simplifying Fractions** A fraction is in simplest terms when the numerator and denominator have no common factor.	p. 3
$\dfrac{2}{3} \cdot \dfrac{5}{6} = \dfrac{10}{18} = \dfrac{5}{9}$	**Multiplying Fractions** To multiply two fractions, multiply the numerators, then multiply the denominators. Simplification can be done before or after the multiplication.	p. 4
$\dfrac{3}{5} \div \dfrac{2}{7} = \dfrac{3}{5} \cdot \dfrac{7}{2} = \dfrac{21}{10}$	**Dividing Fractions** To divide two fractions, invert the divisor (the second fraction), then multiply the fractions.	p. 5
$\dfrac{2}{5} + \dfrac{5}{8} = \dfrac{16}{40} + \dfrac{25}{40} = \dfrac{41}{40}$	**Adding Fractions** To add two fractions, find the LCD (least common denominator), rewrite the fractions with this denominator, then add the numerators.	p. 5
$\dfrac{2}{3} - \dfrac{1}{4} = \dfrac{8}{12} - \dfrac{3}{12} = \dfrac{5}{12}$	**Subtracting Fractions** To subtract two fractions, find the LCD, rewrite the fractions with this denominator, then subtract the numerators.	p. 6
	Signed Numbers—The Terms	0.2
Negative numbers Positive numbers $-3\ -2\ -1\ \ 0\ \ 1\ \ 2\ \ 3$ Zero is neither positive nor negative.	**Positive Numbers** Numbers used to name points to the right of 0 on the number line.	p. 10
	Negative Numbers Numbers used to name points to the left of 0 on the number line.	p. 10
	Signed Numbers A set containing both positive and negative numbers and zero.	p. 10

Summary for Chapter 0

The opposite of 5 is −5. (5 units each direction from 0)	**Opposites** Two numbers are opposites if the points name the same distance from 0 on the number line, but in opposite directions.	p. 11
	The opposite of a positive number is negative.	
The opposite of −3 is 3. (3 units each direction from 0)	The opposite of a negative number is positive.	
	0 is its own opposite.	
The absolute value of a number a is written $\lvert a \rvert$. $\lvert 7 \rvert = 7 \quad \lvert -8 \rvert = 8$	**Absolute Value** The distance on the number line between the point named by a number and 0. The absolute value of a number is always positive or 0.	p. 12
The integers are $\{ \ldots, -3, -2, -1, 0, 1, 2, 3, \ldots \}$	**Integers** The set consisting of the natural numbers, their opposites, and 0.	p. 13
	Operations on Signed Numbers	0.3–0.4
$5 + 8 = 13$ $-3 + (-7) = -10$ $5 + (-3) = 2$ $7 + (-9) = -2$	**To Add Signed Numbers** 1. If two numbers have the same sign, add their absolute values. Give the sum the sign of the original numbers. 2. If two numbers have different signs, subtract the smaller absolute value from the larger. Give the result the sign of the number with the larger absolute value.	p. 17
$4 - (-2) = 4 + 2 = 6$ *The opposite of −2.*	**To Subtract Signed Numbers** To subtract signed numbers, add the first number and the opposite of the number being subtracted.	p. 22
$5 \cdot 7 = 35$ $(-4)(-6) = 24$ $(8)(-7) = -56$	**To Multiply Signed Numbers** To multiply signed numbers, multiply the absolute values of the numbers. Then attach a sign to the product according to the following rules: 1. If the numbers have different signs, the product is negative. 2. If the numbers have the same sign, the product is positive.	p. 28
$\dfrac{-8}{-2} = 4$ $27 \div (-3) = -9$ $\dfrac{-16}{8} = -2$	**To Divide Signed Numbers** To divide signed numbers, divide the absolute values of the numbers. Then attach a sign to the quotient according to the following rules: 1. If the numbers have the same sign, the quotient is positive. 2. If the numbers have different signs, the quotient is negative.	p. 36

	The Properties of Addition and Multiplication	0.3–0.4
	The Commutative Properties If a and b are any numbers,	p. 19
$3 + 4 = 4 + 3$ $7 = 7$	1. $a + b = b + a$ 2. $a \cdot b = b \cdot a$	
	The Associative Properties If a, b, and c are any numbers,	p. 19
$3 \cdot (4 \cdot 5) = (3 \cdot 4) \cdot 5$ $3 \cdot (20) = (12) \cdot 5$ $60 = 60$	1. $a + (b + c) = (a + b) + c$ 2. $a \cdot (b \cdot c) = (a \cdot b) \cdot c$	
$2(5 + 3) = 2 \cdot 5 + 2 \cdot 3$ $2(8) = 10 + 6$ $16 = 16$	**The Distributive Property** If a, b, and c are any numbers, $a(b + c) = a \cdot b + a \cdot c$	p. 34
	Exponents and the Order of Operations	0.5
	The Notation	p. 43
$5^3 = 5 \cdot 5 \cdot 5$ $= 125$ $a^2 b^3 = a \cdot a \cdot b \cdot b \cdot b$ $6m^2 = 6 \cdot m \cdot m$	Exponent $a^4 = \underbrace{a \cdot a \cdot a \cdot a}_{\text{4 factors}}$ Base The number or letter used as a factor, here a, is called the *base*. The *exponent*, which is written above and to the right of the base, tells us how many times the base is used as a factor.	
Operate inside grouping symbols. $5 + 3(6-4)^2$ Evaluate the power. $5 + 3 \cdot 2^2$ Multiply. $= 5 + 3 \cdot 4$ Add. $= 5 + 12$ $= 17$	**The Order of Operations** 1. Do any operations within grouping symbols first. 2. Evaluate all expressions containing exponents. 3. Do any multiplication or division in order, working from left to right. 4. Do any addition or subtraction in order, working from left to right.	p. 45

Self-Test for Chapter 0

1. $\dfrac{1}{3}$
2. $\dfrac{3}{11}$
3. $\dfrac{9}{11}$
4. $\dfrac{25}{16}$
5. $\dfrac{29}{30}$
6. $\dfrac{21}{80}$
7. $\dfrac{9}{44}$
8. $\dfrac{14}{5}$
9. 2
10. −12
11. 0
12. −3
13. −12
14. −14
15. 11
16. 15
17. −35
18. 54
19. 3
20. 36
21. 0
22. 7
23. −4
24. Undefined
25. 15
26. 6
27. <
28. >
29. 7
30. 86

Each self-test in the bridge section of this text allows you to assess your mastery of the material in a particular chapter. After you have reviewed the summary material for the chapter, try taking the self-test. When you complete the test, check your answers against those in the back of the text. If you miss any questions, go back to the appropriate section of the text and review the material in greater depth than the summary permitted. Once you have completed all of the self-tests, you will be ready for the final exam. Good luck.

In Exercises 1 to 4, simplify each fraction.

1. $\dfrac{12}{36}$
2. $\dfrac{27}{99}$
3. $\dfrac{18}{22}$
4. $\dfrac{100}{64}$

5. Find the sum $\dfrac{4}{15} + \dfrac{7}{10}$
6. Find the difference $\dfrac{9}{16} - \dfrac{3}{10}$

7. Find the product $\dfrac{3}{8} \cdot \dfrac{18}{33}$
8. Find the quotient $\dfrac{4}{5} \div \dfrac{2}{7}$

In Exercises 9 to 16, add or subtract as indicated.

9. $5 + (-3)$
10. $(-5) + 0 + (-7)$
11. $17 + (-17)$
12. $13 + (-11) + (-5)$
13. $23 - 35$
14. $-6 - 8$
15. $7 - (-4)$
16. $-3 - (-18)$

In Exercises 17 to 24, multiply or divide as indicated.

17. $(-7)(5)$
18. $(-9)(-6)$
19. $(-2.5)(-1.2)$
20. $(-6)^2$
21. $0 \div (-5)$
22. $-28 \div (-4)$
23. $(-40) \div 10$
24. $20 \div 0$

Use the distributive property to remove parentheses and simplify.

25. $5(-8 + 11)$
26. $-3(7 - 9)$

Complete the following statements using the symbol <, >, or =.

27. -7 _____ -5
28. $|-8|$ _____ -8

Evaluate each expression.

29. $23 - 4 \cdot 12 \div 3$
30. $4 \cdot 5^2 - 35 + 21$

Summary for Chapter 1

Example	Topic	Reference
	From Arithmetic to Algebra	1.1
The sum of x and 5 is $x + 5$. 7 more than a is $a + 7$. b increased by 3 is $b + 3$.	**Addition** $x + y$ means the **sum** of x **and** y or x **plus** y. Some other words indicating addition are "more than" and "increased by."	p. 56
The difference of x and 3 is $x - 3$. 5 less than p is $p - 5$. a decreased by 4 is $a - 4$.	**Subtraction** $x - y$ means the **difference** of x **and** y or x **minus** y. Some other words indicating subtraction are "less than" and "decreased by."	p. 57
The product of m and n is mn. The product of 2 and the sum of a and b is $2(a + b)$.	**Multiplication** $\left. \begin{array}{c} x \cdot y \\ (x)(y) \\ xy \end{array} \right\}$ These all mean the *product* of x and y or x *times* y.	p. 58
n divided by 5 is $\frac{n}{5}$. The sum of a and b, divided by 3, is $\frac{a+b}{3}$.	**Division** $\frac{x}{y}$ means x divided by y or the *quotient* when x is divided by y.	p. 60
	Evaluating Algebraic Expressions	1.2
Evaluate $$\frac{4a - b}{2c}$$ if $a = -6$, $b = 8$, and $c = -4$. $\frac{4a - b}{2c} = \frac{4(-6) - 8}{2(-4)}$ $= \frac{-24 - 8}{-8}$ $= \frac{-32}{-8} = 4$	**To evaluate an algebraic expression:** 1. Replace each variable by the given number value. 2. Do the necessary arithmetic operations. (Be sure to follow the rules for the order of operations.)	p. 65
	Adding and Subtracting Algebraic Expressions	1.3
$3x^2y$ is a term.	**Term** A number, or the product of a number and one or more variables, raised to a power.	p. 75

■ Summary for Chapter 1   453

$4a^2$ and $3a^2$ are like terms. $5x^2$ and $2xy^2$ are not like terms.	**Like Terms** Terms that contain exactly the same variables raised to the same powers.	p. 76
$5a + 3a = 8a$ $7xy - 3xy = 4xy$	**Combining Like Terms** 1. Add or subtract the numerical coefficients. 2. Attach the common variables.	p. 78
	Sets	**1.4**
$A = \{2, 3, 4, 5\}$ is a set.	**Set** A set is a collection of objects classified together.	p. 86
2 is an element of set A.	**Elements** The elements are the objects in a set.	p. 86
$S = \{2, 4, 6, 8\}$ is in roster form.	**Roster Form** A set is said to be in roster form if the elements are listed and enclosed in braces.	p. 86
$\{x\vert x < 4\}$ is written in set-builders notation.	**Set-Builder Notation** $\{x\vert x > a\}$ is read "the set of all x, where x is greater than a." $\{x\vert x < a\}$ is read "the set of all x, where x is less than a." $\{x\vert a < x < b\}$ is read "the set of all x, where x is greater than a and less than b."	p. 87
$\{x\vert x < 4\}$ ←——(——→ −4 0 4 The parenthesis indicates every number below the marked value (here it is 4).	**Plotting the Elements of a Set on a Number Line** $\{x\vert x < a\}$ indicates the set of all points on the number line to the left of a. We plot those points by using a parenthesis at a (indicating that a is not included), then a bold line to the left.	p. 88
$\{x\vert x \geq -3\}$ ←[———→ −3 0 3 The bracket indicates every number at or above the indicated value (−3).	$\{x\vert x \geq a\}$ indicates the set of all points on the number line to the right of, and including, a. We plot those points by using a bracket at a (indicating that a is included), then a bold line to the right.	
$\{x\vert 3 \leq x < 10\}$ ←—[———)—→ 0 3 10 This notation indicates every number between 3 and 10, including 3 but not including 10.	$\{x\vert a < x < b\}$ indicates the set of all points on the number line between a and b including a. We plot those points by using an opening bracket at a and a closing parenthesis at b then a bold line in between.	

Self-Test for Chapter 1

Each self-test in the bridge section of this text allows you to assess your mastery of the material in a particular chapter. After you have reviewed the summary material for the chapter, try taking the self-test. When you complete the test, check your answers against those in the back of the text. If you miss any questions, go back to the appropriate section of the text and review the material in greater depth than the summary permitted. Once you have completed all of the self-tests, you will be ready for the final exam. Good luck.

Write in symbols.

1. The sum of x and y
2. The difference m minus n
3. The product of a and b
4. The quotient when p is divided by q
5. 7 more than a
6. 5 less than c
7. The product of 3 and the quantity a plus b
8. 3 times the difference of m and n

Evaluate where $x = -3$.

9. $-4x - 12$ 10. $3x^2 + 2x - 4$

In Exercises 11 to 14, evaluate each expression if $a = -2$, $b = 6$, and $c = -4$.

11. $4a - c$ 12. $6(2b - 3c)$ 13. $\dfrac{3a - 4b}{a + c}$ 14. $a^2 - 3c$

In Exercises 15 to 17, combine like terms.

15. $8a - 3b - 5a + 2b$ 16. $8x - (-3y - 2x)$

17. $7x^2 - 3x + 2 - (5x^2 - 3x - 6)$

18. Use the roster method to list the elements of the set that consists of all negative integers greater than -5.

---

1. $x + y$
2. $m - n$
3. $a \cdot b$
4. $\dfrac{p}{q}$
5. $a + 7$
6. $c - 5$
7. $3(a + b)$
8. $3(m - n)$
9. 0
10. 17
11. -4
12. 144
13. 5
14. 16
15. $3a - b$
16. $10x + 3y$
17. $2x^2 + 8$
18. $\{-4, -3, -2, -1\}$

- Self-Test for Chapter 1

19. $\{x \mid -2 \leq x \leq 4\}$
20. ←|+|+|+|+|+|+|+|]→
    −5   0   3
21. $\{x \mid -3 < x \leq 2\}$
22. $2x - 8$
23. $2w + 4$
24. $85 - 8 \cdot 7 = \$29$
25. $\dfrac{184 - 160}{12} = \dfrac{24}{12} = 2$ lb/wk

19. Use set-builder notation to represent the set of all real numbers between -2 and 4 inclusive.

20. Plot the elements of the set $\{x \mid -5 < x \leq 3\}$ on a number line.

21. Use set-builder notation to describe the set ←|+|(|+|+|]|+|+|+|→
    −3   0   2

22. Tom is 8 years younger than twice Christopher's age. Write an expression for Tom's age.

23. The length of a rectangle is 4 more than twice its width. Write an expression for the length of the rectangle.

24. Patrick worked all day mowing lawns and was paid \$7 per hour. If he had \$85 at the end of an 8-hour day, how much did he have before he started working?

25. A woman weighed 184 pounds when she started her diet. At the end of 12 weeks, she weighed 160 pounds. How much did she lose each week, if weight loss was *equal* each week?

Summary for Chapter 2

Example	Topic	Reference
	Algebraic Equations	2.1–2.3
$3x - 5 = 7$ is an equation.	**Equation** A statement that two expressions are equal.	p. 99
4 is a solution for the equation because $$3 \cdot 4 - 5 = 7$$ $$12 - 5 = 7$$ $$7 = 7 \quad \text{(True)}$$	**Solution** A value for the variable that will make an equation a true statement.	p. 99
	Equivalent Equations Equations that have exactly the same solutions.	p. 102

If $x = y + 3$, then $x + 2 = y + 3 + 2$ $5x = 20$ and $x = 4$ are equivalent equations.	**Writing Equivalent Equations** There are two basic properties that will yield equivalent equations.
	
	1. If $a = b$, then $a + c = b + c$.
	
	   Adding (or subtracting) the same quantity on each side of an equation gives an equivalent equation.
	
	2. If $a = b$, then $ac = bc$, $c \neq 0$.
	
	   Multiplying (or dividing) both sides of an equation by the same number gives an equivalent equation.
		p. 102

Solve: $3(x - 2) + 4x = 3x + 14$ $3x - 6 + 4x = 3x + 14$ $7x - 6 = 3x + 14$ $\underline{ + 6 + 6}$ $7x = 3x + 20$ $\underline{-3x -3x}$ $4x = 20$ $\dfrac{4x}{4} = \dfrac{20}{4}$ $x = 5$	**Solving Linear Equations** We say that an equation is "solved" when we have an equivalent equation of the form $x = \Box$ or $\Box = x$ where the $\Box$ is some number The steps of solving a linear equation are as follows: 1. Use the distributive property to remove any grouping symbols that appear. Then simplify by combining any like terms. 2. Add or subtract the same term on both sides of the equation until the term containing the variable is on one side and a number is on the other. 3. Multiply or divide both sides of the equation by the same nonzero number so that the variable is alone on one side of the equation. 4. Check the solution in the original equation.	p. 103

Solving Literal Equations — 2.4

$a = \dfrac{2b+c}{3}$

is a literal equation.

Literal Equation An equation that involves more than one letter or variable. — p. 141

Solve for b:

$$a = \dfrac{2b+c}{3}$$

$$3a = 3\left(\dfrac{2b+c}{3}\right)$$

$$3a = 2b + c$$

$$\underline{\;-c\qquad -c\;}$$

$$3a - c = 2b$$

$$\dfrac{3a-c}{2} = b$$

Solving Literal Equations — p. 142

1. Multiply both sides of the equation by the lowest common denominator (LCD) to clear of fractions.
2. Add or subtract the same term on both sides of the equation so that all terms containing the variable you are solving for are on one side.
3. Divide both sides by any numbers or letters multiplying the variable that you are solving for.

Applying Equations — 2.4

Using Equations to Solve Word Problems Follow these steps. — p. 145

1. Read the problem carefully. Then reread it to decide what you are asked to find.
2. Choose a letter to represent one of the unknowns in the problem. Then represent each of the unknowns with an expression that uses the same letter.
3. Translate the problem to the language of algebra to form an equation.
4. Solve the equation and answer the question of the original problem.
5. Check your solution by returning to the original problem.

Inequalities—An Introduction — 2.5

$4 < 9$
$-1 > -6$
$2 \leq 2$
$3 \geq -4$

Inequality A statement that one quantity is less than (or greater than) another. Four symbols are used: — p. 156

$\qquad a < b \qquad\qquad a > b$
$\quad$ ↑ $\qquad\qquad\quad$ ↑
a is less than b. $\quad a$ is greater than b.

$\qquad a \leq b \qquad\qquad a \geq b$
$\quad$ ↑ $\qquad\qquad\quad$ ↑
a is less than $\quad\; a$ is greater than
or equal to b. $\quad\;$ or equal to b.

$x < 6$ is graphed [graph: 0 to 6, open at 6, arrow left] $x \geq 5$ [graph: 0 to 5, closed at 5, arrow right]	**Graphing Inequalities** To graph $x < a$, we use a parenthesis and an arrow pointing left. [graph with a] To graph $x \geq b$, we use a bracket and an arrow pointing right. [graph with b]	p. 157
$\begin{aligned} 2x - 3 &> 5x + 6 \\ +3 & \quad\quad +3 \\ \hline 2x &> -5x + 9 \\ -5x & \quad\quad -5x \\ \hline -3x &> 9 \\ \frac{-3x}{-3} &< \frac{9}{-3} \\ x &< -3 \end{aligned}$ [graph: -3 to 0, open at -3, arrow left]	**Solving Inequalities** An inequality is "solved" when it is in the form $x < \square$ or $x > \square$. Proceed as in solving equations by using the following properties. **1.** If $a < b$, then $a + c < b + c$. Adding (or subtracting) the same quantity to both sides of an inequality gives an equivalent inequality. **2.** If $a < b$, then $ac < bc$ when $c > 0$ and $ac > bc$ when $c < 0$. Multiplying both sides of an inequality by the same *positive number* gives an equivalent inequality. When both sides of an inequality are multiplied by the same *negative number*, you must *reverse the sense* of the inequality to give an equivalent inequality.	p. 158
	Absolute Value Equations	2.6
$\|2x - 5\| = 7$ is equivalent to $2x - 5 = -7$ or $2x - 5 = 7$ so $x = -1$ or $x = 6$	**Property 1** For any positive number p, if $$\|x\| = p$$ then $$x = -p \quad \text{or} \quad x = p$$	p. 174
$\|x\| = \|2x - 6\|$ $x = 2x - 6$ or $x = -(2x - 6)$ $x = 6$ or $x = 2$	**Property 2** If $\|x\| = \|y\|$ then $$x = y \quad \text{or} \quad x = -y$$	p. 176

	Absolute Value Inequalities	2.6
$\|3x - 5\| < 7$ is equivalent to $-7 < 3x - 5 < 7$ This yields $-2 < 3x < 12$ $-\dfrac{2}{3} < x < 4$	**Property 1** For any positive number p, if $$\|x\| < p$$ then $$-p < x < p$$ To solve this form of inequality, translate to the equivalent compound inequality and solve as before.	p. 180
$\|2 - 5x\| \geq 12$ is equivalent to $2 - 5x \leq -12$ or $2 - 5x \geq 12$ This yields $x \geq \dfrac{14}{5}$ or $x \leq -2$	**Property 2** For any positive number p, if $$\|x\| > p$$ then $$x < -p \quad \text{or} \quad x > p$$ To solve this form of inequality, translate to the equivalent compound inequality and solve as before.	p. 181

Self-Test for Chapter 2

1. No
2. Yes
3. $\{x = 11\}$
4. $\{x = 12\}$
5. $\{x = 7\}$
6. $\{7\}$
7. $\{-12\}$
8. $\{25\}$
9. $\{4\}$
10. $\left\{-\dfrac{2}{3}\right\}$
11. $\{-5\}$
12. $\left\{\dfrac{4}{5}\right\}$
13. $\{2\}$
14. $\left\{\dfrac{5}{2}\right\}$
15. $\left\{\dfrac{14}{5}\right\}$
16. $\dfrac{C}{2\pi}$
17. $\dfrac{3V}{B}$

Each self-test in the bridge section of this text allows you to assess your mastery of the material in a particular chapter. After you have reviewed the summary material for the chapter, try taking the self-test. When you complete the test, check your answers against those in the back of the text. If you miss any questions, go back to the appropriate section of the text and review the material in greater depth than the summary permitted. Once you have completed all of the self-tests, you will be ready for the final exam. Good luck.

Tell whether the number shown in parentheses is a solution for the given equation.

1. $7x - 3 = 25$ (5)
2. $8x - 3 = 5x + 9$ (4)

Solve the following equations and check your results.

3. $x - 7 = 4$
4. $7x - 12 = 6x$

5. $9x - 2 = 8x + 5$

Solve the following equations and check your results. Express your answer in set notation.

6. $7x = 49$
7. $\dfrac{1}{4}x = -3$

8. $\dfrac{4}{5}x = 20$
9. $10 - 3x = -2$

10. $7x - 3 = 4x - 5$
11. $2x - 7 = 5x + 8$

12. $7 - 5x = 3$
13. $5x - 3(x - 5) = 19$

14. $6x - (-4x - 5) = -2(-4x - 5)$
15. $8x - [5 - (4x - 9)] = 7x$

Solve for the indicated variable.

16. $C = 2\pi r$ (for r)
17. $V = \dfrac{1}{3}Bh$ (for h)

Solve and graph the solution sets for the following inequalities.

18. $x - 5 \leq 9$ **19.** $5 - 3x > 17$

20. $|x + 6| < 4$ **21.** $|x - 4| > 2$

Solve the following word problems. Be sure to show the equation you used for the solution.

22. 5 times a number, decreased by 7, is 28. What is the number?

23. The sum of three consecutive integers is 66. Find the three integers.

24. Jan is twice as old as Juwan, while Rick is 5 years older than Jan. If the sum of their ages is 35 years, find each of their ages.

25. The perimeter of a rectangle is 62 inches (in.). If the length of the rectangle is 1 in. more than twice its width, what are the dimensions of the rectangle?

26. At 10 A.M., Sandra left her house on a business trip and drove an average of 45 mi/h. One hour later, Adam discovered that Sandra had left her briefcase behind, and he began driving at 55 mi/h along the same route. When will Adam catch up with Sandra?

Solve each of the following inequalities.

27. $6 \leq x + 5 \leq 9$ **28.** $3x + 7 < -8$ or $-4x + 5 < -11$

Solve the following inequalities.

29. $|x - 1| < 4$ **30.** $|2x + 1| \geq 7$

18. $\{x | x \leq 14\}$

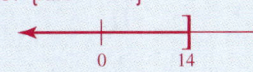

19. $\{x | x < -4\}$

20. $\{x | -10 < x < -2\}$

21. $\{x | x < 2 \text{ or } x > 6\}$

22. 7

23. 21, 22, 23

24. 6, 12, 17

25. 10 in. by 21 in.

26. 3:30 P.M.

27. $\{x | 1 \leq x \leq 4\}$

28. $\{x | x < -5 \text{ or } x > 4\}$

29. $\{x | -3 < x < 5\}$

30. $\{x | x \leq -4 \text{ or } x \geq 3\}$

Summary for Chapter 3

Example	Topic	Reference
	Equations in Two Variables	3.1
If $2x - y = 10$, $(6, 2)$ is a solution for the equation, because substituting 6 for x and 2 for y gives a true statement.	**Solutions of Linear Equations** A pair of values that satisfy the equation. Solutions for linear equations in two variables are written as *ordered pairs*. An ordered pair has the form (x, y) x coordinate, y coordinate	p. 193
	The Rectangular Coordinate System	3.2
(graph showing y axis, x axis, and Origin)	**The Rectangular Coordinate System** A system formed by two perpendicular axes that intersect at a point called the **origin**. The horizontal line is called the **x axis**. The vertical line is called the **y axis**.	p. 201
To graph the point corresponding to $(2, 3)$: (graph showing point (2,3), 3 units up and 2 units right)	**Graphing Points from Ordered Pairs** The coordinates of an ordered pair allow you to associate a point in the plane with every ordered pair. To graph a point in the plane, 1. Start at the origin. 2. Move right or left according to the value of the x coordinate: to the right if x is positive or to the left if x is negative. 3. Then move up or down according to the value of the y coordinate: up if y is positive and down if y is negative.	p. 202

Summary for Chapter 3

Graphing Linear Equations — 3.3

Linear Equation An equation that can be written in the form

$$Ax + By = C$$

where A and B are not both 0.

p. 214

$2x - 3y = 4$ is a linear equation.

Graphing Linear Equations

1. Find at least three solutions for the equation, and put your results in tabular form.
2. Graph the solutions found in step 1.
3. Draw a straight line through the points determined in step 2 to form the graph of the equation.

p. 216

Graph of $x - y = 6$ passing through $(0, -6)$, $(3, -3)$, and $(6, 0)$.

x	y
0	-6
3	-3
6	0

The Slope of a Line — 3.4

Slope The slope of a line gives a numerical measure of the steepness of the line. The slope m of a line containing the distinct points in the plane $P(x_1, y_1)$ and $Q(x_2, y_2)$ is given by

$$m = \frac{y_2 - y_1}{x_2 - x_1} \quad \text{where } x_2 \neq x_1$$

p. 241

To find the slope of the line through $(-2, -3)$ and $(4, 6)$,

$$m = \frac{6 - (-3)}{4 - (-2)} = \frac{6 + 3}{4 + 2} = \frac{9}{6} = \frac{3}{2}$$

Forms of Linear Equations — 3.5

Slope-Intercept Form The slope-intercept form for the equation of a line is

$$y = mx + b$$

where the line has slope m and y intercept b.

p. 246

For the equation

$$y = \frac{2}{3}x - 3$$

the slope m is $\frac{2}{3}$ and b, the y intercept, is -3.

Parallel Lines Two lines are parallel if they have the same slope.

p. 249

$y = 3x - 7$ and $y = 3x + \frac{2}{5}$ are parallel lines.

Perpendicular Lines Two lines are perpendicular if the product of the two slopes is -1. The slopes are said to be **negative reciprocals**.

p. 259

$y = 2x + 1$ and $y = -\frac{1}{2}x + 5$ are perpendicular lines.

$$(2)\left(-\frac{1}{2}\right) = -1$$

Self-Test for Chapter 3

1. (4, 0), (5, 4)
2. (3, 0), (0, 4), $\left(\dfrac{3}{4}, 3\right)$
3. (4, 2)
4. (−4, 6)
5. (0, −7)
6., 7., 8.

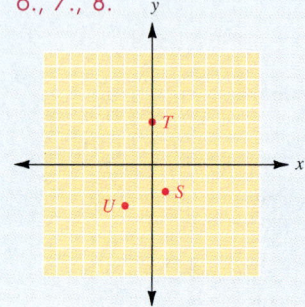

Each self-test in the bridge section of this text allows you to assess your mastery of the material in a particular chapter. After you have reviewed the summary material for the chapter, try taking the self-test. When you complete the test, check your answers against those in the back of the text. If you miss any questions, go back to the appropriate section of the text and review the material in greater depth than the summary permitted. Once you have completed all of the self-tests, you will be ready for the final exam. Good luck.

1. Determine which of the ordered pairs are solutions for the given equation.

$$4x - y = 16 \qquad (4, 0), (3, -1), (5, 4)$$

2. Complete the ordered pairs so that each is a solution for the given equation.

$$4x + 3y = 12 \qquad (3,), (, 4), (, 3)$$

Give the coordinates of the points graphed below.

3. A
4. B
5. C

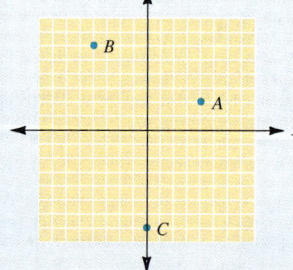

Plot points with the coordinates shown.

6. S(1, −2)
7. T(0, 3)
8. U(−2, −3)

Graph each of the following equations.

9. $x + y = 4$

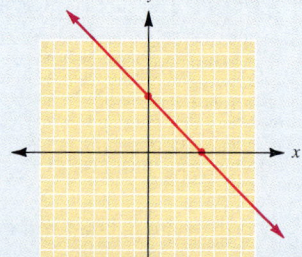

10. $y = 3x$

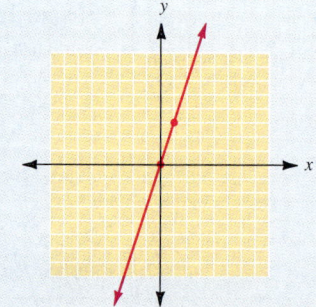

■ Self-Test for Chapter 3 **465**

15. 1

16. $\frac{3}{4}$

17. $y = -3x + 6$

18. $y = \frac{2}{5}x - 3$

19. Slope: -5;
    y intercept: -9

20. Slope: $-\frac{6}{5}$;
    y intercept: 6

21. Slope: 0; y intercept: 5

22. $y = 5x - 2$

23. $y = -4x - 16$

24. $y = 4x + 3$

25. $y = -\frac{5}{2}x - 17$

11. $y = \frac{3}{4}x - 4$

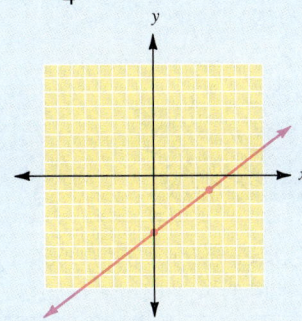

12. $x + 3y = 6$

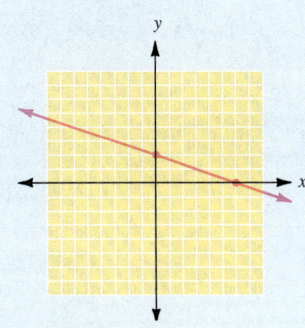

13. $2x + 5y = 10$

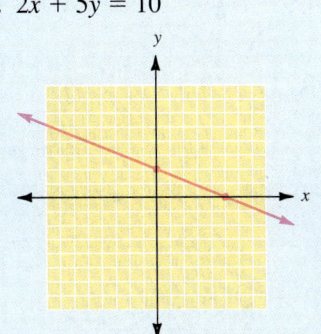

14. $y = -4$

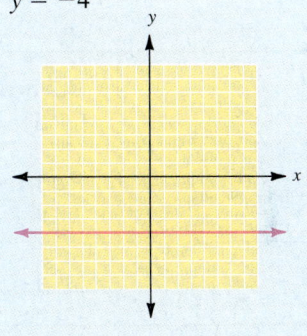

Find the slope of the line through the following pairs of points.

15. $(-3, 5)$ and $(2, 10)$ **16.** $(-2, 6)$ and $(2, 9)$

Write the equation of the line with the given slope and y intercept. Then graph each line, *using* the slope and y intercept.

17. $m = -3, b = 6$

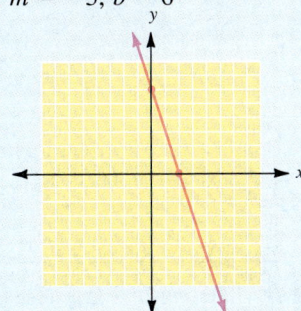

18. $m = \frac{2}{5}, b = -3$

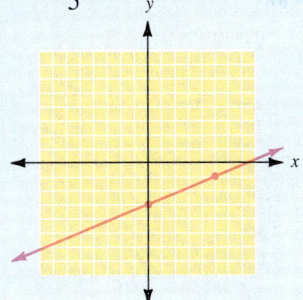

Find the slope and y intercept of the line represented by each of the given equations.

19. $y = -5x - 9$ **20.** $6x + 5y = 30$ **21.** $y = 5$

In 22–25, write the equation of the line L satisfying the given set of conditions.

22. L has slope 5 and y intercept -2.   **23.** L passes through $(-5, 4)$ and $(-2, -8)$.

24. L has y intercept 3 and is parallel to the line with the equation $4x - y = 9$.

25. L passes through the point $(-6, -2)$ and is perpendicular to the line with equation $2x - 5y = 10$.

Summary for Chapter 4

Example	Topic	Reference
	Ordered Pairs and Relations	4.1
(1, 4) is an ordered pair.	**Ordered Pair** Given two related values, x and y, we write the pair of values as (x, y).	p. 280
The set {(1, 4), (2, 5), (1, 6)} is a relation.	**Relation** A relation is a set of ordered pairs.	p. 280
The domain is {1, 2}.	**Domain** The domain is the set of all first elements of a relation.	p. 281
The range is {4, 5, 6}.	**Range** The range is the set of all second elements of a relation.	p. 281
	An Introduction to Functions	4.2–4.3
	Graph The graph of a relation is the set of points in the plane that correspond to the ordered pairs of the relation.	p. 290
{(1, 2), (2, 3), (3, 4)} is a function. {(1, 2), (2, 3), (2, 4)} is *not* a function.	**Function** A function is a set of ordered pairs (a relation) in which no two first elements are equal.	p. 296
A relation–*not* a function	**Vertical-Line Test** The vertical line test is used to determine, from the graph, whether a relation is a function. If a vertical line meets the graph of a relation in two or more points, the relation is *not* a function. If no vertical line passes through two or more points on the graph of a relation, it is the graph of a function.	p. 297

■ Summary for Chapter 4   **467**

	Reading Values from Graphs	**4.4**
	For a specific value of x, let's call it a, we can find $f(a)$ with the following algorithm:	p. 309
	1. Draw a vertical line through a on the x axis.	
	2. Find the point of intersection of that line with the graph.	
	3. Draw a horizontal line through the graph at that point.	
	4. Find the intersection of the horizontal line with the y axis.	
	5. $f(a)$ is that y value.	
	If given the function value, one finds the x value associated with it as follows:	p. 312
	1. Find the given function value on the y axis.	
	2. Draw a horizontal line through that point.	
	3. Find every point on the graph that intersects the horizontal line.	
	4. Draw a vertical line through each of those points of intersection.	
	5. The x value(s) are each point of intersection of the vertical lines and the x axis.	
	Operations on Functions	**4.5**
Let $f(x) = 2x + 1$ and $g(x) = 3x^2$ $(f + g)(x) = (2x + 1) + (3x^2)$ $= 3x^2 + 2x + 1$	The **sum of two functions** f and g is written $f + g$. It is defined as $$f + g: (f + g)(x) = f(x) + g(x)$$	p. 321
$(f - g)(x) = (2x + 1) - (3x^2)$ $= -3x^2 + 2x + 1$	The **difference of two functions** f and g is written $f - g$. It is defined as $$f - g: (f - g)(x) = f(x) - g(x)$$	p. 321
$(f \cdot g)(x) = (2x + 1)(3x^2)$	The **product of two functions** f and g is written $f \cdot g$. It is defined as $$f \cdot g: (f \cdot g)(x) = f(x) \cdot g(x)$$	p. 323
$(f \div g)(x) = (2x + 1) \div (3x^2)$ $= \dfrac{(2x + 1)}{3x^2}$	The **quotient of two functions** f and g is written $f \div g$. It is defined as $$f \div g: (f \div g)(x) = f(x) \div g(x)$$	p. 323

Self-Test for Chapter 4

Each self-test in the bridge section of this text allows you to assess your mastery of the material in a particular chapter. After you have reviewed the summary material for the chapter, try taking the self-test. When you complete the test, check your answers against those in the back of the text. If you miss any questions, go back to the appropriate section of the text and review the material in greater depth than the summary permitted. Once you have completed all of the self-tests, you will be ready for the final exam. Good luck.

1. Identify which of the following are ordered pairs.

   (a) $\{-1, 3\}$ (b) $(1, 4)$ (c) $[2, 4]$ (d) $5, 6$ (e) $\{(2, 1), (4, 5)\}$

2. For each of the following sets of ordered pairs, identify the domain and range.

   (a) $\{(1, 6), (-3, 5), (2, 1), (4, -2), (3, 0)\}$ (b) $\{(x, y) \mid 4x + 5y = 20\}$

In each of the following, evaluate $f(x)$ for the value given.

3. $f(x) = x^2 - 4x - 5$; find $f(0)$ and $f(-2)$.
4. $f(x) = -x^3 + 5x - 3x^2 - 8$; find $f(-1)$ and $f(1)$.
5. $f(x) = -7x - 15$; find $f(0)$ and $f(-3)$.
6. $f(x) = 3x - 25$; find $f(a)$ and $f(x - 1)$.

In each of the following, determine which relations are functions.

7. $\{(1, 2), (3, 2), (-1, 4), (2, 4)\}$ 8. $\{(3, 1), (2, 0), (-1, 4), (2, 4)\}$

9.
x	y
1	3
4	2
−1	−3
5	1

10.
x	y
1	2
0	3
−1	

Use the vertical line test to determine whether the following graphs represent functions.

11.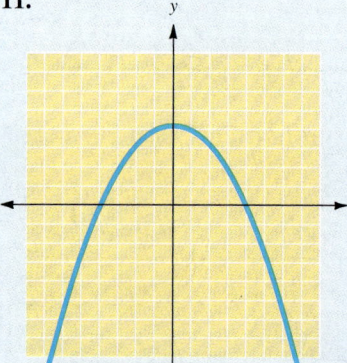

12.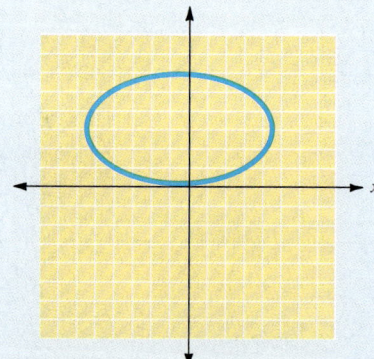

Answers (margin):

1. b
2. Dom = $\{-3, 1, 2, 3, 4\}$; range = $\{-2, 0, 1, 5, 6\}$
   Domain = $\mathbb{R}$; range = $\mathbb{R}$
3. −5, 7
4. −15, −7
5. −15, 6
6. $3a - 25$, $3(x - 1) - 25 = 3x - 28$
7. Function
8. Not a function
9. Function
10. Function
11. Function
12. Not a function

Self-Test for Chapter 4

13. Rewrite each equation as a function of x.

(a) $y = -6x + 9$ (b) $4x - 7y = 35$

Graph each of the following functions.

14. $f(x) = 9x - 2$ **15.** $f(x) = -7x + 3$ **16.** $f(x) = -3x - 2$

Find the coordinates of the labeled points.

17.
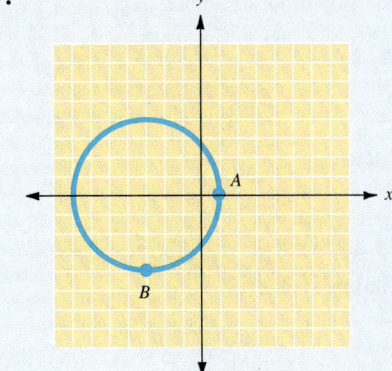

18.
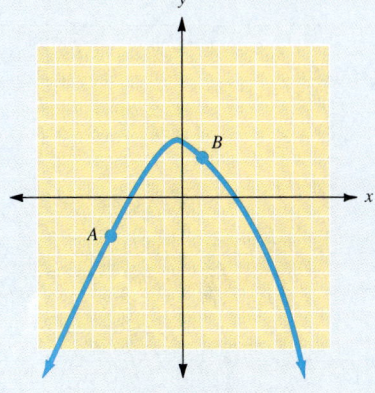

In the following, use the graph to estimate (a) $f(-1)$, (b) $f(0)$, and (c) $f(1)$.

19.
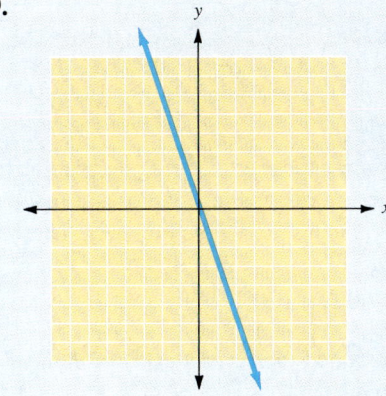

20.
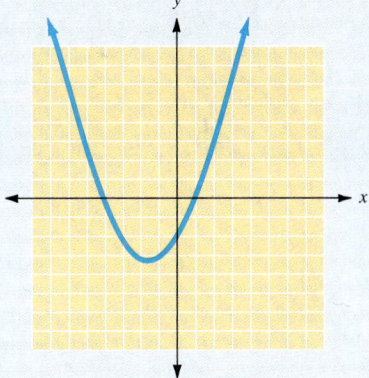

For Exercises 21 to 24, use the functions $f(x) = 3x - 4$ and $g(x) = 2x + 1$.

21. Find $f(4) + g(4)$.

22. Add the functions $f(x) + g(x)$.

23. Find $f(-2) - g(-2)$.

24. Find the difference $f(x) - g(x)$.

---

13. $f(x) = -6x + 9$

$f(x) = \dfrac{4x - 35}{7}$

14.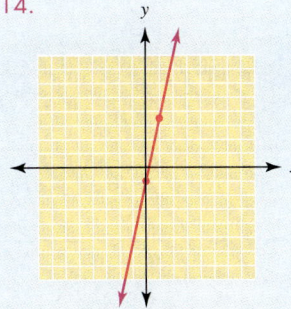

15.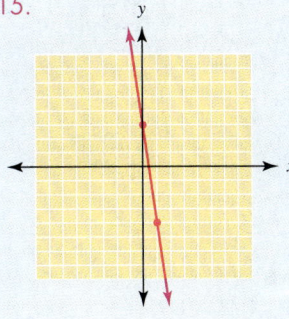

16.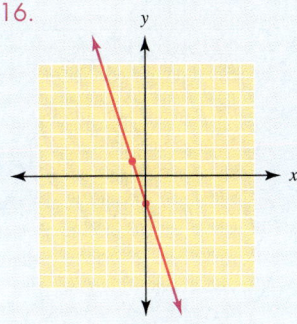

17. $A(1, 0)$; $B(-3, -4)$
18. $A(-4, -2)$; $B(1, 2)$
19. (a) 3; (b) 0; (c) -3
20. (a) -3; (b) -2; (c) 0
21. 17
22. $5x - 3$
23. -7
24. $x - 5$

Summary for Chapter 5

Example	Topic	Reference
	Positive Integer Exponents	5.1
	Properties of Exponents For any nonzero real numbers a and b and integers m and n:	p. 335
$x^5 \cdot x^7 = x^{5+7} = x^{12}$	**Product Rule** $$a^m \cdot a^n = a^{m+n}$$	p. 336
$\dfrac{x^7}{x^5} = x^{7-5} = x^2$	**Quotient Rule** $$\dfrac{a^m}{a^n} = a^{m-n},\ m > n$$	p. 338
$(2y)^3 = 2^3 y^3 = 8y^3$	**Product-Power Rule** $$(ab)^n = a^n b^n$$	p. 339
$\left(\dfrac{2}{3}\right)^2 = \dfrac{2^2}{3^2} = \dfrac{4}{9}$	**Quotient-Power Rule** $$\left(\dfrac{a}{b}\right)^m = \dfrac{a^m}{b^m}$$	p. 340
$(2^3)^4 = 2^{12}$	**Power Rule** $$(a^m)^n = a^{mn}$$	p. 340
	Polynomials	5.2
$4x^3 - 3x^2 + 5x$ is a polynomial.	**Polynomial** An algebraic expression made up of terms in which the exponents are whole numbers. These terms are connected by plus or minus signs. Each sign (+ or −) is attached to the term following that sign.	p. 346
The terms of $4x^3 - 3x^2 + 5x$ are $4x^3$, $-3x^2$, and $5x$.	**Term** A number, or the product of a number and variables, raised to a power.	p. 346
The coefficients of $4x^3 - 3x^2$ are 4 and −3.	**Coefficient** In each term of a polynomial, the number that is multiplied by the variable(s) is called the *numerical coefficient* or, more simply, the *coefficient* of that term.	p. 346
$2x^3$ is a monomial. $3x^2 - 7x$ is a binomial. $5x^5 - 5x^3 + 2$ is a trinomial.	**Types of Polynomials** A polynomial can be classified according to the number of terms it has. A *monomial* has exactly one term. A *binomial* has exactly two terms. A *trinomial* has exactly three terms.	p. 347

■ Summary for Chapter 5

The degree of $4x^5 - 5x^3 + 3x$ is 5.	**Degree of a Polynomial with Only One Variable** The highest power of the variable appearing in any one term.	p. 347
$4x^5 - 5x^3 + 3x$ is written in descending-exponent form.	**Descending-Exponent Form** The form of a polynomial when it is written with the highest-degree term first, the next highest-degree term second, and so on.	p. 348
	Adding and Subtracting Polynomials	5.3
	Removing Signs of Grouping	p. 352
$+(3x - 5) = -3x - 5$	1. If a plus sign $(+)$ or no sign at all appears in front of parentheses, just remove the parentheses. No other changes are necessary.	
$-(3x - 5) = -3x + 5$	2. If a minus sign $(-)$ appears in front of parentheses, the parentheses can be removed by changing the sign of each term inside the parentheses.	
$(2x + 3) + (3x - 5)$ $= 2x + 3 + 3x - 5$ $= 5x - 2$	**Adding Polynomials** Remove the signs of grouping. Then collect and combine any like terms.	p. 353
$(3x^2 + 2x) - (2x^2 + 3x - 1)$ $= 3x^2 + 2x - 2x^2 - 3x + 1$ ↖↑↑ Sign changes $= 3x^2 - 2x^2 + 2x - 3x + 1$ $= x^2 - x + 1$	**Subtracting Polynomials** Remove the signs of grouping by changing the sign of each term in the polynomial being subtracted. Then combine any like terms.	p. 355
	Multiplying Polynomials	5.4
$2x(x^2 + 4)$ $= 2x^3 + 8x$	**To Multiply a Polynomial by a Monomial** Multiply each term of the polynomial by the monomial, and simplify the results.	p. 363
$(2x - 3)(3x + 5)$ $= 6x^2 + 10x - 9x - 15$ F O I L $= 6x^2 + x - 15$	**To Multiply a Binomial by a Binomial** Use the FOIL method: $$\text{F}\text{O}\text{I}\text{L}$$ $$(a + b)(c + d) = a \cdot c + a \cdot d + b \cdot c + b \cdot d$$	p. 365

$\begin{array}{r} x^2 - 3x + 5 \\ 2x - 3 \\ \hline -3x^2 + 9x - 15 \\ 2x^3 - 6x^2 + 10x \\ \hline 2x^3 - 9x^2 + 19x - 15 \end{array}$	**To Multiply a Polynomial by a Polynomial** Arrange the polynomials vertically. Multiply each term of the upper polynomial by each term of the lower polynomial, and combine the results.	p. 367
	Special Products	5.4
$\begin{aligned} &(2x - 5)^2 \\ &= 4x^2 + 2 \cdot 2x \cdot (-5) + 25 \\ &= 4x^2 - 20x + 25 \end{aligned}$	**The Square of a Binomial** $$(a + b)^2 = a^2 + 2ab + b^2$$ 1. The first term of the square is the square of the first term of the binomial. 2. The middle term is twice the product of the two terms of the binomial. 3. The last term is the square of the last term of the binomial.	p. 368
$\begin{aligned} &(2x - 5y)(2x + 5y) \\ &= (2x)^2 - (5y)^2 \\ &= 4x^2 - 25y^2 \end{aligned}$	**The Product of Binomials That Differ Only in Sign** Subtract the square of the second term from the square of the first term. $$(a + b)(a - b) = a^2 - b^2$$	p. 370

Self-Test for Chapter 5

Each self-test in the bridge section of this text allows you to assess your mastery of the material in a particular chapter. After you have reviewed the summary material for the chapter, try taking the self-test. When you complete the test, check your answers against those in the back of the text. If you miss any questions, go back to the appropriate section of the text and review the material in greater depth than the summary permitted. Once you have completed all of the self-tests, you will be ready for the final exam. Good luck.

Simplify each expression, using the properties of exponents.

1. $(3x^2y)(-2xy^3)$
2. $\left(\dfrac{8m^2n^5}{2p^3}\right)^2$
3. $(x^4y^5)^2$
4. $\dfrac{9c^5d^3}{18c^2d^2}$
5. $(3x^2y)^3(-2xy^2)^2$
6. $(-2xy)^3(4x^2y)^2$
7. Given $f(x) = 3x^2 - 4x + 5$, find $f(-2)$.

Classify each of the following polynomials as a monomial, binomial, or trinomial.

8. $6x^2 + 7x$
9. $5x^2 + 8x - 8$

Arrange in descending-exponent form, and give the coefficients and degree.

10. $-3x^2 + 8x^4 - 7$

Add.

11. $3x^2 - 7x + 2$ and $7x^2 - 5x - 9$
12. $7a^2 - 3a$ and $7a^3 + 4a^2$

Subtract.

13. $5x^2 - 2x + 5$ from $8x^2 + 9x - 7$
14. $2b^2 + 5$ from $3b^2 - 7b$
15. $5a^2 + a$ from the sum of $3a^2 - 5a$ and $9a^2 - 4a$

Use the vertical method for 16 and 17.

16. Add $x^2 + 3$, $5x - 9$, and $3x^2$
17. Subtract $3x^2 - 5$ from $5x^2 - 7x$

Multiply.

18. $5ab(3a^2b - 2ab + 4ab^2)$
19. $(x - 2)(3x + 7)$
20. $(a - 7b)(a + 7b)$
21. $(x + 3y)(4x - 5y)$
22. $(3m + 2n)^2$
23. $(2x + y)(x^2 + 3xy - 2y^2)$
24. $(2x + 5y)(2x - 5y)$
25. $(x - 5y)(x - 5y)$

---

1. $-6x^3y^4$
2. $\dfrac{16m^4n^{10}}{p^6}$
3. x^8y^{10}
4. $\dfrac{c^3d}{2}$
5. $108x^8y^7$
6. $-128x^7y^5$
7. 25
8. Binomial
9. Trinomial
10. $8x^4 - 3x^2 - 7$; $8, -3, -7$; 4
11. $10x^2 - 12x - 7$
12. $7a^3 + 11a^2 - 3a$
13. $3x^2 + 11x - 12$
14. $b^2 - 7b - 5$
15. $7a^2 - 10a$
16. $4x^2 + 5x - 6$
17. $2x^2 - 7x + 5$
18. $15a^3b^2 - 10a^2b^2 + 20a^2b^3$
19. $3x^2 + x - 14$
20. $a^2 - 49b^2$
21. $4x^2 + 7xy - 15y^2$
22. $9m^2 + 12mn + 4n^2$
23. $2x^3 + 7x^2y - xy^2 - 2y^3$
24. $4x^2 - 25y^2$
25. $x^2 - 10xy + 25y^2$

Summary for Chapter 6

Example	Topic	Reference
	Common-Term Factoring	**6.1**
$4x^2$ is the greatest common monomial factor of $8x^4 - 12x^3 + 16x^2$.	**Common Monomial Factor** A single term that is a factor of every term of the polynomial. The greatest common factor (GCF) is the common monomial factor that has the largest possible numerical coefficient and the largest possible exponents.	p. 384
$8x^4 - 12x^3 + 16x^2$ $= 4x^2(2x^2 - 3x + 4)$	**Factoring a Monomial from a Polynomial** 1. Determine the greatest common factor. 2. Apply the distributive law in the form $$ab + ac = a(b + c)$$ ↑ The greatest common factor	p. 385
	The Difference of Squares	**6.2**
To factor: $16x^2 - 25y^2$: Think: $(4x)^2 - (5y)^2$ so $16x^2 - 25y^2$ $= (4x + 5y)(4x - 5y)$	**Factoring a Difference of Squares** Use the following form: $$a^2 - b^2 = (a + b)(a - b)$$	p. 393
	Factoring Trinomials	**6.3**
$4x^2 - 6x + 10x - 15$ $= 2x(2x - 3) + 5(2x - 3)$ $= (2x - 3)(2x + 5)$	**Factoring by Grouping** When there are four terms of a polynomial, factor the first pair and factor the last pair. If these two pairs have a common binomial factor, factor that out. The result will be the product of two binomials.	p. 400
$x^2 + 3x - 28$ $ac = -28; b = 3$ $mn = -28; m + n = 3$ $m = 7, n = -4$ $x^2 + 7x - 4x - 28$ $= x(x + 7) - 4(x + 7)$ $= (x + 7)(x + 4)$	**Factoring Trinomials** To factor a trinomial, first use the *ac* test to determine factorability. If the trinomial is factorable, the *ac* test will yield two terms (which have as their sum the middle term of the trinomial) that allow the factoring to be completed by using the grouping method.	p. 402

Dividing Polynomials — 6.4

$\dfrac{9x^4 + 6x^3 - 15x^2}{3x}$ $= 3x^3 + 2x^2 - 5x$	**To Divide a Polynomial by a Monomial** Divide each term of the polynomial by the monomial. Then combine the results.	p. 422
$\begin{array}{r} x + 5 \\ x - 3 \overline{\smash{)}\, x^2 + 2x - 7} \\ \underline{x^2 - 3x} \\ 5x - 17 \\ \underline{5x - 15} \\ 8 \end{array}$ The result is $x + 5 + \dfrac{8}{x-3}$	**To Divide a Polynomial by a Polynomial** Use the long division method.	p. 424

Solving Quadratic Equations by Factoring — 6.5

To solve: $x^2 + 7x = 30$ $x^2 + 7x - 30 = 0$ $(x + 10)(x - 3) = 0$ $x + 10 = 0$ or $x - 3 = 0$ $x = -10$ and $x = 3$ are solutions.	1. Add or subtract the necessary terms on both sides of the equation so that the equation is in standard form (set equal to 0). 2. Factor the quadratic expression. 3. Set each factor equal to 0. 4. Solve the resulting equations to find the solutions. 5. Check each solution by substituting in the original equation.	p. 432

Self-Test for Chapter 6

Each self-test in the bridge section of this text allows you to assess your mastery of the material in a particular chapter. After you have reviewed the summary material for the chapter, try taking the self-test. When you complete the test, check your answers against those in the back of the text. If you miss any questions, go back to the appropriate section of the text and review the material in greater depth than the summary permitted. Once you have completed all of the self-tests, you will be ready for the final exam. Good luck.

Factor each of the following polynomials.

1. $12b + 18$
2. $9p^3 - 12p^2$
3. $5x^2 - 10x + 20$
4. $6a^2b - 18ab + 12ab^2$

Factor each of the following polynomials completely.

5. $a^2 - 25$
6. $64m^2 - n^2$
7. $49x^2 - 16y^2$
8. $32a^2b - 50b^3$
9. $8x^3 - 27y^3$
10. $128x^4y + 250xy^4$
11. $a^2 - 5a - 14$
12. $b^2 + 8b + 15$
13. $x^2 - 11x + 28$
14. $y^2 + 12yz + 20z^2$
15. $x^2 + 2x - 5x - 10$
16. $6x^2 + 2x - 9x - 3$
17. $2x^2 + 15x - 8$
18. $3w^2 + 10w + 7$
19. $8x^2 - 2xy - 3y^2$
20. $6x^3 + 3x^2 - 30x$
21. $6x^2 + 4x - 15x - 10$
22. $4x^2 + 4x + 5x + 5$

Perform the indicated long division.

23. $\dfrac{3x^2 - 2x - 4}{3x + 1}$
24. $\dfrac{4x^3 - 5x^2 + 7x - 9}{x - 2}$
25. $\dfrac{3x^4 - 2x^2 - 5}{x^2 + 1}$

Solve each equation for the variable x.

26. $x^2 - 2x - 3 = 0$
27. $x^2 - 5x - 14 = 0$
28. $x^2 - 11x = -30$
29. $2x^2 + 16x + 30 = 0$
30. $6x^2 - 7x = 3$

---

1. $6(2b + 3)$
2. $3p^2(3p - 4)$
3. $5(x^2 - 2x + 4)$
4. $6ab(a - 3 + 2b)$
5. $(a + 5)(a - 5)$
6. $(8m - n)(8m + n)$
7. $(7x - 4y)(7x + 4y)$
8. $2b(4a + 5b)(4a - 5b)$
9. $(2x - 3y)(4x^2 + 6xy + 9y^2)$
10. $2xy(4x + 5y)(16x^2 - 20xy + 25y^2)$
11. $(a - 7)(a + 2)$
12. $(b + 3)(b + 5)$
13. $(x - 7)(x - 4)$
14. $(y + 10z)(y + 2z)$
15. $(x + 2)(x - 5)$
16. $(3x + 1)(2x - 3)$
17. $(2x - 1)(x + 8)$
18. $(3w + 7)(w + 1)$
19. $(4x - 3y)(2x + y)$
20. $3x(2x + 5)(x - 2)$
21. $(3x + 2)(2x - 5)$
22. $(x + 1)(4x + 5)$
23. $x - 1 + \dfrac{-3}{3x + 1}$
24. $4x^2 + 3x + 13 + \dfrac{17}{x - 2}$
25. $3x^2 - 5$
26. $\{-1, 3\}$
27. $\{-2, 7\}$
28. $\{5, 6\}$
29. $\{-3, -5\}$
30. $\left\{-\dfrac{1}{3}, \dfrac{3}{2}\right\}$

Final Exam for Chapters 0-6

Answers (left column):

1. 27
2. −1
3. 6
4. $a + 6b$
5. $2x^2y + 3x + 2xy$
6. $x^2 + 9x$
7. $a^2 − 9b^2$
8. $x^2 − 4xy + 4y^2$
9. $x^2 + 3x − 10$
10. $a^2 + a − 12$
11. $3x + 2$
12. $3x + 3 + \dfrac{1}{x − 1}$
13. 180
14. 118
15. −16
16. {11}
17. {−33}
18. $\left\{\dfrac{8}{5}\right\}$
19. {1}
20. {−3, 15}
21. {−10, 1}
22. {x|x > −6}
23. $\left\{x \mid \dfrac{1}{2} < x < \dfrac{13}{2}\right\}$
24. {x|x < −8 or x > −2}
25. $\left\{x \mid x \geq \dfrac{14}{3} \text{ or } x \leq −2\right\}$
26. $4(3x + 5)$
27. $(5x + 6y)(5x − 6y)$
28. $(4x − 5)(3x + 2)$
29. $(2x − 3)(x − 5)$
30. Function
31. Not a function

Simplify the expression.

1. $3(5 − 2)^2$
2. $|3| − |−4|$
3. $35 − 5(7) + 6$
4. $3a + 5b − 2a + b$
5. $7x^2y + 3x − 5x^2y + 2xy$
6. $(5x^2 + 4x − 3) − (4x^2 − 5x − 3)$

Multiply or divide as indicated.

7. $(a − 3b)(a + 3b)$
8. $(x − 2y)^2$
9. $(x − 2)(x + 5)$
10. $(a − 3)(a + 4)$
11. $(9x^2 + 12x + 4) \div (3x + 2)$
12. $(3x^2 − 2) \div (x − 1)$

Evaluate each expression where $x = 3$ and $y = −5$.

13. $−12xy$
14. $x^5 + y^3$
15. $(x + y)(x − y)$

Solve each equation and check your result.

16. $7a − 3 = 6a + 8$
17. $\dfrac{2}{3}x = −22$
18. $12 − 5x = 4$
19. $5x − 2(x − 3) = 9$
20. $|x − 6| = 9$
21. $|2x + 9| − 5 = 6$

Solve each inequality.

22. $2x − 7 < 5 + 4x$
23. $|2x − 7| < 6$
24. $|x + 5| > 3$
25. $|3x − 4| \geq 10$

Factor each expression.

26. $12x + 20$
27. $25x^2 − 36y^2$
28. $12x^2 − 15x + 8x − 10$
29. $2x^2 − 13x + 15$

Determine whether each of the following represents a function.

30. {(1, 1), (2, 1), (3, 1), (4, 1)}
31. {(1, 1), (2, 2), (1, −1)}

32. Not a function

33. Function

38. $124 = 2(x + 1) + 2x$; 30.5 in. by 31.5 in.

39. $x + (x + 5) = 141$; biology: $73, math: $68

40. $\{-4, 7\}$

41. $\{-2, 5\}$

42. $\left\{-\dfrac{3}{7}, \dfrac{3}{7}\right\}$

43. $\left\{-8, \dfrac{1}{2}\right\}$

32.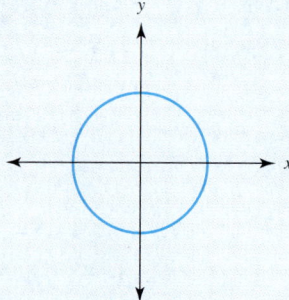

33.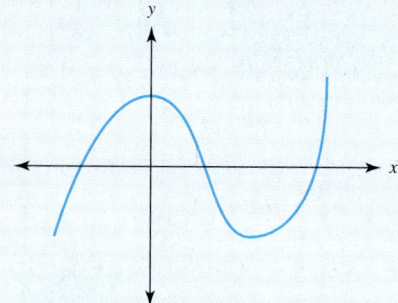

Graph each of the following functions.

34. $f(x) = 3x$

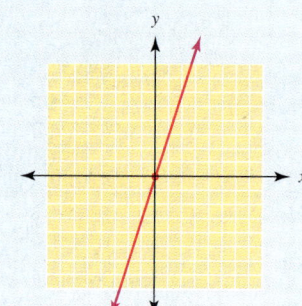

35. $f(x) = x + 3$

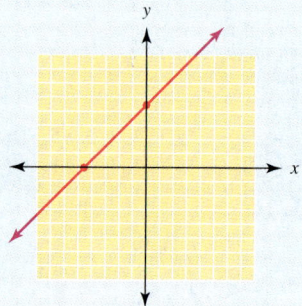

36. $f(x) = -3x - 2$

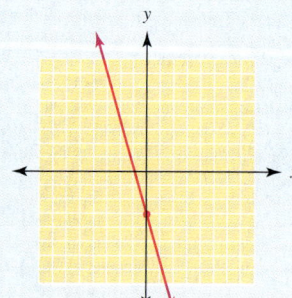

37. $f(x) = 3x + 3$

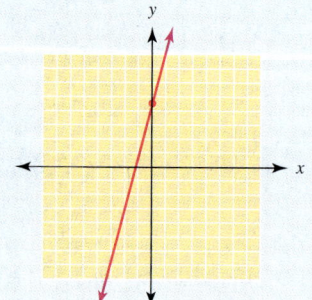

Solve the following word problems. Show the equation used.

38. The perimeter of a rectangle is 124 inches. If the length of the rectangle is one inch more than its width, what are the dimensions of the rectangle?

39. Juan's biology text cost five dollars more than his mathematics text. Together they cost $141. Find the cost of each text.

Solve each equation by factoring.

40. $x^2 - 3x = 28$

41. $x^2 - 10 = 3x$

42. $49x^2 - 9 = 0$

43. $2x^2 + 15x = 8$

CHAPTER 7
Rational Expressions

LIST OF SECTIONS

7.1 Simplifying Rational Expressions

7.2 Multiplication and Division of Rational Expressions

7.3 Addition and Subtraction of Rational Expressions

7.4 Complex Fractions

7.5 Rational Equations and Inequalities in One Variable

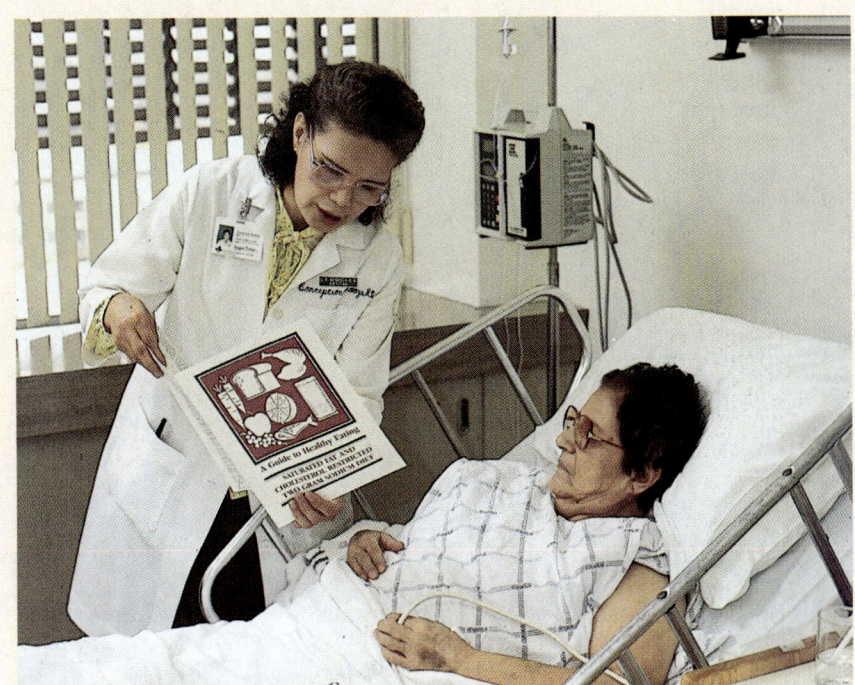

In the United States, disorders of the heart and circulatory system kill more people than all other causes combined. The major risk factors for heart disease are smoking, high blood pressure, obesity, cholesterol over 240, and a family history of heart problems. Although nothing can be done about family history, everyone can affect the first four risk factors by diet and exercise.

One quick way to check your risk of heart problems is to compare your waist and hip measurements. Measure around your waist at the naval and around your hips at the largest point. These measures may be in inches or centimeters. Use the ratio $\frac{w}{h}$ to assess your risk. For women, $\frac{w}{h} \geq 0.8$ indicates an increased health risk, and for men, $\frac{w}{h} \geq 0.95$ is the indicator of an increased risk.

SECTION 7.1 Simplifying Rational Expressions

7.1 OBJECTIVES

1. Simplify rational expressions
2. Identify rational functions
3. Simplify rational functions
4. Graph rational functions

The word "rational" comes from "ratio."

Our work in this chapter will expand your experience with algebraic expressions to include algebraic fractions or **rational expressions.** We consider the four basic operations of addition, subtraction, multiplication, and division in the next sections. Fortunately, you will observe many parallels to your previous work with arithmetic fractions.

Evaluating Rational Expressions

First, let's define what we mean by a rational expression. Recall that a rational number is the ratio of two integers. Similarly, a rational expression can be written as the ratio of two polynomials, in which the denominator cannot have the value 0.

> A *rational expression* is the ratio of two polynomials. It can be written as
>
> $$\frac{P}{Q}$$
>
> where P and Q are polynomials and Q cannot have the value 0.

$$\frac{x-3}{x+1} \qquad \frac{x^2+5}{x-3} \quad \text{and} \quad \frac{x^2-2x}{x^2+3x+1}$$

are all rational expressions.

Example 1

Precluding Division by Zero

(a) For what values of x is the following expression undefined?

$$\frac{x}{x-5}$$

A fraction is undefined when its denominator is equal to 0.

To answer this question, we must find where the denominator is 0. Set

$$x - 5 = 0$$
$$x = 5$$

Note that when $x = 5$, $\frac{x}{x-5}$ becomes $\frac{5}{5-5}$, or $\frac{5}{0}$.

The expression $\frac{x}{x-5}$ is undefined for $x = 5$.

480

(b) For what values of x is the following expression undefined?

$$\frac{3}{x+5}$$

Again, set the denominator equal to 0:

$$x + 5 = 0$$

or

$$x = -5$$

The expression $\dfrac{3}{x+5}$ is undefined for $x = -5$.

✓ CHECK YOURSELF 1

For what values of the variable are the following expressions undefined?

(a) $\dfrac{1}{r+7}$ (b) $\dfrac{5}{2x-9}$

Scientific calculators are often used to evaluate rational expressions for values of the variable. The *parentheses keys* help in this process.

Example 2 Evaluating a Rational Expression

Using a calculator, evaluate the following expressions for the given value of the variable.

(a) $\dfrac{3x}{2x-5}$ for $x = 4$

Caution
Be sure to use the parentheses keys before the 2 and after the 5.

Enter the expression in your calculator as follows:

$$3 \;\boxed{\times}\; 4 \;\boxed{\div}\; \boxed{(}\; 2 \;\boxed{\times}\; 4 \;\boxed{-}\; 5 \;\boxed{)}\; \boxed{=}$$

The display will read the value 4.

(b) $\dfrac{2x + 7}{4x - 11}$ for $x = 4$

Enter the expression as follows:

$\boxed{(}\ 2\ \boxed{\times}\ 4\ \boxed{+}\ 7\ \boxed{)}\ \boxed{\div}\ \boxed{(}\ 4\ \boxed{\times}\ 4\ \boxed{-}\ 11\ \boxed{)}\ \boxed{=}$

The display will read the value 3.

Because the numerator has more than one term, it must be enclosed in parentheses.

✓ **CHECK YOURSELF 2**

Using a scientific calculator, evaluate each of the following.

(a) $\dfrac{5x}{3x - 2}$ for $x = 4$ (b) $\dfrac{2x + 9}{3x - 4}$ for $x = 3$

Simplifying Rational Expressions

Generally, we want to write rational expressions in the simplest possible form. To begin our discussion of simplifying rational expressions, let's review for a moment. As we pointed out previously, there are many parallels to your work with arithmetic fractions. Recall that

$$\dfrac{3}{5} = \dfrac{3 \cdot 2}{5 \cdot 2} = \dfrac{6}{10}$$

so

$$\dfrac{3}{5} \quad \text{and} \quad \dfrac{6}{10}$$

name equivalent fractions. In a similar fashion,

$$\dfrac{10}{15} = \dfrac{5 \cdot 2}{5 \cdot 3} = \dfrac{2}{3}$$

so

$$\dfrac{10}{15} \quad \text{and} \quad \dfrac{2}{3}$$

name equivalent fractions.

Section 7.1 ■ Simplifying Rational Expressions

We can always multiply or divide the numerator and denominator of a fraction by the same nonzero number. The same pattern is true in algebra.

> **Fundamental Principle of Rational Expressions**
>
> For polynomials P, Q, and R,
>
> $$\frac{P}{Q} = \frac{PR}{QR} \quad \text{where } Q \neq 0 \quad \text{and} \quad R \neq 0$$

This principle can be used in two ways. We can multiply or divide the numerator and denominator of a rational expression by the same nonzero polynomial. The result will always be an expression that is equivalent to the original one.

In simplifying arithmetic fractions, we used this principle to divide the numerator and denominator by all common factors. With arithmetic fractions, those common factors are generally easy to recognize. Given rational expressions where the numerator and denominator are polynomials, we must determine those factors as our first step. The most important tools for simplifying expressions are the factoring techniques in Chapter 6.

In fact, you will see that most of the methods in this chapter depend on factoring polynomials.

Example 3

Simplifying Rational Expressions

Simplify each rational expression. Assume denominators are not 0.

We find the common factors 4, x, and y in the numerator and denominator. We divide the numerator and denominator by the common factor $4xy$. Note that

$$\frac{4xy}{4xy} = 1$$

(a) $\dfrac{4x^2y}{12xy^2} = \dfrac{4xy \cdot x}{4xy \cdot 3y}$

$= \dfrac{x}{3y}$

We have divided the numerator and denominator by the common factor $x - 2$. Again note that

$$\frac{x-2}{x-2} = 1$$

(b) $\dfrac{3x - 6}{x^2 - 4} = \dfrac{3(x - 2)}{(x + 2)(x - 2)}$ Factor the numerator and the denominator.

We can now divide the numerator and denominator by the common factor $x - 2$:

$$\frac{3(x - 2)}{(x + 2)(x - 2)} = \frac{3}{x + 2}$$

and the rational expression is in simplest form.

484 Chapter 7 ■ Rational Expressions

Caution
Pick any value other than 0 for the variable x, and substitute. You will quickly see that
$$\frac{x+2}{x+3} \neq \frac{2}{3}$$

Be Careful! Given the expression

$$\frac{x+2}{x+3}$$

students are often tempted to divide by variable x, as in

$$\frac{x+2}{x+3} \stackrel{?}{=} \frac{2}{3}$$

This is not a valid operation. We can only divide by common *factors*, and in the expression above the variable x is a *term* in both the numerator and the denominator. The numerator and denominator of a rational expression must be factored *before* common factors are divided out. Therefore,

$$\frac{x+2}{x+3}$$

is in its simplest possible form.

✓ **CHECK YOURSELF 3**

Simplify each expression.

(a) $\dfrac{36a^3b}{9ab^2}$ (b) $\dfrac{x^2-25}{4x+20}$

The same techniques are used when trinomials need to be factored.

Example 4 **Simplifying Rational Expressions**

Simplify each rational expression.

Divide by the common factor $x+1$, using the fact that
$$\frac{x+1}{x+1} = 1$$
where $x \neq -1$

(a) $\dfrac{5x^2-5}{x^2-4x-5} = \dfrac{5(x-1)(x+1)}{(x-5)(x+1)}$

$\phantom{(a) \dfrac{5x^2-5}{x^2-4x-5}} = \dfrac{5(x-1)}{x-5}$

(b) $\dfrac{2x^2+x-6}{2x^2-x-3} = \dfrac{(x+2)(2x-3)}{(x+1)(2x-3)}$

$\phantom{(b) \dfrac{2x^2+x-6}{2x^2-x-3}} = \dfrac{x+2}{x+1}$

In part (c) we factor by grouping in the numerator and use the sum of cubes in the denominator. Note that

$x^3 + 2x^2 - 3x - 6$
$= x^2(x + 2) - 3(x + 2)$
$= (x + 2)(x^2 - 3)$

(c) $\dfrac{x^3 + 2x^2 - 3x - 6}{x^3 + 8} = \dfrac{(x + 2)(x^2 - 3)}{(x + 2)(x^2 - 2x + 4)} = \dfrac{x^2 - 3}{x^2 - 2x + 4}$

✓ **CHECK YOURSELF 4**

Simplify each rational expression.

(a) $\dfrac{x^2 - 5x + 6}{3x^2 - 6x}$ (b) $\dfrac{3x^2 + 14x - 5}{3x^2 + 2x - 1}$

Simplifying certain algebraic expressions involves recognizing a particular pattern. Verify for yourself that

$$3 - 9 = -(9 - 3)$$

In general, it is true that

$$a - b = -(-a + b) = -(b - a) = -1(b - a)$$

or, by dividing both sides of the equation by $b - a$,

|Note that

$\dfrac{a - b}{a - b} = 1$

but

$\dfrac{a - b}{b - a} = -1$

$$\dfrac{a - b}{b - a} = \dfrac{-(b - a)}{b - a} = -1, \ a \neq b$$

Example 5 makes use of this result.

Example 5 — Simplifying Rational Expressions

Simplify each rational expression.

Note that

$\dfrac{x - 2}{2 - x} = -1$

(a) $\dfrac{2x - 4}{4 - x^2} = \dfrac{\overset{-1}{2(x - 2)}}{(2 + x)(2 - x)}$

$= \dfrac{2(-1)}{2 + x} = \dfrac{-2}{2 + x}$

(b) $\dfrac{9 - x^2}{x^2 + 2x - 15} = \dfrac{(3 + x)\overset{-1}{(3 - x)}}{(x + 5)(x - 3)}$

$= \dfrac{(3 + x)(-1)}{x + 5} = \dfrac{-x - 3}{x + 5}$

486 Chapter 7 ■ Rational Expressions

> **Simplifying Rational Expressions**
> 1. Completely factor both the numerator and denominator of the expression.
> 2. Divide the numerator and denominator by *all* common factors.
> 3. The resulting expression will be in simplest form (or in lowest terms).

✓ **CHECK YOURSELF 5**

Simplify each rational expression.

(a) $\dfrac{5x - 20}{16 - x^2}$ (b) $\dfrac{x^2 - 6x - 27}{81 - x^2}$

> A **rational function** is a function that is defined by a rational expression. It can be written as
>
> $$f(x) = \frac{P}{Q}$$
>
> where P and Q are polynomials and Q does not have a value of zero.

Example 6

Identifying Rational Functions

Which of the following are rational functions?

(a) $f(x) = 3x^3 - 2x + 5$ This is a rational function; it could be written over the denominator 1, and 1 is a polynomial.

(b) $f(x) = \dfrac{3x^2 - 5x + 2}{2x - 1}$ This is a rational function; it is the ratio of two polynomials.

Recall from Chapter 5 that there are no square roots of variables in a polynomial.

(c) $f(x) = 3x^3 + 3\sqrt{x}$ This is not a rational function; it is not the ratio of two polynomials.

CHECK YOURSELF 6

Which of the following are rational functions?

(a) $f(x) = x^5 - 2x^4 - 1$ (b) $f(x) = \dfrac{x^2 - x + 7}{\sqrt{x} - 1}$ (c) $f(x) = \dfrac{3x^3 + 3x}{2x + 1}$

Simplifying Rational Functions

When we simplify a rational function, it is important that we note the x values that need to be excluded, particularly when we are trying to draw the graph of a function. The set of ordered pairs of the simplified function will be exactly the same as the set of ordered pairs of the original function. If we plug the excluded value(s) for x into the simplified expression, we get a set of ordered pairs that represent "holes" in the graph. These holes are breaks in the curve. We use an open circle to designate them on a graph.

Example 7 Simplifying a Rational Function

Given the function

$$f(x) = \frac{x^2 + 2x + 1}{x + 1}$$

(a) Rewrite the function in simplified form.

$$f(x) = \frac{x^2 + 2x + 1}{x + 1} = \frac{(x + 1)(x + 1)}{(x + 1)}$$

$$= (x + 1) \qquad x \neq -1$$

(b) Find the ordered pair associated with the hole in the graph of the original function.

Plugging -1 into the simplified function yields the ordered pair $(-1, 0)$. This represents the hole in the graph of the function $f(x)$.

CHECK YOURSELF 7

Given the function

$$f(x) = \frac{5x^2 - 10x}{5x}$$

(a) Rewrite the function in simplified form.

(b) Find the ordered pair associated with the hole in the graph of the original function.

Example 8 Graphing a Rational Function

Graph the following function.

$$f(x) = \frac{x^2 + 2x + 1}{x + 1}$$

From Example 7, we know that

$$\frac{x^2 + 2x + 1}{x + 1} = x + 1 \qquad x \neq -1$$

Therefore,

$$f(x) = x + 1 \qquad x \neq -1$$

The graph will be the graph of the line $f(x) = x + 1$, with an open circle at the point $(-1, 0)$.

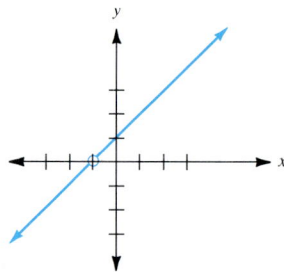

✓ CHECK YOURSELF 8

Graph the function $f(x) = \dfrac{5x^2 - 10x}{5x}$.

✓ CHECK YOURSELF ANSWERS

1. (a) $r = -7$; (b) $x = \dfrac{9}{2}$.   2. (a) 2; (b) 3.   3. (a) $\dfrac{4a^2}{b}$; (b) $\dfrac{x-5}{4}$.
4. (a) $\dfrac{x-3}{3x}$; (b) $\dfrac{x+5}{x+1}$.   5. (a) $\dfrac{-5}{x+4}$; (b) $\dfrac{-x-3}{x+9}$.   6. (a) A rational function, (b) not a rational function, and (c) a rational function.
7. (a) $f(x) = x - 2$, $x \neq 0$; (b) $(0, -2)$.   8.

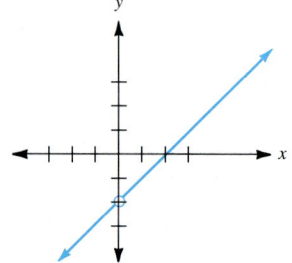

Exercises • 7.1

Answers (left column):
1. 5
2. −7
3. Never undefined
4. Never undefined
5. $\frac{1}{2}$
6. $-\frac{1}{3}$
7. 0
8. 0
9. −2
10. $\frac{7}{3}$
11. 0
12. $-\frac{1}{9}$
13. 2
14. 2
15. 3
16. 3
17. $\frac{2}{3}$
18. $\frac{3}{5}$
19. $\frac{2x^3}{3}$
20. $\frac{5x^4}{4}$
21. $\frac{2xy^3}{5}$
22. $\frac{3}{4a^2}$
23. $\frac{-3x^2}{y^2}$
24. $\frac{3x^2y}{4}$
25. $\frac{a^3b^2}{3c^2}$
26. $\frac{-4p^2}{3q^2}$
27. $\frac{6}{x+4}$
28. $\frac{x+5}{3}$
29. $\frac{x+1}{6}$
30. $\frac{5y}{y+3}$
31. $\frac{x+2}{x+7}$
32. $\frac{m+7}{2m+3}$
33. $3b+1$
34. $\frac{a-3b}{a+5b}$
35. $\frac{y-z}{y+3z}$
36. $\frac{2x+1}{x-1}$
37. $\frac{x^2+4x+16}{x+4}$

In Exercises 1 to 12, for what values of the variable is each rational expression undefined?

1. $\dfrac{x}{x-5}$
2. $\dfrac{y}{y+7}$
3. $\dfrac{x+5}{3}$
4. $\dfrac{x-6}{4}$
5. $\dfrac{2x-3}{2x-1}$
6. $\dfrac{3x-2}{3x+1}$
7. $\dfrac{2x+5}{x}$
8. $\dfrac{3x-7}{x}$
9. $\dfrac{x(x+1)}{x+2}$
10. $\dfrac{x+2}{3x-7}$
11. $\dfrac{4-x}{x}$
12. $\dfrac{2x+7}{3x+\frac{1}{3}}$

In Exercises 13 to 16, evaluate each expression, using a calculator.

13. $\dfrac{3x}{2x-1}$ for $x = 2$
14. $\dfrac{5x}{4x-3}$ for $x = 2$
15. $\dfrac{2x+3}{x+3}$ for $x = -6$
16. $\dfrac{4x-7}{2x-1}$ for $x = -2$

In Exercises 17 to 48, simplify each expression. Assume the denominators are not 0.

17. $\dfrac{14}{21}$
18. $\dfrac{45}{75}$
19. $\dfrac{4x^5}{6x^2}$
20. $\dfrac{25x^6}{20x^2}$
21. $\dfrac{10x^2y^5}{25xy^2}$
22. $\dfrac{18a^2b^3}{24a^4b^3}$
23. $\dfrac{-42x^3y}{14xy^3}$
24. $\dfrac{-15x^3y^3}{-20xy^2}$
25. $\dfrac{28a^5b^3c^2}{84a^2bc^4}$
26. $\dfrac{-52p^5q^3r^2}{39p^3q^5r^2}$
27. $\dfrac{6x-24}{x^2-16}$
28. $\dfrac{x^2-25}{3x-15}$
29. $\dfrac{x^2+2x+1}{6x+6}$
30. $\dfrac{5y^2-10y}{y^2+y-6}$
31. $\dfrac{x^2-5x-14}{x^2-49}$
32. $\dfrac{2m^2+11m-21}{4m^2-9}$
33. $\dfrac{3b^2-14b-5}{b-5}$
34. $\dfrac{a^2-9b^2}{a^2+8ab+15b^2}$
35. $\dfrac{2y^2+3yz-5z^2}{2y^2+11yz+15z^2}$
36. $\dfrac{6x^2-x-2}{3x^2-5x+2}$
37. $\dfrac{x^3-64}{x^2-16}$

489

38. $\dfrac{r-3s}{r^2-2rs+4s^2}$

39. $\dfrac{(a^2+9)(a-3)}{a+2}$

40. $\dfrac{(c^2+4)(c-2)}{c-5}$

41. $\dfrac{y-2}{x+5}$

42. $\dfrac{c+5}{d-4}$

43. $\dfrac{x+6}{x^2-2}$

44. $\dfrac{y+7}{y-3}$

45. $\dfrac{-2}{m+5}$

46. $\dfrac{-5}{x+4}$

47. $\dfrac{-x-7}{2x+1}$

48. $\dfrac{-2x+1}{x+3}$

49. Rational

50. Not rational

51. Rational

52. Not rational

53. Not rational

54. Rational

55. (a) $f(x) = x - 2$;
    (b) $(-1, -3)$

56. (a) $f(x) = x - 3$;
    (b) $(-4, -7)$

57. (a) $f(x) = 3x - 1$;
    (b) $(-2, -7)$

58. (a) $f(x) = x - 1$;
    (b) $\left(\dfrac{5}{2}, \dfrac{3}{2}\right)$

59. (a) $f(x) = \dfrac{x+2}{5}$;
    (b) $(-2, 0)$

38. $\dfrac{r^2-rs-6s^2}{r^3+8s^3}$

39. $\dfrac{a^4-81}{a^2+5a+6}$

40. $\dfrac{c^4-16}{c^2-3c-10}$

41. $\dfrac{xy-2x+3y-6}{x^2+8x+15}$

42. $\dfrac{cd-3c+5d-15}{d^2-7d+12}$

43. $\dfrac{x^2+3x-18}{x^3-3x^2-2x+6}$

44. $\dfrac{y^2+2y-35}{y^2-5y-3y+15}$

45. $\dfrac{2m-10}{25-m^2}$

46. $\dfrac{5x-20}{16-x^2}$

47. $\dfrac{49-x^2}{2x^2-13x-7}$

48. $\dfrac{2x^2-7x+3}{9-x^2}$

In Exercises 49 to 54, identify which functions are rational functions.

49. $f(x) = 4x^2 - 5x + 6$

50. $f(x) = \dfrac{x^3 - 2x^2 + 7}{\sqrt{x} + 2}$

51. $f(x) = \dfrac{x^2 - x - 1}{x + 2}$

52. $f(x) = \dfrac{\sqrt{x} - x + 3}{x - 2}$

53. $f(x) = 5x^2 - \sqrt[3]{x}$

54. $f(x) = \dfrac{x^2 - x + 5}{x}$

For the given functions in Exercises 55 to 60, (a) rewrite the function in simplified form and (b) find the ordered pair associated with the hole in the graph of the original function.

55. $f(x) = \dfrac{x^2 - x - 2}{x + 1}$

56. $f(x) = \dfrac{x^2 + x - 12}{x + 4}$

57. $f(x) = \dfrac{3x^2 + 5x - 2}{x + 2}$

58. $f(x) = \dfrac{2x^2 - 7x + 5}{2x - 5}$

59. $f(x) = \dfrac{x^2 + 4x + 4}{5(x + 2)}$

60. $f(x) = \dfrac{x^2 - 6x + 9}{7(x - 3)}$

60. (a) $f(x) = \dfrac{x-3}{7}$;
    (b) $(3, 0)$

Section 7.1 ■ Simplifying Rational Expressions   491

61.–62.

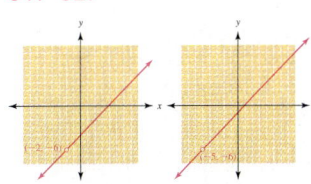

In Exercises 61 to 66, graph the rational functions. Indicate the coordinates of the hole in the graph.

61. $f(x) = \dfrac{x^2 - 2x - 8}{x + 2}$ **62.** $f(x) = \dfrac{x^2 + 4x - 5}{x + 5}$ **63.** $f(x) = \dfrac{x^2 + 4x + 3}{x + 1}$

64. $f(x) = \dfrac{x^2 + 7x + 10}{x + 2}$ **65.** $f(x) = \dfrac{x^2 - 4x + 3}{x - 1}$ **66.** $f(x) = \dfrac{x^2 - 6x + 8}{x - 4}$

67. Explain why the following statement is false.

$$\dfrac{6m^2 + 2m}{2m} = 6m^2 + 1$$

68. State and explain the fundamental principle of rational expressions.

69. The rational expression $\dfrac{x^2 - 4}{x + 2}$ can be simplified to $x - 2$. Is this reduction true for all values of x? Explain.

70. What is meant by a rational expression in lowest terms?

In Exercises 71 to 76, simplify.

63.–64.

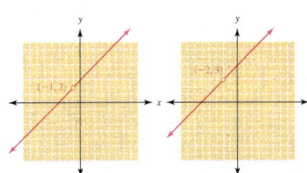

71. $\dfrac{2(x + h) - 2x}{(x + h) - x}$ **72.** $\dfrac{-3(x + h) - (-3x)}{(x - h) - x}$

73. $\dfrac{3(x + h) - 3 - (3x - 3)}{(x + h) - x}$ **74.** $\dfrac{2(x + h) + 5 - (2x + 5)}{(x + h) - x}$

75. $\dfrac{(x + h)^2 - x^2}{(x + h) - x}$ **76.** $\dfrac{(x + h)^3 - x^3}{(x + h) - x}$

Given $f(x) = \dfrac{P(x)}{Q(x)}$, if the graphs of $P(x)$ and $Q(x)$ intersect at $(a, 0)$, then $x - a$ is a factor of both $P(x)$ and $Q(x)$. Use a graphing calculator to find the common factor for the expressions in Exercises 77 and 78.

65.–66.

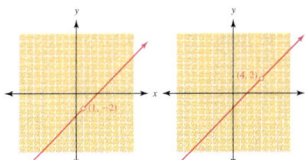

77. $f(x) = \dfrac{x^2 + 4x - 5}{x^2 + 3x - 10}$ **78.** $f(x) = \dfrac{2x^2 + 11x - 21}{2x^2 + 15x + 7}$

79. Revenue. The total revenue from the sale of a popular video is approximated by the rational function

$$R(x) = \dfrac{300x^2}{x^2 + 9}$$

71. 2   72. −3
73. 3   74. 2
75. $2x + h$
76. $3x^2 + 3xh + h^2$
77. $x + 5$
78. $x + 7$

where x is the number of months since the video has been released and $R(x)$ is the total revenue in hundreds of dollars.

79. $3000

$9231

$15,000

$6231

80. $R(x) = \dfrac{3500 + 8.75x}{x}$,

$78.75

81. (a) 3

(a) Find the total revenue generated by the end of the first month.

(b) Find the total revenue generated by the end of the second month.

(c) Find the total revenue generated by the end of the third month.

(d) Find the revenue in the second month only.

80. Cost. A company has a set-up cost of $3500 for the production of a new product. The cost to produce a single unit is $8.75.

(a) Write a rational expression that gives the average cost per unit when x units are produced.

(b) Find the average cost when 50 units are produced.

81. Besides holes, we sometimes encounter a different sort of "break" in the graph of a rational function. Consider the rational function

$$f(x) = \dfrac{1}{x-3}$$

(a) For what value(s) of x is the function undefined?

(b) Complete the following table.

x	$f(x)$
4	1
3.1	10
3.01	100
3.001	1,000
3.0001	10,000

(c) What do you observe concerning $f(x)$ as x is chosen close to 3 (but slightly larger than 3)?

(d) Complete the table.

x	$f(x)$
2	-1
2.9	-10
2.99	-100
2.999	-1000
2.9999	$-10,000$

(e) What do you observe concerning $f(x)$ as x is chosen close to 3 (but slightly smaller than 3)?

(f) Graph the function on your graphing calculator. Describe the behavior of the graph of f near $x = 3$.

SECTION 7.2 Multiplication and Division of Rational Expressions

7.2 OBJECTIVES

1. Multiply and divide rational expressions
2. Multiply and divide two rational functions

For all problems with rational expressions, assume denominators are not 0.

Once again, let's turn to an example from arithmetic to begin our discussion of multiplying rational expressions. Recall that to multiply two fractions, we multiply the numerators and multiply the denominators. For instance,

$$\frac{2}{5} \cdot \frac{3}{7} = \frac{2 \cdot 3}{5 \cdot 7} = \frac{6}{35}$$

In algebra, the pattern is exactly the same.

> **Multiplying Rational Expressions**
>
> For polynomials P, Q, R, and S,
>
> $$\frac{P}{Q} \cdot \frac{R}{S} = \frac{PR}{QS} \qquad \text{where } Q \neq 0 \quad \text{and} \quad S \neq 0$$

Example 1

Multiplying Rational Expressions

Multiply.

$$\frac{2x^3}{5y^2} \cdot \frac{10y}{3x^2} = \frac{20x^3y}{15x^2y^2}$$

$$= \frac{5x^2y \cdot 4x}{5x^2y \cdot 3y} \qquad \text{Divide by the common factor } 5x^2y \text{ to simplify.}$$

$$= \frac{4x}{3y}$$

✓ **CHECK YOURSELF 1**

Multiply.

$$\frac{9a^2b^3}{5ab^4} \cdot \frac{20ab^2}{27ab^3}$$

Chapter 7 ■ Rational Expressions

The factoring methods in Chapter 6 are used to simplify rational expressions.

Generally, you will find it best to divide by any common factors before you multiply, as Example 2 illustrates.

Example 2 Multiplying Rational Expressions

Multiply as indicated.

(a) $\dfrac{x}{x^2 - 3x} \cdot \dfrac{6x - 18}{9x}$ *Factor.*

$= \dfrac{\cancel{x}}{\cancel{x}(\cancel{x - 3})} \cdot \dfrac{\overset{2}{\cancel{6}}(\cancel{x - 3})}{\underset{3}{\cancel{9}\cancel{x}}}$ *Divide by the common factors of 3, x, and x − 3.*

$= \dfrac{2}{3x}$

(b) $\dfrac{x^2 - y^2}{5x^2 - 5xy} \cdot \dfrac{10xy}{x^2 + 2xy + y^2}$ *Factor and divide by the common factors of 5, x, x − y, and x + y.*

$= \dfrac{(x + y)\cancel{(x - y)}}{\cancel{5}x\cancel{(x - y)}} \cdot \dfrac{\overset{2}{\cancel{10}}xy}{\cancel{(x + y)}(x + y)}$

$= \dfrac{2y}{x + y}$

Note that

$\dfrac{2 - x}{x - 2} = -1$

(c) $\dfrac{4}{x^2 - 2x} \cdot \dfrac{10x - 5x^2}{8x + 24}$

$= \dfrac{\cancel{4}}{x\cancel{(x - 2)}} \cdot \dfrac{\overset{-1}{5x\cancel{(2 - x)}}}{\underset{2}{\cancel{8}(x + 3)}}$

$= \dfrac{-5}{2(x + 3)}$

✓ CHECK YOURSELF 2

Multiply as indicated.

(a) $\dfrac{x^2 - 5x - 14}{4x^2} \cdot \dfrac{8x + 56}{x^2 - 49}$ (b) $\dfrac{x}{2x - 6} \cdot \dfrac{3x - x^2}{2}$

The following algorithm summarizes our work in multiplying rational expressions.

Multiplying Rational Expressions

Step 1 Write each numerator and denominator in completely factored form.
Step 2 Divide by any common factors appearing in both the numerator and denominator.
Step 3 Multiply as needed to form the desired product.

Dividing Rational Expressions

In dividing rational expressions, you can again use your experience from arithmetic. Recall that

We invert the divisor (the second fraction) and multiply.

$$\frac{3}{5} \div \frac{2}{3} = \frac{3}{5} \cdot \frac{3}{2} = \frac{9}{10}$$

Once more, the pattern in algebra is identical.

Dividing Rational Expressions

For polynomials P, Q, R, and S,

$$\frac{P}{Q} \div \frac{R}{S} = \frac{P}{Q} \cdot \frac{S}{R} = \frac{PS}{QR}$$

where $Q \neq 0$, $R \neq 0$, and $S \neq 0$.

To divide rational expressions, invert the divisor and multiply as before, as Example 3 illustrates.

Example 3 Dividing Rational Expressions

Divide as indicated.

Invert the divisor and multiply.

Caution
Be Careful! Invert the divisor, then factor.

(a) $\dfrac{3x^2}{8x^3y} \div \dfrac{9x^2y^2}{4y^4} = \dfrac{3x^2}{8x^3y} \cdot \dfrac{4y^4}{9x^2y^2} = \dfrac{y}{6x^3}$

(b) $\dfrac{2x^2 + 4xy}{9x - 18y} \div \dfrac{4x + 8y}{3x - 6y} = \dfrac{2x^2 + 4xy}{9x - 18y} \cdot \dfrac{3x - 6y}{4x + 8y}$

$= \dfrac{\overset{}{2x(x + 2y)}}{\underset{3}{9(x - 2y)}} \cdot \dfrac{\overset{}{3(x - 2y)}}{\underset{2}{4(x + 2y)}} = \dfrac{x}{6}$

(c) $\dfrac{2x^2 - x - 6}{4x^2 + 6x} \div \dfrac{x^2 - 4}{4x} = \dfrac{2x^2 - x - 6}{4x^2 + 6x} \cdot \dfrac{4x}{x^2 - 4}$

$= \dfrac{(2x+3)(x-2)}{2x(2x+3)} \cdot \dfrac{4x^2}{(x+2)(x-2)} = \dfrac{2x}{x+2}$

✓ CHECK YOURSELF 3

Divide and simplify.

(a) $\dfrac{5xy}{7x^3} \div \dfrac{10y^2}{14x^3}$

(b) $\dfrac{3x - 9y}{2x + 10y} \div \dfrac{x^2 - 3xy}{4x^2 + 20xy}$

(c) $\dfrac{x^2 - 9}{x^3 - 27} \div \dfrac{x^2 - 2x - 15}{2x^2 - 10x}$

We summarize our work in dividing fractions with the following algorithm.

Dividing Rational Expressions

Step 1 Invert the divisor (the *second* rational expression) to write the problem as one of multiplication.

Step 2 Proceed as in the algorithm for the multiplication of rational expressions.

Multiplying Rational Functions

The product of two rational functions is always a rational function. Given two rational functions, $f(x)$ and $g(x)$, we can rename the product, so

$$h(x) = f(x) \cdot g(x)$$

This will always be true for values of x for which both f and g are defined. So, for example, $h(1) = f(1) \cdot g(1)$ as long as both $f(1)$ and $g(1)$ exist.

Section 7.2 ▪ Multiplication and Division of Rational Expressions

If
$$h(x) = f(x) \cdot g(x)$$
then the set of ordered pairs
$$(x, f(x) \cdot g(x)) = (x, h(x))$$

Example 4 illustrates this concept.

Example 4 Multiplying Rational Functions

Given the rational functions

$$f(x) = \frac{x^2 - 3x - 10}{x + 1} \quad \text{and} \quad g(x) = \frac{x^2 - 4x - 5}{x - 5}$$

find the following.

(a) $f(0) \cdot g(0)$

If $f(0) = -10$ and $g(0) = 1$, then $f(0) \cdot g(0) = (-10)(1) = -10$.

(b) $f(5)g(5)$

Although we can find $f(5)$, $g(5)$ is undefined. 5 is excluded from the domain of the function. Therefore, $f(5)g(5)$ is undefined.

(c) $h(x) = f(x) \cdot g(x)$

$$h(x) = f(x) \cdot g(x)$$
$$= \frac{x^2 - 3x - 10}{x + 1} \cdot \frac{x^2 - 4x - 5}{x - 5}$$
$$= \frac{(x - 5)(x + 2)}{\cancel{(x + 1)}} \cdot \frac{\cancel{(x + 1)}(x - 5)}{\cancel{(x - 5)}}$$
$$= (x - 5)(x + 2) \qquad x \neq -1, x \neq 5$$

(d) $h(0)$

$$h(0) = (0 - 5)(0 + 2) = -10$$

(e) $h(5)$

Although the temptation is to substitute 5 for x in part c, notice that the function is undefined when x is -1 or 5. As was true in part b, the function is undefined at that point.

✓ CHECK YOURSELF 4

Given the rational functions

$$f(x) = \frac{x^2 - 2x - 8}{x + 2} \quad \text{and} \quad g(x) = \frac{x^2 - 3x - 10}{x - 4}$$

find the following.

(a) $f(0)g(0)$ (b) $f(4)g(4)$ (c) $h(x) = f(x)g(x)$ (d) $h(0)$ (e) $h(4)$

Dividing Rational Functions

When we divide two rational functions to create a third rational function, we must be certain to exclude values for which the polynomial in the denominator is equal to zero, as Example 5 illustrates.

Example 5   Dividing Polynomial Functions

Given the rational functions

$$f(x) = \frac{x^3 - 2x^2}{x + 2} \quad \text{and} \quad g(x) = \frac{x^2 - 3x + 2}{x - 4}$$

complete the following.

(a) Find $\dfrac{f(0)}{g(0)}$.

If $f(0) = 0$ and $g(0) = -\dfrac{1}{2}$, then

$$\frac{f(0)}{g(0)} = \frac{0}{-\dfrac{1}{2}} = 0$$

(b) Find $\dfrac{f(1)}{g(1)}$.

Although we can find both $f(1)$ and $g(1)$, $g(1) = 0$, so division is undefined. 1 is excluded from the domain of the quotient.

(c) Find $h(x) = \dfrac{f(x)}{g(x)}$.

$h(x) = \dfrac{f(x)}{g(x)}$ Note that -2 is excluded from the domain of f and 4 is excluded from the domain of g.

$= \dfrac{\dfrac{x^3 - 2x^2}{x+2}}{\dfrac{x^2 - 3x + 2}{x - 4}}$ Invert and multiply.

$= \dfrac{x^3 - 2x^2}{x+2} \cdot \dfrac{x-4}{x^2 - 3x + 2}$

$= \dfrac{x^2(x-2)}{x+2} \cdot \dfrac{x-4}{(x-1)(x-2)}$ Because $(x-1)(x-2)$ is part of the denominator, 1 and 2 are excluded from the domain of h.

$= \dfrac{x^2(x-4)}{(x+2)(x-1)}$ $x \neq -2, 1, 2, 4$

(d) For which values of x is $h(x)$ undefined?

$h(x)$ will be undefined for any value of x that would cause division by zero. $h(x)$ is undefined for the values -2, 1, 2, and 4.

✓ CHECK YOURSELF 5

Given the rational functions

$$f(x) = \dfrac{x^2 - 2x + 1}{x + 3} \quad \text{and} \quad g(x) = \dfrac{x^2 - 5x + 4}{x - 2}$$

(a) Find $\dfrac{f(0)}{g(0)}$.   (b) Find $\dfrac{f(1)}{g(1)}$.   (c) Find $h(x) = \dfrac{f(x)}{g(x)}$.

(d) For which values of x is $h(x)$ undefined?

✓ CHECK YOURSELF ANSWERS

1. $\dfrac{4a}{3b^2}$.   2. (a) $\dfrac{2(x+2)}{x^2}$; (b) $\dfrac{-x^2}{4}$.   3. (a) $\dfrac{x}{y}$; (b) 6; (c) $\dfrac{2x}{x^2 + 3x + 9}$.
4. (a) -10; (b) undefined; (c) $h(x) = (x-5)(x+2)$, $x \neq -2$, $x \neq 4$; (d) -10; and (e) undefined.
5. (a) $-\dfrac{1}{6}$; (b) undefined; (c) $h(x) = \dfrac{(x-1)(x-2)}{(x+3)(x-4)}$; and (d) $x \neq -3, 1, 2, 4$.

Calculating Probability. Probability, or the study of chance, measures the relative likelihood of events. Suppose a situation has N equally likely possible outcomes, and an event includes E of these. Let P be the probability that the event will occur. Then

$$P = \frac{E}{N}$$

From this definition, it is clear that any probability will always be between 0 and 1: $0 \leq P \leq 1$. The probability of an event not happening is $1 - P$.

The probability of two separate, independent events, event 1 and event 2, *both* happening is the *product* of the probability of the first event times the probability of the second event: $P(E_1 \text{ and } E_2) = P(E_1) \cdot P(E_2)$. The events must be independent, which means the first event cannot have any effect on the outcome of the second.

Work with a partner to complete the following.

What is the probability that two people in your class have the same birthday? Surprisingly, it is pretty common in a class of 30 people to find at least two people who have the same birthday. You can use the probability that two people will *not* have the same birthday to figure this out.

Begin with one person from the class. This person can have a birthday on any day. What is the probability that the second person will *not* have the same birthday? (How many days are in the year?) Let us say that the second person does not have the same birthday as the first person. Now there are two days in the year when no one else in the class can have a birthday. Next we find the probability that the third person does not have the same birthday. If these two probabilities are multiplied, we get the probability that three people do not have the same birthday. Continue until all 30 people have been accounted for. What is the probability that none of them have the same birthday?

$P(\text{no two birthdays are the same among } n \text{ people}) = 1 - (P_2)(P_3)(P_4) \ldots (P_n)$

where

$$P_i = \frac{365 - (i - 1)}{365}$$

is the probability that the *i*th person does not have the same birthday as anyone who came before.

Exercises • 7.2

In Exercises 1 to 36, multiply or divide as indicated. Express your result in simplest form.

1. $\dfrac{2}{x}$
2. $\dfrac{-3}{2y^2}$
3. $\dfrac{3}{a^4}$
4. $\dfrac{-3p^4}{2}$
5. $\dfrac{5}{12x}$
6. $\dfrac{-x^2}{6y^3}$
7. $\dfrac{16b^3}{3a}$
8. $\dfrac{-3y}{2}$
9. $5mn$
10. $\dfrac{8cd^3}{3}$
11. $\dfrac{15x}{2}$
12. $\dfrac{4a^2}{3}$
13. $\dfrac{9b}{8}$
14. $\dfrac{21m^2}{5}$
15. $x^2 + 2x$
16. $\dfrac{3y^2}{y+8}$
17. $\dfrac{3(c-2)}{5}$
18. $\dfrac{4m(m-7)}{3}$
19. $\dfrac{5x}{2(x-2)}$
20. $\dfrac{2}{y-2}$
21. $\dfrac{5d}{4(d-3)}$
22. $\dfrac{2}{b-2}$
23. $\dfrac{x-5}{2x+3}$
24. $\dfrac{2p-1}{3p-2}$
25. $\dfrac{2a+1}{2a}$
26. $\dfrac{2x-7}{5(2x-3)}$
27. $\dfrac{-6}{w+2}$
28. $\dfrac{-12}{y+3}$

1. $\dfrac{x^2}{3} \cdot \dfrac{6x}{x^4}$

2. $\dfrac{-y^3}{10} \cdot \dfrac{15y}{y^6}$

3. $\dfrac{a}{7a^3} \div \dfrac{a^2}{21}$

4. $\dfrac{p^5}{8} \div \dfrac{-p^2}{12p}$

5. $\dfrac{4xy^2}{15x^3} \cdot \dfrac{25xy}{16y^3}$

6. $\dfrac{3x^3y}{10xy^3} \cdot \dfrac{5xy^2}{-9xy^3}$

7. $\dfrac{8b^3}{15ab} \div \dfrac{2ab^2}{20ab^3}$

8. $\dfrac{4x^2y^2}{9x^3} \div \dfrac{-8y^2}{27xy}$

9. $\dfrac{m^3n}{2mn} \cdot \dfrac{6mn^2}{m^3n} \div \dfrac{3mn}{5m^2n}$

10. $\dfrac{4cd^2}{5cd} \cdot \dfrac{3c^3d}{2c^2d} \div \dfrac{9cd}{20cd^3}$

11. $\dfrac{5x+15}{3x} \cdot \dfrac{9x^2}{2x+6}$

12. $\dfrac{a^2-3a}{5a} \cdot \dfrac{20a^2}{3a-9}$

13. $\dfrac{3b-15}{6b} \div \dfrac{4b-20}{9b^2}$

14. $\dfrac{7m^2+28m}{4m} \div \dfrac{5m+20}{12m^2}$

15. $\dfrac{x^2-3x-10}{5x} \cdot \dfrac{15x^2}{3x-15}$

16. $\dfrac{y^2-8y}{4y} \cdot \dfrac{12y^2}{y^2-64}$

17. $\dfrac{c^2+2c-8}{6c} \div \dfrac{5c+20}{18c}$

18. $\dfrac{m^2-49}{5m} \div \dfrac{3m+21}{20m^2}$

19. $\dfrac{x^2-2x-8}{4x-16} \cdot \dfrac{10x}{x^2-4}$

20. $\dfrac{y^2+7y+10}{y^2+5y} \cdot \dfrac{2y}{y^2-4}$

21. $\dfrac{d^2-3d-18}{16d-96} \div \dfrac{d^2-9}{20d}$

22. $\dfrac{b^2+6b+8}{b^2+4b} \div \dfrac{b^2-4}{2b}$

23. $\dfrac{2x^2-x-3}{3x^2+7x+4} \cdot \dfrac{3x^2-11x-20}{4x^2-9}$

24. $\dfrac{4p^2-1}{2p^2-9p-5} \cdot \dfrac{3p^2-13p-10}{9p^2-4}$

25. $\dfrac{a^2-9}{2a^2-6a} \div \dfrac{2a^2+5a-3}{4a^2-1}$

26. $\dfrac{2x^2-5x-7}{4x^2-9} \div \dfrac{5x^2+5x}{2x^2+3x}$

27. $\dfrac{2w-6}{w^2+2w} \cdot \dfrac{3w}{3-w}$

28. $\dfrac{3y-15}{y^2+3y} \cdot \dfrac{4y}{5-y}$

501

502 Chapter 7 ■ Rational Expressions

29. $\dfrac{-a}{6}$ 30. $\dfrac{-3}{4x}$

31. $\dfrac{2}{x}$ 32. $\dfrac{a}{2b}$

33. $\dfrac{3}{m}$ 34. $\dfrac{y}{4}$

35. $\dfrac{5}{x}$ 36. $3a$

37. (a) -4; (b) undefined; (c) $(x + 1)(x - 4)$, $x \neq -2$, $x \neq 4$; (e) undefined

38. (a) 0; (b) undefined; (c) $(x - 1)(x + 2)$, $x \neq -5$, $x \neq 3$; (d) 0; (e) undefined

39. (a) -6; (b) undefined; (c) $(2x - 5)(3x - 1)$, $x \neq -2$, $x \neq -1$; (d) -6; (e) undefined

40. (a) 15; (b) undefined; (c) $(x + 1)(x + 3)$, $x \neq 1$, $x \neq 3$; (d) 15; (e) undefined

41. (a) $\dfrac{-4}{5}$; (b) $\dfrac{5}{4}$; (c) $\dfrac{(3x - 2)(x + 4)}{(x - 2)(x - 5)}$; (d) $2, -4, -1, 5$

42. (a) 0; (b) $\dfrac{-3}{2}$; (c) $\dfrac{x(x + 1)}{(x - 3)(x + 2)}$; (d) $-2, 3, 5$

43. $\dfrac{x^2 + 2}{4}$ 44. $\dfrac{3}{5a}$

45. $\dfrac{x + 2}{x(x + 3)}$ 46. $\dfrac{1}{w(w - 1)}$

29. $\dfrac{a - 7}{2a + 6} \div \dfrac{21 - 3a}{a^2 + 3a}$

30. $\dfrac{x - 4}{x^2 + 2x} \div \dfrac{16 - 4x}{3x + 6}$

31. $\dfrac{x^2 - 9y^2}{2x^2 - xy - 15y^2} \cdot \dfrac{4x + 10y}{x^2 + 3xy}$

32. $\dfrac{2a^2 - 7ab - 15b^2}{2ab - 10b^2} \cdot \dfrac{2a^2 - 3ab}{4a^2 - 9b^2}$

33. $\dfrac{3m^2 - 5mn + 2n^2}{9m^2 - 4n^2} \div \dfrac{m^3 - m^2n}{9m^2 + 6mn}$

34. $\dfrac{2x^2y - 5xy^2}{4x^2 - 25y^2} \div \dfrac{4x^2 + 20xy}{2x^2 + 15xy + 25y^2}$

35. $\dfrac{x^3 + 8}{x^2 - 4} \cdot \dfrac{5x - 10}{x^3 - 2x^2 + 4x}$

36. $\dfrac{a^3 - 27}{a^2 - 9} \div \dfrac{a^3 + 3a^2 + 9a}{3a^3 + 9a^2}$

37. Let $f(x) = \dfrac{x^2 - 3x - 4}{x + 2}$ and $g(x) = \dfrac{x^2 - 2x - 8}{x - 4}$. Find (a) $f(0) \cdot g(0)$, (b) $f(4) \cdot g(4)$, (c) $h(x) = f(x) \cdot g(x)$, (d) $h(0)$, and (e) $h(4)$.

38. Let $f(x) = \dfrac{x^2 - 4x + 3}{x + 5}$ and $g(x) = \dfrac{x^2 + 7x + 10}{x - 3}$. Find (a) $f(1) \cdot g(1)$, (b) $f(3) \cdot g(3)$, (c) $h(x) = f(x) \cdot g(x)$, (d) $h(1)$, and (e) $h(3)$.

39. Let $f(x) = \dfrac{2x^2 - 3x - 5}{x + 2}$ and $g(x) = \dfrac{3x^2 + 5x - 2}{x + 1}$. Find (a) $f(1) \cdot g(1)$, (b) $f(-2) \cdot g(-2)$, (c) $h(x) = f(x) \cdot g(x)$, (d) $h(1)$, and (e) $h(-2)$.

40. Let $f(x) = \dfrac{x^2 - 1}{x - 3}$ and $g(x) = \dfrac{x^2 - 9}{x - 1}$. Find (a) $f(2) \cdot g(2)$, (b) $f(3) \cdot g(3)$, (c) $h(x) = f(x) \cdot g(x)$, (d) $h(2)$, and (e) $h(3)$.

41. Let $f(x) = \dfrac{3x^2 + x - 2}{x - 2}$ and $g(x) = \dfrac{x^2 - 4x - 5}{x + 4}$. Find (a) $\dfrac{f(0)}{g(0)}$, (b) $\dfrac{f(1)}{g(1)}$, (c) $h(x) = \dfrac{f(x)}{g(x)}$, and (d) the values of x for which $h(x)$ is undefined.

42. Let $f(x) = \dfrac{x^2 + x}{x - 5}$ and $g(x) = \dfrac{x^2 - x - 6}{x - 5}$. Find (a) $\dfrac{f(0)}{g(0)}$, (b) $\dfrac{f(2)}{g(2)}$, (c) $h(x) = \dfrac{f(x)}{g(x)}$, and (d) the values of x for which $h(x)$ is undefined.

The results from multiplying and dividing rational expressions can be checked by using a graphing calculator. To do this, define one expression in Y_1 and the other in Y_2. Then define the operation in Y_3 as $Y_1 \cdot Y_2$ or $Y_1 \div Y_2$. Put your simplified result in Y_4 (sorry, you still must simplify algebraically). Deselect the graphs for Y_1 and Y_2. If you have correctly simplified the expression, the graphs of Y_3 and Y_4 will be identical. Use this technique in Exercises 43 to 46.

43. $\dfrac{x^3 - 3x^2 + 2x - 6}{x^2 - 9} \cdot \dfrac{5x^2 + 15x}{20x}$

44. $\dfrac{3a^3 + a^2 - 9a - 3}{15a^2 + 5a} \cdot \dfrac{3a^2 + 9}{a^4 - 9}$

45. $\dfrac{x^4 - 16}{x^2 + x - 6} \div (x^3 + 4x)$

46. $\dfrac{w^3 + 27}{w^2 + 2w - 3} \div (w^3 - 3w^2 + 9w)$

SECTION 7.3 Addition and Subtraction of Rational Expressions

7.3 OBJECTIVES

1. Add and subtract rational expressions
2. Add and subtract rational functions

Adding and Subtracting Rational Expressions

Recall that adding or subtracting two arithmetic fractions with the same denominator is straightforward. The same is true in algebra. To add or subtract two rational expressions with the same denominator, we add or subtract their numerators and then write that sum or difference over the common denominator.

Adding or Subtracting Rational Expressions

$$\frac{P}{R} + \frac{Q}{R} = \frac{P+Q}{R}$$

and

$$\frac{P}{R} - \frac{Q}{R} = \frac{P-Q}{R}$$

where $R \neq 0$.

Example 1

Adding and Subtracting Rational Expressions

Perform the indicated operations.

Since we have common denominators, we simply perform the indicated operations on the numerators.

$$\frac{3}{2a^2} - \frac{1}{2a^2} + \frac{5}{2a^2} = \frac{3 - 1 + 5}{2a^2}$$

$$= \frac{7}{2a^2}$$

✓ **CHECK YOURSELF 1**

Perform the indicated operations.

$$\frac{5}{3y^2} + \frac{4}{3y^2} - \frac{7}{3y^2}$$

The sum or difference of rational expressions should always be expressed in simplest form. Consider Example 2.

Example 2 — Adding and Subtracting Rational Expressions

Add or subtract as indicated.

(a) $\dfrac{5x}{x^2 - 9} + \dfrac{15}{x^2 - 9}$ Add the numerators.

$= \dfrac{5x + 15}{x^2 - 9}$

$= \dfrac{5(x + 3)}{(x - 3)(x + 3)} = \dfrac{5}{x - 3}$ Factor and divide by the common factor.

(b) $\dfrac{3x + y}{2x} - \dfrac{x - 3y}{2x} = \dfrac{(3x + y) - (x - 3y)}{2x}$ Be sure to *enclose the second numerator* in parentheses.

$= \dfrac{3x + y - x + 3y}{2x}$ Remove the parentheses by *changing each sign*.

$= \dfrac{2x + 4y}{2x} = \dfrac{2(x + 2y)}{2x}$ Factor and divide by the common factor of 2.

$= \dfrac{x + 2y}{x}$

✓ CHECK YOURSELF 2

Perform the indicated operations.

(a) $\dfrac{6a}{a^2 - 2a - 8} + \dfrac{12}{a^2 - 2a - 8}$ (b) $\dfrac{5x - y}{3y} - \dfrac{2x - 4y}{3y}$

By **inspection**, we mean you look at the denominators and find that the LCD is obvious (as in Example 2).

Now, what if our rational expressions *do not* have common denominators? In that case, we must use the least common denominator (LCD). The **least common denominator** is the simplest polynomial that is divisible by each of the individual denominators. Each expression in the desired sum or difference is then "built up" to an equivalent expression having that LCD as a denominator. We can then add or subtract as before.

Although in many cases we can find the LCD by inspection, we can state an algorithm for finding the LCD that is similar to the one used in arithmetic.

Section 7.3 ■ Addition and Subtraction of Rational Expressions

Finding the Least Common Denominator

Step 1 Write each of the denominators in completely factored form.
Step 2 Write the LCD as the product of each prime factor to the highest power to which it appears in the factored form of any individual denominators.

Again, we see the key role that factoring plays in the process of working with rational expressions.

Example 3 illustrates the procedure.

Example 3

Finding the LCD for Two Rational Expressions

Find the LCD for each of the following pairs of rational expressions.

(a) $\dfrac{3}{4x^2}$ and $\dfrac{5}{6xy}$

Factor the denominators.

$$4x^2 = 2^2 \cdot x^2$$
$$6y = 2 \cdot 3 \cdot x \cdot y$$

You may very well be able to find this LCD by inspecting the numerical coefficients and the variable factors.

The LCD must have the factors

$$2^2 \cdot 3 \cdot x^2 \cdot y$$

and so $12x^2y$ is the desired LCD.

(b) $\dfrac{7}{x-3}$ and $\dfrac{2}{x+5}$

Here, neither denominator can be factored. The LCD must have the factors $x - 3$ and $x + 5$. So the LCD is

$$(x - 3)(x + 5)$$

It is generally best to leave the LCD in this factored form.

✓ **CHECK YOURSELF 3**

Find the LCD for the following pairs of rational expressions.

(a) $\dfrac{3}{8a^3}$ and $\dfrac{5}{6a^2}$ (b) $\dfrac{4}{x+7}$ and $\dfrac{3}{x-5}$

Example 4

Finding the LCD for Two Rational Expressions

Find the LCD for the following pairs of rational expressions.

(a) $\dfrac{2}{x^2 - x - 6}$ and $\dfrac{1}{x^2 - 9}$

Factoring, we have

$$x^2 - x - 6 = (x + 2)(x - 3)$$

and

$$x^2 - 9 = (x + 3)(x - 3)$$

The LCD of the given expressions is then

The LCD must contain each of the factors appearing in the original denominators.

$$(x + 2)(x - 3)(x + 3)$$

(b) $\dfrac{5}{x^2 - 4x + 4}$ and $\dfrac{3}{x^2 + 2x - 8}$

Again, we factor:

$$x^2 - 4x + 4 = (x - 2)^2$$
$$x^2 + 2x - 8 = (x - 2)(x + 4)$$

The LCD must contain $(x - 2)^2$ as a factor since $x - 2$ appears twice as a factor in the first denominator.

The LCD is then

$$(x - 2)^2(x + 4)$$

✓ CHECK YOURSELF 4

Find the LCD for the following pairs of rational expressions.

(a) $\dfrac{3}{x^2 - 2x - 15}$ and $\dfrac{5}{x^2 - 25}$

(b) $\dfrac{5}{y^2 + 6y + 9}$ and $\dfrac{3}{y^2 - y - 12}$

Section 7.3 ■ Addition and Subtraction of Rational Expressions

Let's look at Example 5, in which the concept of the LCD is applied in adding or subtracting rational expressions.

Example 5

Adding and Subtracting Rational Expressions

Add or subtract as indicated.

(a) $\dfrac{5}{4xy} + \dfrac{3}{2x^2}$

The LCD for $2x^2$ and $4xy$ is $4x^2y$. We rewrite each of the rational expressions with the LCD as a denominator.

Note that in each case we are multiplying by 1: $\dfrac{x}{x}$ in the first fraction and $\dfrac{2y}{2y}$ in the second fraction, which is why the resulting fractions are equivalent to the original ones.

$\dfrac{5}{4xy} + \dfrac{3}{2x^2} = \dfrac{5 \cdot x}{4xy \cdot x} + \dfrac{3 \cdot 2y}{2x^2 \cdot 2y}$

$= \dfrac{5x}{4x^2y} + \dfrac{6y}{4x^2y} = \dfrac{5x + 6y}{4x^2y}$

Multiply the first rational expression by $\dfrac{x}{x}$ and the second by $\dfrac{2y}{2y}$ to form the LCD of $4x^2y$.

(b) $\dfrac{3}{a-3} - \dfrac{2}{a}$

The LCD for a and $a - 3$ is $a(a - 3)$. We rewrite each of the rational expressions with that LCD as a denominator.

$$\dfrac{3}{a-3} - \dfrac{2}{a}$$

$$= \dfrac{3a}{a(a-3)} - \dfrac{2(a-3)}{a(a-3)} \quad \text{Subtract the numerators.}$$

$$= \dfrac{3a - 2(a-3)}{a(a-3)} \quad \text{Remove the parentheses, and combine like terms.}$$

$$= \dfrac{3a - 2a + 6}{a(a-3)} = \dfrac{a+6}{a(a-3)}$$

✓ **CHECK YOURSELF 5**

Perform the indicated operations.

(a) $\dfrac{3}{2ab} + \dfrac{4}{5b^2}$ (b) $\dfrac{5}{y+2} - \dfrac{3}{y}$

Example 6 — Adding and Subtracting Rational Expressions

Add or subtract as indicated.

(a) $\dfrac{-5}{x^2 - 3x - 4} + \dfrac{8}{x^2 - 16}$

We first factor the two denominators.

$$x^2 - 3x - 4 = (x + 1)(x - 4)$$
$$x^2 - 16 = (x + 4)(x - 4)$$

We see that the LCD must be

$$(x + 1)(x + 4)(x - 4)$$

Again, rewriting the original expressions with factored denominators gives

We use the facts that

$\dfrac{x + 4}{x + 4} = 1$ and

$\dfrac{x + 1}{x + 1} = 1$

$\dfrac{-5}{(x + 1)(x - 4)} + \dfrac{8}{(x - 4)(x + 4)}$

$= \dfrac{-5(x + 4)}{(x + 1)(x - 4)(x + 4)} + \dfrac{8(x + 1)}{(x - 4)(x + 4)(x + 1)}$

$= \dfrac{-5(x + 4) + 8(x + 1)}{(x + 1)(x - 4)(x + 4)}$ Now add the numerators.

$= \dfrac{-5x - 20 + 8x + 8}{(x + 1)(x - 4)(x + 4)}$ Combine like terms in the numerator.

$= \dfrac{3x - 12}{(x + 1)(x - 4)(x + 4)}$ Factor.

$= \dfrac{3(x - 4)}{(x + 1)(x - 4)(x + 4)}$ Divide by the common factor $x - 4$.

$= \dfrac{3}{(x + 1)(x + 4)}$

(b) $\dfrac{5}{x^2 - 5x + 6} - \dfrac{3}{4x - 12}$

Again, factor the denominators.

$$x^2 - 5x + 6 = (x - 2)(x - 3)$$
$$4x - 12 = 4(x - 3)$$

The LCD is $4(x - 2)(x - 3)$, and proceeding as before, we have

$$\frac{5}{(x-2)(x-3)} - \frac{3}{4(x-3)}$$

$$= \frac{5 \cdot 4}{4(x-2)(x-3)} - \frac{3(x-2)}{4(x-2)(x-3)}$$

$$= \frac{20 - 3(x-2)}{4(x-2)(x-3)}$$

$$= \frac{20 - 3x + 6}{4(x-2)(x-3)} = \frac{-3x + 26}{4(x-2)(x-3)} \quad \text{Simplify the numerator and combine like terms.}$$

✓ CHECK YOURSELF 6

Add or subtract as indicated.

(a) $\dfrac{-4}{x^2 - 4} + \dfrac{7}{x^2 - 3x - 10}$ (b) $\dfrac{5}{3x - 9} - \dfrac{2}{x^2 - 9}$

Example 7 looks slightly different from those you have seen thus far, but the reasoning involved in performing the subtraction is exactly the same.

Example 7 Subtracting Rational Expressions

Subtract.

$$3 - \frac{5}{2x - 1}$$

To perform the subtraction, remember that 3 is equivalent to the fraction $\dfrac{3}{1}$, so

$$3 - \frac{5}{2x - 1} = \frac{3}{1} - \frac{5}{2x - 1}$$

The LCD for 1 and $2x - 1$ is just $2x - 1$. We now rewrite the first expression with that denominator.

$$3 - \frac{5}{2x - 1} = \frac{3(2x - 1)}{2x - 1} - \frac{5}{2x - 1}$$

$$= \frac{3(2x - 1) - 5}{2x - 1} = \frac{6x - 8}{2x - 1}$$

✓ CHECK YOURSELF 7

Subtract.

$$\frac{4}{3x+1} - 3$$

Example 8 uses an observation from Section 7.1. Recall that

$$a - b = -(b - a)$$
$$= -1(b - a)$$

Example 8 — Adding and Subtracting Rational Expressions

Add.

$$\frac{x^2}{x-5} + \frac{3x+10}{5-x}$$

Your first thought might be to use a denominator of $(x - 5)(5 - x)$. However, we can simplify our work considerably if we multiply the numerator and denominator of the second fraction by -1 to find a common denominator.

$$\frac{x^2}{x-5} + \frac{3x+10}{5-x}$$

$$= \frac{x^2}{x-5} + \frac{(-1)(3x+10)}{(-1)(5-x)}$$

$$= \frac{x^2}{x-5} + \frac{-3x-10}{x-5}$$

$$= \frac{x^2 - 3x - 10}{x-5} = \frac{(x+2)(x-5)}{x-5}$$

$$= x + 2$$

Use

$$\frac{-1}{-1} = 1$$

Note that

$$(-1)(5 - x) = x - 5$$

The fractions now have a common denominator, and we can add as before.

CHECK YOURSELF 8

Add.

$$\frac{x^2}{x-7} + \frac{10x-21}{7-x}$$

Adding Rational Functions

The sum of two rational functions is always a rational function. Given two rational functions, $f(x)$ and $g(x)$, we can rename the sum, so $h(x) = f(x) + g(x)$. This will always be true for values of x for which both f and g are defined. So, for example, $h(-2) = f(-2) + g(-2)$, so long as both $f(-2)$ and $g(-2)$ exist.

> If
> $$h(x) = f(x) + g(x)$$
> then the set of ordered pairs
> $$(x, f(x) + g(x)) = (x, h(x))$$

Example 9 — Adding Two Rational Functions

Given

$$f(x) = \frac{3x}{x+5} \quad \text{and} \quad g(x) = \frac{x}{x-4}$$

complete the following.

(a) Find $f(1) + g(1)$.

If $f(1) = \frac{1}{2}$ and $g(1) = -\frac{1}{3}$, then

$$f(1) + g(1) = \frac{1}{2} + \left(-\frac{1}{3}\right)$$
$$= \frac{3}{6} + \left(-\frac{2}{6}\right)$$
$$= \frac{1}{6}$$

(b) Find $h(x) = f(x) + g(x)$.

$$h(x) = f(x) + g(x)$$
$$= \frac{3x}{x+5} + \frac{x}{x-4}$$
$$= \frac{3x(x-4) + x(x+5)}{(x+5)(x-4)} = \frac{3x^2 - 12x + x^2 + 5x}{(x+5)(x-4)}$$
$$= \frac{4x^2 - 7x}{(x+5)(x-4)} \quad x \neq -5, 4$$

(c) Find the ordered pair $(1, h(1))$.

$$h(1) = \frac{-3}{-18} = \frac{1}{6}$$

The ordered pair is $\left(1, \frac{1}{6}\right)$.

✓ CHECK YOURSELF 9

Given

$$f(x) = \frac{x}{2x - 5} \quad \text{and} \quad g(x) = \frac{2x}{3x - 1}$$

complete the following.

(a) Find $f(1) + g(1)$.

(b) Find $h(x) = f(x) + g(x)$.

(c) Find the ordered pair $(1, h(1))$.

Subtracting Rational Functions

When subtracting rational functions, one must take particular care with the signs in the numerator of the expression being subtracted.

Example 10 Subtracting Rational Functions

Given

$$f(x) = \frac{3x}{x + 5} \quad \text{and} \quad g(x) = \frac{x - 2}{x - 4}$$

complete the following.

(a) Find $f(1) - g(1)$.

If $f(1) = \frac{1}{2}$ and $g(1) = \frac{1}{3}$, then

$$f(1) + g(1) = \frac{1}{2} - \left(\frac{1}{3}\right)$$

$$= \frac{3}{6} - \left(\frac{2}{6}\right)$$

$$= \frac{1}{6}$$

(b) Find $h(x) = f(x) - g(x)$.

$$h(x) = \frac{3x}{x+5} - \frac{x-2}{x-4}$$

$$= \frac{3x(x-4) - (x-2)(x+5)}{(x+5)(x-4)}$$

$$= \frac{(3x^2 - 12x) - (x^2 + 3x - 10)}{(x+5)(x-4)}$$

$$= \frac{2x^2 - 15x + 10}{(x+5)(x-4)} \qquad x \neq -5, 4$$

(c) Find the ordered pair $(1, h(1))$.

$$h(1) = \frac{-3}{-18} = \frac{1}{6}$$

The ordered pair is $\left(1, \dfrac{1}{6}\right)$.

✓ CHECK YOURSELF 10

Given

$$f(x) = \frac{x}{2x-5} \quad \text{and} \quad g(x) = \frac{2x-1}{3x-1}$$

complete the following.

(a) Find $f(1) - g(1)$. (b) Find $h(x) = f(x) - g(x)$.

(c) Find the ordered pair $(1, h(1))$.

✓ CHECK YOURSELF ANSWERS

1. $\dfrac{2}{3y^2}$. **2.** (a) $\dfrac{6}{a-4}$; (b) $\dfrac{x+y}{y}$. **3.** (a) $24a^3$; (b) $(x+7)(x-5)$.

4. (a) $(x-5)(x+5)(x+3)$; (b) $(y+3)^2(y-4)$. **5.** (a) $\dfrac{8a+15b}{10ab^2}$;

(b) $\dfrac{2y-6}{y(y+2)}$. **6.** (a) $\dfrac{3}{(x-2)(x-5)}$; (b) $\dfrac{5x+9}{3(x+3)(x-3)}$. **7.** $\dfrac{-9x+1}{3x+1}$.

8. $x-3$. **9.** (a) $\dfrac{2}{3}$; (b) $h(x) = \dfrac{7x^2 - 11x}{(2x-5)(3x-1)}$ $x \neq \dfrac{5}{2}, \dfrac{1}{3}$; (c) $\left(1, \dfrac{2}{3}\right)$.

10. (a) $-\dfrac{5}{6}$ (b) $h(x) = \dfrac{-x^2 + 11x - 5}{(2x-5)(3x-1)}$ $x \neq \dfrac{5}{2}, \dfrac{1}{3}$; (c) $\left(1, -\dfrac{5}{6}\right)$.

Probability and Pari-Mutuel Betting. In most gambling games, payoffs are determined by the **odds**. At horse and dog tracks, the odds (D) are a ratio that is calculated by taking into account the total amount wagered (A), the amount wagered on a particular animal (a), and the government share, called the take-out (f). The ratio is then rounded down to a comparison of integers like 99 to 1, 3 to 1, or 5 to 2. Below is the formula that tracks use to find odds.

$$(D) = \frac{A(1-f)}{a} - 1$$

Work with a partner to complete the following.

1. Assume that the government takes 10%, and simplify the expression for D. Use this formula to compute the odds on each horse if a total of $10,000 were bet on all the horses and the amounts were distributed as shown in the table.

Horse	Total Amount Wagered on This Horse to Win	Odds: Amount Paid on Each Dollar Bet if Horse Wins
1	$5000	
2	$1000	
3	$2000	
4	$1500	
5	$ 500	

2. Odds can be used as a guide in determining the chance that a given horse will win. The probability of a horse winning is related to many variables, such as track condition, how the horse is feeling, and weather. However, the odds do reflect the consensus opinion of racing fans and can be used to give some idea of the probability. The relationship between odds and probability is given by the equations

   $$P(\text{win}) = \frac{1}{D+1}$$

   and $P(\text{loss}) = 1 - P(\text{win})$

   or $P(\text{loss}) = 1 - \frac{1}{D+1}$

   Solve this equation for D, the odds against the horse winning. Do the probabilities for each horse winning all add up to 1? Should they add to 1?

Exercises • 7.3

Answers (left column, red):

1. $\dfrac{6}{x^2}$
2. $\dfrac{3}{b^3}$
3. $\dfrac{7}{3a+7}$
4. $\dfrac{3}{5x+3}$
5. 2
6. 7
7. $\dfrac{y-1}{2}$
8. $\dfrac{x+3}{4}$
9. 2
10. 4
11. $\dfrac{3}{x+2}$
12. $\dfrac{2}{x-3}$
13. $\dfrac{19}{6x}$
14. $\dfrac{1}{20w}$
15. $\dfrac{3(2a+1)}{a^2}$
16. $\dfrac{3p-7}{p^2}$
17. $\dfrac{2(n-m)}{mn}$
18. $\dfrac{3(x+y)}{xy}$
19. $\dfrac{9b-20}{12b^3}$
20. $\dfrac{8-15x}{10x^3}$
21. $\dfrac{a-4}{a(a-2)}$
22. $\dfrac{7c+4}{c(c+1)}$
23. $\dfrac{5x+7}{(x+1)(x+2)}$
24. $\dfrac{2(3y+5)}{(y+3)(y-1)}$

In Exercises 1 to 36, perform the indicated operations. Express your results in simplest form.

1. $\dfrac{7}{2x^2} + \dfrac{5}{2x^2}$
2. $\dfrac{11}{3b^3} - \dfrac{2}{3b^3}$
3. $\dfrac{5}{3a+7} + \dfrac{2}{3a+7}$
4. $\dfrac{6}{5x+3} - \dfrac{3}{5x+3}$
5. $\dfrac{2x}{x-3} - \dfrac{6}{x-3}$
6. $\dfrac{7w}{w+3} + \dfrac{21}{w+3}$
7. $\dfrac{y^2}{2y+8} + \dfrac{3y-4}{2y+8}$
8. $\dfrac{x^2}{4x-12} - \dfrac{9}{4x-12}$
9. $\dfrac{4m-7}{m-5} - \dfrac{2m+3}{m-5}$
10. $\dfrac{3b-8}{b-6} + \dfrac{b-16}{b-6}$
11. $\dfrac{x-7}{x^2-x-6} + \dfrac{2x-2}{x^2-x-6}$
12. $\dfrac{5x-12}{x^2-8x+15} - \dfrac{3x-2}{x^2-8x+15}$
13. $\dfrac{5}{3x} + \dfrac{3}{2x}$
14. $\dfrac{4}{5w} - \dfrac{3}{4w}$
15. $\dfrac{6}{a} + \dfrac{3}{a^2}$
16. $\dfrac{3}{p} - \dfrac{7}{p^2}$
17. $\dfrac{2}{m} - \dfrac{2}{n}$
18. $\dfrac{3}{x} + \dfrac{3}{y}$
19. $\dfrac{3}{4b^2} - \dfrac{5}{3b^3}$
20. $\dfrac{4}{5x^3} - \dfrac{3}{2x^2}$
21. $\dfrac{2}{a} - \dfrac{1}{a-2}$
22. $\dfrac{4}{c} + \dfrac{3}{c+1}$
23. $\dfrac{2}{x+1} + \dfrac{3}{x+2}$
24. $\dfrac{4}{y-1} + \dfrac{2}{y+3}$

25. $\dfrac{4(y+2)}{(y-3)(y+1)}$

26. $\dfrac{x-19}{(x-1)(x+5)}$

27. $\dfrac{w(3w-11)}{(w-7)(w-2)}$

28. $\dfrac{n(4n-7)}{(n+5)(n-4)}$

29. $\dfrac{7x}{(3x-2)(2x+1)}$

30. $\dfrac{c(20c-17)}{(2c-3)(5c-1)}$

31. $\dfrac{4}{m-7}$

32. $\dfrac{8}{a-5}$

33. $\dfrac{2x+11}{(x+4)(x-4)}$

34. $\dfrac{2y+11}{(y+2)(y+3)}$

35. $\dfrac{3m+1}{(m-1)(m-2)}$

36. $\dfrac{-x-2}{(x+1)(x-1)}$

37. $\dfrac{15}{y-5}$

38. $\dfrac{12}{a-6}$

39. $\dfrac{-12}{x-4}$

40. $\dfrac{6}{p+2}$

41. $\dfrac{5z+14}{(z+2)(z-2)(z+4)}$

42. $\dfrac{7x+29}{(x+5)(x-5)(x+2)}$

---

25. $\dfrac{5}{y-3} - \dfrac{1}{y+1}$

26. $\dfrac{4}{x+5} - \dfrac{3}{x-1}$

27. $\dfrac{2w}{w-7} + \dfrac{w}{w-2}$

28. $\dfrac{3n}{n+5} + \dfrac{n}{n-4}$

29. $\dfrac{3x}{3x-2} - \dfrac{2x}{2x+1}$

30. $\dfrac{5c}{5c-1} + \dfrac{2c}{2c-3}$

31. $\dfrac{6}{m-7} + \dfrac{2}{7-m}$

32. $\dfrac{5}{a-5} - \dfrac{3}{5-a}$

33. $\dfrac{3}{x^2-16} + \dfrac{2}{x-4}$

34. $\dfrac{5}{y^2+5y+6} + \dfrac{2}{y+2}$

35. $\dfrac{4m}{m^2-3m+2} - \dfrac{1}{m-2}$

36. $\dfrac{x}{x^2-1} - \dfrac{2}{x-1}$

As we saw in Section 7.2 exercises, the graphing calculator can be used to check our work. In Exercises 37 to 42, enter the first rational expression in Y_1 and the second in Y_2. In Y_3, you will enter either $Y_1 + Y_2$ or $Y_1 - Y_2$. Enter your algebraically simplified rational expression in Y_4. The graphs of Y_3 and Y_4 will be identical if you have correctly simplified the expression.

37. $\dfrac{6y}{y^2-8y+15} + \dfrac{9}{y-3}$

38. $\dfrac{8a}{a^2-8a+12} + \dfrac{4}{a-2}$

39. $\dfrac{6x}{x^2-10x+24} - \dfrac{18}{x-6}$

40. $\dfrac{21p}{p^2-3p-10} - \dfrac{15}{p-5}$

41. $\dfrac{2}{z^2-4} + \dfrac{3}{z^2+2z-8}$

42. $\dfrac{5}{x^2-3x-10} + \dfrac{2}{x^2-25}$

Section 7.3 • Addition and Subtraction of Rational Expressions

In Exercises 43 to 46, find (a) $f(1) + g(1)$, (b) $h(x) = f(x) + g(x)$, and (c) the ordered pair $(1, h(1))$.

43. $f(x) = \dfrac{3x}{x+1}$ and $g(x) = \dfrac{2x}{x-3}$

44. $f(x) = \dfrac{4x}{x-4}$ and $g(x) = \dfrac{x+4}{x+1}$

45. $f(x) = \dfrac{x}{x+1}$ and $g(x) = \dfrac{1}{x^2+2x+1}$

46. $f(x) = \dfrac{x+2}{x-4}$ and $g(x) = \dfrac{x+3}{x+4}$

In Exercises 47 to 50, find (a) $f(1) - g(1)$, (b) $h(x) = f(x) - g(x)$, and (c) the ordered pair $(1, h(1))$.

47. $f(x) = \dfrac{x+5}{x-5}$ and $g(x) = \dfrac{x-5}{x+5}$

48. $f(x) = \dfrac{2x}{x-4}$ and $g(x) = \dfrac{3x}{x+7}$

49. $f(x) = \dfrac{x+9}{4x-36}$ and $g(x) = \dfrac{x-9}{x^2-18x+81}$

50. $f(x) = \dfrac{4x+1}{x+5}$ and $g(x) = -\dfrac{2}{x}$

In Exercises 51 to 60, evaluate each expression at the given variable value(s).

51. $\dfrac{5x+5}{x^2+3x+2} - \dfrac{x-3}{x^2+5x-6}$, $x = -4$

52. $\dfrac{y-3}{y^2-6y+8} + \dfrac{2y-6}{y^2-4}$, $y = 3$

53. $\dfrac{2m+2n}{m^2-n^2} + \dfrac{m-2n}{m^2+2mn+n^2}$, $m=3, n=2$

54. $\dfrac{w-3z}{w^2-2wz+z^2} - \dfrac{w+2z}{w^2-z^2}$, $w=2, z=1$

55. $\dfrac{1}{a-3} - \dfrac{1}{a+3} + \dfrac{2a}{a^2-9}$, $a=4$

56. $\dfrac{1}{m+1} + \dfrac{1}{m-3} - \dfrac{4}{m^2-2m-3}$, $m=-2$

57. $\dfrac{3w^2+16w-8}{w^2+2w-8} + \dfrac{w}{w+4} - \dfrac{w-1}{w-2}$, $w=3$

58. $\dfrac{4x^2-7x-45}{x^2-6x+5} - \dfrac{x+2}{x-1} - \dfrac{x}{x-5}$, $x=-3$

59. $\dfrac{a^2-9}{2a^2-5a-3} \cdot \left(\dfrac{1}{a-2} + \dfrac{1}{a+3}\right)$, $a=-3$

60. $\dfrac{m^2-2mn+n^2}{m^2+2mn-3n^2} \cdot \left(\dfrac{2}{m-n} - \dfrac{1}{m+n}\right)$, $m=4, n=-3$

---

43. (a) $\dfrac{1}{2}$; (b) $\dfrac{5x^2-7x}{(x+1)(x-3)}$; (c) $\left(1, \dfrac{1}{2}\right)$

44. (a) $\dfrac{7}{6}$; (b) $\dfrac{5x^2+4x-16}{(x-4)(x+1)}$; (c) $\left(1, \dfrac{7}{6}\right)$

45. (a) $\dfrac{3}{4}$; (b) $\dfrac{x^2+x+1}{(x+1)^2}$; (c) $\left(1, \dfrac{3}{4}\right)$

46. (a) $\dfrac{-1}{5}$; (b) $\dfrac{2x^2+5x-4}{x^2-16}$; (c) $\left(1, \dfrac{-1}{5}\right)$

47. (a) $-\dfrac{5}{6}$; (b) $\dfrac{20x}{(x-5)(x+5)}$; (c) $\left(1, \dfrac{5}{6}\right)$

48. (a) $\dfrac{-25}{24}$; (b) $\dfrac{-x^2+26x}{(x-4)(x+7)}$; (c) $\left(1, \dfrac{-25}{24}\right)$

49. (a) $-\dfrac{3}{16}$; (b) $\dfrac{(x+5)}{4(x-9)}$; (c) $\left(1, -\dfrac{3}{16}\right)$

50. (a) $\dfrac{17}{6}$; (b) $\dfrac{4x^2+3x+10}{x(x+5)}$; (c) $\left(1, \dfrac{17}{6}\right)$

51. 1   **52.** 0
53. $\dfrac{49}{25}$ **54.** $-\dfrac{7}{3}$
55. 2   **56.** −2
57. 8   **58.** $\dfrac{-1}{4}$
59. Undefined
60. 1

Complex Fractions

7.4 OBJECTIVES

1. Use the fundamental principle to simplify complex fractions
2. Use division to simplify complex fractions

Fundamental principle:

$$\frac{P}{Q} = \frac{PR}{QR}$$

where $Q \neq 0$ and $R \neq 0$.

Again, we are multiplying by $\frac{10}{10}$ or 1.

Our work in this section deals with two methods for simplifying complex fractions. We begin with a definition. A **complex fraction** is a fraction that has a fraction in its numerator or denominator (or both). Some examples are

$$\frac{\dfrac{5}{6}}{\dfrac{3}{4}} \qquad \frac{\dfrac{4}{x}}{\dfrac{3}{x+1}} \qquad \text{and} \qquad \frac{1 + \dfrac{1}{x}}{1 - \dfrac{1}{x}}$$

Two methods can be used to simplify complex fractions. Method 1 involves the fundamental principle, and Method 2 involves inverting and multiplying.

Method 1 for Simplifying Complex Fractions

Recall that by the *fundamental principle* we can always multiply the numerator and denominator of a fraction by the same nonzero quantity. In simplifying a complex fraction, we multiply the numerator and denominator by the LCD of all fractions that appear within the complex fraction.

Here the denominators are 5 and 10, so we can write

$$\frac{\dfrac{3}{5}}{\dfrac{7}{10}} = \frac{\dfrac{3}{5} \cdot 10}{\dfrac{7}{10} \cdot 10} = \frac{6}{7}$$

Method 2 for Simplifying Complex Fractions

Our second approach interprets the complex fraction as indicating division and applies our earlier work in dividing fractions in which we *invert and multiply*.

$$\frac{\dfrac{3}{5}}{\dfrac{7}{10}} = \frac{3}{5} \div \frac{7}{10} = \frac{3}{5} \cdot \frac{10}{7} = \frac{6}{7} \qquad \text{Invert and multiply.}$$

Which method is better? The answer depends on the expression you are trying to simplify. Both approaches are effective, and you should be familiar with both. With practice you will be able to tell which method may be easier to use in a particular situation.

Section 7.4 ■ Complex Fractions

Let's look at the same two methods applied to the simplification of an algebraic complex fraction.

Example 1 **Simplifying Complex Fractions**

Simplify.

$$\frac{1 + \dfrac{2x}{y}}{2 - \dfrac{x}{y}}$$

Method 1 The LCD of 1, $\dfrac{2x}{y}$, 2, and $\dfrac{x}{y}$ is y. So we multiply the numerator and denominator by y.

$$\frac{1 + \dfrac{2x}{y}}{2 - \dfrac{x}{y}} = \frac{\left(1 + \dfrac{2x}{y}\right) \cdot y}{\left(2 - \dfrac{x}{y}\right) \cdot y} \quad \text{Distribute } y \text{ over the numerator and denominator.}$$

$$= \frac{1 \cdot y + \dfrac{2x}{y} \cdot y}{2 \cdot y - \dfrac{x}{y} \cdot y} \quad \text{Simplify.}$$

$$= \frac{y + 2x}{2y - x}$$

Method 2 In this approach, we must *first work separately* in the numerator and denominator to form single fractions.

Make sure you understand the steps in forming a single fraction in the numerator and denominator.

$$\frac{1 + \dfrac{2x}{y}}{2 - \dfrac{x}{y}} = \frac{\dfrac{y}{y} + \dfrac{2x}{y}}{\dfrac{2y}{y} - \dfrac{x}{y}} = \frac{\dfrac{y + 2x}{y}}{\dfrac{2y - x}{y}}$$

$$= \frac{y + 2x}{y} \cdot \frac{y}{2y - x} \quad \text{Invert the divisor and multiply.}$$

$$= \frac{y + 2x}{2y - x}$$

CHECK YOURSELF 1

Simplify.

$$\frac{\dfrac{x}{y} - 1}{\dfrac{2x}{y} + 2}$$

Again, simplifying a complex fraction means writing an equivalent simple fraction in lowest terms, as Example 2 illustrates.

Example 2 Simplifying Complex Fractions

Simplify.

$$\frac{1 - \dfrac{2y}{x} + \dfrac{y^2}{x^2}}{1 - \dfrac{y^2}{x^2}}$$

We choose the first method of simplification in this case. The LCD of all the fractions that appear is x^2. So we multiply the numerator and denominator by x^2.

$$\frac{1 - \dfrac{2y}{x} + \dfrac{y^2}{x^2}}{1 - \dfrac{y^2}{x^2}} = \frac{\left(1 - \dfrac{2y}{x} + \dfrac{y^2}{x^2}\right) \cdot x^2}{}$$

Distribute x^2 over the numerator and denominator, and simplify.

$$= \frac{x^2 - 2xy + y^2}{x^2 - y^2}$$

Factor the numerator and denominator.

$$= \frac{(x - y)(x - y)}{(x + y)(x - y)} = \frac{x - y}{x + y}$$

Divide by the common factor $x - y$.

✓ CHECK YOURSELF 2

Simplify.

$$\frac{1 + \dfrac{5}{x} + \dfrac{6}{x^2}}{1 - \dfrac{9}{x^2}}$$

In Example 3, we will illustrate the second method of simplification for purposes of comparison.

Example 3 — Simplifying Complex Fractions

Simplify.

$$\frac{1 - \dfrac{1}{x+2}}{x - \dfrac{2}{x-1}}$$

Again, take time to make sure you understand how the numerator and denominator are rewritten as single fractions.

Note: Method 2 is probably the more efficient in this case. The LCD of the denominators would be $(x + 2)(x - 1)$, leading to a somewhat more complicated process if method 1 were used.

$$\frac{1 - \dfrac{1}{x+2}}{x - \dfrac{2}{x-1}} = \frac{\dfrac{x+2}{x+2} - \dfrac{1}{x+2}}{\dfrac{x(x-1)}{x-1} - \dfrac{2}{x-1}} = \frac{\dfrac{x+1}{x+2}}{\dfrac{x^2 - x - 2}{x-1}}$$

$$= \frac{x+1}{x+2} \cdot \frac{x-1}{x^2 - x - 2}$$

$$= \frac{x+1}{x+2} \cdot \frac{x-1}{(x-2)(x+1)}$$

$$= \frac{x-1}{(x+2)(x-2)}$$

✓ CHECK YOURSELF 3

Simplify.

$$\frac{2 + \dfrac{5}{x-3}}{x - \dfrac{1}{2x+1}}$$

The following algorithm summarizes our work with complex fractions.

> **Simplifying Complex Fractions**
>
> **Method 1**
>
> 1. Multiply the numerator and denominator of the complex fraction by the LCD of all the fractions that appear within the numerator and denominator.
>
> 2. Simplify the resulting rational expression, writing the expression in lowest terms.
>
> **Method 2**
>
> 1. Write the numerator and denominator of the complex fraction as single fractions, if necessary.
>
> 2. Invert the denominator and multiply as before, writing the result in lowest terms.

✓ **CHECK YOURSELF ANSWERS**

1. $\dfrac{x-y}{2x+2y}$. 2. $\dfrac{x+2}{x-3}$. 3. $\dfrac{2x+1}{(x-3)(x+1)}$.

Exercises • 7.4

In Exercises 1 to 39, simplify each complex fraction.

Answers (left column):

1. $\dfrac{8}{9}$ 2. $\dfrac{5}{4}$

3. $\dfrac{14}{5}$ 4. 2

5. $\dfrac{5}{6}$ 6. $\dfrac{14}{15}$

7. $\dfrac{1}{2x}$ 8. $\dfrac{3}{2a}$

9. $\dfrac{m}{2}$ 10. $\dfrac{3x}{4}$

11. $\dfrac{2(y+1)}{y-1}$

12. $\dfrac{x+3}{2(x-3)}$

13. $\dfrac{3b}{a^2}$ 14. $\dfrac{2n}{m^2}$

15. $\dfrac{x}{(x+2)(x-3)}$

16. $\dfrac{x+6}{x(x-5)}$

17. $\dfrac{2x-1}{2x+1}$

18. $\dfrac{3b+1}{3b-1}$

19. $y-x$

20. $\dfrac{1}{b+a}$ 21. $\dfrac{x-y}{y}$

22. $\dfrac{n}{m-2n}$

23. $\dfrac{a+4}{a+3}$ 24. $\dfrac{x-4}{x-3}$

25. $\dfrac{x^2 y(x+y)}{(x-y)}$

26. $\dfrac{ab(a-b)}{a-2b}$

27. $\dfrac{x}{x-2}$

Exercises:

1. $\dfrac{\frac{2}{3}}{\frac{6}{8}}$ 2. $\dfrac{\frac{5}{6}}{\frac{10}{15}}$ 3. $\dfrac{\frac{2}{3}+\frac{1}{2}}{\frac{3}{4}-\frac{1}{3}}$

4. $\dfrac{\frac{3}{4}+\frac{1}{2}}{\frac{7}{8}-\frac{1}{4}}$ 5. $\dfrac{2+\frac{1}{3}}{3-\frac{1}{5}}$ 6. $\dfrac{1+\frac{3}{4}}{2-\frac{1}{8}}$

7. $\dfrac{\frac{x}{8}}{\frac{x^2}{4}}$ 8. $\dfrac{\frac{a^2}{10}}{\frac{a^3}{15}}$ 9. $\dfrac{\frac{3}{m}}{\frac{6}{m^2}}$

10. $\dfrac{\frac{15}{x^2}}{\frac{20}{x^3}}$ 11. $\dfrac{\frac{y+1}{y}}{\frac{y-1}{2y}}$ 12. $\dfrac{\frac{x+3}{4x}}{\frac{x-3}{2x}}$

13. $\dfrac{\frac{a+2b}{3a}}{\frac{a^2+2ab}{9b}}$ 14. $\dfrac{\frac{m-3n}{4m}}{\frac{m^2-3mn}{8n}}$ 15. $\dfrac{\frac{x-2}{x^2-9}}{\frac{x^2-4}{x^2+3x}}$

16. $\dfrac{\frac{x+5}{x^2-6x}}{\frac{x^2-25}{x^2-36}}$ 17. $\dfrac{2-\frac{1}{x}}{2+\frac{1}{x}}$ 18. $\dfrac{3+\frac{1}{b}}{3-\frac{1}{b}}$

19. $\dfrac{\frac{1}{x}-\frac{1}{y}}{\frac{1}{xy}}$ 20. $\dfrac{\frac{1}{ab}}{\frac{1}{a}+\frac{1}{b}}$ 21. $\dfrac{\frac{x^2}{y^2}-1}{\frac{x}{y}+1}$

22. $\dfrac{\frac{m}{n}+2}{\frac{m^2}{n^2}-4}$ 23. $\dfrac{1+\frac{3}{a}-\frac{4}{a^2}}{1+\frac{2}{a}-\frac{3}{a^2}}$ 24. $\dfrac{1-\frac{2}{x}-\frac{8}{x^2}}{1-\frac{1}{x}-\frac{6}{x^2}}$

25. $\dfrac{\frac{x^2}{y}+2x+y}{\frac{1}{y^2}-\frac{1}{x^2}}$ 26. $\dfrac{\frac{a}{b}+1-\frac{2b}{a}}{\frac{1}{b^2}-\frac{4}{a^2}}$ 27. $\dfrac{1+\frac{1}{x-1}}{1-\frac{1}{x-1}}$

524 Chapter 7 ▪ Rational Expressions

28. $\dfrac{2m-5}{2m-3}$

29. $\dfrac{y+2}{(y-1)(y+4)}$

30. $\dfrac{x-3}{(x+2)(x-6)}$

31. $\dfrac{x}{3}$ 32. $\dfrac{3m-8}{m-4}$

33. 1

34. $\dfrac{y-1}{3}$

35. $\dfrac{2a}{a^2+1}$

36. $\dfrac{4x}{x^2+4}$

37. $\dfrac{2x+1}{x+1}$

38. $\dfrac{2y-1}{y-1}$

39. $\dfrac{3x+2}{2x+1}$

40. $1 + \dfrac{1}{1 + \dfrac{1}{1 + \dfrac{1}{1 + \dfrac{1}{x}}}}$

41. $\dfrac{5x+3}{3x+2}$

---

28. $\dfrac{2 - \dfrac{1}{m-2}}{2 + \dfrac{1}{m-2}}$

29. $\dfrac{1 - \dfrac{1}{y-1}}{y - \dfrac{8}{y+2}}$

30. $\dfrac{1 + \dfrac{1}{x+2}}{x - \dfrac{18}{x-3}}$

31. $\dfrac{\dfrac{1}{x-3} + \dfrac{1}{x+3}}{\dfrac{1}{x-3} - \dfrac{1}{x+3}}$

32. $\dfrac{\dfrac{2}{m-2} + \dfrac{1}{m-3}}{\dfrac{2}{m-2} - \dfrac{1}{m-3}}$

33. $\dfrac{\dfrac{x}{x+1} + \dfrac{1}{x-1}}{\dfrac{x}{x-1} - \dfrac{1}{x+1}}$

34. $\dfrac{\dfrac{y}{y-4} + \dfrac{1}{y+2}}{\dfrac{4}{y-4} - \dfrac{1}{y+2}}$

35. $\dfrac{\dfrac{a+1}{a-1} - \dfrac{a-1}{a+1}}{\dfrac{a+1}{a-1} + \dfrac{a-1}{a+1}}$

36. $\dfrac{\dfrac{x+2}{x-2} - \dfrac{x-2}{x+2}}{\dfrac{x+2}{x-2} + \dfrac{x-2}{x+2}}$

37. $1 + \dfrac{1}{1 + \dfrac{1}{x}}$

38. $1 + \dfrac{1}{1 - \dfrac{1}{y}}$

39. $1 + \dfrac{1}{1 + \dfrac{1}{1 + \dfrac{1}{x}}}$

40. Extend the "continued fraction" patterns in Exercises 37 and 39 to write the next complex fraction.

41. Simplify the complex fraction in Exercise 40.

42. Outline the two different methods used to simplify a complex fraction. What are the advantages of each method?

43. Can the expression $\dfrac{x^2+y^2}{x+y}$ be written as $\dfrac{x^2}{x} + \dfrac{y^2}{y}$? If not, is there a correct simplified form?

44. Write and simplify a complex fraction that is the reciprocal of $x + \dfrac{6}{x-1}$.

45. Let $f(x) = \dfrac{3}{x}$. Write and simplify a complex fraction whose numerator is $f(3+h) - f(3)$ and whose denominator is h.

46. Write and simplify a complex fraction that is the arithmetic mean of $\dfrac{1}{x}$ and $\dfrac{1}{x-1}$.

47. Compare your results in Exercises 37, 39, and 41. Could you have predicted the result?

Suppose you drive at 40 mi/h from city A to city B. You then return along the same route from city B to city A at 50 mi/h. What is your average rate for the round trip? Your obvious guess would be 45 mi/h, but you are in for a surprise.

Suppose that the cities are 200 mi apart. Your time from city A to city B is the distance divided by the rate, or

$$\dfrac{200 \text{ mi}}{40 \text{ mi/h}} = 5 \text{ h}$$

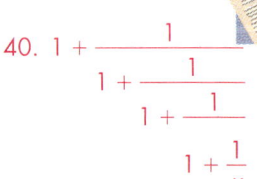
44. $\dfrac{1}{x + \dfrac{6}{x-1}} = \dfrac{x-1}{x^2 - x + 6}$

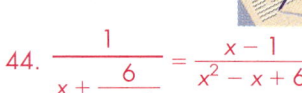
45. $\dfrac{\dfrac{3}{3+h} - 1}{h} = \dfrac{-1}{3+h}$

46. $\dfrac{\dfrac{1}{x} + \dfrac{1}{x-1}}{2} = \dfrac{2x-1}{2x(x-1)}$

49. $44\frac{4}{9}$ mi/h

50. $54\frac{6}{11}$ mi/h

51. $54\frac{6}{9}$ mi/h

Similarly, your time from city B to city A is

$$\frac{200 \text{ mi}}{50 \text{ mi/h}} = 4 \text{ h}$$

The total time is then 9 h, and now using *rate equals distance divided by time,* we have

$$\frac{400 \text{ mi}}{9 \text{ h}} = \frac{400}{9} \text{ mi/h} = 44\frac{4}{9} \text{ mi/h}$$

Note that the rate for the round trip is independent of the distance involved. For instance, try the same computations above if cities A and B are 400 mi apart.

The answer to the problem above is the complex fraction

$$R = \frac{2}{\frac{1}{R_1} + \frac{1}{R_2}}$$

where

R_1 = rate going

R_2 = rate returning

R = rate for round trip

Use this information to solve Exercises 48 to 51.

48. Verify that if $R_1 = 40$ mi/h and $R_2 = 50$ mi/h, then $R = 44\frac{4}{9}$ mi/h, by simplifying the complex fraction *after* substituting those values.

49. Simplify the given complex fraction first. *Then* substitute 40 for R_1 and 50 for R_2 to calculate R.

50. Repeat Exercise 48, where $R_1 = 50$ mi/h and $R_2 = 60$ mi/h.

51. Use the procedure in Exercise 49 with the above values for R_1 and R_2.

52. Mathematicians have shown that there are situations in which the method used to determine the number of U.S. representatives each state gets may not be fair, and a state may not get its basic quota of representatives. They give the table below of a hypothetical seven states and their populations as an example.

State	Population	Exact Quota	Number of Reps.
A	325	1.625	2
B	788	3.940	4
C	548	2.740	3
D	562	2.810	3
E	4,263	21.315	21
F	3,219	16.095	15
G	295	1.475	2
Total	10,000	50	50

In this case, the total population of all states is 10,000, and there are 50 representatives in all, so there should be no more than 10,000/50 or 200 people per representative. The quotas are found by dividing the population by 200. Whether a state, A, should get an additional representative before another state, E, should get one is decided in this method by using the simplified inequality below. If the ratio

$$\frac{A}{\sqrt{a(a+1)}} > \frac{E}{\sqrt{e(e+1)}}$$

is true, then A gets an extra representative before E does.

(a) If you go through the process of comparing the inequality above for each pair of states, state F loses a representative to state G. Do you see how this happens? Will state F complain?

(b) Alexander Hamilton, one of the signers of the Constitution, proposed that the extra representative positions be given one at a time to states with the largest remainder until all the "extra" positions were filled. How would this affect the table? Do you agree or disagree?

53. In Italy in the 1500s, Pietro Antonio Cataldi expressed square roots as infinite, continued fractions. It is not a difficult process to follow. For instance, if you want the square root of 5, then let

$$x + 1 = \sqrt{5}$$

Squaring both sides gives

$$(x + 1)^2 = 5 \quad \text{or} \quad x^2 + 2x + 1 = 5$$

which can be written

$$x(x + 2) = 4$$

$$x = \frac{4}{x + 2}$$

One can continue replacing x with $\frac{4}{x+2}$:

$$x = \cfrac{4}{2 + \cfrac{4}{2 + \cfrac{4}{2 + \cfrac{4}{2 + \cdots}}}}$$

to obtain

$$\sqrt{5} - 1$$

(a) Evaluate the complex fraction from the previous page (ignore the three dots) and then add 1, and see how close it is to the square root of 5. What should you put where the ellipses (. . .) are? Try a number you feel is close to $\sqrt{5}$. How far would you have to go to get the square root correct to the nearest hundredth?

(b) Develop an infinite complex fraction for $\sqrt{10}$.

In Exercises 54 and 55, use the table utility on your graphing calculator to complete the table. Comment on the equivalence of the two expressions.

54.

x	-3	-2	-1	0	1	2	3
$\dfrac{1-\dfrac{2}{x}}{1-\dfrac{4}{x^2}}$							
$\dfrac{x}{x+2}$							

55.

x	-3	-2	-1	0	1	2	3
$\dfrac{-8+\dfrac{20}{x}}{4-\dfrac{25}{x^2}}$							
$\dfrac{-4x}{2x+5}$							

56. Here is yet another method for simplifying a complex fraction. Suppose we want to simplify

$$\dfrac{\dfrac{3}{5}}{\dfrac{7}{10}}$$

Multiply the numerator and denominator of the complex fraction by $\dfrac{10}{7}$.

(a) What principle allows you to do this?

(b) Why was $\dfrac{10}{7}$ chosen?

(c) When learning to divide fractions, you may have heard the saying "Yours is not to reason why . . . just invert and multiply." How does this method serve to explain the "reason why" we invert and multiply?

SECTION 7.5 Rational Equations and Inequalities in One Variable

7.5 OBJECTIVES

1. Solve rational equations in one variable algebraically
2. Solve literal equations involving a rational expression
3. Solve rational inequalities in one variable algebraically

"One person's constant is another person's variable."

–Susan Gerhart

Rational Equations

Applications of your work in algebra will often result in equations involving rational expressions. Our objective in this section is to develop methods to find solutions for such equations.

The usual technique for solving such equations is to multiply both sides of the equation by the lowest common denominator (LCD) of all the rational expressions appearing in the equation. The resulting equation will be cleared of fractions, and we can then proceed to solve the equation as before. Example 1 illustrates the process.

Example 1

Clearing Equations of Fractions

Solve.

$$\frac{2x}{3} + \frac{x}{5} = 13$$

The LCD for 3 and 5 is 15. Multiplying both sides of the equation by 15, we have

$$15\left(\frac{2x}{3} + \frac{x}{5}\right) = 15 \cdot 13 \qquad \text{Distribute 15 on the left.}$$

$$15 \cdot \frac{2x}{3} + 15 \cdot \frac{x}{5} = 15 \cdot 13$$

$$10x + 3x = 195 \qquad \text{Simplify. The equation is now cleared of fractions.}$$

$$13x = 195$$

$$x = 15$$

The solution set is {15}.

528

Section 7.5 ■ Rational Equations and Inequalities in One Variable

To check, substitute 15 in the original equation.

$$\frac{2 \cdot 15}{3} + \frac{15}{5} \stackrel{?}{=} 13$$

$$10 + 3 \stackrel{?}{=} 13$$

$$13 = 13 \quad \text{A true statement.}$$

So 15 is the solution for the equation.

Caution

Be Careful! A common mistake is to confuse an *equation* such as

$$\frac{2x}{3} + \frac{x}{5} = 13$$

and an *expression* such as

$$\frac{2x}{3} + \frac{x}{5}$$

Let's compare.

Equation: $\quad \dfrac{2x}{3} + \dfrac{x}{5} = 13$

Here we want to *solve the equation for x*, as in Example 1. We multiply both sides by the LCD to clear fractions and proceed as before.

Expression: $\quad \dfrac{2x}{3} + \dfrac{x}{5}$

Here we want to find *a third fraction* that is equivalent to the given expression. We write each fraction as an equivalent fraction with the LCD as a common denominator.

$$\frac{2x}{3} + \frac{x}{5} = \frac{2x \cdot 5}{3 \cdot 5} + \frac{x \cdot 3}{5 \cdot 3}$$

$$= \frac{10x}{15} + \frac{3x}{15} = \frac{10x + 3x}{15}$$

$$= \frac{13x}{15}$$

✓ **CHECK YOURSELF 1**

Solve.

$$\frac{3x}{2} - \frac{x}{3} = 7$$

Example 2 Solving an Equation Involving Rational Expressions

Solve.

We assume that x cannot have the value 0. Do you see why?

$$\frac{7}{4x} - \frac{3}{x^2} = \frac{1}{2x^2}$$

The LCD of $4x$, x^2, and $2x^2$ is $4x^2$. So, multiplying both sides by $4x^2$, we have

$$4x^2\left(\frac{7}{4x} - \frac{3}{x^2}\right) = 4x^2 \cdot \frac{1}{2x^2} \quad \text{Distribute } 4x^2 \text{ on the left side.}$$

$$4x^2 \cdot \frac{7}{4x} - 4x^2 \cdot \frac{3}{x^2} = 4x^2 \cdot \frac{1}{2x^2} \quad \text{Simplify.}$$

$$7x - 12 = 2$$

$$7x = 14$$

$$x = 2$$

The solution set is {2}.

We leave the check of the solution, $x = 2$, to you. Be sure to return to the original equation and substitute 2 for x.

 CHECK YOURSELF 2

Solve.

$$\frac{5}{2x} - \frac{4}{x^2} = \frac{7}{2x^2}$$

Example 3 illustrates the same solution process when there are binomials in the denominators.

Example 3 Solving an Equation Involving Rational Expressions

Solve.

Here we assume that x cannot have the value −2 or 3.

$$\frac{4}{x+2} + 3 = \frac{3x}{x-3}$$

Section 7.5 ■ Rational Equations and Inequalities in One Variable

The LCD is $(x + 2)(x - 3)$. Multiplying by that LCD, we have

$$(x + 2)(x - 3)\left(\frac{4}{x + 2}\right) + (x + 2)(x - 3)(3) = (x + 2)(x - 3)\left(\frac{3x}{x - 3}\right)$$

Note that multiplying *each term* by the LCD is the same as multiplying both sides of the equation by the LCD.

Or, simplifying each term, we have

$$4(x - 3) + 3(x + 2)(x - 3) = 3x(x + 2)$$

We now clear the parentheses and proceed as before.

$$4x - 12 + 3x^2 - 3x - 18 = 3x^2 + 6x$$
$$3x^2 + x - 30 = 3x^2 + 6x$$
$$x - 30 = 6x$$
$$-5x = 30$$
$$x = -6$$

The solution set is $\{-6\}$.

Again, we leave the check of this solution to you.

 CHECK YOURSELF 3

Solve.

$$\frac{5}{x - 4} + 2 = \frac{2x}{x - 3}$$

Factoring plays an important role in solving equations containing rational expressions.

Example 4 **Solving an Equation Involving Rational Expressions**

Solve.

$$\frac{3}{x - 3} - \frac{7}{x + 3} = \frac{2}{x^2 - 9}$$

In factored form, the denominator on the right side is $(x - 3)(x + 3)$, which forms the LCD, and we multiply each term by that LCD.

$$(x - 3)(x + 3)\left(\frac{3}{x - 3}\right) - (x - 3)(x + 3)\left(\frac{7}{x + 3}\right) = (x - 3)(x + 3)\left[\frac{2}{(x - 3)(x + 3)}\right]$$

Again, simplifying each term on the right and left sides, we have

$$3(x + 3) - 7(x - 3) = 2$$
$$3x + 9 - 7x + 21 = 2$$
$$-4x = -28$$
$$x = 7$$

The solution set is {7}.

Be sure to check this result by substitution in the original equation.

✓ **CHECK YOURSELF 4**

Solve $\dfrac{4}{x - 4} - \dfrac{3}{x + 1} = \dfrac{5}{x^2 - 3x - 4}$.

Whenever we multiply both sides of an equation by an expression containing a variable, there is the possibility that a proposed solution may make that multiplier 0. As we pointed out earlier, multiplying by 0 does not give an equivalent equation, and therefore verifying solutions by substitution serves not only as a check of our work but also as a check for extraneous solutions. Consider Example 5.

Example 5 — Solving an Equation Involving Rational Expressions

Solve.

Note that we must assume that $x \neq 2$.

$$\dfrac{x}{x - 2} - 7 = \dfrac{2}{x - 2}$$

The LCD is $x - 2$, and multiplying, we have

Note that each of the three terms gets multiplied by $(x - 2)$.

$$\left(\dfrac{x}{x - 2}\right)(x - 2) - 7(x - 2) = \left(\dfrac{2}{x - 2}\right)(x - 2)$$

Simplifying yields

$$x - 7(x - 2) = 2$$
$$x - 7x + 14 = 2$$
$$-6x = -12$$
$$x = 2$$

Caution
Because division by 0 is undefined, we conclude that 2 is *not a solution* for the original equation. It is an extraneous solution. The original equation has no solution.

To check this result, by substituting 2 for x, we have

$$\frac{2}{2-2} - 7 \stackrel{?}{=} \frac{2}{2-2}$$

$$\frac{2}{0} - 7 \stackrel{?}{=} \frac{2}{0}$$

The solution set is empty. The set is written $\{\}$ or $\varnothing$.

✓ CHECK YOURSELF 5

Solve $\dfrac{x-3}{x-4} = 4 + \dfrac{1}{x-4}$.

Equations involving rational expressions may also lead to quadratic equations, as illustrated in Example 6.

Example 6

Solving an Equation Involving Rational Expressions

Solve.

Assume $x \neq 3$ and $x \neq 4$.

$$\frac{x}{x-4} = \frac{15}{x-3} - \frac{2x}{x^2 - 7x + 12}$$

After factoring the trinomial denominator on the right, the LCD of $x - 3$, $x - 4$, and $x^2 - 7x + 12$ is $(x - 3)(x - 4)$. Multiplying by that LCD, we have

$$(x-3)(x-4)\left(\frac{x}{x-4}\right) = (x-3)(x-4)\left(\frac{15}{x-3}\right) - (x-3)(x-4)\left[\frac{2x}{(x-3)(x-4)}\right]$$

Simplifying yields

$x(x - 3) = 15(x - 4) - 2x$ Remove the parentheses.

$x^2 - 3x = 15x - 60 - 2x$ Write in standard form and factor.

$x^2 - 16x + 60 = 0$

$(x - 6)(x - 10) = 0$

So

$$x = 6 \quad \text{or} \quad x = 10$$

Verify that 6 and 10 are both solutions for the original equation. The solution set is $\{6, 10\}$.

✓ CHECK YOURSELF 6

Solve $\dfrac{3x}{x+2} - \dfrac{2}{x+3} = \dfrac{36}{x^2+5x+6}$.

The following algorithm summarizes our work in solving equations containing rational expressions.

> **Solving Equations Containing Rational Expressions**
>
> **Step 1** Clear the equation of fractions by multiplying both sides of the equation by the LCD of all the fractions that appear.
> **Step 2** Solve the equation resulting from step 1.
> **Step 3** Check all solutions by substitution in the original equation.

The method in this section may also be used to solve certain literal equations for a specified variable. Consider Example 7.

Example 7 Solving a Literal Equation

A parallel electric circuit. The symbol for a resistor is

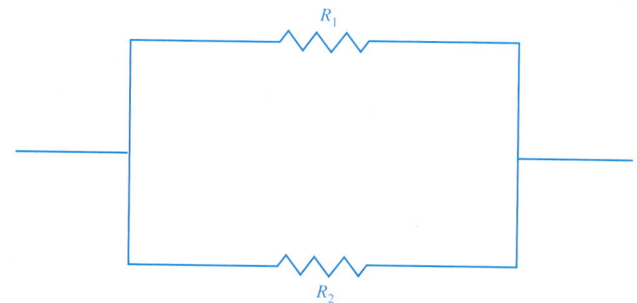

If two resistors with resistances R_1 and R_2 are connected in parallel, the combined resistance R can be found from

Recall that the numbers 1 and 2 are *subscripts*. We read R_1 as "R sub 1" and R_2 as "R sub 2."

$$\frac{1}{R} = \frac{1}{R_1} + \frac{1}{R_2}$$

Solve the formula for R.
First, the LCD is RR_1R_2, and we multiply:

$$RR_1R_2 \cdot \frac{1}{R} = RR_1R_2 \cdot \frac{1}{R_1} + RR_1R_2 \cdot \frac{1}{R_2}$$

Simplifying yields

$$R_1R_2 = RR_2 + RR_1 \quad \text{Factor out } R \text{ on the right.}$$
$$R_1R_2 = R(R_2 + R_1) \quad \text{Divide by } R_2 + R_1 \text{ to isolate } R.$$
$$\frac{R_1R_2}{R_2 + R_1} = R \quad \text{or} \quad R = \frac{R_1R_2}{R_1 + R_2}$$

✓ CHECK YOURSELF 7

Solve for D_1.

Note: This formula involves the focal length of a convex lens.

$$\frac{1}{F} = \frac{1}{D_1} + \frac{1}{D_2}$$

The Distance Formula

$$d = r \cdot t \quad \text{Distance equals rate times time.}$$

Sometimes use of this formula leads to a rational equation, as in Example 8.

Example 8

Solving a Motion Problem

A boat can travel 16 miles per hour in still water. If the boat can travel 5 miles downstream in the same time it takes to travel 3 miles upstream, what is the rate of the river's current?

Step 1 We want to find the rate of the current.

Step 2 Let c be the rate of the current. Then $16 + c$ is the rate of the boat going downstream and $16 - c$ is the rate going upstream.

Step 3 We'll use a chart to help us set up an appropriate equation.

	d	r	t
Downstream	5 miles	$16 + c$	$\dfrac{5}{16 + c}$
Upstream	3 miles	$16 - c$	$\dfrac{3}{16 - c}$

We found the time by dividing distance by rate. If $d = r \cdot t$, then $t = \dfrac{d}{r}$. The key to finding our equation is noting that the time is the same upstream and downstream, so

$$\frac{5}{16 + c} = \frac{3}{16 - c}$$

Step 4 Multiplying both sides by the LCD $(16 + c)(16 - c)$ yields

$$5(16 - c) = 3(16 + c)$$
$$80 - 5c = 48 + 3c$$
$$32 = 8c$$
$$c = 4$$

The current is moving at 4 miles per hour.

Step 5 To check, verify that $\dfrac{5}{16 + 4} = \dfrac{3}{16 - 4}$.

✓ CHECK YOURSELF 8

A plane flew 540 miles into a steady 30 mi/h wind. The pilot then returned along the same route with a tailwind. If the entire trip took seven and one-half hours, what would his speed have been in still air?

Another application that frequently results in rational equations is the work problem. Example 9 illustrates.

Example 9 — Solving a Work Problem

One computer printer can print a company's paychecks in 40 minutes (min). A second printer can print them in 80 min. If both printers are working, how long will it take to print the paychecks?

Before we learn how to solve work problems, we should look at a couple of common errors made in attempting to solve such problems.

Caution

Error 1: Students sometimes try adding the two times together. If we add the 40 min and the 80 min we get 120 min. Is this a reasonable answer? Certainly not! If one printer does the job in 40 min, why would it take 120 min for two printers to do it? It wouldn't.

Caution

Error 2: A more reasonable approach would be to give half the job to each printer. The first printer would finish its half of the job in 20 min. The second would finish its half in 40 min. The first printer would be idle for the final 20 min, so we know the job could have been finished faster.

Although error 2 does not solve the problem, it does give us guidelines for a reasonable answer. When the printers work together, it will take them somewhere between 20 and 40 min to finish the job. To find the exact amount of time, we use the work principle.

The Work Principle: Given an object, A, which completes a task in time a, and an object, B, which completes the same task in time b, we find the amount of time it takes them to complete the task together by solving the following equation for variable t.

$$\frac{t}{a} + \frac{t}{b} = 1$$

Now we can solve the problem.

Step 1   We are looking for the time it takes to print the paychecks.

Step 2   Our variable, t is the time it will take the printers to complete the task.

Step 3   We have the equation $\frac{t}{40} + \frac{t}{80} = 1$

Step 4   Multiply by LCD 80.

$$2t + t = 80$$
$$3t = 80$$
$$t = \frac{80}{3} \text{ min}$$
$$= 26\frac{2}{3} \text{ min}$$
$$= 26 \text{ min } 40 \text{ sec}$$

Step 5   To verify this answer, we find what fraction of the job each printer does in this time. The first printer does $\frac{\frac{80}{3}}{40} = \frac{2}{3}$ of the job. The second printer does $\frac{\frac{80}{3}}{80} = \frac{1}{3}$ of the job. Together, they do the entire job $\left(\frac{2}{3} + \frac{1}{3} = 1\right)$.

✓ CHECK YOURSELF 9

It would take Sasha 48 days to paint the house. Natasha could do it in 36 days. How long would it take them to paint the house if they worked together?

Example 10

Finding the Zeros of a Function

Find the zeros of

$$f(x) = \frac{1}{x} - \frac{3}{7x} - \frac{4}{21}$$

Set the function equal to 0, and solve the resulting equation for x.

$$f(x) = \frac{1}{x} - \frac{3}{7x} - \frac{4}{21} = 0$$

The LCD for x, $7x$, and 21 is $21x$. Multiplying both sides by $21x$, we have

$$21x\left(\frac{1}{x} - \frac{3}{7x} - \frac{4}{21}\right) = 21x \cdot 0$$

$21 - 9 - 4x = 0$ Distribute $21x$ on the left side.

$12 - 4x = 0$ Simplify.

$12 = 4x$

$3 = x$

So 3 is the value of x for which $f(x) = 0$, that is, 3 is a zero of $f(x)$.

✓ CHECK YOURSELF 10

Find the zeros of the function.

$$f(x) = \frac{5x + 2}{x - 5} - \frac{11}{4}$$

Rational Inequalities

To solve inequalities involving rational expressions, we need some properties of division over the real numbers. Recall that

1. The quotient of two positive numbers is always positive.
2. The quotient of two negative numbers is always positive.
3. The quotient of a positive number and a negative number is always negative.

We solve rational inequalities by using sign graphs, as Example 11 illustrates.

Example 11

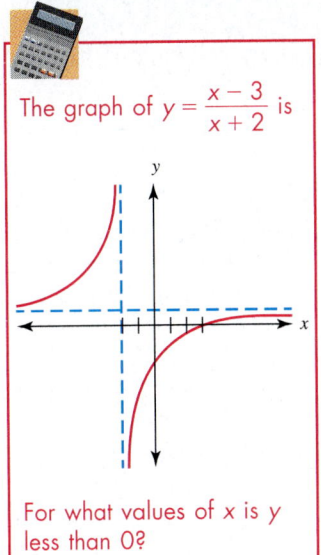

The graph of $y = \dfrac{x-3}{x+2}$ is

For what values of x is y less than 0?

Solving a Rational Inequality

Solve.

$$\dfrac{x-3}{x+2} < 0$$

The inequality states that the quotient of $x - 3$ and $x + 2$ must be negative (less than 0). This means that the numerator and denominator must have opposite signs.

We start by finding the critical points. These are points where either the numerator or denominator is 0. In this case, the critical points are 3 and -2.

The solution depends on determining whether the numerator and denominator are positive or negative. To visualize the process, start with a number line and label it as shown below.

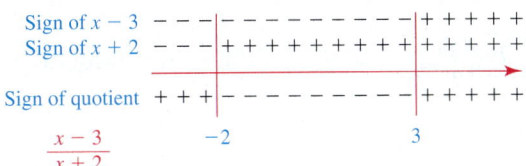

Examining the sign of the numerator and denominator, we see that

For any x less than -2, the quotient is positive (quotient of two negatives).

For any x between -2 and 3, the quotient is negative (quotient of a negative and a positive).

For any x greater than 3, the quotient is positive (quotient of two positives).

We return to the original inequality

$$\dfrac{x-3}{x+2} < 0$$

This inequality is true only when the quotient is negative, that is, when x is between -2 and 3. This solution set can be written as $\{x \mid -2 < x < 3\}$ and represented on a graph as follows.

✓ CHECK YOURSELF 11

Solve and graph the solution set.

$$\frac{x-4}{x+2} > 0$$

The solution process illustrated in Example 11 is valid only when the rational expression is isolated on one side of the inequality and is related to 0. If this is not the case, we must write an equivalent inequality as the first step, as Example 12 illustrates.

Example 12   Solving a Rational Inequality

Solve.

$$\frac{2x-3}{x+1} \geq 1$$

Since the rational expression is not related to 0, we use the following procedure.

We have subtracted 1 from both sides.

$\dfrac{2x-3}{x+1} - 1 \geq 0$ Form a common denominator on the left side.

$\dfrac{2x-3}{x+1} - \dfrac{x+1}{x+1} \geq 0$ Combine the expressions on the left side.

$\dfrac{2x-3-(x+1)}{x+1} \geq 0$ Simplify.

$\dfrac{x-4}{x+1} \geq 0$

We can now proceed as before since the rational expression is related to 0. The critical points are 4 and -1, and the sign graph is formed as shown below.

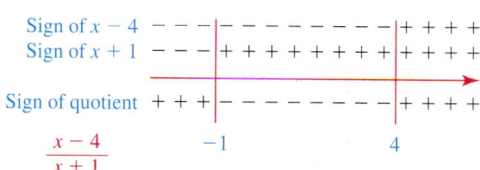

Section 7.5 ■ Rational Equations and Inequalities in One Variable **541**

We want a *positive* quotient.

From the sign graph the solution is

$$\{x|x < -1 \text{ or } x \geq 4\}$$

The graph is shown below.

Note that 4 is included, but -1 cannot be included in the solution set. Why?

✓ CHECK YOURSELF 12

Solve and graph the solution set.

$$\frac{2x - 3}{x - 2} \leq 1$$

✓ CHECK YOURSELF ANSWERS

1. $\{6\}$. **2.** $\{3\}$. **3.** $\{9\}$. **4.** $\{-11\}$. **5.** No solution or $\varnothing$.

6. $\left\{-5, \dfrac{8}{3}\right\}$. **7.** $\dfrac{FD_2}{D_2 - F}$. **8.** 150 mi/h. **9.** $20\dfrac{4}{7}$ days. **10.** 7.

11. $\{x|x < -2 \text{ or } x > 4\}$

12. $\{x|1 \leq x < 2\}$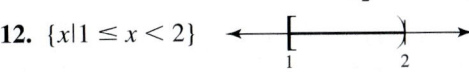

Exercises • 7.5

Answers (left column):

1. Equation, {36}
2. Equation, {28}
3. Expression, $\dfrac{3x}{10}$
4. Expression, $\dfrac{x}{24}$
5. Equation, {5}
6. Expression, $\dfrac{21x - 25}{20}$
7. Equation, {3}
8. Expression, $\dfrac{7x - 2}{6}$
9. {5}   10. {−7}
11. {8}   12. {6}
13. $\left\{\dfrac{3}{2}\right\}$ 14. {2}
15. {−1} 16. {−3}
17. $\left\{-\dfrac{9}{5}\right\}$ 18. $\left\{-\dfrac{2}{3}\right\}$
19. $\left\{-\dfrac{2}{3}\right\}$ 20. {−3}
21. No solution or ∅
22. {7}   23. {−23}
24. {5}   25. {6}
26. x is any real except $x = 0$ and $x = 2$
27. {4}   28. {6}
29. {8}   30. {5}
31. {4}   32. {−4}

In Exercises 1 to 8, decide whether each of the following is an expression or an equation. If it is an equation, find a solution. If it is an expression, write it as a single fraction.

1. $\dfrac{x}{2} - \dfrac{x}{3} = 6$ 2. $\dfrac{x}{4} - \dfrac{x}{7} = 3$ 3. $\dfrac{x}{2} - \dfrac{x}{5}$

4. $\dfrac{x}{6} - \dfrac{x}{8}$ 5. $\dfrac{3x + 1}{4} = x - 1$ 6. $\dfrac{3x - 1}{2} - \dfrac{x}{5} - \dfrac{x + 3}{4}$

7. $\dfrac{x}{4} = \dfrac{x}{12} + \dfrac{1}{2}$ 8. $\dfrac{2x - 1}{3} + \dfrac{x}{2}$

In Exercises 9 to 50, solve each equation.

9. $\dfrac{x}{3} + \dfrac{3}{2} = \dfrac{x}{6} + \dfrac{7}{3}$ 10. $\dfrac{x}{10} - \dfrac{1}{5} = \dfrac{x}{5} + \dfrac{1}{2}$ 11. $\dfrac{4}{x} + \dfrac{3}{4} = \dfrac{10}{x}$

12. $\dfrac{3}{x} = \dfrac{5}{3} - \dfrac{7}{x}$ 13. $\dfrac{5}{4x} - \dfrac{1}{2} = \dfrac{1}{2x}$ 14. $\dfrac{7}{6x} - \dfrac{1}{3} = \dfrac{1}{2x}$

15. $\dfrac{3}{x + 4} = \dfrac{2}{x + 3}$ 16. $\dfrac{5}{x - 2} = \dfrac{4}{x + 1}$ 17. $\dfrac{9}{x} + 2 = \dfrac{2x}{x + 3}$

18. $\dfrac{6}{x} + 3 = \dfrac{3x}{x + 1}$ 19. $\dfrac{3}{x + 2} - \dfrac{5}{x} = \dfrac{13}{x + 2}$ 20. $\dfrac{7}{x} - \dfrac{2}{x - 3} = \dfrac{6}{x}$

21. $\dfrac{3}{2} + \dfrac{2}{2x - 4} = \dfrac{1}{x - 2}$ 22. $\dfrac{2}{x - 1} + \dfrac{5}{2x - 2} = \dfrac{3}{4}$

23. $\dfrac{x}{3x + 12} + \dfrac{x - 1}{x + 4} = \dfrac{5}{3}$ 24. $\dfrac{x}{4x - 12} - \dfrac{x - 4}{x - 3} = \dfrac{1}{8}$

25. $\dfrac{x - 1}{x + 3} - \dfrac{x - 3}{x} = \dfrac{3}{x^2 + 3x}$ 26. $\dfrac{x + 1}{x - 2} - \dfrac{x + 3}{x} = \dfrac{6}{x^2 - 2x}$

27. $\dfrac{1}{x - 2} - \dfrac{2}{x + 2} = \dfrac{2}{x^2 - 4}$ 28. $\dfrac{1}{x + 4} + \dfrac{1}{x - 4} = \dfrac{12}{x^2 - 16}$

29. $\dfrac{7}{x + 5} - \dfrac{1}{x - 5} = \dfrac{x}{x^2 - 25}$ 30. $\dfrac{2}{x - 2} = \dfrac{3}{x + 2} + \dfrac{x}{x^2 - 4}$

31. $\dfrac{11}{x + 2} - \dfrac{5}{x^2 - x - 6} = \dfrac{1}{x - 3}$ 32. $\dfrac{5}{x - 4} = \dfrac{1}{x + 2} - \dfrac{2}{x^2 - 2x - 8}$

Section 7.5 ▪ Rational Equations and Inequalities in One Variable

33. $\left\{\dfrac{3}{2}\right\}$ 34. $\left\{\dfrac{11}{2}\right\}$

35. No solution or $\varnothing$

36. No solution or $\varnothing$

37. $\{7\}$ 38. $\{3\}$

39. $\{5\}$ 40. $\{-3\}$

41. $\left\{-\dfrac{5}{2}\right\}$

42. No solution or $\varnothing$

43. $\left\{-\dfrac{1}{2}, 6\right\}$

44. $\{-4, 3\}$

45. $\left\{-\dfrac{1}{2}\right\}$ 46. $\left\{\dfrac{2}{3}\right\}$

47. $\left\{-\dfrac{1}{3}, 7\right\}$

48. $\left\{-\dfrac{1}{2}, 6\right\}$

49. $\{-8, 9\}$

50. $\left\{-\dfrac{9}{2}, 10\right\}$

51. $\dfrac{ab}{b-a}$ 52. $\dfrac{bx}{b-x}$

53. $\dfrac{RR_2}{R_2-R}$ 54. $\dfrac{D_1 F}{D_1-F}$

55. $\dfrac{y+1}{y-1}$ 56. $\dfrac{2y-3}{y-1}$

57. $\dfrac{A}{1+rt}$ 58. $\dfrac{IR}{E-Ir}$

33. $\dfrac{5}{x-2} - \dfrac{3}{x+3} = \dfrac{24}{x^2+x-6}$

34. $\dfrac{3}{x+1} - \dfrac{5}{x+6} = \dfrac{2}{x^2+7x+6}$

35. $\dfrac{x}{x-3} - 2 = \dfrac{3}{x-3}$

36. $\dfrac{x}{x-5} + 2 = \dfrac{5}{x-5}$

37. $\dfrac{2}{x^2-3x} - \dfrac{1}{x^2+2x} = \dfrac{2}{x^2-x-6}$

38. $\dfrac{2}{x^2-x} - \dfrac{4}{x^2+5x-6} = \dfrac{3}{x^2+6x}$

39. $\dfrac{2}{x^2-4x+3} - \dfrac{3}{x^2-9} = \dfrac{2}{x^2+2x-3}$

40. $\dfrac{2}{x^2-4} - \dfrac{1}{x^2+x-2} = \dfrac{3}{x^2-3x+2}$

41. $\dfrac{7}{x-5} - \dfrac{3}{x+5} = \dfrac{40}{x^2-25}$

42. $\dfrac{3}{x-3} - \dfrac{18}{x^2-9} = \dfrac{5}{x+3}$

43. $\dfrac{2x}{x-3} + \dfrac{2}{x-5} = \dfrac{3x}{x^2-8x+15}$

44. $\dfrac{x}{x-4} = \dfrac{5x}{x^2-x-12} - \dfrac{3}{x+3}$

45. $\dfrac{2x}{x+2} = \dfrac{5}{x^2-x-6} - \dfrac{1}{x-3}$

46. $\dfrac{3x}{x-1} = \dfrac{2}{x-2} - \dfrac{2}{x^2-3x+2}$

47. $\dfrac{7}{x-2} + \dfrac{16}{x+3} = 3$

48. $\dfrac{5}{x-2} + \dfrac{6}{x+2} = 2$

49. $\dfrac{11}{x-3} - 1 = \dfrac{10}{x+3}$

50. $\dfrac{17}{x-4} - 2 = \dfrac{10}{x+2}$

In Exercises 51 to 58, solve each equation for the indicated variable.

51. $\dfrac{1}{x} = \dfrac{1}{a} - \dfrac{1}{b}$ for x

52. $\dfrac{1}{x} = \dfrac{1}{a} + \dfrac{1}{b}$ for a

53. $\dfrac{1}{R} = \dfrac{1}{R_1} + \dfrac{1}{R_2}$ for R_1

54. $\dfrac{1}{F} = \dfrac{1}{D_1} + \dfrac{1}{D_2}$ for D_2

55. $y = \dfrac{x+1}{x-1}$ for x

56. $y = \dfrac{x-3}{x-2}$ for x

57. $t = \dfrac{A-P}{Pr}$ for P

58. $I = \dfrac{nE}{R+nr}$ for n

59. $\{x|-1 < x < 2\}$

60. $\{x|x < -3 \text{ or } x > 2\}$

61. $\{x|x < 2 \text{ or } x > 4\}$

62. $\{x|-6 < x < 3\}$

63. $\{x|-3 < x \leq 5\}$

64. $\{x|-3 \leq x < 2\}$

65. $\left\{x \mid x < -3 \text{ or } x \geq \dfrac{1}{2}\right\}$

66. $\left\{x \mid \dfrac{2}{3} \leq x < 4\right\}$

67. $\{x|-2 \leq x < 3\}$

68. $\{x|x < -5 \text{ or } x > 3\}$

69. $\{x|-3 < x \leq 4\}$

70. $\{x|x \leq 2 \text{ or } x > 5\}$

71. $\{x|x < 2 \text{ or } x > 3\}$

In Exercises 59 to 72, solve each inequality, and graph the solution set.

59. $\dfrac{x-2}{x+1} < 0$

60. $\dfrac{x+3}{x-2} > 0$

61. $\dfrac{x-4}{x-2} > 0$

62. $\dfrac{x+6}{x-3} < 0$

63. $\dfrac{x-5}{x+3} \leq 0$

64. $\dfrac{x+3}{x-2} \leq 0$

65. $\dfrac{2x-1}{x+3} \geq 0$

66. $\dfrac{3x-2}{x-4} \leq 0$

67. $\dfrac{x}{x-3} + \dfrac{2}{x-3} \leq 0$

68. $\dfrac{x}{x+5} - \dfrac{3}{x+5} > 0$

69. $\dfrac{x}{x+3} \leq \dfrac{4}{x+3}$

70. $\dfrac{x}{x-5} \geq \dfrac{2}{x-5}$

71. $\dfrac{2x-5}{x-2} > 1$

72. $\dfrac{2x+3}{x+4} \geq 1$

In Exercises 73 to 80, find the zeros of each function.

73. $f(x) = \dfrac{x}{10} - \dfrac{12}{5}$

74. $f(x) = \dfrac{4x}{3} - \dfrac{x}{6}$

75. $f(x) = \dfrac{12}{x+5} - \dfrac{5}{x}$

76. $f(x) = \dfrac{1}{x-2} - \dfrac{3}{x}$

72. $\{x|x < -4 \text{ or } x \geq 1\}$

73. 24

74. 0

75. $\dfrac{25}{7}$

76. 3

77. $\dfrac{3}{4}$

78. 1

79. 3, 13

80. −9, 8

81. 6, 24

82. 4, 6

83. 9

84. 5

85. 4 mi/h

86. 4 mi/h

87. 150 mi/h

88. 40 mi/h

89. 36 min

90. 30 h

Section 7.5 ▪ Rational Equations and Inequalities in One Variable   545

77. $f(x) = \dfrac{1}{x-3} + \dfrac{2}{x} - \dfrac{5}{3x}$

78. $f(x) = \dfrac{2}{x} - \dfrac{1}{x+1} - \dfrac{3}{x^2+x}$

79. $f(x) = 1 + \dfrac{39}{x^2} - \dfrac{16}{x}$

80. $f(x) = x - \dfrac{72}{x} + 1$

Solve the following problems.

81. **Number analysis.** One number is 4 times another number. The sum of the reciprocals of the numbers is $\dfrac{5}{24}$. Find the two numbers.

82. **Number analysis.** The sum of the reciprocals of two consecutive even integers is equal to 10 times the reciprocal of the product of those integers. Find the two integers.

83. **Number analysis.** If the same number is subtracted from the numerator and denominator of $\dfrac{11}{15}$, the result is $\dfrac{1}{3}$. Find that number.

84. **Number analysis.** The numerator of $\dfrac{8}{9}$ is multiplied by a number. That same number is subtracted from the denominator, and the result is 10. What is that number?

85. **Motion.** A motorboat can travel 20 mi/h in still water. If the boat can travel 3 mi downstream on a river in the same time it takes to travel 2 mi upstream, what is the rate of the river's current?

86. **Motion.** Janet and Michael took a canoeing trip, traveling 6 mi upstream along a river, against a 2 mi/h current. They then returned downstream to the starting point of their trip. If their entire trip took 4 h, what was their rate in still water?

87. **Motion.** A plane flew 720 mi with a steady 30-mi/h tailwind. The pilot then returned to the starting point, flying against the same wind. If the round-trip flight took 10 h, what was the plane's airspeed?

88. **Motion.** A small jet has an airspeed (the rate in still air) of 300 mi/h. During one day's flights, the pilot noted that the plane could fly 85 mi with a tailwind in the same time it took to fly 65 mi against the same wind. What was the rate of the wind?

89. **Work.** One computer printer can print a company's weekly payroll checks in 60 min. A second printer would take 90 min to complete the job. How long would it take the two printers, operating together, to print the checks?

90. **Work.** An electrician can wire a house in 20 h. If she works with an apprentice, the same job can be completed in 12 h. How long would it take the apprentice, working alone, to wire the house?

91. What special considerations must be made when an equation contains rational expressions with variables in the denominator?

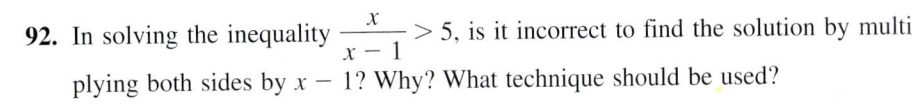

92. In solving the inequality $\dfrac{x}{x-1} > 5$, is it incorrect to find the solution by multiplying both sides by $x - 1$? Why? What technique should be used?

Summary Exercises · 7

Answers (left column):
1. Never undefined
2. $y \neq 0$
3. $x \neq 5$
4. $x \neq \dfrac{5}{2}$
5. $\dfrac{3x^2}{4}$
6. $\dfrac{-3m^2}{n}$
7. 8
8. $\dfrac{5}{x+4}$
9. $-\dfrac{x+3}{x+5}$
10. $\dfrac{3w-7}{2w+3}$
11. $\dfrac{2a-b}{3a-b}$
12. $\dfrac{3}{4w^2+2wz+z^2}$

This summary exercise set is provided to give you practice with each of the objectives in the chapter. Each exercise is keyed to the appropriate chapter section.

[7.1] For what value of the variable will each of the following rational expressions be defined?

1. $\dfrac{x}{2}$ 2. $\dfrac{3}{y}$ 3. $\dfrac{2}{x-5}$ 4. $\dfrac{3x}{2x-5}$

[7.1] Simplify each of the following rational expressions.

5. $\dfrac{18x^5}{24x^3}$ 6. $\dfrac{15m^3n}{-5mn^2}$ 7. $\dfrac{8y-64}{y-8}$

8. $\dfrac{5x-20}{x^2-16}$ 9. $\dfrac{9-x^2}{x^2+2x-15}$

10. $\dfrac{3w^2+8w-35}{2w^2+13w+15}$ 11. $\dfrac{6a^2-ab-b^2}{9a^2-b^2}$ 12. $\dfrac{6w-3z}{8w^3-z^3}$

[7.1] Graph the following rational functions. Indicate the coordinates of the hole in the graph.

13. $f(x) = \dfrac{x^2-3x-4}{x+1}$

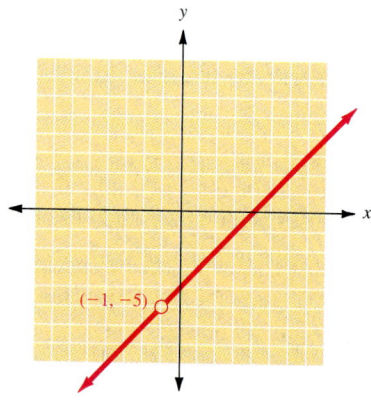

14. $f(x) = \dfrac{x^2+x-6}{x-2}$

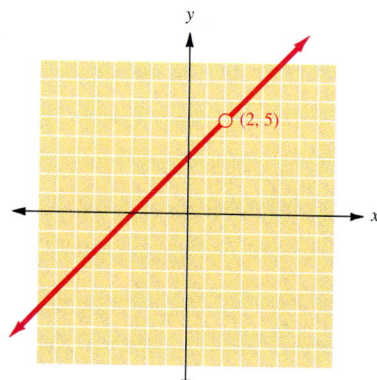

546

- Summary Exercises   **547**

15. $\dfrac{5x^2}{6}$ 16. $3a^2$

17. $\dfrac{4}{3y}$ 18. $\dfrac{m}{m+5}$

19. $\dfrac{3}{2a}$ 20. $\dfrac{r-s}{5r}$

21. $\dfrac{x+2y}{x-5y}$

22. $\dfrac{1}{(w^2-2)(w^2-3w+9)}$

23. (a) -8; (b) $x^2 + x - 20$; (c) -8

24. (a) 0; (b) $\dfrac{x^2-2x-3}{x+2}$; (c) 0

25. $\dfrac{3x^2-16x-8}{(x+4)(x-4)}$

26. $\dfrac{10x+9}{12x^2}$

27. $\dfrac{x+5}{x(x-5)}$

28. $\dfrac{5y+23}{(y+5)(y+4)}$

29. $\dfrac{-11}{6(m-1)}$

30. $\dfrac{12}{x-3}$

31. $\dfrac{15x+5}{4(x+1)(x-1)}$

32. $\dfrac{10}{a-5}$

33. $\dfrac{-4}{s+2}$

34. $\dfrac{x-13}{(x+3)(x-3)(x-1)}$

35. $\dfrac{3x-1}{x+2}$

36. $\dfrac{2}{w-z}$

[7.2] Multiply or divide as indicated. Express your results in simplest form.

15. $\dfrac{x^5}{24} \cdot \dfrac{20}{x^3}$

16. $\dfrac{a^3 b}{4ab^2} \div \dfrac{ab}{12ab^2}$

17. $\dfrac{6y-18}{9y} \cdot \dfrac{10}{5y-15}$

18. $\dfrac{m^2-3m}{m^2-5m+6} \cdot \dfrac{m^2-4}{m^2+7m+10}$

19. $\dfrac{a^2-2a}{a^2-4} \div \dfrac{2a^2}{3a+6}$

20. $\dfrac{r^2+2rs}{r^3-r^2 s} \div \dfrac{5r+10s}{r^2-2rs+s^2}$

21. $\dfrac{x^2-2xy-3y^2}{x^2-xy-2y^2} \cdot \dfrac{x^2-4y^2}{x^2-8xy+15y^2}$

22. $\dfrac{w^3+3w^2+2w+6}{w^4-4} \div (w^3+27)$

23. Let $f(x) = \dfrac{x^2-16}{x-5}$ and $g(x) = \dfrac{x^2-25}{x+4}$. Find (a) $f(3) \cdot g(3)$, (b) $h(x) = f(x) \cdot g(x)$, and (c) $h(3)$.

24. Let $f(x) = \dfrac{2x^2-5x-3}{x-4}$ and $g(x) = \dfrac{x^2-3x-4}{2x^2+5x+2}$. Find (a) $f(3) \cdot g(3)$, (b) $h(x) = f(x) \cdot g(x)$, and (c) $h(3)$.

[7.3] Perform the indicated operations. Express your results in simplified form.

25. $\dfrac{5x+7}{x+4} - \dfrac{2x-5}{x-4}$

26. $\dfrac{3}{4x^2} + \dfrac{5}{6x}$

27. $\dfrac{2}{x-5} - \dfrac{1}{x}$

28. $\dfrac{2}{y+5} + \dfrac{3}{y+4}$

29. $\dfrac{2}{3m-3} - \dfrac{5}{2m-2}$

30. $\dfrac{7}{x-3} - \dfrac{5}{3-x}$

31. $\dfrac{5}{4x+4} + \dfrac{5}{2x-2}$

32. $\dfrac{2a}{a^2-9a+20} + \dfrac{8}{a-4}$

33. $\dfrac{2}{s-1} - \dfrac{6s}{s^2+s-2}$

34. $\dfrac{4}{x^2-9} - \dfrac{3}{x^2-4x+3}$

35. $\dfrac{x^2-14x-8}{x^2-2x-8} + \dfrac{2x}{x-4} - \dfrac{3}{x+2}$

36. $\dfrac{w^2+2wz+z^2}{w^2-wz-2z^2} \cdot \left(\dfrac{3}{w+z} - \dfrac{1}{w-z} \right)$

37. (a) 8; (b) $\dfrac{3x^2 - 8x}{x^2 - 5x + 6}$; (c) (4, 8)

38. (a) 4; (b) $\dfrac{2x^2 - 6x - 16}{x^2 - 9x + 14}$; (c) (3, 4)

39. $\dfrac{2}{3x}$

40. $\dfrac{1}{y + 2}$

41. $\dfrac{b + a}{b - a}$

42. $\dfrac{y}{2y + x}$

43. $\dfrac{rs}{s + r}$

44. $\dfrac{x + 1}{x + 3}$

45. $\dfrac{x - 4}{(x - 1)(x - 1)}$

46. $\dfrac{w^2 - 2w - 1}{w^2 + 2w - 1}$

47. $-y + 2$

48. $\dfrac{x}{2x - 1}$

49. $\dfrac{x + 2}{(x - 1)(x + 4)}$

50. $2 + y$ 51. {5}

52. {2}   53. {6}

54. No solution

55. {3}

56. {5}

---

37. Let $f(x) = \dfrac{2x}{x - 2}$ and $g(x) = \dfrac{x}{x - 3}$. Find (a) $f(4) + g(4)$, (b) $h(x) = f(x) + g(x)$, and (c) the ordered pair $(4, h(4))$.

38. Let $f(x) = \dfrac{x + 2}{x - 2}$ and $g(x) = \dfrac{x + 1}{x - 7}$. Find (a) $f(3) + g(3)$, (b) $h(x) = f(x) + g(x)$, and (c) the ordered pair $(3, h(3))$.

[7.4] Simplify each of the following complex fractions.

39. $\dfrac{\dfrac{x^2}{12}}{\dfrac{x^3}{8}}$

40. $\dfrac{\dfrac{y - 1}{y^2 - 4}}{\dfrac{y^2 - 1}{y^2 - y - 2}}$

41. $\dfrac{1 + \dfrac{a}{b}}{1 - \dfrac{a}{b}}$

42. $\dfrac{2 - \dfrac{x}{y}}{4 - \dfrac{x^2}{y^2}}$

43. $\dfrac{\dfrac{1}{r} - \dfrac{1}{s}}{\dfrac{1}{r^2} - \dfrac{1}{s^2}}$

44. $\dfrac{1 - \dfrac{1}{x + 2}}{1 + \dfrac{1}{x + 2}}$

45. $\dfrac{1 - \dfrac{2}{x - 1}}{x + \dfrac{3}{x - 4}}$

46. $\dfrac{\dfrac{w}{w + 1} - \dfrac{1}{w - 1}}{\dfrac{w}{w - 1} + \dfrac{1}{w + 1}}$

47. $1 - \dfrac{1}{1 - \dfrac{1}{y - 1}}$

48. $1 - \dfrac{1}{1 + \dfrac{1}{1 - \dfrac{1}{x}}}$

49. $\dfrac{1 - \dfrac{1}{x - 1}}{x - \dfrac{8}{x + 2}}$

50. $1 - \dfrac{1}{1 + \dfrac{1}{y + 1}}$

[7.5] Solve each of the following equations.

51. $\dfrac{1}{2x} + \dfrac{1}{3x} = \dfrac{1}{6}$

52. $\dfrac{5}{2x^2} - \dfrac{1}{4x} = \dfrac{1}{x}$

53. $\dfrac{x}{x - 2} + 1 = \dfrac{x + 4}{x - 2}$

54. $\dfrac{2x - 1}{x - 3} - \dfrac{5}{x - 3} = 1$

55. $\dfrac{2}{3x + 1} = \dfrac{1}{x + 2}$

56. $\dfrac{5}{x + 1} + \dfrac{1}{x - 2} = \dfrac{7}{x + 1}$

Summary Exercises

57. $\dfrac{4}{x-1} - \dfrac{5}{3x-7} = \dfrac{3}{x-1}$

58. $\dfrac{7}{x} - \dfrac{1}{x-3} = \dfrac{9}{x^2-3x}$

59. $\dfrac{2}{x-3} - \dfrac{11}{x^2-9} = \dfrac{3}{x+3}$

60. $\dfrac{5}{x+3} + \dfrac{1}{x-5} = \dfrac{1}{x+3}$

61. $\dfrac{2}{x-4} = \dfrac{x}{x-2} - \dfrac{x+4}{x^2-6x+8}$

62. $\dfrac{x}{x-5} = \dfrac{3x}{x^2-7x+10} + \dfrac{8}{x-2}$

[7.5] Solve each of the following inequalities.

63. $\dfrac{x-2}{x+1} < 0$

64. $\dfrac{x+3}{x-2} > 0$

65. $\dfrac{x-4}{x-2} > 0$

66. $\dfrac{x+6}{x+3} < 0$

67. $\dfrac{x-5}{x+3} \le 0$

68. $\dfrac{x+3}{x-2} \le 0$

69. $\dfrac{2x-1}{x+3} \ge 0$

70. $\dfrac{3x-2}{x-4} \le 0$

71. $\dfrac{x}{x-3} + \dfrac{2}{x-3} \le 0$

72. $\dfrac{x}{x+5} - \dfrac{3}{x+5} > 0$

73. $\dfrac{x}{x+3} \le \dfrac{4}{x+3}$

74. $\dfrac{x}{x-5} \ge \dfrac{2}{x-5}$

75. $\dfrac{2x-5}{x-2} > 1$

76. $\dfrac{2x+3}{x+4} \ge 1$

77. **Number analysis.** The sum of the reciprocals of two consecutive integers is equal to 11 times the reciprocal of the product of those two integers. What are the two integers?

78. **Motion.** Karl drove 224 mi on the expressway for a business meeting. On his return, he decided to use a shorter route of 200 mi, but road construction slowed his speed by 6 mi/h. If the trip took the same time each way, what was his average speed in each direction?

79. **Work.** An electrician can wire a certain model home in 20 h while it would take her apprentice 30 h to wire the same model. How long would it take the two of them, working together, to wire the house?

80. **Motion.** A light plane took 1 h longer to fly 540 mi on the first portion of a trip than to fly 360 mi on the second. If the rate was the same for each portion, what was the flying time for each leg of the trip?

---

57. $\{-1\}$

58. $\{5\}$

59. $\{4\}$

60. $\left\{\dfrac{17}{5}\right\}$

61. $\{0, 7\}$

62. $\{8\}$

63. $\{x \mid -1 < x < 2\}$

64. $\{x \mid x < -3 \text{ or } x > 2\}$

65. $\{x \mid x < 2 \text{ or } x > 4\}$

66. $\{x \mid -6 < x < -3\}$

67. $\{x \mid -3 < x \le 5\}$

68. $\{x \mid -3 \le x < 2\}$

69. $\left\{x \mid x < -3 \text{ or } x \ge \dfrac{1}{2}\right\}$

70. $\left\{x \mid \dfrac{2}{3} \le x < 4\right\}$

71. $\{x \mid -2 \le x < 3\}$

72. $\{x \mid x < -5 \text{ or } x > 3\}$

73. $\{x \mid -3 < x \le 4\}$

74. $x \le 2 \text{ or } x > 5$

75. $\{x \mid x < 2 \text{ or } x > 3\}$

76. $\{x \mid x < -4 \text{ or } x \ge 1\}$

77. 5, 6

78. Going: 56 mi/h; coming 50 mi/h

79. 12 h

80. First portion: 3 h; second portion: 2 h

Self-Test 7

Answers:

1. $\dfrac{-3x^4}{4y^2}$

2. $\dfrac{w+1}{w-2}$

3. $\dfrac{x+3}{x-2}$

5. $\dfrac{4a^2}{7b}$

6. $\dfrac{m-4}{4}$

7. $\dfrac{2}{x-1}$

8. $\dfrac{2x+y}{x(x+3y)}$

9. $\dfrac{-5(3x+1)}{(3x-1)}$

The purpose of this self-test is to help you check your progress and to review for a chapter test in class. Allow yourself about 1 hour to take the test. When you are done, check your answers in the back of the book. If you missed any answers, be sure to go back and review the appropriate sections in the chapter and the exercises that are provided.

Simplify each of the following rational expressions.

1. $\dfrac{-21x^5y^3}{28xy^5}$
2. $\dfrac{3w^2 + w - 2}{3w^2 - 8w + 4}$
3. $\dfrac{x^3 + 2x^2 - 3x}{x^3 - 3x^2 + 2x}$

4. Graph the following.

$$f(x) = \dfrac{x^2 - 5x + 4}{x - 4}$$

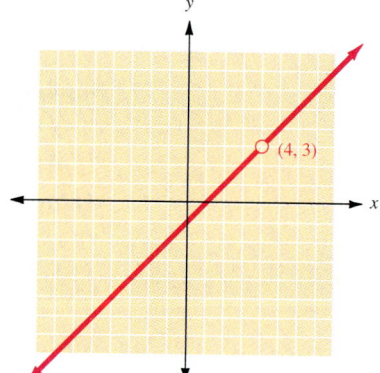

Multiply or divide as indicated.

5. $\dfrac{3ab^2}{5ab^3} \cdot \dfrac{20a^2b}{21b}$

6. $\dfrac{m^2 - 3m}{m^2 - 9} \div \dfrac{4m}{m^2 - m - 12}$

7. $\dfrac{x^2 - 3x}{5x^2} \cdot \dfrac{10x}{x^2 - 4x + 3}$

8. $\dfrac{x^2 + 3xy}{2x^3 - x^2y} \div \dfrac{x^2 + 6xy + 9y^2}{4x^2 - y^2}$

9. $\dfrac{9x^2 - 9x - 4}{6x^2 - 11x + 3} \cdot \dfrac{15 - 10x}{3x - 4}$

550

Self-Test 7

10. $\dfrac{2(2x+1)}{x(x-2)}$

11. $\dfrac{2}{x-3}$ 12. $\dfrac{4}{x-2}$

13. $\dfrac{8x+17}{(x-4)(x+1)(x+4)}$

14. $\dfrac{y}{3y+x}$ 15. $\dfrac{z-1}{2(z+3)}$

16. $\dfrac{1}{xy(x-y)}$

17. $\{2, 6\}$

18. $\{x \mid -4 \leq x < 3\}$

$\longleftarrow\underset{-4}{[}\rule{1cm}{0.4pt}\underset{3}{)}\longrightarrow$

19. $\{x \mid -3 < x \leq -1\}$

$\longleftarrow\underset{-3}{(}\rule{1cm}{0.4pt}\underset{-1}{]}\longrightarrow$

20. 3

Add or subtract as indicated.

10. $\dfrac{5}{x-2} - \dfrac{1}{x}$ 11. $\dfrac{2}{x+3} + \dfrac{12}{x^2-9}$

12. $\dfrac{6x}{x^2-x-2} - \dfrac{2}{x+1}$ 13. $\dfrac{3}{x^2-3x-4} + \dfrac{5}{x^2-16}$

Simplify each of the following complex fractions.

14. $\dfrac{3 - \dfrac{x}{y}}{9 - \dfrac{x^2}{y^2}}$ 15. $\dfrac{1 - \dfrac{10}{z+3}}{2 - \dfrac{12}{z-1}}$ 16. $\dfrac{\dfrac{1}{x} + \dfrac{1}{y}}{x^2 - y^2}$

Solve the following equation.

17. $\dfrac{5}{x} - \dfrac{x-3}{x+2} = \dfrac{22}{x^2+2x}$

Solve the following and graph the solution set.

18. $\dfrac{x+4}{x-3} \leq 0$ 19. $\dfrac{x+7}{x+3} \geq 3$

20. The numerator of $\dfrac{4}{7}$ is multiplied by a number. That same number is added to the denominator, and the result is $\dfrac{6}{5}$. What was that number?

Cumulative Test ■ 0-7

This test is provided to help you in the process of reviewing previous chapters. Answers are provided in the back of the book. If you missed any answers, be sure to go back and review the appropriate chapter sections.

1. Solve the equation $4x - 2(x + 1) = 3(5 - x) - 7$.
2. If $f(x) = 4x^3 - 5x^2 + 7x - 11$, find $f(-2)$.
3. Find the x and y intercepts and graph the equation $2x + 3y = 12$.
4. Find the equation of the line that passes through the point $(-1, -2)$ and is perpendicular to the line $4x + 5y = 15$.
5. Simplify the expression $(4x^3y^2)^3(-2x^2y^3)^2$.
6. If $P(x) = 4x^5 + 7x^4 - 3x^2 - 5x + 1$, find $P(-1)$.
7. Find the domain and range of the relation $4x - 3y = 15$.
8. Solve the following equation.

$$x^2 + 5x - 6 = 0$$

Simplify each of the following polynomials.

9. $(2x^2 - 3x + 4) - (3x^2 + 2x - 5)$
10. $(2x + 3)(5x - 4)$

Factor each of the following completely.

11. $3x^3 - x^2 - 2x$
12. $16x^2 - 25y^2$
13. $3x^2 - 3xy + x - y$
14. If $f(x) = -5x + 1$ and $g(x) = 8x + 6$, find (a) $f + g$, (b) $f - g$, (c) $f \cdot g$, and (d) $\dfrac{f}{g}$.

Solve the following equations.

15. $-6x - 6(2x - 9) = -3(x + 7)$
16. $|2x + 46| = 10$
17. $4(3x - 6) = -8(x - 2)$

Solve the following inequalities.

18. $4(3 + x) \geq 8$
19. $|x + 9| \leq 4$
20. $-3|x + 5| > -3$

---

1. $\{2\}$
2. -77
3. x intercept: 6; y intercept: 4
4. $y = \dfrac{5}{4}x - \dfrac{3}{4}$
5. $256x^{10}y^{12}$
6. 6
7. Domain: all reals; range: all reals
8. $\{-6, 1\}$
9. $x^2 - 5x + 9$
10. $10x^2 + 7x - 12$
11. $x(3x + 2)(x - 1)$
12. $(4x + 5y)(4x - 5y)$
13. $(x - y)(3x + 1)$
14. (a) $3x + 7$;
    (b) $-13x - 5$;
    (c) $-40x^2 - 22x + 6$;
    (d) $\dfrac{-5x + 1}{8x + 6}$
15. $\{5\}$
16. $\{-18, -28\}$
17. $\{2\}$
18. $\{x | x \geq -1\}$
19. $\{x | -13 \leq x \leq -5\}$
20. $\{x | -6 < x < -4\}$

CHAPTER 8
Systems of Linear Equations and Inequalities

LIST OF SECTIONS

8.1 Solving Systems of Linear Equations by Graphing

8.2 Systems of Equations in Two Variables with Applications

8.3 Systems of Linear Equations in Three Variables

8.4 Graphing Linear Inequalities in Two Variables

8.5 Systems of Linear Inequalities in Two Variables

Successful businesses juggle many factors, including workers' schedules, available machine time, and costs of raw material and storage. Getting the most efficient mix of these factors is crucial if the business is to make money. Systems of linear equations can be used to find the most efficient ways to combine these costs.

For example, the owner of a small bakery must decide how much of several kinds of bread to make based on the time it takes to produce each kind and the profit to be made on each one. The owner knows the bakery requires 0.75 hours of oven time and 1.25 hours of preparation time for every 8 loaves of bread, and that a coffee cake takes 1 hour of oven time and 1.25 hours of preparation time for every 6 coffee cakes. In 1 day the bakery has 12 hours of oven time and 16 hours of preparation time available. The owner knows that she clears a profit of $0.50 for every loaf of bread sold and $1.75 for every coffee cake sold. But, she has found that on a regular day she sells no more than 12 coffee cakes. Given these constraints, how many of each type of product can be made in a day? Which combination of products would give the highest profit if all the products are sold?

These questions can be answered by graphing a system of equations that model the constraints, where b is the number of loaves of bread, and c is the number of coffee cakes.

Section 8.1 Solving Systems of Linear Equations by Graphing

8.1 OBJECTIVES

1. Recognize the solution of a linear system of equations
2. Solve a system graphically
3. Use slopes to identify consistent systems

In Section 2.1, we defined a solution set as "the set of all values for the variable that makes an equation a true statement." If we have the equation

$$2x - 3x + 5 = x - 7$$

the solution set is $\{6\}$. This tells us that 6 is the only value for the variable x that makes the equation a true statement.

When we studied equations in two variables in Chapter 3, we found that a solution to a two-variable equation was an ordered pair. Given the equation

$$y = 2x - 5$$

one possible solution to the equation is the ordered pair $(1, -3)$. There are an infinite number of other possible solutions, so we don't list them in a solution set.

In this section, we are introducing a topic that has many applications in chemistry, business, economics, and physics. Each of these areas has occasion to solve a system of equations.

> A **system of equations** is a set of two or more related equations.

Our goal in this chapter is to solve linear systems of equations.

> A **solution** for a linear system of equations in two variables is an ordered pair of real numbers (x, y) that satisfies all of the equations in the system.

Over the next couple of sections, we will look at different ways in which a linear system of equations can be solved. Our first look will be at a graphical method of solving a system.

Example 1

Solving a System by Graphing

Solve the system by graphing.

$$2x + y = 4$$
$$x - y = 5$$

Solve each equation for y and then graph.

$$y = -2x + 4$$
$$\text{and}$$
$$y = x - 5$$

We can *approximate* the solution by tracing the curves near their intersection.

We graph the lines corresponding to the two equations of the system.

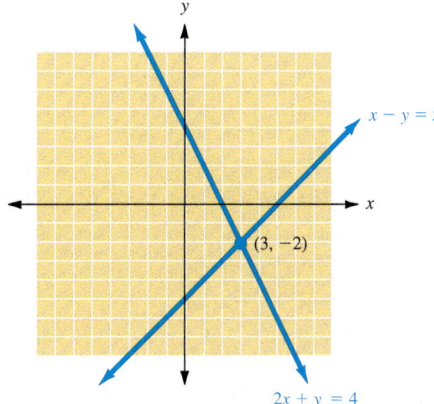

Each equation has an infinite number of solutions (ordered pairs) corresponding to points on a line. The point of intersection, here $(3, -2)$, is the *only* point lying on both lines, and so $(3, -2)$ is the only ordered pair satisfying both equations, and $(3, -2)$ is the solution for the system.

 CHECK YOURSELF 1

Solve the system by graphing.

$$3x - y = 2$$
$$x + y = 6$$

In Example 1, the two lines are nonparallel and intersect at only one point. The system has a unique solution corresponding to that point. Such a system is called a **consistent system.** In Example 2, we examine a system representing two lines that have no point of intersection.

Example 2 Solving a System by Graphing

Solve the system by graphing.

$$2x - y = 4$$
$$6x - 3y = 18$$

The lines corresponding to the two equations are graphed below.

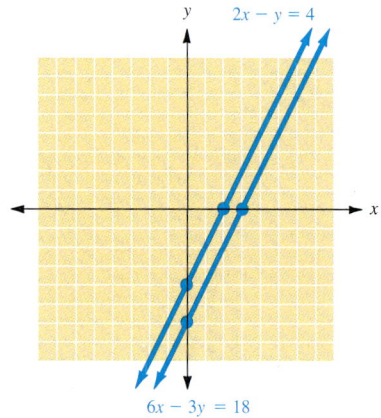

The lines are distinct and parallel. There is no point at which they intersect, so the system has no solution. We call such a system an **inconsistent system.**

✓ CHECK YOURSELF 2

Solve the system, if possible.

$$3x - y = 1$$
$$6x - 2y = 3$$

Sometimes the equations in a system have the same graph.

Example 3 Solving a System by Graphing

Solve the system by graphing.

$$2x - y = 2$$
$$4x - 2y = 4$$

The equations are graphed, as follows.

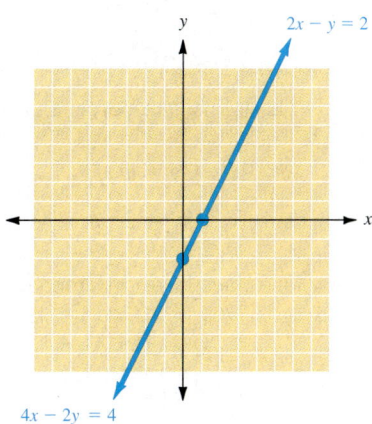

The lines have the same graph, so they have an infinite number of solutions in common. We call such a system a **dependent system.**

 CHECK YOURSELF 3

Solve the system by graphing.

$$6x - 3y = 12$$
$$y = 2x - 4$$

We've now seen the three possible types of solutions to a system of two linear equations. There will be either a single solution (a consistent system), an infinite number of solutions (a dependent system), or no solution (an inconsistent system).

Note that, for both the dependent system and the inconsistent system, the slopes of the two lines in the system must be the same. (Do you see why that is true?) Given any two lines with different slopes, they will intersect at exactly one point. This idea is used in Example 4.

Example 4

Identifying Consistent Systems

For each system, determine the number of solutions.

(a) $y = 2x - 5$
   $y = 2x + 9$

There is no solution. Both lines have a slope of 2, but different y intercepts. We have two distinct parallel lines, and therefore an inconsistent system.

(b) $y = 3x + 7$

$y = -\dfrac{1}{3}x + 2$

These lines are perpendicular. There is one solution. The system is consistent.

The slopes are $\dfrac{2}{3}$ and $-\dfrac{3}{5}$.

(c) $2x - 3y = 7$

$3x + 5y = 2$

The lines have different slopes. There is a single solution. The system is consistent.

(d) $y = \dfrac{2}{3}x - 6$

$2x - 3y = 12$

Both lines have a slope of $\dfrac{2}{3}$, but different y intercepts. There are no solutions. The system is not consistent.

✓ CHECK YOURSELF 4

For each system, determine the number of solutions.

(a) $y = 2x - 1$ (b) $y = -3x - 2$

$y = 3x + 7$ $y = -\dfrac{1}{3}x + 4$

(c) $6x - 3y = 4$ (d) $y = \dfrac{1}{2}x - 4$

$-2x + y = 9$ $x - y = 6$

✓ CHECK YOURSELF ANSWERS

1.   2.   3.

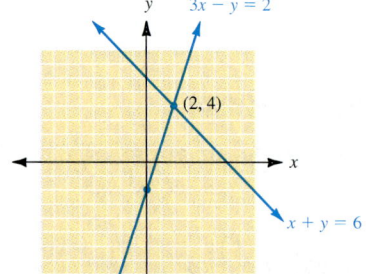
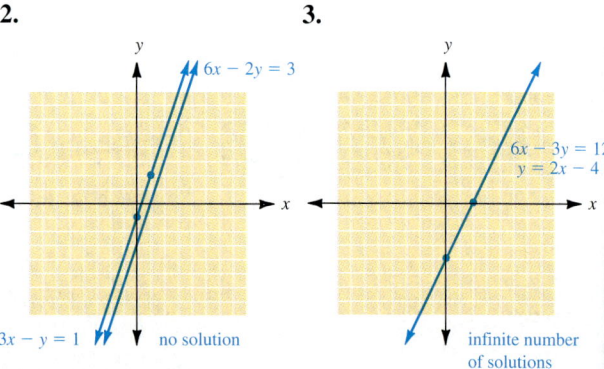

4. (a) One solution; (b) one solution; (c) no solutions; (d) one solution.

Exercises • 8.1

1. The solution is (5, 1).
2. The solution is (5, −3).
3. The solution is (1, 4).
4. The solution is (2, 5).
5. The solution is (2, 1).
6. The solution is (1, 3).
7. The solution is (2, 4).
8. The solution is (2, 2).
9. The solution is (6, 2).
10. The system is inconsistent.
11. The solution is (2, 3).
12. The solution is (2, −3).
13. There are infinitely many solutions. The system is dependent.
14. The solution is (2, −4).
15. The solution is (4, 2).
16. The solution is (−6, −2).
17. The solution is (4, 3).
18. The solution is (6, −4).
19. No solution, the system is inconsistent.
20. There are infinitely many solutions. The system is dependent.
21. The solution is $\left(0, \frac{3}{2}\right)$.
22. The solution is (−6, 0).
23. The solution is (4, −6).
24. The solution is (−3, 5).
25. (3.18, 28.58)
26. (61.65, 7.34)

Solve each of the following systems by graphing.

1. $x + y = 6$
   $x - y = 4$

2. $x - y = 8$
   $x + y = 2$

3. $-x + y = 3$
   $x + y = 5$

4. $x + y = 7$
   $-x + y = 3$

5. $x + 2y = 4$
   $x - y = 1$

6. $3x + y = 6$
   $x + y = 4$

7. $2x + y = 8$
   $2x - y = 0$

8. $x - 2y = -2$
   $x + 2y = 6$

9. $x + 3y = 12$
   $2x - 3y = 6$

10. $2x - y = 4$
    $2x - y = 6$

11. $3x + 2y = 12$
    $y = 3$

12. $x - 2y = 8$
    $3x - 2y = 12$

13. $x - y = 4$
    $2x - 2y = 8$

14. $2x - y = 8$
    $x = 2$

15. $x - 4y = -4$
    $x + 2y = 8$

16. $x - 6y = 6$
    $-x + y = 4$

17. $3x - 2y = 6$
    $2x - y = 5$

18. $4x + 3y = 12$
    $x + y = 2$

19. $3x - y = 3$
    $3x - y = 6$

20. $3x - 6y = 9$
    $x - 2y = 3$

21. $2y = 3$
    $x - 2y = -3$

22. $x + y = -6$
    $-x + 2y = 6$

23. $x = 4$
    $y = -6$

24. $x = -3$
    $y = 5$

Use a graphing calculator to solve each of the following exercises. Estimate your answer to the nearest hundredth. You may need to adjust the viewing window to see the point of intersection.

25. $88x + 57y = 1909$
    $95x + 48y = 1674$

26. $32x + 45y = 2303$
    $29x - 38y = 1509$

27. $25x - 65y = 5312$
    $-21x + 32y = 1256$

28. $-27x + 76y = 1676$
    $56x - 2y = -678$

For each system, determine whether it is consistent, inconsistent, or dependent.

29. $y = 3x + 7$
    $y = 7x - 2$

30. $y = 2x - 5$
    $y = -2x + 9$

31. $y = 7x - 1$
    $y = 7x + 8$

32. $y = -5x + 9$
    $y = -5x - 11$

27. (−445.35, −253.01)
28. (−11.47, 17.98)
29. Consistent
30. Consistent
31. Inconsistent
32. Inconsistent

559

33. Consistent

34. Inconsistent

35. Dependent

36. Consistent

37. $m = 2; b = 5$

38. $m = -6; b = 13$

39. Answers will vary.

40. Answers will vary.

41. Answers will vary.

42. Answers will vary.

43. (a) $\dfrac{-A}{B}$

    (b) $\dfrac{-D}{E}$

    (c) $AE - BD \neq 0$

33. $3x + 4y = 12$
$9x - 5y = 10$

34. $2x - 4y = 11$
$-8x + 16y = 15$

35. $7x - 2y = 5$
$14x - 4y = 10$

36. $3x + 2y = 8$
$6x - 4y = -12$

37. Find values for m and b in the following system so that the solution to the system is $(1, 2)$.

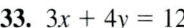

$$mx + 3y = 8$$
$$-3x + 4y = b$$

38. Find values for m and b in the following system so that the solution to the system is $(-3, 4)$.

$$5x + 7y = b$$
$$mx + y = 22$$

39. Complete the following statements in your own words:

"To solve an equation means to . . . ."

"To solve a system of equations means to . . . ."

40. A system of equations such as the one below is sometimes called a "2-by-2" system of linear equations.

$$3x + 4y = 1$$
$$x - 2y = 6$$

Explain this term.

41. Complete this statement in your own words: "All the points on the graph of the equation $2x + 3y = 6$ . . . ." Exchange statements with other students. Do you agree with other students' statements?

42. Does a system of linear equations always have a solution? How can you tell without graphing that a system of two equations will be graphed as two parallel lines? Give some examples to explain your reasoning.

43. Suppose we have the following linear system:

$$Ax + By = C \quad (1)$$
$$Dx + Ey = F \quad (2)$$

(a) Write the slope of the line determined by equation (1).

(b) Write the slope of the line determined by equation (2).

(c) What must be true about the given coefficients in order to guarantee that the system is consistent?

SECTION 8.2 Systems of Equations in Two Variables with Applications

8.2 OBJECTIVES

1. Solve a system by the addition method
2. Solve a system by the substitution method
3. Use a system of equations to solve an application

Graphical solutions to linear systems are excellent for seeing and estimating solutions. The drawback comes in precision. No matter how carefully one graphs the lines, the displayed solution rarely leads to one that is exact. This problem is exaggerated when the solution includes fractional values. In this section, we will look at a couple of methods that result in exact solutions.

One method for solving systems of linear equations in two variables is by the **addition method.** This method of solving systems is based on finding equivalent systems. Two systems are equivalent if they have the same solution set.

An equivalent system is formed whenever

1. One of the equations is multiplied by a nonzero number.
2. One of the equations is replaced by the sum of a constant multiple of another equation and that equation.

Example 1 illustrates the addition method of solution.

Example 1 Solving a System by the Addition Method

Solve the system by the addition method.

$$5x - 2y = 12 \qquad (1)$$
$$3x + 2y = 12 \qquad (2)$$

The addition method is sometimes called **solution by elimination** for this reason.

In this case, adding the equations will eliminate variable y, and we have

$$8x = 24$$
$$x = 3 \qquad (3)$$

Now equation (3) can be paired with either of the original equations to form an equivalent system. We let $x = 3$ in equation (1):

The solution should be checked by substituting these values into equation (2). Here

$$3(3) + 2\left(\frac{3}{2}\right) \stackrel{?}{=} 12$$
$$9 + 3 \stackrel{?}{=} 12$$
$$12 = 12$$

is a true statement.

$$5(3) - 2y = 12$$
$$15 - 2y = 12$$
$$-2y = -3$$
$$y = \frac{3}{2}$$

and $\left(3, \frac{3}{2}\right)$ is the solution for our system.

✓ CHECK YOURSELF 1

Solve the system by the addition method.

$$4x - 3y = 19$$
$$-4x + 5y = -25$$

Remember that multiplying one or both of the equations by a nonzero constant produces an equivalent system.

Example 1 and Check Yourself 1 were straightforward in that adding the equations of the system immediately eliminated one of the variables. Example 2 illustrates a common situation in which we must multiply one or both of the equations by a nonzero constant before the addition method is applied.

Example 2

All these solutions can be approximated by graphing the lines and tracing near the intersection. This is particularly useful when the solutions are not integers (the technical term for such solutions is

Solving a System by the Addition Method

Solve the system by the addition method.

$$3x - 5y = 19 \quad (4)$$
$$5x + 2y = 11 \quad (5)$$

It is clear that adding the equations of the given system will *not* eliminate one of the variables. Therefore, we must use multiplication to form an equivalent system. The choice of multipliers depends on which variable we decide to eliminate. Here we have decided to eliminate *y*. We multiply equation (4) by 2 and equation (5) by 5. We then have a system that is equivalent to the original system, but easier to solve.

Note that the coefficients of *y* are now *opposites* of each other.

$$6x - 10y = 38$$
$$25x + 10y = 55$$

Adding now eliminates *y* and yields

$$31x = 93$$
$$x = 3 \quad (6)$$

Pairing equation (6) with equation (4) gives an equivalent system, and we can substitute 3 for x in equation (4):

$$3 \cdot 3 - 5y = 19$$
$$9 - 5y = 19$$
$$-5y = 10$$
$$y = -2$$

Again, the solution should be checked in both equations by substitution in equation (5).

The solution for the system is $(3, -2)$.

✓ CHECK YOURSELF 2

Solve the system by the addition method.

$$2x + 3y = -18$$
$$3x - 5y = 11$$

The following algorithm summarizes the addition method of solving linear systems of two equations in two variables.

Solving by the Addition Method

Step 1 If necessary, multiply one or both of the equations by a constant so that one of the variables can be eliminated by addition.
Step 2 Add the equations of the equivalent system formed in step 1.
Step 3 Solve the equation found in step 2.
Step 4 Substitute the value found in step 3 into either of the equations of the original system to find the corresponding value of the remaining variable. The ordered pair formed is the solution to the system.
Step 5 Check the solution by substituting the pair of values found in step 4 into the other equation of the original system.

Example 3 illustrates two special situations you may encounter while applying the addition method.

Example 3 — Solving a System by the Addition Method

Solve each system by the addition method.

(a) $4x + 5y = 20$ (7)
    $8x + 10y = 19$ (8)

Multiply equation (7) by -2. Then

If these equations are graphed, we have two parallel lines.

$$-8x - 10y = -40$$
$$8x + 10y = 19$$
$$\overline{0 = -21}$$

We add the two left sides to get 0 and the two right sides to get -21.

The result $0 = -21$ is a *false* statement, which means that there is no point of intersection. Therefore, the system is inconsistent, and there is no solution.

(b) $\quad 5x - 7y = 9 \hfill (9)$

$\quad 15x - 21y = 27 \hfill (10)$

Multiply equation (9) by -3. We then have

$$-15x + 21y = -27$$
$$15x - 21y = 27$$
$$\overline{0 = 0}$$

We add the two equations.

The solution set could be written $\{(x, y) | 5x - 7y = 9\}$. This means the set of all ordered pairs (x, y) that make $5x - 7y = 9$ a true statement.

Both variables have been eliminated, and the result is a *true* statement. If the graphs of the two lines coincide, then there are an infinite number of solutions, one for each point on that line. Recall that this a *dependent system*.

✓ CHECK YOURSELF 3

Solve each system by the addition method, if possible.

(a) $\quad 3x + 2y = 8 \qquad$ (b) $\quad x - 2y = 8$

$\quad\quad 9x + 6y = 11 \qquad\qquad\quad 3x - 6y = 24$

The results of Example 3 can be summarized as follows.

When a system of two linear equations is solved:

1. If a false statement such as $3 = 4$ is obtained, then the system is inconsistent and has no solution.

2. If a true statement such as $8 = 8$ is obtained, then the system is dependent and has an infinite number of solutions.

The Substitution Method

A third method for finding the solutions of linear systems in two variables is called the **substitution method.** You may very well find the substitution method more difficult to apply in solving certain systems than the addition method, particularly when the equations involved in the substitution lead to fractions. However, the substitution method does have important extensions to systems involving higher-degree equations, as you will see in later mathematics classes.

To outline the technique, we solve one of the equations from the original system for one of the variables. That expression is then substituted into the *other* equation of the system to provide an equation in a single variable. That equation is solved, and the corresponding value for the other variable is found as before, as Example 4 illustrates.

Example 4 Solving a System by the Substitution Method

(a) Solve the system by the substitution method.

$$2x - 3y = -3 \quad \textbf{(11)}$$
$$y = 2x - 1 \quad \textbf{(12)}$$

Since equation (12) is already solved for y, we substitute $2x - 1$ for y in equation (11).

$$2x - 3(2x - 1) = -3$$

We now have an equation in the single variable x.

Solving for x gives

$$2x - 6x + 3 = -3$$
$$-4x = -6$$
$$x = \frac{3}{2}$$

We now substitute $\frac{3}{2}$ for x in equation (12).

$$y = 2\left(\frac{3}{2}\right) - 1$$
$$= 3 - 1 = 2$$

To check this result, we substitute these values in both equation (11) and (12) and have

$$2\left(\frac{3}{2}\right) - 3 \cdot 2 \stackrel{?}{=} -3$$
$$3 - 6 \stackrel{?}{=} -3$$
$$-3 = -3$$

A true statement!

The solution for our system is $\left(\frac{3}{2}, 2\right)$.

(b) Solve the system by the substitution method.

$$2x + 3y = 16 \quad \textbf{(13)}$$
$$3x - y = 2 \quad \textbf{(14)}$$

| Why did we choose to solve for *y* in equation (14)? We could have solved for *x*, so that

$$x = \frac{y+2}{3}$$

We simply chose the easier case to avoid fractions. | We start by solving equation (14) for *y*.

$$3x - y = 2$$
$$-y = -3x + 2$$
$$y = 3x - 2 \qquad (15)$$

Substituting in equation (13) yields

$$2x + 3(3x - 2) = 16$$
$$2x + 9x - 6 = 16$$
$$11x = 22$$
$$x = 2$$

We now substitute 2 for *x* in equation (15).

$$y = 3 \cdot 2 - 2$$
$$= 6 - 2 = 4$$ |

The solution should be checked in *both* equations of the original system.

The solution for the system is (2, 4). We leave the check of this result to you.

✓ CHECK YOURSELF 4

Solve each system by the substitution method.

(a) $2x + 3y = 6$
    $x = 3y + 6$

(b) $3x + 4y = -3$
    $x + 4y = -1$

The following algorithm summarizes the substitution method for solving linear systems of two equations in two variables.

Solving by the Substitution Method

Step 1 If necessary, solve one of the equations of the original system for one of the variables.

Step 2 Substitute the expression obtained in step 1 into the *other* equation of the system to write an equation in a single variable.

Step 3 Solve the equation found in step 2.

Step 4 Substitute the value found in step 3 into the equation derived in step 1 to find the corresponding value of the remaining variable. The ordered pair formed is the solution for the system.

Step 5 Check the solution by substituting the pair of values found in step 4 into *both* equations of the original system.

Section 8.2 ■ Systems of Equations in Two Variables with Applications

A natural question at this point is, How do you decide which solution method to use? First, the graphical method can generally provide only approximate solutions. When exact solutions are necessary, one of the algebraic methods must be applied. Which method to use depends totally on the given system.

If you can easily solve for a variable in one of the equations, the substitution method should work well. However, if solving for a variable in either equation of the system leads to fractions, you may find the addition approach more efficient.

Solving Applications

We are now ready to apply our equation-solving skills to solving various applications or word problems. Being able to extend these skills to problem solving is an important goal, and the procedures developed here are used throughout the rest of the book.

Although we consider applications from a variety of areas in this section, all are approached with the same five-step strategy presented here to begin the discussion.

> **Solving Applications**
>
> Step 1 Read the problem carefully to determine the unknown quantities.
> Step 2 Choose variables to represent the unknown quantities.
> Step 3 Translate the problem to the language of algebra to form a system of equations.
> Step 4 Solve the system of equations, and answer the question of the original problem.
> Step 5 Verify your solution by returning to the original problem.

Example 5

Solving a Mixture Problem

A coffee merchant has two types of coffee beans, one selling for \$3 per pound and the other for \$5 per pound. The beans are to be mixed to provide 100 lb of a mixture selling for \$4.50 per pound. How much of each type of coffee bean should be used to form 100 lb of the mixture?

Step 1 The unknowns are the amounts of the two types of beans.

Step 2 We use two variables to represent the two unknowns. Let x be the amount of \$3 beans and y the amount of \$5 beans.

Step 3 We now want to establish a system of two equations. One equation will be based on the *total amount* of the mixture, the other on the mixture's *value*.

Since we use two variables, we must form two equations.

$$x + y = 100 \quad \text{The mixture must weigh 100 lb.} \quad (16)$$

$$3x + 5y = 450 \quad (17)$$

The total value of the mixture comes from:
$100(4.50) = 450$

Value of \$3 beans   Value of \$5 beans   Total value

Step 4 An easy approach to the solution of the system is to multiply equation (16) by -3 and add to eliminate x.

$$-3x - 3y = -300$$
$$3x + 5y = 450$$
$$2y = 150$$
$$y = 75 \text{ lb}$$

By substitution in equation (16), we have

$$x = 25 \text{ lb}$$

$3(25) + 5(75) = 75 + 375$
$= 450$

We should use 25 lbs of $3 beans and 75 lbs of $5 beans.

Step 5 To check the result, show that the value of the $3 beans, added to the value of the $5 beans, equals the desired value of the mixture.

✓ CHECK YOURSELF 5

Peanuts, which sell for $2.40 per pound, and cashews, which sell for $6 per pound, are to be mixed to form a 60-lb mixture selling for $3 per pound. How much of each type of nut should be used?

A related problem is illustrated in Example 6.

Example 6

Solving a Mixture Problem

A chemist has a 25% and a 50% acid solution. How much of each solution should be used to form 200 mL of a 35% acid solution?

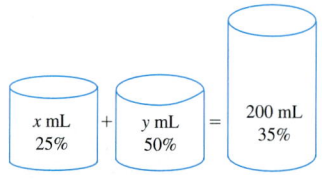

Step 1 The unknowns in this case are the amounts of the 25% and 50% solutions to be used in forming the mixture.

Drawing a sketch of a problem is often a valuable part of the problem-solving strategy.

Step 2 Again we use two variables to represent the two unknowns. Let x be the amount of the 25% solution and y the amount of the 50% solution. Let's draw a picture before proceeding to form a system of equations.

Step 3 Now, to form our two equations, we want to consider two relationships: the *total amounts* combined and the *amounts of acid* combined.

Total amounts combined.

$$x + y = 200 \tag{18}$$

Amounts of acid combined.

$$0.25x + 0.50y = 0.35(200) \tag{19}$$

Step 4 Now, clear equation (19) of decimals by multiplying equation (19) by 100. The solution then proceeds as before, with the result

$$x = 120 \text{ mL} \quad (25\% \text{ solution})$$
$$y = 80 \text{ mL} \quad (50\% \text{ solution})$$

We need 120 mL of the 25% solution and 80 mL of the 50% solution.

Step 5 To check, show that the amount of acid in the 25% solution, (0.25)(120), added to the amount in the 50% solution, (0.50)(80), equals the correct amount in the mixture, (0.35)(200). We leave that to you.

✓ CHECK YOURSELF 6

A pharmacist wants to prepare 300 mL of a 20% alcohol solution. How much of a 30% solution and a 15% solution should be used to form the desired mixture?

Applications that involve a constant rate of travel, or speed, require the use of the distance formula.

$$d = rt$$

where

$$d = \text{distance traveled}$$
$$r = \text{rate or speed}$$
$$t = \text{time}$$

Example 7 illustrates this approach.

Example 7

Solving a Distance-Rate-Time Problem

A boat can travel 36 mi downstream in 2 h. Coming back upstream, the boat takes 3 h. What is the rate of the boat in still water? What is the rate of the current?

Step 1 We want to find the two rates.

Step 2 Let x be the rate of the boat in still water and y the rate of the current.

Step 3 To form a system, think about the following. Downstream, the rate of the boat is *increased* by the effect of the current. Upstream, the rate is *decreased*.

In many applications, it helps to lay out the information in tabular form. Let's try that strategy here.

Downstream the rate is then

$$x + y$$

Upstream, the rate is

$$x - y$$

	d	r	t
Downstream	36	$x + y$	2
Upstream	36	$x - y$	3

Since $d = rt$, from the table we can easily form two equations:

$$36 = (x + y)(2) \qquad (20)$$
$$36 = (x - y)(3) \qquad (21)$$

Step 4 We clear equations (20) and (21) of parentheses and simplify, to write the equivalent system

$$x + y = 18$$
$$x - y = 12$$

Solving, we have

$$x = 15 \text{ mi/h}$$
$$y = 3 \text{ mi/h}$$

The rate of the current is 3 mi/h and the rate of the boat in still water is 15 mi/h.

Step 5 To check, verify the $d = rt$ equation in *both* the upstream and the downstream cases. We leave that to you.

✓ Check Yourself 7

A plane flies 480 mi in an easterly direction, with the wind, in 4 h. Returning westerly along the same route, against the wind, the plane takes 6 h. What is the rate of the plane in still air? What is the rate of the wind?

Systems of equations in problem solving have many applications in a business setting. Example 8 illustrates one such application.

Example 8 — Solving a Business-Based Application

A manufacturer produces a standard model and a deluxe model of a 13-inch (in.) television set. The standard model requires 12 h of labor to produce, while the deluxe model requires 18 h. The company has 360 h of labor available per week. The plant's capacity is a total of 25 sets per week. If all the available time and capacity are to be used, how many of each type of set should be produced?

Step 1 The unknowns in this case are the number of standard and deluxe models that can be produced.

Section 8.2 ■ Systems of Equations in Two Variables with Applications **571**

The choices for x and y could have been reversed.

Step 2 Let x be the number of standard models and y the number of deluxe models.

Step 3 Our system will come from the two given conditions that fix the total number of sets that can be produced and the total labor hours available.

$$x + y = 25 \quad \leftarrow \text{Total number of sets}$$
$$12x + 18y = 360 \quad \leftarrow \text{Total labor hours available}$$

Labor hours— standard sets   Labor hours— deluxe sets

Step 4 Solving the system in step 3, we have

$$x = 15 \quad \text{and} \quad y = 10$$

which tells us that to use all the available capacity, the plant should produce 15 standard sets and 10 deluxe sets per week.

Step 5 We leave the check of this result to the reader.

✓ CHECK YOURSELF 8

A manufacturer produces standard cassette players and compact disc players. Each cassette player requires 2 h of electronic assembly, and each CD requires 3 h. The cassette players require 4 h of case assembly and the CDs 2 h. The company has 120 h of electronic assembly time available per week and 160 h of case assembly time. How many of each type of unit can be produced each week if all available assembly time is to be used?

Let's look at one final application that leads to a system of two equations.

Example 9 Solving a Business-Based Application

Two car rental agencies have the following rate structures for a subcompact car. Company A charges $20 per day plus 15¢ per mile. Company B charges $18 per day plus 16¢ per mile. If you rent a car for 1 day, for what number of miles will the two companies have the same total charge?

Letting c represent the total a company will charge and m the number of miles driven, we calculate the following.

For company A:

You first saw this type of linear model in exercises in Section 4.2.

$$c(m) = 20 + 0.15m \tag{22}$$

For company B:

$$c(m) = 18 + 0.16m \tag{23}$$

The system can be solved most easily by substitution. Substituting $18 + 0.16m$ for $c(m)$ in equation (22) gives

$$18 + 0.16m = 20 + 0.15m$$
$$0.01m = 2$$
$$m = 200 \text{ mi}$$

The graph of the system is shown below.

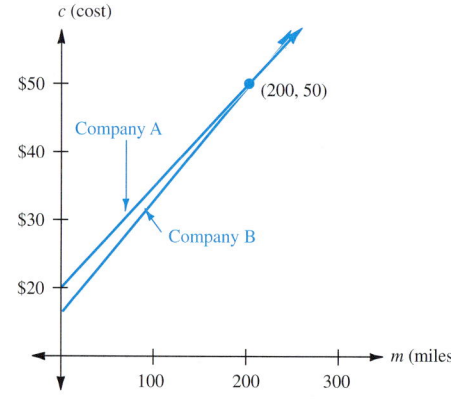

From the graph, how would you make a decision about which agency to use?

✓ CHECK YOURSELF 9

For a compact car, the same two companies charge $27 per day plus 20¢ per mile and $24 per day plus 22¢ per mile. For a 2-day rental, for what mileage will the charges be the same? What is the total charge?

✓ CHECK YOURSELF ANSWERS

1. $\left\{\left(\dfrac{5}{2}, -3\right)\right\}$.   2. $\{(-3, -4)\}$.
3. (a) Inconsistent system: no solution; (b) dependent system: an infinite number of solutions.   4. (a) $\left\{\left(4, \dfrac{-2}{3}\right)\right\}$; (b) $\left\{\left(-2, \dfrac{3}{4}\right)\right\}$.   5. 50 lb of peanuts and 10 lb of cashews.   6. 100 mL of the 30% and 200 mL of the 15%.   7. 100 mi/h plane and 20 mi/h wind.   8. 30 cassette players and 20 CDs.
9. At 300 mi, $114 charge.

Exercises ■ 8.2

1. (2, 3) 2. (6, 2)
3. $\left(-5, \frac{3}{2}\right)$ 4. (5, −3)
5. (2, 1) 6. (3, 5)
7. Dependent
8. (6, 4) 9. (5, −3)
10. Inconsistent
11. Inconsistent
12. (−4, 3) 13. (−8, −2)
14. Dependent
15. (5, −2) 16. (10, 6)
17. (−4, −3)
18. Inconsistent
19. Dependent
20. (4, −2)
21. $\left(\frac{1}{3}, 2\right)$
22. $\left(5, \frac{3}{5}\right)$
23. (10, 1)
24. (−2, −1)
25. Inconsistent
26. (3, −1)
27. $\left(-2, -\frac{8}{3}\right)$
28. $\left(\frac{3}{5}, -1\right)$
29. (3, 4)
30. $\left(-2, \frac{3}{2}\right)$
31. (4, −5)
32. (−6, −3)

In Exercises 1 to 14, solve each system by the addition method. If a unique solution does not exist, state whether the system is inconsistent or dependent.

1. $\begin{aligned} 2x - y &= 1 \\ -2x + 3y &= 5 \end{aligned}$

2. $\begin{aligned} x + 3y &= 12 \\ 2x - 3y &= 6 \end{aligned}$

3. $\begin{aligned} x + 2y &= -2 \\ 3x + 2y &= -12 \end{aligned}$

4. $\begin{aligned} 2x + 3y &= 1 \\ 5x + 3y &= 16 \end{aligned}$

5. $\begin{aligned} x + y &= 3 \\ 3x - 2y &= 4 \end{aligned}$

6. $\begin{aligned} x - y &= -2 \\ 2x + 3y &= 21 \end{aligned}$

7. $\begin{aligned} 2x + y &= 8 \\ -4x - 2y &= -16 \end{aligned}$

8. $\begin{aligned} 3x - 4y &= 2 \\ 4x - y &= 20 \end{aligned}$

9. $\begin{aligned} 5x - 2y &= 31 \\ 4x + 3y &= 11 \end{aligned}$

10. $\begin{aligned} 2x - y &= 4 \\ 6x - 3y &= 10 \end{aligned}$

11. $\begin{aligned} 3x - 2y &= 7 \\ -6x + 4y &= -15 \end{aligned}$

12. $\begin{aligned} 3x + 4y &= 0 \\ 5x - 3y &= -29 \end{aligned}$

13. $\begin{aligned} -2x + 7y &= 2 \\ 3x - 5y &= -14 \end{aligned}$

14. $\begin{aligned} 5x - 2y &= 3 \\ 10x - 4y &= 6 \end{aligned}$

In Exercises 15 to 26, solve each system by the substitution method. If a unique solution does not exist, state whether the system is inconsistent or dependent.

15. $\begin{aligned} x - y &= 7 \\ y &= 2x - 12 \end{aligned}$

16. $\begin{aligned} x - y &= 4 \\ x &= 2y - 2 \end{aligned}$

17. $\begin{aligned} 3x + 2y &= -18 \\ x &= 3y + 5 \end{aligned}$

18. $\begin{aligned} 3x - 18y &= 4 \\ x &= 6y + 2 \end{aligned}$

19. $\begin{aligned} 10x - 2y &= 4 \\ y &= 5x - 2 \end{aligned}$

20. $\begin{aligned} 4x + 5y &= 6 \\ y &= 2x - 10 \end{aligned}$

21. $\begin{aligned} 3x + 4y &= 9 \\ y &= 3x + 1 \end{aligned}$

22. $\begin{aligned} 6x - 5y &= 27 \\ x &= 5y + 2 \end{aligned}$

23. $\begin{aligned} x - 7y &= 3 \\ 2x - 5y &= 15 \end{aligned}$

24. $\begin{aligned} 4x + 3y &= -11 \\ 5x + y &= -11 \end{aligned}$

25. $\begin{aligned} 4x - 12y &= 5 \\ -x + 3y &= -1 \end{aligned}$

26. $\begin{aligned} 5x - 6y &= 21 \\ x - 2y &= 5 \end{aligned}$

In Exercises 27 to 32, solve each system by any method discussed in this section.

27. $\begin{aligned} 2x - 3y &= 4 \\ x &= 3y + 6 \end{aligned}$

28. $\begin{aligned} 5x + y &= 2 \\ 5x - 3y &= 6 \end{aligned}$

29. $\begin{aligned} 4x - 3y &= 0 \\ 5x + 2y &= 23 \end{aligned}$

30. $\begin{aligned} 7x - 2y &= -17 \\ x + 4y &= 4 \end{aligned}$

31. $\begin{aligned} 3x - y &= 17 \\ 5x + 3y &= 5 \end{aligned}$

32. $\begin{aligned} 7x + 3y &= -51 \\ y &= 2x + 9 \end{aligned}$

33. (12, −6)

34. (10, 4)

35. (9, −15)

36. (−8, 4)

37. d

38. f

39. g

40. a

41. h

In Exercises 33 to 36, solve each system by any method discussed in this section. (*Hint*: You should multiply to clear fractions as your first step.)

33. $\dfrac{1}{2}x - \dfrac{1}{3}y = 8$

$\dfrac{1}{3}x + y = -2$

34. $\dfrac{1}{5}x - \dfrac{1}{2}y = 0$

$x - \dfrac{3}{2}y = 4$

35. $\dfrac{2}{3}x + \dfrac{3}{5}y = -3$

$\dfrac{1}{3}x + \dfrac{2}{5}y = -3$

36. $\dfrac{3}{8}x - \dfrac{1}{2}y = -5$

$\dfrac{1}{4}x + \dfrac{3}{2}y = 4$

Each application in Exercises 37 to 44 can be solved by the use of a system of linear equations. Match the application with the appropriate system below.

(a) $12x + 5y = 116$
$8x + 12y = 112$

(b) $x + y = 8000$
$0.06x + 0.09y = 600$

(c) $x + y = 200$
$0.20x + 0.60y = 90$

(d) $x + y = 36$
$y = 3x - 4$

(e) $2(x + y) = 36$
$3(x - y) = 36$

(f) $x + y = 200$
$5.50x + 4y = 980$

(g) $L = 2W + 3$
$2L + 2W = 36$

(h) $x + y = 120$
$2.20x + 5.40y = 360$

37. One number is 4 less than 3 times another. If the sum of the numbers is 36, what are the two numbers?

38. Suppose a movie theater sold 200 adult and student tickets for a showing with a revenue of $980. If the adult tickets were $5.50 and the student tickets were $4, how many of each type of ticket were sold?

39. The length of a rectangle is 3 cm more than twice its width. If the perimeter of the rectangle is 36 cm, find the dimensions of the rectangle.

40. An order of 12 dozen roller-ball pens and 5 dozen ballpoint pens cost $116. A later order for 8 dozen roller-ball pens and 12 dozen ballpoint pens cost $112. What was the cost of 1 dozen of each type of pen?

41. A candy merchant wants to mix peanuts selling at $2.20 per pound with cashews selling at $5.40 per pound to form 120 lb of a mixed-nut blend that will sell for $3 per pound. What amount of each type of nut should be used?

42. b

43. c

44. e

45. 550 adult tickets; 200 student tickets

46. 300 main-floor tickets; 200 balcony tickets

47. 27 in. by 15 in.

48. 8 cm by 29 cm

49. Mulch: $1.80; fertilizer: $3.20

50. $1.20 per disk; $3.50 per package of paper

51. 105 lb of $4 beans, 45 lb of $6.50 beans

52. 50 lb of jelly beans; 150 lb of gumdrops

42. Donald has investments totaling $8000 in two accounts—one a savings account paying 6% interest and the other a bond paying 9%. If the annual interest from the two investments was $600, how much did he have invested at each rate?

43. A chemist wants to combine a 20% alcohol solution with a 60% solution to form 200 mL of a 45% solution. How much of each solution should be used to form the mixture?

44. Xian was able to make a downstream trip of 36 mi in 2 h. Returning upstream, he took 3 h to make the trip. How fast can his boat travel in still water? What was the rate of the river's current?

In Exercises 45 to 66, solve by choosing a variable to represent each unknown quantity and writing a system of equations.

45. **Mixture problem.** Suppose 750 tickets were sold for a concert with a total revenue of $5300. If adult tickets were $8 and students tickets were $4.50, how many of each type of ticket were sold?

46. **Mixture problem.** Theater tickets sold for $7.50 on the main floor and $5 in the balcony. The total revenue was $3250, and there were 100 more main-floor tickets sold than balcony tickets. Find the number of each type of ticket sold.

47. **Geometry.** The length of a rectangle is 3 in. less than twice its width. If the perimeter of the rectangle is 84 in., find the dimensions of the rectangle.

48. **Geometry.** The length of a rectangle is 5 cm more than 3 times its width. If the perimeter of the rectangle is 74 cm, find the dimensions of the rectangle.

49. **Mixture problem.** A garden store sold 8 bags of mulch and 3 bags of fertilizer for $24. The next purchase was for 5 bags of mulch and 5 bags of fertilizer. The cost of that purchase was $25. Find the cost of a single bag of mulch and a single bag of fertilizer.

50. **Mixture problem.** The cost of an order for 10 computer disks and 3 packages of paper was $22.50. The next order was for 30 disks and 5 packages of paper, and its cost was $53.50. Find the price of a single disk and a single package of paper.

51. **Mixture problem.** A coffee retailer has two grades of decaffeinated beans—one selling for $4 per pound and the other for $6.50 per pound. She wishes to blend the beans to form a 150-lb mixture that will sell for $4.75 per pound. How many pounds of each grade of bean should be used in the mixture?

52. **Mixture problem.** A candy merchant sells jelly beans at $3.50 per pound and gumdrops at $4.70 per pound. To form a 200-lb mixture that will sell for $4.40 per pound, how many pounds of each type of candy should be used?

Answers (left column)

53. $7000 time deposit; $5000 bond
54. $4000: savings; $6000: mutual
55. 100 mL of 10%; 300 mL of 50%
56. 200 mL of 70%; 300 mL of 20%
57. 15 mi/h boat; 3 mi/h current
58. Jet: 525 mi/h; air: 75 mi/h
59. 26
60. 82
61. 15 battery-powered calculators; 20 solar models
62. 18 standard; 12 cordless

Problems

53. Investment. Cheryl decided to divide $12,000 into two investments—one a time deposit that pays 8% annual interest and the other a bond that pays 9%. If her annual interest was $1010, how much did she invest at each rate?

54. Investment. Miguel has $2000 more invested in a mutual fund paying 10% interest than in a savings account paying 7%. If he received $880 in interest for 1 year, how much did he have invested in the two accounts?

55. Science. A chemist mixes a 10% acid solution with a 50% acid solution to form 400 mL of a 40% solution. How much of each solution should be used in the mixture?

56. Science. A laboratory technician wishes to mix a 70% saline solution and a 20% saline solution to prepare 500 mL of a 40% solution. What amount of each solution should be used?

57. Motion. A boat traveled 36 mi up a river in 3 h. Returning downstream, the boat took 2 h. What is the boat's rate in still water, and what is the rate of the river's current?

58. Motion. A jet flew east a distance of 1800 mi with the jetstream in 3 h. Returning west, against the jetstream, the jet took 4 h. Find the jet's speed in still air and the rate of the jetstream.

59. Number problem. The sum of the digits of a two-digit number is 8. If the digits are reversed, the new number is 36 more than the original number. Find the original number. (*Hint*: If u represents the units digit of the number and t the tens digit, the original number can be represented by $10t + u$.)

60. Number problem. The sum of the digits of a two-digit number is 10. If the digits are reversed, the new number is 54 less than the original number. What was the original number?

61. Business. A manufacturer produces a battery-powered calculator and a solar model. The battery-powered model requires 10 min of electronic assembly and the solar model 15 min. There are 450 min of assembly time available per day. Both models require 8 min for packaging, and 280 min of packaging time are available per day. If the manufacturer wants to use all the available time, how many of each unit should be produced per day?

62. Business. A small tool manufacturer produces a standard model and a cordless model power drill. The standard model takes 2 h of labor to assembly and the cordless model 3 h. There are 72 h of labor available per week for the drills. Material costs for the standard drill are $10, and for the cordless drill they are $20. The company wishes to limit material costs to $420 per week. How many of each model drill should be produced in order to use all the available resources?

Section 8.2 ■ Systems of Equations in Two Variables with Applications

63. 15
64. 10
65. 100 mi
66. 800 sq ft
67. $\left(\dfrac{2}{3}, \dfrac{2}{5}\right)$
68. $\left(\dfrac{3}{2}, 9\right)$
69. $\left(\dfrac{1}{3}, -\dfrac{3}{2}\right)$
70. (8, 2)

63. Economics. In economics, a demand equation gives the quantity D that will be demanded by consumers at a given price p, in dollars. Suppose that $D = 210 - 4p$ for a particular product.

A supply equation gives the supply S that will be available from producers at price p. Suppose also that for the same product $S = 10p$.

The equilibrium point is that point where the supply equals the demand (here, where $S = D$). Use the given equations to find the equilibrium point.

64. Economics. Suppose the demand equation for a product is $D = 150 - 3p$ and the supply equation is $S = 12p$. Find the equilibrium point for the product.

65. Consumer affairs. Two car rental agencies have the following rate structure for compact cars.

Company A: $30/day and 22¢/mi.

Company B: $28/day and 26¢/mi.

For a 2-day rental, at what number of miles will the charges be the same?

66. Construction. Two construction companies submit the following bid.

Company A: $5000 plus $15/square foot of building.

Company B: $7000 plus $12.50/square foot of building.

For what number of square feet of building will the bids of the two companies be the same?

Certain systems that are not linear can be solved with the methods of this section if we first substitute to change variables. For instance, the system

$$\dfrac{1}{x} + \dfrac{1}{y} = 4$$

$$\dfrac{1}{x} - \dfrac{3}{y} = -6$$

can be solved by the substitutions $u = \dfrac{1}{x}$ and $v = \dfrac{1}{y}$. That gives the system $u + v = 4$ and $u - 3v = -6$. The system is then solved for u and v, and the corresponding values for x and y are found. Use this method to solve the systems in Exercises 67 to 70.

67. $\dfrac{1}{x} + \dfrac{1}{y} = 4$

$\dfrac{1}{x} - \dfrac{3}{y} = -6$

68. $\dfrac{1}{x} + \dfrac{3}{y} = 1$

$\dfrac{4}{x} + \dfrac{3}{y} = 3$

69. $\dfrac{2}{x} + \dfrac{3}{y} = 4$

$\dfrac{2}{x} - \dfrac{6}{y} = 10$

70. $\dfrac{4}{x} - \dfrac{3}{y} = -1$

$\dfrac{12}{x} - \dfrac{1}{y} = 1$

71. $y = \frac{3}{2}x - 2$

72. $y = -\frac{2}{3}x + 5$

73. $(1.3, -0.5)$

74. $(-1.5, 0.6)$

75. $(5.8, 1.7)$

76. $(4.7, 10.7)$

77. Answers will vary.

78. Answers will vary.

79. $y = \dfrac{AF - CD}{AE - BD}$

$AE - BD \neq 0$

$x = \dfrac{CE - BF}{AE - BD}$

$AE - BD \neq 0$

Writing the equation of a line through two points can be done by the following method. Given the coordinates of two points, substitute each pair of values into the equation $y = mx + b$. This gives a system of two equations in variables m and b, which can be solved as before.

In Exercises 71 and 72, write the equation of the line through each of the following pairs of points, using the method outlined above.

71. $(2, 1)$ and $(4, 4)$ **72.** $(-3, 7)$ and $(6, 1)$

In Exercises 73 and 74, use your calculator to approximate the solution to each system. Express each coordinate to the nearest tenth.

73. $y = 2x - 3$ **74.** $3x - 4y = -7$

$2x + 3y = 1$ $2x + 3y = -1$

For Exercises 75 and 76, adjust the viewing window on your calculator so that you can see the point of intersection for the two lines representing the equations in the system. Then approximate the solution, expressing each coordinate to the nearest tenth.

75. $5x - 12y = 8$ **76.** $9x - 3y = 10$

$7x + 2y = 44$ $x + 5y = 58$

77. We have discussed three different methods of solving a system of two linear equations in two unknowns: the graphical method, the addition method, and the substitution method. Discuss the strengths and weaknesses of each method.

78. Determine a system of two linear equations for which the solution is $(3, 4)$. Are there other systems which have the same solution? If so, determine at least one more and explain why this can be true.

79. Suppose we have the following linear system:

$$Ax + By = C \tag{1}$$

$$Dx + Ey = F \tag{2}$$

(a) Multiply equation (1) by $-D$, multiply equation (2) by A and add. This will allow you to eliminate x. Solve for y and indicate what must be true about the coefficients in order for a unique value for y to exist.

(b) Now return to the original system and eliminate y instead of x. (*Hint*: try multiplying equation (1) by E and equation (2) by $-B$.) Solve for x and again indicate what must be true about the coefficients for a unique value for x to exist.

SECTION 8.3: Systems of Linear Equations in Three Variables

8.3 OBJECTIVES

1. Find ordered triples associated with each equation
2. Solve by the addition method
3. Look at a graphic interpretation of a solution
4. Solve an application of a system with three variables

Suppose an application involves three quantities that we want to label x, y, and z. A typical equation used for the solution might be

$$2x + 4y - z = 8$$

This is called a **linear equation in three variables.** A solution for such an equation is an **ordered triple** (x, y, z) of real numbers that satisfies the equation. For example, the ordered triple $(2, 1, 0)$ is a solution for the equation above since substituting 2 for x, 1 for y, and 0 for z results in the following true statement.

$$2 \cdot 2 + 4 \stackrel{?}{=} 8$$
$$4 + 4 \stackrel{?}{=} 8$$
$$8 = 8 \quad \text{True}$$

Of course, other solutions, in fact infinitely many, exist. You might want to verify that $(1, 1, -2)$ and $(3, 1, 2)$ are also solutions. To extend the concepts of the last section, we want to consider systems of three linear equations in three variables such as

$$x + y + z = 5$$
$$2x - y + z = 9$$
$$x - 2y + 3z = 16$$

For a unique solution to exist, when *three variables* are involved, we must have *three equations*.

The choice of which variable to eliminate is yours. Generally, you should pick the variable that allows the easiest computation.

The solution for such a system is the set of all ordered triples that satisfy each equation of the system. In this case, you should verify that $(2, -1, 4)$ is a solution for the system since that ordered triple makes each equation a true statement.

Let's turn now to the solution process itself. In this section, we will consider the addition method. We will then apply what we have learned to solving applications.

The Addition Method

The central idea is to choose *two pairs* of equations from the system and, by the addition method, to eliminate the *same variable* from each of those pairs. The method is best illustrated by example. So let's proceed to see how the solution for the previous system was determined.

579

Example 1 — Solving a Linear System in Three Variables

Solve the system.

$$x + y + z = 5 \quad (1)$$
$$2x - y + z = 9 \quad (2)$$
$$x - 2y + 3z = 16 \quad (3)$$

First we choose two of the equations and the variable to eliminate. Variable y seems convenient in this case. Pairing equations (1) and (2) and then adding, we have

Any pair of equations could have been selected.

$$\begin{array}{r} x + y + z = 5 \\ 2x - y + z = 9 \\ \hline 3x \phantom{{}+y} + 2z = 14 \end{array} \quad (4)$$

We now want to choose a different pair of equations to eliminate y. Using equations (1) and (3) this time, we multiply equation (1) by 2 and then add the result to equation (3):

$$\begin{array}{r} 2x + 2y + 2z = 10 \\ x - 2y + 3z = 16 \\ \hline 3x \phantom{{}+2y} + 5z = 26 \end{array} \quad (5)$$

We now have equations (4) and (5) in variables x and z.

$$3x + 2z = 14$$
$$3x + 5z = 26$$

Since we are now dealing with a system of two equations in two variables, any of the methods of the previous section apply. We have chosen to multiply equation (4) by -1 and then add that result to equation (5). This yields

$$3z = 12$$
$$z = 4$$

Substituting $z = 4$ in equation (4) gives

$$3x + 2 \cdot 4 = 14$$
$$3x + 8 = 14$$
$$3x = 6$$
$$x = 2$$

Any of the original equations could have been used.

To check, substitute these values into the other equations of the original system.

Finally, letting $x = 2$ and $z = 4$ in equation (1) gives

$$2 + y + 4 = 5$$
$$y = -1$$

and $(2, -1, 4)$ is shown to be the solution for the system.

 CHECK YOURSELF 1

Solve the system.

$$x - 2y + z = 0$$
$$2x + 3y - z = 16$$
$$3x - y - 3z = 23$$

One or more of the equations of a system may already have a missing variable. The elimination process is simplified in that case, as Example 2 illustrates.

Example 2

Solving a Linear System in Three Variables

Solve the system.

$$2x + y - z = -3 \quad (6)$$
$$y + z = 2 \quad (7)$$
$$4x - y + z = 12 \quad (8)$$

Noting that equation (7) involves only y and z, we must simply find another equation in those same two variables. Multiply equation (6) by -2 and add the result to equation (8) to eliminate x.

$$-4x - 2y + 2z = 6$$
$$\underline{4x - y + z = 12}$$
$$-3y + 3z = 18$$
$$y - z = -6 \quad (9)$$

We now have a *second* equation in y and z.

We now form a system consisting of equations (7) and (9) and solve as before.

$$y + z = 2$$
$$\underline{y - z = -6} \quad \text{Adding eliminates } z.$$
$$2y = -4$$
$$y = -2$$

From equation (7), if $y = -2$,

$$-2 + z = 2$$
$$z = 4$$

and from equation (6), if $y = -2$ and $z = 4$,

$$2x - 2 - 4 = -3$$
$$2x = 3$$
$$x = \frac{3}{2}$$

The solution for the system is

$$\left(\frac{3}{2}, -2, 4\right)$$

✓ CHECK YOURSELF 2

Solve the system.

$$x + 2y - z = -3$$
$$x - y + z = 2$$
$$x - z = 3$$

The following algorithm summarizes the procedure for finding the solutions for a linear system of three equations in three variables.

> **Solving a System of Three Equations in Three Unknowns**
>
> **Step 1** Choose a pair of equations from the system, and use the addition method to eliminate one of the variables.
>
> **Step 2** Choose a *different* pair of equations, and eliminate the *same* variable.
>
> **Step 3** Solve the system of two equations in two variables determined in steps 1 and 2.
>
> **Step 4** Substitute the values found above into one of the original equations, and solve for the remaining variable.
>
> **Step 5** The solution is the ordered triple of values found in steps 3 and 4. It can be checked by substituting into the other equations of the original system.

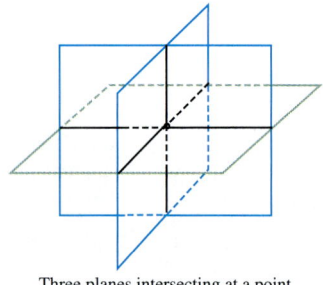

Three planes intersecting at a point

Systems of three equations in three variables may have (1) exactly one solution, (2) infinitely many solutions, or (3) no solution. Before we look at an algebraic approach in the second and third cases, let's discuss the geometry involved.

The graph of a linear equation in three variables is a plane (a flat surface) in three dimensions. Two distinct planes either will be parallel or will intersect in a line.

If three distinct planes intersect, that intersection will be either a single point (as in our first example) or a line (think of three pages in an open book—they intersect along the binding of the book).

Let's look at an example of how the solution proceeds in these cases.

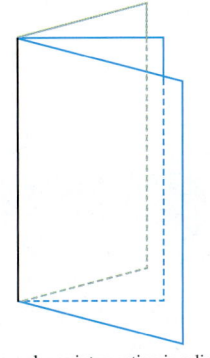

Three planes intersecting in a line

Example 3

Solving a Dependent Linear System in Three Variables

Solve the system.

$$x + 2y - z = 5 \quad (10)$$
$$x - y + z = -2 \quad (11)$$
$$-5x - 4y + z = -11 \quad (12)$$

We begin as before by choosing two pairs of equations from the system and eliminating the same variable from each of the pairs. Adding equations (10) and (11) gives

$$2x + y = 3 \quad (13)$$

Adding equations (10) and (12) gives

$$-4x - 2y = -6 \quad (14)$$

Now consider the system formed by equations (13) and (14). We multiply equation (13) by 2 and add again:

$$\begin{aligned} 4x + 2y &= 6 \\ -4x - 2y &= -6 \\ \hline 0 &= 0 \end{aligned}$$

There are ways of representing the solutions, as you will see in later courses.

This true statement tells us that the system has an infinite number of solutions (lying along a straight line). Again, such a system is dependent.

✓ CHECK YOURSELF 3

Solve the system.

$$2x - y + 3z = 3$$
$$-x + y - 2z = 1$$
$$y - z = 5$$

There is a third possibility for the solutions in three variables, as Example 4 illustrates.

Example 4 Solving an Inconsistent Linear System in Three Variables

Solve the system.

$$3x + y - 3z = 1 \quad (15)$$
$$-2x - y + 2z = 1 \quad (16)$$
$$-x - y + z = 2 \quad (17)$$

This time we eliminate variable *y*. Adding equations (15) and (16), we have

$$x - z = 2 \quad (18)$$

Adding equations (15) and (17) gives

$$2x - 2z = 3 \quad (19)$$

Now, multiply equation (18) by -2 and add the result to equation (19).

$$-2x + 2z = -4$$
$$\underline{2x - 2z = 3}$$
$$0 = -1$$

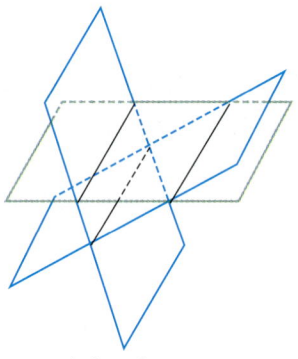

An inconsistent system

All the variables have been eliminated, and we have arrived at a contradiction, $0 = -1$. This means that the system is *inconsistent* and has no solutions. There is *no* point common to all three planes.

CHECK YOURSELF 4

Solve the system.

$$x - y - z = 0$$
$$-3x + 2y + z = 1$$
$$3x - y + z = -1$$

As a closing note, we have by no means illustrated all possible types of inconsistent and dependent systems. Other possibilities involve either distinct parallel planes or planes that coincide. The solution techniques in these additional cases are, however, similar to those illustrated above.

Solving Applications

In many instances, if an application involves three unknown quantities, you will find it useful to assign three variables to those quantities and then build a system of three equations from the given relationships in the problem. The extension of our problem solving strategy is natural, as Example 5 illustrates.

Example 5

Solving a Number Problem

The sum of the digits of a three-digit number is 12. The tens digit is 2 less than the hundreds digit, and the units digit is 4 less than the sum of the other two digits. What is the number?

Step 1 The three unknowns are, of course, the three digits of the number.

Sometimes it helps to choose variable letters that relate to the words as is done here.

Step 2 We now want to assign variables to each of three digits. Let u be the units digit, t be the tens digit, and h be the hundreds digit.

Take a moment now to go back to the original problem and pick out those conditions. That skill is a crucial part of the problem-solving strategy.

Step 3 There are three conditions given in the problem that allow us to write the necessary three equations. From those conditions

$$h + t + u = 12$$
$$t = h - 2$$
$$u = h + t - 4$$

Step 4 There are various ways to approach the solution. To use addition, write the system in the equivalent form

$$h + t + u = 12$$
$$-h + t = -2$$
$$-h - t + u = -4$$

and solve by our earlier methods. The solution, which you can verify, is $h = 5$, $t = 3$, and $u = 4$. The desired number is 534.

Step 5 To check, you should show that the digits of 534 meet each of the conditions of the original problem.

✓ CHECK YOURSELF 5

The sum of the measures of the angles of a triangle is 180°. In a given triangle, the measure of the second angle is twice the measure of the first. The measure of the third angle is 30° less than the sum of the measures of the first two. Find the measure of each angle.

Let's continue with a slightly different application that will lead to a system of three equations.

Example 6 Solving an Investment Application

Monica decided to divide a total of $42,000 into three investments: a savings account paying 5% interest, a time deposit paying 7%, and a bond paying 9%. Her total annual interest from the three investments was $2600, and the interest from the savings account was $200 less than the total interest from the other two investments. How much did she invest at each rate?

Step 1 The three amounts are the unknowns.

Again, we choose letters that suggest the unknown quantities—s for savings, t for time deposit, and b for bond.

Step 2 We let s be the amount invested at 5%, t the amount at 7%, and b the amount at 9%. Note that the interest from the savings account is then $0.05s$, and so on.

A table will help with the next step.

For 1 year, the interest formula is

$$I = Pr$$

(interest equals principal times rate).

	5%	7%	9%
Principal	s	t	b
Interest	$0.05s$	$0.07t$	$0.09b$

Step 3 Again there are three conditions in the given problem. By using the table above, they lead to the following equations.

Total invested.

$$s + t + b = 42{,}000$$

Total interest.

$$0.05s + 0.07t + 0.09b = 2600$$

The savings interest was $200 *less than* that from the other two investments.

$$0.05s = 0.07t + 0.09b - 200$$

Step 4 We clear of decimals and solve as before, with the result

$$s = \$24{,}000 \qquad t = \$11{,}000 \qquad b = \$7000$$

Find the interest earned from each investment, and verify that the conditions of the problem are satisfied.

Step 5 We leave the check of these solutions to you.

✓ CHECK YOURSELF 6

Glenn has a total of $11,600 invested in three accounts: a savings account paying 6% interest, a stock paying 8%, and a mutual fund paying 10%. The annual interest from the stock and mutual fund is twice that from the savings account, and the mutual fund returned $120 more than the stock. How much did Glenn invest in each account?

✓ CHECK YOURSELF ANSWERS

1. $\{(5, 1, -3)\}$. 2. $\{(1, -3, -2)\}$.
3. The system is dependent (three are an infinite number of solutions).
4. The system is inconsistent (there are no solutions).
5. The three angles are 35°, 70°, and 75°.
6. $5000 in savings, $3000 in stocks, and $3600 in mutual funds.

Exercises · 8.3

Answers (left column):

1. $(1, 2, 4)$
2. Inconsistent
3. $(-2, 1, 2)$
4. $(2, -1, -3)$
5. $(-4, 3, 2)$
6. $(3, -1, -5)$
7. Infinite number of solutions
8. $\left(-4, 4, \dfrac{2}{3}\right)$
9. $\left(3, \dfrac{1}{2}, -\dfrac{7}{2}\right)$
10. Inconsistent
11. $(3, 2, -5)$
12. $\left(\dfrac{1}{2}, -\dfrac{1}{2}, \dfrac{11}{2}\right)$
13. $(2, 0, -3)$
14. $(30, -13, 15)$
15. $\left(4, -\dfrac{1}{2}, \dfrac{3}{2}\right)$
16. $\left(-\dfrac{3}{2}, \dfrac{11}{4}, \dfrac{19}{4}\right)$
17. Inconsistent
18. $(3, -1, -2)$
19. $\left(2, \dfrac{5}{2}, -\dfrac{3}{2}\right)$
20. Infinite number of solutions

In Exercises 1 to 20, solve each system of equations. If a unique solution does not exist, state whether the system is inconsistent or has an infinite number of solutions.

1. $\begin{aligned} x - y + z &= 3 \\ 2x + y + z &= 8 \\ 3x + y - z &= 1 \end{aligned}$

2. $\begin{aligned} x - y - z &= 2 \\ 2x + y + z &= 8 \\ x + y + z &= 6 \end{aligned}$

3. $\begin{aligned} x + y + z &= 1 \\ 2x - y + 2z &= -1 \\ -x - 3y + z &= 1 \end{aligned}$

4. $\begin{aligned} -x - y - z &= 6 \\ -x + 3y + 2z &= -11 \\ 3x + 2y + z &= 1 \end{aligned}$

5. $\begin{aligned} x + y + z &= 1 \\ -2x + 2y + 3z &= 20 \\ 2x - 2y - z &= -16 \end{aligned}$

6. $\begin{aligned} x + y + z &= 3 \\ 3x + y - z &= 13 \\ 3x + y - 2z &= 18 \end{aligned}$

7. $\begin{aligned} 2x + y - z &= 2 \\ -x - 3y + z &= -1 \\ -4x + 3y + z &= -4 \end{aligned}$

8. $\begin{aligned} x + 4y - 6z &= 8 \\ 2x - y + 3z &= -10 \\ 3x - 2y + 3z &= -18 \end{aligned}$

9. $\begin{aligned} 3x - y + z &= 5 \\ x + 3y + 3z &= -6 \\ x + 4y - 2z &= 12 \end{aligned}$

10. $\begin{aligned} 2x - y + 3z &= 2 \\ x - 2y + 3z &= 1 \\ 4x - y + 5z &= 5 \end{aligned}$

11. $\begin{aligned} x + 2y + z &= 2 \\ 2x + 3y + 3z &= -3 \\ 2x + 3y + 2z &= 2 \end{aligned}$

12. $\begin{aligned} x - 4y - z &= -3 \\ x + 2y + z &= 5 \\ 3x - 7y - 2z &= -6 \end{aligned}$

13. $\begin{aligned} x + 3y - 2z &= 8 \\ 3x + 2y - 3z &= 15 \\ 4x + 2y + 3z &= -1 \end{aligned}$

14. $\begin{aligned} x + y - z &= 2 \\ 3x + 5y - 2z &= -5 \\ 5x + 4y - 7z &= -7 \end{aligned}$

15. $\begin{aligned} x + y - z &= 2 \\ x \phantom{{}+y} - 2z &= 1 \\ 2x - 3y - z &= 8 \end{aligned}$

16. $\begin{aligned} x + y + z &= 6 \\ x - 2y \phantom{{}+z} &= -7 \\ 4x + 3y + z &= 7 \end{aligned}$

17. $\begin{aligned} x - 3y + 2z &= 1 \\ 16y - 9z &= 5 \\ 4x + 4y - z &= 8 \end{aligned}$

18. $\begin{aligned} x - 4y + 4z &= -1 \\ y - 3z &= 5 \\ 3x - 4y + 6z &= 1 \end{aligned}$

19. $\begin{aligned} x + 2y - 4z &= 13 \\ 3x + 4y - 2z &= 19 \\ 3x \phantom{{}+4y} + 2z &= 3 \end{aligned}$

20. $\begin{aligned} x + 2y - z &= 6 \\ -3x - 2y + 5z &= -12 \\ x \phantom{{}+2y} - 2z &= 3 \end{aligned}$

21. 3, 5, 8

22. 5, 7, 12

23. 3 nickels; 5 dimes; 17 quarters

24. 150 adult tickets; 80 student tickets; 48 children's tickets

25. 4 cm, 7 cm, 8 cm

26. (35°, 50°, 95°)

27. $3000 savings; $9000 bond; $5000 money market

28. $3000 savings; $5000 time deposit; $2000 bond

29. 243

30. 261

31. 30

32. 36

Solve Exercises 21 to 36 by choosing a variable to represent each unknown quantity and writing a system of equations.

21. **Number problem.** The sum of three numbers is 16. The largest number is equal to the sum of the other two, and 3 times the smallest number is 1 more than the largest. Find the three numbers.

22. **Number problem.** The sum of three numbers is 24. Twice the smallest number is 2 less than the largest number, and the largest number is equal to the sum of the other two. What are the three numbers?

23. **Coin problem.** A cashier has 25 coins consisting of nickels, dimes, and quarters with a value of $4.90. If the number of dimes is 1 less than twice the number of nickels, how many of each type of coin does she have?

24. **Recreation.** A theater has tickets at $6 for adults, $3.50 for students, and $2.50 for children under 12 years old. A total of 278 tickets were sold for one showing with a total revenue of $1300. If the number of adult tickets sold was 10 less than twice the number of student tickets, how many of each type of ticket were sold for the showing?

25. **Geometry.** The perimeter of a triangle is 19 cm. If the length of the longest side is twice that of the shortest side and 3 cm less than the sum of the lengths of the other two sides, find the lengths of the three sides.

26. **Geometry.** The measure of the largest angle of a triangle is 10° more than the sum of the measures of the other two angles and 10° less than 3 times the measure of the smallest angle. Find the measures of the three angles of the triangle.

27. **Investments.** Jovita divides $17,000 into three investments: a savings account paying 6% annual interest, a bond paying 9%, and a money market fund paying 11%. The annual interest from the three accounts is $1540, and she has 3 times as much invested in the bond as in the savings account. What amount does she have invested in each account?

28. **Investments.** Adrienne has $10,000 invested in a savings account paying 5%, a time deposit paying 7%, and a bond paying 10%. She has $1000 less invested in the bond than in her savings account, and she earned $700 in annual interest. What has she invested in each account?

29. **Number problem.** The sum of the digits of a three-digit number is 9, and the tens digit of the number is twice the hundreds digit. If the digits are reversed in order, the new number is 99 more than the original number. What is the original number?

30. **Number problem.** The sum of the digits of a three-digit number is 9. The tens digit is 3 times the hundreds digit. If the digits are reversed in order, the new number is 99 less than the original number. Find the original three-digit number.

31. **Business.** A manufacturer can produce and sell x items per week at a cost of $C = 20x + 3600$. The revenue from selling those items is given by $R = 140x$. Find the break-even point for this product.

32. **Business.** If the cost for a second product is given by $C = 30x + 3600$ and the revenue by $R = 130x$, find the break-even point for that product.

33. 110 single; 230 carpool

34. 160 single; 260 carpool

35. Roy 8 mi; Sally 16 mi; Jeff 26 mi

36. 20 motorcycles; 135 cars; 25 vans

37. $(1, 2, -1, -2)$

38. $(3, 2, 0, 1)$

39. $(7, 2)$

40. No solution

33. Consumer affairs. To encourage carpooling, a city charges $10 per single driver or $4 per person for carpools of two or more people in its city parking lots. If one parking lot took in $2020 and 340 commuters used that lot for 1 day, how many of each type of commuter—single driver or carpool rider—used that lot that day?

34. Consumer affairs. To encourage carpooling, a city charges $12 per single driver or $5 per person for carpools of two or more people in its city parking lots. If one parking lot took in $3220 and 420 commuters used that lot for 1 day, how many of each type of commuter—single driver or carpool rider—used that lot that day?

35. Motion. Roy, Sally, and Jeff drive a total of 50 mi to work each day. Sally drives twice as far as Roy, and Jeff drives 10 mi farther than Sally. Use a system of three equations in three unknowns to find how far each person drives each day.

36. Consumer affairs. A parking lot has spaces reserved for motorcycles, cars, and vans. There are 5 more spaces reserved for vans than for motorcycles. There are 3 times as many car spaces as van and motorcycle spaces combined. If the parking lot has 180 total reserved spaces, how many of each type are there?

The solution process illustrated in this section can be extended to solving systems of more than three variables in a natural fashion. For instance, if four variables are involved, eliminate one variable in the system and then solve the resulting system in three variables as before. Substituting those three values into one of the original equations will provide the value for the remaining variable and the solution for the system.

In Exercises 37 and 38, use this procedure to solve the system.

37. $\begin{aligned} x + 2y + 3z + w &= 0 \\ -x - y - 3z + w &= -2 \\ x - 3y + 2z + 2w &= -11 \\ -x + y - 2z + w &= 1 \end{aligned}$

38. $\begin{aligned} x + y - 2z - w &= 4 \\ x - y + z + 2w &= 3 \\ 2x + y - z - w &= 7 \\ x - y + 2z + w &= 2 \end{aligned}$

In some systems of equations there are more equations than variables. We can illustrate this situation with a system of three equations in two variables. To solve this type of system, pick any two of the equations and solve this system. Then substitute the solution obtained into the third equation. If a true statement results, the solution used is the solution to the entire system. If a false statement occurs, the system has no solution.

In Exercises 39 and 40, use this procedure to solve each system.

39. $\begin{aligned} x - y &= 5 \\ 2x + 3y &= 20 \\ 4x + 5y &= 38 \end{aligned}$

40. $\begin{aligned} 3x + 2y &= 6 \\ 5x + 7y &= 35 \\ 7x + 9y &= 8 \end{aligned}$

41. $T = 20$, $C = 25$, $B = 40$

42. $T = 30$, $C = 16$, $R = 95$

43. (a) $y = 2x^2 - x + 4$

    (b) $y = 3x^2 - 2x + 1$

41. Experiments have shown that cars (C), trucks (T), and buses (B) emit different amounts of air pollutants. In one such experiment, a truck emitted 1.5 pounds (lb) of carbon dioxide (CO_2) per passenger-mile and 2 grams (g) of nitrogen oxide (NO) per passenger-mile. A car emitted 1.1 lb of CO_2 per passenger-mile and 1.5 g of NO per passenger-mile. A bus emitted 0.4 lb of CO_2 per passenger-mile and 1.8 g of NO per passenger-mile. A total of 85 mi was driven by the three vehicles, and 73.5 lb of CO_2 and 149.5 g of NO were collected. Use the following system of equations to determine the miles driven by each vehicle.

$$T + C + B = 85.0$$
$$1.5T + 1.1C + 0.4B = 73.5$$
$$2T + 1.5C + 1.8B = 149.5$$

42. Experiments have shown that cars (C), trucks (T), and trains (R) emit different amounts of air pollutants. In one such experiment, a truck emitted 0.8 lb of carbon dioxide per passenger-mile and 1 g of nitrogen oxide per passenger-mile. A car emitted 0.7 lb of CO_2 per passenger-mile and 0.9 g of NO per passenger-mile. A train emitted 0.5 lb of CO_2 per passenger-mile and 4 g of NO per passenger-mile. A total of 141 mi was driven by the three vehicles, and 82.7 lb of CO_2 and 424.4 g of NO were collected. Use the following system of equations to determine the miles driven by each vehicle.

$$T + C + R = 141.0$$
$$0.8T + 0.7C + 0.5R = 82.7$$
$$T + 0.9C + 4R = 424.4$$

43. In Chapter 11 you will learn about quadratic functions and their graphs. A quadratic function has the form $y = ax^2 + bx + c$, where a, b, and c are specific numbers and $a \neq 0$. Three distinct points on the graph are enough to determine the equation.

(a) Suppose that $(1, 5)$, $(2, 10)$, and $(3, 19)$ are on the graph of $y = ax^2 + bx + c$. Substituting the pair $(1, 5)$ into this equation (that is, let $x = 1$ and $y = 5$) yields $5 = a + b + c$. Substituting each of the other ordered pairs yields: $10 = 4a + 2b + c$ and $19 = 9a + 3b + c$. Solve the resulting system of equations to determine the values of a, b, and c. Then write the equation of the function.

(b) Repeat the work of part a using the following three points: $(1, 2)$, $(2, 9)$, and $(3, 22)$.

SECTION 8.4 Graphing Linear Inequalities in Two Variables

8.4 OBJECTIVES

1. Graph linear inequalities in two variables
2. Graph a region defined by linear inequalities

What does the solution set look like when we are faced with an inequality in two variables? We will see that it is a set of ordered pairs best represented by a shaded region. The general form for a linear inequality in two variables is

$$ax + by < c$$

where a and b cannot both be 0. The symbol $<$ can be replaced with $>$, $\leq$, or $\geq$. Some examples are

$$y < -2x + 6 \qquad x - 2y \leq 4 \qquad \text{or} \qquad 2x - 3y \geq x + 5y$$

As was the case with an equation, the solution set of a linear inequality is a set of ordered pairs of real numbers. However, in the case of the linear inequalities, we will find that the solution sets will be all the points in an entire region of the plane, called a **half plane.**

To determine such a solution set, let's start with the first inequality listed above. To graph the solution set of

$$y < -2x + 6$$

we begin by writing the corresponding linear equation

$$y = -2x + 6$$

First, note that the graph of $y = -2x + 6$ is simply a straight line.

Now, to graph the solution set of $y < -2x + 6$, we must include all ordered pairs that satisfy that inequality. For instance if $x = 1$, we have

$$y < -2 \cdot 1 + 6$$

$$y < 4$$

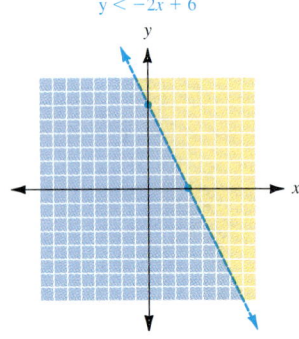

The line is dashed to indicate that the equality is *not* included.

So we want to include all points of the form $(1, y)$, where $y < 4$. Of course, since $(1, 4)$ is *on* the corresponding line, this means that we want all points *below* the line along the vertical line $x = 1$. The result will be similar for any choice of x, and our solution set will then contain all points below the line $y = -2x + 6$. We can then graph the solution set as the shaded region shown. We have the following definition.

We call the graph of the equation

$ax + by = c$

the **boundary line** of the half planes.

In general, the solution set of an inequality of the form

$ax + by < c$ or $ax + by > c$

will be a half plane either above or below the corresponding line determined by

$ax + by = c$

How do we decide which half plane represents the desired solution set? The use of a **test point** provides an easy answer. Choose any point *not* on the line. Then substitute the coordinates of that point into the given inequality. If the coordinates satisfy the inequality (result in a true statement), then shade the region or half plane that includes the test point; if not, shade the opposite half plane. Example 1 illustrates the process.

Example 1

Graphing a Linear Inequality

Graph the linear inequality

$$x - 2y < 4$$

First, we graph the corresponding equation

$$x - 2y = 4$$

to find the boundary line. Now to decide on the appropriate half plane, we need a test point *not* on the line. As long as the line *does not pass through the origin,* we can always use (0, 0) as a test point. It provides the easiest computation.

Here letting $x = 0$ and $y = 0$, we have

$$0 - 2 \cdot 0 < 4$$

$$0 < 4$$

Since this a true statement, we proceed to shade the half plane including the origin (the test point), as shown.

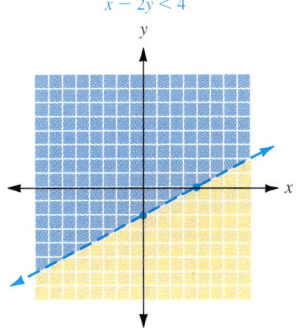

✓ CHECK YOURSELF 1

Graph the solution set of $3x + 4y > 12$.

The graphs of some linear inequalities will include the boundary line. That will be the case whenever equality is included with the inequality statement, as illustrated in Example 2.

Example 2 Graphing a Linear Inequality

Graph the inequality

$$2x + 3y \geq 6$$

First, we graph the boundary line, here corresponding to $2x + 3y = 6$. Note that we use a solid line in this case since equality is included in the original statement.

Again, we choose a convenient test point not on the line. As before, the origin will provide the simplest computation.

Substituting $x = 0$ and $y = 0$, we have

$$2 \cdot 0 + 3 \cdot 0 \geq 6$$

$$0 \geq 6$$

A solid boundary line means that equality is included.

This is a *false* statement. Hence the graph will consist of all points on the *opposite* side from the origin. The graph will then be the upper half plane, as shown.

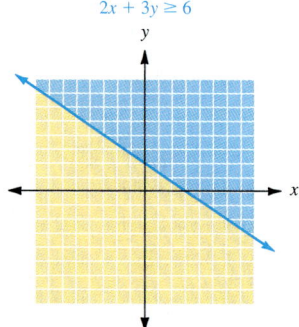

✓ CHECK YOURSELF 2

Graph the solution set of $x - 3y \leq 6$.

Example 3 Graphing a Linear Inequality

Graph the solution set of

$$y \leq 2x$$

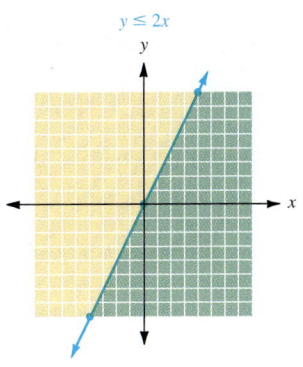

$y \leq 2x$

The choice of (1, 1) is arbitrary. We simply want *any* point *not* on the line.

We proceed as before by graphing the boundary line (it is solid since equality is included). The only difference between this and previous examples is that we *cannot use the origin* as a test point. Do you see why?

Choosing (1, 1) as our test point gives the statement

$$1 \leq 2 \cdot 1$$
$$1 \leq 2$$

Since the statement is *true*, we shade the half plane *including* the test point (1, 1).

✓ **CHECK YOURSELF 3**

Graph the solution set of $3x + y > 0$.

Let's consider a special case of graphing linear inequalities in the rectangular coordinate system.

Example 4 Graphing a Linear Inequality

Graph the solution set of $x > 3$.

Here we specify the rectangular coordinate system to indicate we want a two-dimensional graph.

First, we draw the boundary line (a dashed line since equality is not included) corresponding to

$$x = 3$$

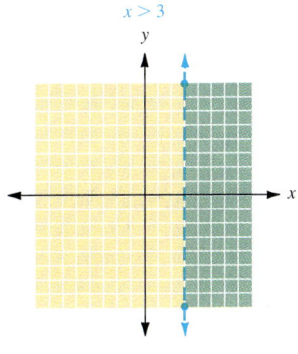

$x > 3$

We can choose the origin as a test point in this case, and that results in the false statement

$$0 > 3$$

We then shade the half plane *not* including the origin. In this case, the solution set is represented by the half plane to the right of the vertical boundary line.

As you may have observed, in this special case choosing a test point is not really necessary. Since we want values of x that are *greater than* 3, we want those ordered pairs that are to the *right* of the boundary line.

CHECK YOURSELF 4

Graph the solution set of

$$y \leq 2$$

in the rectangular coordinate system.

Example 5 — Graphing a Region Defined by Linear Inequalities

Graph the region satisfying the following conditions.

$$3x + 4y \leq 12$$
$$x \geq 0$$
$$y \geq 0$$

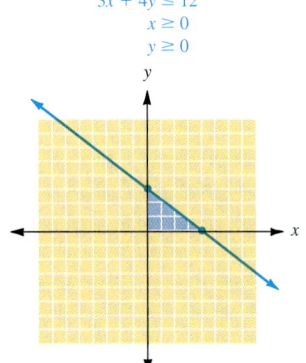

The solution set in this case must satisfy *all three conditions*. As before, the solution set of the first inequality is graphed as the half plane *below* the boundary line. The second and third inequalities mean that x and y must also be nonnegative. Therefore, our solution set is restricted to the first quadrant (and the appropriate segments of the x and y axes), as shown.

✓ CHECK YOURSELF 5

Graph the region satisfying the following conditions.

$$3x + 4y < 12$$
$$x \geq 0$$
$$y \geq 0$$

The following algorithm summarizes our work in graphing linear inequalities in two variables.

To Graph a Linear Inequality

1. Replace the inequality symbol with an equality symbol to form the equation of the boundary line of the solution set.

2. Graph the boundary line. Use a dashed line if equality is not included ($<$ or $>$). Use a solid line if equality is included ($\leq$ or $\geq$).

3. Choose any convenient test point *not* on the boundary line.

4. If the inequality is *true* for the test point, shade the half plane *including* the test point. If the inequality is *false* for the test point, shade the half plane *not including* the test point.

✓ CHECK YOURSELF ANSWERS

1.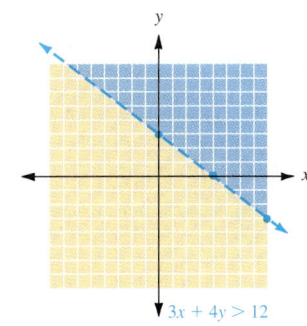
$3x + 4y > 12$

2.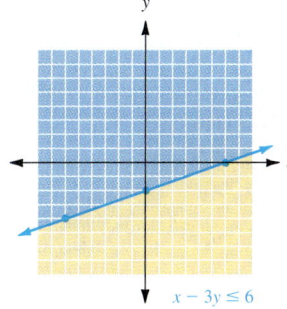
$x - 3y \leq 6$

3.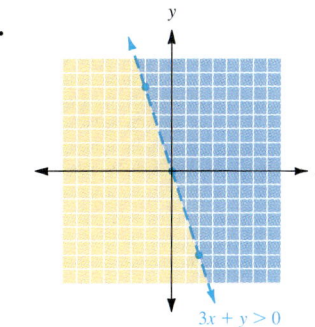
$3x + y > 0$

4.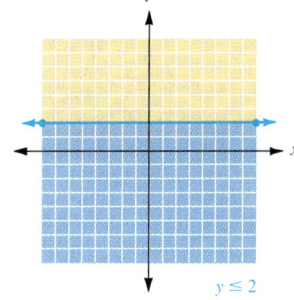
$y \leq 2$

5.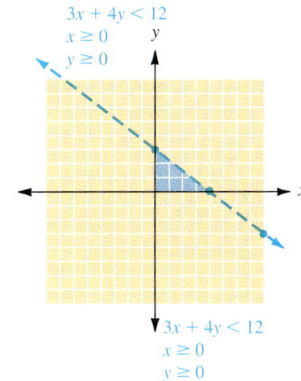
$3x + 4y < 12$
$x \geq 0$
$y \geq 0$

Exercises 8.4

In Exercises 1–24, graph the solution sets of the linear inequalities.

1. $x + y < 4$

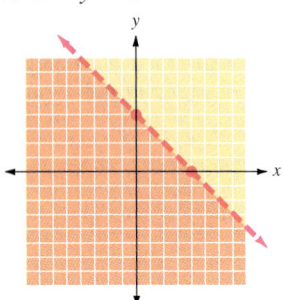

2. $x + y \geq 6$

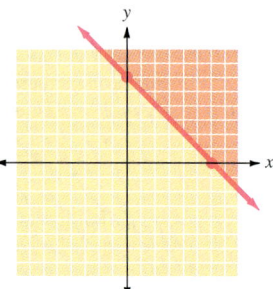

3. $x - y \geq 3$

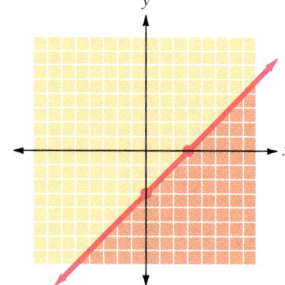

4. $x - y < 5$

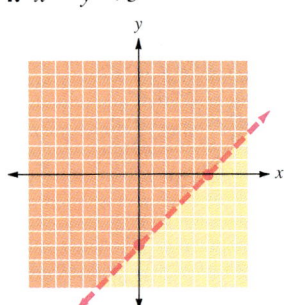

5. $y \geq 2x + 1$

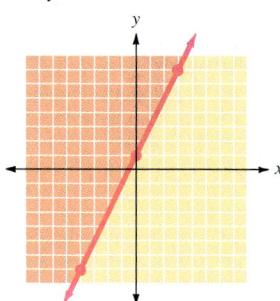

6. $y < 3x - 4$

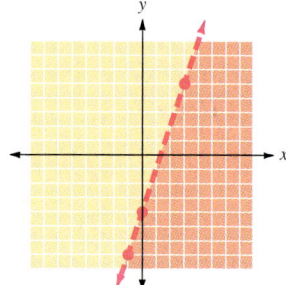

7. $2x + 3y < 6$

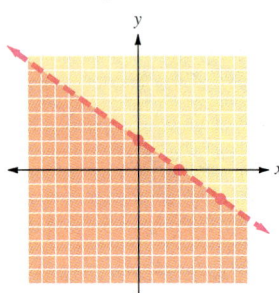

8. $3x - 4y \geq 12$

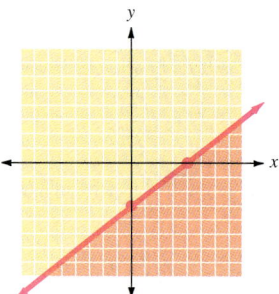

9. $x - 4y > 8$

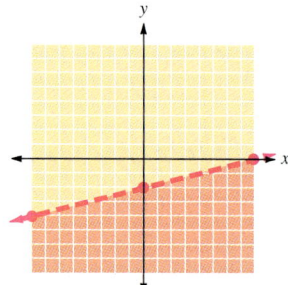

10. $2x + 5y \leq 10$

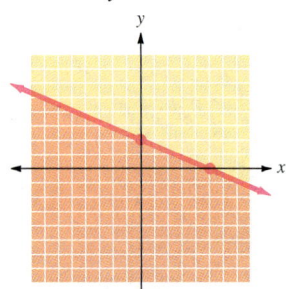

11. $y \geq 3x$

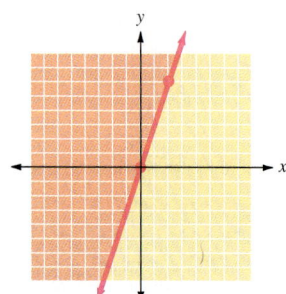

12. $y \leq -2x$

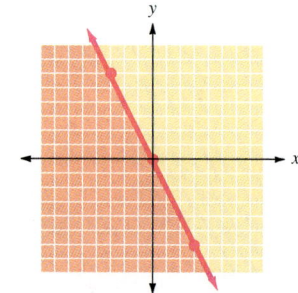

Section 8.4 ■ Graphing Linear Inequalities in Two Variables   **599**

13. $x - 2y > 0$

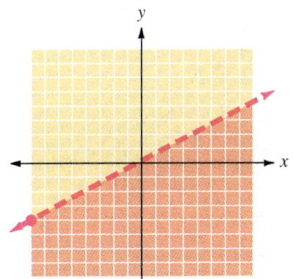

14. $x + 4y \leq 0$

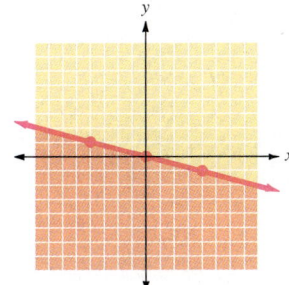

15. $x < 3$

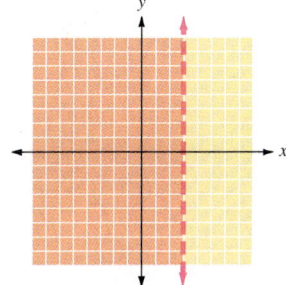

16. $y < -2$

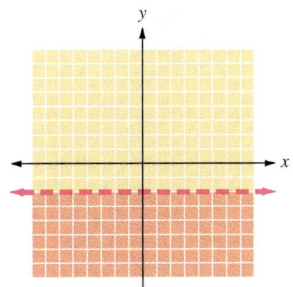

17. $y > 3$

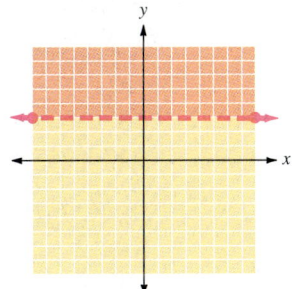

18. $x \leq -4$

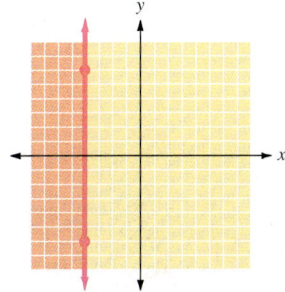

19. $3x - 6 \leq 0$

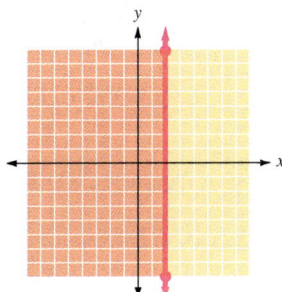

20. $-2y > 6$

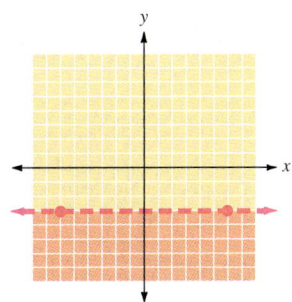

21. $0 < x < 1$

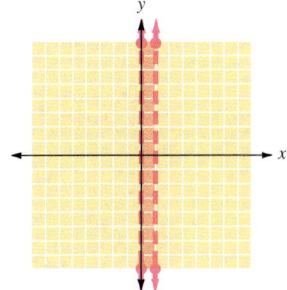

22. $-2 \leq y \leq 1$

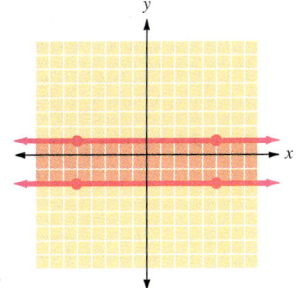

23. $1 \leq x \leq 3$

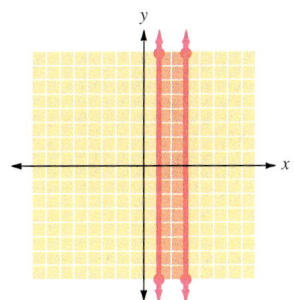

24. $1 < y < 5$

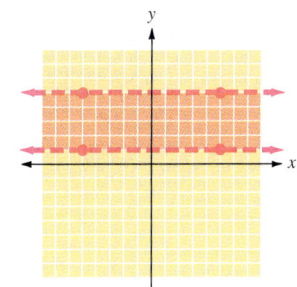

31.
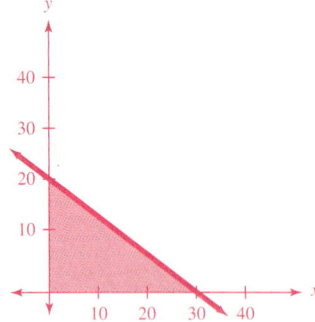

32.
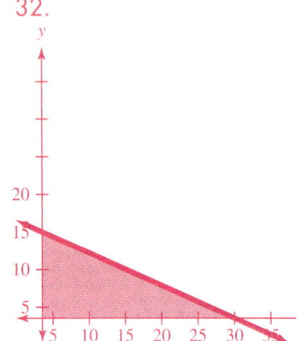

In Exercises 25 to 28, graph the region satisfying each set of conditions.

25. $0 \leq x \leq 3$
$2 \leq y \leq 4$

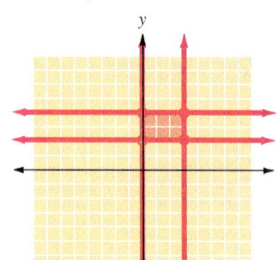

26. $1 \leq x \leq 5$
$0 \leq y \leq 3$

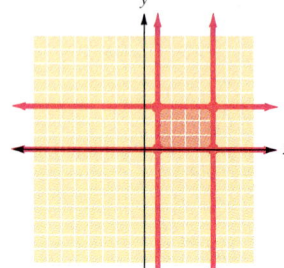

27. $x + 2y \leq 4$
$x \geq 0$
$y \geq 0$

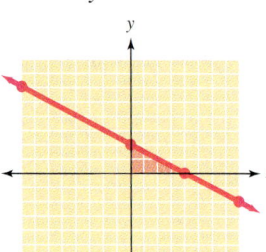

28. $2x + 3y \leq 6$
$x \geq 0$
$y \geq 0$

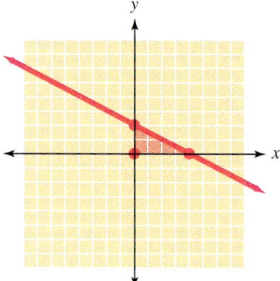

29. Assume that you are working only with the variable x. Describe the solution to the statement $x > -1$.

30. Now, assume that you are working in two variables, x and y. Describe the solution to the statement $x > -1$.

31. Manufacturing. A manufacturer produces a standard model and a deluxe model of a 13-in. television set. The standard model requires 12 h to produce, while the deluxe model requires 18 h. The labor available is limited to 360 h per week.

If x represents the number of standard-model sets produced per week and y represents the number of deluxe models, draw a graph of the region representing the feasible values for x and y. Keep in mind that the values for x and y must be nonnegative since they represent a quantity of items. (This will be the solution set for the system of inequalities.)

32. Manufacturing. A manufacturer produces standard record turntables and CD players. The turntables require 10 h of labor to produce while CD players require 20 h. Let x represent the number of turntables produced and y the number of CD players.

If the labor hours available are limited to 300 h per week, graph the region representing the feasible values for x and y.

33.

$3x + 4y \leq 1000$,
$x \geq 0$,
$y \geq 0$

34.

$3 \leq x \leq 7$
$0 \leq y < 4$

35. $y \geq -x + 4$

36. $y \leq -\dfrac{3}{2}x + 3$

37. $y < \dfrac{1}{2}x - 3$

38. $y > -\dfrac{5}{6}x - 5$

33. **Serving capacity.** A hospital food service can serve at most 1000 meals per day. Patients on a normal diet receive 3 meals per day and patients on a special diet receive 4 meals per day. Write a linear inequality that describes the number of patients that can be served per day and draw its graph.

34. **Time on job.** The movie and TV critic for the local radio station spends 3 to 7 hours daily reviewing movies and fewer than 4 hours reviewing TV shows. Let x represent the hours watching movies and y represent the time spent watching TV. Write two inequalities that model the situation, and graph their intersection.

In Exercises 35 to 38 write an inequality for the shaded region shown in the figure.

35.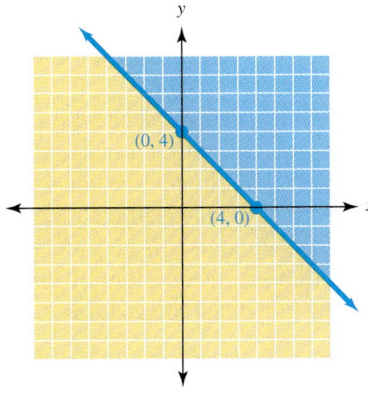

36.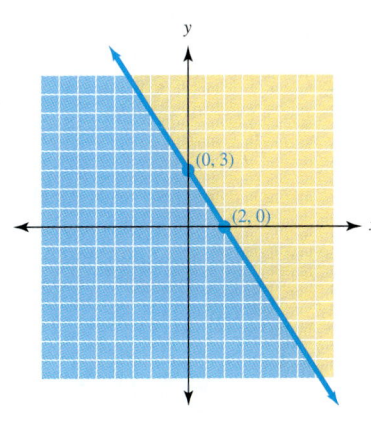

37.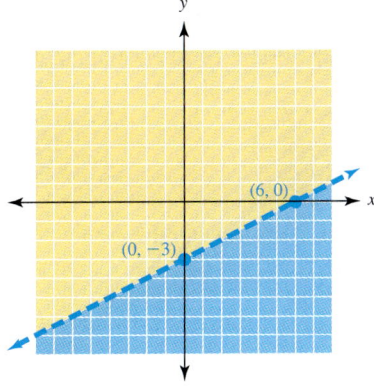

38.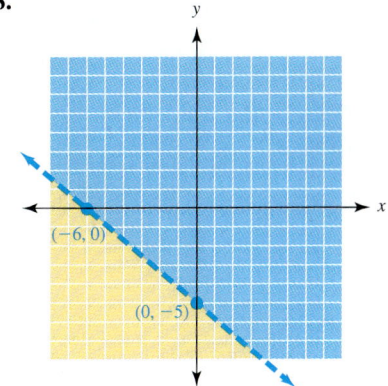

SECTION 8.5
Systems of Linear Inequalities in Two Variables

8.5 OBJECTIVES

1. Graph systems of linear inequalities
2. Find the solution to an application of a system of linear inequalities

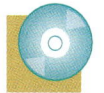

"When I'm working on a problem, I never think about beauty. I think only how to solve the problem. But when I have finished, if the solution is not beautiful, I know it is wrong."

–Richard Buckminster Fuller

Our previous work in this chapter dealt with finding the solution set of a system of linear equations. That solution set represented the points of intersection of the graphs of the equations in the system. In this section, we extend the work we did in 8.4 in which we solved systems of linear inequalities.

In this case, the solution set is all ordered pairs that satisfy each inequality. **The graph of the solution set of a system of linear inequalities** is then the intersection of the graphs of the individual inequalities. Let's look at an example.

Example 1   Solving a System by Graphing

Solve the following system of linear inequalities by graphing

$$x + y > 4$$
$$x - y < 2$$

We start by graphing each inequality separately. The boundary line is drawn, and using $(0, 0)$ as a test point, we see that we should shade the half plane above the line in both graphs.

Note that the boundary line is dashed to indicate it is *not* included in the graph.

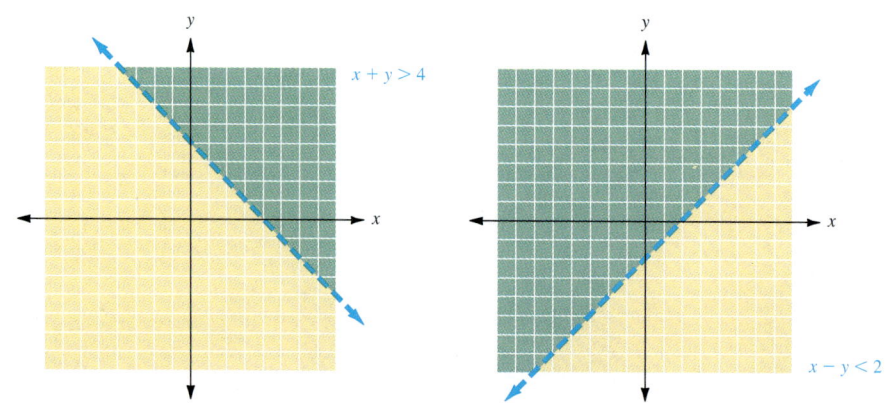

602

In practice, the graphs of the two inequalities are combined on the same set of axes, as is shown below. The graph of the solution set of the original system is the intersection of the graphs drawn above.

Points on the lines are not included in the solution.

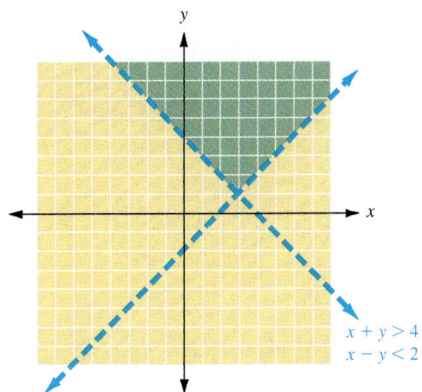

$x + y > 4$
$x - y < 2$

✓ CHECK YOURSELF 1

Solve the following system of linear inequalities by graphing.

$$2x - y < 4$$
$$x + y < 3$$

Most applications of systems of linear inequalities lead to **bounded regions.** This requires a system of three or more inequalities, as shown in Example 2.

Example 2 Solving a System by Graphing

Solve the following system of linear inequalities by graphing.

$$x + 2y \leq 6$$
$$x + y \leq 5$$
$$x \geq 2$$
$$y \geq 0$$

On the same set of axes, we graph the boundary line of each of the inequalities. We then choose the appropriate half planes (indicated by the arrow that is perpendicular to the line) in each case, and we locate the intersection of those regions for our graph.

The vertices of the shaded region are given because they have particular significance in later applications of this concept. Can you see how the coordinates of the vertices were determined?

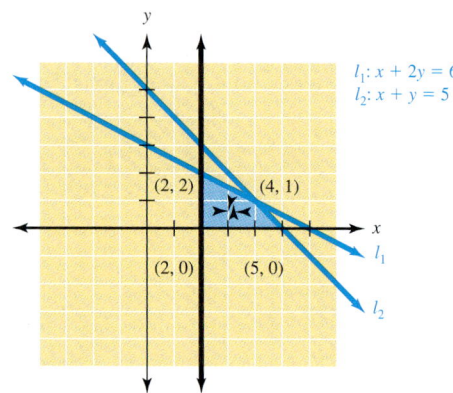

✓ CHECK YOURSELF 2

Solve the following system of linear inequalities by graphing.

$$2x - y \le 8 \qquad x \ge 0$$
$$x + y \le 7 \qquad y \ge 0$$

Let's expand on Example 8, Section 8.2 to see an application of our work with systems of linear inequalities. Consider Example 3.

Example 3 Solving a Business-Based Application

A manufacturer produces a standard model and a deluxe model of a 13-in. television set. The standard model requires 12 h of labor to produce, while the deluxe model requires 18 h. The labor available is limited to 360 h per week. Also, the plant capacity is limited to producing a total of 25 sets per week. Draw a graph of the region representing the number of sets that can be produced, given these conditions.

As suggested earlier, we let x represent the number of standard-model sets produced and y the number of deluxe-model sets. Since the labor is limited to 360 h, we have

The total labor is limited to (or less than or equal to) 360 h.

$$12x + 18y \le 360 \qquad (1)$$

↑ 12 h per standard set   ↑ 18 h per deluxe set

The total production, here $x + y$ sets, is limited to 25, so we can write

$$x + y \le 25 \qquad (2)$$

We have $x \geq 0$ and $y \geq 0$ since the number of sets produced cannot be negative.

For convenience in graphing, we divide both members of inequality (1) by 6, to write the equivalent system

$$2x + 3y \leq 60$$
$$x + y \leq 25$$
$$x \geq 0$$
$$y \geq 0$$

We now graph the system of inequalities as before. The shaded area represents all possibilities in terms of the number of sets that can be produced.

The shaded area is called the **feasible region.** All points in the region meet the given conditions of the problem and represent possible production options.

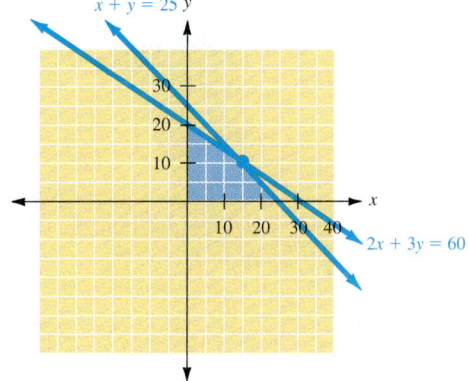

✓ CHECK YOURSELF 3

A manufacturer produces standard record turntables and CD players. The turntables require 10 h of labor to produce while the CD players require 20 h. The labor hours available are limited to 300 h per week. Existing orders require that at least 10 turntables and at least 5 CD players be produced per week. Draw a graph of the region representing the possible production options.

✓ CHECK YOURSELF ANSWERS

1. $2x - y < 4$
   $x + y < 3$

2. $2x - y \leq 8$
   $x + y \leq 7$
   $x \geq 0$
   $y \geq 0$

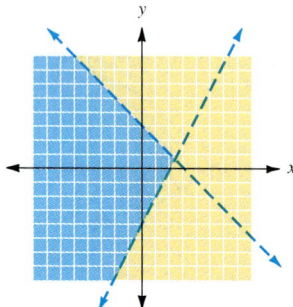 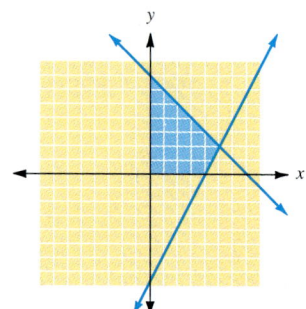

3. Let x be the number of turntables and y be the number of CD players. The system is

$$10x + 20y \leq 300$$
$$x \geq 10$$
$$y \geq 5$$

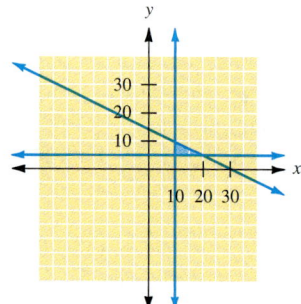

Exercises • 8.5

In Exercises 1 to 18, solve each system of linear inequalities graphically.

1. $x + 2y \leq 4$
   $x - y \geq 1$

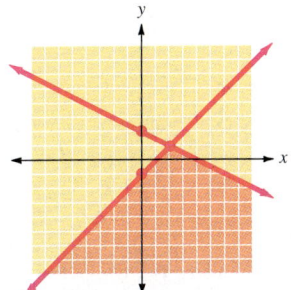

2. $3x - y > 6$
   $x + y < 6$

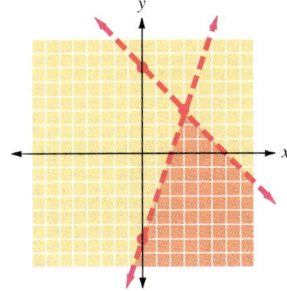

3. $3x + y < 6$
   $x + y > 4$

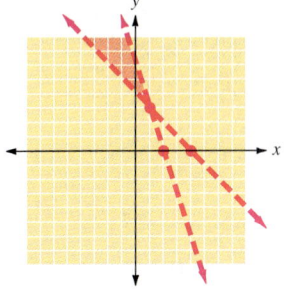

4. $2x + y \geq 8$
   $x + y \geq 4$

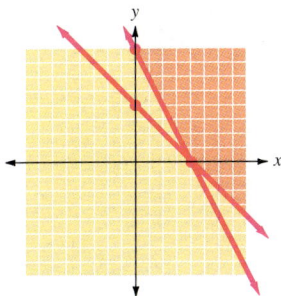

5. $x + 3y \leq 12$
   $2x - 3y \leq 6$

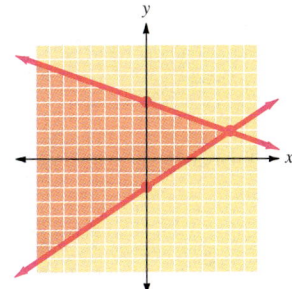

6. $x - 2y > 8$
   $3x - 2y > 12$

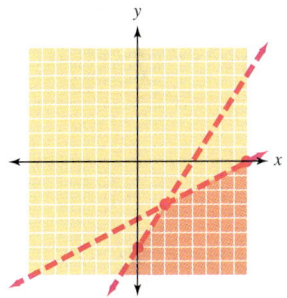

7. $3x + 2y \leq 12$
   $x \geq 2$

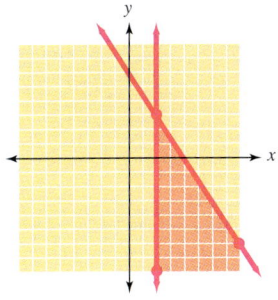

8. $2x + y \leq 6$
   $y \geq 1$

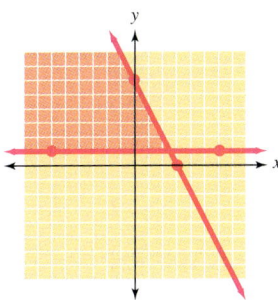

9. $2x + y \leq 8$
   $x > 1$
   $y > 2$

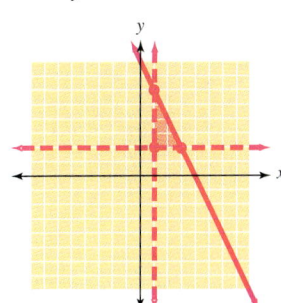

10. $3x - y \leq 6$
$x \geq 1$
$y \leq 3$

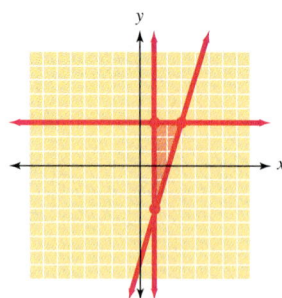

11. $x + 2y \leq 8$
$2 \leq x \leq 6$
$y \geq 0$

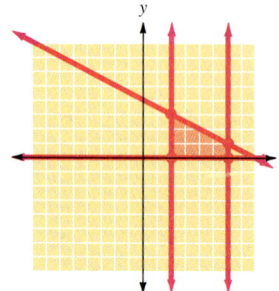

12. $x + y < 6$
$0 \leq y \leq 3$
$x \geq 1$

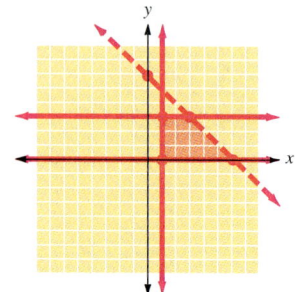

13. $3x + y \leq 6$
$x + y \leq 4$
$x \geq 0$
$y \geq 0$

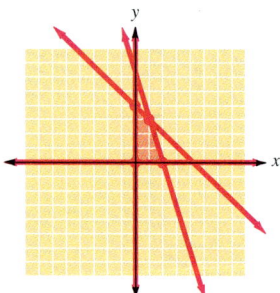

14. $x - 2y \geq -2$
$x + 2y \leq 6$
$x \geq 0$
$y \geq 0$

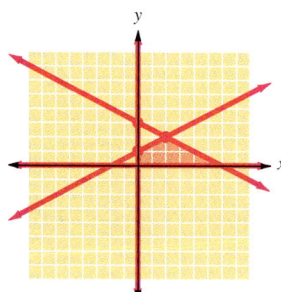

15. $4x + 3y \leq 12$
$x + 4y \leq 8$
$x \geq 0$
$y \geq 0$

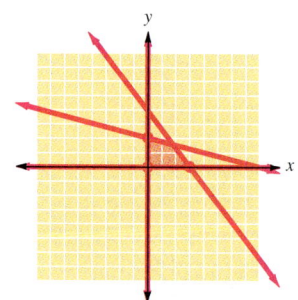

16. $2x + y \leq 8$
$x + y \geq 3$
$x \geq 0$
$y \geq 0$

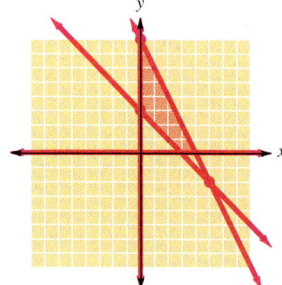

17. $x - 4y \leq -4$
$x + 2y \leq 8$
$x \geq 2$

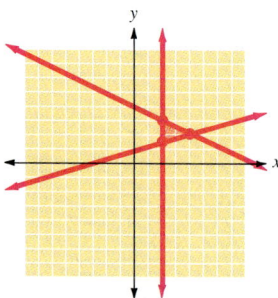

18. $x - 3y \geq -6$
$x + 2y \geq 4$
$x \leq 4$

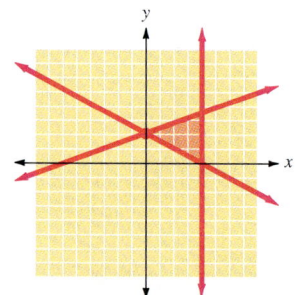

Section 8.5 ■ Systems of Linear Inequalities in Two Variables **609**

19.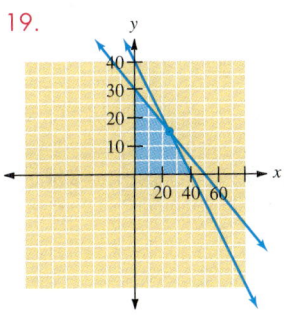

In Exercises 19 and 20, draw the appropriate graph.

19. **Manufacturing.** A manufacturer produces both two-slice and four-slice toasters. The two-slice toaster takes 6 h of labor to produce and the four-slice toaster 10 h. The labor available is limited to 300 h per week, and the total production capacity is 40 toasters per week. Draw a graph of the feasible region, given these conditions, where x is the number of two-slice toasters and y is the number of four-slice toasters.

20. **Production.** A small firm produces both AM and AM/FM car radios. The AM radios take 15 h to produce, and the AM/FM radios take 20 h. The number of production hours is limited to 300 h per week. The plant's capacity is limited to a total of 18 radios per week, and existing orders require that at least 4 AM radios and at least 3 AM/FM radios be produced per week. Draw a graph of the feasible region, given these conditions, where x is the number of AM radios and y the number of AM/FM radios.

21. When one solves a system of linear inequalities, it is often easier to shade the region that is not part of the solution, rather than the region that is. Try this method, then describe its benefits.

22. Describe a system of linear inequalities for which there is no solution.

23. Write the system of inequalities whose graph is the shaded region.

20.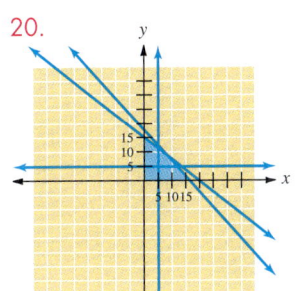

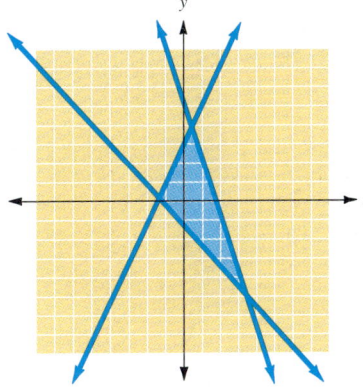

24. Write the system of inequalities whose graph is the shaded region.

23. $y < 2x + 3$

$y < -3x + 5$

$y > -x - 1$

24. $x \leq 0$

$y < 3x + 4$

$y > -2x - 1$

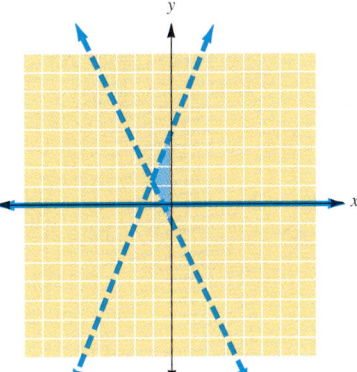

Summary Exercises • 8

This summary exercise set will give you practice with each of the objectives in the chapter.

[8.1] Solve each of the following systems by graphing.

1. $x + y = 8$
   $x - y = 4$

2. $x + 2y = 8$
   $x - y = 5$

3. $2x + 3y = 12$
   $2x + y = 8$

4. $x + 4y = 8$
   $y = 1$

[8.2] Solve each of the following systems by the addition method. If a unique solution does not exist, state whether the given system is inconsistent or dependent.

5. $x + 2y = 7$
   $x - y = 1$

6. $x + 3y = 14$
   $4x + 3y = 29$

7. $3x - 5y = 5$
   $-x + y = -1$

8. $x - 4y = 12$
   $2x - 8y = 24$

9. $6x + 5y = -9$
   $-5x + 4y = 32$

10. $3x + y = -17$
    $5x - 3y = -19$

11. $3x + y = 8$
    $-6x - 2y = -10$

12. $5x - y = -17$
    $4x + 3y = -6$

13. $7x - 4y = 27$
    $5x + 6y = 6$

14. $4x - 3y = 1$
    $6x + 5y = 30$

15. $x - \dfrac{1}{2}y = 8$
    $\dfrac{2}{3}x + \dfrac{3}{2}y = -2$

16. $\dfrac{1}{5}x - 2y = 4$
    $\dfrac{3}{5}x + \dfrac{2}{3}y = -8$

[8.2] Solve each of the following systems by the substitution method. If a unique solution does not exist, state whether the given system is inconsistent or dependent.

17. $2x + y = 23$
    $x = y + 4$

18. $x - 5y = 26$
    $y = x - 10$

19. $3x + y = 7$
    $y = -3x + 5$

20. $2x - 3y = 13$
    $x = 3y + 9$

21. $5x - 3y = 13$
    $x - y = 3$

22. $4x - 3y = 6$
    $x + y = 12$

23. $3x - 2y = -12$
    $6x + y = 1$

24. $x - 4y = 8$
    $-2x + 8y = -16$

---

1. Solution: (6, 2)
2. Solution: (6, 1)
3. Solution: (3, 2)
4. Solution: (4, 1)
5. (3, 2)
6. (5, 3)
7. (0, −1)
8. Dependent
9. (−4, 3)
10. (−5, −2)
11. Inconsistent
12. (−3, 2)
13. $\left(3, -\dfrac{3}{2}\right)$
14. $\left(\dfrac{5}{2}, 3\right)$
15. (6, −4)
16. (−10, −3)
17. (9, 5)
18. (6, −4)
19. Inconsistent
20. $\left(4, -\dfrac{5}{3}\right)$
21. (2, −1)
22. (6, 6)
23. $\left(-\dfrac{2}{3}, 5\right)$
24. Dependent

25. 7, 23

26. 28 $5 bills; 50 $10 bills

27. 800 adult tickets; 400 student tickets

28. Cassettes: $2.50; videotapes: $4.00

29. 20 cm, 12 cm

30. 96 lb of peanuts; 24 lb of cashews

31. $8000 in savings; $9000 in time deposit

32. 200 mL of 20%; 400 mL of 50%

33. Jet 500 mi/h, wind 50 mi/h

34. 27

35. 15 $5\frac{1}{4}$ in.; 8 $3\frac{1}{2}$ in.

36. (15, 195)

[8.2] Solve each of the following problems by choosing a variable to represent each unknown quantity. Then, write a system of equations that will allow you to solve for each variable.

25. Number problem. One number is 2 more than 3 times another. If the sum of the two numbers is 30, find the two numbers.

26. Money value. Suppose that a cashier has 78 $5 and $10 dollar bills with a value of $640. How many of each type of bill does she have?

27. Ticket sales. Tickets for a basketball game sold at $7 for an adult ticket and $4.50 for a student ticket. If the revenue from 1200 tickets was $7400, how many of each type of ticket were sold?

28. Purchase price. A purchase of 8 blank cassette tapes and 4 blank videotapes costs $36. A second purchase of 4 cassette tapes and 5 videotapes costs $30. What is the price of a single cassette tape and of a single videotape?

29. Rectangles. The length of a rectangle is 4 cm less than twice its width. If the perimeter of the rectangle is 64 cm, find the dimensions of the rectangle.

30. Mixture. A grocer in charge of bulk foods wishes to combine peanuts selling for $2.25 per pound and cashews selling for $6 per pound. What amount of each nut should be used to form a 120-lb mixture selling for $3 per pound?

31. Investments. Reggie has two investments totaling $17,000—one a savings account paying 6%, the other a time deposit paying 8%. If his annual interest is $1200, what does he have invested in each account?

32. Mixtures. A pharmacist mixes a 20% alcohol solution and a 50% alcohol solution to form 600 mL of a 40% solution. How much of each solution should she use in forming the mixture?

33. Motion. A jet flying east, with the wind, makes a trip of 2200 mi in 4 h. Returning, against the wind, the jet can travel only 1800 mi in 4 h. What is the plane's rate in still air? What is the rate of the wind?

34. Number problem. The sum of the digits of a two-digit number is 9. If the digits are reversed, the new number is 45 more than the original number. What was the original number?

35. Work. A manufacturer produces $5\frac{1}{4}$-in. computer disk drives and $3\frac{1}{2}$-in. drives. The $5\frac{1}{4}$-in. drives require 20 min of component assembly time; the $3\frac{1}{2}$-in. drives, 25 min. The manufacturer has 500 min of component assembly time available per day. Each drive requires 30 min for packaging and testing, and 690 min of that time is available per day. How many of each of the drives should be produced daily to use all the available time?

36. Equilibrium price. If the demand equation for a product is $D = 270 - 5p$ and the supply equation is $S = 13p$, find the equilibrium point.

612 Chapter 8 ▪ Systems of Linear Equations and Inequalities

37. 300 mi

38. $\left(3, \dfrac{8}{3}, -\dfrac{1}{3}\right)$

39. (6, 1, −2)

40. Inconsistent

41. (5, 2, 1)

42. (4, 1, −3)

43. Dependent

44. $\left(\dfrac{1}{2}, \dfrac{2}{3}, 4\right)$

45. 2, 5, 8

46. 394

47. 200 orchestra;
    40 balcony;
    120 box-seats

48. 30°, 45°, 105°

49. $6000 savings;
    $2000 stock;
    $4000 mutual fund

---

37. Rental charges. Two car rental agencies have the following rates for the rental of a compact automobile:

Company A: $18 per day plus 12¢ per mile.

Company B: $20 per day plus 10¢ per mile.

For a 3-day rental, at what number of miles will the charges from the two companies be the same?

[8.3] Solve each of the following systems by the addition method. If a unique solution does not exist, state whether the given system is inconsistent or dependent.

38. $\begin{aligned} x - y + z &= 0 \\ x + 4y - z &= 14 \\ x + y - z &= 6 \end{aligned}$

39. $\begin{aligned} x - y + z &= 3 \\ 3x + y + 2z &= 15 \\ 2x - y + 2z &= 7 \end{aligned}$

40. $\begin{aligned} x - y - z &= 2 \\ -2x + 2y + z &= -5 \\ -3x + 3y + z &= -10 \end{aligned}$

41. $\begin{aligned} x - y &= 3 \\ 2y + z &= 5 \\ x + 2z &= 7 \end{aligned}$

42. $\begin{aligned} x + y + z &= 2 \\ x + 3y - 2z &= 13 \\ y - 2z &= 7 \end{aligned}$

43. $\begin{aligned} x + y - z &= -1 \\ x - y + 2z &= 2 \\ -5x - y - z &= -1 \end{aligned}$

44. $\begin{aligned} 2x + 3y + z &= 7 \\ -2x - 9y + 2z &= 1 \\ 4x - 6y + 3z &= 10 \end{aligned}$

[8.3] Solve each of the following problems by choosing a variable to represent each unknown quantity.

45. Number problem. The sum of three numbers is 15. The largest number is 4 times the smallest number, and it is also 1 more than the sum of the other two numbers. Find the three numbers.

46. Number problem. The sum of the digits of a three-digit number is 16. The tens digit is 3 times the hundreds digit, and the units digit is 1 more than the hundreds digit. What is the number?

47. Tickets sold. A theater has orchestra tickets at $10, box-seat tickets at $7, and balcony tickets at $5. For one performance, a total of 360 tickets were sold, and the total revenue was $3040. If the number of orchestra tickets sold was 40 more than that of the other two types combined, how many of each type of ticket were sold for the performance?

48. Triangles. The measure of the largest angle of a triangle is 15° less than 4 times the measure of the smallest angle and 30° more than the sum of the measures of the other two angles. Find the measures of the three angles of the triangle.

49. Investments. Rachel divided $12,000 into three investments: a savings account paying 5%, a stock paying 7%, and a mutual fund paying 9%. Her annual interest from the investments was $800, and the amount that she had invested at 5% was equal to the sum of the amounts invested in the other accounts. How much did she have invested in each type of account?

Summary Exercises 8 **613**

50. 19, 22

51. 51, 93

52. 120 ft, 80 ft

53. (40, 4000)

50. Number problem. The difference of two positive numbers is 3, and the sum of those numbers is 41. Find the two numbers.

51. Number problem. The sum of two integers is 144, and the difference is 42. What are the two integers?

52. Rectangle. A rectangular building lot is $1\frac{1}{2}$ times as wide as it is long. The perimeter of the lot is 400 ft. Find the length and width of the lot.

53. Break-even analysis. A manufacturer's cost for producing x units of a product is given by

$$C = 10x + 3600$$

The revenue from selling x units of that product is given by

$$R = 100x$$

Find the break-even point for this product.

[8.4] Graph the solution set for each of the following linear inequalities.

54. $y < 2x + 1$

55. $y \geq -2x + 3$

56. $3x + 2y \geq 6$

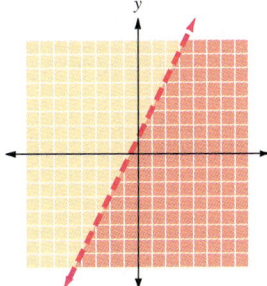 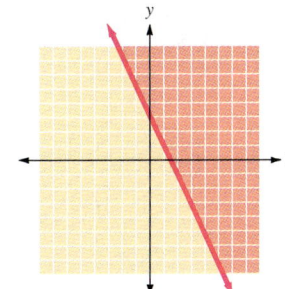 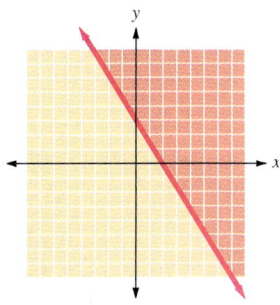

57. $3x - 5y < 15$

58. $y < -2x$

59. $4x - y \geq 0$

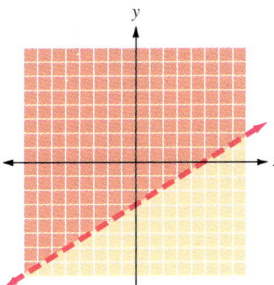 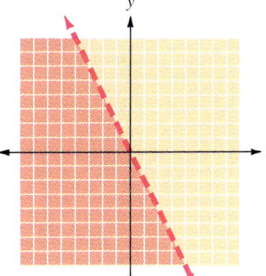 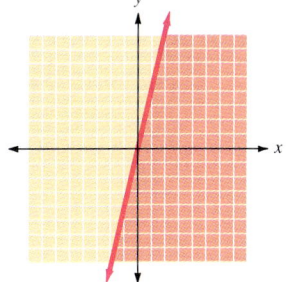

60. $y \geq -3$

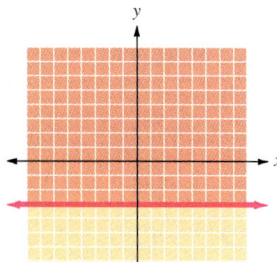

61. $x < 4$

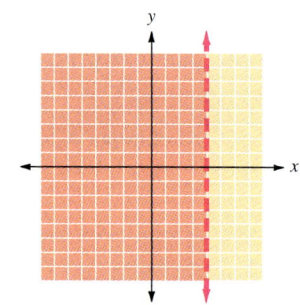

[8.5] Solve each of the following systems of linear inequalities graphically.

62. $x - y < 7$
$x + y > 3$

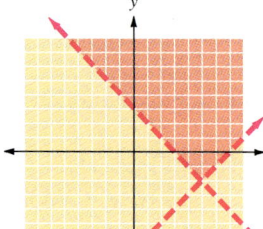

63. $x - 2y \leq -2$
$x + 2y \leq 6$

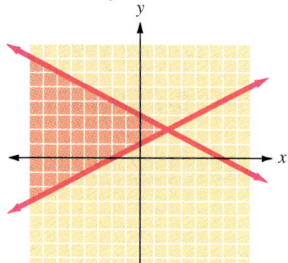

64. $x - 6y < 6$
$-x + 6y < 4$

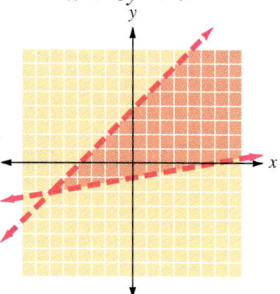

65. $2x + y \leq 8$
$x \geq 1$
$y \geq 0$

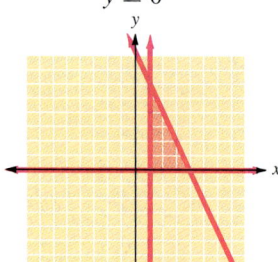

66. $2x + y \leq 6$
$x \geq 1$
$y \geq 0$

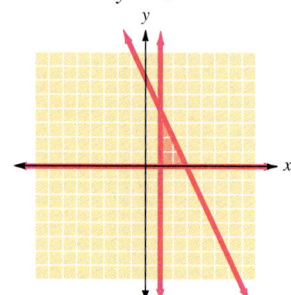

67. $4x + y \leq 8$
$x \geq 0$
$y \geq 2$

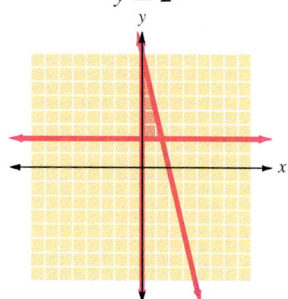

68. $4x + 2y \leq 8$
$x + y \leq 3$
$x \geq 0$
$y \geq 0$

69. $3x + y \leq 6$
$x + y \leq 4$
$x \geq 0$
$y \geq 0$

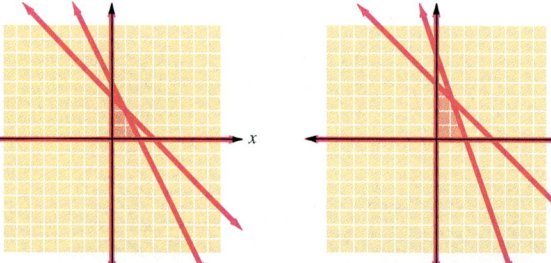

Self-Test 8

1. (−3, 4)
2. Dependent
3. Inconsistent
4. (−2, −5)
5. (5, 0)
6. $\left(3, -\frac{5}{3}\right)$
7. (−1, 2, 4)
8. $\left(2, -3, -\frac{1}{2}\right)$
9. Disks: $2.50; ribbons; $6
10. 60 lb jawbreakers; 40 lb licorice
11. Four 5-in. sets; six 12-in. sets
12. $8000 savings; $4000 bond; $2000 mutual fund
13. 50 ft by 80 ft

The purpose of this self-test is to help you check your progress and to review for a chapter test in class. Allow yourself about 1 hour to take the test. When you are done, check your answers in the back of the book. If you missed any answers, be sure to go back and review the appropriate sections in the chapter and the exercises that are provided.

Solve each of the following systems. If a unique solution does not exist, state whether the given system is inconsistent or dependent.

1. $\begin{aligned} 3x + y &= -5 \\ 5x - 2y &= -23 \end{aligned}$

2. $\begin{aligned} 4x - 2y &= -10 \\ y &= 2x + 5 \end{aligned}$

3. $\begin{aligned} 9x - 3y &= 4 \\ -3x + y &= -1 \end{aligned}$

4. $\begin{aligned} 5x - 3y &= 5 \\ 3x + 2y &= -16 \end{aligned}$

5. $\begin{aligned} x - 2y &= 5 \\ 2x + 5y &= 10 \end{aligned}$

6. $\begin{aligned} 5x - 3y &= 20 \\ 4x + 9y &= -3 \end{aligned}$

Solve each of the following systems.

7. $\begin{aligned} x - y + z &= 1 \\ -2x + y + z &= 8 \\ x + 5z &= 19 \end{aligned}$

8. $\begin{aligned} x + 3y - 2z &= -6 \\ 3x - y + 2z &= 8 \\ -2x + 3y - 4z &= -11 \end{aligned}$

Solve each of the following problems by choosing a variable to represent each unknown quantity. Then write a system of equations that will allow you to solve for each variable.

9. An order for 30 computer disks and 12 printer ribbons totaled $147. A second order for 12 more disks and 6 additional ribbons cost $66. What was the cost per individual disk and ribbon?

10. A candy dealer wants to combine jawbreakers selling for $2.40 per pound and licorice selling for $3.90 per pound to form a 100-lb mixture that will sell for $3 per pound. What amount of each type of candy should be used?

11. A small electronics firm assembles 5-in. portable television sets and 12-in. models. The 5-in. set requires 9 h of assembly time; the 12-in. set, 6 h. Each unit requires 5 h for packaging and testing. If 72 h of assembly time and 50 h of packaging and testing time are available per week, how many of each type of set should be finished if the firm wishes to use all its available capacity?

12. Hans decided to divide $14,000 into three investments: a savings account paying 6% annual interest, a bond paying 9%, and a mutual fund paying 13%. His annual interest from the three investments was $1100, and he had twice as much invested in the bond as in the mutual fund. What amount did he invest in each type?

13. The fence around a rectangular yard requires 260 ft of fencing. The length is 20 ft less than twice the width. Find the dimensions of the yard.

Graph the solution set in each of the following.

14. $5x + 6y \leq 30$ **15.** $x + 3y > 6$

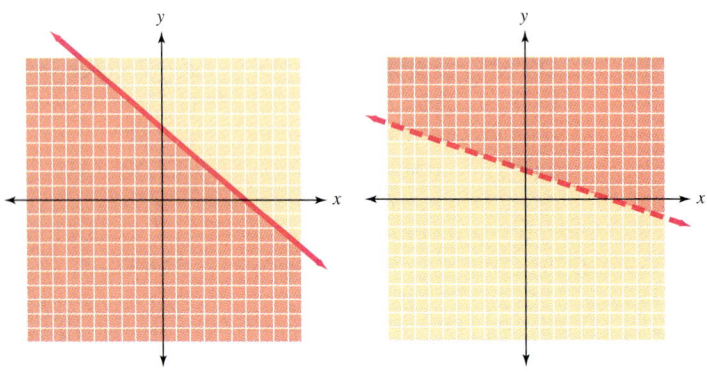

16. $4x - 8 \leq 0$ **17.** $2y + 4 > 0$

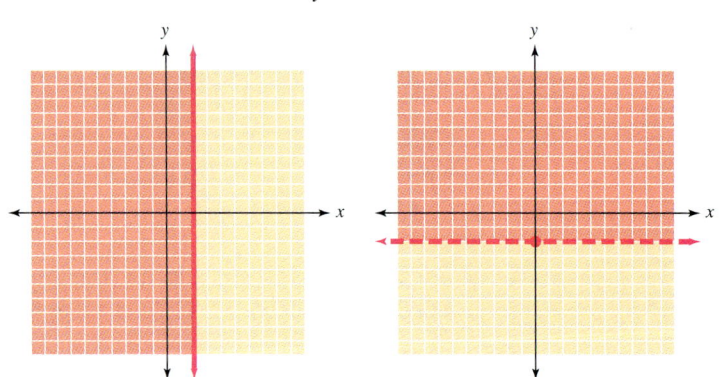

Solve each of the following systems of linear inequalities graphically.

18. $x - 2y < 6$
  $x + y < 3$

19. $3x + 4y \geq 12$
  $x \geq 1$

20. $x + 2y \leq 8$
  $x + y \leq 6$
  $x \geq 0$
  $y \geq 0$

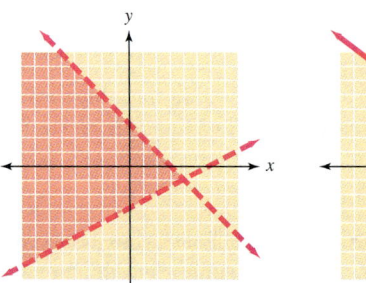 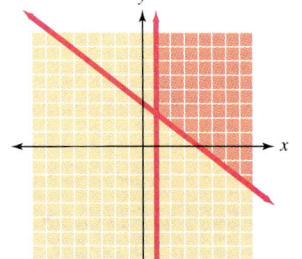 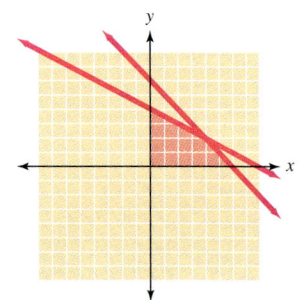

Cumulative Test 1-8

1. $\frac{22}{4}$ 2. $x > -2$
3. $4, -1$
4. $-4 \leq x \leq \frac{2}{3}$
5. $x < -\frac{17}{5}$ or $x > 5$
6. 2
7.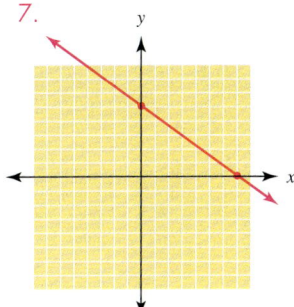
8.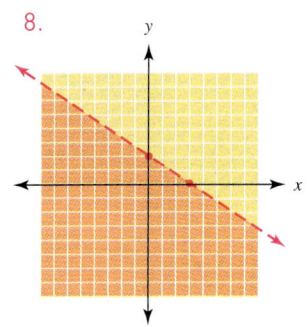
9. $\frac{P - P_0}{IT}$
10. 7
11. $y = -x + 3$
12. $2x^2 - 5x - 3$
13. $9x^2 - 12x + 4$
14. $(x - 3)(x^2 - 5)$
15. $y = \frac{4}{5}x - \frac{2}{5}$
16. 90

This test is provided to help you in the process of reviewing the previous chapters. Answers are provided in the back of the book. If you missed any answers, be sure to go back and review the appropriate sections.

Solve each of the following.

1. $3x - 2(x + 5) = 12 - 3x$
2. $2x - 7 < 3x - 5$
3. $|2x - 3| = 5$
4. $|3x + 5| \leq 7$
5. $|5x - 4| > 21$
6. $2x + 3(x - 2) = -4(x + 1) + 16$

Graph each of the following.

7. $5x + 7y = 35$
8. $2x + 3y < 6$
9. Solve the equation $P = P_0 + IRT$ for R.
10. Find the slope of the line connecting $(4, 6)$ and $(3, -1)$.
11. Write the equation of the line that passes through the points $(-1, 4)$ and $(5, -2)$.

Simplify the following expressions.

12. $(2x + 1)(x - 3)$
13. $(3x - 2)^2$
14. Completely factor the expression $x^3 - 3x^2 - 5x + 15$.
15. Write the equation of the line passing through the point $(3, 2)$ and parallel to the line $4x - 5y = 20$.
16. Find $f(-5)$ if $f(x) = 3x^2 - 4x - 5$.

Solve each of the following systems of equations.

17. $2x + 3y = 6$
    $5x + 3y = -24$
18. $x + y + z = 3$
    $2x - y + 2z = 0$
    $-x - 3y + z = -9$

Solve each of the following applications.

19. The length of a rectangle is 3 cm more than twice its width. If the perimeter of the rectangle is 54 cm, find the dimensions of the rectangle.
20. The sum of the digits of a two-digit number is 10. If the digits are reversed, the number is 36 less than the original number. What was the original number?

17. $\left(-10, \frac{26}{3}\right)$

18. $(2, 2, -1)$

19. 8 cm by 19 cm

20. 73

CHAPTER 9
Graphical Solutions

LIST OF SECTIONS

9.1 Graphical Solutions to Equations in One Variable

9.2 Graphing Linear Inequalities in One Variable

9.3 Absolute Value Functions

9.4 Absolute Value Inequalities

Quality control is exercised in nearly all manufacturing processes. In the pharmaceutical-making process, great caution must be exercised to ensure that the medicines and drugs are pure and contain precisely what is indicated on the label. Guaranteeing such purity is a task the quality control division of the pharmaceutical company assumes.

A lab technician working in quality control must run a series of tests on samples of every ingredient, even simple ingredients such as salt (NaCl). One such test is a measure of how much weight is lost as a sample is dried. The technician must set up a 3-hour procedure that involves cleaning and drying bottles and stoppers and then weighing them while they are empty and again when they contain samples of the substance to be heated and dried. At the end of the procedure, to compute the percentage of weight loss from drying, the technician uses the formula

$$L = \frac{W_g - W_f}{W_g - T} \cdot 100$$

where L = percentage loss in drying
W_g = weight of container and sample
W_f = weight of container and sample after drying process completed
T = weight of empty container

The pharmaceutical company may have a standard of acceptability for this substance. For instance, the substance may not be acceptable if the loss of weight from drying is greater than 10%. The technician would then use the following inequality to calculate acceptable weight loss:

$$10 \geq \frac{W_g - W_f}{W_g - T} \cdot 100$$

Such inequalities are more useful when solved for one of the variables, here W_f or T. In this chapter, you will learn how to solve such an inequality.

SECTION 9.1 Graphical Solutions to Equations in One Variable

9.1 OBJECTIVES

1. Rewrite a linear equation in one variable as $f(x) = g(x)$
2. Find the point of intersection of $f(x)$ and $g(x)$
3. Interpret the point of intersection of $f(x)$ and $g(x)$
4. Solve a linear equation in one variable by writing it as the functional equality $f(x) = g(x)$

"Descartes commanded the future from his study more than Napolean from the throne."

–Oliver Wendell Holmes

In Section 2.3, you learned to solve linear equations in one variable algebraically. In this section, we will look at graphical solutions for the same type of equations.

Recall that a solution to a linear equation is a value for the variable that makes the equation a true statement. In this section, we will look at a graphical method for finding the solution to a linear equation. This method will be particularly useful in the next section, when we discuss linear inequalities.

In our first example, we will look at the solution to a simple linear equation. The method may seem unnecessarily complicated, but remember that we are learning the method so we can solve more complex equations and inequalities later.

Example 1 A Graphical Solution to a Linear Equation

Graphically solve the following equation.

$$2x - 6 = 0$$

Step 1 Let each side of the equation represent a function of x.

$$f(x) = 2x - 6$$
$$g(x) = 0$$

Step 2 Graph the two functions on the same set of axes.

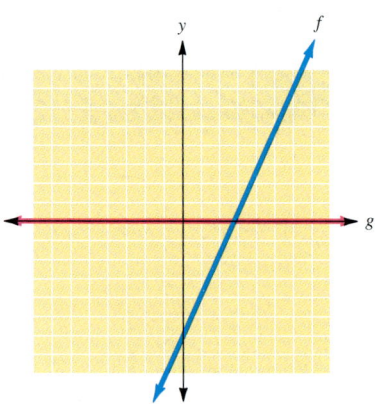

Step 3 Find the point of intersection of the two graphs. The *x* coordinate of this point represents the solution to the original equation.

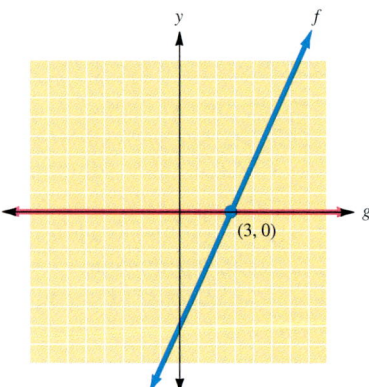

The two lines intersect on the *x* axis at the point (3, 0). We are looking for the *x* value at the point of intersection, which is 3.

✓ CHECK YOURSELF 1

Graphically solve the following equation.

$$-3x + 6 = 0$$

The same three-step process is used for solving any equation. In Example 2, we look for a point of intersection that is *not* on the *x* axis.

Example 2   **A Graphical Solution to a Linear Equation**

Graphically solve the following equation.

$$2x - 6 = -3x + 4$$

Step 1   Let each side of the equation represent a function of x.

$$f(x) = 2x - 6$$
$$g(x) = -3x + 4$$

Step 2   Graph the two functions on the same set of axes.

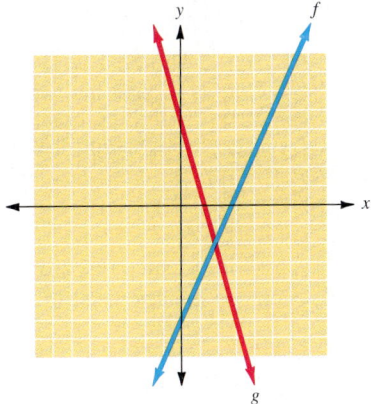

Step 3   Find the point of intersection of the two graphs. The x coordinate of this point represents the solution to the original equation.

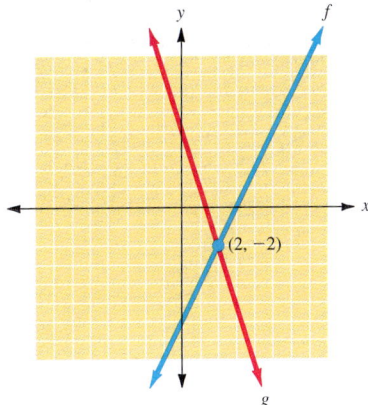

If the solution was an ordered pair, the solution set would be $\{(2, -2)\}$, but we are looking for only the x value.

The two lines intersect at the point $(2, -2)$. Again, the solution is the x value of the ordered pair, so the solution is $x = 2$ and the solution set is $\{2\}$. But what is the significance of the y value of the ordered pair? Note that, when you substitute 2 for x in the two functions, both yield -2.

✓ CHECK YOURSELF 2

Graphically display the solution to the following equation.

$$-3x - 4 = 2x - 1$$

The following algorithm summarizes our work in finding a graphical solution for an equation.

Finding a Graphical Solution for an Equation

Step 1 Let each side of the equation represent a function of x.
Step 2 Graph the two functions on the same set of axes.
Step 3 Find the point of intersection of the two graphs. The x value at this point represents the solution to the original equation.

Linear equations are often first solved by algebraic means. The graph of the equation can then be used to check the solution. This concept is illustrated in Example 3.

Example 3 Solving Linear Equations Algebraically and Graphically

Solve the linear equation algebraically, then graphically display the solution.

$$2(x + 3) = -3x - 4$$

Begin by using the distributive property to rid the left side of parentheses.

$$2x + 6 = -3x - 4$$
$$5x + 6 = -4$$
$$5x = -10$$
$$x = -2$$

The solution set is $\{-2\}$.
To graphically display the solution, let

$$f(x) = 2(x + 3)$$
$$g(x) = -3x - 4$$

Graphing both lines, we get

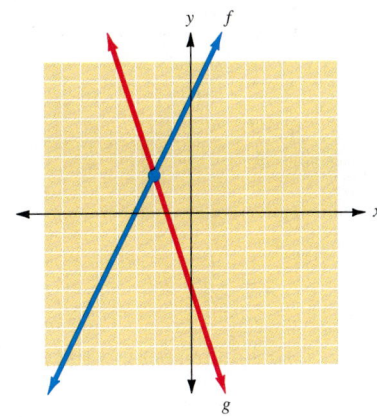

The point of intersection appears to be $(-2, 2)$, which confirms that -2 is a reasonable solution to the equation

$$2(x + 3) = -3x - 4$$

✓ CHECK YOURSELF 3

First solve the linear equation algebraically, then graphically display the solution.

$$3(x - 2) = -4x + 1$$

In Example 4, we turn to a business application.

Example 4 Solving a Business Application

A manufacturer can produce and sell x items per week at the following cost in dollars.

$$C(x) = 30x + 800$$

The revenue from selling those items is given by

$$R(x) = 110x$$

Find the break-even point, which is the number of units at which the revenue equals the cost.

We form a linear system from the given equations.

$$C(x) = 30x + 800 \qquad (1)$$
$$R(x) = 110x \qquad (2)$$

Finding the break-even point requires that revenue equals cost, or $R(x) = C(x)$. From equation (2), we can substitute $110x$ for $C(x)$ in equation (1). We then have

$$110x = 30x + 800$$

or

$$110x - 30x = 800$$
$$80x = 800$$
$$x = 10$$

The solution set is $\{10\}$.

For $x = 10$ units, the cost (and the revenue) is $1100. This system is illustrated below.

Note that to the right of the break-even point, the revenue line is above the cost line.

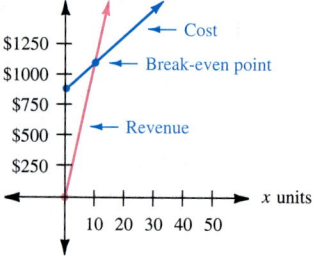

Note that if the company sells more than 10 units, it makes a profit since the revenue exceeds the cost.

✓ CHECK YOURSELF 4

A manufacturer can produce and sell x items per week at a cost

$$C(x) = 30x + 1800$$

The revenue from selling those items is given by

$$R(x) = 120x$$

Find the break-even point.

✓ CHECK YOURSELF ANSWERS

1. $f(x) = -3x + 6$
$g(x) = 0$

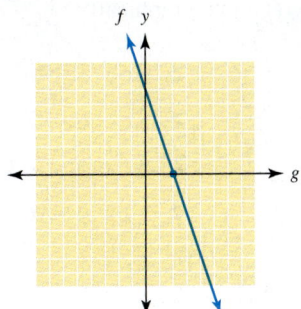

Solution $\{2\}$

2. $f(x) = -3x - 4$
$g(x) = 2x - 1$

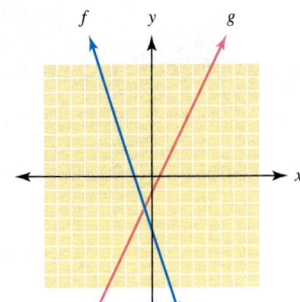

Solution $\left\{-\dfrac{3}{5}\right\}$

3. $3(x - 2) = -4x + 1$
$3x - 6 = -4x + 1$
$7x = 7$
$x = 1$

Graphically,

$f(x) = 3(x - 2)$
$g(x) = -4x + 1$

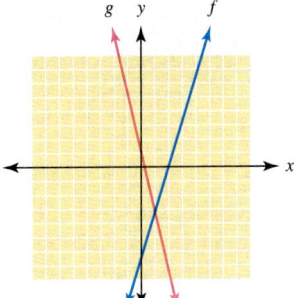

4. 20 items.

Exercises • 9.1

1. {4}
2. {−3}
3. {1}
4. {3}
5. {2}
6. {−2}
7. {5}
8. {−1}
9. {3}
10. {1}
11. {5}
12. {−1}
13. {2}
14. {3}
15. {5}
16. {−1}
17. Always use the lower graph to determine the cheaper cost.
18. 800 flashlights
19. 250 sets

Solve the following equations graphically.

1. $2x - 8 = 0$
2. $4x + 12 = 0$
3. $7x - 7 = 0$
4. $2x - 6 = 0$
5. $5x - 8 = 2$
6. $4x + 5 = -3$
7. $2x - 3 = 7$
8. $5x + 9 = 4$
9. $4x - 2 = 3x + 1$
10. $6x + 1 = x + 6$
11. $\frac{7}{5}x - 3 = \frac{3}{10}x + \frac{5}{2}$
12. $2x - 3 = 3x - 2$
13. $3(x - 1) = 4x - 5$
14. $2(x + 1) = 5x - 7$
15. $7\left(\frac{1}{5}x - \frac{1}{7}\right) = x + 1$
16. $2(3x - 1) = 12x + 4$

17. The following graph represents the rates that two different car rental agencies charge. The x axis represents the number of miles driven (in hundreds of miles), and the y axis represents the total charge. How would you use this graph to decide which agency to use?

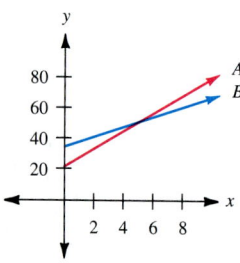

18. **Business.** A firm producing flashlights finds that its fixed cost is $2400 per week, and its variable cost is $4.50 per flashlight. The revenue is $7.50 per flashlight, so the cost and revenue equations are, respectively,

$$C(x) = 4.50x + 2400 \quad \text{and} \quad R(x) = 7.50x$$

Find the break-even point for the firm (the point at which the revenue equals the cost).

19. **Business.** A company that produces portable television sets determines that its fixed cost is $8750 per month. The variable cost is $70 per set, and the revenue is $105 per set. The cost and revenue equations, respectively, are given by

$$C(x) = 70x + 8750 \quad \text{and} \quad R(x) = 105x$$

Find the number of sets the company must produce and sell in order to break even.

Section 9.1 ■ Graphical Solutions to Equations in One Variable   **629**

20. Graphs can be used to solve distance, time, and rate problems because graphs make pictures of the action.

    (a) Consider this earlier exercise: "Robert left on a trip, traveling at 45 mi/h. One-half hour later, Laura discovered that Robert forgot his luggage and so she left along the same route, traveling at 54 mi/h, to catch up with him. When did Laura catch up with Robert?" How could drawing a graph help solve this problem? If you graph Robert's distance as a function of time and Laura's distance as a function of time, what does the slope of each line correspond to in the problem?

    (b) Use a graph to solve this problem: Marybeth and Sam left her mother's house to drive home to Minneapolis along the interstate. They drove an average of 60 mi/h. After they had been gone for $\frac{1}{2}$ h, Marybeth's mother realized they had left their laptop computer. She grabbed it, jumped into her car, and pursued the two at 70 mi/h. Marybeth and Sam also noticed the missing computer, but not until 1 h after they had left. When they noticed that it was missing, they slowed to 45 mi/h while they considered what to do. After driving for another $\frac{1}{2}$ h, they turned around and drove back toward the home of Marybeth's mother at 65 mi/h. Where did they meet? How long had Marybeth's mother been driving then they met?

    (c) Now that you have become experts at this, try solving this problem by drawing a graph. It will require that you think about the slope and perhaps make several guesses when drawing the graphs. If you ride your new bicycle to class, it takes you 1.2 h. If you drive, it takes you 40 min. If you drive in traffic an average of 15 mi/h faster than you can bike, how far way from school do you live? Write an explanation of how you solved this problem by using a graph.

21. **Graphing.** The family next door to you is trying to decide which health maintenance organization (HMO) to join. One parent has a job with health benefits for the employee only, but the rest of the family can be covered if the employee agrees to a payroll deduction. The choice is between The Empire Group, which would cost the family $185 per month for coverage and $25.50 for each office visit, and Group Vitality, which costs $235 per month and $4.00 for each office visit.

    (a) Write an equation showing total yearly costs for each HMO. Graph the cost per year as a function of the number of visits, and put both graphs on the same axes.

    (b) Write a note to the family explaining when The Empire Group would be better and when Group Vitality would be better. Explain how they can use your data and graph to help make a good decision. What other issues might be of concern to them?

SECTION 9.2 Graphing Linear Inequalities in One Variable

9.2 OBJECTIVES

1. Solve linear inequalities in one variable graphically

In Section 9.1, we looked at the graphical approach to solving a linear equation. In this section, we will use the graphs of linear functions to determine the solutions of a linear inequality.

Linear inequalities in one variable, x are obtained from linear equations by replacing the symbol for equality (=) with one of the inequality symbols ($<, >, \leq, \geq$).

The general form for a linear inequality in one variable is

$$x < a$$

where the symbol $<$ can be replaced with $>$, $\leq$, or $\geq$. Examples of linear inequalities in one variable include

$$x \geq -3 \qquad 2x + 5 > 7 \qquad 2x - 3 \leq 5x + 6$$

Recall that the solution set for an equation is the set of all values for the variable(s) that make the equation a true statement. Similarly, the solution set for an inequality is the set of all values that make the inequality a true statement. Example 1 looks at the two-dimensional graph of an equation in one variable.

Example 1

Graphing the Solution Set to an Inequality

Use a graph to find the solution set to the inequality

$$2x + 5 > 7$$

First, rewrite the inequality as a comparison of two functions. Here $f(x) > g(x)$, where $f(x) = 2x + 5$ and $g(x) = 7$.

Now graph the two functions on a single set of axes.

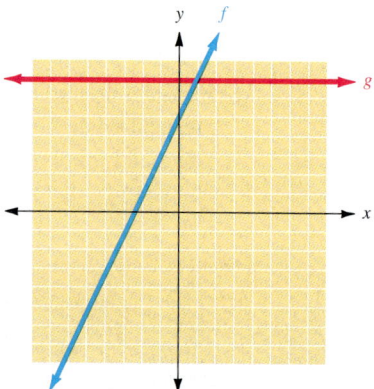

Section 9.2 ■ Graphing Linear Inequalities in One Variable   **631**

Next, draw a vertical dotted line through each point of intersection of the two functions. In this case, there will be a vertical line through the point (1, 7).

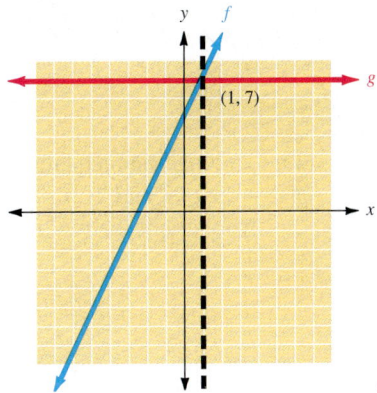

The solution set is every x value that results in $f(x)$ being greater than $g(x)$. The graph of $f(x)$ is above the graph of $g(x)$ for all x values greater than 1. The solution set is everything to the right of the dotted line at $x = 1$.

The solution set will be all the x values that make the original statement, $2x + 5 > 7$, true.

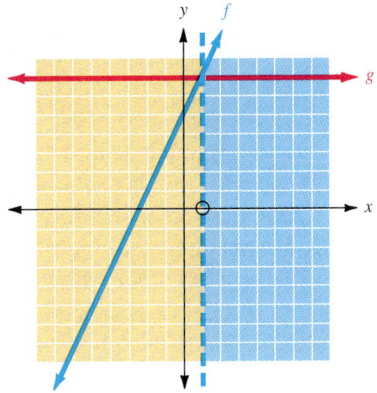

Finally, we express the solution set in set notation

$$\{x | x > 1\}$$

which is read as, "The set of all x such that x is greater than 1."

 CHECK YOURSELF 1

Graph the solution set to the inequality $3x - 2 < 4$.

In Example 1, the function $g(x) = 7$ resulted in a horizontal line. In Example 2, we see that the same method works when comparing any two functions.

Example 2 — Graphing the Solution Set to an Inequality

Graph the solution set to the inequality

$$2x - 3 \geq 5x$$

First, rewrite the inequality as a comparison of two functions. Here, $f(x) \geq g(x)$, where $f(x) = 2x - 3$ and $g(x) = 5x$.

Now graph the two functions on a single set of axes.

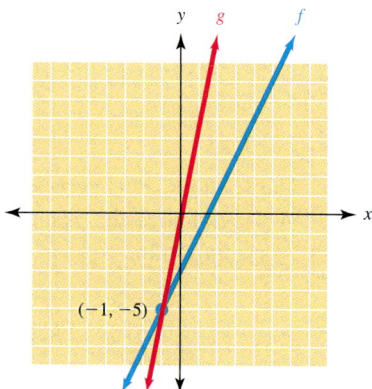

As in Example 1, draw a vertical line through each point of intersection of the two functions. In this case, there will be a vertical line through the point $(-1, -5)$. In this case, the line is included (greater than or *equal to*), so the line is solid, not dotted.

Again, we need to fill in every x value that makes the statement true. In this case, that is every x for which the line representing $f(x)$ is above or intersects the line representing $g(x)$. That is the region in which $f(x)$ is greater than or equal to $g(x)$. We mark the region to the left of the vertical line, but we also want to include the x value on the vertical line.

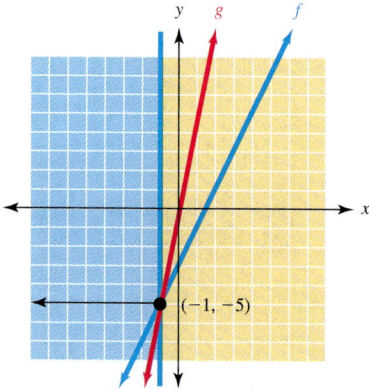

Finally, we express the solution in set notation. We see that the solution set consists of all x values such that x is -1, or less, so we write

$$\{x | x \leq -1\}$$

✓ CHECK YOURSELF 2

Graph the solution set to the inequality

$$3x + 2 \geq -2x - 8$$

The following algorithm summarizes our work in this section.

Finding the Solution for an Inequality in One Variable

Step 1 Rewrite the inequality as a comparison of two functions.

$$f(x) < g(x) \quad f(x) > g(x) \quad f(x) \leq g(x) \quad f(x) \geq g(x)$$

Step 2 Graph the two functions on a single set of axes.

Step 3 Draw a vertical line through each point of intersection of the two functions. Use a dotted line if equality is not included ($<$ or $>$). Use a solid line if equality is included ($\leq$ or $\geq$).

Step 4 Mark the x values that make the inequality a true statement.

Step 5 Write the solution in set notation.

✓ CHECK YOURSELF ANSWERS

1.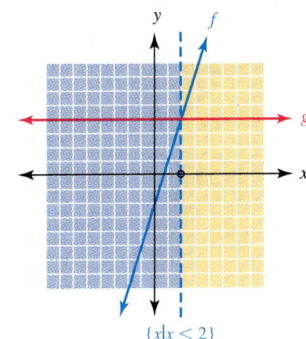
   $\{x \mid x < 2\}$

2.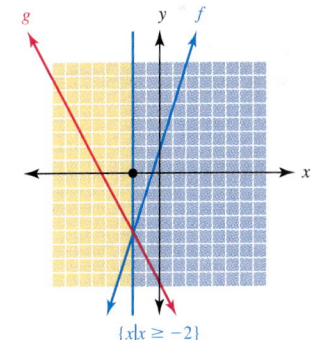
   $\{x \mid x \geq -2\}$

Exercises · 9.2

In Exercises 1 to 16, solve the linear inequalities graphically.

1. $2x < 8$

2. $-x < 4$

3. $\dfrac{x+3}{2} < -1$

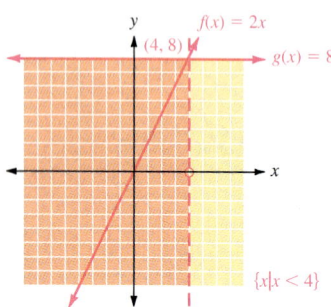
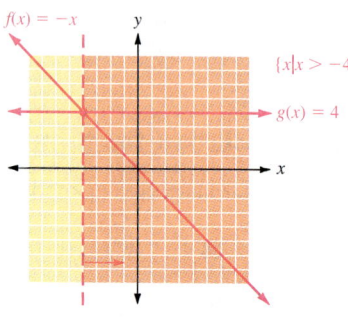
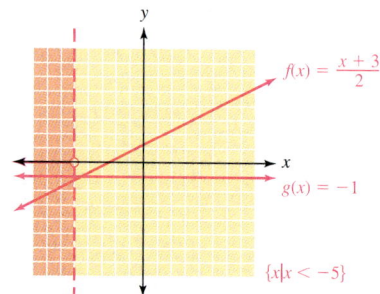

4. $\dfrac{-3x+3}{4} > -3$

5. $6x \geq 6$

6. $-3x \leq 6$

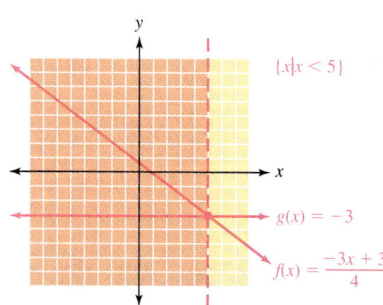
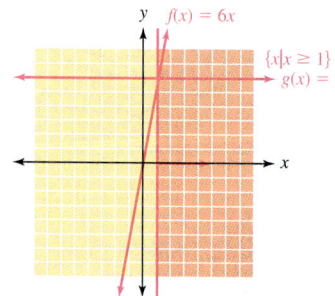
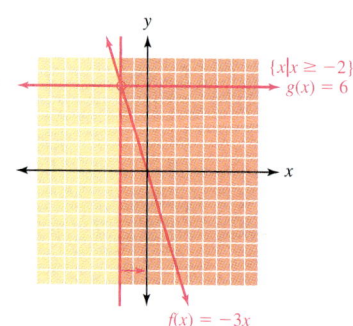

7. $7x - 7 < -2x + 2$

8. $7x + 2 > x - 4$

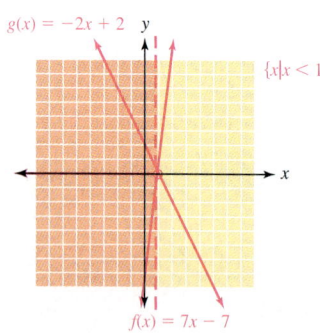
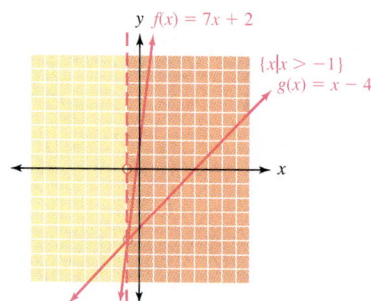

9. $\dfrac{14x + 4}{3} > 2(4x - 1)$

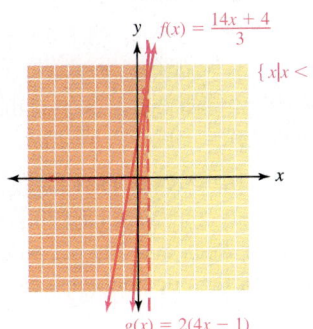

10. $2(3x + 1) < 4(x + 1)$

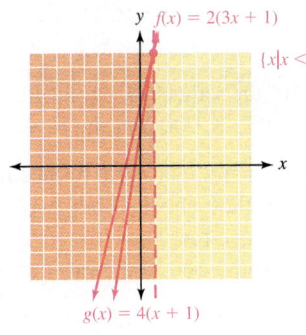

11. $6(1 + x) \geq 2(3x - 5)$

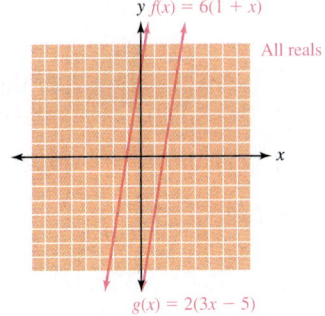

12. $2(x - 5) \geq 2x - 1$

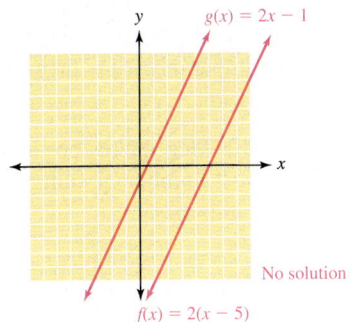

13. $7x > \dfrac{9x - 5}{2}$

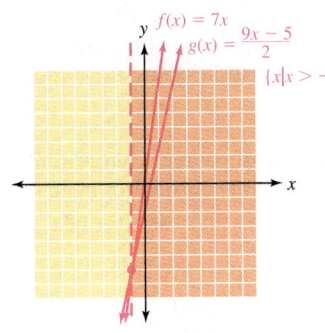

14. $-4x - 12 < x + 8$

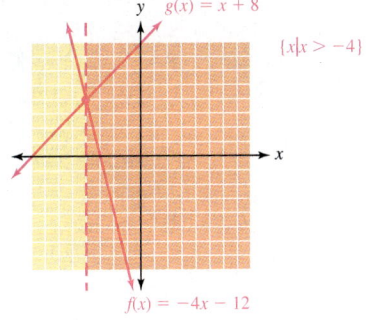

15. $4x - 6 \leq 2x - 2(5x - 12)$

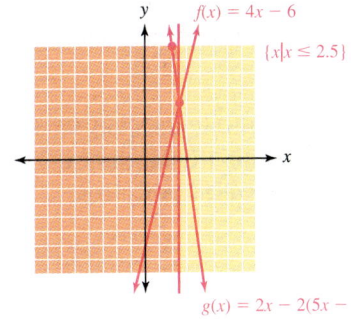

16. $5x + 3 > 2(4 - x) + 7x$

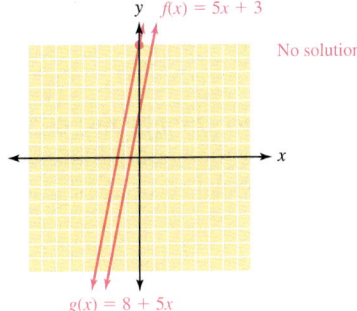

17. $x \geq 500$

18. $x \geq 45$

19. If miles are under 700, Downtown Edsel; if over 700, Wheels, Inc.; $W = \$28 \times 7 = \$196$; $DE = 98 + 0.14x$ (x is number of miles)

20. Less than 29,655 miles

21. 110 people

22. 4

In Exercises 17 to 22, solve the following applications.

17. **Business.** The cost to produce x units of wire is $C(x) = 50x + 5000$, and the revenue generated is $R(x) = 60x$. Find all values of x for which the product will at least break even.

18. **Business.** Find the values of x for which a product will at least break even if the cost is $C(x) = 85x + 900$ and the revenue is given by $R(x) = 105x$.

19. **Car Rental.** Tom and Jean went to Salem, Massachusetts, for 1 week. They needed to rent a car, so they checked out two rental firms. Wheels, Inc. wanted $28 per day with no mileage fee. Downtown Edsel wanted $98 per week and 14¢ per mile. Set up equations to express the rates of the two firms, and then decide when each deal should be taken.

20. **Mileage.** A fuel company has a fleet of trucks. The annual operating cost per truck is $C(x) = 0.58x + 7800$, where x is the number of miles traveled by a truck per year. What number of miles will yield an operating cost that is less than $25,000?

21. **Wedding.** Amanda and Joe are having their wedding reception at the Richland Fire Hall. They can spend at the most $3000 for the reception. If the hall charges a $250 cleanup fee plus $25 per person, find the greatest number of people they can invite.

22. **Tuition.** A nearby college charges annual tuition of $6440. Meg makes no more than $1610 per year in her summer job. What is the fewest number of summers that she must work in order to make enough for 1 year's tuition?

23. **Graphing.** Explain to a relative how a graph is helpful in solving each inequality below. Be sure to include the significance of the point at which the lines meet (or what happens if the lines do not meet).

   (a) $3x - 2 < 5$ (b) $3x - 2 \leq 4 - x$ (c) $4(x - 1) \geq 2 + 4x$

24. **College.** Look at the data here about enrollment in college. Assume that the changes occurred at a constant rate over the years. Make one linear graph for men and one for women, but on the same set of axes. What conclusions could you draw from reading the graph?

Year	Number (in millions) of Men in the United States Enrolled in College	Number (in millions) of Women in the United States Enrolled in College
1960	2.3	1.2
1998	6.4	7.8

SECTION 9.3 Absolute Value Functions

  OBJECTIVES

1. Graph an absolute value function
2. Solve absolute value equations in one variable graphically

Graph the function
$y = |x|$ as
$Y_1 = \text{abs}(x)$

In Section 2.6 we learned to solve absolute value equations algebraically. In this section, we will examine a graphical method for solving similar equations.

To demonstrate the graphical method, we will first look at the graph of an absolute value function. We will start by looking at the graph of the function $f(x) = |x|$. All other graphs of absolute value functions are variations of this graph.

The graph can be found using a graphing calculator (most graphing calculators use abs to represent the absolute value). We will develop the graph from a table of values.

x	$f(x) = |x|$
-3	3
-2	2
-1	1
0	0
1	1
2	2

Plotting these ordered pairs, we see a pattern emerge. The graph is like a large V that has its vertex at the origin. The slope of the line to the right of 0 is 1, and the slope of the line to the left of 0 is -1.

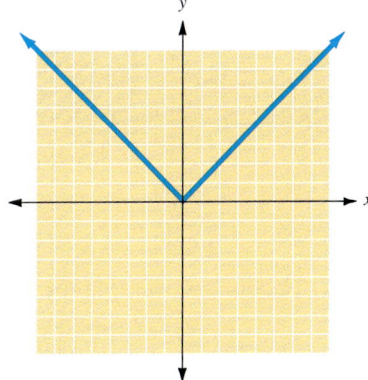

Let us now see what happens to the graph when we add or subtract some constant inside the absolute value bars.

637

Example 1 Graphing an Absolute Value Function

$f(x) = |x - 3|$
Would be entered as
$Y_1 = abs(x - 3)$

Graph each function.

On a TI calculator, (a) $f(x) = |x - 3|$

Again, we start with a table of values.

x	$f(x)$
-2	5
-1	4
0	3
1	2
2	1
3	0
4	1
5	2

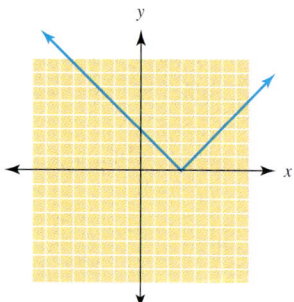

Then, we plot the points associated with the set of ordered pairs. The graph is shown to the left.

The graph of the function $f(x) = |x - 3|$ is the same shape as the graph of the function $f(x) = |x|$; it has just shifted to the right 3 units.

(b) $f(x) = |x + 1|$

We begin with a table of values.

x	$f(x)$
-2	1
-1	0
0	1
1	2
2	3
3	4

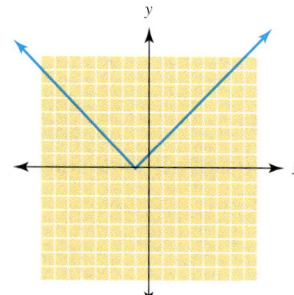

Again, you will find the graph in the margin.

Note that the graph of $f(x) = |x + 1|$ is the same shape as the graph of the function $f(x) = |x|$, except that it has shifted 1 unit to the left.

✓ CHECK YOURSELF 1

Graph each function.

(a) $f(x) = |x - 2|$ (b) $f(x) = |x + 3|$

We can summarize what we have discovered about the horizontal shift of the graph of an absolute value function.

> The graph of the function $f(x) = |x - a|$ will be the same shape as the graph of $f(x) = |x|$ except that the graph will be shifted a units
>
> to the right if a is positive.
>
> to the left if a is negative.

If a is negative, $x - a$ will be x plus some positive number.

We will now use these methods to solve equations that contain an absolute value expression.

Example 2 — Solving an Absolute Value Equation Graphically

Graphically find the solution set for the equation.

$$|x - 3| = 4$$

We graph the function associated with each side of the equation.

$$f(x) = |x - 3| \quad \text{and} \quad g(x) = 4$$

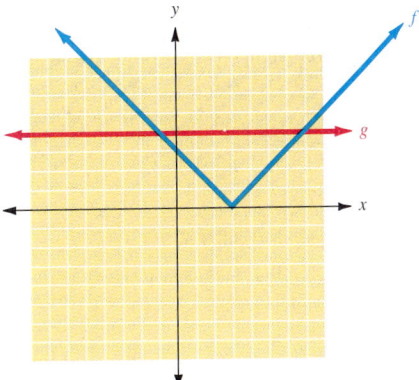

Then, we draw a vertical line through each of the intersection points.

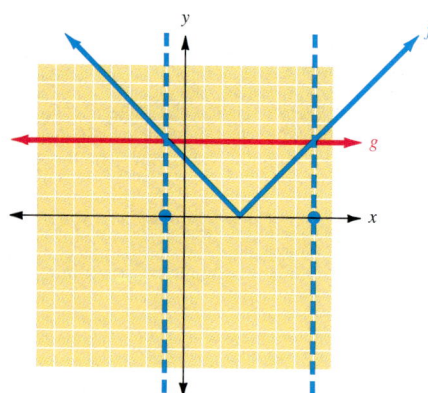

Looking at the x values of the two vertical lines, we find the solutions to the original equation. There are two x values that make the statement true: -1 and 7. The solution set is $\{-1, 7\}$.

✓ CHECK YOURSELF 2

Graphically find the solution set for the equation.

$$|x - 2| = 3$$

✓ CHECK YOURSELF ANSWERS

1. (a)

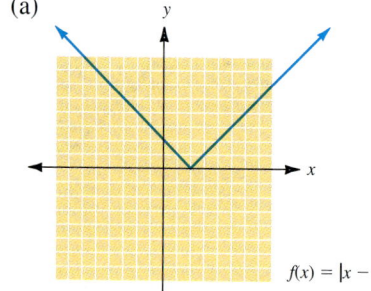

$f(x) = |x - 2|$

(b)

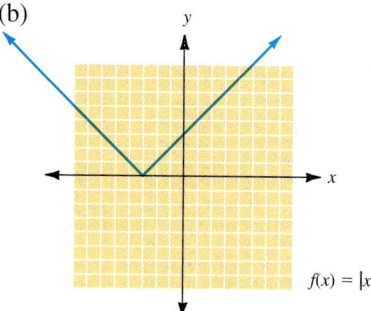

$f(x) = |x + 3|$

2.

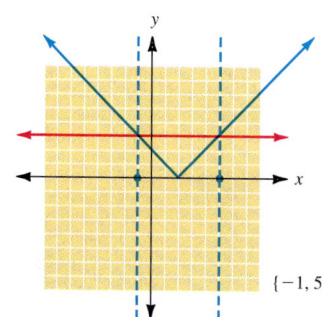

$\{-1, 5\}$

Assessing Piston Design. Combustion engines get their power from the force exerted by burning fuel on a piston inside a chamber. The piston is forced down out of the cylinder by the force of a small explosion caused by burning fuel mixed with air. The piston in turn moves a piston rod, which transfers the motion to the work of the engine. The rod is attached to a flywheel, which pushes the piston back into the cylinder to begin the process all over. Cars usually have four to eight of these cylinders and pistons. It is crucial that the piston and the cylinder fit well together, with just a thin film of oil separating the sides of the piston and the sides of the cylinder. When these are manufactured, the measurements for each part must be accurate. But, there is always some error. How much error is a matter for the engineers to set and for the quality control department to check.

Suppose the diameter of the cylinder is meant to be 7.6 cm, and the engineer specifies that this part must be manufactured to within 0.1 mm of that measurement. This figure is called the **tolerance**. As parts come off the assembly line, someone in quality control takes samples and measures the cylinders and the pistons. Given this information, complete the following.

1. Write an absolute value statement about the diameter, d_c, of the cylinder.

2. If the diameter of the piston is to be 7.59 cm with a tolerance of 0.1 mm, write an absolute value statement about the diameter, d_p, of the piston.

3. Investigate all the possible ways these two parts will fit together. If the two parts have to be within 0.1 mm of each other for the engine to run well, is there a problem with the way the parts may be paired together? Write your answer and use a graph to explain.

4. Accuracy in machining the parts is expensive, so the tolerance should be close enough to make sure the engine runs correctly, but not so close that the cost is prohibitive. If you think a tolerance of 0.1 mm is too large, find another that you think would work better. If it is too small, how much can it be enlarged and still have the engine run according to design? (That is, so $|d_c - d_p| \leq 0.1$ mm.) Write the tolerance using absolute value signs. Explain your reasoning if you think a tolerance of 0.1 mm is not workable.

5. After you have decided on the appropriate tolerance for these parts, think about the quality control engineer's job. Hazard a few educated opinions to answer these questions: How many parts should be pulled off the line and measured? How often? How many parts can reasonably be expected to be outside the expected tolerance before the whole line is shut down and the tools corrected?

Exercises · 9.3

In Exercises 1 to 6, graph each function.

1. $f(x) = |x - 3|$

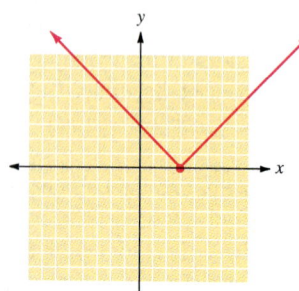

2. $f(x) = |x + 2|$

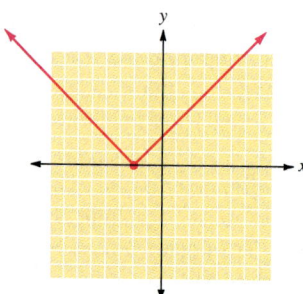

3. $f(x) = |x + 3|$

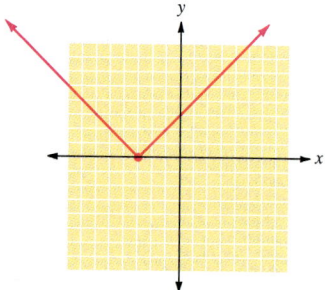

4. $f(x) = |x - 4|$

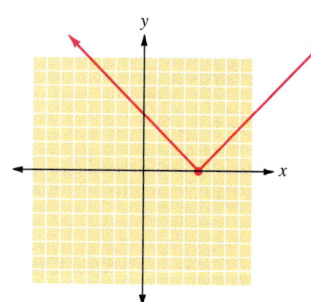

5. $f(x) = |x - (-3)|$

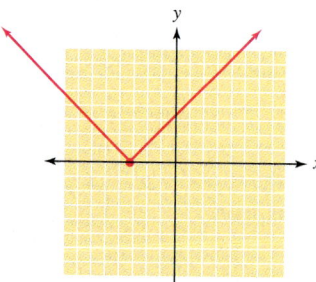

6. $f(x) = |x - (-5)|$

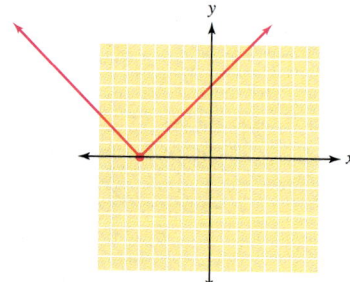

In Exercises 7 to 12, solve the equations graphically.

7. $|x| = 3$

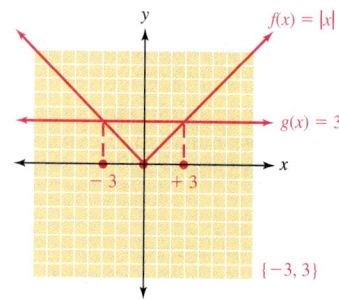

$\{-3, 3\}$

8. $|x| = 5$

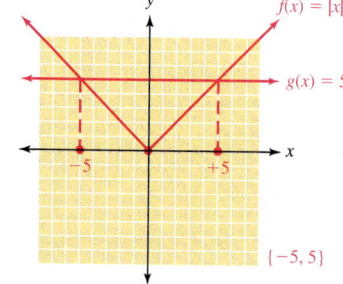

$\{-5, 5\}$

9. $|x - 2| = 5$

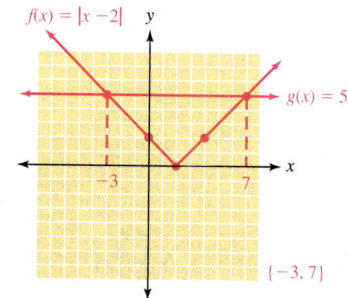

$\{-3, 7\}$

642

10. $|x - 5| = 3$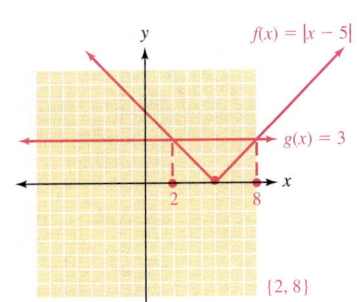

11. $|x + 2| = 4$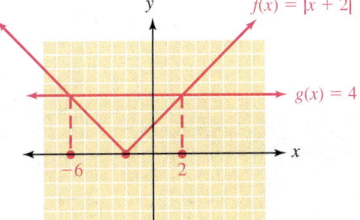

12. $|x + 4| = 2$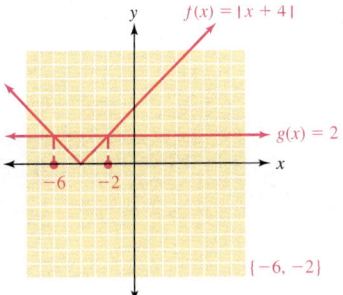

In Exercises 13 to 16, determine the function represented by each graph.

13. $f(x) = |x - 2|$

14. $f(x) = |x + 3|$

15. $f(x) = |x - 3|$

16. $f(x) = |x + 2|$

13.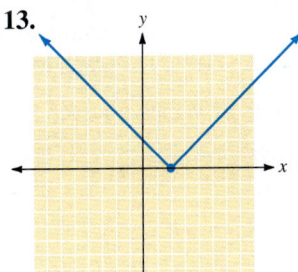

14.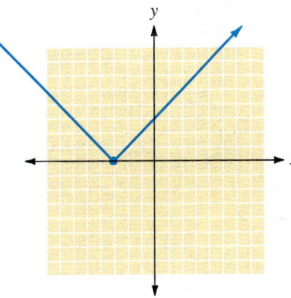

15.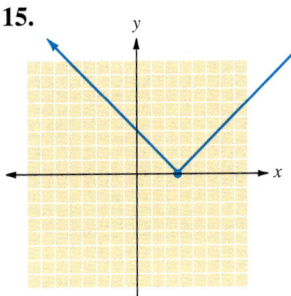

16.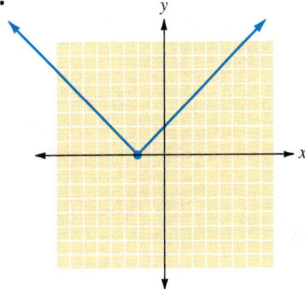

In Exercises 17 to 19, use a table of values for each function to sketch its graph.

17. $f(x) = -|x|$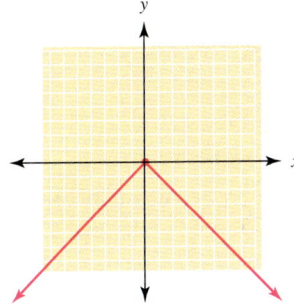

18. $f(x) = -|x - 2|$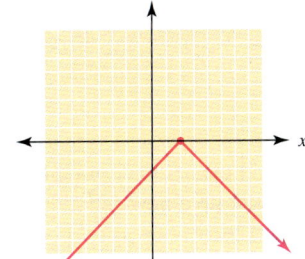

19. $f(x) = -|x + 3|$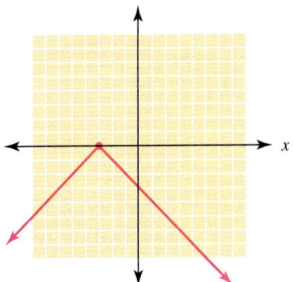

 20. What observations can you make concerning the graphs in Exercises 17 to 19? Based on these observations, make up a new function similar to those, predict the shape and position of the graph, and then create the graph to check your predictions.

In Exercises 21 to 24, use a table of values for each function to sketch its graph.

21. $f(x) = |x| - 3$ **22.** $f(x) = |x| + 2$ **23.** $f(x) = |x + 1| + 4$ **24.** $f(x) = |x + 2| - 3$

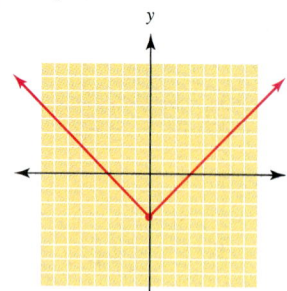 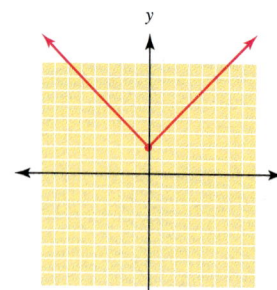 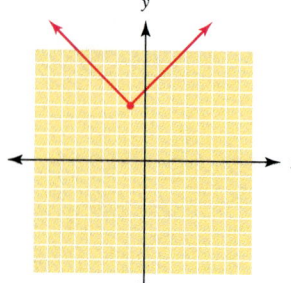 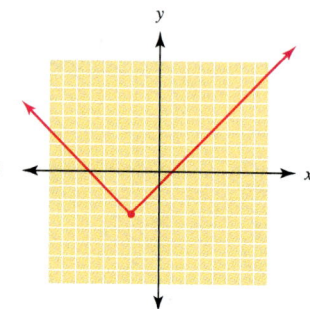

25. What observations can you make concerning the graphs in Exercises 21 to 24? Based on these observations, make up a new function similar to those, predict the shape and position of the graph, and then create the graph to check your predictions.

In Exercises 26 to 30, use a table of values for each function to sketch its graph.

26. $f(x) = 2|x|$ **27.** $f(x) = 3|x|$ **28.** $f(x) = \frac{1}{2}|x|$

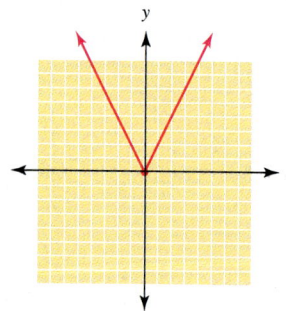 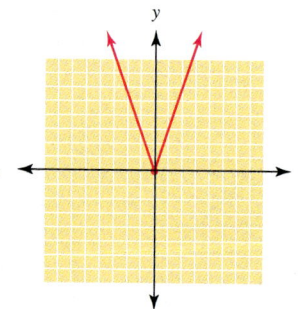 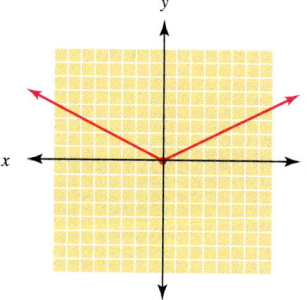

29. $f(x) = -2|x|$ **30.** $f(x) = -\frac{1}{2}|x|$

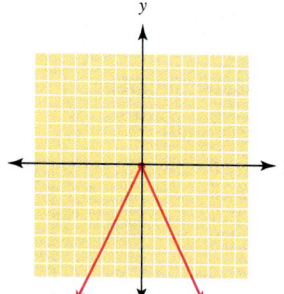 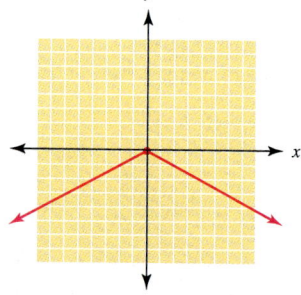

31. What observations can you make concerning the graphs in Exercises 26 to 30? Based on these observations, make up a new function similar to those, predict the shape and position of the graph, and then create the graph to check your predictions.

In Exercises 32 to 35 use graphing techniques to solve each equation.

32. $|x - 2| = |x + 4|$ **33.** $|x - 3| = |x + 5|$

34. $|x - 1| = |x + 5|$ **35.** $|x - 3| = |x + 1|$

32. $\{-1\}$
33. $\{-1\}$
34. $\{-2\}$
35. $\{1\}$

SECTION 9.4 Absolute Value Inequalities

9.4 OBJECTIVES

1. Solve absolute value inequalities in one variable graphically
2. Solve absolute value inequalities in one variable algebraically

In Section 9.3, we looked at a graphical method for solving an absolute value equation. In this section, we will look at a graphical method for solving absolute value inequalities.

Absolute value inequalities in one variable, x, are obtained from absolute value equations by replacing the symbol for equality $(=)$ with one of the inequality symbols $(<, >, \leq, \geq)$.

The general form for an absolute value inequality in one variable is

$$|x - a| < b$$

where the symbol $<$ can be replaced with $>$, $\leq$, or $\geq$. Examples of absolute value inequalities in one variable include

$$|x| < 6 \qquad |x - 4| \geq 2 \qquad |3x - 5| \leq 8$$

Example 1

Solving an Absolute Value Inequality Graphically

Graphically solve

$$|x| < 6$$

As we did in previous sections, we begin by letting each side of the inequality represent a function. Here

$$f(x) = |x| \qquad \text{and} \qquad g(x) = 6$$

Now we graph both functions on the same set of axes.

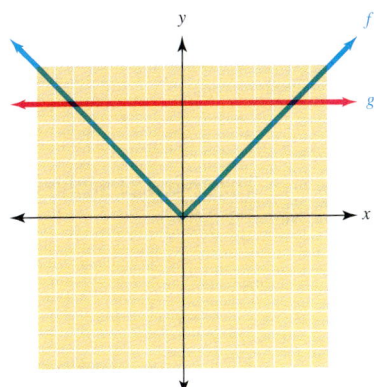

645

We next draw a vertical dotted line (equality is not included) through the points of intersection of the two graphs.

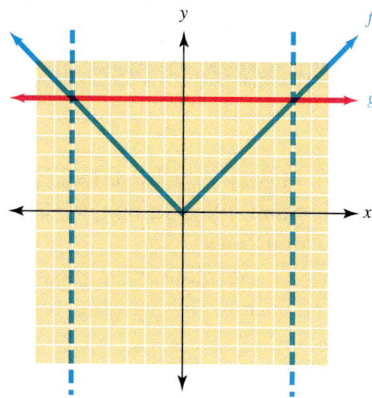

The solution set is any value of x for which the graph of $f(x)$ is below the graph of $g(x)$.

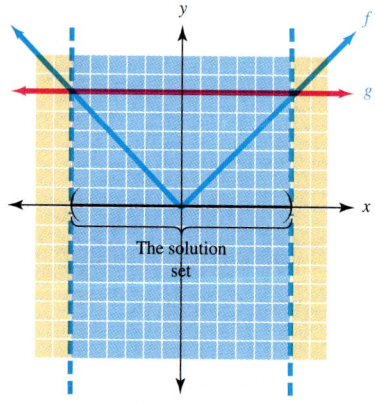

In set notation, we write $\{x \mid -6 < x < 6\}$.

✓ CHECK YOURSELF 1

Graphically solve the inequality

$$|x| < 3$$

The graphical method of Example 1 relates to the following general statement from Section 2.6.

Absolute Value Inequalities, Property 1

For any positive number p, if

$$|x| < p$$

then

$$-p < x < p$$

Let's look at an application of Property 1 in solving an absolute value inequality.

Example 2 Solving Absolute Value Inequalities

Solve the following inequality and graph the solution set on a number line.

$$|x - 3| < 5$$

From Property 1, we know that the given absolute value inequality is equivalent to the compound inequality

$$-5 < x - 3 < 5$$

Solve as before.

$$-5 < x - 3 < 5 \qquad \text{Add 3 to all three parts.}$$
$$-2 < x < 8$$

With Property 1 we can translate *an absolute value inequality to an inequality* not *containing an absolute value, which can be solved by our earlier methods.*

The solution set is

$$\{x \mid -2 < x < 8\}$$

The graph is shown below.

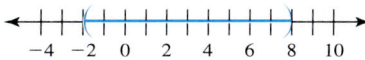

Note that the solution is an open interval on the number line.

✓ CHECK YOURSELF 2

Solve the following inequality and graph the solution set on a number line.

$$|x - 4| \leq 8$$

In example 3, we will look at a graphical method for solving an inequality.

Example 3 **Finding a Graphical Solution for an Absolute Value Inequality**

Use a graph to solve the inequality

$$|x - 3| < 5$$

Let $f(x) = |x - 3|$ and $g(x) = 5$, and graph both functions on the same set of axes.

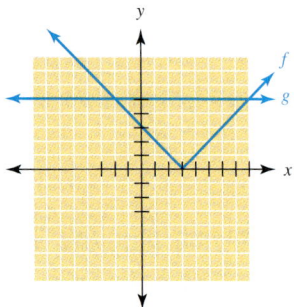

Drawing a vertical dotted line through the intersection points, we find the set of x values for which $f(x) < g(x)$.

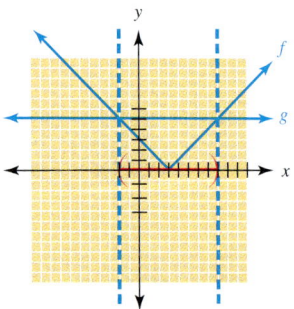

We have confirmed that the solution set is $\{x | -2 < x < 8\}$

 CHECK YOURSELF 3

Graphically solve the inequality

$$|x - 4| \leq 8$$

We know the solution set for the absolute value inequality

$$|x| < 4$$

consists of those numbers whose distance from the origin is *less than 4*. Now, what about the solution set for $|x| > 4$?

Example 4

Solving Absolute Value Inequalities

Graphically solve

$$|x| > 4$$

Again, we graph both functions on the same set of axes. We also draw vertical dotted lines through the points of intersection of the graphs.

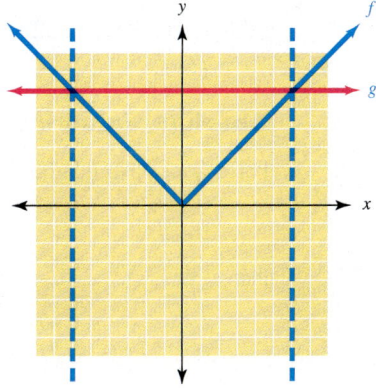

The solution set is every value of x for which the graph of $f(x)$ is *above* the graph of $g(x)$.

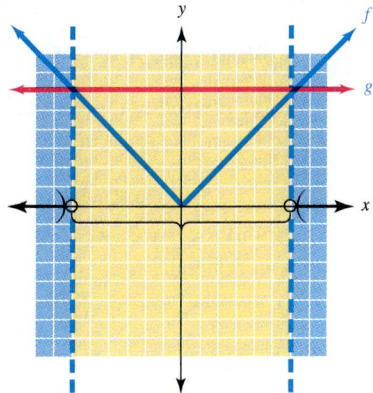

In set notation, we write $\{x|x < -4 \textbf{ or } x > 4\}$

✓ **CHECK YOURSELF 4**

Graphically solve the inequality

$$|x| > 3$$

The solution set for Example 4 consists of those numbers whose distance from the origin is *greater than 4*. The solution set is pictured below.

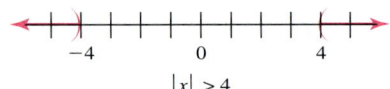

$|x| > 4$

The solution set can be described by the compound inequality

$$x < -4 \quad \text{or} \quad x > 4$$

and this relates to the following general statement from Section 2.6:

Absolute Value Inequalities, Property 2

For any positive number p, if

$$|x| > p$$

then

$$x < -p \quad \text{or} \quad x > p$$

Let's apply Property 2 to the solution of an absolute value inequality

Example 5

Solving Absolute Value Inequalities

Solve the following inequality and graph the solution set on a number line.

$$|2 - x| > 8$$

From Property 2, we know that the given absolute inequality is equivalent to the compound inequality.

$$2 - x < -8 \quad \text{or} \quad 2 - x > 8$$

Again we translate the absolute value inequality to the compound inequality not *containing an absolute value.*

Solving as before, we have

$$2 - x < -8 \quad \text{or} \quad 2 - x > 8$$
$$-x < -10 \qquad\qquad -x > 6$$
$$x > 10 \qquad\qquad x < -6$$

When we divide by a negative number, we reverse the inequality.

The solution set is $\{x \mid x < -6 \text{ or } x > 10\}$, and the graph is shown below.

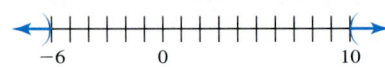

✓ CHECK YOURSELF 5

Solve the following inequality and graph the solution set on a number line.

$$|3 - x| \geq 4$$

Our final example looks at a graphical method for solving the inequality in Example 5.

Example 6 — Finding a Graphical Solution for an Absolute Value Inequality

Graphically solve the inequality

$$|2 - x| > 8$$

Let $f(x) = |2 - x|$ and $g(x) = 8$, and graph both functions on the same set of axes.

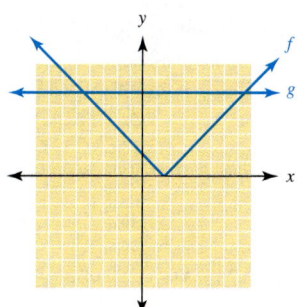

Drawing a vertical dotted line through the intersection points, we find the set of x values for which $f(x) > g(x)$.

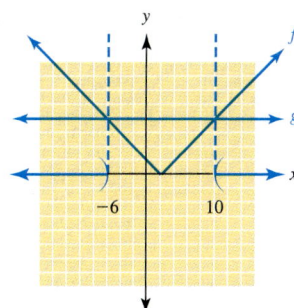

We have confirmed that the solution set is $\{x | x < -6 \text{ or } x > 10\}$.

✓ CHECK YOURSELF 6

Graphically solve the inequality

$$|3 - x| \geq 4$$

✓ CHECK YOURSELF ANSWERS

1. $\{x | -3 < x < 3\}$

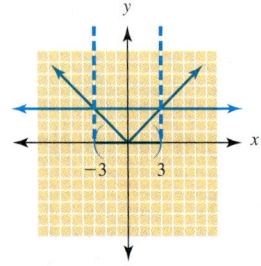

2. $\{x | -4 \leq x \leq 12\}$

3. $\{x | -4 \leq x \leq 12\}$

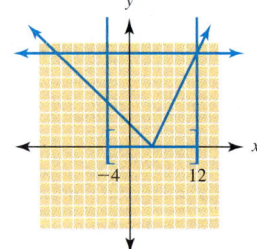

4. $\{x | x < -3 \text{ or } x > 3\}$

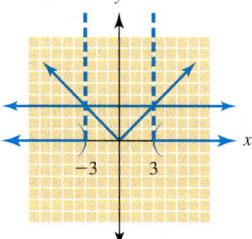

5. $\{x | x \leq -1 \text{ or } x \geq 7\}$

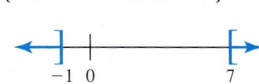

6. $\{x | x \leq -1 \text{ or } x \geq 7\}$

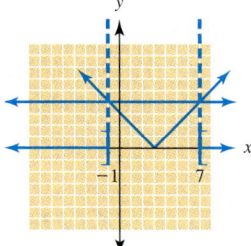

Determining Heating Costs. Ed and Sharon's heating system has just broken down. They can replace their current electrical heat pump system with another electrical heat pump system or replace it with an oil heating system or a gas system. Ed knows that in order to install an oil heating system, he will need a new air conditioner and extensive duct remodeling in his house. He also knows that a gas heating system will require gas lines be laid on his street and then connected to the house. Both Ed and Sharon are concerned about installing another heat pump system because it does not seem to work well in their cold Northeast climate. They both want to install a system that will be economical in the long run. Ed did some research and collected installation costs and yearly operational costs. These data appear in the following chart.

Type of Heating	Installation Costs	Operational Costs (per year)
Reinstall an electric system	$4,200	$900
Convert to oil system	$8,500	$550
Convert to gas system	$13,000	$425

1. Use 5-year intervals to construct a table that shows the total cost of heating over a 35-year period for the three heating systems.

2. Determine the average yearly heating costs for each of the three systems after (a) 5 years, (b) 10 years, (c) 15 years, (d) 20 years, (e) 25 years, (f) 30 years, and (g) 35 years.

3. The total cost of each system is a function composed of the operational costs multiplied by the number of years of operation added to the installation costs. For each of the three systems, develop a function that represents the total cost.

4. Graph the three straight lines developed in step 3 on the same coordinate axis.

5. Using the graphs in step 4, estimate the following:

    (a) When does the oil system become cheaper than the electric system?

    (b) When does the gas system become cheaper than the electric system?

    (c) When does the gas system become cheaper than the oil system?

6. Using the functions developed in step 3, determine the following:

    (a) When does the oil system become cheaper than the electric system?

    (b) When does the gas system become cheaper than the electric system?

    (c) When does the gas system become cheaper than the oil system?

7. Do the following:

    (a) Contact your local electric and gas utility and obtain the current rate schedule as well as the cost of installing a new heating system that would be adequate for a 2000-square ft home with 8-ft ceilings.

    (b) Contact a local oil company and obtain the current prices as well as installation costs of a new heating system that would be adequate for a 2000-square ft home with 8-ft ceilings.

    (c) Use the information from (a) and (b) and construct a table similar to that developed in step 1.

    (d) Develop a function that represents the total heating costs for each of the three systems based on the information you obtained in (a) and (b).

    (e) Based solely on the costs associated with the heating systems, which one would you choose to install in your house?

8. In addition to cost, what other factors would you consider in choosing a new heating system for your home?

9. Considering all the factors involved, decide what type of heating system you would pick for a new home you are building, and write a paragraph explaining your decision.

Exercises • 9.4

In Exercises 1 to 12, solve each inequality graphically.

1. $|x| < 4$

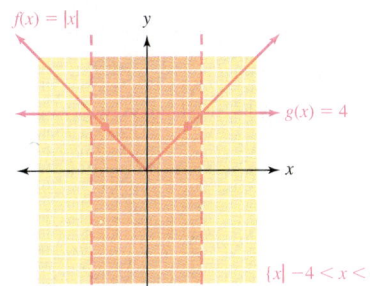

2. $|x| < 6$

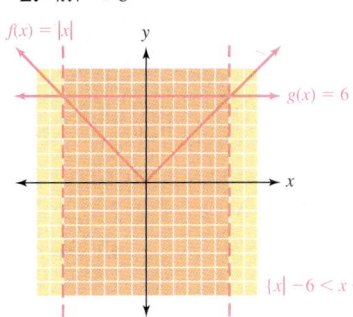

3. $|x| \geq 5$

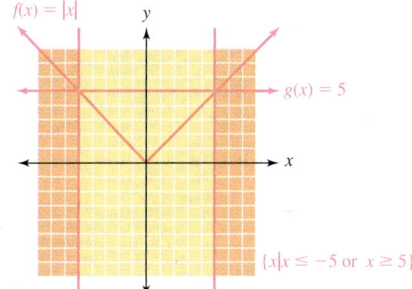

4. $|x| \geq 2$

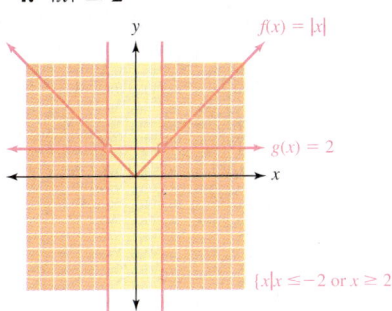

5. $|x - 3| < 4$

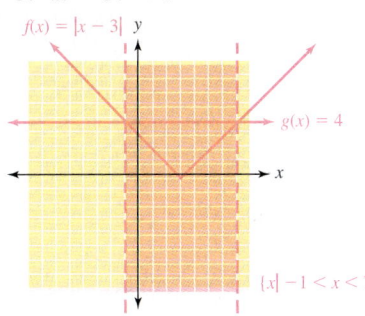

6. $|x - 1| < 5$

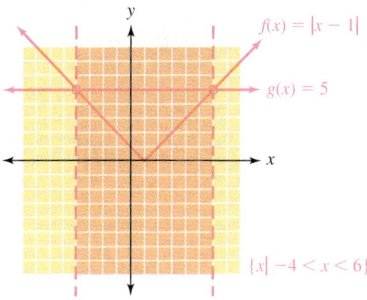

7. $|x - 2| \geq 5$

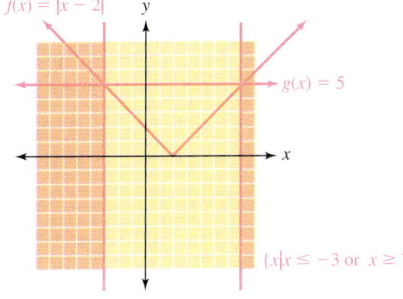

8. $|x + 2| > 4$

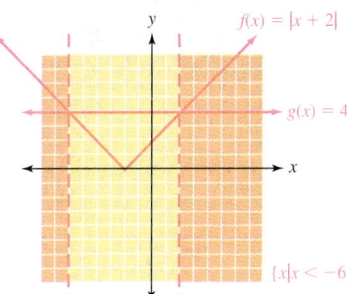

9. $|x + 1| \leq 5$

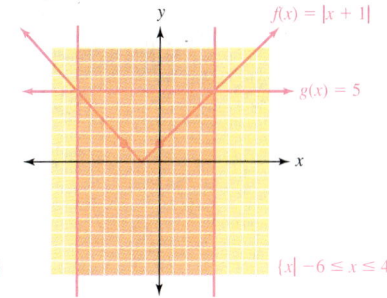

10. $|x + 4| > 1$

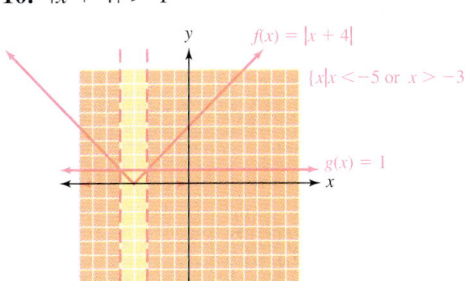

11. $|x + 2| \geq -2$

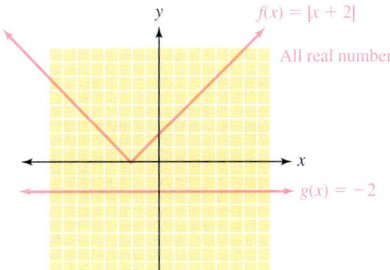

12. $|x - 4| > -1$

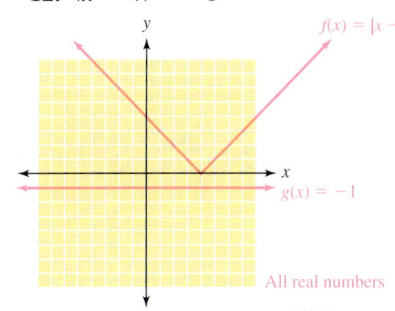

In Exercises 13 to 32, solve each inequality. Graph the solution set.

13. $|x| < 5$

14. $|x| > 3$

15. $|x| \geq 7$

16. $|x| \leq 4$

17. $|x - 4| > 2$

18. $|x + 5| < 3$

19. $|x + 6| \leq 4$

20. $|x - 7| > 5$

21. $|3 - x| > 5$

22. $|5 - x| < 3$

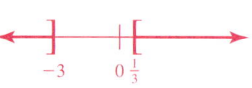

23. $|x - 7| < 0$

No solution

24. $|x + 5| \geq 0$

25. $|2x - 5| < 3$

26. $|3x - 1| > 8$

27. $|3x + 4| \geq 5$

28. $|2x + 3| \leq 9$

29. $|5x - 3| > 7$

30. $|6x - 5| < 13$

31. $|2 - 3x| \leq 11$

32. $|3 - 2x| > 11$

In Exercises 33 to 40, use absolute value notation to write an inequality that represents each sentence.

33. $|x| < 3$

33. x is within 3 units of 0 on the number line.

34. $|x| < 4$

34. x is within 4 units of 0 on the number line.

35. $|x| \geq 5$

35. x is at least 5 units from 0 on the number line.

36. $|x| \geq 2$

36. x is at least 2 units from 0 on the number line.

37. $|x - (-2)| < 7$

37. x is less than 7 units from -2 on the number line.

38. $|x - 4| > 6$

38. x is more than 6 units from 4 on the number line.

39. $|x - (-4)| \geq 3$

39. x is at least 3 units from -4 on the number line.

40. $|x - 4| \leq 3$

40. x is at most 3 units from 4 on the number line.

Summary Exercises ■ 9

This summary exercise set is provided to give you practice with each of the objectives in the chapter. Each exercise is keyed to the appropriate chapter section.

[9.1] Solve the following equations graphically.

1. $3x - 6 = 0$

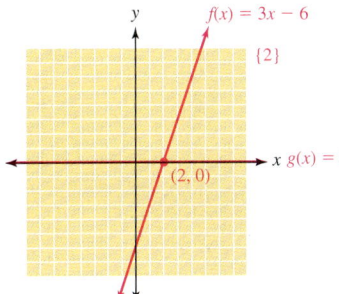

2. $4x + 3 = 7$

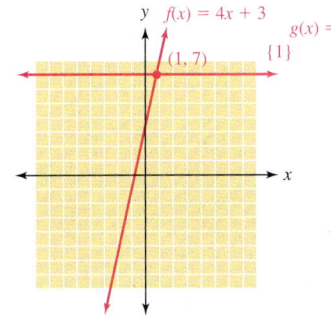

3. $3x + 5 = x + 7$

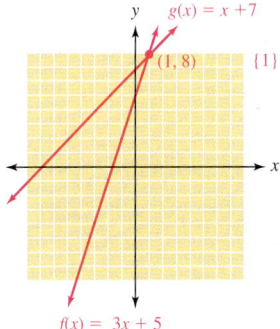

4. $4x - 3 = x - 6$

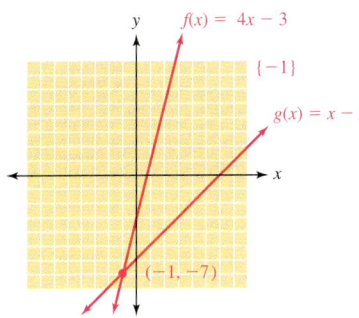

5. $\dfrac{6x - 1}{2} = 2(x - 1)$

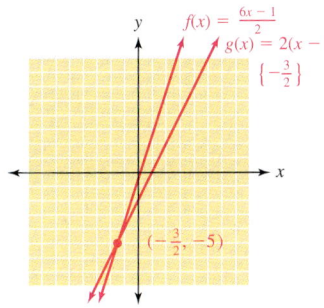

6. $3x + 2 = 2x - 1$

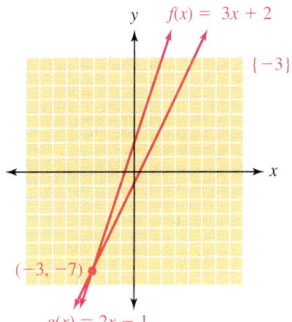

7. $3(x - 2) = 2(x - 1)$

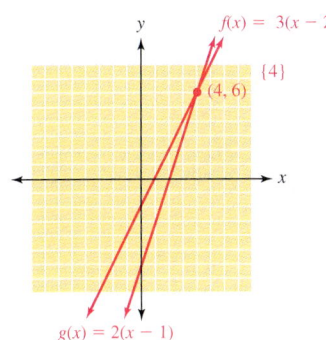

8. $3(x + 1) + 3 = -7(x + 2)$

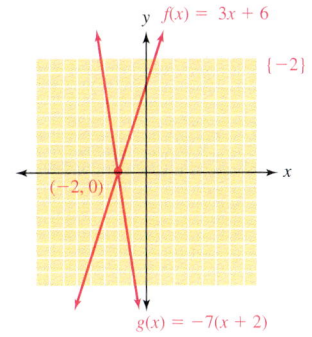

657

658 Chapter 9 ■ Graphical Solutions

[9.2] Solve the following inequalities graphically.

9. $5 < 2x + 1$

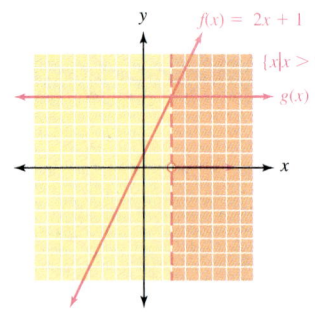

10. $-3 \geq -2x + 3$

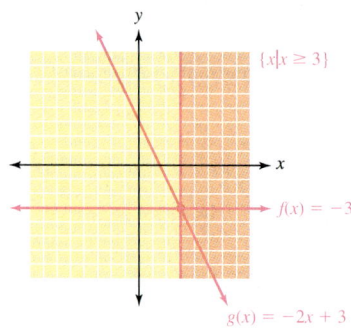

11. $3x + 2 \geq 6 - x$

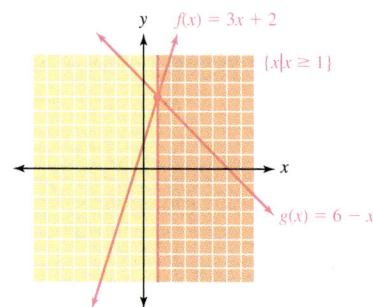

12. $3x - 5 < 15 - 2x$

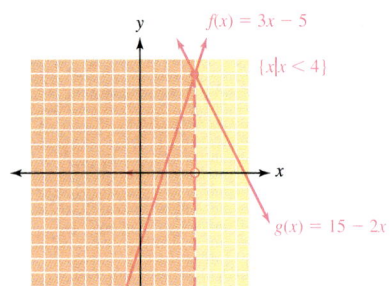

13. $x + 6 < -2x - 3$

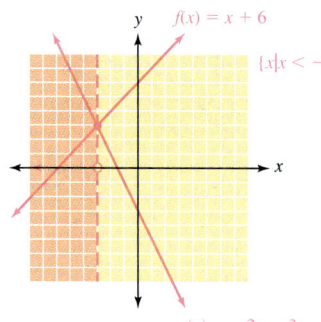

14. $4x - 3 \geq 7 - x$

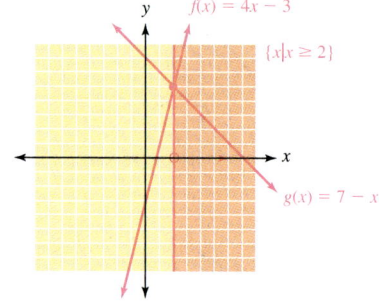

15. $x \geq -3 - 5x$

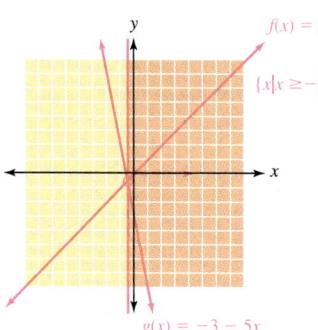

16. $x < 4 + 2x$

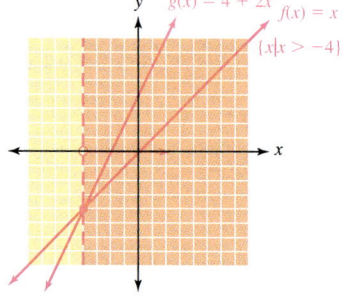

17. $x \geq 15$

18. 4

■ Summary Exercises

[9.2] Solve each of the following applications.

17. The cost to produce x units of a product is $C = 100x + 6000$, and the revenue is $R = 500x$. Find all values of x for which the product will at least break even.

18. For a particular line of lamps, a store has average monthly costs of $C = 55x + 180$ and corresponding revenue of $R = 100x$, where x is the number of units sold. How many units must be sold for the store to break even on the lamps for the month?

[9.3] Solve the following equations graphically.

19. $|x + 3| = 5$

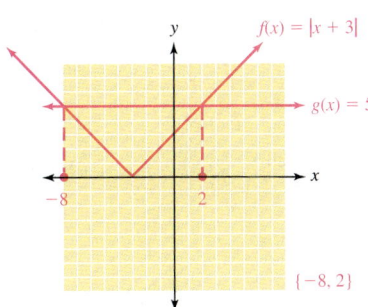

$\{-8, 2\}$

20. $|x - 2| = 7$

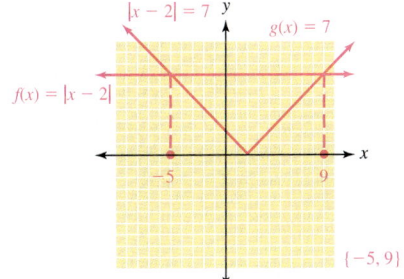

$\{-5, 9\}$

21. $|2 - x| = 3$

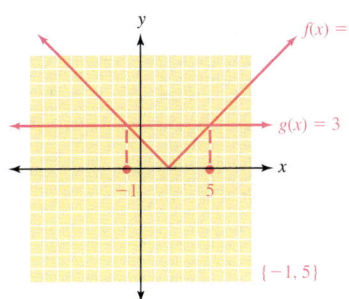

$\{-1, 5\}$

22. $|4 - x| = 2$

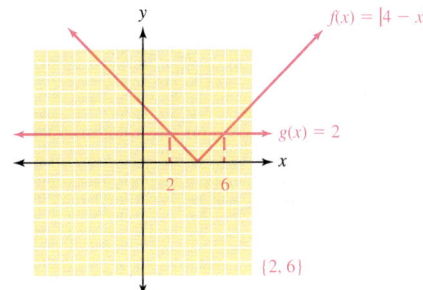

$\{2, 6\}$

[9.4] Solve the following inequalities graphically.

23. $|x| < 7$

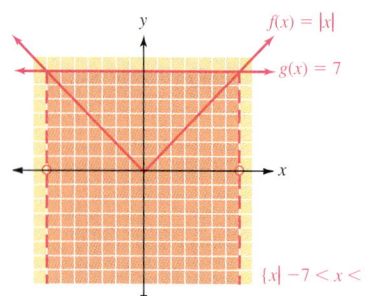

$\{x|-7 < x < 7\}$

24. $|x| \leq 9$

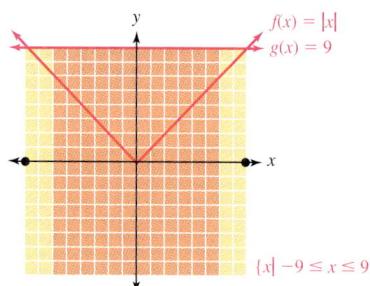

$\{x|-9 \leq x \leq 9\}$

25. $|x - 1| > 6$

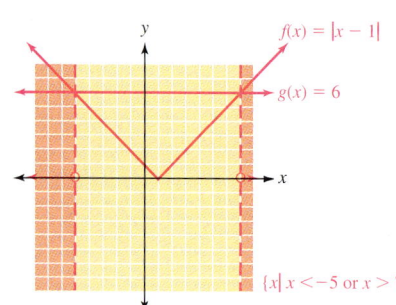

$\{x| x < -5 \text{ or } x > 7\}$

26. $|x + 3| \leq 4$

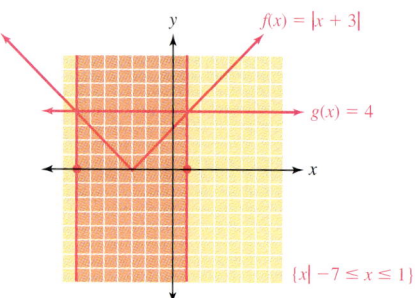

$\{x|-7 \leq x \leq 1\}$

27. $|x + 5| \geq 2$

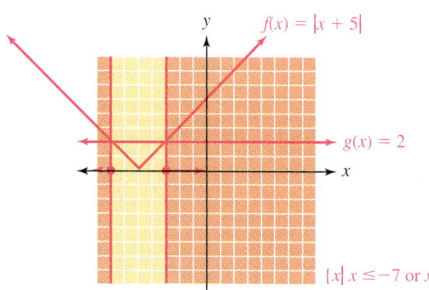

$\{x| x \leq -7 \text{ or } x \geq -3\}$

28. $|x - 1| < 8$

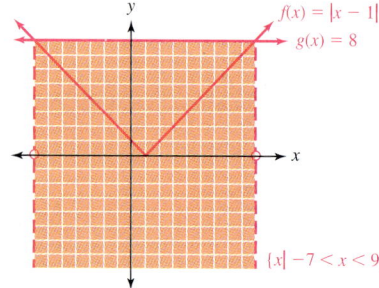

$\{x|-7 < x < 9\}$

Self-Test ■ 9

The purpose of this self-test is to help you check your progress and to review for a chapter test in class. Allow yourself about 1 hour to take the test. When you are done, check your answers in the back of the book. If you missed any answers, be sure to go back and review the appropriate sections in the chapter and the exercises that are provided.

Solve the following equations graphically.

1. $4x - 7 = 5$

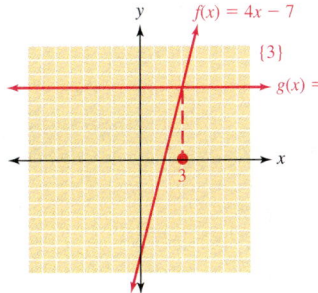

2. $6 - x = 4(x - 1)$

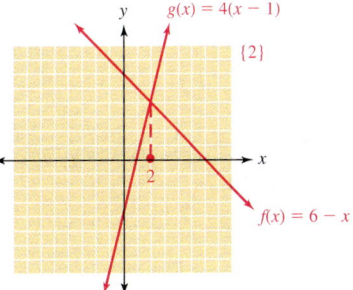

3. $-8x + 11 = 2x - 9$

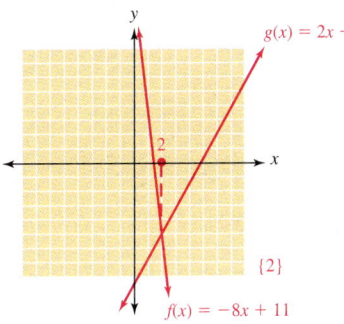

4. $6(x - 1) = -3(x - 4)$

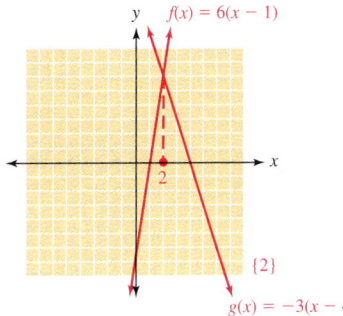

Solve the following inequalities graphically.

5. $5x - 3 < 7$

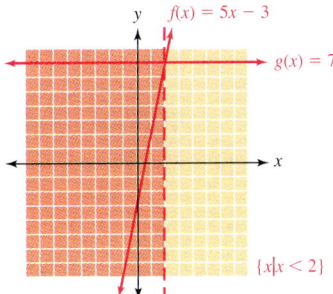

6. $2x - 1 \leq 3(x - 1)$

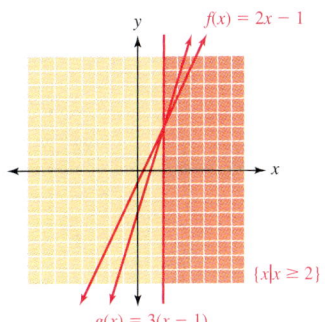

661

662 Chapter 9 ■ Graphical Solutions

Solve the following equations graphically.

7. $|x + 3| = 7$

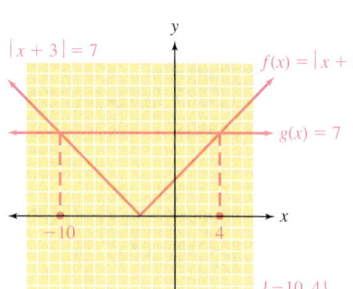

8. $|x - 5| = 9$

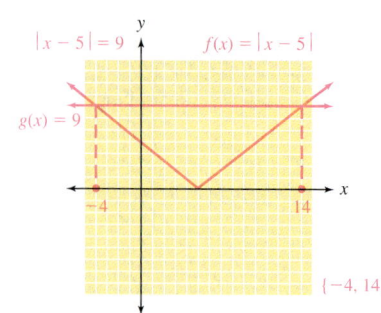

Solve the following inequalities graphically.

9. $|x| \leq 3$

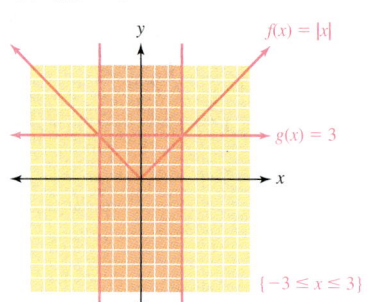

10. $|x + 1| \geq 4$

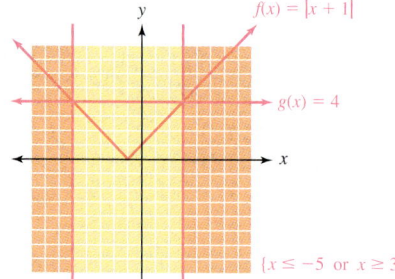

11. $|x - 1| > 5$

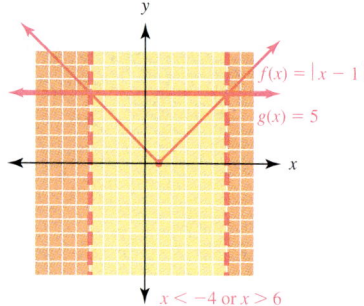

12. $|x + 2| < 4$

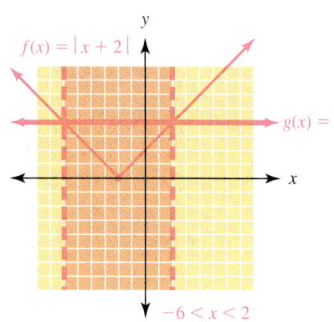

13. $|x - 1| < 4$

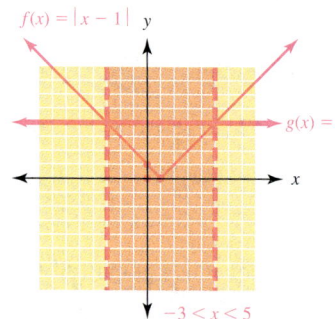

14. $|3 + x| \geq 2$

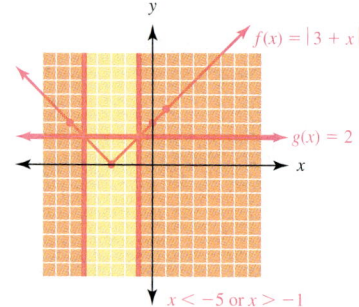

15. $|3 - x| \leq 5$

16. $|2 - x| > 5$

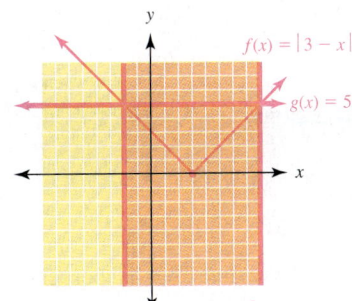

$-2 \leq x \leq 8$

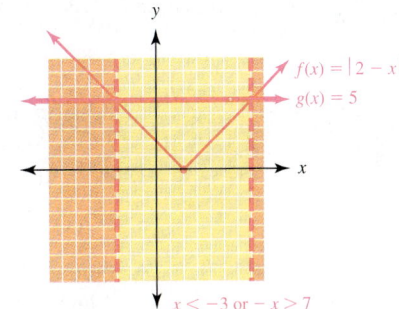

$x < -3$ or $-x > 7$

Solve the following applications.

17. 150 mi at 30 mph and 80 mi at 40 mph

18. 20,000 boxes

19. $\left\{ x \mid -\dfrac{9}{2} < x < \dfrac{7}{2} \right\}$

20. $\left\{ x \mid x \leq -\dfrac{7}{3} \text{ or } x \geq \dfrac{11}{3} \right\}$

17. By traveling 30 mph for one period of time and 40 mph for another, Maria traveled 230 miles. Had she gone 10 mph faster throughout, she would have traveled 300 miles. How many miles did she travel at each rate?

18. A company that manufactures boxed mints sells each box for $3. The company incurs a cost of $1.50 per box with a total fixed cost of $30,000. How many boxes of mints must be sold for the company to break even?

Solve the following inequalities for x.

19. $|2x + 1| < 8$

20. $|3x - 2| \geq 9$

Cumulative Test 1-9

Answers (left column):

1. $-\dfrac{4}{3}$
2. $\{-2, 8\}$
3. -21
4. $y = \dfrac{2}{3}x - 2$
5. $4x^2 - 5x - 3$
6. $x(x+3)(x-3)$
7. $(2x+1)(x-3)$
8. $(x-2)(x+5)$
9. $\dfrac{1}{x+1}$
10. $\dfrac{-x^2 + x + 5}{(x+3)(x-2)}$
11. $\dfrac{1}{(x-4)(x+1)}$
12. $(2, -2)$
13. x intercept: 3; y intercept: 6
14. $\{x \mid x \geq -8\}$
15. $\{x \mid -3 \leq x \leq 9\}$
16. $\{x \mid x < -3 \text{ or } x > 7\}$
17. $\dfrac{1}{x-5} + x + 6$
18. $\dfrac{1}{(x-5)(x+6)}$

Solve the following equations.

1. $4x - 2(x+1) = -5 - (x+1)$
2. $|x - 3| = 5$
3. If $f(x) = 4x^3 - 3x^2 + 5x - 9$, find $f(-1)$.
4. Find the equation of the line that is perpendicular to $3x + 2y = 6$ and has a y intercept of -2.
5. Simplify $4x - 3(x+1) + 6x(x-1) - 2x^2$.

Factor completely.

6. $x^3 - 9x$ 7. $2x^2 - 5x - 3$ 8. $x^2 - 2x - 10 + 5x$

Simplify.

9. $\dfrac{4}{x+1} - \dfrac{3}{x+1}$ 10. $\dfrac{x+1}{x^2 + x - 6} - \dfrac{x+2}{x+3}$

11. $\dfrac{2x-3}{x^2 - 16} \div \dfrac{2x^2 - x - 3}{x+4}$

12. Solve the following system.

$$3x + 2y = 2$$
$$x + 3y = -4$$

13. Find the x and y intercepts of the equation $4x + 2y = 12$.

Solve the following inequalities.

14. $x - 3 \leq 2x + 5$ 15. $|x - 3| \leq 6$ 16. $-3|x+1| < -6$

If $f(x) = \dfrac{1}{x-5}$ and $g(x) = x + 6$, find the following.

17. $f + g$ 18. $f \div g$

Graph the following functions.

19. $f(x) = -\dfrac{2}{3}x + 5$ 20. $f(x) = |x + 4|$

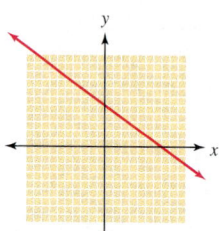 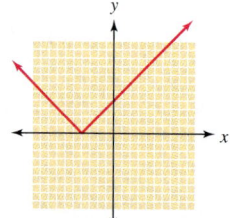

664

CHAPTER 10
Radicals and Exponents

LIST OF SECTIONS

10.1 Zero and Negative Integer Exponents and Scientific Notation

10.2 Evaluating Radical Expressions

10.3 Rational Exponents

10.4 Complex Numbers

As we have all experienced firsthand, consumer goods increase in price from year to year. This increase is usually measured by the Consumer Price Index (CPI), which measures a change in the prices of such everyday goods and services as energy, food, shelter, apparel, transportation, medical care, and utilities. The percent change in the CPI is a reflection of the purchasing power of the dollar and indicates the rate of inflation.

Since many labor contracts and government benefits programs such as Social Security increase or decrease along with the CPI, the method used to calculate this index is hotly debated by economists and statisticians. Beginning in April 1997, the Bureau of Labor Statistics began releasing an experimental CPI that uses a **geometric mean formula.** This new method may more accurately reflect the true cost-of-living increase or decrease because it takes into consideration that consumers' buying habits change as prices fluctuate. For instance, consumers may switch from romaine lettuce to iceberg lettuce or spinach if the price of romaine lettuce is too high.

665

SECTION 10.1 Zero and Negative Exponents and Scientific Notation

10.1 OBJECTIVES

1. Define the zero exponent
2. Simplify expressions with negative exponents
3. Write a number in scientific notation
4. Solve an application of scientific notation

In Section 5.1, we examined properties of exponents, but all of the exponents were positive integers. In this section, we look at zero and negative exponents. First we extend the quotient rule so that we can define an exponent of zero.

Recall that, in the quotient rule, to divide two expressions that have the same base, we keep the base and subtract the exponents.

$$\frac{a^m}{a^n} = a^{m-n}$$

Now, suppose that we allow m to equal n. We then have

$$\frac{a^m}{a^m} = a^{m-m} = a^0 \qquad (1)$$

But we know that it is also true that

$$\frac{a^m}{a^m} = 1 \qquad (2)$$

Comparing equations (1) and (2), we see that the following definition is reasonable.

The Zero Exponent

For any real number a where $a \neq 0$,

$$a^0 = 1$$

We must have $a \neq 0$. The form 0^0 is called **indeterminate** and is considered in later mathematics classes.

Example 1

The Zero Exponent

Use the above definition to simplify each expression.

(a) $17^0 = 1$
(b) $(a^3 b^2)^0 = 1$
(c) $6x^0 = 6 \cdot 1 = 6$
(d) $-3y^0 = -3$

Note that in $6x^0$ the exponent 0 applied *only* to x.

Section 10.1 ■ Zero and Negative Exponents and Scientific Notation

✓ **CHECK YOURSELF 1**

Simplify each expression.

(a) 25^0 (b) $(m^4n^2)^0$ (c) $8s^0$ (d) $-7t^0$

Recall that, in the product rule, to multiply expressions with the same base, keep the base and add the exponents.

$$a^m \cdot a^n = a^{m+n}$$

Now, what if we allow one of the exponents to be negative and apply the product rule? Suppose, for instance, that $m = 3$ and $m = -3$. Then

$$a^m \cdot a^n = a^3 \cdot a^{-3} = a^{3+(-3)}$$
$$= a^0 = 1$$

so

$$a^3 \cdot a^{-3} = 1$$

John Wallis (1616–1702), an English mathematician, was the first to fully discuss the meaning of 0, negative, and rational exponents (which we discuss in Section 10.3).

> **Negative Integer Exponents**
>
> For any nonzero real number a and whole number n,
>
> $$a^{-n} = \frac{1}{a^n}$$
>
> and a^{-n} is the **multiplicative inverse** of a^n.

Example 2 illustrates this definition.

Example 2

From this point on, to *simplify* will mean to write the expression with *positive exponents only*.

Also, we will restrict all variables so that they represent nonzero real numbers.

Using Properties of Exponents

Simplify the following expressions.

(a) $y^{-5} = \dfrac{1}{y^5}$

(b) $4^{-2} = \dfrac{1}{4^2} = \dfrac{1}{16}$

(c) $(-3)^{-3} = \dfrac{1}{(-3)^3} = \dfrac{1}{-27} = -\dfrac{1}{27}$

(d) $\left(\dfrac{2}{3}\right)^{-3} = \dfrac{1}{\left(\dfrac{2}{3}\right)^3} = \dfrac{1}{\dfrac{8}{27}} = \dfrac{27}{8}$

✓ CHECK YOURSELF 2

Simplify each of the following expressions.

(a) a^{-10} (b) 2^{-4} (c) $(-4)^{-2}$ (d) $\left(\dfrac{5}{2}\right)^{-2}$

Example 3 illustrates the case where coefficients are involved in an expression with negative exponents. As will be clear, some caution must be used.

Example 3 — Using Properties of Exponents

Simplify each of the following expressions.

(a) $2x^{-3} = 2 \cdot \dfrac{1}{x^3} = \dfrac{2}{x^3}$

The exponent -3 applies only to the variable x, and *not* to the coefficient 2.

Caution
The expressions
$4w^{-2}$ and $(4w)^{-2}$
are *not* the same.
Do you see why?

(b) $4w^{-2} = 4 \cdot \dfrac{1}{w^2} = \dfrac{4}{w^2}$

(c) $(4w)^{-2} = \dfrac{1}{(4w)^2} = \dfrac{1}{16w^2}$

✓ CHECK YOURSELF 3

Simplify each of the following expressions.

(a) $3w^{-4}$ (b) $10x^{-5}$ (c) $(2y)^{-4}$ (d) $-5t^{-2}$

Section 10.1 ■ Zero and Negative Exponents and Scientific Notation **669**

Suppose that a variable with a negative exponent appears in the denominator of an expression. Our previous definition can be used to write a complex fraction that can then be simplified. For instance,

$$\frac{1}{a^{-2}} = \frac{1}{\frac{1}{a^2}} = 1 \cdot \frac{a^2}{1} = a^2 \longleftarrow \text{Positive exponent in numerator}$$

Negative exponent in denominator

To divide, we invert and multiply.

To avoid the intermediate steps, we can write that, in general,

> For any nonzero real number a and integer n,
>
> $$\frac{1}{a^{-n}} = a^n$$

Example 4 Using Properties of Exponents

Simplify each of the following expressions.

(a) $\dfrac{1}{y^{-3}} = y^3$

(b) $\dfrac{1}{2^{-5}} = 2^5 = 32$

(c) $\dfrac{3}{4x^{-2}} = \dfrac{3x^2}{4}$ The exponent -2 applies only to x, not to 4.

(d) $\dfrac{a^{-3}}{b^{-4}} = \dfrac{b^4}{a^3}$

✓ **CHECK YOURSELF 4**

Simplify each of the following expressions.

(a) $\dfrac{1}{x^{-4}}$ (b) $\dfrac{1}{3^{-3}}$ (c) $\dfrac{2}{3a^{-2}}$ (d) $\dfrac{c^{-5}}{d^{-7}}$

To review these properties, return to Section 5.1.

The product and quotient rules for exponents apply to expressions that involve any integral exponent—positive, negative, or 0. Example 5 illustrates this concept.

Example 5 — Using Properties of Exponents

Simplify each of the following expressions, and write the result, using positive exponents only.

(a) $x^3 \cdot x^{-7} = x^{3+(-7)}$ Add the exponents by the product rule.
$$= x^{-4} = \frac{1}{x^4}$$

(b) $\dfrac{m^{-5}}{m^{-3}} = m^{-5-(-3)} = m^{-5+3}$ Subtract the exponents by the quotient rule.
$$= m^{-2} = \frac{1}{m^2}$$

(c) $\dfrac{x^5 x^{-3}}{x^{-7}} = \dfrac{x^{5+(-3)}}{x^{-7}} = \dfrac{x^2}{x^{-7}} = x^{2-(-7)} = x^9$ We apply first the product rule and then the quotient rule.

Note that m^{-5} in the numerator becomes m^5 in the denominator, and m^{-3} in the denominator becomes m^3 in the numerator. We then simplify as before.

In simplifying expressions involving negative exponents, there are often alternate approaches. For instance, in Example 5, part b, we could have made use of our earlier work to write

$$\frac{m^{-5}}{m^{-3}} = \frac{m^3}{m^5} = m^{3-5} = m^{-2} = \frac{1}{m^2}$$

✓ CHECK YOURSELF 5

Simplify each of the following expressions.

(a) $x^9 \cdot x^{-5}$ (b) $\dfrac{y^{-7}}{y^{-3}}$ (c) $\dfrac{a^{-3} a^2}{a^{-5}}$

The properties of exponents can be extended to include negative exponents. One of these properties, the *quotient-power rule,* is particularly useful when rational expressions are raised to a negative power. Let's look at the rule and apply it to negative exponents.

Quotient-Power Rule

$$\left(\frac{a}{b}\right)^n = \frac{a^n}{b^n}$$

Section 10.1 ■ Zero and Negative Exponents and Scientific Notation

> **Raising Quotients to a Negative Power**
>
> $$\left(\frac{a}{b}\right)^{-n} = \frac{a^{-n}}{b^{-n}} = \frac{b^n}{a^n} = \left(\frac{b}{a}\right)^n \qquad a \neq 0, b \neq 0$$

Example 6 Extending the Properties of Exponents

Simplify each expression.

(a) $\left(\dfrac{s^3}{t^2}\right)^{-2} = \left(\dfrac{t^2}{s^3}\right)^2 = \dfrac{t^4}{s^6}$

(b) $\left(\dfrac{m^2}{n^{-2}}\right)^{-3} = \left(\dfrac{n^{-2}}{m^2}\right)^3 = \dfrac{n^{-6}}{m^6} = \dfrac{1}{m^6 n^6}$

✓ **CHECK YOURSELF 6**

Simplify each expression.

(a) $\left(\dfrac{3t^2}{s^3}\right)^{-3}$ (b) $\left(\dfrac{x^5}{y^{-2}}\right)^{-3}$

Example 7 Using Properties of Exponents

Simplify each of the following expressions.

(a) $\left(\dfrac{3}{q^5}\right)^{-2} = \left(\dfrac{q^5}{3}\right)^2$

$= \dfrac{q^{10}}{9}$

(b) $\left(\dfrac{x^3}{y^4}\right)^{-3} = \left(\dfrac{y^4}{x^3}\right)^3$

$= \dfrac{(y^4)^3}{(x^3)^3} = \dfrac{y^{12}}{x^9}$

✓ CHECK YOURSELF 7

Simplify each of the following expressions.

(a) $\left(\dfrac{r^4}{5}\right)^{-2}$ (b) $\left(\dfrac{a^4}{b^3}\right)^{-3}$

As you might expect, more complicated expressions require the use of more than one of the properties, for simplification. Example 8 illustrates such cases.

Example 8 — Using Properties of Exponents

Simplify each of the following expressions.

(a) $\dfrac{(a^2)^{-3}(a^3)^4}{(a^{-3})^3} = \dfrac{a^{-6} \cdot a^{12}}{a^{-9}}$ Apply the power rule to each factor.

$= \dfrac{a^{-6+12}}{a^{-9}} = \dfrac{a^6}{a^{-9}}$ Apply the product rule.

$= a^{6-(-9)} = a^{6+9} = a^{15}$ Apply the quotient rule.

It may help to separate the problem into three fractions, one for the coefficients and one for each of the variables.

(b) $\dfrac{8x^{-2}y^{-5}}{12x^{-4}y^3} = \dfrac{8}{12} \cdot \dfrac{x^{-2}}{x^{-4}} \cdot \dfrac{y^{-5}}{y^3}$

$= \dfrac{2}{3} \cdot x^{-2-(-4)} \cdot y^{-5-3}$

$= \dfrac{2}{3} \cdot x^2 \cdot y^{-8} = \dfrac{2x^2}{3y^8}$

(c) $\left(\dfrac{pr^3s^{-5}}{p^3r^{-3}s^{-2}}\right)^{-2} = (p^{1-3}r^{3-(-3)}s^{-5-(-2)})^{-2}$

$= (p^{-2}r^6s^{-3})^{-2}$ Apply the quotient rule inside the parentheses.

$= (p^{-2})^{-2}(r^6)^{-2}(s^{-3})^{-2}$ Apply the rule for a product to a power.

$= p^4r^{-12}s^6 = \dfrac{p^4s^6}{r^{12}}$ Apply the power rule.

Caution

Be Careful! Another possible first step (and generally an efficient one) is to rewrite an expression by using our earlier definitions.

$$a^{-n} = \dfrac{1}{a^n} \quad \text{and} \quad \dfrac{1}{a^{-n}} = a^n$$

For instance, in Example 8, part b, we would *correctly* write

$$\dfrac{8x^{-2}y^{-5}}{12x^{-4}y^3} = \dfrac{8x^4}{12x^2y^3y^5}$$

Section 10.1 ■ Zero and Negative Exponents and Scientific Notation   **673**

Caution

A *common error* is to write

$$\frac{8x^{-2}y^{-5}}{12x^{-4}y^3} = \frac{12x^4}{8x^2y^3y^5} \qquad \text{This is } not \text{ correct.}$$

The coefficients should not have been moved along with the factors in *x*. Keep in mind that the negative exponents apply *only* to the variables. The coefficients remain *where they were* in the original expression when the expression is rewritten using this approach.

✓ CHECK YOURSELF 8

Simplify each of the following expressions.

(a) $\dfrac{(x^5)^{-2}(x^2)^3}{(x^{-4})^3}$ (b) $\dfrac{12a^{-3}b^{-2}}{16a^{-2}b^3}$ (c) $\left(\dfrac{xy^{-3}z^{-5}}{x^{-4}y^{-2}z^3}\right)^{-3}$

Let us now take a look at an important use of exponents, scientific notation.

We begin the discussion with a calculator exercise. On most calculators, if you multiply 2.3 times 1000, the display will read

$$2300$$

Multiply by 1000 a second time. Now you will see

$$2300000.$$

Multiplying by 1000 a third time will result in the display

This must equal
2,300,000,000.

$$2.3 \quad 09 \qquad \text{or} \qquad 2.3 \quad \text{E09}$$

And multiplying by 1000 again yields

Consider the following table:

$$2.3 \quad 12 \qquad \text{or} \qquad 2.3 \quad \text{E12}$$

$2.3 = 2.3 \times 10^0$
$23 = 2.3 \times 10^1$
$230 = 2.3 \times 10^2$
$2300 = 2.3 \times 10^3$
$23{,}000 = 2.3 \times 10^4$
$230{,}000 = 2.3 \times 10^5$

Can you see what is happening? This is the way calculators display very large numbers. The number on the left is always between 1 and 10, and the number on the right indicates the number of places the decimal point must be moved to the right to put the answer in standard (or decimal) form.

This notation is used frequently in science. It is not uncommon in scientific applications of algebra to find yourself working with very large or very small numbers. Even in the time of Archimedes (287–212 B.C.), the study of such numbers was not unusual. Archimedes estimated that the universe was 23,000,000,000,000,000 m in diameter, which is the approximate distance light travels in $2\dfrac{1}{2}$ years. By comparison, Polaris (the North Star) is 680 light-years from the earth. Example 10 will discuss the idea of light-years.

In scientific notation, his estimate for the diameter of the universe would be

$$2.3 \times 10^{16} \text{ m}$$

In general, we can define scientific notation as follows.

> **Scientific Notation**
>
> Any number written in the form
>
> $$a \times 10^n$$
>
> where $1 \leq a < 10$ and n is an integer, is written in scientific notation.

Example 9 Using Scientific Notation

Write each of the following numbers in scientific notation.

Note the pattern for writing a number in scientific notation.

(a) $120{,}000. = 1.2 \times 10^5$
    5 places — The power is 5.

(b) $88{,}000{,}000. = 8.8 \times 10^7$
    7 places — The power is 7.

The exponent on 10 shows the number of places *we must move the decimal point so that the multiplier will be a number between 1 and 10. A positive exponent tells us to move right, while a negative exponent indicates to move left.*

(c) $520{,}000{,}000. = 5.2 \times 10^8$
    8 places

(d) $4{,}000{,}000{,}000. = 4 \times 10^9$
    9 places

(e) $0.0005 = 5 \times 10^{-4}$ If the decimal point is to be moved to the left,
    4 places   the exponent will be negative.

Note: To convert back to standard or decimal form, the process is simply reversed.

(f) $0.0000000081 = 8.1 \times 10^{-9}$
    9 places

✓ CHECK YOURSELF 9

Write in scientific notation.

(a) 212,000,000,000,000,000   (b) 0.00079

(c) 5,600,000   (d) 0.0000007

Section 10.1 ■ Zero and Negative Exponents and Scientific Notation   **675**

Example 10 — An Application of Scientific Notation

(a) Light travels at a speed of 3.05×10^8 meters per second (m/s). There are approximately 3.15×10^7 s in a year. How far does light travel in a year?

We multiply the distance traveled in 1 s by the number of seconds in a year. This yields

$$(3.05 \times 10^8)(3.15 \times 10^7) = (3.05 \cdot 3.15)(10^8 \cdot 10^7)$$ Multiply the coefficients, and add the exponents.

$$= 9.6075 \times 10^{15}$$

Note that $9.6075 \times 10^{15} \approx 10 \times 10^{15} = 10^{16}$.

For our purposes we round the distance light travels in 1 year to 10^{16} m. This unit is called a **light-year,** and it is used to measure astronomical distances.

(b) The distance from earth to the star Spica (in Virgo) is 2.2×10^{18} m. How many light-years is Spica from earth?

We divide the distance (in meters) by the number of meters in 1 light-year.

$$\frac{2.2 \times 10^{18}}{10^{16}} = 2.2 \times 10^{18-16}$$

$$= 2.2 \times 10^2 = 220 \text{ light-years}$$

✓ CHECK YOURSELF 10

The farthest object that can be seen with the unaided eye is the Andromeda galaxy. This galaxy is 2.3×10^{22} m from earth. What is this distance in light-years?

✓ CHECK YOURSELF ANSWERS

1. (a) 1; (b) 1; (c) 8; (d) -7.   **2.** (a) $\frac{1}{a^{10}}$; (b) $\frac{1}{16}$; (c) $\frac{1}{16}$; (d) $\frac{4}{25}$.
3. (a) $\frac{3}{w^4}$; (b) $\frac{10}{x^5}$; (c) $\frac{1}{16y^4}$; (d) $-\frac{5}{t^2}$.   **4.** (a) x^4; (b) 27; (c) $\frac{2a^2}{3}$; (d) $\frac{d^7}{c^5}$.
5. (a) x^4; (b) $\frac{1}{y^4}$; (c) a^4.   **6.** (a) $\frac{s^9}{27t^6}$; (b) $\frac{1}{x^{15}y^6}$.
7. (a) $\frac{25}{r^8}$; (b) $\frac{b^9}{a^{12}}$.   **8.** (a) x^8; (b) $\frac{3}{4ab^5}$; (c) $\frac{y^3 z^{24}}{x^{15}}$.   **9.** (a) 2.12×10^{17};
(b) 7.9×10^{-4}; (c) 5.6×10^6; (d) 7×10^{-7}.   **10.** 2,300,000 light-years.

Exercises ▪ 10.1

1. $\dfrac{1}{x^5}$ 2. $\dfrac{1}{27}$
3. $\dfrac{1}{25}$ 4. $\dfrac{1}{x^8}$
5. $\dfrac{1}{25}$ 6. $-\dfrac{1}{27}$
7. $-\dfrac{1}{8}$ 8. $\dfrac{1}{16}$
9. $\dfrac{27}{8}$ 10. $\dfrac{16}{9}$
11. $\dfrac{3}{x^2}$ 12. $\dfrac{4}{x^3}$
13. $\dfrac{-5}{x^4}$ 14. $\dfrac{1}{16x^4}$
15. $\dfrac{1}{9x^2}$ 16. $\dfrac{-5}{x^2}$
17. x^3 18. x^5
19. $\dfrac{2x^3}{5}$ 20. $\dfrac{3x^4}{4}$
21. $\dfrac{y^4}{x^3}$ 22. $\dfrac{y^3}{x^5}$
23. x^2 24. y
25. $\dfrac{1}{a^3}$ 26. $\dfrac{1}{w^2}$
27. $\dfrac{1}{z^{10}}$ 28. $\dfrac{1}{b^8}$
29. 1 30. 1
31. $\dfrac{1}{x^3}$ 32. x^3

In Exercises 1 to 22, simplify each expression.

1. x^{-5} 2. 3^{-3} 3. 5^{-2} 4. x^{-8}

5. $(-5)^{-2}$ 6. $(-3)^{-3}$ 7. $(-2)^{-3}$ 8. $(-2)^{-4}$

9. $\left(\dfrac{2}{3}\right)^{-3}$ 10. $\left(\dfrac{3}{4}\right)^{-2}$ 11. $3x^{-2}$ 12. $4x^{-3}$

13. $-5x^{-4}$ 14. $(-2x)^{-4}$ 15. $(-3x)^{-2}$ 16. $-5x^{-2}$

17. $\dfrac{1}{x^{-3}}$ 18. $\dfrac{1}{x^{-5}}$ 19. $\dfrac{2}{5x^{-3}}$ 20. $\dfrac{3}{4x^{-4}}$

21. $\dfrac{x^{-3}}{y^{-4}}$ 22. $\dfrac{x^{-5}}{y^{-3}}$

In Exercises 23 to 32, use the properties of exponents to simplify expressions.

23. $x^5 \cdot x^{-3}$ 24. $y^{-4} \cdot y^5$ 25. $a^{-9} \cdot a^6$ 26. $w^{-5} \cdot w^3$

27. $z^{-2} \cdot z^{-8}$ 28. $b^{-7} \cdot b^{-1}$ 29. $a^{-5} \cdot a^5$ 30. $x^{-4} \cdot x^4$

31. $\dfrac{x^{-5}}{x^{-2}}$ 32. $\dfrac{x^{-3}}{x^{-6}}$

Section 10.1 ■ Zero and Negative Exponents and Scientific Notation

In Exercises 33 to 58, use the properties of exponents to simplify the following.

33. x^{15}
34. w^{24}
35. $2x^5$
36. $9p^{10}$
37. $3a$
38. $10y^4$
39. $x^{10}y$
40. $r^{10}s^7$
41. $a^{17}b^8c^{13}$
42. $p^4q^7r^2$
43. $\dfrac{1}{x^{15}}$
44. x^6
45. b^8
46. $\dfrac{1}{b^{12}}$
47. $\dfrac{x^{10}}{y^6}$
48. $\dfrac{p^6}{q^4}$
49. $x^{12}y^6$
50. $\dfrac{27}{x^6y^6}$
51. $\dfrac{x^{15}}{32}$
52. $\dfrac{b^4}{a^6}$
53. $\dfrac{y^4}{x^2}$
54. x^9y^6
55. $\dfrac{y^2}{x^4}$
56. $\dfrac{18}{x^{13}}$
57. $\dfrac{48}{x^8}$
58. $\dfrac{x}{20{,}000}$
59. $16x^{26}$
60. $27x^{16}$
61. $\dfrac{72}{x^3}$
62. x^8y^{12}
63. $x^{42}y^{33}z^{25}$
64. $x^5y^5z^4$
65. $75x^2$
66. $4a^6$
67. $144w^2$
68. $288x^{26}$
69. $\dfrac{3x^3}{2y^4}$
70. $2x^5y^3$
71. $-567x^{22}y^{25}$
72. $\dfrac{2x^7z^3}{3w^3y}$
73. $\dfrac{6}{x^2y^5}$
74. $\dfrac{-10a^3}{b^4}$
75. $\dfrac{y^5}{x^3}$
76. $\dfrac{x^5}{4y^2}$
77. $\dfrac{3xy^5}{4z^6}$
78. $\dfrac{2z^4}{3x^3y^6}$
79. $\dfrac{1}{x^5y^3}$

33. $(x^5)^3$
34. $(w^4)^6$
35. $(2x^{-3})(x^2)^4$
36. $(p^4)(3p^3)^2$
37. $(3a^{-4})(a^3)(a^2)$
38. $(5y^{-2})(2y)(y^5)$
39. $(x^4y)(x^2)^3(y^3)^0$
40. $(r^4)^2(r^2s)(s^3)^2$
41. $(ab^2c)(a^4)^4(b^2)^3(c^3)^4$
42. $(p^2qr^2)(p^2)(q^3)^2(r^2)^0$
43. $(x^5)^{-3}$
44. $(x^{-2})^{-3}$
45. $(b^{-4})^{-2}$
46. $(a^0b^{-4})^3$
47. $(x^5y^{-3})^2$
48. $(p^{-3}q^2)^{-2}$
49. $(x^{-4}y^{-2})^{-3}$
50. $(3x^{-2}y^{-2})^3$
51. $(2x^{-3}y^0)^{-5}$
52. $\dfrac{a^{-6}}{b^{-4}}$
53. $\dfrac{x^{-2}}{y^{-4}}$
54. $\left(\dfrac{x^{-3}}{y^2}\right)^{-3}$
55. $\dfrac{x^{-4}}{y^{-2}}$
56. $\dfrac{(3x^{-4})^2(2x^2)}{x^6}$
57. $(4x^{-2})^2(3x^{-4})$
58. $(5x^{-4})^{-4}(2x^3)^{-5}$

In Exercises 59 to 90, simplify each expression.

59. $(2x^5)^4(x^3)^2$
60. $(3x^2)^3(x^2)^4(x^2)$
61. $(2x^{-3})^3(3x^3)^2$
62. $(x^2y^3)^4(xy^3)^0$
63. $(xy^5z)^4(xyz^2)^8(x^6yz)^5$
64. $(x^2y^2z^2)^0(xy^2z)^2(x^3yz^2)$
65. $(3x^{-2})(5x^2)^2$
66. $(2a^3)^2(a^0)^5$
67. $(2w^3)^4(3w^{-5})^2$
68. $(3x^3)^2(2x^4)^5$
69. $\dfrac{3x^6}{2y^9} \cdot \dfrac{y^5}{x^3}$
70. $\dfrac{x^8}{y^6} \cdot \dfrac{2y^9}{x^3}$
71. $(-7x^2y)(-3x^5y^6)^4$
72. $\left(\dfrac{2w^5z^3}{3x^3y^9}\right)\left(\dfrac{x^5y^4}{w^4z^0}\right)^2$
73. $(2x^2y^{-3})(3x^{-4}y^{-2})$
74. $(-5a^{-2}b^{-4})(2a^5b^0)$
75. $\dfrac{(x^{-3})(y^2)}{y^{-3}}$
76. $\dfrac{6x^3y^{-4}}{24x^{-2}y^{-2}}$
77. $\dfrac{15x^{-3}y^2z^{-4}}{20x^{-4}y^{-3}z^2}$
78. $\dfrac{24x^{-5}y^{-3}z^2}{36x^{-2}y^3z^{-2}}$
79. $\dfrac{x^{-5}y^{-7}}{x^0y^{-4}}$

80. $\dfrac{z^{12}}{x^8 y^{10}}$ **81.** $\dfrac{y^8}{x^7}$

82. $x^6 y^9$ **83.** x^{5n}

84. x^{4n+1} **85.** x^2

86. $\dfrac{1}{x^3}$ **87.** y^{3n^2}

88. x^{n^2+n} **89.** x^2

90. x^5 **91.** 9.3×10^7

92. 2.1×10^{-5} m

93. 1.3×10^{11}

94. 6.02×10^{23}

95. 28   **96.** 18

97. 0.008

98. 0.0000075

99. 0.000028

100. 0.000521

101. 5×10^{-4}

102. 3×10^{-6}

103. 3.7×10^{-4}

104. 5.1×10^{-5}

105. 8×10^{-8}

106. 6×10^{-4}

107. 3×10^5

108. 5×10^{-6}

109. 8×10^9

110. 7.5×10^{12}

111. 2×10^2

112. 3×10^5

113. 6×10^{16}

114. 4×10^{11}

80. $\left(\dfrac{xy^3 z^{-4}}{x^{-3} y^{-2} z^2}\right)^{-2}$ **81.** $\dfrac{x^{-2} y^2}{x^3 y^{-2}} \cdot \dfrac{x^{-4} y^2}{x^{-2} y^{-2}}$ **82.** $\left(\dfrac{x^{-3} y^3}{x^{-4} y^2}\right)^3 \cdot \left(\dfrac{x^{-2} y^{-2}}{xy^4}\right)^{-1}$

83. $x^{2n} \cdot x^{3n}$ **84.** $x^{n+1} \cdot x^{3n}$ **85.** $\dfrac{x^{n+3}}{x^{n+1}}$

86. $\dfrac{x^{n-4}}{x^{n-1}}$ **87.** $(y^n)^{3n}$ **88.** $(x^{n+1})^n$

89. $\dfrac{x^{2n} \cdot x^{n+2}}{x^{3n}}$ **90.** $\dfrac{x^n \cdot x^{3n+5}}{x^{4n}}$

In Exercises 91 to 94, express each number in scientific notation.

91. The distance from the earth to the sun: 93,000,000 mi.

92. The diameter of a grain of sand: 0.000021 m.

93. The diameter of the sun: 130,000,000,000 cm.

94. The number of molecules in 22.4 L of a gas: 602,000,000,000,000,000,000,000 (Avogadro's number)

95. The mass of the sun is approximately 1.98×10^{30} kg. If this were written in standard or decimal form, how many 0s would follow the digit 8?

96. Archimedes estimated the universe to be 2.3×10^{19} millimeters (mm) in diameter. If this number were written in standard or decimal form, how many 0s would follow the digit 3?

In Exercises 97 to 100, write each expression in standard notation.

97. 8×10^{-3} **98.** 7.5×10^{-6} **99.** 2.8×10^{-5} **100.** 5.21×10^{-4}

In Exercises 101 to 104, write each of the following in scientific notation.

101. 0.0005 **102.** 0.000003 **103.** 0.00037 **104.** 0.000051

In Exercises 105 to 108, compute the expressions using scientific notation, and write your answer in that form.

105. $(4 \times 10^{-3})(2 \times 10^{-5})$ **106.** $(1.5 \times 10^{-6})(4 \times 10^2)$

107. $\dfrac{9 \times 10^3}{3 \times 10^{-2}}$ **108.** $\dfrac{7.5 \times 10^{-4}}{1.5 \times 10^2}$

In Exercises 109 to 114, perform the indicated calculations. Write your result in scientific notation.

109. $(2 \times 10^5)(4 \times 10^4)$ **110.** $(2.5 \times 10^7)(3 \times 10^5)$ **111.** $\dfrac{6 \times 10^9}{3 \times 10^7}$

112. $\dfrac{4.5 \times 10^{12}}{1.5 \times 10^7}$ **113.** $\dfrac{(3.3 \times 10^{15})(6 \times 10^{15})}{(1.1 \times 10^8)(3 \times 10^6)}$ **114.** $\dfrac{(6 \times 10^{12})(3.2 \times 10^8)}{(1.6 \times 10^7)(3 \times 10^2)}$

Section 10.1 ■ Zero and Negative Exponents and Scientific Notation   679

115. 66 years

116. 210 years

117. 1.55×10^{23}; 2.92×10^{13}

118. 4.66×10^{15} L

119. 6.5×10^{14}

115. Megrez, the nearest of the Big Dipper stars, is 6.6×10^{17} m from earth. Approximately how long does it take light, traveling at 10^{16} m/year, to travel from Megrez to earth?

116. Alkaid, the most distant star in the Big Dipper, is 2.1×10^{18} m from earth. Approximately how long does it take light to travel from Alkaid to earth?

117. The number of liters of water on earth is 15,500 followed by 19 zeros. Write this number in scientific notation. Then use the number of liters of water on earth to find out how much water is available for each person on earth. The population of earth is 5.3 billion.

118. If there are 5.3×10^9 people on earth and there is enough freshwater to provide each person with 8.79×10^5 L, how much freshwater is on earth?

119. The United States uses an average of 2.6×10^6 L of water per person each year. The United States has 2.5×10^8 people. How many liters of water does the United States use each year?

120. Can $(a + b)^{-1}$ be written as $\dfrac{1}{a} + \dfrac{1}{b}$ by using the properties of exponents? If not, why not? Explain.

121. Write a short description of the difference between $(-4)^{-2}$, -4^{-3}, $(-4)^3$, and -4^3. Are any of these equal?

122. If $n > 0$, which of the following expressions are negative?

$$-n^{-3},\ n^{-3},\ (-n)^{-3},\ (-n)^3,\ -n^3$$

If $n < 0$, which of these expressions are negative? Explain what effect a negative in the exponent has on the sign of the result when an exponential expression is simplified.

123. Take the best offer. You are offered a 28-day job in which you have a choice of two different pay arrangements. Plan 1 offers a flat $4,000,000 at the end of the 28th day on the job. Plan 2 offers 1¢ the first day, 2¢ the second day, 4¢ the third day, and so on, with the amount doubling each day. Make a table to decide which offer is the best. Write a formula for the amount you make on the nth day and a formula for the total after n days. Which pay arrangement should you take? Why?

SECTION 10.2 Evaluating Radical Expressions

10.2 OBJECTIVES

1. Evaluate radicals
2. Simplify radical expressions
3. Multiply radical expressions
4. Approximate radical values with a scientific calculator

In Section 10.1, we discussed the properties of integer exponents. In this and the next section we extend these properties. To achieve that objective, we first develop an idea that "reverses" the power process.

A statement such as

$$x^2 = 9$$

is read as "*x* squared equals nine."

In this section, we are concerned with the relationship between the base *x* and the number 9. Equivalently, we can say that "*x* is the square root of 9."

We know from experience that *x* must equal 3 (since $3^2 = 9$) or -3 [since $(-3)^2 = 9$]. We see that 9 has two square roots, 3 and -3. In fact, every positive number has two square roots, one positive and one negative. In general, if $x^2 = a$, we say that *x* is the *square root* of *a*.

We also know that $3^3 = 27$, and we similarly call 3 the *cube root* of 27. Here 3 is the *only* real number with that property. Every real number (positive or negative) has exactly one real cube root. This brings us to the following definition.

Given

$$x^n = a$$

we say *x* is the **nth root of a.**

The symbol $\sqrt{}$ is called the *radical*. $\sqrt{a}$ is used to designate the *principal (positive) square root of a*. For example,

$$\sqrt{9} = 3$$

indicates that 3 is the principal square root of 9. In some applications, we will want to indicate the negative square root; to do so we write

$$-\sqrt{9} = -3$$

If both square roots need to be indicated, we write

$$\pm\sqrt{9} = \pm 3$$

Section 10.2 ■ Evaluating Radical Expressions   **681**

The symbol $\sqrt[n]{a}$ is used to designate the principal *n*th root of *a*. We call *n* the *index* of the radical, and the expression *a* is called the *radicand*. Anytime the index is an even number, the radicand must be positive to produce a real value. Example 1 illustrates this idea.

Example 1

Evaluating Radicals

Evaluate each radical.

(a) $\sqrt[3]{64}$ is the cube root of 64. It has a value of 4 since $4 \cdot 4 \cdot 4 = 64$.

(b) $\sqrt[4]{16}$ is the principal fourth root of 16. It has a value of 2 since $2^4 = 16$.

(c) $\sqrt[3]{-27} = -3$.

(d) $\sqrt[4]{-16}$ is not defined as a real value. No real number, taken to an even power, will produce a negative result.

(e) $\sqrt{0} = 0$.

✓ **CHECK YOURSELF 1**

Evaluate each radical.

(a) $\sqrt[3]{125}$ (b) $\sqrt[6]{64}$ (c) $\sqrt[3]{-1000}$ (d) $\sqrt[4]{-2500}$

For an expression to be written in simplest radical form, several conditions must be satisfied. The first is that no factor in the radicand can be raised to a power equal to or greater than the index. The product theorem allows us to remove such factors from the radicand.

> **The Product Theorem**
> $$\sqrt[n]{ab} = \sqrt[n]{a} \cdot \sqrt[n]{b}$$

Example 2

Simplify each radical expression. Assume all variables are positive.

(a) $\sqrt{32} = \sqrt{2^5} = \sqrt{2^2 \cdot 2^2 \cdot 2} = \sqrt{2^2} \cdot \sqrt{2^2} \cdot \sqrt{2} = 2 \cdot 2 \cdot \sqrt{2} = 4\sqrt{2}$

(b) $\sqrt{75a^3 b} = \sqrt{25a^2 \cdot 3ab} = \sqrt{5^2 a^2 \cdot 3ab} = \sqrt{5^2 a^2} \cdot \sqrt{3ab} = 5a\sqrt{3ab}$

(c) $\sqrt[3]{8x^3 y \cdot 2xy^2} = \sqrt[3]{8x^3 y^3 \cdot 2x} = \sqrt[3]{2^3 x^3 y^3} \cdot \sqrt[3]{2x} = 2xy\sqrt[3]{2x}$

CHECK YOURSELF 2

Write in simplest radical form.

(a) $\sqrt{50}$ (b) $\sqrt{32a^3}$ (c) $\sqrt[3]{-16x^4y^5}$

We can use the product theorem to multiply expressions containing radicals. The product of two radical expressions is the radical of the product of the radicands. The result should always be expressed in simplest radical form.

Example 3 — Multiplying Radicals

Multiply.

(a) $\sqrt{5} \cdot \sqrt{3} = \sqrt{15}$

(b) $\sqrt{3ab} \cdot \sqrt{6a^2b^3} = \sqrt{18a^3b^4} = \sqrt{9a^2b^4 \cdot 2a} = 3ab^2\sqrt{2a}$

CHECK YOURSELF 3

Multiply.

(a) $\sqrt{2x} \cdot \sqrt{6}$ (b) $\sqrt{3a^2b} \cdot \sqrt{12ab^3}$

A second theorem allows us to rewrite expressions that have fractions inside the radical.

The Quotient Theorem

$$\sqrt[n]{\frac{a}{b}} = \frac{\sqrt[n]{a}}{\sqrt[n]{b}}$$

Example 4 — Removing Fractions From the Radicand

Simplify the following expression.

$$\sqrt{\frac{3a}{16a^2}} = \frac{\sqrt{3a}}{\sqrt{16a^2}} = \frac{\sqrt{3a}}{4a}$$

Section 10.2 ■ Evaluating Radical Expressions   **683**

✓ **CHECK YOURSELF 4**

Simplify the following expression.

$$\sqrt{\frac{5x}{36y^2}}$$

Whenever the radicand is a real number, we can use a scientific calculator to approximate the value of a radical. This is particularly useful when we are trying to graph an equation or solve an application problem.

Example 5   **Approximating the Value of a Radical Expression**

For each expression, use a calculator to approximate its value. Express your answer to the nearest tenth.

(a) $\sqrt{7x^3}$, $x = 8$

$\sqrt{7(8)^3} \approx 59.9$

(b) $\sqrt{3x} + 4\sqrt{2x^3} - \sqrt{7x}$, $x = 3$

$\sqrt{3(3)} + 4\sqrt{2(3)^3} - \sqrt{7(3)} \approx 27.8$

If your calculator has a MATH menu, you will probably find the command for the *n*th root in that menu.

(c) $\sqrt[5]{7x} + \sqrt[3]{15x^5} - \sqrt[4]{127x}$, $x = 5$

$\sqrt[5]{7(5)} + \sqrt[3]{15(5)^5} - \sqrt[4]{127(5)} \approx 33.1$

✓ **CHECK YOURSELF 5**

For each expression, use a calculator to approximate the value at the given *x*.

(a) $\sqrt{15x^5}$, $x = 2$

(b) $\sqrt{7x^3} + 3\sqrt{2x} - \sqrt{17x}$, $x = 5$

(c) $\sqrt[3]{9x} - \sqrt[5]{7x^2} - \sqrt[3]{12x^4}$, $x = 4$

✓ **CHECK YOURSELF ANSWERS**

1. (a) 5; (b) 2; (c) -10; (d) no real value.   **2.** (a) $5\sqrt{2}$; (b) $4a\sqrt{2a}$; (c) $-2xy\sqrt[3]{2xy^2}$.   **3.** (a) $2\sqrt{3x}$; (b) $6ab^2\sqrt{a}$.   **4.** $\dfrac{\sqrt{5x}}{6y}$.   **5.** (a) ≈ 21.9; (b) ≈ 29.8; (c) ≈ -6.7.

The Average or Mean of Two or More Numbers. If b is the mean of two numbers, a and c, where $a < c$, then, according to the Pythagoreans, a group of individuals who studied mathematics, music, and mysticism about 500 B.C., there are 10 different ways to define this mean. Two of the most well-known are the geometric and the arithmetic means, defined by the Pythagoreans as follows:

$$\text{Arithmetic mean:} \quad \frac{b-a}{c-b} = \frac{a}{a}$$

$$\text{Geometric mean:} \quad \frac{b-a}{c-b} = \frac{a}{b}$$

Work with a partner to complete the following.

1. Solve each equation for b to find the formula for computing each mean. The arithmetic mean is commonly called the *average,* although this is not the only way to define "average."

2. Try computing the geometric and arithmetic mean of several pairs of numbers. When is the geometric mean greater than the arithmetic mean? When are they equal?

3. In a country in economic crisis, inflation usually soars. If one year the price of basic food items doubled, then the next year they tripled, and the third year they dropped in price by half, what was the *average* amount that prices were multiplied by each year? Compute the arithmetic and geometric mean of the increases each year. Which is a better measure of the average in this case? Explain.

Exercises ▪ 10.2

Answers (left column):

1. 7
2. 6
3. −6
4. −9
5. Not a real number
6. Not a real number
7. 4
8. −4
9. −6
10. 6
11. 3
12. Not a real number
13. 2
14. −2
15. −2
16. 2
17. $\frac{2}{3}$
18. $\frac{3}{5}$
19. $\frac{2}{3}$
20. $\frac{3}{4}$
21. 6
22. r
23. 9
24. 5
25. −5
26. 5
27. $2\sqrt{3}$
28. $2\sqrt{6}$
29. $-6\sqrt{3}$
30. $-4\sqrt{6}$
31. $4\sqrt{2}$
32. $5\sqrt{10}$
33. $-2\sqrt[3]{6}$
34. $-3\sqrt[3]{2}$
35. $2\sqrt[4]{6}$
36. $3\sqrt[4]{3}$
37. $3x^2\sqrt{7}$
38. $3w^2\sqrt{6}$
39. $5a^2\sqrt{3a}$
40. $7m\sqrt{2m}$
41. $4xy\sqrt{5y}$
42. $6p^2q\sqrt{3p}$
43. $5x^2\sqrt[3]{2x^2}$
44. $4r^2\sqrt[3]{2s^2}$
45. $2x^2yz\sqrt[3]{7y^2z}$
46. $-5ab^5c^3\sqrt[3]{2a}$

In Exercises 1 to 26, evaluate each radical.

1. $\sqrt{49}$
2. $\sqrt{36}$
3. $-\sqrt{36}$
4. $-\sqrt{81}$
5. $\sqrt{-49}$
6. $\sqrt{-25}$
7. $\sqrt[3]{64}$
8. $\sqrt[3]{-64}$
9. $-\sqrt[3]{216}$
10. $-\sqrt[3]{-216}$
11. $\sqrt[4]{81}$
12. $\sqrt[4]{-81}$
13. $\sqrt[5]{32}$
14. $\sqrt[5]{-32}$
15. $-\sqrt[5]{32}$
16. $-\sqrt[5]{-32}$
17. $\sqrt{\frac{4}{9}}$
18. $\sqrt{\frac{9}{25}}$
19. $\sqrt[3]{\frac{8}{27}}$
20. $\sqrt[3]{\frac{27}{64}}$
21. $\sqrt{6^2}$
22. $\sqrt[3]{r^3}$
23. $\sqrt{(-9)^2}$
24. $\sqrt{(-5)^2}$
25. $\sqrt[3]{(-5)^3}$
26. $\sqrt[3]{5^3}$

In Exercises 27 to 58, simplify each radical expression.

27. $\sqrt{12}$
28. $\sqrt{24}$
29. $-\sqrt{108}$
30. $-\sqrt{96}$
31. $\sqrt{32}$
32. $\sqrt{250}$
33. $\sqrt[3]{-48}$
34. $\sqrt[3]{-54}$
35. $\sqrt[4]{96}$
36. $\sqrt[4]{243}$
37. $\sqrt{63x^4}$
38. $\sqrt{54w^4}$
39. $\sqrt{75a^5}$
40. $\sqrt{98m^3}$
41. $\sqrt{80x^2y^3}$
42. $\sqrt{108p^5q^2}$
43. $\sqrt[3]{250x^8}$
44. $\sqrt[3]{128r^6s^2}$
45. $\sqrt[3]{56x^6y^5z^4}$
46. $-\sqrt[3]{250a^4b^{15}c^9}$

685

Answers (left column)

47. $3y^3\sqrt[4]{2}$
48. $3a^3\sqrt[4]{3a^3}$
49. $2\sqrt[5]{2w}$
50. $2ab^2\sqrt[5]{3b^2}$
51. $\dfrac{\sqrt{5}}{4}$
52. $\dfrac{\sqrt{11}}{6}$
53. $\dfrac{x^2}{5}$
54. $\dfrac{a^3}{7}$
55. $\dfrac{\sqrt{5}}{3y^2}$
56. $\dfrac{\sqrt{7}}{5x}$
57. $\dfrac{\sqrt[3]{3}}{4}$
58. $\dfrac{\sqrt[3]{4x^2}}{3}$
59. $\sqrt{42}$
60. $\sqrt{30}$
61. $\sqrt{42}$
62. $\sqrt{105}$
63. $\sqrt[3]{36}$
64. $\sqrt[3]{35}$
65. 6
66. 10
67. $4x\sqrt{2x}$
68. $6w^2\sqrt{2}$
69. $5x\sqrt[3]{2x}$
70. $3rs\sqrt[3]{6r}$
71. 3.9
72. 5.4
73. 14.6
74. 12.5
75. Not a real number
76. Not a real number
77. 9.1
78. 9.8
79. -3.9
80. Not a real number
81. 2.1
82. 2.2
83. 8.7
84. 11.3
85. 2.2
86. 2.8
87. 1.9
88. 2.4
89. 1.9
90. 2.0

Exercises

47. $\sqrt[4]{162y^{12}}$ 48. $\sqrt[4]{243a^{15}}$ 49. $\sqrt[5]{64w}$ 50. $\sqrt[5]{96a^5b^{12}}$

51. $\sqrt{\dfrac{5}{16}}$ 52. $\sqrt{\dfrac{11}{36}}$ 53. $\sqrt{\dfrac{x^4}{25}}$ 54. $\sqrt{\dfrac{a^6}{49}}$

55. $\sqrt{\dfrac{5}{9y^4}}$ 56. $\sqrt{\dfrac{7}{25x^2}}$ 57. $\sqrt[3]{\dfrac{3}{64}}$ 58. $\sqrt[3]{\dfrac{4x^2}{27}}$

In Exercises 59 to 70, combine and simplify each expression.

59. $\sqrt{7} \cdot \sqrt{6}$ 60. $\sqrt{3} \cdot \sqrt{10}$ 61. $\sqrt{3} \cdot \sqrt{7} \cdot \sqrt{2}$

62. $\sqrt{5} \cdot \sqrt{7} \cdot \sqrt{3}$ 63. $\sqrt[3]{4} \cdot \sqrt[3]{9}$ 64. $\sqrt[3]{5} \cdot \sqrt[3]{7}$

65. $\sqrt{3} \cdot \sqrt{12}$ 66. $\sqrt{5} \cdot \sqrt{20}$ 67. $\sqrt{8x^2} \cdot \sqrt{4x}$

68. $\sqrt{12w^3} \cdot \sqrt{6w}$ 69. $\sqrt[3]{25x^2} \cdot \sqrt[3]{10x^2}$ 70. $\sqrt[3]{18r^2s^2} \cdot \sqrt[3]{9r^2s}$

In Exercises 71 to 90, use a calculator to evaluate the expressions. Round each answer to the nearest tenth.

71. $\sqrt{15}$ 72. $\sqrt{29}$ 73. $\sqrt{213}$ 74. $\sqrt{156}$

75. $\sqrt{-15}$ 76. $\sqrt{-79}$ 77. $\sqrt{83}$ 78. $\sqrt{97}$

79. $-\sqrt{15}$ 80. $\sqrt{-29}$ 81. $\sqrt{\dfrac{23}{5}}$ 82. $\sqrt{\dfrac{39}{8}}$

83. $\sqrt{\dfrac{1124}{15}}$ 84. $\sqrt{\dfrac{896}{7}}$ 85. $\sqrt[4]{\dfrac{236}{10}}$ 86. $\sqrt[4]{\dfrac{715}{11}}$

87. $\sqrt[5]{\dfrac{2110}{85}}$ 88. $\sqrt[5]{\dfrac{1376}{19}}$ 89. $\sqrt[8]{\dfrac{6432}{38}}$ 90. $\sqrt[7]{\dfrac{4123}{31}}$

91. False

92. False

93. True

94. False

95. False

96. False

97. No

In Exercises 91 to 96, label the following as True or False.

91. $\sqrt{16x^{16}} = 4x^4$

92. $\sqrt{(x-4)^2} = x - 4$

93. $\sqrt{16x^{-4}y^{-4}}$ is a real number

94. $\sqrt{x^2 + y^2} = x + y$

95. $\dfrac{\sqrt{x^2 - 25}}{x - 5} = \sqrt{x+5}$

96. $\sqrt{2} + \sqrt{6} = \sqrt{8}$

97. Is there any prime number whose square root is an integer? Explain your answer.

98. Explain the difference between the conjugate, in which the middle sign is changed, of a binomial and the opposite of a binomial. To illustrate, use $4 - \sqrt{7}$.

99. Determine two consecutive integers whose square roots are also consecutive integers.

100. Determine the missing binomial in the following: $(\sqrt{3} - 2)($ 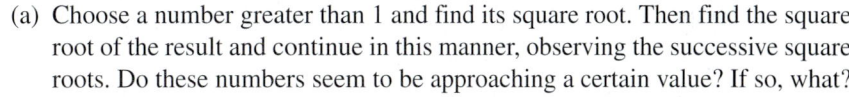 $) = -1$.

101. Try the following using your calculator.

(a) Choose a number greater than 1 and find its square root. Then find the square root of the result and continue in this manner, observing the successive square roots. Do these numbers seem to be approaching a certain value? If so, what?

98. Conjugate $4 + \sqrt{7}$; opposite $-4 + \sqrt{7}$

99. 0, 1

100. $\sqrt{3} + 2$

104. (a) 1.1s; (b) 1.4s; (c) 1.7s; (d) 1.9s; (e) 2.0s

(b) Choose a number greater than 0 but less than 1 and find its square root. Then find the square root of the result, and continue in this manner, observing successive square roots. Do these numbers seem to be approaching a certain value? If so, what?

102. (a) Can a number be equal to its own square root?

(b) Other than the number(s) found in part a, is a number always greater than its square root? Investigate.

103. Let a and b be positive numbers. If a is greater than b, is it always true that the square root of a is greater than the square root of b? Investigate.

104. Suppose that a weight is attached to a string of length L, and the other end of the string is held fixed. If we pull the weight and then release it, allowing the weight to swing back and forth, we can observe the behavior of a simple pendulum. The period, T, is the time required for the weight to complete a full cycle, swinging forward and then back. The following formula may be used to describe the relationship between T and L.

$$T = 2\pi \sqrt{\dfrac{L}{g}}$$

If L is expressed in centimeters, then $g = 980$ cm/s². For each of the following string lengths, calculate the corresponding period. Round to the nearest tenth of a second.

(a) 30 cm (b) 50 cm (c) 70 cm (d) 90 cm (e) 110 cm

SECTION 10.3 Rational Exponents

10.3 OBJECTIVES

1. Define rational exponents
2. Simplify expressions with rational exponents
3. Estimate the value of an expression using a scientific calculator
4. Write expressions in radical or exponential form

In Section 10.2, we discussed radical notation along with the concept of roots. In this section, we use that concept to develop a new idea, using exponents to provide an alternative way of writing these roots.

This new idea involves **rational numbers as exponents.** To begin the development, we extend all of the previous properties to include rational exponents.

Given that extension, suppose that

$$a = 4^{1/2} \tag{1}$$

Squaring both sides of the equation yields

$$a^2 = (4^{1/2})^2$$
$$a^2 = 4^{(1/2)(2)}$$
$$a^2 = 4^1$$
$$a^2 = 4 \tag{2}$$

From equation (2) we see that a is the number whose square is 4; that is, a is the principal square root of 4. Using our earlier notation, we can write

$$a = \sqrt{4}$$

But from equation (1)

$$a = 4^{1/2}$$

and to be consistent, we must have

$$4^{1/2} = \sqrt{4}$$

This argument can be repeated for any exponent of the form $\frac{1}{n}$, so it seems reasonable to make the following definition.

We will see later in this chapter that the property $(x^m)^n = x^{mn}$ holds for rational numbers m and n.

$4^{1/2}$ indicates the *principal square root* of 4.

> If a is any real number and n is a positive integer ($n > 1$), then
> $$a^{1/n} = \sqrt[n]{a}$$
> We restrict a so that a is nonnegative when n is even. In words, $a^{1/n}$ indicates the principal nth root of a.

Example 3

Writing Expressions in Radical Form

Write each expression in radical form and then simplify.

(a) $25^{1/2} = \sqrt{25} = 5$

$27^{1/3}$ is the *cube root* of 27.

(b) $27^{1/3} = \sqrt[3]{27} = 3$

(c) $-36^{1/2} = -\sqrt{36} = -6$

(d) $(-36)^{1/2} = \sqrt{-36}$ is not a real number.

$32^{1/5}$ is the *fifth root* of 32.

(e) $32^{1/5} = \sqrt[5]{32} = 2$

✓ CHECK YOURSELF 1

Write each expression in radical form and simplify.

(a) $8^{1/3}$ (b) $-64^{1/2}$ (c) $81^{1/4}$

We are now ready to extend our exponent notation to allow *any* rational exponent, again assuming that our previous exponent properties must still be valid. Note that

This is because

$$\frac{m}{n} = (m)\left(\frac{1}{n}\right) = \left(\frac{1}{n}\right)(m)$$

$$a^{m/n} = (a^{1/n})^m = (a^m)^{1/n}$$

From our earlier work, we know that $a^{1/n} = \sqrt[n]{a}$, and combining this with the above observation, we offer the following definition for $a^{m/n}$.

The two radical forms for $a^{m/n}$ are equivalent, and the choice of which form to use generally depends on whether we are evaluating numerical expressions or rewriting expressions containing variables in radical form.

> For any positive number a and positive integers m and n with $n > 1$,
>
> $$a^{m/n} = (\sqrt[n]{a})^m = \sqrt[n]{a^m}$$

This new extension of our rational exponent notation is applied in Example 2.

Example 2

Simplifying Expressions with Rational Exponents

Simplify each expression.

(a) $9^{3/2} = (9^{1/2})^3 = (\sqrt{9})^3$

$\phantom{(a)\ 9^{3/2}} = 3^3 = 27$

(b) $\left(\dfrac{16}{81}\right)^{3/4} = \left(\left(\dfrac{16}{81}\right)^{1/4}\right)^3 = \left(\sqrt[4]{\dfrac{16}{81}}\right)^3$

$= \left(\dfrac{2}{3}\right)^3 = \dfrac{8}{27}$

(c) $(-8)^{2/3} = ((-8)^{1/3})^2 = (\sqrt[3]{-8})^2$

$= (-2)^2 = 4$

This illustrates why we use $(\sqrt[n]{a})^m$ for $a^{m/n}$ when evaluating numerical expressions. The numbers involved will be smaller and easier to work with.

In part a we could also have evaluated the expression as

$$9^{3/2} = \sqrt{9^3} = \sqrt{729}$$
$$= 27$$

✓ **CHECK YOURSELF 2**

Simplify each expression.

(a) $16^{3/4}$ (b) $\left(\dfrac{8}{27}\right)^{2/3}$ (c) $(-32)^{3/5}$

Now we want to extend our rational exponent notation. Using the definition of negative exponents, we can write

$$a^{-m/n} = \dfrac{1}{a^{m/n}}$$

Example 3 illustrates the use of negative rational exponents.

Example 3

Simplifying Expressions with Rational Exponents

Simplify each expression.

(a) $16^{-1/2} = \dfrac{1}{16^{1/2}} = \dfrac{1}{4}$

(b) $27^{-2/3} = \dfrac{1}{27^{2/3}} = \dfrac{1}{(\sqrt[3]{27})^2} = \dfrac{1}{3^2} = \dfrac{1}{9}$

✓ **CHECK YOURSELF 3**

Simplify each expression.

(a) $16^{-1/4}$ (b) $81^{-3/4}$

Section 10.3 ■ Rational Exponents **691**

Calculators can be used to evaluate expressions that contain rational exponents by using the y^x key and the parentheses keys.

Example 4   Estimating Powers Using a Calculator

Using a graphing calculator, evaluate each of the following. Round all answers to three decimal places.

(a) $45^{2/5}$

If you are using a scientific calculator, try using the y^x key in place of the $\wedge$ key.

Enter 45 and press the $\wedge$ key. Then use the following keystrokes:

$$ (\; 2 \; \div \; 5 \;) $$

Press ENTER, and the display will read 4.584426407. Rounded to three decimal places, the result is 4.584.

(b) $38^{-2/3}$

Enter 38 and press the $\wedge$ key. Then use the following keystrokes:

$$ (\; (-) \; 2 \; \div \; 3 \;) $$

The $(-)$ key changes the sign of the exponent to minus.

Press $=$, and the display will read 0.088473037. Rounded to three decimal places, the result is 0.088.

✓ CHECK YOURSELF 4

Evaluate each of the following by using a calculator. Round each answer to three decimal places.

(a) $23^{3/5}$ (b) $18^{-4/7}$

As we mentioned earlier in this section, we assume that all our previous exponent properties will continue to hold for rational exponents. Those properties are restated here.

Chapter 10 ■ Radicals and Exponents

> **Properties of Exponents**
>
> For any nonzero real numbers a and b and rational numbers m and n,
>
> 1. Product rule $a^m \cdot a^n = a^{m+n}$
> 2. Quotient rule $\dfrac{a^m}{a^n} = a^{m-n}$
> 3. Power rule $(a^m)^n = a^{mn}$
> 4. Product-power rule $(ab)^m = a^m b^m$
> 5. Quotient-power rule $\left(\dfrac{a}{b}\right)^m = \dfrac{a^m}{b^m}$
>
> We restrict a and b to being nonnegative real numbers when m or n indicates an even root.

Example 5 illustrates the use of our extended properties to simplify expressions involving rational exponents. Here, we assume that all variables represent positive real numbers.

Example 5

Simplifying Expressions

Simplify each expression.

Product rule—add the exponents.

(a) $x^{2/3} \cdot x^{1/2} = x^{2/3 + 1/2}$
$= x^{4/6 + 3/6} = x^{7/6}$

Quotient rule—subtract the exponents.

(b) $\dfrac{w^{3/4}}{w^{1/2}} = w^{3/4 - 1/2}$
$= w^{3/4 - 2/4} = w^{1/4}$

Power rule—multiply the exponents.

(c) $(a^{2/3})^{3/4} = a^{(2/3)(3/4)}$
$= a^{1/2}$

✓ **CHECK YOURSELF 5**

Simplify each expression.

(a) $z^{3/4} \cdot z^{1/2}$ (b) $\dfrac{x^{5/6}}{x^{1/3}}$ (c) $(b^{5/6})^{2/5}$

Section 10.3 ■ Rational Exponents 693

As you would expect from your previous experience with exponents, simplifying expressions often involves using several exponent properties.

Example 6

Simplifying Expressions

Simplify each expression.

(a) $(x^{2/3} \cdot y^{5/6})^{3/2}$

$= (x^{2/3})^{3/2} \cdot (y^{5/6})^{3/2}$ Product-power rule

$= x^{(2/3)(3/2)} \cdot y^{(5/6)(3/2)} = xy^{5/4}$ Power rule

(b) $\left(\dfrac{r^{-1/2}}{s^{1/3}}\right)^6 = \dfrac{(r^{-1/2})^6}{(s^{1/3})^6}$ Quotient-power rule

$= \dfrac{r^{-3}}{s^2} = \dfrac{1}{r^3 s^2}$ Power rule

(c) $\left(\dfrac{4a^{-2/3} \cdot b^2}{a^{1/3} \cdot b^{-4}}\right)^{1/2} = \left(\dfrac{4b^2 \cdot b^4}{a^{1/3} \cdot a^{2/3}}\right)^{1/2} = \left(\dfrac{4b^6}{a}\right)^{1/2}$ We simplify inside the parentheses as the first step.

$= \dfrac{(4b^6)^{1/2}}{a^{1/2}} = \dfrac{4^{1/2}(b^6)^{1/2}}{a^{1/2}}$

$= \dfrac{2b^3}{a^{1/2}}$

✓ **CHECK YOURSELF 6**

Simplify each expression.

(a) $(a^{3/4} \cdot b^{1/2})^{2/3}$ (b) $\left(\dfrac{w^{1/2}}{z^{-1/4}}\right)^4$ (c) $\left(\dfrac{8x^{-3/4}y}{x^{1/4} \cdot y^{-5}}\right)^{1/3}$

We can also use the relationships between rational exponents and radicals to write expressions involving rational exponents as radicals and vice versa.

Example 7

Writing Expressions in Radical Form

Note: Here we use $a^{m/n} = \sqrt[n]{a^m}$, which is generally the preferred form in this situation.

Write each expression in radical form.

(a) $a^{3/5} = \sqrt[5]{a^3}$

(b) $(mn)^{3/4} = \sqrt[4]{(mn)^3}$
$= \sqrt[4]{m^3 n^3}$

Note that the exponent applies only to the variable y.

(c) $2y^{5/6} = 2\sqrt[6]{y^5}$

Now the exponent applies to 2y because of the parentheses.

(d) $(2y)^{5/6} = \sqrt[6]{(2y)^5}$
$= \sqrt[6]{32y^5}$

✓ CHECK YOURSELF 7

Write each expression in radical form.

(a) $(ab)^{2/3}$ (b) $3x^{3/4}$ (c) $(3x)^{3/4}$

Example 8

Writing Expressions in Exponential Form

Using rational exponents, write each expression and simplify.

(a) $\sqrt[3]{5x} = (5x)^{1/3}$

(b) $\sqrt{9a^2 b^4} = (9a^2 b^4)^{1/2}$
$= 9^{1/2}(a^2)^{1/2}(b^4)^{1/2} = 3ab^2$

(c) $\sqrt[4]{16w^{12} z^8} = (16w^{12} z^8)^{1/4}$
$= 16^{1/4}(w^{12})^{1/4}(z^8)^{1/4} = 2w^3 z^2$

✓ CHECK YOURSELF 8

Using rational exponents, write each expression and simplify.

(a) $\sqrt{7a}$ (b) $\sqrt[3]{27p^6 q^9}$ (c) $\sqrt[4]{81x^8 y^{16}}$

✓ CHECK YOURSELF ANSWERS

1. (a) 2; (b) -8; (c) 3.   2. (a) 8; (b) $\frac{4}{9}$; (c) -8.   3. (a) $\frac{1}{2}$; (b) $\frac{1}{27}$.
4. (a) 6.562; (b) 0.192.   5. (a) $z^{5/4}$; (b) $x^{1/2}$; (c) $b^{1/3}$.   6. (a) $a^{1/2} b^{1/3}$; (b) $w^2 z$; (c) $\frac{2y^2}{x^{1/3}}$.   7. (a) $\sqrt[3]{a^2 b^2}$; (b) $3\sqrt[4]{x^3}$; (c) $\sqrt[4]{27x^3}$.   8. (a) $(7a)^{1/2}$; (b) $3p^2 q^3$; (c) $3x^2 y^4$.

Exercises • 10.3

Answers (left column):

1. 6
2. 10
3. −5
4. Not a real number
5. Not a real number
6. −7
7. 3
8. −4
9. 3
10. −2
11. $\frac{2}{3}$
12. $\frac{3}{2}$
13. 9
14. 64
15. 16
16. 25
17. 4
18. −27
19. 729
20. −27
21. $\frac{4}{9}$
22. $\frac{27}{8}$
23. $\frac{1}{5}$
24. $\frac{1}{3}$
25. $\frac{1}{3}$
26. $\frac{1}{11}$
27. $\frac{1}{27}$
28. $\frac{1}{8}$
29. $\frac{1}{32}$
30. $\frac{1}{64}$
31. $\frac{5}{2}$
32. $\frac{4}{9}$
33. x
34. a
35. $y^{4/5}$
36. $m^{3/2}$
37. $b^{13/6}$
38. $p^{3/2}$
39. $x^{1/3}$
40. $a^{2/3}$
41. s
42. z^3
43. $w^{3/4}$
44. $b^{1/2}$
45. x
46. y
47. $a^{3/5}$

In Exercises 1 to 12, use the definition of $a^{1/n}$ to evaluate each expression.

1. $36^{1/2}$
2. $100^{1/2}$
3. $-25^{1/2}$
4. $(-64)^{1/2}$
5. $(-49)^{1/2}$
6. $-49^{1/2}$
7. $27^{1/3}$
8. $(-64)^{1/3}$
9. $81^{1/4}$
10. $-32^{1/5}$
11. $\left(\frac{4}{9}\right)^{1/2}$
12. $\left(\frac{27}{8}\right)^{1/3}$

In Exercises 13 to 22, use the definition of $a^{m/n}$ to evaluate each expression.

13. $27^{2/3}$
14. $16^{3/2}$
15. $(-8)^{4/3}$
16. $125^{2/3}$
17. $32^{2/5}$
18. $-81^{3/4}$
19. $81^{3/2}$
20. $(-243)^{3/5}$
21. $\left(\frac{8}{27}\right)^{2/3}$
22. $\left(\frac{9}{4}\right)^{3/2}$

In Exercises 23 to 32, use the definition of $a^{-m/n}$ to evaluate the following expression. Use your calculator to check each answer.

23. $25^{-1/2}$
24. $27^{-1/3}$
25. $81^{-1/4}$
26. $121^{-1/2}$
27. $9^{-3/2}$
28. $16^{-3/4}$
29. $64^{-5/6}$
30. $16^{-3/2}$
31. $\left(\frac{4}{25}\right)^{-1/2}$
32. $\left(\frac{27}{8}\right)^{-2/3}$

In Exercises 33 to 76, use the properties of exponents to simplify each expression. Assume all variables represent positive real numbers.

33. $x^{1/2} \cdot x^{1/2}$
34. $a^{2/3} \cdot a^{1/3}$
35. $y^{3/5} \cdot y^{1/5}$
36. $m^{1/4} \cdot m^{5/4}$
37. $b^{2/3} \cdot b^{3/2}$
38. $p^{5/6} \cdot p^{2/3}$
39. $\dfrac{x^{2/3}}{x^{1/3}}$
40. $\dfrac{a^{5/6}}{a^{1/6}}$
41. $\dfrac{s^{7/5}}{s^{2/5}}$
42. $\dfrac{z^{9/2}}{z^{3/2}}$
43. $\dfrac{w^{5/4}}{w^{1/2}}$
44. $\dfrac{b^{7/6}}{b^{2/3}}$
45. $(x^{3/4})^{4/3}$
46. $(y^{4/3})^{3/4}$
47. $(a^{2/5})^{3/2}$

695

696 Chapter 10 ■ Radicals and Exponents

48. $p^{1/2}$
49. $\dfrac{1}{y^6}$
50. $\dfrac{1}{w^4}$
51. $a^4 b^9$
52. $p^3 q^{10}$
53. $32xy^3$
54. $81 m^3 n^5$
55. $st^{1/3}$
56. $xy^{2/7}$
57. $4pq^{5/3}$
58. $8a^{1/4} b^{1/2}$
59. $x^{2/5} y^{1/2} z$
60. $p^{1/2} q^{2/5} r$
61. $a^{1/2} b^{1/4}$
62. $x^{1/6} y^{1/4}$
63. $\dfrac{s^2}{r^4}$
64. $\dfrac{1}{w^4 z^2}$
65. $\dfrac{x^3}{y^2}$
66. $\dfrac{p^3}{q^2}$
67. $\dfrac{1}{mn^2}$
68. $r^2 s^5$
69. $\dfrac{s^3}{r^2 t}$
70. $\dfrac{a^2 c}{b}$
71. $\dfrac{2xz^3}{y^2}$
72. $\dfrac{rp^2}{4q^3}$
73. $\dfrac{2n}{m^{1/5}}$
74. $\dfrac{3x^{2/3}}{y}$
75. $xy^{3/4}$
76. pr
77. $\sqrt[4]{a^3}$
78. $\sqrt[6]{m^5}$
79. $2\sqrt[3]{x^2}$
80. $\dfrac{3}{\sqrt[5]{m^2}}$
81. $3\sqrt[5]{x^2}$
82. $\dfrac{2}{\sqrt[4]{y^3}}$
83. $\sqrt[5]{9x^2}$
84. $\dfrac{1}{\sqrt[4]{8y^3}}$
85. $(7a)^{1/2}$
86. $5w^2$
87. $2m^2 n^3$
88. $2r^2 s^3$
89. 9.946
90. 2.449
91. 0.370
92. 0.068

48. $(p^{3/4})^{2/3}$
49. $(y^{-3/4})^8$
50. $(w^{-2/3})^6$
51. $(a^{2/3} \cdot b^{3/2})^6$
52. $(p^{3/4} \cdot q^{5/2})^4$
53. $(2x^{1/5} \cdot y^{3/5})^5$
54. $(3m^{3/4} \cdot n^{5/4})^4$
55. $(s^{3/4} \cdot t^{1/4})^{4/3}$
56. $(x^{5/2} \cdot y^{5/7})^{2/5}$
57. $(8p^{3/2} \cdot q^{5/2})^{2/3}$
58. $(16a^{1/3} \cdot b^{2/3})^{3/4}$
59. $(x^{3/5} \cdot y^{3/4} \cdot z^{3/2})^{2/3}$
60. $(p^{5/6} \cdot q^{2/3} \cdot r^{5/3})^{3/5}$
61. $\dfrac{a^{5/6} \cdot b^{3/4}}{a^{1/3} \cdot b^{1/2}}$
62. $\dfrac{x^{2/3} \cdot y^{3/4}}{x^{1/2} \cdot y^{1/2}}$
63. $\dfrac{(r^{-1} \cdot s^{1/2})^3}{r \cdot s^{-1/2}}$
64. $\dfrac{(w^{-2} \cdot z^{-1/4})^6}{w^{-8} z^{1/2}}$
65. $\left(\dfrac{x^{12}}{y^8}\right)^{1/4}$
66. $\left(\dfrac{p^9}{q^6}\right)^{1/3}$
67. $\left(\dfrac{m^{-1/4}}{n^{1/2}}\right)^4$
68. $\left(\dfrac{r^{1/5}}{s^{-1/2}}\right)^{10}$
69. $\left(\dfrac{r^{-1/2} \cdot s^{3/4}}{t^{1/4}}\right)^4$
70. $\left(\dfrac{a^{1/3} \cdot b^{-1/6}}{c^{-1/6}}\right)^6$
71. $\left(\dfrac{8x^3 \cdot y^{-6}}{z^{-9}}\right)^{1/3}$
72. $\left(\dfrac{16p^{-4} \cdot q^6}{r^2}\right)^{-1/2}$
73. $\left(\dfrac{16m^{-3/5} \cdot n^2}{m^{1/5} \cdot n^{-2}}\right)^{1/4}$
74. $\left(\dfrac{27x^{5/6} \cdot y^{-4/3}}{x^{-7/6} \cdot y^{5/3}}\right)^{1/3}$
75. $\left(\dfrac{x^{3/2} \cdot y^{1/2}}{z^2}\right)^{1/2} \left(\dfrac{x^{3/4} \cdot y^{3/2}}{z^{-3}}\right)^{1/3}$
76. $\left(\dfrac{p^{1/2} \cdot q^{4/3}}{r^{-4}}\right)^{3/4} \left(\dfrac{p^{15/8} \cdot q^{-3}}{r^6}\right)^{1/3}$

In Exercises 77 to 84, write each expression in radical form. Do not simplify.

77. $a^{3/4}$
78. $m^{5/6}$
79. $2x^{2/3}$
80. $3m^{-2/5}$
81. $3x^{2/5}$
82. $2y^{-3/4}$
83. $(3x)^{2/5}$
84. $(2y)^{-3/4}$

In Exercises 85 to 88, write each expression using rational exponents, and simplify where necessary.

85. $\sqrt{7a}$
86. $\sqrt{25w^4}$
87. $\sqrt[3]{8m^6 n^9}$
88. $\sqrt[5]{32 r^{10} s^{15}}$

In Exercises 89 to 92, evaluate each expression, using a calculator. Round each answer to three decimal places.

89. $46^{3/5}$
90. $23^{2/7}$
91. $12^{-2/5}$
92. $36^{-3/4}$

93. Describe the difference between x^{-2} and $x^{1/2}$.

Section 10.3 ■ Rational Exponents

Answers:

95. $a^2 + a^{5/4}$
96. $6x - 10$
97. $a - 4$
98. $w^{2/3} - 9$ 99. $m - n$
100. $x^{2/3} - y^{2/3}$
101. $x + 4x^{1/2} + 4$
102. $a^{2/3} - 6a^{1/3} + 9$
103. $r + 2r^{1/2}s^{1/2} + s$
104. $p - 2p^{1/2}q^{1/2} + q$
105. $(x^{1/3} + 1)(x^{1/3} + 3)$
106. $(y^{1/5} - 4)(y^{1/5} + 2)$
107. $(a^{2/5} - 3)(a^{2/5} - 4)$
108. $(w^{2/3} + 5)(w^{2/3} - 2)$
109. $(x^{2/3} - 2)(x^{2/3} + 2)$
110. $(x^{1/5} + 4)(x^{1/5} - 4)$
111. x^{5n} 112. p^4
113. y^{4n} 114. a^{9n}
115. r^2 116. w^3
117. $a^{6n}b^{4n}$
118. $c^{12m}d^{6m}$
119. x
120. b
121. $\sqrt[4]{x}$
122. $\sqrt[6]{a}$
123. $\sqrt[8]{y}$
124. $\sqrt[6]{w}$
125. 2×10^4
126. 2×10^2
127. 2×10^{-3}
128. 3×10^{-2}
129. 4×10^{-4}
130. 8×10^{-6}

94. Some rational exponents, like $\dfrac{1}{2}$, can easily be rewritten as terminating decimals (0.5). Others, like $\dfrac{1}{3}$, cannot. What is it that determines which rational numbers can be rewritten as terminating decimals?

In Exercises 95 to 104, apply the appropriate multiplication patterns. Then simplify your result.

95. $a^{1/2}(a^{3/2} + a^{3/4})$ **96.** $2x^{1/4}(3x^{3/4} - 5x^{-1/4})$

97. $(a^{1/2} + 2)(a^{1/2} - 2)$ **98.** $(w^{1/3} - 3)(w^{1/3} + 3)$

99. $(m^{1/2} + n^{1/2})(m^{1/2} - n^{1/2})$ **100.** $(x^{1/3} + y^{1/3})(x^{1/3} - y^{1/3})$

101. $(x^{1/2} + 2)^2$ **102.** $(a^{1/3} - 3)^2$

103. $(r^{1/2} + s^{1/2})^2$ **104.** $(p^{1/2} - q^{1/2})^2$

As is suggested by several of the preceding exercises, certain expressions containing rational exponents are factorable. For instance, to factor $x^{2/3} - x^{1/3} - 6$, let $u = x^{1/3}$. Note that $x^{2/3} = (x^{1/3})^2 = u^2$.

Substituting, we have $u^2 - u - 6$, and factoring yields $(u - 3)(u + 2)$ or $(x^{1/3} - 3)(x^{1/3} + 2)$.

In Exercises 105 to 110, use this technique to factor each expression.

105. $x^{2/3} + 4x^{1/3} + 3$ **106.** $y^{2/5} - 2y^{1/5} - 8$ **107.** $a^{4/5} - 7a^{2/5} + 12$

108. $w^{4/3} + 3w^{2/3} - 10$ **109.** $x^{4/3} - 4$ **110.** $x^{2/5} - 16$

In Exercises 111 to 120, perform the indicated operations. Assume that n represents a positive integer and that the denominators are not zero.

111. $x^{3n} \cdot x^{2n}$ **112.** $p^{1-n} \cdot p^{n+3}$ **113.** $(y^2)^{2n}$ **114.** $(a^{3n})^3$

115. $\dfrac{r^{n+2}}{r^n}$ **116.** $\dfrac{w^n}{w^{n-3}}$ **117.** $(a^3 \cdot b^2)^{2n}$ **118.** $(c^4 \cdot d^2)^{3m}$

119. $\left(\dfrac{x^{n+2}}{x^n}\right)^{1/2}$ **120.** $\left(\dfrac{b^n}{b^{n-3}}\right)^{1/3}$

In Exercises 121 to 124, write each expression in exponent form, simplify, and give the result as a single radical.

121. $\sqrt{\sqrt{x}}$ **122.** $\sqrt[3]{\sqrt{a}}$ **123.** $\sqrt[4]{\sqrt{y}}$ **124.** $\sqrt{\sqrt[3]{w}}$

In Exercises 125 to 130, simplify each expression. write your answer in scientific notation.

125. $(4 \times 10^8)^{1/2}$ **126.** $(8 \times 10^6)^{1/3}$ **127.** $(16 \times 10^{-12})^{1/4}$

128. $(9 \times 10^{-4})^{1/2}$ **129.** $(16 \times 10^{-8})^{1/2}$ **130.** $(16 \times 10^{-8})^{3/4}$

131. 40

132. 4.5

131. While investigating rainfall runoff in a region of semiarid farmland, a researcher encounters the following expression:

$$t = C\left(\frac{L}{xy^2}\right)^{1/3}$$

Evaluate t when $C = 20$, $L = 600$, $x = 3$, and $y = 5$.

132. The average velocity of water in an open irrigation ditch is given by the formula

$$V = \frac{1.5x^{2/3}y^{1/2}}{z}$$

Evaluate V when $x = 27$, $y = 16$, and $z = 12$.

133. Use the properties of exponents to decide what x should be to make each statement true. Explain your choices regarding which properties of exponents you decide to use.

(a) $(a^{2/3})^x = a$

(b) $(a^{5/6})^x = \dfrac{1}{a}$

(c) $a^{2x} \cdot a^{3/2} = 1$

(d) $(\sqrt{a^{2/3}})^x = a$

134. The geometric mean is used to measure average inflation rates or interest rates. If prices increased by 15% over 5 years, then the average *annual* rate of inflation is obtained by taking the 5th root of 1.15:

$$(1.15)^{1/5} = 1.0283 \quad \text{or} \quad \sim 2.8\%$$

The 1 is added to 0.15 because we are taking the original price and adding 15% of that price. We could write that as

$$P + 0.15P$$

Factoring, we get

$$P + 0.15P = P(1 + 0.15)$$
$$= P(1.15)$$

From December 1990 through February 1997, the Bureau of Labor Statistics computed an inflation rate of 16.2%, which is equivalent to an annual growth rate of 2.46%. From December 1990 through February 1997 is 75 months. To what exponent was 1.162 raised to obtain this average annual growth rate?

135. (a) 81;
  (b) Not defined

135. On your calculator, try evaluating $(-9)^{4/2}$ in the following two ways:

(a) $((-9)^4)^{1/2}$ (b) $((-9)^{1/2})^4$

Discuss the results.

SECTION 10.4 Complex Numbers

10.4 OBJECTIVES

1. Define a complex number
2. Add and subtract complex numbers
3. Multiply and divide complex numbers

$i = \sqrt{-1}$ is called the **imaginary unit.**

Radicals such as

$$\sqrt{-4} \text{ and } \sqrt{-49}$$

are not real numbers since no real number squared produces a negative number. Our work in this section will extend our number system to include these **imaginary numbers,** which will allow us to consider radicals such as those mentioned above.

First we offer a definition.

> The number i is defined as
> $$i = \sqrt{-1}$$
> Note that this means that
> $$i^2 = -1$$

This definition of the number i gives us an alternate means of indicating the square root of a negative number.

> When a is a positive real number,
> $$\sqrt{-a} = \sqrt{a(-1)} = \sqrt{a} \cdot \sqrt{-1} = \sqrt{a}\,i \quad \text{or} \quad i\sqrt{a}$$

In each case, the i *must* be *visibly* clear of the radical.
Note that

> $\sqrt{a}\sqrt{b} = \sqrt{ab}$ as long as a and b are not *both* negative.

Example 1

Using the Number i

Write each expression as a multiple of i.

(a) $\sqrt{-4} = \sqrt{4}\, i = 2i$

(b) $-\sqrt{-9} = -\sqrt{9}\, i = -3i$

> We simplify $\sqrt{8}$ as $2\sqrt{2}$. Note that we write the i in front of the radical to make it clear that i is *not part* of the radicand.

(c) $\sqrt{-8} = \sqrt{8}\, i = 2\sqrt{2}\, i$ or $2i\sqrt{2}$

(d) $\sqrt{-7} = \sqrt{7}\, i$ or $i\sqrt{7}$

✓ **CHECK YOURSELF 1**

Write each radical as a multiple of i.

(a) $\sqrt{-25}$ (b) $\sqrt{-24}$

We are now ready to define complex numbers in terms of the number i.

> The term "imaginary number" was introduced by René Descartes in 1637. Euler used i to indicate $\sqrt{-1}$ in 1748, but it was not until 1832 that Gauss used the term "complex number."

A **complex number** is any number that can be written in the form

$$a + bi$$

where a and b are real numbers and

$$i = \sqrt{-1}$$

> The first application of these numbers was made by Charles Steinmetz (1865–1923) in explaining the behavior of electric circuits.

The form $a + bi$ is called the **standard form** of a complex number. We call a the **real part** of the complex number and b the **imaginary part.** Some examples follow.

$3 + 7i$ is an example of a complex number with real part 3 and imaginary part 7.

> Also, $5i$ is called a **pure imaginary** number.

$5i$ is also a complex number since it can be written as $0 + 5i$.

-3 is a complex number since it can be written as $-3 + 0i$.

> The real numbers can be considered a subset of the set of complex numbers.

The basic operations of addition and subtraction on complex numbers are defined here.

Adding and Subtracting Complex Numbers

For the complex numbers $a + bi$ and $c + di$,

$$(a + bi) + (c + di) = (a + c) + (b + d)i$$
$$(a + bi) - (c + di) = (a - c) + (b - d)i$$

In words, we add or subtract the real parts and the imaginary parts of the complex numbers.

Example 2 illustrates the use of these definitions.

Example 2

Adding and Subtracting Complex Numbers

Perform the indicated operations.

(a) $(5 + 3i) + (6 - 7i) = (5 + 6) + (3 - 7)i$
$= 11 - 4i$

(b) $5 + (7 - 5i) = (5 + 7) + (-5i)$
$= 12 - 5i$

(c) $(8 - 2i) - (3 - 4i) = (8 - 3) + [-2 - (-4)]i$
$= 5 + 2i$

✓ CHECK YOURSELF 2

Perform the indicated operations.

(a) $(4 - 7i) + (3 - 2i)$ (b) $-7 + (-2 + 3i)$ (c) $(-4 + 3i) - (-2 - i)$

We now consider the basic operation of multiplication on complex numbers.

Multiplying Complex Numbers

For the complex numbers $a + bi$ and $c + di$,

$$(a + bi)(c + di) = ac + adi + bci + bdi^2$$
$$= ac + adi + bci - bd$$
$$= (ac - bd) + (ad + bi)i$$

This formula for the general product of two complex numbers can be memorized. However, you will find it much easier to get used to the multiplication pattern as it is applied to complex numbers than to memorize this formula.

Example 3

Multiplying Complex Numbers

Multiply.

$$(2 + 3i)(3 - 4i)$$
$$= 2 \cdot 3 + 2 \cdot (-4i) + 3i \cdot 3 + (3i)(-4i)$$
$$= 6 + (-8i) + 9i + (-12)\,i^2$$
$$= 6 + (-8i) + 9i + (-12)(-1)$$
$$= 6 - 8i + 9i + 12$$
$$= (6 + 12) + (-8 + 9)\,i$$
$$= 18 + i$$

We can replace i^2 with -1 because of the definition of i, and we usually do so because of the resulting simplification.

✓ CHECK YOURSELF 3

Multiply $(2 - 5i)(3 - 2i)$.

There is one particular product form that is very important. We call $a + bi$ and $a - bi$ **complex conjugates.** For instance,

$$3 + 2i \quad \text{and} \quad 3 - 2i$$

are complex conjugates.

Consider the product

$$(3 + 2i)(3 - 2i) = 3^2 - (2i)^2$$
$$= 9 - 4i^2 = 9 - 4(-1)$$
$$= 9 + 4 = 13$$

The product of $3 + 2i$ and $3 - 2i$ is a real number. In general, we can write the product of two complex conjugates as

$$(a + bi)(a - bi) = a^2 + b^2$$

The fact that this product is always a real number will be very useful when we consider the division of complex numbers later in this section.

Example 4 — Multiplying Complex Numbers

Multiply.

We could get the same result by applying the formula above with $a = 7$ and $b = 4$.

$$(7 - 4i)(7 + 4i) = 7^2 - (4i)^2$$
$$= 7^2 - 4^2(-1)$$
$$= 7^2 + 4^2$$
$$= 49 + 16$$
$$= 65$$

✓ CHECK YOURSELF 4

Multiply $(5 + 3i)(5 - 3i)$.

We are now ready to discuss the division of complex numbers. Generally, we find the quotient by multiplying the numerator and denominator by the conjugate of the denominator, as Example 5 illustrates.

Example 5 — Dividing Complex Numbers

Divide.

Think of $3i$ as $0 + 3i$ and of its conjugate as $0 - 3i$, or $-3i$.

(a) $\dfrac{6 + 9i}{3i}$

$$\frac{6 + 9i}{3i} = \frac{(6 + 9i)(-3i)}{(3i)(-3i)}$$

The conjugate of $3i$ is $-3i$, and so we multiply the numerator and denominator by $-3i$.

$$= \frac{-18i - 27i^2}{-9i^2}$$

Note: Multiplying the numerator and denominator in the original expression by i would yield the same result. Try it yourself.

$$= \frac{-18i - 27(-1)}{(-9)(-1)}$$

$$= \frac{27 - 18i}{9} = 3 - 2i$$

We multiply by $\dfrac{3 - 2i}{3 - 2i}$, or 1.

(b) $\dfrac{3 - i}{3 + 2i} = \dfrac{(3 - i)(3 - 2i)}{(3 + 2i)(3 - 2i)}$

$$= \frac{9 - 6i - 3i + 2i^2}{9 - 4i^2}$$

$$= \frac{9 - 9i - 2}{9 + 4}$$

To write a complex number in standard form, we separate the real component from the imaginary.

$$= \frac{7 - 9i}{13} = \frac{7}{13} - \frac{9}{13}i$$

(c) $\dfrac{2+i}{4-5i} = \dfrac{(2+i)(4+5i)}{(4-5i)(4+5i)}$

$= \dfrac{8+4i+10i+5i^2}{16-25i^2}$

$= \dfrac{8+14i-5}{16+25}$

$= \dfrac{3+14i}{41} = \dfrac{3}{41} + \dfrac{14}{41}i$

✓ CHECK YOURSELF 5

Divide.

(a) $\dfrac{5+i}{5-3i}$ (b) $\dfrac{4+10i}{2i}$

We conclude this section with the following diagram, which summarizes the structure of the system of complex numbers.

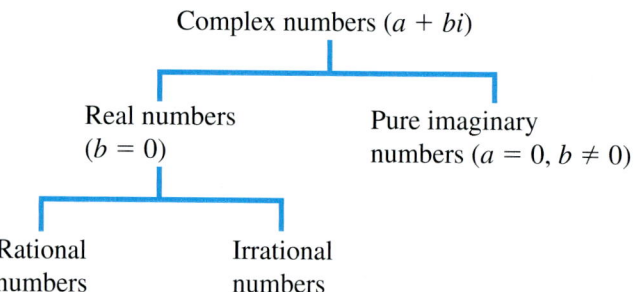

✓ CHECK YOURSELF ANSWERS

1. (a) $5i$; (b) $2i\sqrt{6}$.   2. (a) $7-9i$; (b) $-9+3i$; (c) $-2+4i$.
3. $-4-19i$.   4. 34.   5. (a) $\dfrac{11}{17} + \dfrac{10}{17}i$; (b) $5-2i$.

Exercises • 10.4

Answers (left column):
1. $4i$
2. $6i$
3. $-8i$
4. $-5i$
5. $i\sqrt{21}$
6. $i\sqrt{19}$
7. $2i\sqrt{3}$
8. $2i\sqrt{6}$
9. $-6i\sqrt{3}$
10. $-8i\sqrt{3}$
11. $8 + 3i$
12. $6 + 8i$
13. $1 + 5i$
14. $-7 + 4i$
15. $2 + 2i$
16. $4 + i$
17. $5 - 7i$
18. $9 + 2i$
19. $7 + 11i$
20. $5 + 8i$
21. $3 + 11i$
22. $10 + 3i$
23. $6 - 3i$
24. $2 - 4i$
25. $0 + 0i$ or 0
26. $0 + 0i$ or 0
27. $-15 + 9i$
28. $-6 + 14i$
29. $28 + 12i$
30. $-6 + 12i$
31. $-6 - 8i$
32. $-35 - 10i$
33. $-5 + 4i$
34. $-3 + 2i$
35. $13i$

In Exercises 1 to 10, write each root as a multiple of i. Simplify your results where necessary.

1. $\sqrt{-16}$
2. $\sqrt{-36}$
3. $-\sqrt{-64}$
4. $-\sqrt{-25}$
5. $\sqrt{-21}$
6. $\sqrt{-19}$
7. $\sqrt{-12}$
8. $\sqrt{-24}$
9. $-\sqrt{-108}$
10. $-\sqrt{-192}$

In Exercises 11 to 26, perform the indicated operations.

11. $(3 + i) + (5 + 2i)$
12. $(2 + 3i) + (4 + 5i)$
13. $(3 - 2i) + (-2 + 7i)$
14. $(-5 - 3i) + (-2 + 7i)$
15. $(5 + 4i) - (3 + 2i)$
16. $(7 + 6i) - (3 + 5i)$
17. $(8 - 5i) - (3 + 2i)$
18. $(7 - 3i) - (-2 - 5i)$
19. $(5 + i) + (2 + 3i) + 7i$
20. $(3 - 2i) + (2 + 3i) + 7i$
21. $(2 + 3i) - (3 - 5i) + (4 + 3i)$
22. $(5 - 7i) + (7 + 3i) - (2 - 7i)$
23. $(7 + 3i) - [(3 + i) - (2 - 5i)]$
24. $(8 - 2i) - [(4 + 3i) - (-2 + i)]$
25. $(5 + 3i) + (-5 - 3i)$
26. $(8 - 7i) + (-8 + 7i)$

In Exercises 27 to 42, find each product. Write your answer in standard form.

27. $3i(3 + 5i)$
28. $2i(7 + 3i)$
29. $4i(3 - 7i)$
30. $2i(6 + 3i)$
31. $-2i(4 - 3i)$
32. $-5i(2 - 7i)$
33. $6i\left(\dfrac{2}{3} + \dfrac{5}{6}i\right)$
34. $4i\left(\dfrac{1}{2} + \dfrac{3}{4}i\right)$
35. $(3 + 2i)(2 + 3i)$

Answers (left column)

36. $13 - 11i$
37. $23 + 14i$ 38. $25 - 8i$
39. $18 + i$ 40. $11 + 23i$
41. $21 - 20i$
42. $-40 + 42i$
43. $3 + 2i, 13$
44. $5 - 2i, 29$
45. $2 - 3i, 13$
46. $7 + i, 50$
47. $-3 + 2i, 13$
48. $-5 + 7i, 74$
49. $-5i, 25$
50. $3i, 9$
51. $2 - 3i$
52. $3 + 5i$
53. $-2 - 3i$
54. $-3 + 2i$
55. $\dfrac{6}{29} - \dfrac{15}{29}i$
56. $\dfrac{10}{13} - \dfrac{15}{13}i$
57. $2 - 3i$
58. $-\dfrac{3}{2} + \dfrac{5}{2}i$
59. $\dfrac{17}{25} + \dfrac{6}{25}i$
60. $\dfrac{13}{17} + \dfrac{1}{17}i$
61. $-\dfrac{7}{25} - \dfrac{24}{25}i$
62. $\dfrac{45}{53} + \dfrac{28}{53}i$

Exercises

36. $(5 - 2i)(3 - i)$
37. $(4 - 3i)(2 + 5i)$
38. $(7 + 2i)(3 - 2i)$
39. $(-2 - 3i)(-3 + 4i)$
40. $(-5 - i)(-3 - 4i)$
41. $(5 - 2i)^2$
42. $(3 + 7i)^2$

In Exercises 43 to 50, write the conjugate of each complex number. Then find the product of the given number and the conjugate.

43. $3 - 2i$
44. $5 + 2i$
45. $2 + 3i$
46. $7 - i$
47. $-3 - 2i$
48. $-5 - 7i$
49. $5i$
50. $-3i$

In Exercises 51 to 62, find each quotient, and write your answer in standard form.

51. $\dfrac{3 + 2i}{i}$
52. $\dfrac{5 - 3i}{-i}$
53. $\dfrac{6 - 4i}{2i}$
54. $\dfrac{8 + 12i}{-4i}$
55. $\dfrac{3}{2 + 5i}$
56. $\dfrac{5}{2 - 3i}$
57. $\dfrac{13}{2 + 3i}$
58. $\dfrac{-17}{3 + 5i}$
59. $\dfrac{2 + 3i}{4 + 3i}$
60. $\dfrac{4 - 2i}{5 - 3i}$
61. $\dfrac{3 - 4i}{3 + 4i}$
62. $\dfrac{7 + 2i}{7 - 2i}$

63. The first application of complex numbers was suggested by the Norwegian surveyor Caspar Wessel in 1797. He found that complex numbers could be used to represent distance and direction on a two-dimensional grid. Why would a surveyor care about such a thing?

64. $\mathbb{N}, \mathbb{Z}, \mathbb{Q}, \mathbb{R}$
65. $-\sqrt{35}$
66. $-\sqrt{30}$
67. -6
68. -10
69. $-3\sqrt{10}$
70. $-5\sqrt{6}$
71. -10
72. -11
73. -1
74. i
75. 1
76. $-i$
77. -1
78. 1
79. $-i$
80. i

64. To what sets of numbers does 1 belong?

In this section, we defined $\sqrt{-4} = \sqrt{4}\,i = 2i$ in the process of expressing the square root of a negative number as a multiple of i.

Particular care must be taken with products where two negative radicands are involved. For instance,

$$\sqrt{-3} \cdot \sqrt{-12} = (i\sqrt{3})(i\sqrt{12})$$
$$= i^2\sqrt{36} = (-1)\sqrt{36} = -6$$

is correct. However, if we try to apply the product property for radicals, we have

$$\sqrt{-3} \cdot \sqrt{-12} \stackrel{?}{=} \sqrt{(-3)(-12)} = \sqrt{36} = 6$$

which is *not* correct. The property $\sqrt{a} \cdot \sqrt{b} = \sqrt{ab}$ is not applicable in the case where a and b are both negative. Radicals such as $\sqrt{-a}$ must be written in the standard form $i\sqrt{a}$ *before* multiplying to use the rules for real valued radicals.

In Exercises 65 to 72, find each product.

65. $\sqrt{-5} \cdot \sqrt{-7}$ **66.** $\sqrt{-3} \cdot \sqrt{-10}$ **67.** $\sqrt{-2} \cdot \sqrt{-18}$
68. $\sqrt{-4} \cdot \sqrt{-25}$ **69.** $\sqrt{-6} \cdot \sqrt{-15}$ **70.** $\sqrt{-5} \cdot \sqrt{-30}$
71. $\sqrt{-10} \cdot \sqrt{-10}$ **72.** $\sqrt{-11} \cdot \sqrt{-11}$

Since $i^2 = -1$, the positive integral powers of i form an interesting pattern. Consider the following.

$$i = i$$
$$i^2 = -1$$
$$i^3 = i^2 \cdot i = (-1)i = -i$$
$$i^4 = i^2 \cdot i^2 = (-1)(-1) = 1$$

$$i^5 = i^4 \cdot i = 1 \cdot i = i$$
$$i^6 = i^4 \cdot i^2 = 1(-1) = -1$$
$$i^7 = i^4 \cdot i^3 = 1(-i) = -i$$
$$i^8 = i^4 \cdot i^4 = 1 \cdot 1 = 1$$

Given the pattern above, do you see that any power of i will simplify to i, -1, $-i$, or 1? The easiest approach to simplifying higher powers of i is to write that power in terms of i^4 (because $1^4 = 1$). As an example,

$$i^{18} = i^{16} \cdot i^2 = (i^4)^4 \cdot i^2 = 1^4(-1) = -1$$

In Exercises 73 to 80, use these comments to simplify each power of i.

73. i^{10} **74.** i^9 **75.** i^{20} **76.** i^{15}
77. i^{38} **78.** i^{40} **79.** i^{51} **80.** i^{61}

Summary Exercises • 10

This summary exercise set is provided to give you practice with each of the objectives in the chapter. Each exercise is keyed to the appropriate chapter section.

[10.1] Simplify each expression, using the properties of exponents.

1. $4x^{-5}$
2. $(2w)^{-3}$
3. $\dfrac{3}{m^{-4}}$
4. $\dfrac{a^{-5}}{b^{-4}}$
5. $y^{-5} \cdot y^2$
6. $\dfrac{w^{-7}}{w^{-3}}$
7. $(m^{-6})^{-2}$
8. $(m^3 n^{-5})^{-2}$
9. $\left(\dfrac{a^{-4}}{b^{-2}}\right)^3$
10. $\left(\dfrac{r^{-5}}{s^4}\right)^{-2}$
11. $(5a^2 b^{-3})(2a^{-2} b^{-6})$
12. $\left(\dfrac{m^{-3} n^{-3}}{m^{-4} n^4}\right)^3$

[10.1] Perform the indicated calculation. Write your result in scientific notation.

13. $(3 \times 10^5)(5 \times 10^7)$
14. $\dfrac{(2.4 \times 10^7)(5.1 \times 10^4)}{(1.6 \times 10^3)(3.4 \times 10^5)}$

15. Write 0.0000425 in scientific notation.
16. Write 3.1×10^{-4} in standard notation.

[10.2] Evaluate each of the following roots over the set of real numbers.

17. $-\sqrt{64}$
18. $\sqrt{-81}$
19. $\sqrt[3]{64}$
20. $\sqrt[3]{-64}$
21. $\sqrt[4]{81}$
22. $\sqrt{\dfrac{9}{16}}$
23. $\sqrt[3]{-\dfrac{8}{27}}$
24. $\sqrt{8^2}$

Simplify each of the following expressions. Assume that all variables represent positive real numbers for all subsequent exercises in this exercise set.

25. $\sqrt{4x^2}$
26. $\sqrt{a^4}$
27. $\sqrt{36y^2}$
28. $\sqrt{49w^4 z^6}$
29. $\sqrt[3]{x^9}$
30. $\sqrt[3]{-27b^6}$
31. $\sqrt[3]{8r^3 s^9}$
32. $\sqrt[4]{16x^4 y^8}$
33. $\sqrt[5]{32p^5 q^{15}}$
34. $\sqrt{45}$
35. $-\sqrt{75}$
36. $\sqrt{60x^2}$

---

1. $\dfrac{4}{x^5}$
2. $\dfrac{1}{8w^3}$
3. $3m^4$
4. $\dfrac{b^4}{a^5}$
5. $\dfrac{1}{y^3}$
6. $\dfrac{1}{w^4}$
7. m^{12}
8. $\dfrac{n^{10}}{m^6}$
9. $\dfrac{b^6}{a^{12}}$
10. $r^{10} s^8$
11. $\dfrac{10}{b^9}$
12. $\dfrac{m^3}{n^{21}}$
13. 1.5×10^{13}
14. 2.25×10^3
15. 4.25×10^{-5}
16. 0.00031
17. -8
18. Not a real number
19. 4
20. -4
21. 3
22. $\dfrac{3}{4}$
23. $-\dfrac{2}{3}$
24. 8
25. $2x$
26. a^2
27. $6y$
28. $7w^2 z^3$
29. x^3
30. $-3b^2$
31. $2rs^3$
32. $2xy^2$
33. $2pq^3$
34. $3\sqrt{5}$
35. $-5\sqrt{3}$
36. $2x\sqrt{15}$

Summary Exercises 10

Answer column (left):

37. $6a\sqrt{3a}$ 38. $2\sqrt[3]{4}$
39. $-2wz\sqrt[3]{10w}$
40. $\dfrac{\sqrt{11}}{9}$ 41. $\dfrac{\sqrt{7}}{6}$
42. $\dfrac{y^2}{7}$ 43. $\dfrac{\sqrt{2x}}{3}$
44. $\dfrac{\sqrt{5}}{4x}$
45. $\dfrac{\sqrt[3]{5a^2}}{3}$
46. 3   47. 4
48. 6   49. 2
50. 3   51. 2
52. $\sqrt{21xy}$ 53. $6x\sqrt{3}$
54. $ab\sqrt[3]{4}$
55. $\sqrt{15} + 2\sqrt{5}$
56. $2\sqrt{3}$ 57. $6a\sqrt{5}$
58. $-32 - 2\sqrt{3}$
59. $7 + \sqrt{21} - \sqrt{14} - \sqrt{6}$
60. 1   61. 4
62. $7 + 4\sqrt{3}$
63. $7 - 2\sqrt{10}$
64. 4.1   65. 7.7
66. 3.6   67. 1.8
68. 6.1   69. 12.5
70. 6.5   71. 2.9
72. 17.7
73. Not a real number
74. −4.1   75. 2.0
76. 17.0
77. Not a real number
78. −5.4   79. 1.6

---

37. $\sqrt{108a^3}$ **38.** $\sqrt[3]{32}$ **39.** $\sqrt[3]{-80w^4z^3}$ **40.** $\sqrt{\dfrac{11}{81}}$

41. $\sqrt{\dfrac{7}{36}}$ **42.** $\sqrt{\dfrac{y^4}{49}}$ **43.** $\sqrt{\dfrac{2x}{9}}$ **44.** $\sqrt{\dfrac{5}{16x^2}}$

45. $\sqrt[3]{\dfrac{5a^2}{27}}$ **46.** $\dfrac{\sqrt{27}}{\sqrt{3}}$ **47.** $\dfrac{\sqrt{80}}{\sqrt{5}}$ **48.** $\dfrac{\sqrt{72}}{\sqrt{2}}$

49. $\sqrt[3]{16}$ **50.** $\sqrt[3]{81}$ **51.** $\sqrt[4]{32}$

[10.2] Multiply and simplify each of the following expressions.

52. $\sqrt{3x} \cdot \sqrt{7y}$ **53.** $\sqrt{6x^2} \cdot \sqrt{18}$

54. $\sqrt[3]{4a^2b} \cdot \sqrt[3]{ab^2}$ **55.** $\sqrt{5}(\sqrt{3} + 2)$

56. $\sqrt{6}(\sqrt{8} - \sqrt{2})$ **57.** $\sqrt{a}(\sqrt{5a} + \sqrt{125a})$

58. $(\sqrt{3} + 5)(\sqrt{3} - 7)$ **59.** $(\sqrt{7} - \sqrt{2})(\sqrt{7} + \sqrt{3})$

60. $(\sqrt{5} - 2)(\sqrt{5} + 2)$ **61.** $(\sqrt{7} - \sqrt{3})(\sqrt{7} + \sqrt{3})$

62. $(2 + \sqrt{3})^2$ **63.** $(\sqrt{5} - \sqrt{2})^2$

[10.2] Use a calculator to evaluate the following. Round each answer to the nearest tenth.

64. $\sqrt{17}$ **65.** $\sqrt{59}$ **66.** $\sqrt[3]{45}$ **67.** $\sqrt{\dfrac{37}{11}}$

68. $\sqrt{37}$ **69.** $\sqrt{156}$ **70.** $\sqrt[3]{278}$ **71.** $\sqrt{\dfrac{457}{56}}$

72. $\sqrt{315}$ **73.** $\sqrt{-75}$ **74.** $\sqrt[3]{-69}$ **75.** $\sqrt[4]{\dfrac{567}{36}}$

76. $\sqrt{288}$ **77.** $\sqrt{-36}$ **78.** $\sqrt[3]{-159}$ **79.** $\sqrt[5]{\dfrac{529}{52}}$

710 Chapter 10 ■ Radicals and Exponents

80. x^4

81. $b^{13/6}$

82. r

83. $a^{3/4}$

84. $x^{2/5}$

85. $\dfrac{1}{y^8}$

86. $x^8 y^{15}$

87. $8x^{1/4} y^{1/2}$

88. $\dfrac{x^6}{y^{7/2}}$

89. $\dfrac{3xy}{z^2}$

90. $\sqrt[4]{x^3}$

91. $\sqrt[5]{w^4 z^2}$

92. $3\sqrt[3]{a^2}$

93. $\sqrt[3]{9a^2}$

94. $(7x)^{1/5}$

95. $4w^2$

96. $3pq^3$

97. $2a^2 b^4$

98. $7i$

99. $i\sqrt{13}$

100. $-2i\sqrt{15}$

101. $5 - 2i$

102. $4 - 5i$

103. $3 - 8i$

104. $-3 + 5i$

105. $8 + 28i$

106. $23 + 14i$

107. $-7 - 24i$

108. 13

109. $-3 - i$

110. $\dfrac{6}{5} + \dfrac{8}{5}i$

111. $\dfrac{5}{13} - \dfrac{12}{13}i$

112. $4 + 3i$

[10.3] Use the properties of exponents to simplify each of the following expressions.

80. $x^{3/2} \cdot x^{5/2}$

81. $b^{2/3} \cdot b^{3/2}$

82. $\dfrac{r^{8/5}}{r^{3/5}}$

83. $\dfrac{a^{5/4}}{a^{1/2}}$

84. $(x^{3/5})^{2/3}$

85. $(y^{-4/3})^6$

86. $(x^{4/5} y^{3/2})^{10}$

87. $(16x^{1/3} \cdot y^{2/3})^{3/4}$

88. $\left(\dfrac{x^{-2} y^{-1/6}}{x^{-4} y}\right)^3$

89. $\left(\dfrac{27 y^3 z^{-6}}{x^{-3}}\right)^{1/3}$

[10.3] Write each of the following expressions in radical form.

90. $x^{3/4}$

91. $(w^2 z)^{2/5}$

92. $3a^{2/3}$

93. $(3a)^{2/3}$

[10.3] Write each of the following expressions using rational exponents, and simplify where necessary.

94. $\sqrt[5]{7x}$

95. $\sqrt{16w^4}$

96. $\sqrt[3]{27 p^3 q^9}$

97. $\sqrt[4]{16 a^8 b^{16}}$

[10.4] Write each of the following roots as a multiple of i. Simplify your result.

98. $\sqrt{-49}$

99. $\sqrt{-13}$

100. $-\sqrt{-60}$

[10.4] Perform the indicated operations.

101. $(2 + 3i) + (3 - 5i)$

102. $(7 - 3i) + (-3 - 2i)$

103. $(5 - 3i) - (2 + 5i)$

104. $(-4 + 2i) - (-1 - 3i)$

[10.4] Find each of the following products.

105. $4i(7 - 2i)$

106. $(5 - 2i)(3 + 4i)$

107. $(3 - 4i)^2$

108. $(2 - 3i)(2 + 3i)$

[10.4] Find each of the following quotients, and write your answer in standard form.

109. $\dfrac{5 - 15i}{5i}$

110. $\dfrac{10}{3 - 4i}$

111. $\dfrac{3 - 2i}{3 + 2i}$

112. $\dfrac{5 + 10i}{2 + i}$

Self-Test 10

The purpose of this self-test is to help you check your progress and to review for a chapter test in class. Allow yourself about 1 hour to take the test. When you are done, check your answers in the back of the book. If you missed any answers, be sure to go back and review the appropriate sections in the chapter and the exercises that are provided.

Simplify each expression. Assume that all variables represent positive real numbers in all subsequent problems.

1. $(x^4 y^{-5})^2$ **2.** $\dfrac{9c^{-5}d^3}{18c^{-7}d^{-4}}$

Write the following numbers in scientific notation.

3. 4,230,000,000   **4.** 0.000025

Write each of the following expressions in simplified form.

5. $\sqrt{49a^4}$ **6.** $\sqrt[3]{-27w^6 z^9}$ **7.** $\dfrac{7x}{\sqrt{64y^2}}$

8. $\dfrac{\sqrt{28}}{\sqrt{7}}$ **9.** $\sqrt{7x^3}\sqrt{2x^4}$ **10.** $\sqrt[3]{4x^5}\sqrt[3]{8x^6}$

Use a calculator to evaluate the following. Round each answer to the nearest tenth.

11. $\sqrt{43}$ **12.** $\sqrt[3]{\dfrac{73}{27}}$

Use the properties of exponents to simplify each expression.

13. $(16x^4)^{-3/2}$ **14.** $(27m^{3/2}n^{-6})^{2/3}$ **15.** $\left(\dfrac{16r^{-1/3}s^{5/3}}{rs^{-7/3}}\right)^{3/4}$

Write the expression in radical form and simplify.

16. $(a^7 b^3)^{2/5}$

Write the expression, using rational exponents. Then simplify.

17. $\sqrt[3]{125p^9 q^6}$

Perform the indicated operations.

18. $(-2 + 3i) - (-5 - 7i)$ **19.** $(5 - 3i)(-4 + 2i)$ **20.** $\dfrac{10 - 20i}{3 - i}$

---

1. $\dfrac{x^8}{y^{10}}$
2. $\dfrac{c^2 d^7}{2}$
3. 4.23×10^9
4. 2.5×10^{-5}
5. $7a^2$
6. $-3w^2 z^3$
7. $\dfrac{7x}{8y}$
8. 2
9. $x^3 \sqrt{14x}$
10. $2x^3 \sqrt[3]{4x^2}$
11. 6.6
12. 1.4
13. $\dfrac{1}{64x^6}$
14. $\dfrac{9m}{n^4}$
15. $\dfrac{8s^3}{r}$
16. $a^2 b \sqrt[5]{a^4 b}$
17. $5p^3 q^2$
18. $3 + 10i$
19. $-14 + 22i$
20. $5 - 5i$

Cumulative Test · 1–10

This test is provided to help you in the process of reviewing the previous chapters. Answers are provided in the back of the book. If you missed any answers, be sure to go back and review the appropriate chapter section.

1. Solve the equation $7x - 6(x - 1) = 2(5 + x) + 11$.
2. If $f(x) = 3x^6 - 4x^3 + 9x^2 - 11$, find $f(-1)$.
3. Find the equation of the line that has a y intercept of -6 and is parallel to the line $6x - 4y = 18$.
4. Solve the equation $|3x - 5| = 4$.

Simplify each of the following polynomials.

5. $5x^2 - 8x + 11 - (-3x^2 - 2x + 8) - (-2x^2 - 4x + 3)$
6. $(5x + 3)(2x - 9)$

Factor each of the following completely.

7. $2x^3 + x^2 - 3x$ 8. $9x^4 - 36y^4$ 9. $4x^2 + 8xy - 5x - 10y$

Simplify each of the following rational expressions.

10. $\dfrac{2x^2 + 13x + 15}{6x^2 + 7x - 3}$ 11. $\dfrac{3}{x - 5} - \dfrac{2}{x - 1}$

12. $\dfrac{a^2 - 4a}{a^2 - 6a + 8} \cdot \dfrac{a^2 - 4}{2a^2}$ 13. $\dfrac{a^2 - 9}{a^2 - a - 12} \div \dfrac{a^2 - a - 6}{a^2 - 2a - 8}$

Simplify each of the following radical expressions.

14. $\sqrt{3x^3y}\sqrt{4x^5y^6}$ 15. $(\sqrt{3} - 5)(\sqrt{2} + 3)$

Graph each equation.

16. $y = 3x - 5$ 17. $x = -5$ 18. $2x - 3y = 12$

19. Solve the system of equations.

$$4x - 3y = 15$$
$$x + y = 2$$

20. Solve the following inequalities.

(a) $5x - (2 - 3x) \geq 6 + 10x$ (b) $|x - 2| < 8$

---

1. $\{-15\}$
2. 5
3. $y = \dfrac{3}{2}x - 6$
4. $\left\{3, \dfrac{1}{3}\right\}$
5. $10x^2 - 2x$
6. $10x^2 - 39x - 27$
7. $x(x - 1)(2x + 3)$
8. $9(x^2 + 2y^2)(x^2 - 2y^2)$
9. $(x + 2y)(4x - 5)$
10. $\dfrac{x + 5}{3x - 1}$
11. $\dfrac{x + 7}{(x - 5)(x - 1)}$
12. $\dfrac{a + 2}{2a}$
13. 1
14. $2x^4y^3\sqrt{3y}$
15. $\sqrt{6} - 5\sqrt{2} + 3\sqrt{3} - 15$
16. See answer section
17. See answer section
18. See answer section
19. $\{(3, -1)\}$
20. (a) $\{x | x \leq -4\}$;
    (b) $\{x | -6 < x < 10\}$

CHAPTER 11 Quadratic Functions

LIST OF SECTIONS

11.1 Solving Quadratic Equations by Completing the Square

11.2 The Quadratic Formula

11.3 Solving Quadratic Equations by Graphing

11.4 Solving Quadratic Inequalities

Running power lines from a power plant, wind farm, or hydroelectric plant to a city is a very costly enterprise. Land must be cleared, towers built, and conducting wires strung from tower to tower across miles of countryside. Typical construction designs run from about 300- to 1200-foot spans, with towers about 75 to 200 feet high. Of course, if a lot of towers are needed, and many of them must be tall, the construction costs skyrocket.

Power line construction carries a unique set of problems. Towers must be built tall enough and close enough to keep the conducting lines well above the ground. The sag of these wires (how much they droop from the towers) is a function of the weight of the conductor, the span-length, and the tension in the wires. The amount of this sag, measured in feet, is approximated by the following formula:

$$\text{Sag} = \frac{wS^2}{8T}$$

where w = weight of wires in pounds/foot
S = span length in feet
T = tension in wires measured in pounds

The actual curve of the power lines is called a **catenary curve.** The curve that we use to approximate the sag is a **parabola.**

SECTION 11.1 Solving Quadratic Equations by Completing the Square

 11.1 OBJECTIVES

1. Solve quadratic equations by using the square root method
2. Solve quadratic equations by completing the square

Recall that a quadratic equation is an equation of the form $ax^2 + bx + c = 0$ where a is not equal to zero.

In Section 6.5, we solved quadratic equations by factoring and using the zero-product principle. However, not all equations are solvable by that method. In this section, we will learn another technique that can be used to solve some quadratic equation. This technique is called the **square root method.** Let's begin by solving a special type of equation by factoring.

Example 1 — Solving Equations by Factoring

Solve the quadratic equation $x^2 = 16$ by factoring.

We write the equation in standard form:

$$x^2 - 16 = 0$$

Here, we factor the quadratic member of the equation as a difference of squares.

Factoring, we have

$$(x + 4)(x - 4) = 0$$

Finally, the solutions are

$$x = -4 \quad \text{or} \quad x = 4 \quad \text{or} \quad \{\pm 4\}$$

 CHECK YOURSELF 1

Solve each of the following quadratic equations.

(a) $x^2 = 25$ (b) $5x^2 = 180$

The Square Root Method

The equation in Example 1 could have been solved in an alternative fashion. We could have used what is called the **square root method.** Again, given the equation

$$x^2 = 16$$

we can write the equivalent statement

$$x = \sqrt{16} \quad \text{or} \quad x = -\sqrt{16}$$

Note: Be sure to include *both* the positive and the negative square roots when you use the square root method.

This yields the solutions

$$x = 4 \quad \text{or} \quad x = -4 \quad \text{or} \quad \{\pm 4\}$$

This discussion leads us to the following general result.

> **Square Root Property**
>
> If $x^2 = k$, where k is a complex number, then
>
> $$x = \sqrt{k} \quad \text{or} \quad x = -\sqrt{k}$$

Example 2 further illustrates the use of this property.

Example 2 — Using the Square Root Method

Solve each equation by using the square root method.

(a) $x^2 = 9$

By the square root property,

$$x = \sqrt{9} \quad \text{or} \quad x = -\sqrt{9}$$
$$= 3 \qquad\qquad = -3 \quad \text{or} \quad \{\pm 3\}$$

If a calculator were used, $\sqrt{17} = 4.123$ (rounded to three decimal places).

(b) $x^2 - 17 = 0$

Add 17 to both sides of the equation.

$$x^2 = 17 \quad \text{or} \quad \{\pm\sqrt{17}\} \quad \text{or} \quad \{-\sqrt{17}, \sqrt{17}\}$$

(c) $4x^2 - 3 = 0$

$$4x^2 = 3$$

$$x^2 = \frac{3}{4}$$

$$x = \pm\sqrt{\frac{3}{4}}$$

$$x = \pm\frac{\sqrt{3}}{2} \quad \text{or} \quad \left\{\pm\frac{\sqrt{3}}{2}\right\}$$

In Example 2, part d, we see that complex-number solutions may result.

(d) $x^2 + 1 = 0$

$$x^2 = -1$$

$$x = \pm\sqrt{-1}$$

$$x = \pm i \quad \text{or} \quad \{\pm i\}$$

✓ CHECK YOURSELF 2

Solve each equation.

(a) $x^2 = 5$ (b) $x^2 - 2 = 0$ (c) $9x^2 - 8 = 0$ (d) $x^2 + 9 = 0$

We can also use the approach in Example 2 to solve an equation of the form

$$(x + 3)^2 = 16$$

As before, by the square root property we have

$$x + 3 = \pm 4 \qquad \text{Add } -3 \text{ to both sides of the equation.}$$

Solving for x yields

$$x = -3 \pm 4$$

which means that there are two solutions:

$$x = -3 + 4 \qquad \text{or} \qquad x = -3 - 4$$
$$= 1 \qquad\qquad\qquad\quad = -7 \qquad \text{or} \qquad \{1, -7\}$$

Example 3   Using the Square Root Method

Use the square root method to solve each equation.

(a) $(x - 5)^2 - 5 = 0$

$(x - 5)^2 = 5$

$x - 5 = \pm\sqrt{5}$

$x = 5 \pm \sqrt{5}$ or $\{5 \pm \sqrt{5}\}$

The two solutions $5 + \sqrt{5}$ and $5 - \sqrt{5}$ are abbreviated as $5 \pm \sqrt{5}$.

(b) $9(y + 1)^2 - 2 = 0$

$9(y + 1)^2 = 2$

$(y + 1)^2 = \dfrac{2}{9}$

$y + 1 = \pm\sqrt{\dfrac{2}{9}}$

$y = -1 \pm \dfrac{\sqrt{2}}{3}$

$= \dfrac{-3}{3} \pm \dfrac{\sqrt{2}}{3}$

$= \dfrac{-3 \pm \sqrt{2}}{3}$ or $\left\{\dfrac{-3 \pm \sqrt{2}}{3}\right\}$

✓ CHECK YOURSELF 3

Using the square root method, solve each equation.

(a) $(x - 2)^2 - 3 = 0$ (b) $4(x - 1)^2 = 3$

Completing the Square

Not all quadratic equations can be solved directly by factoring or using the square root method. We must extend our techniques.

The square root method is useful in this process because any quadratic equation can be written in the form

$$(x + h)^2 = k$$

which yields the solution

$$x = -h \pm \sqrt{k}$$

If $(x + h)^2 = k$, then

$x + h = \pm\sqrt{k}$

and

$x = -h \pm \sqrt{k}$

The process of changing an equation in standard form

$$ax^2 + bx + c = 0$$

to the form

$$(x + h)^2 = k$$

is called the method of **completing the square,** and it is based on the relationship between the middle term and the last term of any perfect-square trinomial.

Let's look at three perfect-square trinomials to see whether we can detect a pattern:

$$x^2 + 4x + 4 = (x + 2)^2 \quad (1)$$
$$x^2 - 6x + 9 = (x - 3)^2 \quad (2)$$
$$x^2 + 8x + 16 = (x + 4)^2 \quad (3)$$

Note that this relationship is true *only* if the leading, or x^2 coefficient is 1. That will be important later.

Note that in each case the last (or constant) term is the square of one-half of the coefficient of x in the middle (or linear) term. For example, in equation (2),

$$x^2 - 6x + 9 = (x - 3)^2$$

$\frac{1}{2}$ of this coefficient is -3,
and $(-3)^2 = 9$, the constant.

Verify this relationship for yourself in equation (3). To summarize, in perfect-square trinomials, the constant is always the square of one-half the coefficient of x.

We are now ready to use the above observation in the solution of quadratic equations by completing the square. Consider Example 4.

Example 4 — Completing the Square to Solve an Equation

Solve $x^2 + 8x - 7 = 0$ by completing the square.

First, we rewrite the equation with the constant on the *right-hand side*:

$$x^2 + 8x = 7$$

Our objective is to have a perfect-square trinomial on the left-hand side. We know that we must add the square of one-half of the x coefficient to complete the square. In this case, that value is 16, so now we add 16 to each side of the equation.

$$x^2 + 8x + 16 = 7 + 16$$

When you graph the related function, $y = x^2 + 8x - 7$, you will note that the x values for the x intercepts are just below 1 and just above -9. Be certain that you see how these points relate to the exact solutions,

$-4 + \sqrt{23}$ and
$-4 - \sqrt{23}$.

Section 11.1 • Solving Quadratic Equations by Completing the Square   719

$\frac{1}{2} \cdot 8 = 4$ and $4^2 = 16$

Factor the perfect-square trinomial on the left, and combine like terms on the right to yield

$$(x + 4)^2 = 23$$

Now the square root property yields

$$x + 4 = \pm\sqrt{23}$$

Remember that if $(x + h)^2 = k$, then $x = -h \pm \sqrt{k}$.

Subtracting 4 from both sides of the equation gives

$$x = -4 \pm \sqrt{23} \quad \text{or} \quad \{-4 \pm \sqrt{23}\}$$

 CHECK YOURSELF 4

Solve $x^2 - 6x - 2 = 0$ by completing the square.

Example 5

Completing the Square to Solve an Equation

Solve $x^2 + 5x - 3 = 0$ by completing the square.

$$x^2 + 5x - 3 = 0 \qquad \text{Add 3 to both sides.}$$
$$x^2 + 5x = 3 \qquad \text{Make the left-hand side a perfect square.}$$

Add the square of one-half of the x coefficient to both sides of the equation. Note that

$$\frac{1}{2} \cdot 5 = \frac{5}{2}$$

$$x^2 + 5x + \left(\frac{5}{2}\right)^2 = 3 + \left(\frac{5}{2}\right)^2$$

$$\left(x + \frac{5}{2}\right)^2 = \frac{37}{4} \qquad \text{Take the square root of both sides.}$$

$$x + \frac{5}{2} = \pm\frac{\sqrt{37}}{2} \qquad \text{Solve for } x.$$

$$x = \frac{-5 \pm \sqrt{37}}{2} \quad \text{or} \quad \left\{\frac{-5 \pm \sqrt{37}}{2}\right\}$$

 CHECK YOURSELF 5

Solve $x^2 + 3x - 7 = 0$ by completing the square.

Some equations have nonreal complex solutions, as Example 6 illustrates.

Example 6

Completing the Square to Solve an Equation

Solve $x^2 + 4x + 13 = 0$ by completing the square.

$$x^2 + 4x + 13 = 0 \qquad \text{Subtract 13 from both sides.}$$

Note that the graph of $y = x^2 + 4x + 13$ does not intercept the x axis.

$$x^2 + 4x = -13 \qquad \text{Add } \left[\frac{1}{2}(4)\right]^2 \text{ to both sides.}$$

$$x^2 + 4x + 4 = -13 + 4 \qquad \text{Factor the left-hand side.}$$

$$(x + 2)^2 = -9 \qquad \text{Take the square root of both sides.}$$

$$x + 2 = \pm\sqrt{-9} \qquad \text{Simplify the radical.}$$

$$x + 2 = \pm i\sqrt{9}$$

$$x + 2 = \pm 3i$$

$$x = -2 \pm 3i \qquad \text{or} \qquad \{-2 \pm 3i\}$$

✓ CHECK YOURSELF 6

Solve $x^2 + 10x + 41 = 0$.

Example 7 illustrates a situation in which the leading coefficient of the quadratic member is not equal to 1. As you will see, an extra step will be required.

Example 7

Completing the Square to Solve an Equation

Caution
Before you can complete the square on the left, the coefficient of x^2 must be equal to 1. Otherwise, we must *divide* both sides of the equation by that coefficient.

Solve $4x^2 + 8x - 7 = 0$ by completing the square.

$$4x^2 + 8x - 7 = 0 \qquad \text{Add 7 to both sides.}$$

$$4x^2 + 8x = 7 \qquad \text{Divide both sides by 4.}$$

$$x^2 + 2x = \frac{7}{4} \qquad \text{Now, complete the square on the left.}$$

$$x^2 + 2x + 1 = \frac{7}{4} + 1 \qquad \text{The left side is now a perfect square.}$$

$$(x + 1)^2 = \frac{11}{4}$$

$$x + 1 = \pm\sqrt{\frac{11}{4}}$$

$$x = -1 \pm \sqrt{\frac{11}{4}}$$

$$= -1 \pm \frac{\sqrt{11}}{2}$$

$$= \frac{-2 \pm \sqrt{11}}{2} \qquad \text{or} \qquad \left\{\frac{-2 \pm \sqrt{11}}{2}\right\}$$

Section 11.1 ■ Solving Quadratic Equations by Completing the Square

✓ **CHECK YOURSELF 7**

Solve $4x^2 - 8x + 3 = 0$ by completing the square.

The following algorithm summarizes our work in this section with solving quadratic equations by completing the square.

> **Completing the Square**
>
> Step 1 Isolate the constant on the right side of the equation.
> Step 2 Divide both sides of the equation by the coefficient of the x^2 term if that coefficient is not equal to 1.
> Step 3 Add the square of one-half of the coefficient of the linear term to both sides of the equation. This will give a perfect-square trinomial on the left side of the equation.
> Step 4 Write the left side of the equation as the square of a binomial, and simplify on the right side.
> Step 5 Use the square root property, and then solve the resulting linear equations.

✓ **CHECK YOURSELF ANSWERS**

1. (a) $\{-5, 5\}$; (b) $\{-6, 6\}$.   **2.** (a) $\{\sqrt{5}, -\sqrt{5}\}$; (b) $\{\sqrt{2}, -\sqrt{2}\}$;

(c) $\left\{\dfrac{2\sqrt{2}}{3}, -\dfrac{2\sqrt{2}}{3}\right\}$; and (d) $\{3i, -3i\}$.   **3.** (a) $\{2 \pm \sqrt{3}\}$; (b) $\left\{\dfrac{2 \pm \sqrt{3}}{2}\right\}$.

4. $\{3 \pm \sqrt{11}\}$.   **5.** $\left\{\dfrac{-3 \pm \sqrt{37}}{2}\right\}$.   **6.** $\{-5 \pm 4i\}$.   **7.** $\left\{\dfrac{1}{2}, \dfrac{3}{2}\right\}$.

Exercises • 11.1

Answers (left column):

1. $\{-5, -1\}$ 2. $\{-2, -3\}$
3. $\{-5, 7\}$ 4. $\{8, -3\}$
5. $\left\{-\dfrac{1}{2}, 3\right\}$ 6. $\left\{\dfrac{2}{3}, -4\right\}$
7. $\left\{-\dfrac{1}{2}, \dfrac{2}{3}\right\}$ 8. $\left\{\dfrac{1}{5}, -\dfrac{1}{2}\right\}$
9. $\{\pm 6\}$ 10. $\{-12, 12\}$
11. $\{-\sqrt{7}, \sqrt{7}\}$
12. $\{-3\sqrt{2}, 3\sqrt{2}\}$
13. $\{-\sqrt{6}, \sqrt{6}\}$
14. $\{-\sqrt{22}, \sqrt{22}\}$
15. $\{\pm 2i\}$ 16. $\{\pm 3i\}$
17. $\{-1 \pm 2\sqrt{3}\}$
18. $\left\{\dfrac{3 \pm \sqrt{5}}{2}\right\}$
19. $\left\{\dfrac{-1 \pm \sqrt{3}}{2}\right\}$
20. $\left\{\dfrac{4 \pm 3i}{3}\right\}$
21. 36 22. 49
23. 16 24. 64
25. $\dfrac{9}{4}$ 26. $\dfrac{25}{4}$
27. $\dfrac{1}{4}$ 28. $\dfrac{1}{4}$
29. $\dfrac{1}{16}$ 30. $\dfrac{1}{36}$
31. $\dfrac{1}{100}$ 32. $\dfrac{1}{64}$

In Exercises 1 to 8, solve by factoring or completing the square.

1. $x^2 + 6x + 5 = 0$ **2.** $x^2 + 5x + 6 = 0$ **3.** $z^2 - 2z - 35 = 0$

4. $q^2 - 5q - 24 = 0$ **5.** $2x^2 - 5x - 3 = 0$ **6.** $3x^2 + 10x - 8 = 0$

7. $6y^2 - y - 2 = 0$ **8.** $10z^2 + 3z - 1 = 0$

In Exercises 9 to 20, use the square root method to find solutions for the equations.

9. $x^2 = 36$ **10.** $x^2 = 144$ **11.** $y^2 = 7$

12. $p^2 = 18$ **13.** $2x^2 - 12 = 0$ **14.** $3x^2 - 66 = 0$

15. $2t^2 + 12 = 4$ **16.** $3u^2 - 5 = -32$ **17.** $(x + 1)^2 = 12$

18. $(2x - 3)^2 = 5$ **19.** $(2z + 1)^2 - 3 = 0$ **20.** $(3p - 4)^2 + 9 = 0$

In Exercises 21 to 32, find the constant that must be added to each binomial expression to form a perfect-square trinomial.

21. $x^2 + 12x$ **22.** $r^2 - 14r$ **23.** $y^2 - 8y$

24. $w^2 + 16w$ **25.** $x^2 - 3x$ **26.** $z^2 + 5z$

27. $n^2 + n$ **28.** $x^2 - x$ **29.** $x^2 + \dfrac{1}{2}x$

30. $x^2 - \dfrac{1}{3}x$ **31.** $x^2 + -\dfrac{1}{5}x$ **32.** $y^2 - \dfrac{1}{4}y$

722

Section 11.1 • Solving Quadratic Equations by Completing the Square

In Exercises 33 to 54, solve each equation by completing the square.

33. $x^2 + 12x - 2 = 0$
34. $x^2 - 14x - 7 = 0$
35. $y^2 - 2y = 8$
36. $z^2 + 4z - 72 = 0$
37. $x^2 - 2x - 5 = 0$
38. $x^2 - 2x = 3$
39. $x^2 + 10x + 13 = 0$
40. $x^2 + 3x - 17 = 0$
41. $z^2 - 5z - 7 = 0$
42. $q^2 - 8q + 20 = 0$
43. $m^2 - m - 3 = 0$
44. $y^2 + y - 5 = 0$
45. $x^2 + \frac{1}{2}x = 1$
46. $x^2 - \frac{1}{3}x = 2$
47. $2x^2 + 2x - 1 = 0$
48. $3x^2 - 4x = 1$
49. $3x^2 - 8x = 2$
50. $4x^2 + 8x - 1 = 0$
51. $3x^2 - 2x + 12 = 0$
52. $7y^2 - 2y + 3 = 0$
53. $x^2 + 8x + 20 = 0$
54. $x^2 - 2x + 10 = 0$

55. Why must the leading coefficient of the quadratic member be set equal to 1 before using the technique of completing the square?

56. What relationship exists between the solution(s) of a quadratic equation and the graph of a quadratic function?

In Exercises 57 to 62, find the constant that must be added to each binomial to form a perfect-square trinomial. Let x be the variable; other letters represent constants.

57. $x^2 + 2ax$
58. $x^2 + 2abx$
59. $x^2 + 3ax$
60. $x^2 + abx$
61. $a^2x^2 + 2ax$
62. $a^2x^2 + 4abx$

---

33. $\{-6 \pm \sqrt{38}\}$
34. $\{7 \pm 2\sqrt{14}\}$
35. $\{-2, 4\}$
36. $\{-2 \pm 2\sqrt{19}\}$
37. $\{1 \pm \sqrt{6}\}$
38. $\{-1, 3\}$
39. $\{-5 \pm 2\sqrt{3}\}$
40. $\left\{\dfrac{-3 \pm \sqrt{77}}{2}\right\}$
41. $\left\{\dfrac{5 \pm \sqrt{53}}{2}\right\}$
42. $\{4 \pm 2i\}$
43. $\left\{\dfrac{1 \pm \sqrt{13}}{2}\right\}$
44. $\left\{\dfrac{-1 \pm \sqrt{21}}{2}\right\}$
45. $\left\{\dfrac{-1 \pm \sqrt{17}}{4}\right\}$
46. $\left\{\dfrac{1 \pm \sqrt{73}}{6}\right\}$
47. $\left\{\dfrac{-1 \pm \sqrt{3}}{2}\right\}$
48. $\left\{\dfrac{2 \pm \sqrt{7}}{3}\right\}$
49. $\left\{\dfrac{4 \pm \sqrt{22}}{3}\right\}$
50. $\left\{\dfrac{-2 \pm \sqrt{5}}{2}\right\}$
51. $\left\{\dfrac{1 \pm i\sqrt{35}}{3}\right\}$
52. $\left\{\dfrac{1 \pm 2i\sqrt{5}}{7}\right\}$
53. $\{-4 \pm 2i\}$
54. $\{1 \pm 3i\}$
57. a^2
58. a^2b^2
59. $\dfrac{9}{4}a^2$
60. $\dfrac{a^2b^2}{4}$
61. 1
62. $4b^2$

724 Chapter 11 ▪ Quadratic Functions

63. $\{-a \pm \sqrt{4+a^2}\}$
64. $\{-a \pm \sqrt{a^2+8}\}$

In Exercises 63 and 64, solve each equation by completing the square.

63. $x^2 + 2ax = 4$ **64.** $x^2 + 2ax - 8 = 0$

In Exercises 65 to 68, use your graphing utility to find the graph. Approximate the x intercepts for each graph. (You may have to adjust the viewing window to see both intercepts.)

65.
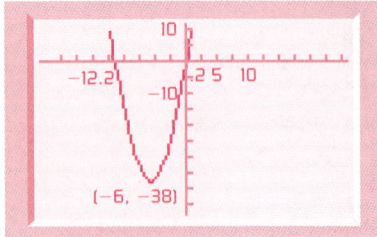

65. $y = x^2 + 12x - 2$ **66.** $y = x^2 - 14x - 7$
67. $y = x^2 - 2x - 8$ **68.** $y = x^2 + 4x - 72$

69. On your graphing calculator, view the graph of $f(x) = x^2 + 1$.

   (a) What can you say about the x intercepts of the graph?

66.
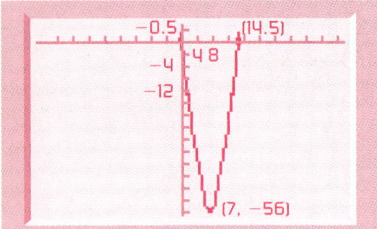

   (b) Determine the zeros of the function, using the square root method.

   (c) How does your answer to part a relate to your answer to part b?

70. Consider the following representation of "completing the square": Suppose we wish to complete the square for $x^2 + 10x$. A square with dimensions x by x has area equal to x^2.

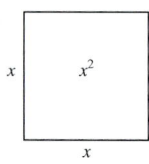

67.
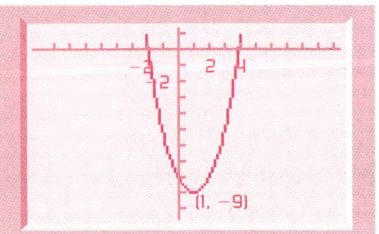

We divide the quantity $10x$ by 2 and get $5x$. If we extend the base x by 5 units, and draw the rectangle attached to the square, the rectangle's dimensions are 5 by x with an area of $5x$.

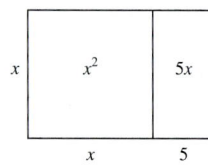

68.
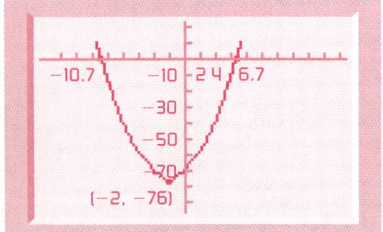

Now we extend the height by 5 units, and draw another rectangle whose area is $5x$.

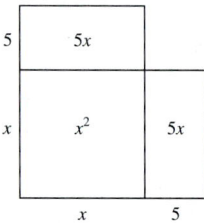

69. (a) There are none; (b) $\pm i$; (c) If the graph of $f(x)$ has no x intercepts, the zeros of the function do not exist.

   (a) What is the total area represented in the figure so far?

   (b) How much area must be added to the figure to "complete the square"?

   (c) Write the area of the completed square as a binomial squared.

71. Repeat the process described in Exercise 70 with $x^2 + 16x$.

70. (a) $x^2 + 5x + 5x$; (b) 25; (c) $x^2 + 10x + 25 = (x + 5)^2$

71. (a) $x^2 + 8x + 8x$; (b) 64; (c) $x^2 + 16x + 64 = (x + 8)^2$

SECTION 11.2 The Quadratic Formula

11.2 OBJECTIVES

1. Solve quadratic equations by using the quadratic formula
2. Determine the nature of the solutions of a quadratic equation by using the discriminant
3. Use the Pythagorean theorem to solve a geometric application

Every quadratic equation can be solved by using the quadratic formula. In this section, we will first describe how the quadratic formula is derived, then we will examine its use. Recall that a quadratic equation is any equation that can be written in the form

$$ax^2 + bx + c = 0 \qquad \text{where } a \neq 0$$

Deriving the Quadratic Formula

Step 1 Isolate the constant on the right side of the equation.
$$ax^2 + bx = -c$$

Step 2 Divide both sides by the coefficient of the x^2 term.
$$x^2 + \frac{b}{a}x = -\frac{c}{a}$$

Step 3 Add the square of one-half the x coefficient to both sides.
$$x^2 + \frac{b}{a}x + \frac{b^2}{4a^2} = -\frac{c}{a} + \frac{b^2}{4a^2}$$

Step 4 Factor the left side as a perfect-square binomial. Then apply the square root property.
$$\left(x + \frac{b}{2a}\right)^2 = \frac{-4ac + b^2}{4a^2}$$
$$x + \frac{b}{2a} = \pm\sqrt{\frac{b^2 - 4ac}{4a^2}}$$

Step 5 Solve the resulting linear equations.
$$x = -\frac{b}{2a} \pm \frac{\sqrt{b^2 - 4ac}}{2a}$$

Step 6 Simplify.
$$= \frac{-b \pm \sqrt{b^2 - 4ac}}{2a}$$

We now use the result derived above to state the **quadratic formula,** a formula that allows us to find the solutions for any quadratic equation.

The Quadratic Formula

Given any quadratic equation in the form

$$ax^2 + bx + c = 0 \qquad \text{where } a \neq 0$$

the two solutions to the equation are found using the formula

$$x = \frac{-b \pm \sqrt{b^2 - 4ac}}{2a}$$

Our first example uses an equation in standard form.

Example 1

Using the Quadratic Formula

Note that the equation is in standard form.

Solve, using the quadratic formula.

$$6x^2 - 7x - 3 = 0$$

First, we determine the values for a, b, and c. Here,

$$a = 6 \qquad b = -7 \qquad c = -3$$

Substituting those values into the quadratic formula, we have

$$x = \frac{-(-7) \pm \sqrt{(-7)^2 - 4(6)(-3)}}{2(6)}$$

Since $b^2 - 4ac = 121$ is a perfect square, the two solutions in this case are rational numbers.

Simplifying inside the radical gives us

$$x = \frac{7 \pm \sqrt{121}}{12}$$

$$= \frac{7 \pm 11}{12}$$

This gives us the solutions

$$x = \frac{3}{2} \quad \text{or} \quad x = -\frac{1}{3} \quad \text{or} \quad \left\{\frac{3}{2}, -\frac{1}{3}\right\}$$

Note that since the solutions for the equation of this example are rational, the original equation could have been solved by our earlier method of factoring.

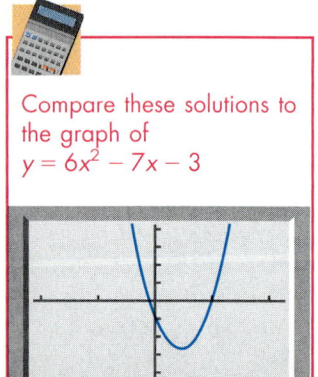

Compare these solutions to the graph of $y = 6x^2 - 7x - 3$

✓ CHECK YOURSELF 1

Solve, using the quadratic formula.

$$3x^2 + 2x - 8 = 0$$

To use the quadratic formula, we often must write the equation in standard form. Example 2 illustrates this approach.

Section 11.2 ■ The Quadratic Formula **727**

Example 2

The equation must be in standard form to determine a, b, and c.

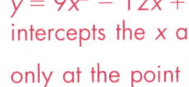

The graph of $y = 9x^2 - 12x + 4$ intercepts the x axis only at the point $\left(\frac{2}{3}, 0\right)$.

Using the Quadratic Formula

Solve by using the quadratic formula.

$$9x^2 = 12x - 4$$

First, we must write the equation in standard form.

$$9x^2 - 12x + 4 = 0$$

Second, we find the values of *a, b,* and *c.* Here,

$$a = 9 \quad b = -12 \quad c = 4$$

Substituting these values into the quadratic formula, we find

$$x = \frac{-(-12) \pm \sqrt{(-12)^2 - 4(9)(4)}}{18}$$

$$= \frac{12 \pm \sqrt{0}}{18}$$

and simplifying yields

$$x = \frac{2}{3} \quad \text{or} \quad \left\{\frac{2}{3}\right\}$$

✓ CHECK YOURSELF 2

Use the quadratic formula to solve the equation.

$$4x^2 - 4x = -1$$

Thus far our examples and exercises have led to rational solutions. That is not always the case, as Example 3 illustrates.

Example 3

Using the Quadratic Formula

Using the quadratic formula, solve

$$x^2 - 3x = 5$$

Once again, to use the quadratic formula, we write the equation in standard form.

$$x^2 - 3x - 5 = 0$$

We now determine values for *a*, *b*, and *c* and substitute.

$$x = \frac{-(-3) \pm \sqrt{(-3)^2 - 4(1)(-5)}}{2}$$

Simplifying as before, we have

$$x = \frac{3 \pm \sqrt{29}}{2} \quad \text{or} \quad \left\{\frac{3 \pm \sqrt{29}}{2}\right\}$$

✓ CHECK YOURSELF 3

Using the quadratic equation, solve $2x^2 = x + 7$.

Example 4 requires some special care in simplifying the solution.

Example 4

Using the Quadratic Formula

Using the quadratic formula, solve

$$3x^2 - 6x + 2 = 0$$

Caution
Students are sometimes tempted to reduce this result to

$$\frac{6 \pm 2\sqrt{3}}{6} \stackrel{?}{=} 1 \pm 2\sqrt{3}$$

This is *not a valid step*. We must divide *each of the terms* in the numerator by 2 when simplifying the expression.

Here, we have $a = 3$, $b = -6$, and $c = 2$. Substituting gives

$$x = \frac{-(-6) \pm \sqrt{(-6)^2 - 4(3)(2)}}{2}$$

$$= \frac{6 \pm \sqrt{12}}{6} \qquad \text{We now look for the largest perfect-square factor of 12, the radicand.}$$

Simplifying, we note that $\sqrt{12}$ is equal to $\sqrt{4 \cdot 3}$, or $2\sqrt{3}$. We can then write the solutions as

$$x = \frac{6 \pm 2\sqrt{3}}{6} = \frac{2(3 \pm \sqrt{3})}{6} = \frac{3 \pm \sqrt{3}}{3}$$

✓ CHECK YOURSELF 4

Solve by using the quadratic formula.

$$x^2 - 4x = 6$$

Example 5

Using the Quadratic Formula

Solve by using the quadratic formula.

$$x^2 - 2x + 2 = 0$$

Labeling the coefficients, we find that

$$a = 1 \quad b = -2 \quad c = 2$$

> The solutions will be non-real any time $b^2 - 4ac$ is negative.

Applying the quadratic formula, we have

$$x = \frac{2 \pm \sqrt{-4}}{2}$$

> The graph of $y = x^2 - 2x + 2$ does not intercept the x axis, so there are no real solutions.

and noting that $\sqrt{-4}$ is $2i$, we can simplify to

$$x = 1 \pm i \quad \text{or} \quad \{1 \pm i\}$$

 CHECK YOURSELF 5

Solve by using the quadratic formula.

$$x^2 - 4x + 6 = 0$$

In attempting to solve a quadratic equation, you should first try the factoring method. If this method does not work, you can apply the quadratic formula or the square root method to find the solution. The following algorithm outlines the steps.

Solving a Quadratic Equation by Using the Quadratic Formula

Step 1 Write the equation in standard form (one side is equal to 0).

$$ax^2 + bx + c = 0$$

Step 2 Determine the values for a, b, and c.
Step 3 Substitute those values into the quadratic formula.

$$x = \frac{-b \pm \sqrt{b^2 - 4ac}}{2a}$$

> Although not necessarily distinct or real, every second-degree equation has two solutions.

Step 4 Simplify.

Given a quadratic equation, the radicand $b^2 - 4ac$ determines the number of real solutions. Because of this, we call the result of substituting a, b, and c into that part of the quadratic formula the discriminant. Because the discriminant is a real number, there are three possibilities, known as the **trichotomy property.**

Graphically, we can see the number of real solutions as the number of times the related quadratic function intercepts the x axis.

The Trichotomy Property

If $b^2 - 4ac \begin{cases} < 0 & \text{there are } no \text{ real solutions, but two non-real solutions.} \\ = 0 & \text{there is } one \text{ real solution (a double solution).} \\ > 0 & \text{there are } two \text{ distinct real solutions.} \end{cases}$

Example 6 — Analyzing the Discriminant

How many real solutions are there for each of the following quadratic equations?

(a) $x^2 + 7x - 15 = 0$

The discriminant $[49 - 4(1)(-15)]$ is 109. This indicates that there are two real solutions.

We could find the two non-real solutions by using the quadratic formula.

(b) $3x^2 - 5x + 7 = 0$

The discriminant, $b^2 - 4ac = -59$, is negative. There are no real solutions.

(c) $9x^2 - 12x + 4 = 0$

The discriminant is 0. There is exactly one real solution (a double solution).

✓ CHECK YOURSELF 6

How many real solutions are there for each of the following quadratic equations?

(a) $2x^2 - 3x + 2 = 0$ (b) $3x^2 + x - 11 = 0$
(c) $4x^2 - 4x + 1 = 0$ (d) $x^2 = -5x - 7$

Frequently, as in Examples 3 and 4, the solutions of a quadratic equation involve square roots. When we are solving algebraic equations, it is generally best to leave solutions in this form. However, if an equation resulting from an application has been solved by the use of the quadratic formula, we will often estimate the root and sometimes accept only positive solutions. Consider the following two applications involving thrown balls that can be solved by using the quadratic formula.

Example 7

Solving a Thrown-Ball Application

If a ball is thrown upward from the ground, the equation to find the height h of such a ball thrown with an initial velocity of 80 ft/s is

$$h(t) = 80t - 16t^2$$

Here h measures the height above the ground, in feet, t seconds (s) after the ball is thrown upward.

Find the time it takes the ball to reach a height of 48 ft.

First we substitute 48 for h, and then we rewrite the equation in standard form.

$$16t^2 - 80t + 48 = 0$$

Note that the result of dividing by 16

$$\frac{0}{16} = 0$$

is 0 on the right.

To simplify the computation, we divide both sides of the equation by the common factor, 16. This yields

$$t^2 - 5t + 3 = 0$$

We solve for t as before, using the quadratic equation, with the result

$$t = \frac{5 \pm \sqrt{13}}{2}$$

There are two solutions because the ball reaches the height twice, once on the way up and once on the way down.

This gives us two solutions, $\frac{5 + \sqrt{13}}{2}$ and $\frac{5 - \sqrt{13}}{2}$. But, because we have specified units of time, we generally estimate the answer to the nearest tenth or hundredth of a second.

In this case, estimating to the nearest tenth of a second gives solutions of 0.7 and 4.3 s.

✓ CHECK YOURSELF 7

The equation to find the height h of a ball thrown with an initial velocity of 64 ft/s is

$$h(t) = 64t - 16t^2$$

Find the time it takes the ball to reach a height of 32 ft, estimating to the nearest tenth of a second.

Example 8

Solving a Thrown-Ball Application

The height, h, of a ball thrown downward from the top of a 240-ft building with an initial velocity of 64 ft/s is given by

$$h(t) = 240 - 64t - 16t^2$$

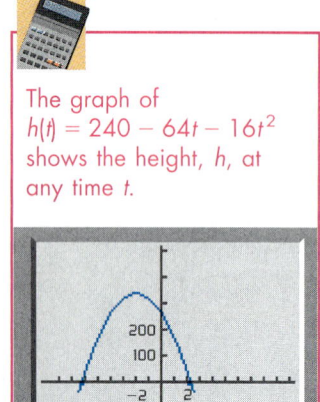

The graph of $h(t) = 240 - 64t - 16t^2$ shows the height, h, at any time t.

At what time will the ball reach a height of 176 ft?

Let $h(t) = 176$, and write the equation in standard form.

$$176 = 240 - 64t - 16t^2$$
$$0 = 64 - 64t - 16t^2$$
$$16t^2 + 64t - 64 = 0$$

Divide both sides of the equation by 16 to simplify the computation.

$$t^2 + 4t - 4 = 0$$

Applying the quadratic formula with $a = 1$, $b = 4$, and $c = -4$ yields

$$t = -2 \pm 2\sqrt{2}$$

The ball has a height of 176 ft at approximately 0.8 s.

Estimating these solutions, we have $t = -4.8$ and $t = 0.8$ s, but of these two values only the *positive value* makes any sense. (To accept the negative solution would be to say that the ball reached the specified height before it was thrown.)

✓ CHECK YOURSELF 8

The height h of a ball thrown upward from the top of a 96-ft building with an initial velocity of 16 ft/s is given by

$$h(t) = 96 + 16t - 16t^2$$

When will the ball have a height of 32 ft? (Estimate your answer to the nearest tenth of a second.)

Another geometric result that generates quadratic equations in applications is the **Pythagorean theorem.** You may recall from earlier algebra courses that the theorem gives an important relationship between the lengths of the sides of a right triangle (a triangle with a 90° angle).

The Pythagorean Theorem

In any right triangle, the square of the longest side (the hypotenuse) is equal to the sum of the squares of the two shorter sides (the legs).

$$c^2 = a^2 + b^2$$

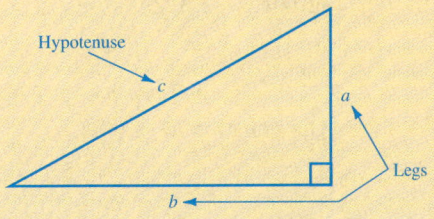

In Example 9, the solution of the quadratic equation contains radicals. Substituting a pair of solutions such as $\dfrac{3 \pm \sqrt{5}}{2}$ is a very difficult process. As in our thrown-ball applications, the emphasis is on checking the "reasonableness" of the answer.

Example 9 A Triangular Application

One leg of a right triangle is 4 cm longer than the other leg. The length of the hypotenuse of the triangle is 12 cm. Find the length of the two legs.

As in any geometric problem, a sketch of the information will help us visualize.

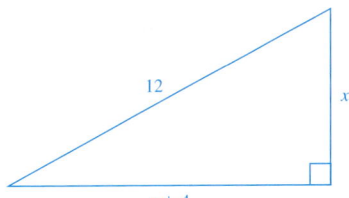

We assign variable x to the shorter leg and $x + 4$ to the other leg.

Remember: The sum of the squares of the legs of the triangle is equal to the square of the hypotenuse.

Now we apply the Pythagorean theorem to write an equation for the solution.

$$x^2 + (x + 4)^2 = (12)^2$$
$$x^2 + x^2 + 8x + 16 = 144$$

or

$$2x^2 + 8x - 128 = 0$$

Dividing both sides by 2, we have the equivalent equation

$$x^2 + 4x - 64 = 0$$

Dividing both sides of a quadratic equation by a common factor is always a prudent step. It simplifies your work with the quadratic formula.

Using the quadratic formula, we get

$$x = -2 + 2\sqrt{17} \quad \text{or} \quad x = -2 - 2\sqrt{17}$$

Now, we check our answers for reasonableness. We can reject $-2 - 2\sqrt{17}$ (do you see why?), but we should still check the reasonableness of the value $-2 + 2\sqrt{17}$. We could substitute $-2 + 2\sqrt{17}$ into the original equation, but it seems more prudent to simply check that it "makes sense" as a solution. Remembering that $\sqrt{16} = 4$, we estimate $-2 + 2\sqrt{17}$ as

$\sqrt{17}$ is just slightly more than $\sqrt{16}$ which is 4.

$$-2 + 2(4) = 6$$

Our equation in step 3,

$$x^2 + (x + 4)^2 = (12)^2$$

where x equals 6, becomes

$$36 + 100 = 144$$

This indicates that our answer is at least reasonable.

✓ CHECK YOURSELF 9

One leg of a right triangle is 2 cm longer than the other. The hypotenuse is 1 cm less than twice the length of the shorter leg. Find the length of each side of the triangle.

✓ CHECK YOURSELF ANSWERS

1. $\left\{-2, \dfrac{4}{3}\right\}$. **2.** $\left\{\dfrac{1}{2}\right\}$. **3.** $\left\{\dfrac{1 \pm \sqrt{57}}{4}\right\}$. **4.** $\{2 \pm \sqrt{10}\}$.
5. $\{2 \pm i\sqrt{2}\}$. **6.** (a) None; (b) two; (c) one; (d) none. **7.** 0.6 and 3.4 s.
8. 2.6 s. **9.** Approximately 4.3, 6.3, and 7.7 cm.

Exercises 11.2

1. $\{-2, 7\}$ 2. $\{-9, 2\}$
3. $\{-13, 5\}$ 4. $\{-13, 10\}$
5. $\left\{-1, \dfrac{1}{5}\right\}$ 6. $\left\{-1, \dfrac{1}{3}\right\}$
7. $\left\{\dfrac{3}{4}\right\}$ 8. $\left\{\dfrac{1}{2}, \dfrac{10}{3}\right\}$
9. $\{1 \pm \sqrt{6}\}$
10. $\{-3 \pm \sqrt{10}\}$
11. $\left\{\dfrac{-3 \pm 3\sqrt{13}}{2}\right\}$
12. $\{-2 \pm \sqrt{11}\}$
13. $\left\{\dfrac{3 \pm \sqrt{15}}{2}\right\}$
14. $\left\{\dfrac{3 \pm \sqrt{7}}{2}\right\}$
15. $\left\{\dfrac{1 \pm \sqrt{3}}{2}\right\}$
16. $\left\{\dfrac{1 \pm i\sqrt{3}}{4}\right\}$
17. $\left\{-\dfrac{2}{3}, 1\right\}$ 18. $\left\{1, \dfrac{3}{2}\right\}$
19. $\left\{\dfrac{1 \pm \sqrt{41}}{4}\right\}$
20. $\left\{-1, \dfrac{1}{3}\right\}$ 21. $\{1, 3\}$
22. $\left\{\dfrac{7 \pm \sqrt{37}}{2}\right\}$
23. $\{4\}$ 24. $\{-10, 3\}$
25. $\left\{\dfrac{1 \pm \sqrt{7}}{2}\right\}$
26. $\left\{\dfrac{3 \pm \sqrt{65}}{4}\right\}$
27. $\left\{\dfrac{1 \pm i\sqrt{2}}{3}\right\}$
28. $\left\{\dfrac{1 \pm \sqrt{11}}{5}\right\}$
29. $\left\{\dfrac{1 \pm \sqrt{161}}{8}\right\}$
30. $\left\{\dfrac{-1 \pm 5\sqrt{13}}{18}\right\}$
31. $\left\{\dfrac{3 \pm \sqrt{39}}{15}\right\}$
32. $\left\{\dfrac{-2 \pm i\sqrt{5}}{6}\right\}$
33. $\left\{\dfrac{-1 \pm \sqrt{41}}{2}\right\}$
34. $\left\{\dfrac{7 \pm \sqrt{93}}{2}\right\}$

In Exercises 1 to 8, solve each quadratic equation first by factoring and then by using the quadratic formula.

1. $x^2 - 5x - 14 = 0$
2. $x^2 + 7x - 18 = 0$
3. $t^2 + 8t - 65 = 0$
4. $q^2 + 3q - 130 = 0$
5. $5x^2 + 4x - 1 = 0$
6. $3x^2 + 2x - 1 = 0$
7. $16t^2 - 24t + 9 = 0$
8. $6m^2 - 23m + 10 = 0$

In Exercises 9 to 20, solve each quadratic equation by (a) completing the square and (b) using the quadratic formula.

9. $x^2 - 2x - 5 = 0$
10. $x^2 + 6x - 1 = 0$
11. $x^2 + 3x - 27 = 0$
12. $t^2 + 4t - 7 = 0$
13. $2x^2 - 6x - 3 = 0$
14. $2x^2 - 6x + 1 = 0$
15. $2q^2 - 2q - 1 = 0$
16. $4r^2 - 2r + 1 = 0$
17. $3x^2 - x - 2 = 0$
18. $2x^2 - 5x + 3 = 0$
19. $2y^2 - y - 5 = 0$
20. $3m^2 + 2m - 1 = 0$

In Exercises 21 to 42, solve each equation by using the quadratic formula.

21. $x^2 - 4x + 3 = 0$
22. $x^2 - 7x + 3 = 0$
23. $p^2 - 8p + 16 = 0$
24. $u^2 + 7u - 30 = 0$
25. $2x^2 - 2x - 3 = 0$
26. $2x^2 - 3x - 7 = 0$
27. $-3s^2 + 2s - 1 = 0$
28. $5t^2 - 2t - 2 = 0$

(*Hint*: Clear each of the following equations of fractions first, or remove grouping symbols, as needed.)

29. $2x^2 - \dfrac{1}{2}x - 5 = 0$
30. $3x^2 + \dfrac{1}{3}x - 3 = 0$
31. $5t^2 - 2t - \dfrac{2}{3} = 0$
32. $3y^2 + 2y + \dfrac{3}{4} = 0$
33. $(x - 2)(x + 3) = 4$
34. $(x + 1)(x - 8) = 3$

735

35. $(t + 1)(2t - 4) - 7 = 0$ **36.** $(2w + 1)(3w - 2) = 1$

37. $3x - 5 = \dfrac{1}{x}$ **38.** $x + 3 = \dfrac{1}{x}$

39. $2t - \dfrac{3}{t} = 3$ **40.** $4p - \dfrac{1}{p} = 6$

41. $\dfrac{5}{y^2} + \dfrac{2}{y} - 1 = 0$ **42.** $\dfrac{6}{x^2} - \dfrac{2}{x} = 1$

In Exercises 43 to 50, for each quadratic equation, find the value of the discriminant and give the number of real solutions.

43. $2x^2 - 5x = 0$ **44.** $3x^2 + 8x = 0$ **45.** $m^2 - 8m + 16 = 0$

46. $4p^2 + 12p + 9 = 0$ **47.** $3x^2 - 7x + 1 = 0$ **48.** $2x^2 - x + 5 = 0$

49. $2w^2 - 5w + 11 = 0$ **50.** $7q^2 - 3q + 1 = 0$

In Exercises 51 to 62, find all the solutions of each quadratic equation. Use any applicable method.

51. $x^2 - 8x + 16 = 0$ **52.** $4x^2 + 12x + 9 = 0$ **53.** $3t^2 - 7t + 1 = 0$

54. $2z^2 - z + 5 = 0$ **55.** $5y^2 - 2y = 0$ **56.** $7z^2 - 6z - 2 = 0$

57. $(x - 1)(2x + 7) = -6$ **58.** $4x^2 - 3 = 0$ **59.** $x^2 + 9 = 0$

60. $(4x - 5)(x + 2) = 1$ **61.** $x - 3 - \dfrac{10}{x} = 0$ **62.** $1 + \dfrac{2}{x} + \dfrac{2}{x^2} = 0$

The equation

$$h(t) = 112t - 16t^2$$

is the equation for the height of an arrow, shot upward from the ground with an initial velocity of 112 ft/s, where t is the time, in seconds, after the arrow leaves the ground. Use this information to solve Exercises 63 and 64. Your answers should be expressed to the nearest tenth of a second.

63. 1.2 or 5.8 s
64. 1.7 or 5.3 s
65. 1.4 s
66. 2.9 s
67. −9, −8 or 8, 9
68. 5, 6
69. 7 by 10 ft
70. 7 cm by 12 cm
71. 5 × 17 cm
72. 6 ft by 9 ft
73. 2.7 cm, 5.4 cm
74. 8.8 and 10.8 ft
75. 5.5, 6.5, 8.5 in.
76. 8.5 cm, 10.5 cm, 13.5 cm
77. 3.2 m, 21.8 m
78. 2.8 cm, 7.2 cm, 7.8 cm

63. Find the time it takes for the arrow to reach a height of 112 ft.

64. Find the time it takes for the arrow to reach a height of 144 ft.

The equation

$$h(t) = 320 - 32t - 16t^2$$

is the equation for the height of a ball, thrown downward from the top of a 320-ft building with an initial velocity of 32 ft/s, where t is the time after the ball is thrown down from the top of the building. Use this information to solve Exercises 65 and 66. Express your results to the nearest tenth of a second.

65. Find the time it takes for the ball to reach a height of 240 ft.

66. Find the time it takes for the ball to reach a height of 96 ft.

67. **Number problem.** The product of two consecutive integers is 72. What are the two integers?

68. **Number problem.** The sum of the squares of two consecutive whole numbers is 61. Find the two whole numbers.

69. **Rectangles.** The width of a rectangle is 3 ft less than its length. If the area of the rectangle is 70 ft², what are the dimensions of the rectangle?

70. **Rectangles.** The length of a rectangle is 5 cm more than its width. If the area of the rectangle is 84 cm², find the dimensions.

71. **Rectangles.** The length of a rectangle is 2 cm more than 3 times its width. If the area of the rectangle is 85 cm², find the dimensions of the rectangle.

72. **Rectangles.** If the length of a rectangle is 3 ft less than twice its width, and the area of the rectangle is 54 ft², what are the dimensions of the rectangle?

73. **Triangles.** One leg of a right triangle is twice the length of the other. The hypotenuse is 6 m long. Find the length of each leg.

74. **Triangles.** One leg of a right triangle is 2 ft longer than the shorter side. If the length of the hypotenuse is 14 ft, how long is each leg?

75. **Triangles.** One leg of a right triangle is 1 in. shorter than the other leg. The hypotenuse is 3 in. longer than the shorter side. Find the length of each side.

76. **Triangles.** The hypotenuse of a given right triangle is 5 cm longer than the shorter leg. The length of the shorter leg is 2 cm less than that of the longer leg. Find the lengths of the three sides.

77. **Triangles.** The sum of the lengths of the two legs of a right triangle is 25 m. The hypotenuse is 22 m long. Find the lengths of the two legs.

78. **Triangles.** The sum of the lengths of one side of a right triangle and the hypotenuse is 15 cm. The other leg is 5 cm shorter than the hypotenuse. Find the length of each side.

Chapter 11 ■ Quadratic Functions

79. (a). 4 s;
    (b). 1 s
80. (a) 6 s;
    (b) 4 s
81. 50 chairs
82. 40 appliances
83. 5.5 s
84. 1 s, 2 s
85. 5 s
86. 0.7 s and 4.3 s

79. Thrown ball. If a ball is thrown vertically upward from the ground, with an initial velocity of 64 ft/s, its height, h, after t s is given by $h(t) = 64t - 16t^2$

(a) How long does it take the ball to return to the ground? (*Hint*: Let $h(t) = 0$.)

(b) How long does it take the ball to reach a height of 48 ft on the way up?

80. Thrown ball. If a ball is thrown vertically upward from the ground, with an initial velocity of 96 ft/s, its height, h, after t s is given by $h(t) = 96t - 16t^2$

(a) How long does it take the ball to return to the ground?

(b) How long does it take the ball to pass through a height of 128 ft on the way back down to the ground?

81. Cost. Suppose that the cost $C(x)$, in dollars, of producing x chairs is given by

$$C(x) = 2400 - 40x + 2x^2$$

How many chairs can be produced for $5400?

82. Profit. Suppose that the profit $T(x)$, in dollars, of producing and selling x appliances is given by

$$T(x) = -3x^2 + 240x - 1800$$

How many appliances must be produced and sold to achieve a profit of $3000?

If a ball is thrown upward from the roof of a building 70 m tall with an initial velocity of 15 m/s, its approximate height, h, after t s is given by

$$h(t) = 70 + 15t - 5t^2$$

Note: The difference between this equation and the one we used in Example 8 has to do with the units used. When we used feet, the t^2 coefficient was -16 (from the fact that the acceleration due to gravity is approximately 32 ft/s²). When we use meters as the height, the t^2 coefficient is -5 (that same acceleration becomes approximately 10 m/s²). Use this information to solve Exercises 83 and 84.

83. Thrown ball. How long does it take the ball to fall back to the ground?

84. Thrown ball. When will the ball reach a height of 80 m?

Changing the initial velocity to 25 m/s will only change the t coefficient. Our new equation becomes

$$h(t) = 70 + 25t - 5t^2$$

85. Thrown ball. How long will it take the ball to return to the thrower?

86. Thrown ball. When will the ball reach a height of 85 m?

The only part of the height equation that we have not discussed is the constant. You have probably noticed that the constant is always equal to the initial height of the ball (70 m in our previous problems). Now, let's have *you* develop an equation.

87. $h(t) = 100 + 20t - 5t^2$
88. 6.9 s
89. 5 s
90. No
91. $h(t) = 100 + 20t - 16t^2$
92. 3.2 s
93. 1.9 s
95. 63 or 27
96. 28, 92
97. $0.94 or $17.06
98. $473.21 or $126.80

A ball is thrown upward from the roof of a 100-m building with an initial velocity of 20 m/s. Use this information to solve Exercises 87 to 90.

87. Thrown ball. Find the equation for the height, h, of the ball after t s.

88. Thrown ball. How long will it take the ball to fall back to the ground?

89. Thrown ball. When will the ball reach a height of 75 m?

90. Thrown ball. Will the ball ever reach a height of 125 m? (*Hint*: Check the discriminant.)

A ball is thrown upward from the roof of a 100-ft building with an initial velocity of 20 ft/s. Use this information to solve Exercises 91 to 94.

91. Thrown ball. Find the height, h, of the ball after t s.

92. Thrown ball. How long will it take the ball to fall back to the ground?

93. Thrown ball. When will the ball reach a height of 80 ft?

94. Thrown ball. Will the ball ever reach a height of 120 ft? Explain.

95. Profit. A small manufacturer's weekly profit in dollars is given by

$$P(x) = -3x^2 + 270x$$

Find the number of items x that must be produced to realize a profit of $5100.

96. Profit. Suppose the profit in dollars is given by

$$P(x) = -2x^2 + 240x$$

Now how many items must be sold to realize a profit of $5100?

97. Equilibrium price. The demand equation for a certain computer chip is given by

$$D = -2p + 14$$

The supply equation is predicted to be

$$S = -p^2 + 16p - 2$$

Find the equilibrium price.

98. Equilibrium price. The demand equation for a certain type of print is predicted to be

$$D = -200p + 36{,}000$$

The supply equation is predicted to be

$$S = -p^2 + 400p - 24{,}000$$

Find the equilibrium price.

101. $\{\pm\sqrt{z^2 - y^2}\}$

102. $\left\{\dfrac{\pm\sqrt{2}}{2yz}\right\}$

103. $\{-6a, 6a\}$

104. $\left\{\dfrac{\pm 3\, b\sqrt{a}}{a}\right\}$

105. $\left\{-3a, \dfrac{a}{2}\right\}$

106. $\left\{\dfrac{b}{3}, 5b\right\}$

107. $\left\{\dfrac{-a \pm a\sqrt{17}}{4}\right\}$

108. $\left\{\dfrac{b \pm b\sqrt{7}}{3}\right\}$

109. $\{-1 \pm \sqrt{6}\}$

110. $\left\{\dfrac{1 \pm \sqrt{3}}{2}\right\}$

111. $\left\{-1, \dfrac{1 \pm i\sqrt{3}}{2}\right\}$

112. $\left\{-\dfrac{1}{2} \pm \dfrac{i\sqrt{3}}{2}\right\}$

113. $\left\{-1, 1, \dfrac{1 \pm i\sqrt{3}}{2}, \dfrac{-1 \pm i\sqrt{3}}{2}\right\}$

114. $\{-2, 2, 1 \pm i\sqrt{3}, -1 \pm i\sqrt{3}\}$

115. (a) $\{-1.2, 4.2\}$;

(b) $-1.2, 4.2$

(c) The solutions to the quadratic equation are the x intercepts of the graph.

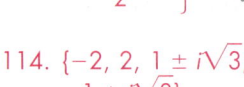

116. (a) $\{3, -1\}$;

(b) $3, -1$

(c) The solutions to the equation are the x coordinates of the points of intersection of the graphs of f and g.

99. Can the solution of a quadratic equation with integer coefficients include one real and one imaginary number? Justify your answer.

100. Explain how the discriminant is used to predict the nature of the solutions of a quadratic equation.

In Exercises 101 to 108, solve each equation for x.

101. $x^2 + y^2 = z^2$

102. $2x^2y^2z^2 = 1$

103. $x^2 - 36a^2 = 0$

104. $ax^2 - 9b^2 = 0$

105. $2x^2 + 5ax - 3a^2 = 0$

106. $3x^2 - 16bx + 5b^2 = 0$

107. $2x^2 + ax - 2a^2 = 0$

108. $3x^2 - 2bx - 2b^2 = 0$

109. Given that the polynomial $x^3 - 3x^2 - 15x + 25 = 0$ has as one of its solutions $x = 5$, find the other two solutions. (*Hint*: If you divide the given polynomial by $x - 5$ the quotient will be a quadratic equation. The remaining solutions will be the solutions for *that* equation.)

110. Given that $2x^3 + 2x^2 - 5x - 2 = 0$ has as one of its solutions $x = -2$, find the other two solutions. (*Hint*: In this case, divide the original polynomial by $x + 2$.)

111. Find all the zeros of the function $f(x) = x^3 + 1$.

112. Find the zeros of the function $f(x) = x^2 + x + 1$.

113. Find all six solutions to the equation $x^6 - 1 = 0$. (*Hint*: Factor the left-hand side of the equation first as the difference of squares, then as the sum and difference of cubes.)

114. Find all six solutions to $x^6 = 64$.

115. (a) Use the quadratic formula to solve $x^2 - 3x - 5 = 0$. For each solution give a decimal approximation to the nearest tenth.

(b) Graph the function $f(x) = x^2 - 3x - 5$ on your graphing calculator. Use a zoom utility and estimate the x intercepts to the nearest tenth.

(c) Describe the connection between parts a and b.

116. (a) Solve the following equation using any appropriate method:

$$x^2 - 2x = 3$$

(b) Graph the following functions on your graphing calculator:

$$f(x) = x^2 - 2x \quad \text{and} \quad g(x) = 3$$

Estimate the points of intersection of the graphs of f and g. In particular note the x coordinates of these points.

(c) Describe the connection between parts a and b.

SECTION 11.3 Solving Quadratic Equations by Graphing

11.3 OBJECTIVES

1. Find an axis of symmetry
2. Find a vertex
3. Graph a parabola
4. Solve quadratic equations by graphing
5. Solve an application involving a quadratic equation

In Section 3.3, we learned to graph a linear equation. We discovered that the graph of every linear equation was a straight line. In this section, we will consider the graph of a quadratic equation.

Consider an equation of the form

$$y = ax^2 + bx + c \qquad a \neq 0$$

This equation is quadratic in x and linear in y. Its graph will always be the curve called the parabola.

In an equation of the form

$$y = ax^2 + bx + c \qquad a \neq 0$$

the parabola opens upward or downward, as follows:

1. If $a > 0$, the parabola opens *upward*.
2. If $a < 0$, the parabola opens *downward*.

$$y = ax^2 + bx + c$$

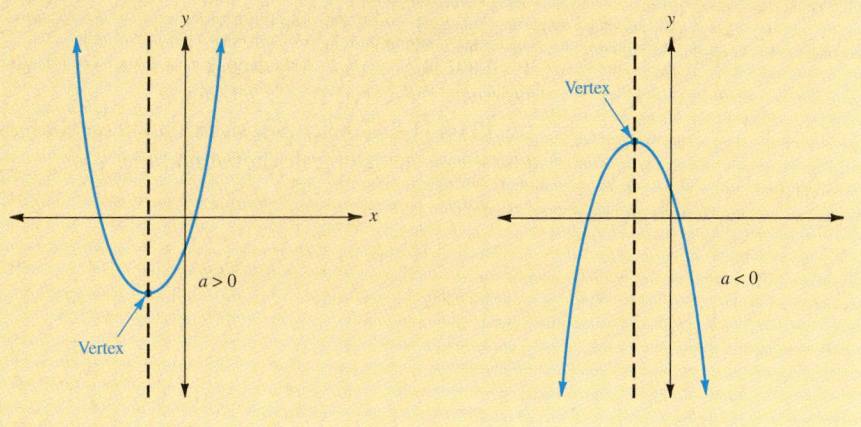

Two concepts regarding the parabola can be made by observation. Consider the above illustrations.

1. There is always a **minimum** (or lowest) point on the parabola if it opens upward. There is always a **maximum** (or highest) point on the parabola if it opens downward. In either case, that maximum or minimum value occurs at the **vertex** of the parabola.

2. Every parabola has an **axis of symmetry.** In the case of parabolas that open upward or downward, that axis of symmetry is a vertical line midway between any pair of symmetric points on the parabola. Also, the point where this axis of symmetry intersects the parabola is the vertex of the parabola.

The following figure summarizes these observations.

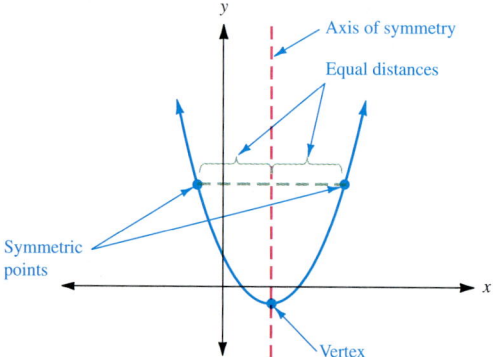

Our objective is to be able to quickly sketch a parabola. This can be done with *as few as three points* if those points are carefully chosen. For this purpose you will want to find the vertex and two symmetric points.

First, let's see how the coordinates of the vertex can be determined from the standard equation

$$y = ax^2 + bx + c \qquad (1)$$

In equation (1), if $x = 0$, then $y = c$, and so $(0, c)$ gives the point where the parabola intersects the y axis (the y intercept).

Look at the sketch. To determine the coordinates of the symmetric point (x_1, c), note that it lies along the horizontal line $y = c$.

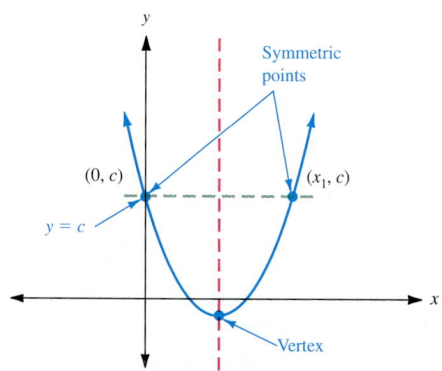

Therefore, let $y = c$ in equation (1):

$$c = ax^2 + bx + c$$
$$0 = ax^2 + bx$$
$$0 = x(ax + b)$$

and

$$x = 0 \quad \text{or} \quad x = -\frac{b}{a}$$

We now know that

$$(0, c) \quad \text{and} \quad \left(-\frac{b}{a}, c\right)$$

are the coordinates of the symmetric points shown. Since the axis of symmetry must be midway between these points, the x value along that axis is given by

$$x = \frac{0 + (-b/a)}{2} = -\frac{b}{2a} \tag{2}$$

Since the vertex for any parabola lies on the axis of symmetry, we can now state the following general result.

Vertex of a Parabola

If

$$y = ax^2 + bx + c \qquad a \neq 0$$

then the x coordinate of the vertex of the corresponding parabola is

$$x = -\frac{b}{2a}$$

Note: The y coordinate of the vertex can be found most easily by substituting the value found for x into the original equation.

We now know how to find the vertex of a parabola, and if two symmetric points can be determined, we are well on our way to the desired graph. Perhaps the simplest case is when the quadratic member of the given equation is factorable. In most cases, the two x intercepts will then give two symmetric points that are very easily found. Example 1 illustrates such a case.

Example 1

Graphing a Parabola

Graph the equation

$$y = x^2 + 2x - 8$$

First, find the axis of symmetry. In this equation, $a = 1$, $b = 2$, and $c = -8$. We then have

$$x = -\frac{b}{2a} = -\frac{2}{2 \cdot 1} = -\frac{2}{2} = -1$$

Sketch the information to help you solve the problem. Begin by drawing—as a dashed line—the axis of symmetry.

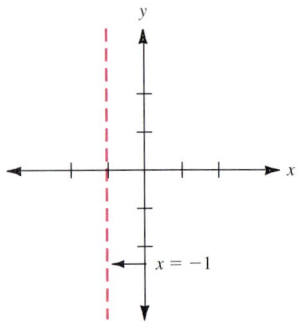

Thus, $x = -1$ is the axis of symmetry.

Second, find the vertex. Since the vertex of the parabola lies on the axis of symmetry, let $x = -1$ in the original equation. If $x = -1$,

$$y = (-1)^2 + 2(-1) - 8 = -9$$

and $(-1, -9)$ is the vertex of the parabola.

Third, find two symmetric points. Note that the quadratic member in this case is factorable, and so setting $y = 0$ in the original equation will quickly give two symmetric points (the x intercepts):

$$0 = x^2 + 2x - 8$$
$$= (x + 4)(x - 2)$$

At this point you can plot the vertex along the axis of symmetry.

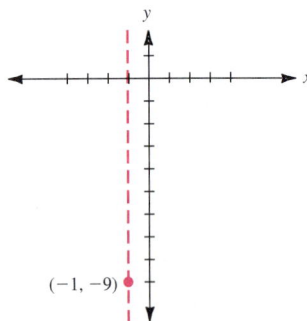

So when $y = 0$,

$$x + 4 = 0 \qquad \text{or} \qquad x - 2 = 0$$
$$x = -4 \qquad\qquad\qquad x = 2$$

and our x intercepts are $(-4, 0)$ and $(2, 0)$.

Fourth, draw a smooth curve connecting the points found above, to form the parabola.

Note: You can choose to find additional pairs of symmetric points at this time if necessary. For instance, the symmetric points $(0, -8)$ and $(-2, -8)$ are easily located.

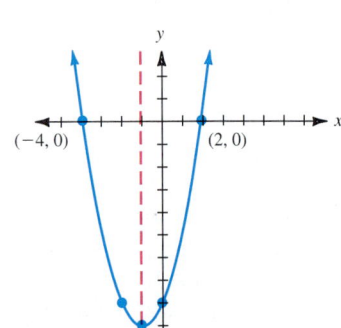

✓ CHECK YOURSELF 1

Graph the equation

$$y = -x^2 - 2x + 3$$

(*Hint*: Since the coefficient of x^2 is negative, the parabola opens downward.)

Section 11.3 ▪ Solving Quadratic Equations by Graphing   **745**

A similar process will work if the quadratic member of the given equation is *not* factorable. In that case, one of two things happens:

1. The *x* intercepts are irrational and therefore not particularly helpful in the graphing process.
2. The *x* intercepts do not exist.

Consider Example 2.

Example 2 — Graphing a Parabola

Graph the function

$$f(x) = x^2 - 6x + 3$$

First, find the axis of symmetry. Here $a = 1$, $b = -6$, and $c = 3$. So

$$x = -\frac{b}{2a} = \frac{-(-6)}{2 \cdot 1} = \frac{6}{2} = 3$$

Thus, $x = 3$ is the axis of symmetry.
Second, find the vertex. If $x = 3$,

$$f(3) = 3^2 - 6 \cdot 3 + 3 = -6$$

and $(3, -6)$ is the vertex of the desired parabola.

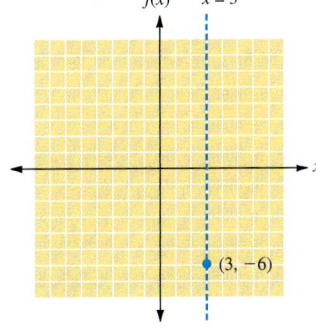

Third, find two symmetric points. Here the quadratic member is not factorable, so we need to find another pair of symmetric points.

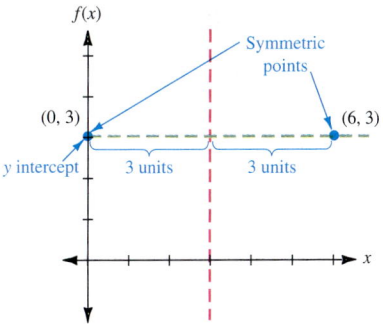

Note that $(0, 3)$ is the *y* intercept of the parabola. We found the axis of symmetry at $x = 3$ in step 1. Note that the symmetric point to $(0, 3)$ lies along the horizontal line through the *y* intercept at the same distance (3 units) from the axis of symmetry. Hence, $(6, 3)$ is our symmetric point.

Fourth, draw a smooth curve connecting the points found above to form the parabola.

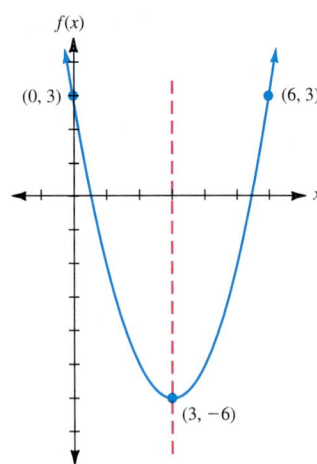

Note: An alternate method is available in step 3. Observing that 3 is the y intercept and that the symmetric point lies along the line $y = 3$, set $f(x) = 3$ in the original equation:

$$3 = x^2 - 6x + 3$$
$$0 = x^2 - 6x$$
$$0 = x(x - 6)$$

so

$$x = 0 \quad \text{or} \quad x - 6 = 0$$
$$x = 6$$

and (0, 3) and (6, 3) are the desired symmetric points.

 CHECK YOURSELF 2

Graph the function

$$f(x) = x^2 + 4x + 5$$

Thus far the coefficient of x^2 has been 1 or -1. The following example shows the effect of different coefficients on the shape of the graph of a quadratic equation.

Example 3 **Graphing a Parabola**

Graph the equation

$$y = 3x^2 - 6x + 5$$

First, find the axis of symmetry.

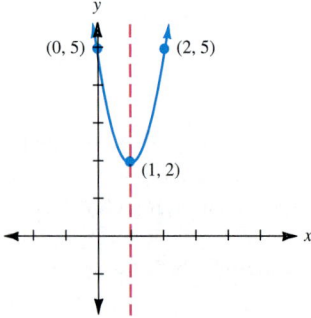

$$x = -\frac{b}{2a} = \frac{-(-6)}{2 \cdot 3} = \frac{6}{6} = 1$$

Second, find the vertex. If $x = 1$,

$$y = 3(1)^2 - 6 \cdot 1 + 5 = 2$$

So (1, 2) is the vertex.

Third, find symmetric points. Again the quadratic member is not factorable, and we use the y intercept (0, 5) and its symmetric point (2, 5).

Fourth, connect the points with a smooth curve to form the parabola. Compare this curve to those in previous examples. Note that the parabola is "tighter" about the axis of symmetry. That is the effect of the larger x^2 coefficient.

 CHECK YOURSELF 3

Graph the equation

$$y = \frac{1}{2}x^2 - 3x - 1$$

The following algorithm summarizes our work thus far in this section.

To Graph a Parabola

Step 1 Find the axis of symmetry.
Step 2 Find the vertex.
Step 3 Determine two symmetric points.
   Note: You can use the x intercepts if the quadratic member of the given equation is factorable. Otherwise use the y intercept and its symmetric point.
Step 4 Draw a smooth curve connecting the points found above to form the parabola. You may choose to find additional pairs of symmetric points at this time.

We have seen that quadratic equations can be solved in three different ways: by factoring (Section 6.5), by completing the square (Section 11.1), or by using the quadratic formula (Section 11.2). We will now look at a fourth technique for solving quadratic equations, a graphical method. Unlike the other methods, the graphical technique may yield only an approximation of the solution(s). On the other hand, it can be a very useful method for checking the reasonableness of exact answers. This is particularly true when technology is used to produce the graph.

We can easily use a graph to identify the number of positive and negative solutions to a quadratic equation. Recall that solutions to the equation

$$0 = ax^2 + bx + c$$

are values for x that make the statement true.

If we have the graph of the parabola

$$y = ax^2 + bx + c$$

then values for which y is equal to zero are solutions to the original equation. We know that $y = 0$ is the equation for the x axis. Solutions for the equation exist wherever the graph crosses the x axis.

Example 4　Identifying the Number of Solutions to a Quadratic Equation

Find the number of positive solutions and the number of negative solutions for each quadratic equation.

(a) $0 = x^2 + 3x - 7$

Using either the methods of this section, or a graphing device, we find the graph of the parabola $y = x^2 + 3x - 7$ first. The axis of symmetry is $x = -\frac{3}{2}$. Two points on the graph could be $(0, -7)$ and $(-3, -7)$.

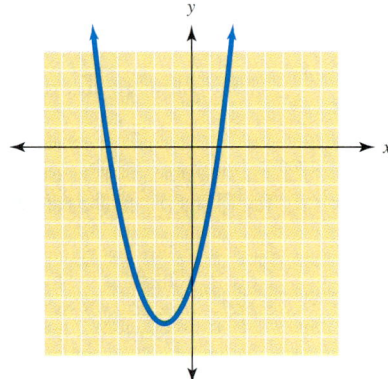

Looking at the graph, we see that the parabola crosses the x axis twice, once to the left of zero and once to the right. There are two real solutions. One is negative and one is positive. Note that due to the basic shape of a parabola, it is not possible for this graph to cross the x axis more than two times.

(b) $0 = 5x^2 - 32x$

Again, we graph the parabola $y = 5x^2 - 32x$ first. The line of symmetry is $x = \frac{16}{5}$. Two points on the parabola are $(0, 0)$ and $\left(\frac{32}{5}, 0\right)$.

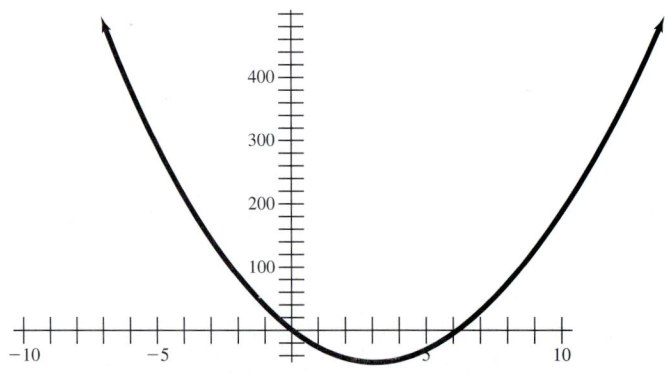

There are two solutions. One is positive and the other appears to be zero. A quick check of the equation confirms that zero is a solution.

(c) $0 = x^2 + 2x + 3$

The axis of symmetry is $x = -1$. Two points on the parabola would be $(0, 3)$ and $(-2, 3)$. Here is the graph of the related parabola.

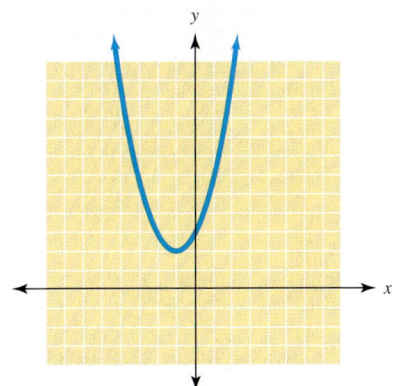

Notice that the graph never touches the x axis. There are no real solutions to the equation $0 = x^2 + 2x + 3$.

✓ CHECK YOURSELF 4

Use the graph of the related parabola to find the number of real solutions to each equation. Identify the number of solutions that are positive and the number that are negative.

(a) $0 = 3x^2 - 16x$ (b) $0 = x^2 + x + 4$ (c) $0 = x^2 - 3x - 9$

We used the x intercepts of the graph to determine the number of solutions for each equation in Example 4. The value of the x coordinate for each intercept is the exact solution. In the next example, we'll use the graphical method to estimate the solutions of a quadratic equation.

Example 5 Solving a Quadratic Equation

Solve each equation.

(a) $0 = x^2 + 3x - 7$

As we did in Example 4, we find the graph of the parabola $y = x^2 + 3x - 7$ first.

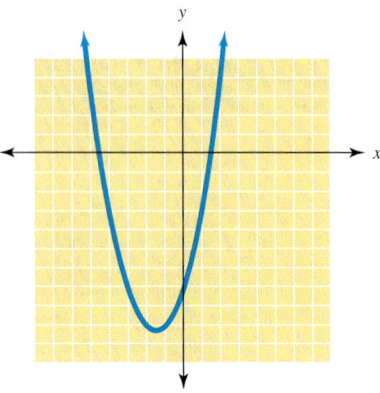

In Example 4, we pointed out that there are two solutions. One solution is between -5 and -4. The other is between 1 and 2. The precision with which we estimate the values depends on the quality of the graph we are examining. From this graph, -4.5 and 1.5 would be reasonable estimates of the solutions.

Using the quadratic formula, we find that the actual solutions are $\dfrac{(-3 - \sqrt{37})}{2}$ and $\dfrac{(-3 + \sqrt{37})}{2}$. Using a calculator, we can round these answers to the nearest hundredth. We get 1.54 and -4.54. Our estimates were quite reasonable.

(b) $0 = 5x^2 - 32x$

Again, we graph the parabola $y = 5x^2 - 32x$ first.

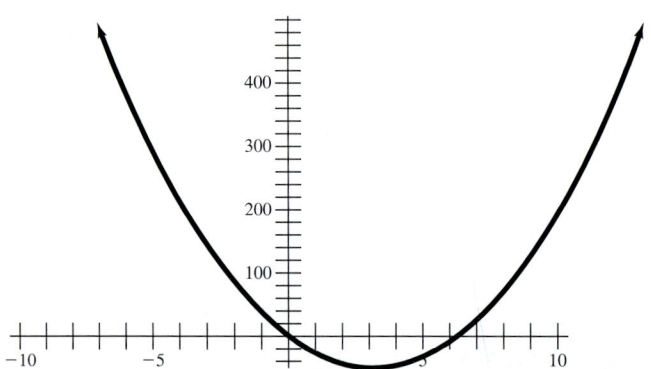

There are two solutions. One is a little more than six and the other appears to be zero. Using the quadratic formula (or factoring) we find that the solutions to this equation are 0 and 6.4.

Section 11.3 ■ Solving Quadratic Equations by Graphing

✓ **CHECK YOURSELF 5**

Use the graph of the related parabola to estimate the solutions to each equation.

(a) $0 = 3x^2 - 16x$ (b) $0 = x^2 - 3x - 9$

From graphs of equations of the form $y = ax^2 + bx + c$, we know that if $a > 0$, then the vertex is the lowest point on the graph (the minimum value). Also, if $a < 0$, then the vertex is the highest point on the graph (the maximum value). We can use this result to solve a variety of problems in which we want to find the maximum or minimum value of a variable. The following are just two of many typical examples.

Example 6

An Application Involving a Quadratic Function

A software company sells a word processing program for personal computers. They have found that their monthly profit in dollars, P, from selling x copies of the program is approximated by

$$P(x) = -0.3x^2 + 90x - 1500$$

Find the number of copies of the program that should be sold in order to maximize the profit.

Since the relating equation is quadratic, the graph must be a parabola. Also since the coefficient of x^2 is negative, the parabola must open downward, and thus the vertex will give the maximum value for the profit, P. To find the vertex,

$$x = -\frac{b}{2a} = \frac{-90}{2(-0.3)} = \frac{-90}{-0.6} = 150$$

The maximum profit must then occur when $x = 150$, and we substitute that value into the original equation:

$$P(x) = -0.3(150)^2 + (90)(150) - 1500$$
$$= \$5250$$

The maximum profit will occur when 150 copies are sold per month, and that profit will be $5250.

✓ CHECK YOURSELF 6

A company that sells portable radios finds that its weekly profit in dollars, P, and the number of radios sold, x, are related by

$$P(x) = -0.2x^2 + 40x - 100$$

Find the number of radios that should be sold to have the largest weekly profit. Also find the amount of that profit.

Example 7 An Application Involving a Quadratic Function

A farmer has 3600 ft of fence and wishes to enclose the largest possible rectangular area with that fencing. Find the largest possible area that can be enclosed.

As usual, when dealing with geometric figures, we start by drawing a sketch of the problem.

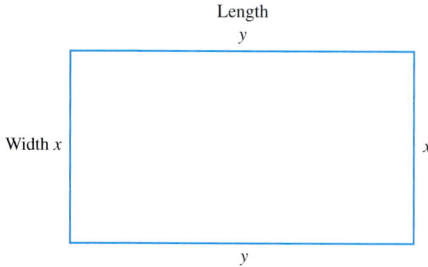

First, we can write the area, A, as

Area = length × width

$$A = xy \qquad (3)$$

Also, since 3600 ft of fence is to be used, we know that

The perimeter of the region is
$2x + 2y$

$$2x + 2y = 3600$$

$$2y = 3600 - 2x \qquad (4)$$

$$y = 1800 - x$$

Substituting for y in equation (3), we have

$$A(x) = x(1800 - x)$$
$$= 1800x - x^2$$
$$= -x^2 + 1800x \qquad (5)$$

Again, the graph for A is a parabola opening downward, and the largest possible area will occur at the vertex. As before,

$$x = \frac{-1800}{2(-1)} = \frac{-1800}{-2} = 900$$

The width x is 900 ft. Since from equation (2)
$$y = 1800 - 900$$
$$= 900 \text{ ft}$$

and the largest possible area is

$$A(x) = -(900)^2 + 1800(900) = 810,000 \text{ ft}^2$$

The length is also 900 ft. The desired region is a square.

✓ CHECK YOURSELF 7

We want to enclose three sides of the largest possible rectangular area by using 900 ft of fence. Assume that an existing wall makes the fourth side. What will be the dimensions of the rectangle?

✓ CHECK YOURSELF ANSWERS

1.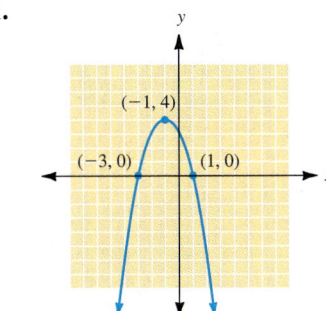

2.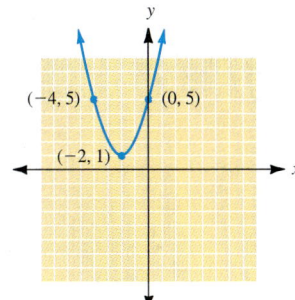

3.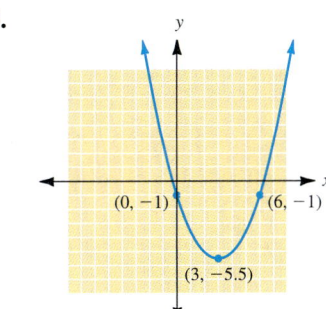

4. (a) Two solutions; one positive, one zero; (b) no real solutions; (c) two solutions; one negative, one positive.   **5.** (a) 0, 5.3; (b) -1.9, 4.9.
6. 100 radios, $1900.   **7.** Width 225 ft, length 450 ft.

Electric Power Costs. You and a partner in a small engineering firm have been asked to help calculate the costs involved in running electric power a distance of 3 miles from a public utility company to a nearby community. You have decided to consider three tower heights: 75 feet, 100 feet, and 120 feet. The 75-ft towers cost $850 each to build and install; the 100-ft towers cost $1110 each; and the 120-ft towers cost $1305 each. Spans between towers typically run 300 to 1200 ft. The conducting wires weigh 1900 pounds per 1000 ft, and the tension on the wires is 6000 pounds. To be safe, the conducting wires should never be closer than 45 ft to the ground.

Work with a partner to complete the following.

1. Develop a plan for the community showing all three scenarios and their costs. Remember that hot weather will expand the wires and cause them to sag more than normally. Therefore, allow for about a 10% margin of error when you calculate the sag amount.

2. Write an accompanying letter to the town council explaining your recommendation. Be sure to include any equations you used to help you make your decision. Remember to look at the formula at the beginning of the chapter, which gives the amount of sag for wires given a certain weight per foot, span length, and tension on the conductors.

Exercises 11.3

In Exercises 1 to 8, match each graph with one of the equations at the left.

(a) $y = x^2 + 2$
(b) $y = 2x^2 - 1$
(c) $y = 2x + 1$
(d) $y = x^2 - 3x$
(e) $y = -x^2 - 4x$
(f) $y = -2x + 1$
(g) $y = x^2 + 2x - 3$
(h) $y = -x^2 + 6x - 8$

1. f
2. d
3. g
4. b
5. h
6. e
7. c
8. a

1.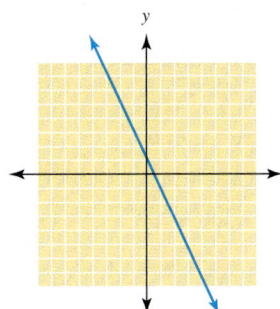

2.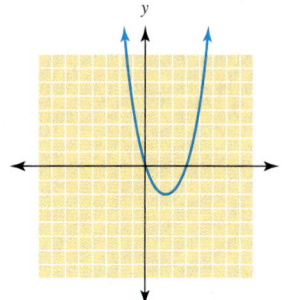

3.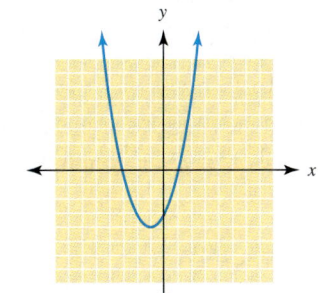

4.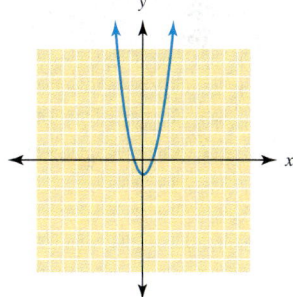

5.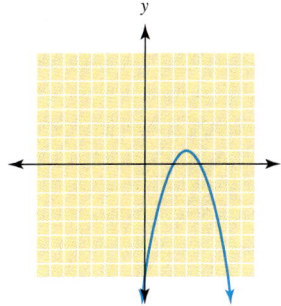

6.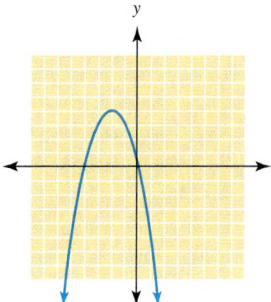

7.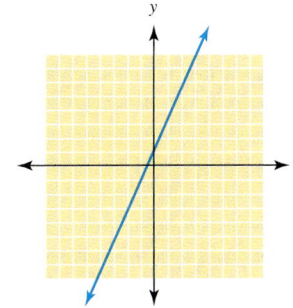

8.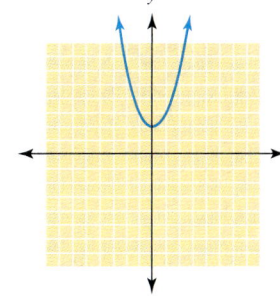

9. (a), (c)

10. (b), (c)

11. (a), (c)

12. (a), (e)

13. (b), (c)

14. (a), (d)

15. $x = 2$; vertex: $(2, -4)$; $(0, 0)$ and $(4, 0)$

16. $x = 0$; vertex: $(0, -1)$; $(-1, 0)$ and $(1, 0)$

17. $x = 0$; vertex: $(0, 4)$; $(-2, 0)$ and $(2, 0)$

18. $x = -1$; vertex: $(-1, -1)$; $(0, 0)$ and $(-2, 0)$

19. $x = -1$; vertex: $(-1, 1)$; $(-2, 0)$ and $(0, 0)$

20. $x = -\frac{3}{2}$; vertex: $\left(-\frac{3}{2}, \frac{9}{4}\right)$; $(0, 0)$ and $(-3, 0)$

21. $x = 3$; vertex: $(3, -4)$; $(1, 0)$ and $(5, 0)$

22. $x = -\frac{1}{2}$; vertex: $\left(-\frac{1}{2}, -\frac{25}{4}\right)$; $(-3, 0)$ and $(2, 0)$

23. $x = \frac{5}{2}$; vertex: $\left(\frac{5}{2}, -\frac{1}{4}\right)$; $(3, 0)$ and $(2, 0)$

24. $x = -3$; vertex: $(-3, -4)$; $(-5, 0)$ and $(-1, 0)$

25. $x = 3$; vertex: $(3, -1)$; $(2, 0)$ and $(4, 0)$

26. $x = -\frac{3}{2}$; vertex: $\left(-\frac{3}{2}, \frac{25}{4}\right)$; $(-4, 0)$ and $(1, 0)$

27. $x = -3$; vertex: $(-3, 4)$; $(-5, 0)$ and $(-1, 0)$

28. $x = 3$; vertex: $(3, 1)$; $(2, 0)$ and $(4, 0)$

In Exercises 9 to 14, which of the given conditions apply to the graphs of the following equations? Note that more than one condition may apply.

(a) The parabola opens upward.

(b) The parabola opens downward.

(c) The parabola has two x intercepts.

(d) The parabola has one x intercept.

(e) The parabola has no x intercept.

9. $y = x^2 - 3$

10. $y = -x^2 + 4x$

11. $y = x^2 - 3x - 4$

12. $y = x^2 - 2x + 2$

13. $y = -x^2 - 3x + 10$

14. $y = x^2 - 8x + 16$

In Exercises 15 to 28, find the equation of the axis of symmetry, the coordinates of the vertex, and the x intercepts. Sketch the graph of each equation.

15. $y = x^2 - 4x$

16. $y = x^2 - 1$

17. $y = -x^2 + 4$

18. $y = x^2 + 2x$

19. $y = -x^2 - 2x$

20. $y = -x^2 - 3x$

21. $y = x^2 - 6x + 5$

22. $y = x^2 + x - 6$

23. $y = x^2 - 5x + 6$

24. $y = x^2 + 6x + 5$

25. $y = x^2 - 6x + 8$

26. $y = -x^2 - 3x + 4$

27. $y = -x^2 - 6x - 5$

28. $y = -x^2 + 6x - 8$

29. $x = 1$; vertex: $(1, -2)$
30. $x = -2$; vertex: $(-2, 2)$
31. $x = 2$; vertex: $(2, -5)$
32. $x = 3$; vertex: $(3, 4)$
33. $x = \frac{3}{2}$; vertex: $\left(\frac{3}{2}, -\frac{3}{4}\right)$
34. $x = -\frac{5}{2}$; vertex: $\left(-\frac{5}{2}, -\frac{13}{4}\right)$
35. $x = -1$; vertex: $(-1, -3)$
36. $x = 1$; vertex: $\left(1, -\frac{3}{2}\right)$
37. $x = \frac{3}{2}$; vertex: $\left(\frac{3}{2}, -\frac{9}{4}\right)$
38. $x = -1$; vertex: $(-1, 1)$
39. $x = -2$; vertex: $(-2, -7)$
40. $x = 1$; vertex: $(1, 4)$
41. 2; 1 positive, 1 negative
42. 2; 2 negative
43. 2; 1 positive
44. 2; 1 positive
45. None
46. 2; 2 negative
47. 2; 1 positive, 1 negative
48. 2; 2 negative
49. 2; 1 positive, 1 negative
50. 2; 2 positive
51. $\{-4, 3\}$
52. $\{-2, -1\}$
53. $\{0, 3.2\}$
54. $\{0, 2.1\}$
55. $\{0.4, -1.8\}$
56. $\{-0.7, -2.3\}$
57. $\{1.8, -3.8\}$
58. $\{1.8, 6.2\}$

In Exercises 29 to 40, find the equation of the axis of symmetry, the coordinates of the vertex, and at least two symmetric points. Sketch the graph of each equation.

29. $y = x^2 - 2x - 1$
30. $y = x^2 + 4x + 6$
31. $y = x^2 - 4x - 1$
32. $y = -x^2 + 6x - 5$
33. $y = -x^2 + 3x - 3$
34. $y = x^2 + 5x + 3$
35. $y = 2x^2 + 4x - 1$
36. $y = \frac{1}{2}x^2 - x - 1$
37. $y = -\frac{1}{3}x^2 + x - 3$
38. $y = -2x^2 - 4x - 1$
39. $y = 3x^2 + 12x + 5$
40. $y = -3x^2 + 6x + 1$

In Exercises 41 to 50, use the graph of the related parabola to find the number of real solutions to each equation. Identify the number of solutions that are positive and the number that are negative.

41. $0 = x^2 + x - 12$
42. $0 = x^2 + 3x + 2$
43. $0 = 6x^2 - 19x$
44. $0 = 7x^2 - 15x$
45. $0 = x^2 - 2x + 5$
46. $0 = x^2 + 6x + 4$
47. $0 = 9x^2 + 12x - 7$
48. $0 = 3x^2 + 9x + 5$
49. $0 = x^2 + 2x - 7$
50. $0 = x^2 - 8x + 11$

In Exercises 51 to 58, use the graph of the related parabola to estimate the solutions to each equation. Round answers to the nearest tenth.

51. $0 = x^2 + x - 12$
52. $0 = x^2 + 3x + 2$
53. $0 = 6x^2 - 19x$
54. $0 = 7x^2 - 15x$
55. $0 = 9x^2 + 12x - 7$
56. $0 = 3x^2 + 9x + 5$
57. $0 = x^2 + 2x - 7$
58. $0 = x^2 - 8x + 11$

758 Chapter 11 ■ Quadratic Functions

59. 100 items, $2600

60. 125 items, $2325

61. 500 ft by 500 ft; 250,000 sq ft

62. 400 ft by 800 ft; 320,000 sq ft

63. 144 ft

64. 64 ft

65. 88 ft/s

66. (a) (3, 1); (b) $y \geq 1$

67. (a) (−4, 2); (b) $y \geq 2$

68. (a) (1, 2); (b) $y \leq 2$

69. (a)(−2, −1); (b) $y \leq -1$

70. (a)(−1, −2); $y \geq -2$

71. (a) (4, 0); (b) $y \leq 0$

73. $-3 \leq x \leq 3$; $-25 \leq y \leq 0$

74. $-1 \leq x \leq 2$; $-8 \leq y \leq 0$

75. $-2 \leq x \leq 4$; $-10 \leq y \leq 0$

76. $-2 \leq x \leq 2$; $0 \leq y \leq 8$

---

59. Profit. A company's weekly profit, P, is related to the number of items sold by $P(x) = -0.3x^2 + 60x - 400$. Find the number of items that should be sold each week in order to maximize the profit. Then find the amount of that weekly profit.

60. Profit. A company's monthly profit, P, is related to the number of items sold by $P(x) = -0.2x^2 + 50x - 800$. How many items should be sold each month to obtain the largest possible profit? What is the amount of that profit?

61. Area. A builder wants to enclose the largest possible rectangular area with 2000 ft of fencing. What should be the dimensions of the rectangle, and what will the area of the rectangle be?

62. Area. A farmer wants to enclose a rectangular area along a river on three sides. If 1600 ft of fencing is to be used, what dimensions will give the maximum enclosed area? Find that maximum area.

63. Motion. A ball is thrown upward into the air with an initial velocity of 96 ft/s. If h gives the height of the ball at time t, then the equation relating h and t is

$$h(t) = -16t^2 + 96t$$

Find the maximum height the ball will attain.

64. Motion. A ball is thrown upward into the air with an initial velocity of 64 ft/s. If h gives the height of the ball at time t, then the equation relating h and t is

$$h(t) = -16t^2 + 64t$$

Find the maximum height the ball will attain.

65. Motion. A ball is thrown upward with an initial velocity v. After 3 s, it attains a height of 120 ft. Find the initial velocity using the equation

$$h(t) = -16t^2 + v \cdot t$$

For each of the following quadratic functions, use your graphing calculator to determine (a) the vertex of the parabola and (b) the range of the function.

66. $f(x) = 2(x - 3)^2 + 1$

67. $g(x) = 3(x + 4)^2 + 2$

68. $f(x) = -(x - 1)^2 + 2$

69. $g(x) = -(x + 2)^2 - 1$

70. $f(x) = 3(x + 1)^2 - 2$

71. $g(x) = -2(x - 4)^2$

72. Explain how to determine the domain and range of the function $f(x) = a(x - h)^2 + k$.

In Exercises 73 to 76, describe a viewing window that would include the vertex and all intercepts for the graph of each function.

73. $f(x) = 3x^2 - 25$

74. $f(x) = 9x^2 - 5x - 7$

75. $f(x) = -2x^2 + 5x - 7$

76. $f(x) = -5x^2 + 2x + 7$

SECTION 11.4 Solving Quadratic Inequalities

11.4 OBJECTIVES

1. Solve a quadratic inequality graphically
2. Solve a quadratic inequality algebraically

A **quadratic inequality** is an inequality that can be written in the form

$$ax^2 + bx + c < 0 \qquad \text{where } a \neq 0$$

Note that the inequality symbol $<$ can be replaced by the symbol $>$, $\leq$, or $\geq$ in the above definition.

In Chapter 8, solutions to linear inequalities such as $4x + 2 < 0$ were analyzed graphically. Recall that, given the graph of the function $f(x) = 4x + 2$

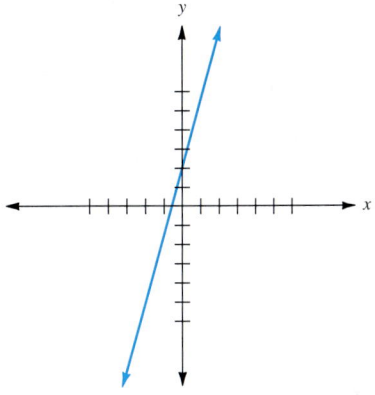

the solution to the inequality was the set of all x values associated with points on the line that were *below* the x axis.

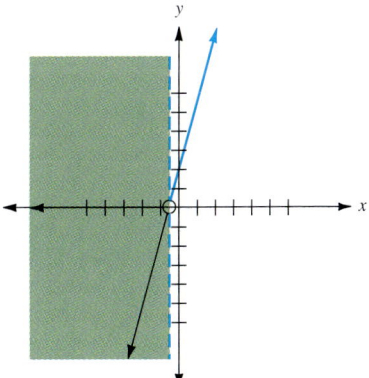

In this case, each x to the left of $-\dfrac{1}{2}$ was associated with a point on the line that was below the x axis. The solution set for the inequality $4x + 2 < 0$ is the set $\left\{x \mid x < -\dfrac{1}{2}\right\}$.

759

Example 1

Solving a Quadratic Inequality Graphically

Solve the inequality

$$x^2 - x - 12 \leq 0$$

First, use the techniques in Section 11.3 to graph the function $f(x) = x^2 - x - 12$.

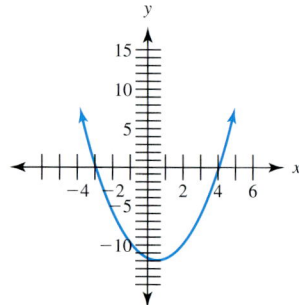

Next, looking at the graph, determine the values of x that make $x^2 - x - 12 < 0$ a true statement. Notice that the graph is below the x axis for values of x between -3 and 4. The graph intercepts the x axis when x is -3 or 4. The solution set for the inequality is $\{x | -3 \leq x \leq 4\}$.

✓ CHECK YOURSELF 1

Use a graph to solve the inequality

$$x^2 - 3x - 10 \geq 0$$

Algebraic methods can also be used to find the exact solution to a quadratic inequality. Subsequent examples in this section will discuss algebraic solutions. When solving an equation or inequality algebraically, it is always a good idea to compare the algebraic solution to a graphical one to ensure that the algebraic solution is reasonable.

Example 2 — Solving a Quadratic Inequality Algebraically

Solve $(x - 3)(x + 1) < 0$.

We start by finding the solutions of the corresponding quadratic equation. So

$$(x - 3)(x + 1) = 0$$

has solutions 3 and -1, called the *critical values*. In fact, the points $(3, 0)$ and $(-1, 0)$ are the points where the graph of the parabola we see in the box to the left passes through the x axis.

Our solution process depends on determining where each factor is positive or negative. To help visualize that process, we start with a number line and label it as shown below. We begin with our first critical point of -1.

$x + 1$ is negative if x is less than -1. $x + 1$ is positive if x is greater than -1.

We now continue in the same manner with the second critical point, 3.

$x - 3$ is negative if x is less than 3. $x - 3$ is positive if x is greater than 3.

In practice, we combine the two steps above for the following result.

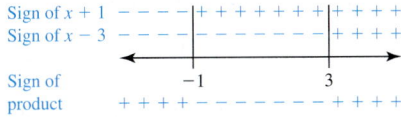

Examining the signs of the factors, we see that in this case,

For any x less than -1, both factors are negative and the product is positive.

For any x between -1 and 3, the factors have opposite signs and the product is negative.

For any x greater than 3, both factors are positive and the product is again positive.

If we expand the left side of statement $(x - 3)(x + 1) < 0$, we get

$$x^2 - 2x - 3 < 0$$

Looking at the graph of

$$y = x^2 - 2x - 3,$$

where is y less than 0 on the graph?

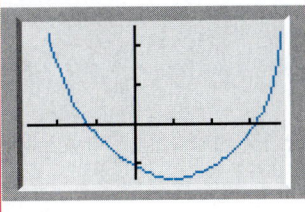

The product of the two binomials must be negative.

We return to the original inequality:

$$(x - 3)(x + 1) < 0$$

We can see that this is true only between -1 and 3. In set notation, the solution can be written as

$$\{x | -1 < x < 3\}$$

On a number line, the graph of the solution is

✓ CHECK YOURSELF 2

Solve and graph the solution set.

$$(x - 2)(x + 4) < 0$$

We now consider an example in which the quadratic member of the inequality must be factored.

Example 3

Solving a Quadratic Inequality Algebraically

Solve $x^2 - 5x + 4 > 0$.

Factoring the quadratic member, we have

$$(x - 1)(x - 4) > 0$$

The critical values are 1 and 4, and we form the sign graph as before.

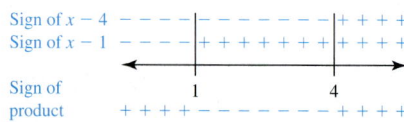

In this case, we want those values of x for which the product is *positive,* and we can see from the sign graph above that the solution is

$$\{x | x < 1 \text{ or } x > 4\}$$

The graph of the solution set is shown below.

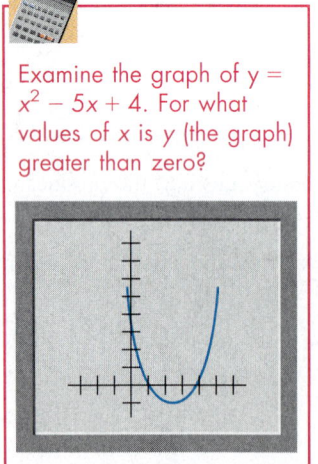

Examine the graph of $y = x^2 - 5x + 4$. For what values of x is y (the graph) greater than zero?

✓ CHECK YOURSELF 3

Solve and graph the solution set.

$$2x^2 - x - 3 > 0$$

The method used in the previous examples works *only* when one side of the inequality is factorable and the other is 0. It is sometimes necessary to rewrite the inequality in an equivalent form in order to attain that form, as Example 4 illustrates.

Example 4 — Solving a Quadratic Inequality Algebraically

Solve $(x + 1)(x - 4) \geq 6$.

First, we multiply to clear the parentheses.

$$x^2 - 3x - 4 \geq 6$$

> Use a calculator to graph both $f(x) = x^2 - 3x - 4$ and $g(x) = 6$. Where is $f(x)$ above $g(x)$? Compare this to the algebraic solution.

Now we subtract 6 from both sides so that the inequality is *related to* 0:

$$x^2 - 3x - 10 \geq 0$$

Factoring the quadratic member, we have

$$(x - 5)(x + 2) \geq 0$$

We can now proceed with the sign graph method as before.

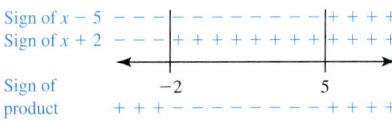

> Both factors are negative if x is less than -2. Both factors are positive if x is greater than 5.

From the graph we see that the solution is

$$\{x \mid x \leq -2 \text{ or } x \geq 5\}$$

The graph is shown below.

✓ CHECK YOURSELF 4

Solve and graph the solution set.

$$(x - 5)(x + 7) \leq -11$$

✓ CHECK YOURSELF ANSWERS

1. $\{x | x \leq -2 \text{ or } x \geq 5\}$. 2. $\{x | -4 < x < 2\}$.

3. $\{x | x < -1 \text{ or } x > \dfrac{3}{2}\}$.

4. $\{x | -6 \leq x \leq 4\}$.

Trajectory and Height. So far in this chapter you have done many exercises involving balls that have been thrown upward with varying velocities. How does trajectory—the angle at which the ball is thrown—affect its height and time in the air?

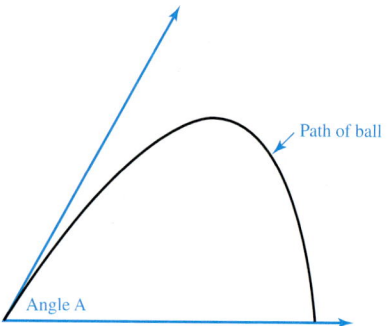

If you throw a ball from ground level with an initial upward velocity of 70 ft per second, the equation

$$h = -16t^2 + \partial(70)t$$

gives you the height, h, in feet t seconds after the ball has been thrown at a certain angle, A. The value of ∂ in the equation depends on the angle A and is given in the accompanying table.

Measure of Angle A in Degrees	Value of ∂
0	0.000
5	0.087
10	0.174
15	0.259
20	0.342
25	0.423
30	0.500
35	0.574
40	0.643
45	0.707
50	0.766
55	0.819
60	0.866
65	0.906
70	0.940
75	0.966
80	0.985
85	0.996
90 (straight up)	1.000

Investigate the following questions and write your conclusions to each one in complete sentences, showing all charts and graphs. Indicate what initial velocity you are using in each case.

1. Suppose an object is thrown from the ground level with an initial upward velocity of 70 ft/s and at an angle of 45 degrees. What will be the height of the ball in 1 s (nearest tenth of a foot)? How long is the ball in the air?

2. Does the ball stay in the air longer if the angle of the throw is greater?

3. If you double the angle of the throw, will the ball stay in the air double the length of time?

4. If you double the angle of the throw, will the ball go twice as high?

5. Is the height of the ball directly related to the angle at which you throw it? That is, does the ball go higher if the angle of the throw is larger?

6. Repeat this exercise using another initial upward velocity.

Exercises · 11.4

1. $\{x \mid -4 < x < 3\}$
2. $\{x \mid x < -5 \text{ or } x > 2\}$
3. $\{x \mid x < -4 \text{ or } x > 3\}$
4. $\{x \mid -5 < x < 2\}$
5. $\{x \mid -4 \leq x \leq 3\}$
6. $\{x \mid x \leq -5 \text{ or } x \geq 2\}$
7. $\{x \mid x \leq -4 \text{ or } x \geq 3\}$
8. $\{x \mid -5 \leq x \leq 2\}$
9. $\{x \mid x < -1 \text{ or } x > 4\}$
10. $\{x \mid -2 < x < 4\}$
11. $\{x \mid -4 \leq x \leq 3\}$
12. $\{x \mid x \leq -3 \text{ or } x \geq 5\}$
13. $\{x \mid x \leq 2 \text{ or } x \geq 3\}$
14. $\{x \mid -5 \leq x \leq -2\}$
15. $\{x \mid -6 \leq x \leq 4\}$
16. $\{x \mid x < -3 \text{ or } x > 6\}$
17. $\{x \mid x < -9 \text{ or } x > 3\}$
18. $\{x \mid 3 \leq x \leq 4\}$
19. $\left\{x \mid -2 \leq x \leq \dfrac{3}{2}\right\}$
20. $\left\{x \mid -\dfrac{2}{3} < x < 4\right\}$
21. $\left\{x \mid -1 < x < \dfrac{3}{4}\right\}$
22. $\left\{x \mid x \leq -\dfrac{2}{5} \text{ or } x \geq 3\right\}$
23. $\{x \mid -4 \leq x \leq 4\}$

In Exercises 1 to 8, solve each inequality, and graph the solution set.

1. $(x - 3)(x + 4) < 0$

2. $(x - 2)(x + 5) > 0$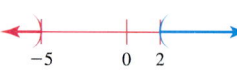

3. $(x - 3)(x + 4) > 0$

4. $(x - 2)(x + 5) < 0$

5. $(x - 3)(x + 4) \leq 0$

6. $(x - 2)(x + 5) \geq 0$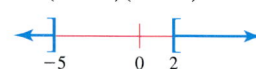

7. $(x - 3)(x + 4) \geq 0$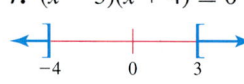

8. $(x - 2)(x + 5) \leq 0$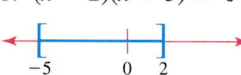

In Exercises 9 to 38, solve each inequality, and graph the solution set.

9. $x^2 - 3x - 4 > 0$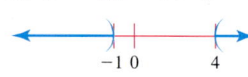

10. $x^2 - 2x - 8 < 0$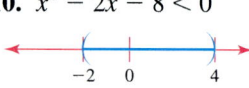

11. $x^2 + x - 12 \leq 0$

12. $x^2 - 2x - 15 \geq 0$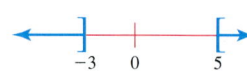

13. $x^2 - 5x + 6 \geq 0$

14. $x^2 + 7x + 10 \leq 0$

15. $x^2 + 2x \leq 24$

16. $x^2 - 3x > 18$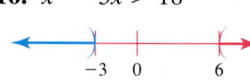

17. $x^2 > 27 - 6x$

18. $x^2 \leq 7x - 12$

19. $2x^2 + x - 6 \leq 0$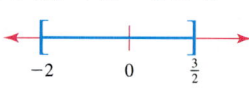

20. $3x^2 - 10x - 8 < 0$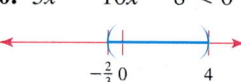

21. $4x^2 + x < 3$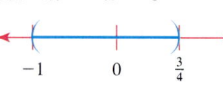

22. $5x^2 - 13x \geq 6$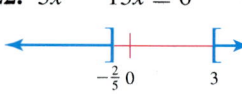

23. $x^2 - 16 \leq 0$

Section 11.4 ■ Solving Quadratic Inequalities   767

24. $\{x | x < -3 \text{ or } x > 3\}$
25. $\{x | x \leq -5 \text{ or } x \geq 5\}$
26. $\{x | -7 < x < 7\}$
27. $\{x | x < -2 \text{ or } x > 2\}$
28. $\{x | -6 \leq x \leq 6\}$
29. $\{x | 0 \leq x \leq 4\}$
30. $\{x | x < -5 \text{ or } x > 0\}$
31. $\{x | x \leq 0 \text{ or } x \geq 6\}$
32. $\{x | 0 < x < 3\}$
33. $\{x | 0 < x < 4\}$
34. $\{x | x \leq 0 \text{ or } x \geq 6\}$
35. $\{x | x = 2\}$
36. All real numbers.
37. $\{x | -4 \leq x \leq 7\}$
38. $\{x | x < -6 \text{ or } x > 7\}$

24. $x^2 - 9 > 0$

25. $x^2 \geq 25$
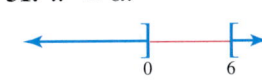

26. $x^2 < 49$
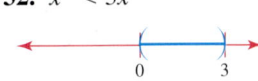

27. $4 - x^2 < 0$
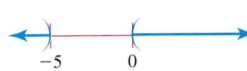

28. $36 - x^2 \geq 0$
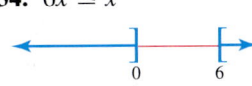

29. $x^2 - 4x \leq 0$
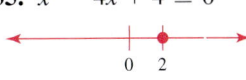

30. $x^2 + 5x > 0$
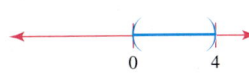

31. $x^2 \geq 6x$

32. $x^2 < 3x$

33. $4x > x^2$
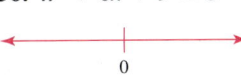

34. $6x \leq x^2$

35. $x^2 - 4x + 4 \leq 0$
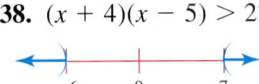

36. $x^2 + 6x + 9 \geq 0$

37. $(x + 3)(x - 6) \leq 10$

38. $(x + 4)(x - 5) > 22$

39. Can a quadratic inequality be solved if the quadratic member of the inequality is not factorable? If so, explain how the solution can be found. If not, explain why not.

40. Is it necessary to relate a quadratic inequality to 0 in order to solve it? Why or why not?

41. $\{x | x < -1 \text{ or } 0 < x < 2\}$
42. $\{x | -3 \leq x \leq 0 \text{ or } x \geq 2\}$

An inequality of the form

$$(x - a)(x - b)(x - c) < 0$$

can be solved by using a sign graph to consider the signs of *all three factors*. In Exercises 41 to 46, use this suggestion to solve each inequality. Then graph the solution set.

41. $x(x - 2)(x + 1) < 0$

42. $x(x + 3)(x - 2) \geq 0$
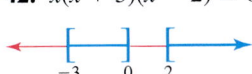

43. $\{x | -2 \leq x \leq 1 \text{ or } x \geq 3\}$

44. $\{x | x < -1 \text{ or } 4 < x < 5\}$

45. $\{x | x \leq -3 \text{ or } 0 \leq x \leq 5\}$

46. $\{x | -6 < x < 0 \text{ or } x > 4\}$

47. $\{x | 50 \leq x \leq 60\}$

48. $\{x | 80 \leq x \leq 100\}$

49. $\{x | 2 \leq t \leq 3\}$

50. $\{x | 1 \leq t \leq 3\}$

51.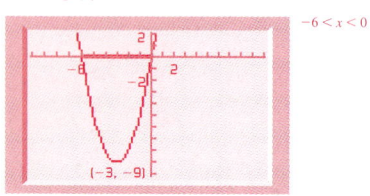

52.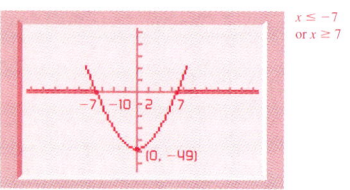

53.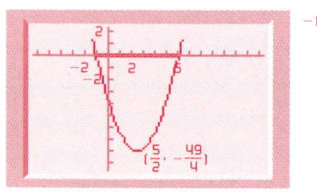

54.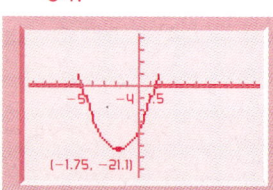

43. $(x - 3)(x + 2)(x - 1) \geq 0$

44. $(x - 5)(x + 1)(x - 4) < 0$

45. $x^3 - 2x^2 - 15x \leq 0$

46. $x^3 + 2x^2 - 24x > 0$

Solve Exercises 47 to 50.

47. A small manufacturer's weekly profit is given by

$$P(x) = -2x^2 + 220x$$

where x is the number of items manufactured and sold. Find the number of items that must be manufactured and sold if the profit is to be greater than or equal to $6000.

48. Suppose that a company's profit is given by

$$P(x) = -2x^2 + 360x$$

How many items must be produced and sold so that the profit will be at least $16,000?

49. If a ball is thrown vertically upward from the ground with an initial velocity of 80 ft/s, its approximate height is given by

$$h(t) = -16t^2 + 80t$$

where t is the time (in seconds) after the ball was released. When will the ball have a height of at least 96 ft?

50. Suppose a ball's height (in meters) is given by

$$h(t) = -5t^2 + 20t$$

When will the ball have a height of at least 15 m?

In Exercises 51 to 54, use your calculator to find the graph for each equation. Use that graph to estimate the solution of the related inequality.

51. $y = x^2 + 6x$; $x^2 + 6x < 0$

52. $y = x^2 - 49$; $x^2 - 49 \geq 0$

53. $y = x^2 - 5x - 6$; $x^2 - 5x - 6 \leq 0$

54. $y = 2x^2 + 7x - 15$; $2x^2 + 7x - 15 > 0$

Summary Exercises • 11

This summary exercise set is provided to give you practice with each of the objectives in the chapter. Each exercise is keyed to the appropriate chapter section.

[11.1] Solve each of the following equations, using the square root method.

1. $x^2 - 8 = 0$
2. $3y^2 - 15 = 0$
3. $(x - 2)^2 = 20$

4. $(2x + 1)^2 - 10 = 0$

[11.1] Find the constant that must be added to each of the following binomials to form a perfect-square trinomial.

5. $x^2 - 12x$
6. $y^2 + 3y$

[11.1] Solve the following equations by completing the square.

7. $x^2 - 4x - 5 = 0$
8. $x^2 + 8x + 12 = 0$
9. $w^2 - 10w - 3 = 0$

10. $y^2 + 3y - 1 = 0$
11. $2x^2 - 6x - 5 = 0$
12. $3x^2 + 4x - 1 = 0$

[11.2] Solve each of the following equations by using the quadratic formula.

13. $x^2 - 5x - 24 = 0$
14. $w^2 + 10w + 25 = 0$
15. $x^2 = 3x + 3$

16. $2y^2 - 5y + 2 = 0$
17. $3y^2 + 4y = 1$
18. $2y^2 + 5y + 4 = 0$

19. $(x - 5)(x + 3) = 13$
20. $\dfrac{1}{x^2} - \dfrac{4}{x} + 1 = 0$
21. $3x^2 + 2x + 5 = 0$

22. $(x - 1)(2x + 3) = -5$

Answers:

1. $\{\pm 2\sqrt{2}\}$
2. $\{\pm\sqrt{5}\}$
3. $\{2 \pm 2\sqrt{5}\}$
4. $\left\{\dfrac{-1 \pm \sqrt{10}}{2}\right\}$
5. 36
6. $\dfrac{9}{4}$
7. $\{-1, 5\}$
8. $\{-6, -2\}$
9. $\{5 \pm 2\sqrt{7}\}$
10. $\left\{\dfrac{-3 \pm \sqrt{13}}{2}\right\}$
11. $\left\{\dfrac{3 \pm \sqrt{19}}{2}\right\}$
12. $\left\{\dfrac{-2 \pm \sqrt{7}}{3}\right\}$
13. $\{-3, 8\}$
14. $\{-5\}$
15. $\left\{\dfrac{3 \pm \sqrt{21}}{2}\right\}$
16. $\left\{\dfrac{1}{2}, 2\right\}$
17. $\left\{\dfrac{-2 \pm \sqrt{7}}{3}\right\}$
18. $\left\{\dfrac{-5 \pm i\sqrt{7}}{4}\right\}$
19. $\{1 \pm \sqrt{29}\}$
20. $\{2 \pm \sqrt{3}\}$
21. $\left\{\dfrac{-1 \pm i\sqrt{14}}{3}\right\}$
22. $\left\{\dfrac{-1 \pm i\sqrt{15}}{4}\right\}$

23. None
24. Two
25. One
26. None
27. 4, 8
28. 8, 10
29. 5
30. Width: 8 ft; length: 10 ft
31. Width: 5 cm; length: 7 cm
32. Width: 11 in.; length: 14 in.
33. $30 \leq x \leq 50$
34. $1 \leq t \leq 3$
35. Width: 4 cm; length: 9 cm
36. 12, 16, 20 in.
37. Width: 8 ft; length: 15 ft

[11.2] For each of the following quadratic equations, use the discriminant to determine the number of real solutions.

23. $x^2 - 3x + 3 = 0$
24. $x^2 + 4x = 2$
25. $4x^2 - 12x + 9 = 0$
26. $2x^2 + 3 = 3x$

27. **Number problem.** The sum of two integers is 12, and their product is 32. Find the two integers.

28. **Number problem.** The product of two consecutive, positive, even integers is 80. What are the two integers?

29. **Number problem.** Twice the square of a positive integer is 10 more than 8 times that integer. Find the integer.

30. **Rectangles.** The length of a rectangle is 2 ft more than its width. If the area of the rectangle is 80 ft^2, what are the dimensions of the rectangle?

31. **Rectangles.** The length of a rectangle is 3 cm less than twice its width. The area of the rectangle is 35 cm^2. Find the length and width of the rectangle.

32. **Rectangles.** An open box is formed by cutting 3-in. squares from each corner of a rectangular piece of cardboard which is 3-in. longer than it is wide. If the box is to have a volume of 120 in.^{3}, what must be the size of the original piece of cardboard?

33. **Profit.** Suppose that a manufacturer's weekly profit P is given by

$$P(x) = -3x^2 + 240x$$

where x is the number of items manufactured and sold. Find the number of items that must be manufactured and sold if the profit is to be at least $4500.

34. **Thrown ball.** If a ball is thrown vertically upward from the ground with an initial velocity of 64 ft/s, its approximate height is given by

$$h(t) = -16t^2 + 64t$$

When will the ball reach a height of at least 48 ft?

35. **Rectangle.** The length of a rectangle is 1 cm more than twice its width. If the length is doubled, the area of the new rectangle is 36 cm^2 more than that of the old. Find the dimensions of the original rectangle.

36. **Triangle.** One leg of a right triangle is 4 in. longer than the other. The hypotenuse of the triangle is 8 in. longer than the shorter leg. What are the lengths of the three sides of the triangle?

37. **Rectangle.** The diagonal of a rectangle is 9 ft longer than the width of the rectangle, and the length is 7 ft more than its width. Find the dimensions of the rectangle.

Summary Exercises

38. Thrown ball. If a ball is thrown vertically upward from the ground, the height, h after t seconds is given by

$$h(t) = 128t - 16t^2$$

(a) How long does it take the ball to return to the ground?

(b) How long does it take the ball to reach a height of 240 ft on the way up?

39. Triangle. One leg of a right triangle is 2 m longer than the other. If the length of the hypotenuse is 8 m, find the length of the other two legs.

40. Thrown ball. Suppose that the height (in meters) of a golf ball, hit off a raised tee, is approximated by

$$h(t) = -5t^2 + 10t + 10$$

t seconds after the ball is hit. When will the ball hit the ground?

[11.1–11.2] Find the zeros of the following functions.

41. $f(x) = x^2 - x - 2$
42. $f(x) = 6x^2 + 7x + 2$
43. $f(x) = -2x^2 - 7x - 6$
44. $f(x) = -x^2 - 1$

[11.3] Find the equation of the axis of symmetry and the coordinates for the vertex of each of the following.

45. $f(x) = x^2$
46. $f(x) = x^2 + 2$
47. $f(x) = x^2 - 5$
48. $f(x) = (x - 3)^2$
49. $f(x) = (x + 2)^2$
50. $f(x) = -(x - 3)^2$
51. $f(x) = (x + 3)^2 + 1$
52. $f(x) = -(x + 2)^2 - 3$
53. $f(x) = -(x - 5)^2 - 2$
54. $f(x) = 2(x - 2)^2 - 5$
55. $f(x) = -x^2 + 2x$
56. $f(x) = x^2 - 4x + 3$
57. $f(x) = -x^2 - x + 6$
58. $f(x) = x^2 + 4x + 5$
59. $f(x) = -x^2 - 6x + 4$

[11.3] Use the graph of the related parabola to find the number of real solutions to each equation. Identify the number of solutions that are positive and the number that are negative.

60. $0 = x^2 - 3x - 10$
61. $0 = x^2 + 5x - 14$
62. $0 = x^2 + 4x + 3$
63. $0 = -3x^2 + 5x + 9$
64. $0 = x^2 - 3x$
65. $0 = 3x^2 + 6x$

---

38. (a) 8 s; (b) 3 s
39. $-1 + \sqrt{31}$, $1 + \sqrt{31}$ or 4.6, 6.6 m
40. $1 + \sqrt{3}$ or 2.7 s
41. $x = 2, -1$
42. $x = -\dfrac{2}{3}, -\dfrac{1}{2}$
43. $x = -\dfrac{3}{2}, -2$
44. $x =$ none
45. $x = 0$; (0, 0)
46. $x = 0$; (0, 2)
47. $x = 0$; (0, −5)
48. $x = 3$; (3, 0)
49. $x = -2$; (−2, 0)
50. $x = 3$; (3, 0)
51. $x = -3$; (−3, 1)
52. $x = -2$; (−2, −3)
53. $x = 5$; (5, −2)
54. $x = 2$; (2, −5)
55. $x = 1$; (1, 1)
56. $x = 2$; (2, −1)
57. $x = -\dfrac{1}{2}$; $\left(-\dfrac{1}{2}, \dfrac{25}{4}\right)$
58. $x = -2$; (−2, 1)
59. $x = -3$; (−3, 13)
60. 2; 1 positive, 1 negative
61. 2; 1 positive, 1 negative
62. 2; 2 negative
63. 2; 1 positive, 1 negative
64. 2; 1 positive
65. 2; 1 negative

66. {−1, 3}

67. {1.2, −3.7}

68. {1.4, 3.6}

69. {0.5, 6.5}

70. {−1, 2.5}

71.–90. See Instructor's Solution Manual for graphs

91. {x | x < −5 or x > 2}

92. {x | 1 < x < 6}

93. {x | −3 ≤ x ≤ −1}

94. {x | x ≤ −4 or x ≥ 5}

95. {x | −3 ≤ x ≤ 8}

96. {x | x ≤ −7 or x ≥ 3}

97. {x | x ≤ −8 or x ≥ 8}

98. {x | x ≤ −5 or x ≥ 0}

99. {x | −3 < x < 7}

100. {x | x ≤ −3 or x ≥ 2}

101. $1 \leq t \leq 3$

102. 50 or less

[11.3] Use the graph of the related parabola to estimate the solutions to each equation. Round answers to the nearest tenth.

66. $0 = x^2 - 2x - 3$
67. $0 = 2x^2 + 5x - 9$
68. $0 = x^2 - 5x + 5$
69. $0 = x^2 - 7x + 5$
70. $0 = 2x^2 - 3x - 5$

[11.3] Graph each of the following functions.

71. $f(x) = x^2$
72. $f(x) = x^2 + 2$
73. $f(x) = x^2 - 5$
74. $f(x) = (x - 3)^2$
75. $f(x) = (x + 2)^2$
76. $f(x) = -(x - 3)^2$
77. $f(x) = (x + 3)^2 + 1$
78. $f(x) = -(x + 2)^2 - 3$
79. $f(x) = x^2 - 4x$
80. $f(x) = -x^2 + 2x$
81. $f(x) = x^2 + 2x - 3$
82. $f(x) = x^2 - 4x + 3$
83. $f(x) = -x^2 - x + 6$
84. $f(x) = -x^2 + 3x + 4$
85. $f(x) = x^2 + 4x + 5$
86. $f(x) = x^2 - 6x + 4$
87. $f(x) = x^2 - 2x + 4$
88. $f(x) = -x^2 + 2x - 2$
89. $f(x) = 2x^2 - 4x + 1$
90. $f(x) = \frac{1}{2}x^2 - 4x$

[11.4] Solve the following inequalities and graph the solution set.

91. $(x - 2)(x + 5) > 0$
92. $(x - 1)(x - 6) < 0$
93. $(x + 1)(x + 3) \leq 0$
94. $(x + 4)(x - 5) \geq 0$
95. $x^2 - 5x - 24 \leq 0$
96. $x^2 + 4x \geq 21$
97. $x^2 \geq 64$
98. $x^2 + 5x \geq 0$
99. $(x + 2)(x - 6) < 9$
100. $(x - 1)(x + 2) \geq 4$

101. If a ball is thrown vertically upward from the ground with an initial velocity of 64 ft/s its approximate height is given by $h(t) = -16t^2 + 64t$. When will the ball reach a height of at least 48 ft?

102. Suppose that the cost, in dollars, of producing x stereo systems is given by the equation $C(x) = 3000 - 60x + 3x^2$. How many systems can be produced if the cost cannot exceed $7500?

Self-Test 11

The purpose of this self-test is to help you check your progress and to review for a chapter test in class. Allow yourself about 1 hour to take the test. When you are done, check your answers in the back of the book. If you missed any answers, be sure to go back and review the appropriate sections in the chapter and the exercises that are provided.

Solve each of the following equations by factoring.

1. $2x^2 + 7x + 3 = 0$
2. $6x^2 = 10 - 11x$
3. $4x^3 - 9x = 0$

Solve each of the following equations, using the square root method.

4. $4w^2 - 20 = 0$
5. $(x - 1)^2 = 10$
6. $4(x - 1)^2 = 23$

Solve each of the following equations by completing the square.

7. $m^2 + 3m - 1 = 0$
8. $2x^2 - 10x + 3 = 0$

Solve each of the following equations, using the quadratic formula.

9. $x^2 - 5x - 3 = 0$
10. $x^2 + 4x = 7$
11. Find the zeros of the function $f(x) = 3x^2 - 10x - 8$.

Solve.

12. The product of two consecutive, positive, odd integers is 63. Find the two integers.

13. Suppose that the height (in feet) of a ball thrown upward from a raised platform is approximated by

$$h(t) = -16t^2 + 32t + 32$$

t s after the ball has been released. How long will it take the ball to hit the ground?

Find the equation of the axis of symmetry and the coordinates of the vertex of each of the following.

14. $y = -3(x + 2)^2 + 1$
15. $y = x^2 - 4x - 5$
16. $y = -2x^2 + 6x - 3$
17. $y = (x - 3)^2 - 2$
18. $y = x^2 - 6x + 2$

---

1. $\left\{-3, -\dfrac{1}{2}\right\}$
2. $\left\{-\dfrac{5}{2}, \dfrac{2}{3}\right\}$
3. $\left\{0, \dfrac{3}{2}, -\dfrac{3}{2}\right\}$
4. $\{\pm\sqrt{5}\}$
5. $\{1 \pm \sqrt{10}\}$
6. $\left\{\dfrac{2 \pm \sqrt{23}}{2}\right\}$
7. $\left\{\dfrac{-3 \pm \sqrt{13}}{2}\right\}$
8. $\left\{\dfrac{5 \pm \sqrt{19}}{2}\right\}$
9. $\left\{\dfrac{5 \pm \sqrt{37}}{2}\right\}$
10. $\{-2 \pm \sqrt{11}\}$
11. $\left\{-\dfrac{2}{3}, 4\right\}$
12. 7, 9
13. 2.7 s
14. $x = -2$; $(-2, 1)$
15. $x = 2$; $(2, -9)$
16. $x = \dfrac{3}{2}$; $\left(\dfrac{3}{2}, \dfrac{3}{2}\right)$
17. $x = 3$; $(3, -2)$
18. $x = 3$; $(3, -7)$

774 Chapter 11 ▪ Quadratic Functions

19.
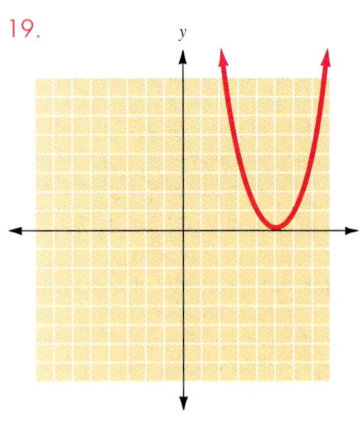

20.

21.

22.
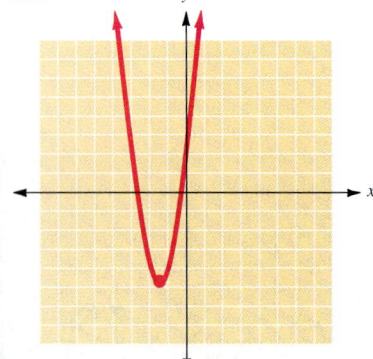

Graph each of the following functions.

19. $f(x) = (x - 5)^2$
20. $f(x) = (x + 2)^2 - 3$
21. $f(x) = -2(x - 3)^2 - 1$
22. $f(x) = 3x^2 + 9x + 2$

Use the graph of the related parabola to estimate the solutions to each equation.

23. $0 = x^2 + 3x - 7$
24. $0 = 4x^2 + 2x - 5$

Solve the following inequalities and graph the solution set.

25. $(x + 3)(x - 1) > 0$

26. $x^2 + 5x - 14 < 0$
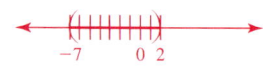

27. $x^2 - 3x \geq 18$

28. $3x^2 + x - 10 \leq 0$
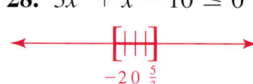

29. $x^2 - 3x \leq 0$
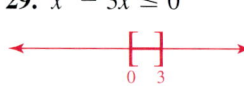

30. $x^2 + 5x > 0$
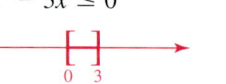

23. $\{-4.5, 1.5\}$

24. $\{-1.4, 0.9\}$

25. $x < -3$ or $x > 1$

26. $-7 < x < 2$

27. $x \leq -3$ or $x \geq 6$

28. $-2 \leq x \leq \dfrac{5}{3}$

29. $0 \leq x \leq 3$

30. $x < -5$ or $x > 0$

Cumulative Test • 1-11

This test is provided to help you in the process of reviewing the previous chapters. Answers are provided in the back of the book. If you missed any answers, be sure to go back and review the appropriate chapter section.

Graph each of the following equations.

1. $2x - 3y = 6$ **2.** $y = -\frac{1}{3}x - 2$ **3.** $y = 4$

Find the slope of the line determined by each set of points.

4. $(-4, 7)$ and $(-3, 4)$ **5.** $(-2, 3)$ and $(-5, -1)$

6. Let $f(x) = 6x^2 - 5x + 1$. Evaluate $f(-2)$.

7. Simplify the function $f(x) = (x^2 - 1)(x + 3)$.

8. Completely factor the expression $x^3 + x^2 - 6x$.

9. Simplify the expression $\dfrac{2}{x + 2} - \dfrac{3x - 2}{x^2 - x - 6}$.

10. Simplify the expression $(\sqrt{7} - \sqrt{2})(\sqrt{3} + \sqrt{6})$.

Solve each equation.

11. $2x - 7 = 0$ **12.** $3x - 5 = 5x + 3$ **13.** $0 = (x - 3)(x + 5)$
14. $x^2 - 3x + 2 = 0$ **15.** $x^2 + 7x - 30 = 0$ **16.** $x^2 - 3x - 3 = 0$
17. $(x - 3)^2 = 5$ **18.** $|x - 2| = 4$ **19.** $\dfrac{x}{3} - \dfrac{4}{9} = \dfrac{5}{18}$

Solve each inequality.

20. $x - 2 \le 7$ **21.** $|x - 1| \ge 8$ **22.** $x^2 - x - 6 \le 0$

Solve the following word problems. Show the equation used for the solution.

23. Five times a number decreased by 7 is -72. Find the number.

24. One leg of a right triangle is 4 ft longer than the shorter leg. If the hypotenuse is 28 ft, how long is each leg?

25. Suppose that a manufacturer's weekly profit P is given by

$$P(x) = -4x^2 + 320x$$

where x is the number of units manufactured and sold. Find the number of items that must be manufactured and sold to guarantee a profit of at least $4956.

1.–3. See graphs in answer section
4. -3
5. $\dfrac{4}{3}$
6. 35
7. $x^3 + 3x^2 - x - 3$
8. $x(x + 3)(x - 2)$
9. $\dfrac{-x - 4}{(x - 3)(x + 2)}$
10. $\sqrt{21} - \sqrt{6} + \sqrt{42} - 2\sqrt{3}$
11. $\left\{\dfrac{7}{2}\right\}$
12. $\{-4\}$
13. $\{-5, 3\}$
14. $\{1, 2\}$
15. $\{-10, 3\}$
16. $\left\{\dfrac{3 \pm \sqrt{21}}{2}\right\}$
17. $\{3 \pm \sqrt{5}\}$
18. $\{-2, 6\}$
19. $\left\{\dfrac{13}{6}\right\}$
20. $\{x | x \le 9\}$
21. $\{x | x \le -7 \text{ or } x \ge 9\}$
22. $\{x | -2 \le x \le 3\}$
23. -13
24. $-2 + 2\sqrt{97}$, $2 + 2\sqrt{97}$ or 17.7 ft by 21.7 ft
25. Between 21 and 59 units

CHAPTER 12 Conic Sections

LIST OF SECTIONS

12.1 More on the Parabola
12.2 The Circle
12.3 The Ellipse
12.4 The Hyperbola

Large cities often commission fireworks artists to choreograph elaborate displays on holidays. Such displays look like beautiful paintings in the sky, in which the fireworks seem to dance to well-known popular and classical music. The displays are feats of engineering and very accurate timing. Suppose the designer wants a second set of rockets of a certain color and shape to be released after the first set of a different color and shape reaches a specific height and explodes. She must know the strength of the initial liftoff and use a quadratic equation to determine the proper time for setting off the second round.

The equation $h = -16t^2 + 100t$ gives the height in feet t seconds after the rockets are shot into the air if the initial velocity is 100 feet per second. Using this equation, the designer knows how high the rocket will ascend and when it will begin to fall. She can time the next round to achieve the effect she wishes. Displays that involve large banks of fireworks in shows that last up to an hour are programmed using computers, but quadratic equations are at the heart of the mechanism that creates the beautiful effects.

SECTION 12.1 More on the Parabola

12.1 OBJECTIVES

1. Recognize the general form for a conic section
2. Graph a horizontal parabola

In Chapter 11, we discussed the graph of a function with the general form

$$y = ax^2 + bx + c \tag{1}$$

Such a graph was called a **parabola.** The parabola is one example from a family of curves that are called conic sections. A **conic section** is a curve formed when a plane cuts through, or forms a section of, a cone. The conic sections comprise four curves—the **parabola, circle, ellipse,** and **hyperbola.** Examples of these curves are shown in the figure.

The inclination of the plane determines which of the sections is formed.

The names *ellipse, parabola,* and *hyperbola* are attributed to Apollonius, a third-century B.C. Greek mathematician and astronomer.

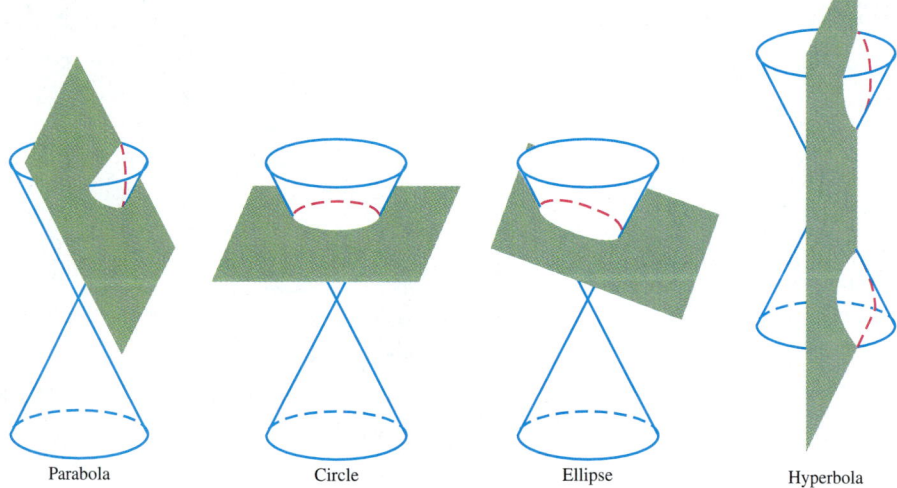

Parabola   Circle   Ellipse   Hyperbola

Algebraically, a conic section is formed from any second-degree equation from the general form

$$ax^2 + by^2 + cxy + dx + ey + f = 0 \tag{2}$$

In the case of the parabola we examined in Section 11.3, the values of b and c are each zero. This leaves the form

$$ax^2 + dx + ey + f = 0 \tag{3}$$

Note that if a, d, e, and f are each given nonzero values, an equation in the form of equation (3) can be rewritten into the form of equation (1). The next example illustrates this.

Example 1

Rewriting Quadratic Equations

Given the standard form for a conic section

$$ax^2 + by^2 + cxy + dx + ey + f = 0$$

and the values $a = -3$, $b = 0$, $c = 0$, $d = 6$, $e = -2$, and $f = -8$, rewrite the equation in standard quadratic form.

Substituting the coefficient values, we have the equation

$$-3x^2 + 6x - 2y - 8 = 0$$

or

$$-2y = 3x^2 - 6x + 8$$

so, dividing by the coefficient of the y term, we have

$$y = -\frac{3}{2}x^2 + 3x - 4$$

✓ CHECK YOURSELF 1

Given the standard form for a conic section

$$ax^2 + by^2 + cxy + dx + ey + f = 0$$

and the values $a = 5$, $b = 0$, $c = 0$, $d = 6$, $e = -3$, and $f = -9$, rewrite the equation in standard quadratic form.

So far we have dealt with equations of the form

$$y = ax^2 + bx + c$$

Suppose we reverse the role of x and y. We then have

$$x = ay^2 + by + c$$

which is quadratic in y but not in x. The graph of such an equation is once again a parabola, but this time the parabola is horizontally oriented.

In an equation of the form

$$x = ay^2 + by + c \qquad a \neq 0$$

the parabola opens leftward or rightward, as follows:

1. If $a > 0$, the parabola opens *rightward*.
2. If $a < 0$, the parabola opens *leftward*.

$$x = ay^2 + by + c$$

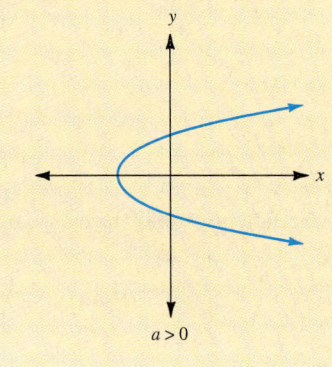
$a > 0$

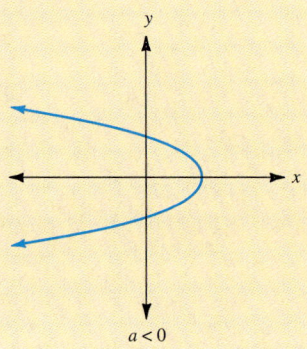
$a < 0$

Much of what we did earlier is easily extended to a horizontally oriented parabola. Example 2 illustrates the changes in the process.

Example 2

Graphing a Parabola

Graph the equation

$$x = y^2 + 4y - 5$$

First, find the axis of symmetry. Now the axis of symmetry is horizontal with the equation

$$y = -\frac{b}{2a}$$

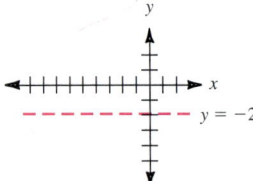

So

$$y = -\frac{b}{2a} = \frac{-4}{2 \cdot 1} = -2$$

Second, find the vertex. If $y = -2$,

$$x = (-2)^2 + 4(-2) - 5 = -9$$

So $(-9, -2)$ is the vertex.

Third, find two symmetric points. Here the quadratic member is factorable, so set $x = 0$ in the original equation. That gives the y intercepts:

$$0 = y^2 + 4y - 5$$
$$= (y + 5)(y - 1)$$

$y + 5 = 0$ or $y - 1 = 0$
$y = -5$ $y = 1$

The y intercepts then are at $(0, -5)$ and $(0, 1)$.

Fourth, draw a smooth curve through the points already found to form the parabola.

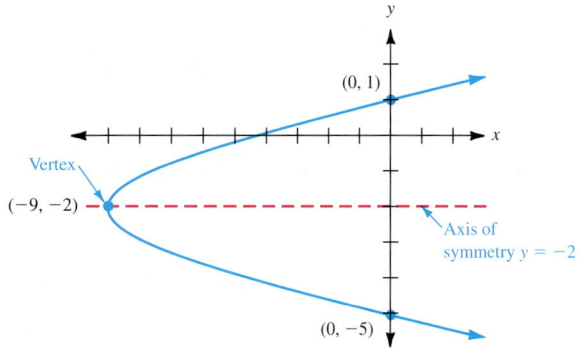

✓ CHECK YOURSELF 2

Graph the equation $x = -y^2 - 2y + 3$.

✓ CHECK YOURSELF ANSWERS

1. $y = \dfrac{5}{3}x^2 + 2x - 3$.   **2.** $x = -y^2 - 2y + 3$

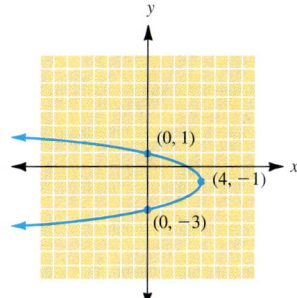

Exercises • 12.1

1. $3x^2 + 2x - 5 = y$
2. $y = 2x^2 + 3x - 4$
3. $y = \frac{1}{2}x^2 - \frac{3}{2}x - \frac{5}{2}$
4. $y = -\frac{4}{3}x^2 + \frac{5}{3}x - \frac{7}{3}$
5. $y = -x^2 + 2x + \frac{3}{2}$
6. $y = -\frac{3}{4}x^2 - \frac{1}{2}x + \frac{5}{4}$

In Exercises 1 to 6, use the given values of a, b, c, d, e, and f to rewrite the standard form for a conic section

$$ax^2 + by^2 + cxy + dx + ey + f = 0$$

into standard quadratic form.

1. $a = 3, b = 0, c = 0, d = 2, e = -1, f = -5$
2. $a = -2, b = 0, c = 0, d = -3, e = 1, f = 4$
3. $a = -1, b = 0, c = 0, d = 3, e = 2, f = 5$
4. $a = 4, b = 0, c = 0, d = -5, e = 3, f = 7$
5. $a = -2, b = 0, c = 0, d = 4, e = -2, f = 3$
6. $a = -3, b = 0, c = 0, d = -2, e = -4, f = 5$

In Exercises 7 to 16, find the equation of the axis of symmetry, the coordinates of the vertex, and at least two symmetric points. Sketch the graph of each equation.

7. $x = y^2 + 4y$

8. $x = y^2 - 4y$

9. $x = y^2 + 8y + 12$

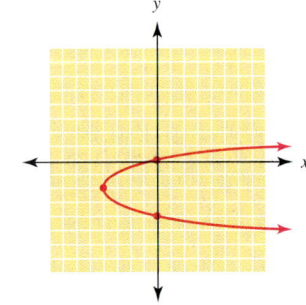
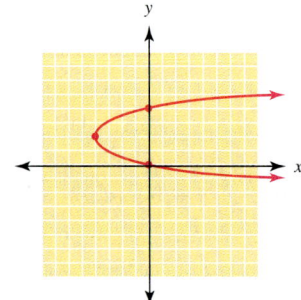
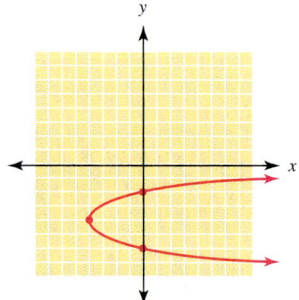

10. $x = y^2 - 2y + 3$

11. $x = -y^2 + 6y - 5$

12. $x = -y^2 - 2y + 2$

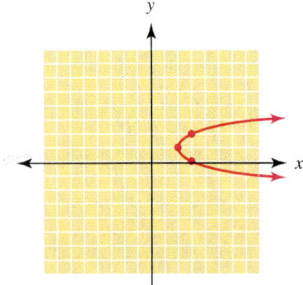
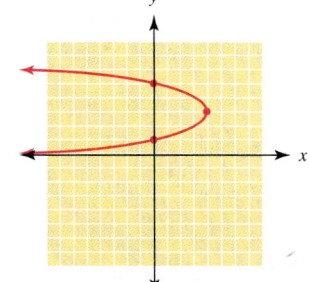
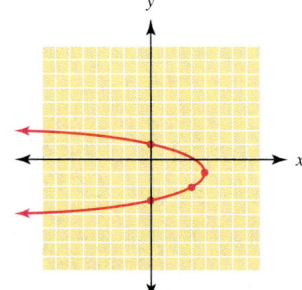

17. $a > 0$ and $h > 0$ or
$a < 0$ and $h < 0$

13. $x = -2y^2 - 4y - 5$

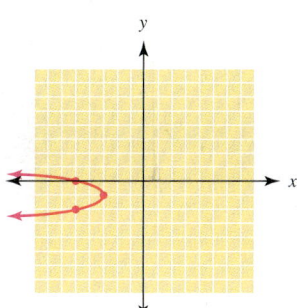

14. $x = -3y^2 - 6y - 5$

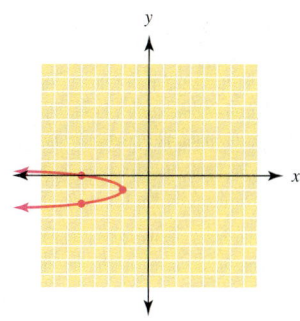

15. $x = 3y^2 - 6y - 1$

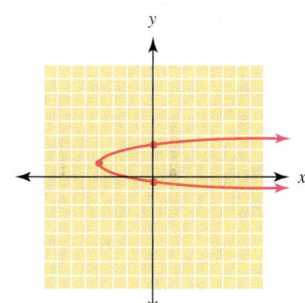

16. $x = 2y^2 - 16y + 37$

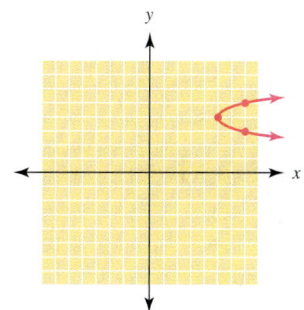

17. Under what conditions will the graph of $x = a(y - k)^2 + h$ have no y intercepts?

18. Discuss similarities and differences between the graphs of $y = x^2 - 3x + 4$ and $x = y^2 - 3y + 4$. Use both graphs in your discussion.

To graph $x = y^2 + 3y + 4$ using a graphing utility, rewrite the equation as a quadratic equation in y

$$y^2 + 3y + (-x + 4) = 0$$

Then use the quadratic formula to solve for y and enter the resulting equations.

$$y_1 = \frac{-3 + \sqrt{9 - 4(-x + 4)}}{2}$$

$$y_2 = \frac{-3 - \sqrt{9 - 4(-x + 4)}}{2}$$

25. {x|x ≥ −9}

26. {x|x ≥ −16}

27. {x|x ≤ 2}

28. {x|x ≤ 7}

In Exercises 19 to 24 use that technique to graph each parabola.

19. $x = y^2 - 4$

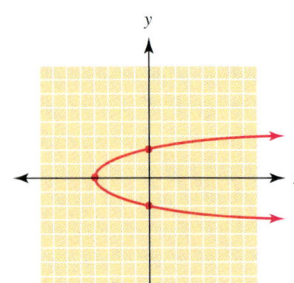

20. $x = y^2 - 9$

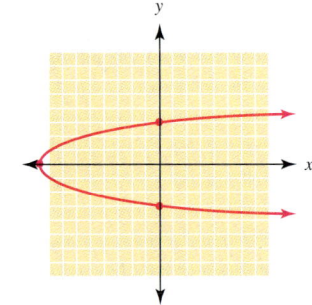

21. $x = -y^2 - 4y + 5$

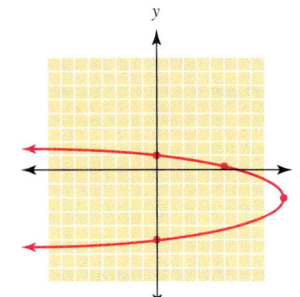

22. $x = -y^2 + 6y - 5$

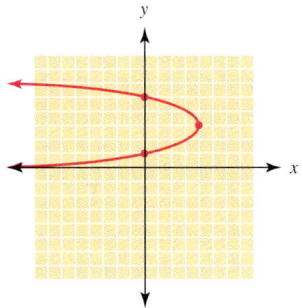

23. $x = 2y^2 + 3y - 2$

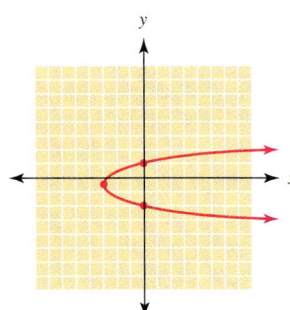

24. $x = 3y^2 + 2y - 5$

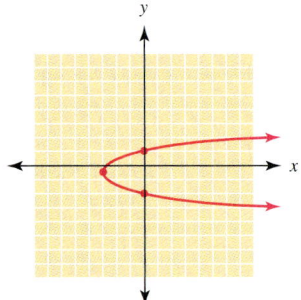

Each equation below defines a relation. Write the domain of each relation. (*Hint*: determine the vertex, and whether the parabola opens to the left or to the right.)

25. $x = y^2 + 6y$

26. $x = y^2 - 8y$

27. $x = -y^2 + 6y - 7$

28. $x = -y^2 - 4y + 3$

SECTION 12.2 The Circle

12.2 OBJECTIVES

1. Identify the graph of an equation as a line, a parabola, or a circle
2. Write the equation of a circle in standard form and graph the circle

The second conic section we look at is the circle. The circle can be described by using the standard form for a conic section,

$$ax^2 + by^2 + cxy + dx + ey + f = 0$$

but we will develop the standard form for a circle through the definition of a circle.

A **circle** is the set of all points in the plane equidistant from a fixed point, called the **center** of the circle. The distance between the center of the circle and any point on the circle is called the **radius** of the circle.

The distance formula is central to any discussion of conic sections.

The Distance Formula

The distance d between two points (x_1, y_1) and (x_2, y_2) is given by

$$d = \sqrt{(x_2 - x_1)^2 + (y_2 - y_1)^2}$$

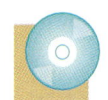

We can use the distance formula to derive the algebraic equation of a circle, given its center and its radius.

Suppose a circle has its center at a point with coordinates (h, k) and radius r. If (x, y) represents any point on the circle, then, by its definition, the distance from (h, k) to (x, y) is r. Applying the distance formula, we have

$$r = \sqrt{(x - h)^2 + (y - k)^2}$$

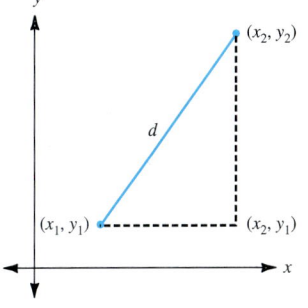

Squaring both sides of the equation gives the equation of the circle

$$r^2 = (x - h)^2 + (y - k)^2$$

In general, we can write the following equation of a circle.

A special case is the circle centered at the origin with radius r. Then $(h, k) = (0, 0)$, and its equation is

$$x^2 + y^2 = r^2$$

Equation of a Circle

The equation of a circle with center (h, k) and radius r is

$$(x - h)^2 + (y - k)^2 = r^2 \qquad (1)$$

Equation (1) can be used in two ways. Given the center and radius of the circle, we can write its equation; or given its equation, we can find the center and radius of a circle.

Example 1

Finding the Equation of a Circle

Find the equation of a circle with center at $(2, -1)$ and radius 3. Sketch the circle.

Let $(h, k) = (2, -1)$ and $r = 3$. Applying equation (1) yields

$$(x - 2)^2 + [y - (-1)]^2 = 3^2$$
$$(x - 2)^2 + (y + 1)^2 = 9$$

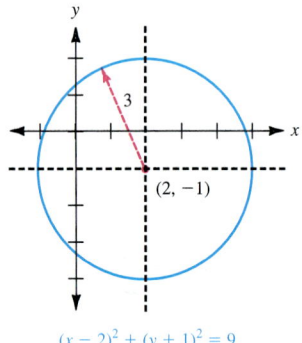

$(x - 2)^2 + (y + 1)^2 = 9$

To sketch the circle, we locate the center of the circle. Then we determine four points 3 units to the right and left and up and down from the center of the circle. Drawing a smooth curve through those four points completes the graph.

✓ **CHECK YOURSELF 1**

Find the equation of the circle with center at $(-2, 1)$ and radius 5. Sketch the circle.

Now, given an equation for a circle, we can also find the radius and center and then sketch the circle. We start with an equation in the special form of equation (1).

Example 2

Finding the Center and Radius of a Circle

Find the center and radius of the circle with equation

$$(x - 1)^2 + (y + 2)^2 = 9$$

Remember, the general form is

$$(x - h)^2 + (y - k)^2 = r^2$$

Our equation "fits" this form when it is written as

Note: $y + 2 = y - (-2)$

$$(x - 1)^2 + [y - (-2)]^2 = 3^2$$

So the center is at $(1, -2)$, and the radius is 3. The graph is shown.

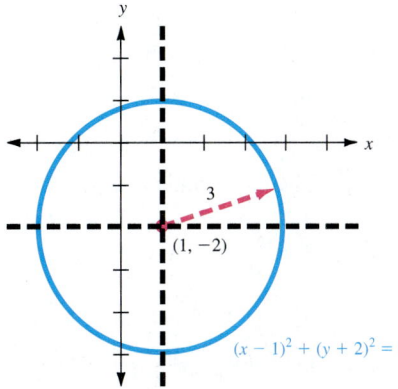

The circle can be graphed on the calculator by solving for y, then graphing both the upper half and lower half of the circle. In this case,

$$(x - 1)^2 + (y + 2)^2 = 9$$
$$(y + 2)^2 = 9 - (x - 1)^2$$
$$(y + 2) = \pm \sqrt{9 - (x - 1)^2}$$
$$y = -2 \pm \sqrt{9 - (x - 1)^2}$$

Now graph the two functions

$$y = -2 + \sqrt{9 - (x - 1)^2}$$

and

$$y = -2 - \sqrt{9 - (x - 1)^2}$$

on your calculator. (The display screen may need to be squared to obtain the shape of a circle.)

✓ **CHECK YOURSELF 2**

Find the center and radius of the circle with equation

$$(x + 3)^2 + (y - 2)^2 = 16$$

Sketch the circle.

To graph the equation of a circle that is not in standard form, we *complete the square*. Let's see how completing the square can be used in graphing the equation of a circle.

Example 3

Finding the Center and Radius of a Circle

To recognize the equation as having the form of a circle, note that the coefficients of x^2 and y^2 are equal.

Find the center and radius of the circle with equation

$$x^2 + 2x + y^2 - 6y = -1$$

Then sketch the circle.

The linear terms in x and y show a translation of the center away from the origin.

We could, of course, simply substitute values of x and try to find the corresponding values for y. A much better approach is to rewrite the original equation so that it matches the standard form.

First, add 1 to both sides to complete the square in x.

$$x^2 + 2x + 1 + y^2 - 6y = -1 + 1$$

Then add 9 to both sides to complete the square in y.

$$x^2 + 2x + 1 + y^2 - 6y + 9 = -1 + 1 + 9$$

We can factor the two trinomials on the left (they are both perfect squares) and simplify on the right.

$$(x + 1)^2 + (y - 3)^2 = 9$$

The equation is now in standard form, and we can see that the center is at $(-1, 3)$ and the radius is 3. The sketch of the circle is shown. Note the "translation" of the center to $(-1, 3)$.

✓ CHECK YOURSELF 3

Find the center and radius of the circle with equation

$$x^2 - 4x + y^2 + 2y = -1$$

Sketch the circle.

✓ CHECK YOURSELF ANSWERS

1. $(x + 2)^2 + (y - 1)^2 = 25$. 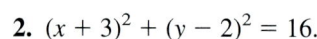 **2.** $(x + 3)^2 + (y - 2)^2 = 16$.

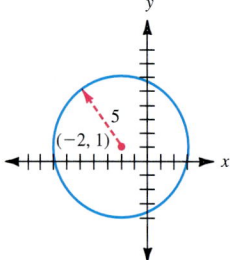 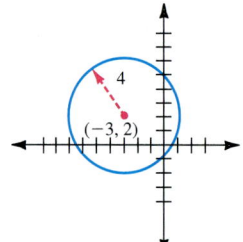

3. $(x - 2)^2 + (y + 1)^2 = 4$.

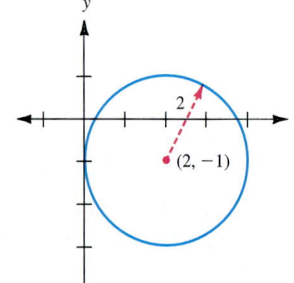

Exercises · 12.2

1. Parabola
2. Circle
3. Line
4. Line
5. Circle
6. Parabola
7. Circle
8. Line
9. None of These
10. Circle
11. Parabola
12. None of These
13. Center: (0, 0); radius: 5
14. Center: (0, 0); radius: $6\sqrt{2}$
15. Center: (3, −1); radius: 4
16. Center: (−3, 0); radius: 9
17. Center: (−1, 0); radius: 4
18. Center: (0, 3); radius: 9
19. Center: (3, −4); radius: $\sqrt{41}$
20. Center: $\left(\frac{5}{2}, \frac{3}{2}\right)$; radius: $\frac{\sqrt{66}}{2}$
21. Center (0, 0); radius: 2
22. Center (0, 0); radius: 5
23. Center (0, 0); radius: 3
24. Center (0, 0); radius: 4
25. Center (1, 0); radius: 3
26. Center (0, −2); radius: 4
27. Center (4, −1); radius: 4
28. Center (−3, −2); radius: 5

In Exercises 1 to 12, decide whether each equation has as its graph a line, a parabola, a circle, or none of these.

1. $y = x^2 − 2x + 5$
2. $y^2 + x^2 = 64$
3. $y = 3x − 2$
4. $2y − 3x = 12$
5. $(x − 3)^2 + (y + 2)^2 = 10$
6. $y + 2(x − 3)^2 = 5$
7. $x^2 + 4x + y^2 − 6y = 3$
8. $4x = 3$
9. $y^2 − 4x^2 = 36$
10. $x^2 + (y − 3)^2 = 9$
11. $y = −2x^2 + 8x − 3$
12. $2x^2 − 3y^2 + 6y = 13$

In Exercises 13 to 20, find the center and the radius for each circle.

13. $x^2 + y^2 = 25$
14. $x^2 + y^2 = 72$
15. $(x − 3)^2 + (y + 1)^2 = 16$
16. $(x + 3)^2 + y^2 = 81$
17. $x^2 + 2x + y^2 = 15$
18. $x^2 + y^2 − 6y = 72$
19. $x^2 − 6x + y^2 + 8y = 16$
20. $x^2 − 5x + y^2 − 3y = 8$

In Exercises 21 to 32, graph each circle by finding the center and the radius.

21. $x^2 + y^2 = 4$
22. $x^2 + y^2 = 25$
23. $4x^2 + 4y^2 = 36$
24. $9x^2 + 9y^2 = 144$
25. $(x − 1)^2 + y^2 = 9$
26. $x^2 + (y + 2)^2 = 16$
27. $(x − 4)^2 + (y + 1)^2 = 16$
28. $(x + 3)^2 + (y + 2)^2 = 25$

29. Center (0, 2); radius: 4

30. Center (3, 0); radius: 3

31. Center (2, −1); radius: 2

32. Center (1, 3); radius: 4

33. Circle with radius zero

35. $10\sqrt{5}$ cm; ≈ 22.4 cm

36. $x^2 + y^2 = 1600$

37. $x^2 + y^2 = \dfrac{4}{9}$

38. $\dfrac{8}{3}$ m

39. $y = \sqrt{36 - x^2}$
    $y = -\sqrt{36 - x^2}$

40. $y = \sqrt{9 - (x - 3)^2}$
    $y = -\sqrt{9 - (x - 3)^2}$

41. $y = \sqrt{36 - (x + 5)^2}$
    $y = -\sqrt{36 - (x + 5)^2}$

42. $y = -1 + \sqrt{25 - (x - 2)^2}$
    $y = -1 - \sqrt{25 - (x - 2)^2}$

43. Domain: $-7 \leq x \leq 1$; range: $-2 \leq y \leq 6$

44. Domain: $-2 \leq x \leq 4$; range: $2 \leq y \leq 8$

45. Domain: $-5 \leq x \leq 5$; range: $-2 \leq y \leq 8$

46. Domain: $-8 \leq x \leq 4$; range: $-6 \leq y \leq 6$

29. $x^2 + y^2 - 4y = 12$

30. $x^2 - 6x + y^2 = 0$

31. $x^2 - 4x + y^2 + 2y = -1$

32. $x^2 - 2x + y^2 - 6y = 6$

33. Describe the graph of $x^2 + y^2 - 2x - 4y + 5 = 0$.

34. Describe how completing the square is used in graphing circles.

35. A solar oven is constructed in the shape of a hemisphere. If the equation

$$x^2 + y^2 = 500$$

describes the circumference of the oven in centimeters, what is its radius?

36. A solar oven in the shape of a hemisphere is to have a diameter of 80 cm. Write the equation that describes the circumference of this oven.

37. A solar water heater is constructed in the shape of a half cylinder, with the water supply pipe at its center. If the water heater has a diameter of $\dfrac{4}{3}$ m, what is the equation that describes its circumference?

38. A solar water heater is constructed in the shape of a half cylinder having a circumference described by the equation

$$9x^2 + 9y^2 - 16 = 0$$

What is its diameter if the units for the equation are meters?

A circle can be graphed on a calculator by plotting the upper and lower semicircles on the same axes. For example, to graph $x^2 + y^2 = 16$, we solve for y:

$$y = \pm\sqrt{16 - x^2}$$

This is then graphed as two separate functions,

$$y = \sqrt{16 - x^2} \quad \text{and} \quad y = -\sqrt{16 - x^2}$$

In Exercises 39 to 42, use that technique to graph each circle.

39. $x^2 + y^2 = 36$

40. $(x - 3)^2 + y^2 = 9$

41. $(x + 5)^2 + y^2 = 36$

42. $(x - 2)^2 + (y + 1)^2 = 25$

Each of the following equations defines a relation. Write the domain and the range of each relation.

43. $(x + 3)^2 + (y - 2)^2 = 16$

44. $(x - 1)^2 + (y - 5)^2 = 9$

45. $x^2 + (y - 3)^2 = 25$

46. $(x + 2)^2 + y^2 = 36$

SECTION 12.3 The Ellipse

12.3 OBJECTIVES

1. Identify an equation whose graph is an ellipse
2. Sketch the graph of an ellipse

Ellipses occur frequently in nature. The planets have elliptical orbits with the sun at one focus.

The reflecting properties of the ellipse are also interesting. Rays from one focus are reflected by the ellipse in such a way that they always pass through the other focus.

Let's turn now to the third conic section, the ellipse. It can be described as an "oval-shaped" curve and has the following geometric description.

An *ellipse* is the set of all points (x, y) such that the sum of the distances from (x, y) to two fixed points, called the *foci* of the ellipse, is constant.

The following sketch illustrates the definition in two particular cases:

1. When the foci are located on the x axis and are symmetric about the origin
2. When the foci are located on the y axis and are symmetric about the origin

An Ellipse

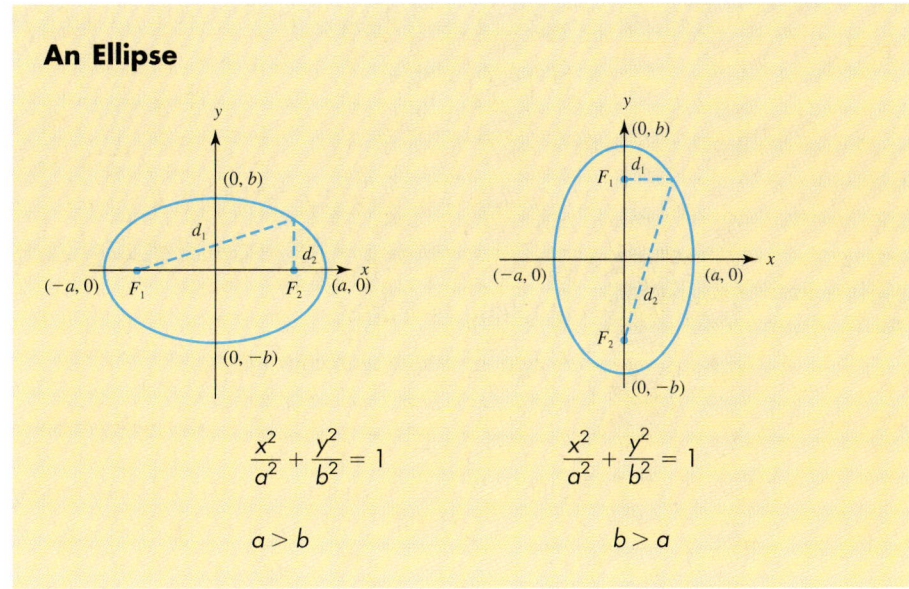

$$\frac{x^2}{a^2} + \frac{y^2}{b^2} = 1$$

$a > b$

$$\frac{x^2}{a^2} + \frac{y^2}{b^2} = 1$$

$b > a$

In either case, $d_1 + d_2$ is constant.

To quickly sketch ellipses of the preceding forms, we need to determine only *four points*, the points where the ellipse intercepts the coordinate axes.

Fortunately those points are easily found when the ellipse is written in standard form:

$$\frac{x^2}{a^2} + \frac{y^2}{b^2} = 1 \qquad (1)$$

Recall:
To find the *x* intercepts, let $y = 0$ and solve for *x*.
To find the *y* intercepts, let $x = 0$ and solve for *y*.

The *x* intercepts are at *a* and at $-a$. The *y* intercepts are at *b* and at $-b$. Let's use this information to sketch an ellipse.

Example 1 — Graphing an Ellipse

Sketch the ellipse

$$\frac{x^2}{9} + \frac{y^2}{4} = 1$$

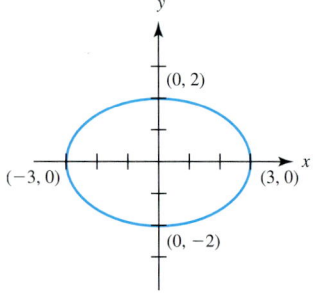

1. The equation is in standard form.
2. Find the *x* intercepts. From equation (1), $a^2 = 9$, and the *x* intercepts are at 3 and at -3.
3. Find the *y* intercepts. From equation (1), $b^2 = 4$, and the *y* intercepts are at 2 and at -2.
4. Plot the intercepts found above, and draw a smooth curve to form the desired ellipse.

✓ **CHECK YOURSELF 1**

Sketch the ellipse.

$$\frac{x^2}{16} + \frac{y^2}{9} = 1$$

Example 2 — Graphing an Ellipse

To recognize the equation as an ellipse in this form, note that the equation has both x^2 and y^2 terms. The coefficients of those terms have the *same* algebraic signs but *different* coefficients.

Sketch the ellipse with equation

$$9x^2 + 4y^2 = 36$$

Step 1 Since this equation is *not* in standard form (the right side is *not* 1), we divide both sides of the equation by the constant 36:

$$\frac{9x^2}{36} + \frac{4y^2}{36} = \frac{36}{36}$$

$$\frac{x^2}{4} + \frac{y^2}{9} = 1$$

We can now proceed as before. Comparing the derived equation with that in standard form, we deduce steps 2 and 3.

Section 12.3 ■ The Ellipse   793

Step 2 The x intercepts are 2 and -2.

Step 3 The y intercepts are 3 and -3.

Step 4 We connect the intercepts with a smooth curve to complete the sketch of the ellipse.

$9x^2 + 4y^2 = 36$

✓ CHECK YOURSELF 2

Sketch the ellipse

$$25x^2 + 4y^2 = 100$$

(*Hint*: First write the equation in standard form by dividing both sides of the equation by 100.)

The following algorithm summarizes our work with graphing ellipses.

Graphing the Ellipse

1. Write the given equation in standard form.
2. From that standard form, determine the x intercepts.
3. Also determine the y intercepts.
4. Plot the four intercepts and connect the points with a smooth curve, to complete the sketch.

✓ CHECK YOURSELF ANSWERS

1. $\dfrac{x^2}{16} + \dfrac{y^2}{9} = 1$ **2.** $\dfrac{x^2}{4} + \dfrac{y^2}{25} = 1$

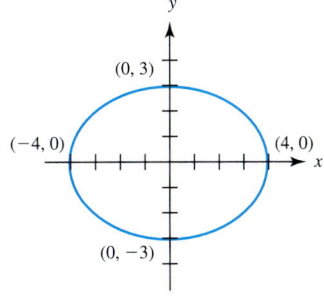
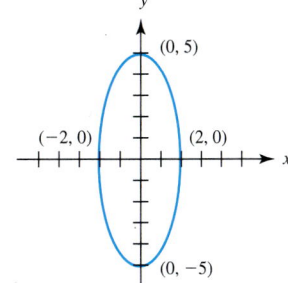

Exercises · 12.3

Graph the following ellipses by finding the *x* and *y* intercepts. If necessary, write the equation in standard form.

1. $\dfrac{x^2}{4} + \dfrac{y^2}{9} = 1$

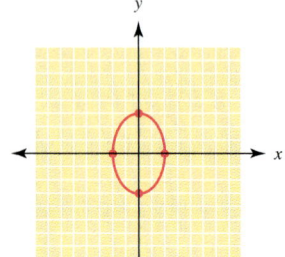

2. $\dfrac{x^2}{16} + \dfrac{y^2}{9} = 1$

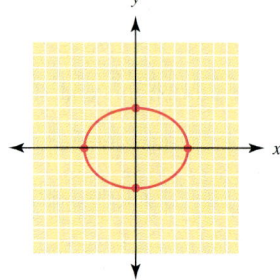

3. $\dfrac{x^2}{9} + \dfrac{y^2}{25} = 1$

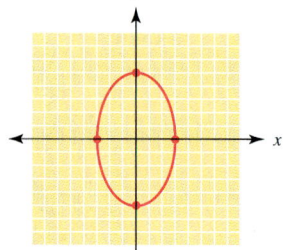

4. $\dfrac{x^2}{36} + \dfrac{y^2}{16} = 1$

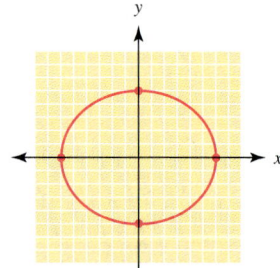

5. $x^2 + 9y^2 = 36$

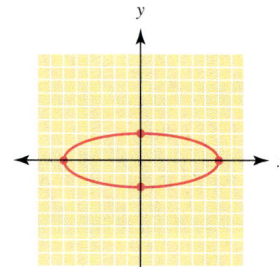

6. $4x^2 + y^2 = 16$

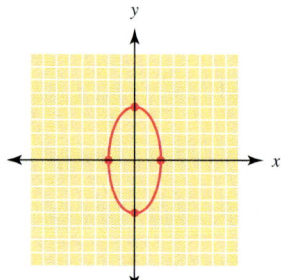

7. $4x^2 + 9y^2 = 36$

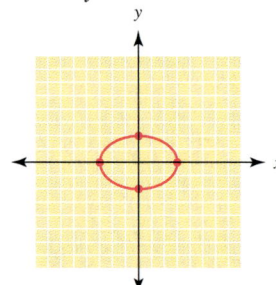

8. $25x^2 + 4y^2 = 100$

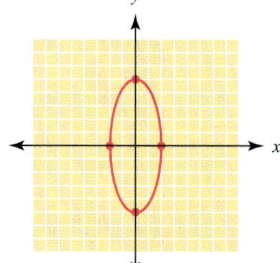

9. $4x^2 + 25y^2 = 100$

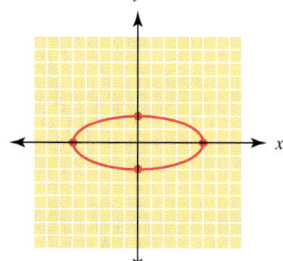

10. $9x^2 + 16y^2 = 144$

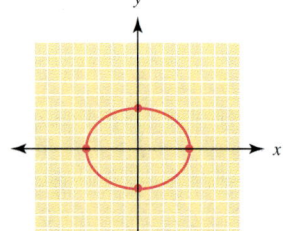

11. $25x^2 + 9y^2 = 225$

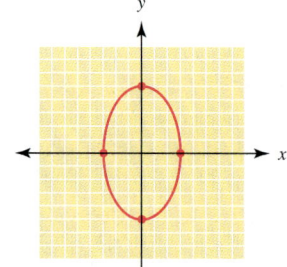

12. $16x^2 + 9y^2 = 144$

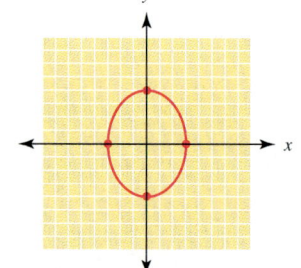

13. Yes

14. Yes

13. A semielliptical archway over a one-way road has a height of 10 feet and a width of 40 feet (see the figure). Will a truck that is 10 feet wide and 9 feet high clear the opening of the highway?

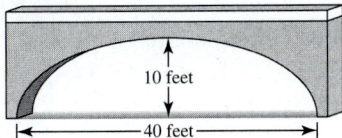

14. A truck that is 8 feet wide is carrying a load that reaches 7 feet above the ground. Will the truck clear a semielliptical arch that is 10 feet high and 30 feet wide?

An ellipse can be graphed on a calculator by plotting the upper and lower halves on the same axes. For example, to graph $9x^2 + 16y^2 = 144$, we solve for y:

$$y_1 = \frac{\sqrt{144 - 9x^2}}{4} \quad \text{and} \quad y_2 = -\frac{\sqrt{144 - 9x^2}}{4}$$

Then graph each equation as a separate function.

In Exercises 15 to 18, use this technique to graph each ellipse.

15. $4x^2 + 16y^2 = 64$

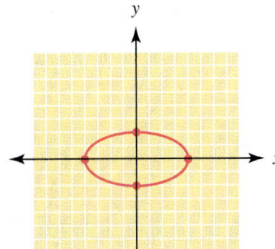

16. $9x^2 + 36y^2 = 324$

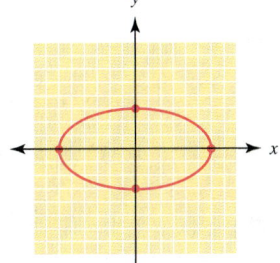

17. $25x^2 + 9y^2 = 225$

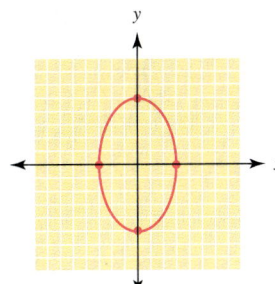

18. $4x^2 + 9y^2 = 36$

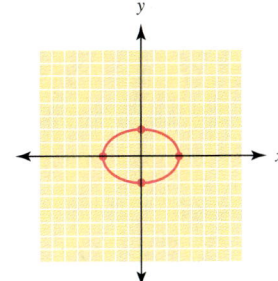

SECTION 12.4 The Hyperbola

12.4 OBJECTIVES

1. Identify an equation whose graph is a hyperbola
2. Sketch the graph of a hyperbola

Our discussion now turns to the last of the conic sections, the hyperbola. As you will see, the geometric description of the hyperbola (and hence the corresponding standard form) is quite similar to that of the ellipse.

> A *hyperbola* is the set of all points (x, y) such that the absolute value of the differences from (x, y) to each of two fixed points, called the *foci* of the hyperbola, is constant.

The following sketch illustrates the definition in the case where the foci are located on the x axis and are symmetric about the origin.

This is the first of two special cases we will investigate in this section.

The difference $|d_1 - d_2|$ remains constant for any point on the hyperbola.

The Hyperbola

$$\frac{x^2}{a^2} - \frac{y^2}{b^2} = 1$$

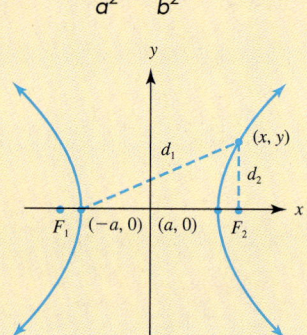

Before we try to sketch a hyperbola from its equation, let's examine the standard form more carefully. For

$$\frac{x^2}{a^2} - \frac{y^2}{b^2} = 1 \tag{1}$$

the graph is a hyperbola that opens to the right and left and is symmetric about the x axis. The points where this hyperbola intercepts the x axis are called the **vertices** of the hyperbola. The vertices of the hyperbola are located at $(a, 0)$ and at $(-a, 0)$.

As we move away from the center of the hyperbola, the *branches* of the hyperbola will approach two straight lines called the **asymptotes** of the hyperbola. The equations of the two asymptotes of the hyperbola are given by

$$y = \frac{b}{a}x \quad \text{and} \quad y = -\frac{b}{a}x$$

These asymptotes prove to be extremely useful aids in sketching the hyperbola. In fact, for most purposes, the vertices and the asymptotes are the only tools that we will need. The following example illustrates.

While we show these equations, you will see an easier method for finding the asymptotes in our first example.

Example 1 Graphing a Hyperbola

Sketch the hyperbola

$$\frac{x^2}{9} - \frac{y^2}{4} = 1$$

Step 1 The equation is in standard form.

Step 2 Find and plot the vertices.

From the standard form we can see that $a^2 = 9$ and $a = 3$ or -3. The vertices of the hyperbola then occur at $(3, 0)$ and $(-3, 0)$.

Step 3 Sketch the asymptotes.

Here is an easy way to sketch the asymptotes. Note again from the standard form (3) that $b^2 = 4$, so $b = 2$ or -2. Plot the points $(0, 2)$ and $(0, -2)$ on the y axis.

Draw (using dashed lines) the rectangle whose sides are parallel to the x and y axes and which pass through the points determined in steps 2 and 3.

Draw the diagonals of the rectangle (again using dashed lines), and then extend those diagonals to form the desired asymptotes.

Step 4 Sketch the hyperbola.

We now complete our task by sketching the hyperbola as two smooth curves, passing through the vertices, and approaching the asymptotes.

It is important to remember that the asymptotes are *not* a part of the graph. They are simply used as aids in sketching the graph as the branches get "closer and closer" to the lines.

The equation of the hyperbola also has both x^2 and y^2 terms. Here the coefficients of those terms have *opposite* signs. If the x^2 coefficient is *positive*, the hyperbola will open *horizontally*.

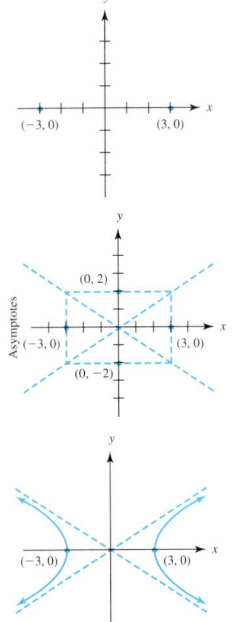

✓ CHECK YOURSELF 1

Sketch the hyperbola

$$\frac{x^2}{16} - \frac{y^2}{9} = 1$$

We now want to consider a second case of the hyperbola and its standard form. Suppose that the foci of the hyperbola are now on the *y* axis and symmetric about the origin. A sketch of such a hyperbola, and the equation that corresponds, follows.

The Hyperbola

$$\frac{y^2}{b^2} - \frac{x^2}{a^2} = 1 \qquad (2)$$

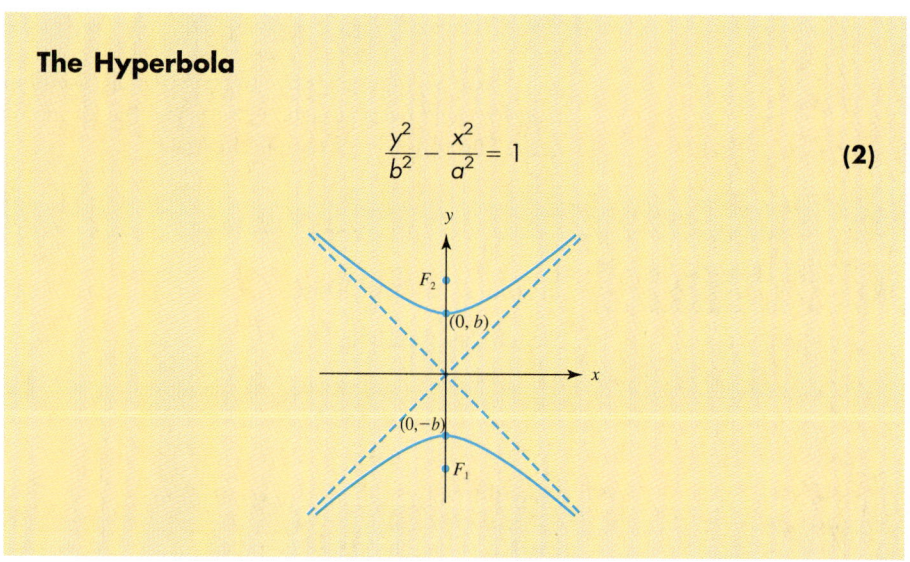

Some observations about this case are in order:

Here the vertices are on the y axis.
The asymptotes are the same as before.

1. The vertices of the hyperbola are now at $(0, b)$ and $(0, -b)$.
2. The asymptotes of the hyperbola have the equations

$$y = \frac{b}{a}x \qquad \text{and} \qquad y = -\frac{b}{a}x$$

The following example illustrates sketching a hyperbola in this case.

Example 2 Graphing a Hyperbola

Sketch the hyperbola

$$4y^2 - 25x^2 = 100$$

You can recognize this equation as corresponding to a hyperbola since the coefficients of the squared terms are *opposite* in sign. Since the y^2 coefficient is *positive*, the hyperbola will open *vertically*.

Step 1 Write the equation in the standard form of equation (2) by dividing both sides by 100:

$$\frac{4y^2}{100} - \frac{25x^2}{100} = \frac{100}{100}$$

$$\frac{y^2}{25} - \frac{x^2}{4} = 1$$

Step 2 Find the vertices.

From the standard form of equation (2), we see that since $b^2 = 25$, $b = 5$, or $b = -5$, so the vertices are at $(0, 5)$ and at $(0, -5)$.

Step 3 Sketch the asymptotes.

Also from the standard form of equation (2), we see that since $a^2 = 4$, $a = 2$ or $a = -2$.

Plot $(2, 0)$ and $(-2, 0)$ on the x axis, and complete the dashed rectangle as before. The diagonals once again extend to form the asymptotes.

Step 4 Sketch the hyperbola.

Draw smooth curves, through the intercepts, that approach the asymptotes to complete the graph.

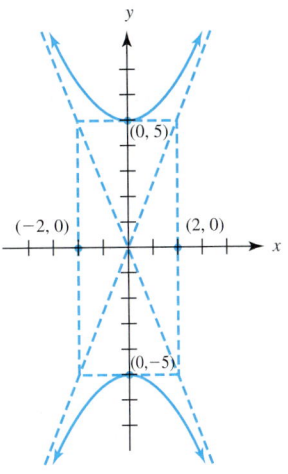

✓ **CHECK YOURSELF 2**

Sketch the hyperbola $9y^2 - 4x^2 = 36$.

The following algorithm summarizes our work with sketching hyperbolas.

Graphing the Hyperbola

1. Write the given equation in standard form.
2. Determine the vertices of the hyperbola.

   If the x^2 coefficient is positive, the vertices are at $(a, 0)$ and $(-a, 0)$ on the x axis.

   If the y^2 coefficient is positive, the vertices are at $(0, b)$ and $(0, -b)$ on the y axis.

> **Graphing the Hyperbola (*continued*)**
>
> 3. Sketch the asymptotes of the hyperbola.
>   Plot points $(a, 0)$, $(-a, 0)$, $(0, b)$, and $(0, -b)$. Form a rectangle from these points. The diagonals (extended) are the asymptotes of the hyperbola.
>
> 4. Sketch the hyperbola.
>   Draw smooth curves, through the intercepts and approaching the asymptotes.

The following chart shows all the equation forms considered in this chapter.

Curve	Example	Recognizing the Curve
Straight line	$4x - 3y = 12$	The equation involves x and/or y to the first power.
Parabola	$y = x^2 - 3x$ or $x = y^2 - 2y + 3$	Only one term, in x or in y, may be squared. The other variable appears to the first power.
Circle	$x^2 + 4x + y^2 = 5$	The equation has both x^2 and y^2 terms. The coefficients of those terms are equal.
Ellipse	$4x^2 + 9y^2 = 36$	The equation has both x^2 and y^2 terms. The coefficients of those terms have the same algebraic sign but different coefficients.
Hyperbola	$4x^2 - 9y^2 = 36$ or $9y^2 - 16x^2 = 144$	The equation has both x^2 and y^2 terms. The coefficients of those terms have different algebraic signs.

✓ CHECK YOURSELF ANSWERS

1. $\dfrac{x^2}{16} - \dfrac{y^2}{9} = 1$

2. $\dfrac{y^2}{4} - \dfrac{x^2}{9} = 1$

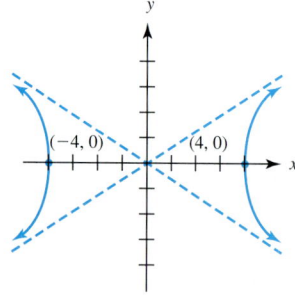

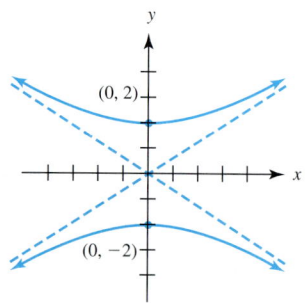

Exercises · 12.4

1. Circle
2. Hyperbola
3. Parabola
4. Ellipse
5. Hyperbola
6. Parabola
7. Hyperbola
8. Parabola
9. Circle
10. Ellipse
11. Hyperbola
12. Parabola

(a) $4x^2 + 25y^2 = 100$
(b) $y = x^2 - 2x - 3$
(c) $x = \dfrac{1}{2}y^2 - 2y$
(d) $\dfrac{x^2}{9} + \dfrac{y^2}{16} = 1$
(e) $\dfrac{y^2}{25} - \dfrac{x^2}{4} = 1$
(f) $16x^2 - 9y^2 = 144$
(g) $(x-2)^2 + (y-2)^2 = 9$
(h) $x^2 + y^2 = 16$

Identify the graph of each of the following equations as one of the conic sections (the parabola, circle, ellipse, or hyperbola).

1. $x^2 + y^2 = 16$
2. $\dfrac{x^2}{4} - \dfrac{y^2}{16} = 1$
3. $y = x^2 - 4$
4. $\dfrac{x^2}{16} + \dfrac{y^2}{9} = 1$
5. $9x^2 - 4y^2 = 36$
6. $x^2 = 4y$
7. $y^2 - 4x^2 = 4$
8. $x = y^2 - 2y + 1$
9. $x^2 - 6x + y^2 + 2x = 2$
10. $4x^2 + 25y^2 = 100$
11. $9y^2 - 16x^2 = 144$
12. $y = x^2 - 6x + 8$

Match each of the curves shown with the appropriate equation on the left.

13. (d)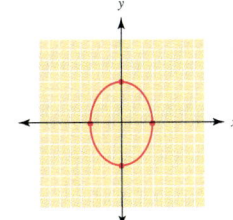

14. (f)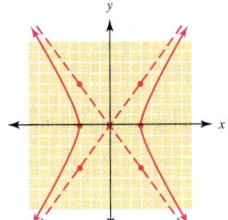

15. (h)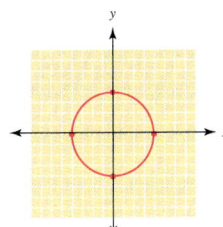

16. (a)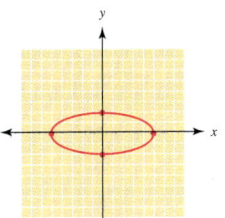

17. (b)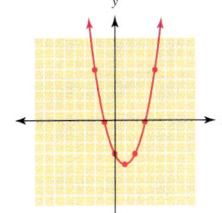

18. (g)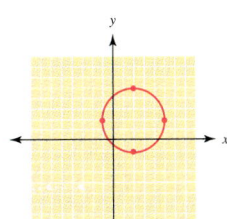

19. (c)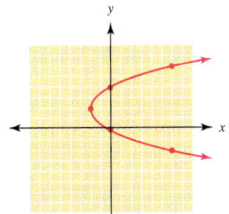

20. (e)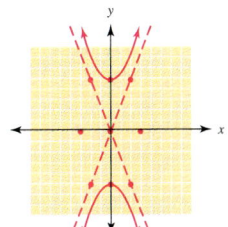

Graph the following hyperbolas by finding the vertices and asymptotes. If necessary, write the equation in standard form.

21. $\dfrac{x^2}{9} - \dfrac{y^2}{9} = 1$

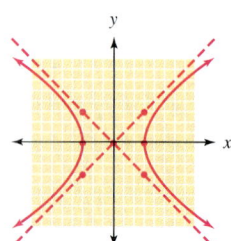

22. $\dfrac{y^2}{9} - \dfrac{x^2}{4} = 1$

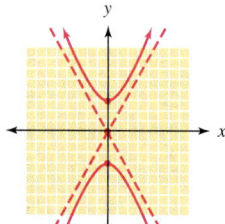

23. $\dfrac{y^2}{16} - \dfrac{x^2}{9} = 1$

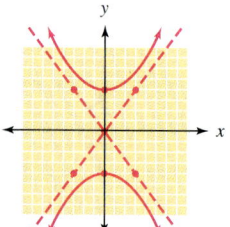

24. $\dfrac{x^2}{25} - \dfrac{y^2}{16} = 1$

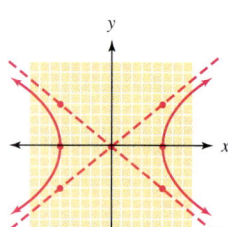

25. $\dfrac{x^2}{36} - \dfrac{y^2}{9} = 1$

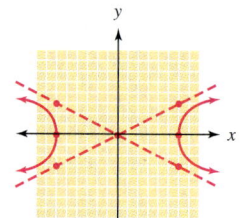

26. $\dfrac{y^2}{25} - \dfrac{x^2}{9} = 1$

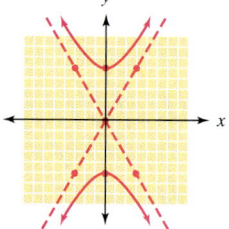

27. $x^2 - 9y^2 = 36$

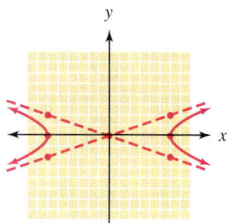

28. $y^2 - 4x^2 = 36$

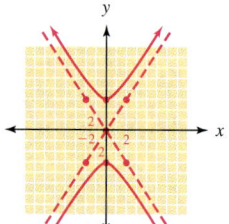

29. $9x^2 - 4y^2 = 36$

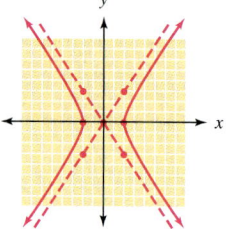

30. $9y^2 - 4x^2 = 36$

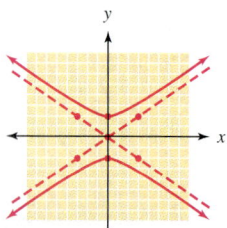

31. $16y^2 - 9x^2 = 144$

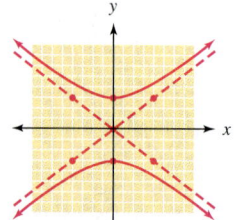

32. $4x^2 - 9y^2 = 36$

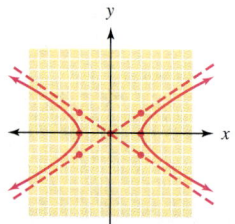

33. $25y^2 - 4x^2 = 100$

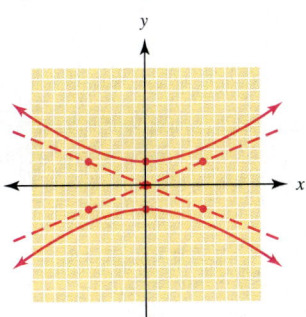

34. $9x^2 - 25y^2 = 225$

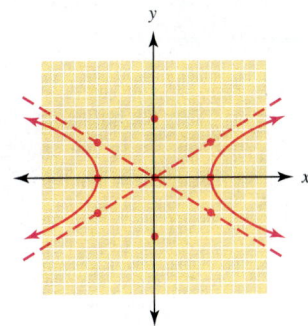

In Exercises 35 to 38, use a graphing utility to graph each equation.

35. $y^2 - 16x^2 = 16$

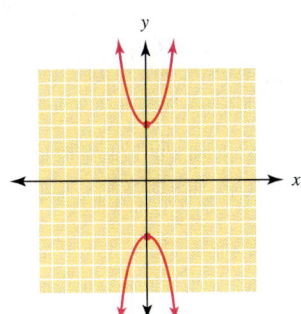

36. $y^2 - 25x^2 = 25$

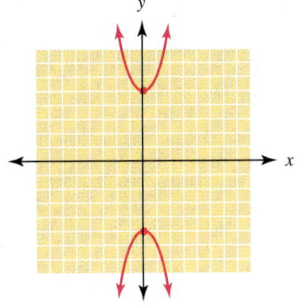

37. $16x^2 - 9y^2 = 144$

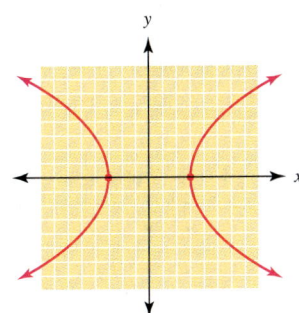

38. $25x^2 - 16y^2 = 400$

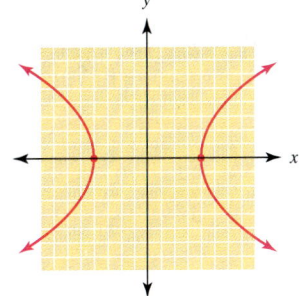

Summary Exercises • 12

This summary exercise set is provided to give you practice with each of the objectives in the chapter. Each chapter is keyed to the appropriate chapter section.

[12.1] In Exercises 1 and 2, use the given values of a, b, c, d, e, and f to rewrite the standard form for a conic section

$$ax^2 + by^2 + cxy + dx + ey + f = 0$$

into standard quadratic form.

1. $a = 2, b = 0, c = 0, d = 3, e = -5, f = -1$

2. $a = -3, b = 0, c = 0, d = -2, e = -3, f = 2$

[12.1] In Exercises 3 to 8, find the equation of the axis of symmetry, the coordinates of the vertex, and at least two symmetric points. Sketch the graph of each equation.

3. $x = -y^2 + 4y$

4. $x = y^2 - 4y$

5. $x = y^2 - 5y + 6$

6. $x = -y^2 + 3y - 2$

7. $x = 2y^2 - 4y + 9$

8. $x = -3y^2 + 5y - 8$

[12.2] Find the center and the radius of the graph of each equation.

9. $x^2 + y^2 = 16$

10. $x^2 + y^2 = 50$

11. $4x^2 + 4y^2 = 36$

12. $3x^2 + 3y^2 = 36$

13. $(x - 3)^2 + y^2 = 36$

14. $(x - 2)^2 + y^2 = 9$

15. $(x - 1)^2 + (y - 2)^2 = 16$

16. $x^2 + 6x + y^2 + 4y = 12$

17. $x^2 + 8x + y^2 + 10y = 23$

18. $x^2 - 6x + y^2 + 6y = 18$

Answers (left margin):

1. $y = \frac{2}{5}x^2 + \frac{3}{5}x - \frac{1}{5}$

2. $y = -x^2 + \frac{2}{3}x + \frac{2}{3}$

3.–8. See Instructor's Solution Manual for graphs.

9. Center: (0, 0); radius: 4

10. Center: (0, 0); radius: $\sqrt{50}$

11. Center: (0, 0); radius: 3

12. Center: (0, 0); radius: $2\sqrt{3}$

13. Center: (3, 0); radius: 6

14. Center: (2, 0); radius: 3

15. Center: (1, 2); radius: 4

16. Center: (−3, −2); radius: 5

17. Center: (−4, −5); radius: 8

18. Center: (3, −3); radius: 6

Summary Exercises

19.–36. See Instructor's Solution Manual for graphs
37. Line
38. Parabola
39. Circle
40. Line
41. Parabola
42. Circle
43. Parabola
44. Parabola
45. Ellipse
46. Circle
47. Hyperbola
48. Hyperbola
49. Ellipse
50. Circle
51. Hyperbola
52. Circle
53. Line
54. Parabola

[12.2] Graph each of the following.

19. $x^2 + y^2 = 16$
20. $4x^2 + 4y^2 = 36$
21. $x^2 + (y + 3)^2 = 25$
22. $(x - 2)^2 + y^2 = 9$
23. $(x - 1)^2 + (y - 2)^2 = 16$
24. $(x + 3)^2 + (y + 3)^2 = 25$
25. $x^2 + y^2 - 4y - 5 = 0$
26. $x^2 - 2x + y^2 - 6y = 6$

[12.3–12.4] Sketch the graph of each of the following equations.

27. $\dfrac{x^2}{36} + \dfrac{y^2}{4} = 1$
28. $\dfrac{x^2}{49} + \dfrac{y^2}{9} = 1$
29. $\dfrac{x^2}{25} + \dfrac{y^2}{9} = 1$
30. $\dfrac{x^2}{4} + \dfrac{y^2}{16} = 1$
31. $9x^2 + 4y^2 = 36$
32. $16x^2 + 9y^2 = 144$
33. $\dfrac{x^2}{9} - \dfrac{y^2}{4} = 1$
34. $\dfrac{y^2}{16} - \dfrac{x^2}{4} = 1$
35. $4x^2 - 9y^2 = 36$
36. $16x^2 - 9y^2 = 144$

[12.1–12.4] For each of the following equations decide whether its graph is a line, parabola, circle, ellipse, or hyperbola.

37. $x + y = 16$
38. $x + y^2 = 5$
39. $4x^2 + 4y^2 = 36$
40. $3x + 3y = 36$
41. $y = (x - 3)^2$
42. $(x - 2)^2 + y^2 = 9$
43. $y = (x - 1)^2 + 1$
44. $x = y^2 + 4y + 4$
45. $\dfrac{x^2}{4} + \dfrac{y^2}{25} = 1$
46. $x^2 - 6x + y^2 + 6y = 18$
47. $\dfrac{x^2}{9} - \dfrac{y^2}{25} = 1$
48. $9x^2 - 4y^2 = 36$
49. $16x^2 + 4y^2 = 64$
50. $x^2 = -y^2 + 18$
51. $\dfrac{x^2}{4} - \dfrac{y^2}{9} = 1$
52. $4x^2 + 4y^2 = 36$
53. $4x - 6y = 12$
54. $3x^2 - y + 4 = 0$

Self-Test 12

1. $y = -x^2 + 3x - \frac{3}{2}$
2. $y = -2$; $(-4, -2)$
3. $y = 3$; $(-1, 3)$
4. $y = 2$; $(6, 2)$
5. Center: $(3, -2)$; radius: 6
6. Center: $(-1, 2)$; radius: 5
7. Center: $(-3, 0)$; radius: 2
8. Center: $(2, -3)$; radius: 3

The purpose of this self-test is to help you check your progress and to review for a chapter test in class. Allow yourself about 1 hour to take the test. When you are done, check your answers in the back of the book. If you missed any answers, be sure to go back and review the appropriate sections in the chapter and the exercises that are provided.

1. Given the values $a = 2$, $b = 0$, $c = 0$, $d = -6$, $e = 2$, and $f = 3$, rewrite the standard form for a conic section ($ax^2 + by^2 + cxy + dx + ey + f = 0$) into standard quadratic form.

For Exercises 2 to 4, find the axis of symmetry, the coordinates of the vertex, and at least two symmetric points. Sketch the graph of each equation.

2. $x = y^2 + 4y$

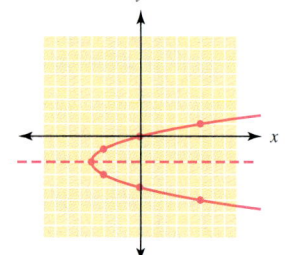

3. $x = 2y^2 - 12y + 17$

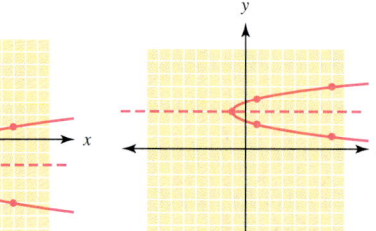

4. $x = -3y^2 + 12y - 6$

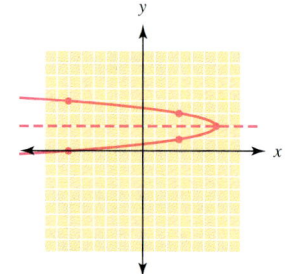

Find the coordinates for the center and the radius and then graph each equation.

5. $(x - 3)^2 + (y + 2)^2 = 36$

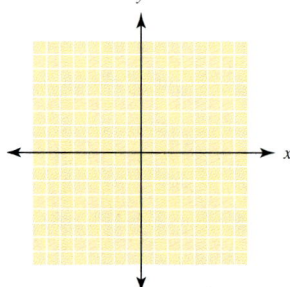

6. $x^2 + 2x + y^2 - 4y - 20 = 0$

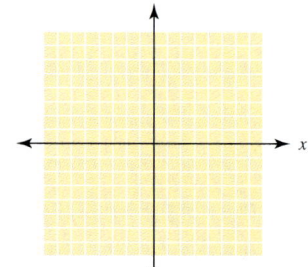

7. $x^2 + 6x + y^2 + 5 = 0$

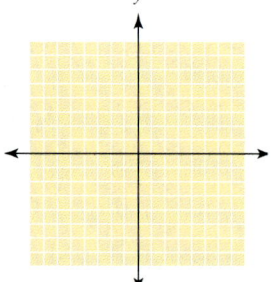

8. $(x - 2)^2 + (y + 3)^2 = 9$

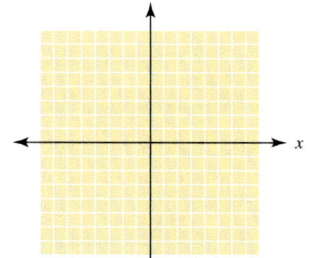

13. Line
14. Parabola
15. Circle
16. Circle
17. Parabola
18. Ellipse
19. Hyperbola
20. Circle

Sketch the graph of each of the following equations.

9. $\dfrac{x^2}{25} + \dfrac{y^2}{9} = 1$

10. $16x^2 + 4y^2 = 64$

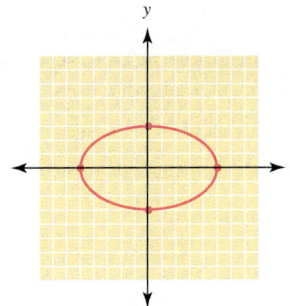
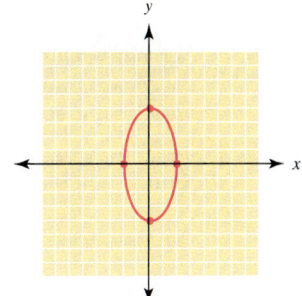

11. $\dfrac{x^2}{9} - \dfrac{y^2}{16} = 1$

12. $4y^2 - 25y^2 = 1$

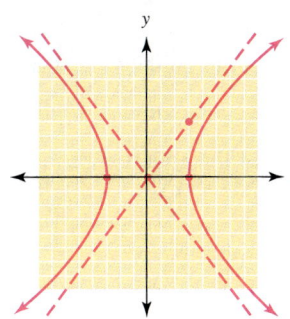
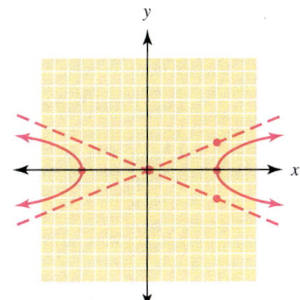

For each of the following equations, decide whether its graph is a line, parabola, circle, ellipse, or hyperbola.

13. $3x - 4y - 1 = 0$

14. $2x^2 + 3y = 9$

15. $5x^2 + 5y^2 = 125$

16. $(x - 4)^2 + y^2 = 36$

17. $4y^2 - 7x = 15$

18. $3x^2 + 5y^2 = 15$

19. $-5x^2 + 6y^2 = 30$

20. $2x^2 + 2y^2 + 4x - 6y = 12$

Cumulative Test 1–12

1. $\left\{\dfrac{11}{2}\right\}$
2. $\{x | x > -2\}$
3. $\{-1, 4\}$
4. $\left\{x \middle| -4 \leq x \leq \dfrac{2}{3}\right\}$
5. $\left\{x \middle| x < -\dfrac{17}{5} \text{ or } x > 5\right\}$
6. $\{8, -3\}$
7.–8. See answers
9. 24.5
10. 7
11. $f(x) = -x + 3$
12. $2x^2 - 5x - 3$
13. $9x^2 - 12x + 4$
14. $(x - 3)(x^2 - 5)$
15.–17. See answers
18. $\left\{\left(-10, \dfrac{26}{3}\right)\right\}$
19. 8 cm by 19 cm
20. 37

This test is provided to help you in the process of reviewing the previous chapters. Answers are provided in the back of the book. If you missed any answers, be sure to go back and review the appropriate sections.

Solve each of the following.

1. $3x - 2(x + 5) = 12 - 3x$
2. $2x - 7 < 3x - 5$
3. $|2x - 3| = 5$
4. $|3x + 5| \leq 7$
5. $|5x - 4| > 21$
6. $x^2 - 5x - 24 = 0$

Graph each of the following.

7. $5x + 7y = 35$
8. $2x + 3y < 6$

9. Find the distance between the points $(-1, 2)$ and $(4, -22)$.

10. Find the slope of the line connecting $(4, 6)$ and $(3, -1)$.

11. Write the function form of the equation of the line that passes through the points $(-1, 4)$ and $(5, -2)$.

Simplify the following polynomials.

12. $(2x + 1)(x - 3)$
13. $(3x - 2)^2$

14. Completely factor the function $f(x) = x^3 - 3x^2 - 5x + 15$.

Graph the following:

15. $y = x^2 - 6x + 5$
16. $(x + 1)^2 + (y - 2)^2 = 25$
17. $\dfrac{x^2}{64} + \dfrac{y^2}{9} = 1$

Solve the following system of equations.

18. $2x + 3y = 6$
    $5x + 3y = -24$

Solve each of the following applications.

19. **Geometry.** The length of a rectangle is 3 cm more than twice its width. If the perimeter of the rectangle is 54 cm, find the dimensions of the rectangle.

20. **Number problem.** The sum of the digits of a two-digit number is 10. If the digits are reversed, the new number is 36 less than the original number. What was the original number?

CHAPTER 13

Exponential and Logarithmic Functions

LIST OF SECTIONS

13.1 Inverse Relations and Functions

13.2 Exponential Functions

13.3 Logarithmic Functions

13.4 Properties of Logarithms

13.5 Logarithmic and Exponential Equations

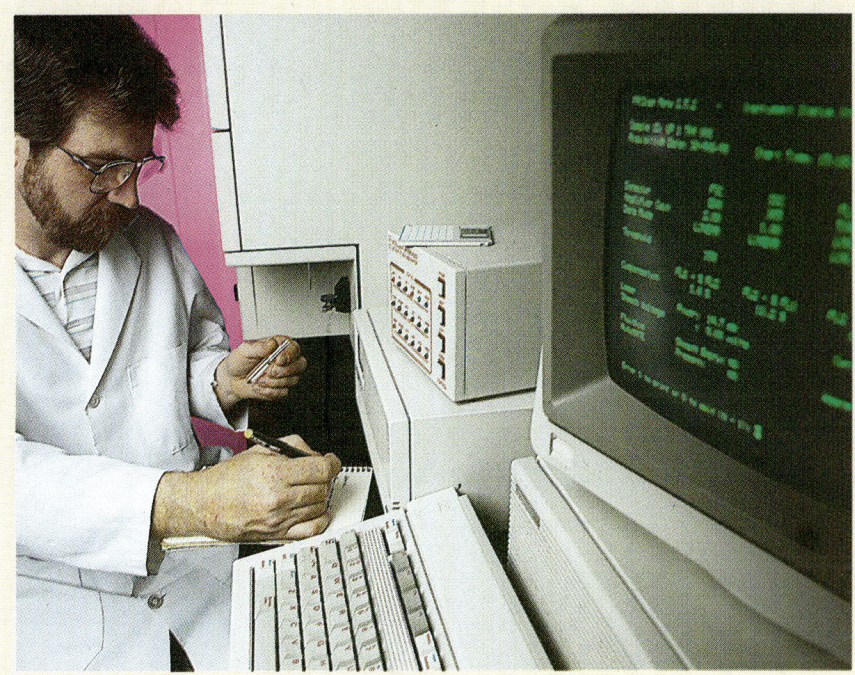

Pharmacologists researching the effects of drugs use exponential and logarithmic functions to model drug absorption and elimination. After a drug is taken orally, it is distributed throughout the body via the circulatory system. Once in the bloodstream, the drug is carried to the body's organs, where it is first absorbed and then eliminated again into the bloodstream. For a medicine or drug to be effective, there must be enough of the substance in the body to achieve the desired effect but not enough to cause harm. This therapeutic level is maintained by taking the proper dosage at timed intervals determined by the rate the body absorbs or eliminates the medicine.

The rate at which the body eliminates the drug is proportional to the amount of the drug present. That is, the more drug there is, the faster the drug is eliminated. The amount of a drug dosage, P, still left after a number of hours, t, is affected by the **half-life** of the drug. In this case, the half-life is how many hours it takes for the body to use up or eliminate half the drug dosage.

If P is the amount of an initial dose, and H is the time it takes the body to eliminate half a dose of a drug, then the amount of the drug still remaining in the system after t units of time is

$$A(t) = Pe^{t\left(\frac{-\ln 2}{H}\right)}$$

If the amount of an initial dose of a drug is 30 mg and if the half-life of the drug in the body is 4 hours, the amount in mg of the drug still in the body t hours after one dose is given by the following formula:

$$A(t) = 30e^{-0.173t}$$

SECTION 13.1 Inverse Relations and Functions

13.1 OBJECTIVES

1. Find the inverse of a relation
2. Graph a relation and its inverse
3. Find the inverse of a function
4. Graph a function and its inverse
5. Identify a one-to-one function

Suppose we are given the relation

$$\{(1, 2), (2, 4), (3, 6)\} \qquad (1)$$

If we *interchange* the first and second components (the x and y values) of each of the ordered pairs in relation (1), we have

$$\{(2, 1), (4, 2), (6, 3)\} \qquad (2)$$

which is another relation. Relations (1) and (2) are called **inverse relations**.

Inverse of a Relation

The *inverse* of a relation is formed by interchanging the components of each of the ordered pairs in the given relation.

We can form an inverse relation by interchanging the roles of x and y in the defining equation. Example 1 illustrates this concept.

Example 1 Finding the Inverse of a Relation

Find the inverse of the relation.

$$f = \{(x, y) | y = 2x - 4\} \qquad (3)$$

First interchange variables x and y to obtain

$$x = 2y - 4$$

Note that x and y have been interchanged from the original equation.

We now solve the defining equation for y.

$$2y = x + 4 \qquad \text{or} \qquad y = \frac{1}{2}x + 2$$

Then, we rewrite the relation in the equivalent form.

$$f^{-1} = \left\{(x, y) \Big| y = \frac{1}{2}x + 2\right\} \qquad (4)$$

Note: The notation f^{-1} has a different meaning from the negative exponent, as in x^{-1} or $\frac{1}{x}$.

The inverse of the original relation (3) is now shown in relation (4) with the defining equation "solved for y." That inverse is denoted f^{-1} (this is read as "f inverse"). We use the notation f^{-1} to indicate the inverse of f when that inverse is *also a function*.

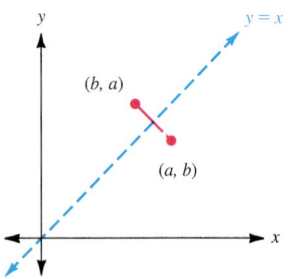

✓ CHECK YOURSELF 1

Write the inverse relation for $g = \{(x, y) | y = 3x + 6\}$.

The graphs of relations and their inverses are related in an interesting way. First, note that the graphs of the ordered pairs (a, b) and (b, a) always have symmetry about the line $y = x$.

Now, with this symmetry in mind, let's consider Example 2.

Example 2 — Graphing a Relation and Its Inverse

Graph the relation f from Example 1 along with its inverse.

Recall that

$$f = \{(x, y) | y = 2x - 4\}$$

and

$$f^{-1} = \left\{(x, y) \,\middle|\, y = \frac{1}{2}x + 2\right\}$$

The graphs of f and f^{-1} are shown below.

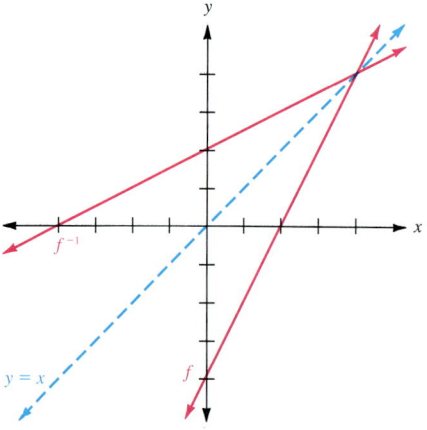

Note that the graphs of f and f^{-1} are symmetric about the line $y = x$. That symmetry follows from our earlier observation about the pairs (a, b) and (b, a) since we simply reversed the roles of x and y in forming the inverse relation.

✓ CHECK YOURSELF 2

Graph the relation g from the Check Yourself 1 exercise along with its inverse.

From our work thus far, it should be apparent that every relation has an inverse. However, that inverse may or may not be a function.

Example 3 — Finding the Inverse of a Function

Find the inverses of the following functions.

(a) $f = \{(1, 3), (2, 4), (3, 9)\}$

Its inverse is

$$\{(3, 1), (4, 2), (9, 3)\}$$

The elements of the ordered pairs have been interchanged.

which is also a function.

(b) $g = \{(1, 3), (2, 6), (3, 6)\}$

Its inverse is

$$\{(3, 1), (6, 2), (6, 3)\}$$

It is not a function because 6 is mapped to both 2 and 3.

which is *not* a function.

✓ CHECK YOURSELF 3

Write the inverses for each of the following relations. Which of the inverses are also functions?

(a) $\{(-1, 2), (0, 3), (1, 4)\}$ (b) $\{(2, 5), (3, 7), (4, 5)\}$

Can we predict in advance whether the inverse of a function will also be a function? The answer is yes.

We already know that for a relation to be a function, no element in its domain can be associated with more than one element in its range.

In addition, if the inverse of a function is to be a function, no element in the range can be associated with more than one element in the domain—that is, no two distinct ordered pairs in the function can have the same second component. A function that satisfies this additional restriction is called a **one-to-one function**.

The function in Example 3, part a,

$$f = \{(1, 3), (2, 4), (3, 9)\}$$

is a one-to-one function and its inverse is also a function. However, the function in Example 3, part b,

$$g = \{(1, 3), (2, 6), (3, 6)\}$$

is *not* a one-to-one function, and its inverse is *not* a function.

From those observations we can state the following general result.

> **Inverse of a Function**
>
> A function f has an inverse f^{-1}, which is also a function, if and only if f is a one-to-one function.

Because the statement is an "if and only if" statement, it can be turned around without changing the meaning. Here we use the same statement as a definition for a one-to-one function.

> **One-to-One Function**
>
> A function f is a *one-to-one function* if and only if it has an inverse f^{-1}, which is also a function.

Our result regarding a one-to-one function and its inverse also has a convenient graphical interpretation, as Example 4 illustrates.

Example 4 — Graphing a Function and Its Inverse

Graph each function and its inverse. State which inverses are functions.

(a) $f = \{(x, y) | y = 4x - 8\}$

Since f is a one-to-one function (no value for y can be associated with more than one value for x), its inverse is also a function. Here,

$$f^{-1} = \left\{(x, y) \,\middle|\, y = \frac{1}{4}x + 2\right\}$$

This is a **linear function** of the form $f = \{(x, y) | y = mx + b\}$. Its graph is a straight line. A linear function, where $m \neq 0$, is always one-to-one.

The graphs of f and f^{-1} are shown in the figure.

> The vertical-line test tells us that *both* f and f^{-1} are functions.

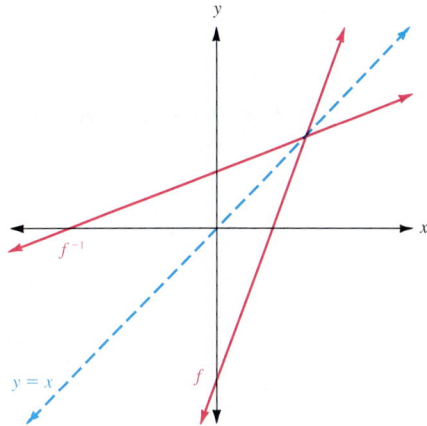

(b) $g = \{(x, y) | y = x^2\}$

This is a **quadratic function** of the form

$$g = \{(x, y) | y = ax^2 + bx + c\} \quad \text{where } a \neq 0$$

Its graph is always a parabola, and a quadratic function is *not* a one-to-one function.

For instance, 4 in the range is associated with both 2 and -2 from the domain. It follows that the inverse of g

$$\{(x, y) | x = y^2\}$$

or

$$\{(x, y) | y = \pm\sqrt{x}\}$$

> By the vertical-line test, we see that the inverse of g is *not* a function because g was *not* one-to-one.

is *not* a function. The graphs of g and its inverse are shown below.

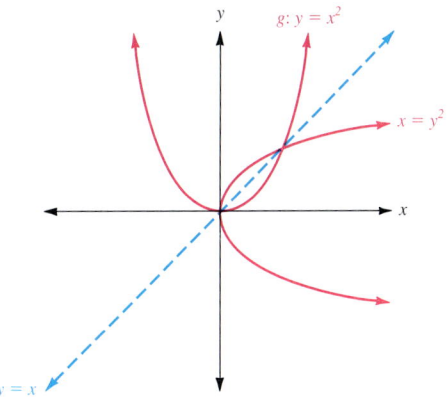

Note: When a function is not one-to-one, as in Example 4, part b, we can restrict the domain of the function so that it will be one-to-one. In this case, if we redefine function g as

$$g = \{(x, y) | y = x^2, x \geq 0\}$$

The domain is now restricted to nonnegative values for x.

it will be one-to-one and its inverse

$$g^{-1} = \{(x, y) | y = \sqrt{x}\}$$

will be a function, as shown in the following graph.

The function g is now one-to-one, and its inverse g^{-1} is also a function.

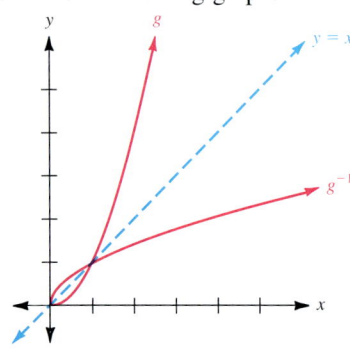

✓ CHECK YOURSELF 4

Graph each function and its inverse. Which inverses are functions?

(a) $f = \{(x, y) | y = 2x - 2\}$ (b) $g = \{(x, y) | y = 2x^2\}$

It is easy to tell from the graph of a function whether that function is one-to-one. If any horizontal line can meet the graph of a function in at most one point, the function is one-to-one. Example 5 illustrates this approach.

Example 5 Identifying a One-to-One Function

Which of the following graphs represents one-to-one functions?

(a)

Since no horizontal line passes through any two points of the graph, f is one-to-one.

One-to-one

(b)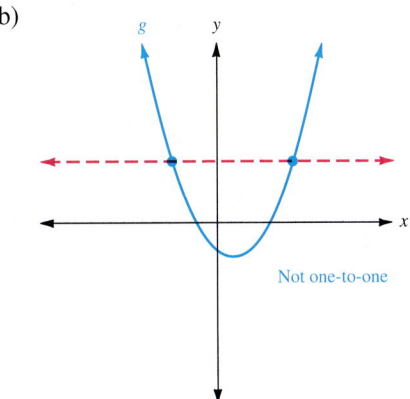

Since a horizontal line can meet the graph of function g at two points, g is *not* a one-to-one function.

 CHECK YOURSELF 5

Consider the graphs of the functions of Check Yourself 4. Which functions are one-to-one?

The following algorithm summarizes our work in this section.

Finding Inverse Relations and Functions

1. Interchange the x and y components of the ordered pairs of the given relation or the roles of x and y in the defining equation.
2. If the relation was described in equation form, solve the defining equation of the inverse for y.
3. If desired, graph the relation and its inverse on the same set of axes. The two graphs will be symmetric about the line $y = x$.

✓ CHECK YOURSELF ANSWERS

1. $g^{-1} = \left\{(x, y) | y = \dfrac{1}{3}x - 2\right\}$ 2.

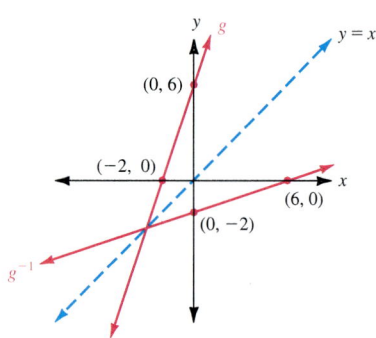

3. (a) $\{(2, -1), (3, 0), (4, 1)\}$, a function; (b) $\{(5, 2), (7, 3), (5, 4)\}$, *not* a function.
4. (a) $f = \{(x, y) | y = 2x - 2\}$, $f^{-1} = \left\{(x, y) | y = \dfrac{1}{2}x + 1\right\}$, the inverse is a function.

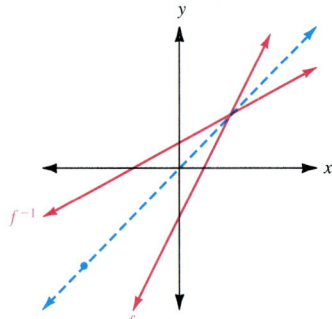

(b) $g = \{(x, y) | y = 2x^2\}$, $g^{-1} = \left\{(x, y) | y = \pm\sqrt{\dfrac{x}{2}}\right\}$, the inverse is not a function.

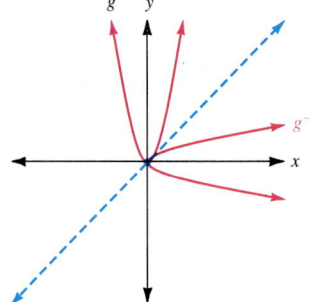

5. (a) Is one-to-one; (b) is not one-to-one.

Exercises · 13.1

1. {(3, 2), (4, 3), (5, 4)}; function
2. {(3, 2), (4, 3), (3, 4)}; not a function
3. {(2, 1), (2, 2), (2, 3)}; not a function
4. {(9, 5) (7, 3), (5, 7)}; function
5. {(4, 2), (9, 3), (16, 4)}; function
6. {(2, −1), (3, 0), (2, 1)}; not a function
7. $y = \frac{1}{2}x - 4$
8. $y = -\frac{1}{2}x - 2$
9. $y = 2x + 1$
10. $y = 3x - 1$
11. $y = \pm \sqrt{x + 1}$ or $x = y^2 - 1$
12. $y = \pm \sqrt{-x + 2}$
13. $4x^2 + y^2 = 36$
14. $x^2 + 4y^2 = 36$
15. $y^2 - x^2 = 9$
16. $4x^2 - y^2 = 4$

In Exercises 1 to 6, write the inverse relation for each function. In each case, decide whether the inverse relation is also a function.

1. {(2, 3), (3, 4), (4, 5)}
2. {(2, 3), (3, 4), (4, 3)}
3. {(1, 2), (2, 2), (3, 2)}
4. {(5, 9), (3, 7), (7, 5)}
5. {(2, 4), (3, 9), (4, 16)}
6. {(−1, 2), (0, 3), (1, 2)}

In Exercises 7 to 16, write an equation for the inverse of the relation defined by each equation.

7. $y = 2x + 8$
8. $y = -2x - 4$
9. $y = \frac{x - 1}{2}$
10. $y = \frac{x + 1}{3}$
11. $y = x^2 - 1$
12. $y = -x^2 + 2$
13. $x^2 + 4y^2 = 36$
14. $4x^2 + y^2 = 36$
15. $x^2 - y^2 = 9$
16. $4y^2 - x^2 = 4$

In Exercises 17 to 22, write an equation for the inverse of the relation defined by each of the following, and graph the relation and its inverse on the same set of axes. Determine which inverse relations are also functions.

17. $y = 3x - 6$

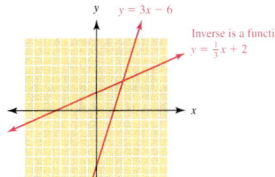

18. $y = 4x + 8$

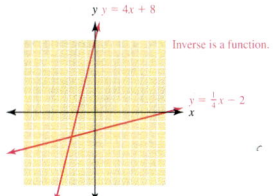

19. $2x - 3y = 6$

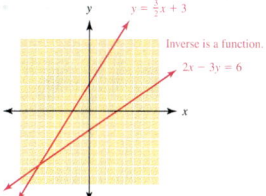

20. $y = 3$

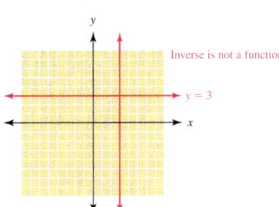

21. $y = x^2 + 1$

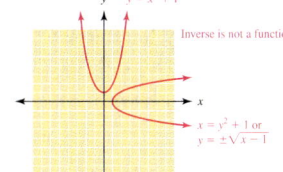

22. $y = -x^2 + 1$

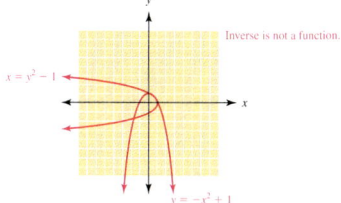

25. 12

26. 4

27. 6

28. 6

29. x

30. x

31. 2

32. 5

33. 3

34. 3

35. x

36. x

37. 16

38. -2

39. 4

40. 4

41. x

42. x

43. 5

44. 4

45. $f^{-1}(x) = \dfrac{x-b}{m}$

46. $\dfrac{1}{3}$

47. $\dfrac{5}{2}$

48. $h(x) = \dfrac{5x+3}{2}$

49. 7

50. x

51. x

52. $f^{-1}(x) = \dfrac{3x+7}{4}$

23. An inverse process is an operation that undoes a procedure. If the procedure is wrapping a present, describe in detail the inverse process.

24. If the procedure is the series of steps that take you from home to your classroom, describe the inverse process.

If $f(x) = 3x - 6$, then $f^{-1}(x) = \dfrac{1}{3}x + 2$. Given these two functions, in Exercises 25 to 30, find each of the following.

25. $f(6)$ 26. $f^{-1}(6)$ 27. $f(f^{-1}(6))$

28. $f^{-1}(f(6))$ 29. $f(f^{-1}(x))$ 30. $f^{-1}(f(x))$

If $g(x) = \dfrac{x+1}{2}$, then $g^{-1}(x) = 2x - 1$. Given these two functions, in Exercises 31 to 36, find each of the following.

31. $g(3)$ 32. $g^{-1}(3)$ 33. $g(g^{-1}(3))$

34. $g^{-1}(g(3))$ 35. $g(g^{-1}(x))$ 36. $g^{-1}(g(x))$

Given $h(x) = 2x + 8$, in Exercises 37 to 42, find each of the following.

37. $h(4)$ 38. $h^{-1}(4)$ 39. $h(h^{-1}(4))$

40. $h^{-1}(h(4))$ 41. $h(h^{-1}(x))$ 42. $h^{-1}(h(x))$

Suppose that f and g are one-to-one functions.

43. If $f(5) = 7$, find $f^{-1}(7)$.   44. If $g^{-1}(4) = 9$, find $g(9)$.

Let f be a linear function; i.e., let $f(x) = mx + b$.

45. Find $f^{-1}(x)$.

46. Based on Exercise 45, if the slope of f is 3, what is the slope of f^{-1}?

47. Based on Exercise 45, if the slope of f is $\dfrac{2}{5}$, what is the slope of f^{-1}?

Consider the function $g(x) = \dfrac{2x-3}{5}$. If you choose a value for x, to find $g(x)$, you first multiply the chosen value by 2, then subtract 3, and finally divide by 5. Imagine a new function h, defined by exactly the opposite of the operations just mentioned, listed in reverse order: multiply by 5, then add 3, and finally divide this result by 2.

48. Write the equation for $h(x)$.

49. Find $g(19)$.   50. Find $g(h(x))$.   51. Find $h(g(x))$.

52. Use the method described to find f^{-1}, where $f(x) = \dfrac{4x-7}{3}$.

SECTION 13.2 Exponential Functions

13.2 OBJECTIVES

1. Graph an exponential function
2. Solve an application of exponential functions
3. Solve an elementary exponential equation

Up to this point in the book, we have worked with polynomial functions and other functions in which the variable was used as a base. We now want to turn to a new classification of functions, the **exponential function.**

Exponential functions are functions whose defining equations involve the variable as an *exponent*. The introduction of these functions will allow us to consider many further applications, including population growth and radioactive decay.

Exponential Functions

An *exponential function* is a function that can be expressed in the form

$$f(x) = b^x$$

where $b > 0$ and $b \neq 1$. We call b the *base* of the exponential function.

The following are examples of exponential functions.

$$f(x) = 2^x \qquad g(x) = 3^x \qquad h(x) = \left(\frac{1}{2}\right)^x$$

As we have done with other new functions, we begin by finding some function values. We then use that information to graph the function.

Example 1

Graphing an Exponential Function

Graph the exponential function

$$f(x) = 2^x$$

First, choose convenient values for x.

Note:

$$2^{-2} = \frac{1}{2^2} = \frac{1}{4}$$

$$f(0) = 2^0 = 1 \qquad f(-1) = 2^{-1} = \frac{1}{2} \qquad f(1) = 2^1 = 2$$

$$f(-2) = 2^{-2} = \frac{1}{4} \qquad f(2) = 2^2 = 4 \qquad f(-3) = 2^{-3} = \frac{1}{8}$$

Next, form a table from these values. Then, plot the corresponding points, and connect them with a smooth curve for the desired graph.

x	$f(x)$
-3	0.125
-2	0.25
-1	0.5
0	1
1	2
2	4
3	8

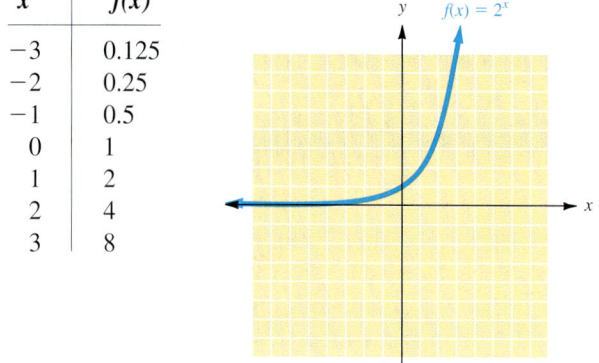

A vertical line will cross the graph at one point at most. The same is true for a horizontal line.

There is no value for x such that
$$2^x = 0$$
so the graph never touches the x axis.

We call $y = 0$ (or the x axis) the **horizontal asymptote**.

Let's examine some characteristics of the graph of the exponential function. First, the vertical-line test shows that this is indeed the graph of a function. Also note that the horizontal-line test shows that the function is one-to-one.

The graph *approaches* the x axis on the left, but it does *not intersect* the x axis. The y intercept is 1 (because $2^0 = 1$ by definition). To the right the functional values get larger. We say that the values *grow without bound*. This same language may be applied to linear or quadratic functions.

✓ CHECK YOURSELF 1

Sketch the graph of the exponential function

$$g(x) = 3^x$$

Let's look at an example in which the base of the function is less than 1.

Example 2 Graphing an Exponential Function

Graph the exponential function

$$f(x) = \left(\frac{1}{2}\right)^x$$

Recall that
$$\left(\frac{1}{2}\right)^x = 2^{-x}$$

First, choose convenient values for x.

$$f(0) = \left(\frac{1}{2}\right)^0 = 1 \qquad f(-1) = \left(\frac{1}{2}\right)^{-1} = 2 \qquad f(1) = \left(\frac{1}{2}\right)^1 = \frac{1}{2}$$

$$f(-2) = \left(\frac{1}{2}\right)^{-2} = 4 \qquad f(2) = \left(\frac{1}{2}\right)^2 = \frac{1}{4} \qquad f(-3) = \left(\frac{1}{2}\right)^{-3} = 8$$

$$f(3) = \left(\frac{1}{2}\right)^3 = \frac{1}{8}$$

Again, form a table of values and graph the desired function.

Again, by the vertical- and horizontal-line tests, this is the graph of a one-to-one function.

x	$f(x)$
-3	8
-2	4
-1	2
0	1
1	0.5
2	0.25
3	0.125

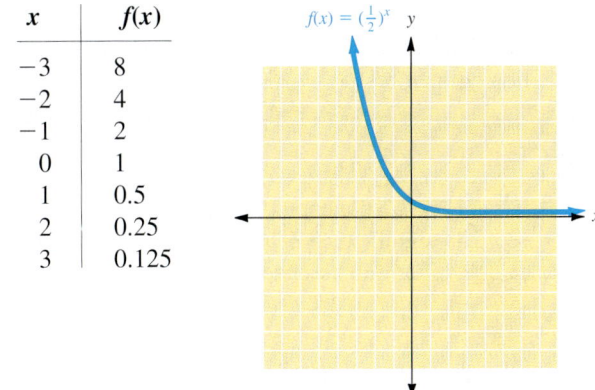

Let's compare this graph and that of Example 1. Clearly, this graph also represents a one-to-one function. As was true in the first example, the graph does not intersect the x axis but approaches that axis, here on the right. The values for the function again grow without bound, but this time on the left. The y intercept for both graphs occurs at 1.

The base of a *growth function* is *greater than* 1.

Note that the graph of Example 1 was *increasing* (going up) as we moved from left to right. That function is an example of a **growth function.**

The base of a *decay function* is *less than* 1 but greater than 0.

The graph of Example 2 was *decreasing* (going down) as we moved from left to right. It is an example of a **decay function.**

✓ CHECK YOURSELF 2

Sketch the graph of the exponential function

$$g(x) = \left(\frac{1}{3}\right)^x$$

The following algorithm summarizes our work thus far in this section.

Graphing an Exponential Function

Step 1 Establish a table of values by considering the function in the form $y = b^x$.
Step 2 Plot points from that table of values and connect them with a smooth curve to form the graph.
Step 3 If $b > 1$, the graph increases from left to right. If $0 < b < 1$, the graph decreases from left to right.
Step 4 All graphs will have the following in common:
(a) The y intercept will be 1.
(b) The graphs will approach, but not touch, the x axis.
(c) The graphs will represent one-to-one functions.

824 Chapter 13 ▪ Exponential and Logarithmic Functions

The use of the letter *e* as a base originated with Leonhard Euler (1707–1783), and *e* is sometimes called *Euler's number* for that reason.

We used bases of 2 and $\frac{1}{2}$ for the exponential functions of our examples because they provided convenient computations. A far more important base for an exponential function is an irrational number named *e*. In fact, when *e* is used as a base, the function defined by

$$f(x) = e^x$$

is called *the* exponential function.

The significance of this number will be made clear in later courses, particularly calculus. For our purposes, *e* can be approximated as

$$e \approx 2.71828$$

Graph $y = e^x$ on your calculator. You may find the $\boxed{e^x}$ key to be the second (or inverse) function to the ln *x* key. Note that e^1 is approximately 2.71828.

The graph of $f(x) = e^x$ is shown below. Of course, it is very similar to the graphs seen earlier in this section.

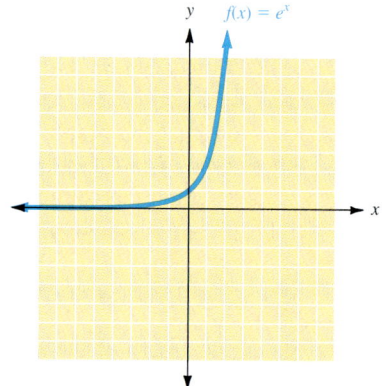

Exponential expressions involving base *e* occur frequently in real-world applications. Example 3 illustrates this approach.

Example 3   A Population Application

Be certain that you enclose the multiplication (0.05 × 5) in parentheses or the calculator will misinterpret your intended order of operation.

(a) Suppose that the population of a city is presently 20,000 and that the population is expected to grow at a rate of 5% per year. The equation

$$P(t) = 20{,}000e^{(0.05)t}$$

gives the town's population after *t* years. Find the population in 5 years.
Let $t = 5$ in the original equation to obtain

$$P(5) = 20{,}000e^{(0.05)(5)} \approx 25{,}681$$

which is the population expected 5 years from now.

Section 13.2 ■ Exponential Functions

Continuous compounding will give the highest accumulation of interest at any rate. However, daily compounding will result in an amount of interest that is only slightly less.

Note that in 9 years the amount in the account is a little more than double *the original principal.*

(b) Suppose $1000 is invested at an annual rate of 8%, compounded continuously. The equation

$$A(t) = 1000e^{0.08t}$$

gives the amount in the account after t years. Find the amount after 9 years.

Let $t = 9$ in the original equation to obtain

$$A(9) = 1000e^{(0.08)(9)} \approx 2054$$

which is the amount in the account after 9 years.

✓ CHECK YOURSELF 3

If $1000 is invested at an annual rate of 6%, compounded continuously, then the equation for the amount in the account after t years is

$$A(t) = 1000e^{0.06t}$$

Use your calculator to find the amount in the account after 12 years.

As we observed, the exponential function is always one-to-one. This yields an important property that can be used to solve certain types of equations involving exponents.

> If $b > 0$ and $b \neq 1$, then
>
> $$b^m = b^n \quad \text{if and only if} \quad m = n \qquad (1)$$

Example 4 Solving an Exponential Equation

(a) Solve $2^x = 8$ for x.

We recognize that 8 is a power of 2, and we can write the equation as

$$2^x = 2^3 \qquad \text{Write with equal bases.}$$

Applying property (1) above, we have

$$x = 3 \qquad \text{Set exponents equal.}$$

and 3 is the solution.

(b) Solve $3^{2x} = 81$ for x.

Since $81 = 3^4$, we can write

$$3^{2x} = 3^4$$
$$2x = 4$$
$$x = 2$$

The answer can easily be checked by substitution. Letting $x = 2$ gives
$$3^{2(2)} = 3^4 = 81$$

We see that 2 is the solution for the equation.

(c) Solve $2^{x+1} = \dfrac{1}{16}$ for x.

Again, we write $\dfrac{1}{16}$ as a power of 2, so that

$$2^{x+1} = 2^{-4}$$

Note:
$$\dfrac{1}{16} = \dfrac{1}{2^4} = 2^{-4}$$

Then

$$x + 1 = -4$$
$$x = -5$$

To verify the solution.
$$2^{-5+1} \stackrel{?}{=} \dfrac{1}{16}$$
$$2^{-4} \stackrel{?}{=} \dfrac{1}{16}$$
$$\dfrac{1}{16} = \dfrac{1}{16}$$

The solution set is $\{-5\}$.

✓ CHECK YOURSELF 4

Solve each of the following equations for x.

(a) $2^x = 16$ (b) $4^{x+1} = 64$ (c) $3^{2x} = \dfrac{1}{81}$

✓ CHECK YOURSELF ANSWERS

1. $y = g(x) = 3^x$.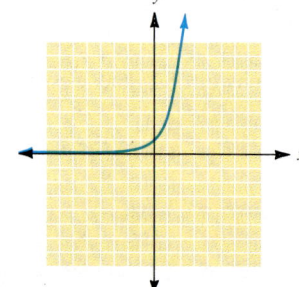

2. $y = \left(\dfrac{1}{3}\right)^x$.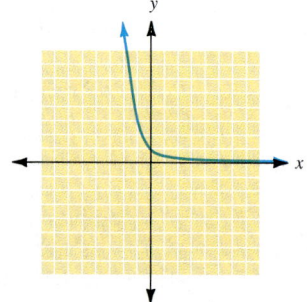

3. $2054.43.   **4.** (a) $\{4\}$; (b) $\{2\}$; (c) $\{-2\}$.

Exercises • 13.2

1. c
2. d
3. b
4. a
5. h
6. e
7. f
8. g

Match the graphs in Exercises 1 to 8 with the appropriate equation.

(a) $y = \left(\dfrac{1}{2}\right)^x$ (b) $y = 2x - 1$ (c) $y = 2^x$ (d) $y = x^2$

(e) $y = 1^x$ (f) $y = 5^x$ (g) $x = 2^y$ (h) $x = y^2$

1.
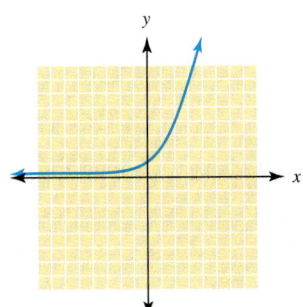

2.
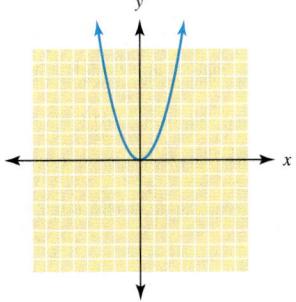

3.
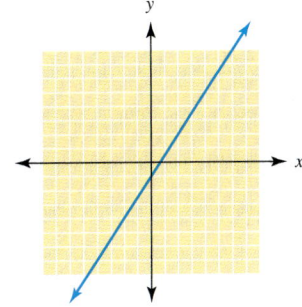

4.
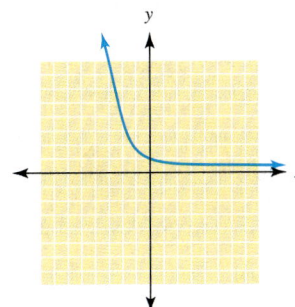

5.
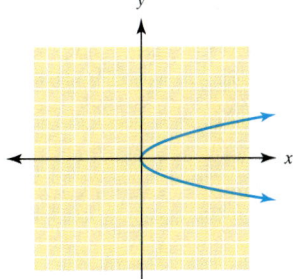

6.
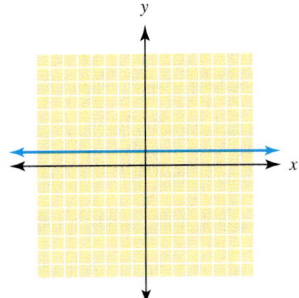

7.
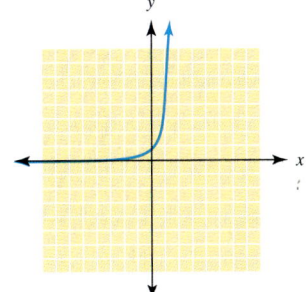

8.
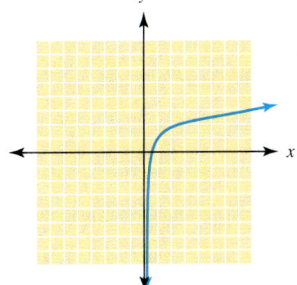

828 Chapter 13 ▪ Exponential and Logarithmic Functions

9. 1
10. 4
11. 16
12. $\dfrac{1}{16}$
13. 4
14. 16
15. 64
16. $\dfrac{1}{4}$
17. 2
18. 5
19. 17
20. $\dfrac{17}{16}$
21. $\dfrac{1}{4}$
22. 4
23. 16
24. $\dfrac{1}{16}$

In Exercises 9 to 12, let $f(x) = 4^x$ and find each of the following.

9. $f(0)$ 10. $f(1)$ 11. $f(2)$ 12. $f(-2)$

In Exercises 13 to 16, let $g(x) = 4^{x+1}$ and find each of the following.

13. $g(0)$ 14. $g(1)$ 15. $g(2)$ 16. $g(-2)$

In Exercises 17 to 20, let $h(x) = 4^x + 1$ and find each of the following.

17. $h(0)$ 18. $h(1)$ 19. $h(2)$ 20. $h(-2)$

In Exercises 21 to 24, let $f(x) = \left(\dfrac{1}{4}\right)^x$ and find each of the following.

21. $f(1)$ 22. $f(-1)$ 23. $f(-2)$ 24. $f(2)$

In Exercises 25 to 36, graph each exponential function.

25. $y = 4^x$ 26. $y = \left(\dfrac{1}{4}\right)^x$ 27. $y = \left(\dfrac{2}{3}\right)^x$

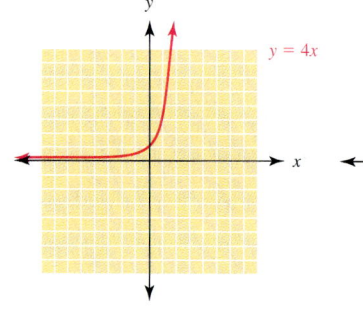
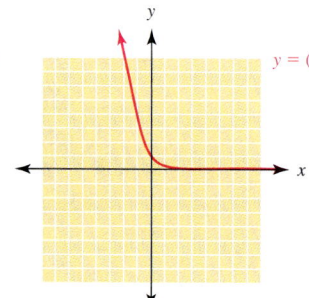
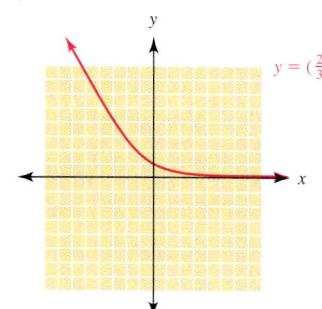

28. $y = \left(\dfrac{3}{2}\right)^x$

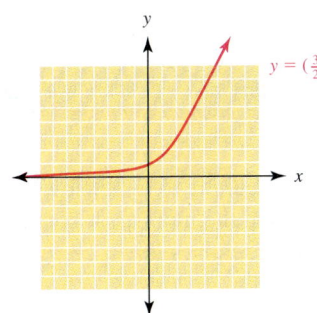

29. $y = 3 \cdot 2^x$

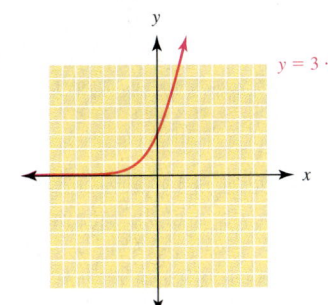

30. $y = 2 \cdot 3^x$

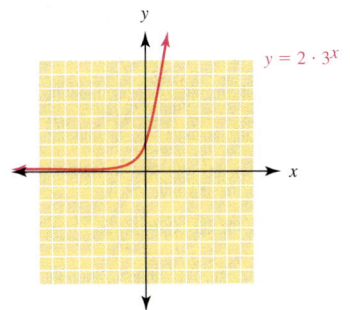

31. $y = 3^x$

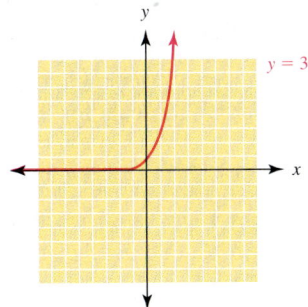

32. $y = 2^{x-1}$

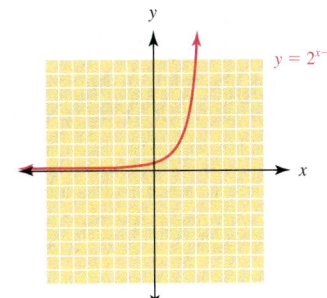

33. $y = 2^{2x}$

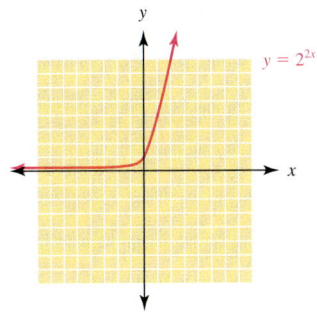

34. $y = \left(\dfrac{1}{2}\right)^{2x}$

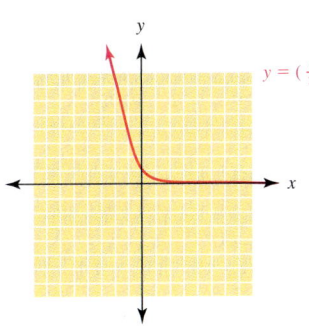

35. $y = e^{-x}$

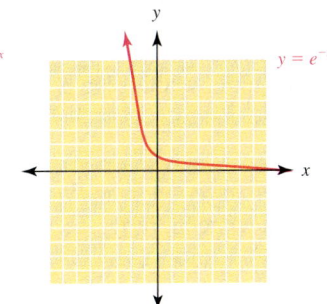

36. $y = e^{2x}$

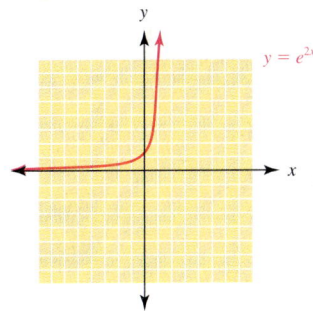

37. {5}
38. {3}
39. {4}
40. {3}
41. {−2}
42. {−4}
43. {3}
44. {2}
45. {5}
46. {3}
47. {−2}
48. {−5}
49. 400
50. 800
51. 3200
52.

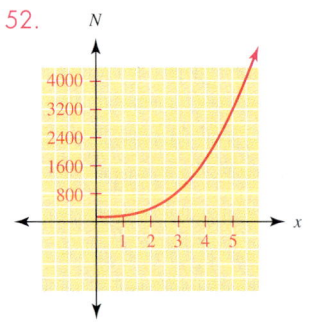

53. 32 g
54. 16 g
55. 8 g
56.

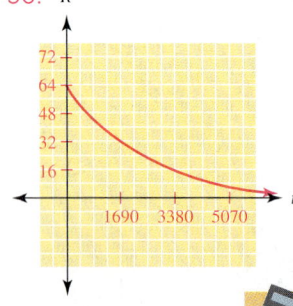

57. $1166.40
58. $1469.33
59. $1999

In Exercises 37 to 48, solve each exponential equation for x.

37. $2^x = 32$ **38.** $4^x = 64$ **39.** $10^x = 10{,}000$ **40.** $5^x = 125$

41. $3^x = \dfrac{1}{9}$ **42.** $2^x = \dfrac{1}{16}$ **43.** $2^{2x} = 64$ **44.** $3^{2x} = 81$

45. $2^{x+1} = 64$ **46.** $4^{x-1} = 16$ **47.** $3^{x-1} = \dfrac{1}{27}$ **48.** $2^{x+2} = \dfrac{1}{8}$

Suppose it takes 1 h for a certain bacterial culture to double by dividing in half. If there are 100 bacteria in the culture to start, then the number of bacteria in the culture after x hours is given by $N(x) = 100 \cdot 2^x$. In Exercises 49 to 52, use this function to find each of the following.

49. The number of bacteria in the culture after 2 h

50. The number of bacteria in the culture after 3 h

51. The number of bacteria in the culture after 5 h

52. Graph the relationship between the number of bacteria in the culture and the number of hours. Be sure to choose an appropriate scale for the N axis.

The half-life of radium is 1690 years. That is, after a 1690-year period, one-half of the original amount of radium will have decayed into another substance. If the original amount of radium was 64 grams (g), the formula relating the amount of radium left after time t is given by $R(t) = 64 \cdot 2^{-\frac{t}{1690}}$. In Exercises 53 to 56, use that formula to find each of the following.

53. The amount of radium left after 1690 years

54. The amount of radium left after 3380 years

55. The amount of radium left after 5070 years

56. Graph the relationship between the amount of radium remaining and time. Be sure to use appropriate scales for the R and t axes.

If $1000 is invested in a savings account with an interest rate of 8%, compounded annually, the amount in the account after t years, is given by $A(t) = 1000(1 + 0.08)^t$. In Exercises 57 to 60, use a calculator to find each of the following.

57. The amount in the account after 2 years

58. The amount in the account after 5 years

59. The amount in the account after 9 years

60. Graph the relationship between the amount in the account and time. Be sure to choose appropriate scales for the A and t axes.

The so-called learning curve in psychology applies to learning a skill, such as typing, in which the performance level progresses rapidly at first and then levels off with time. One can approximate N, the number of words per minute that a person can type after

61. (a) 36; (b) 56; (c) 67

62.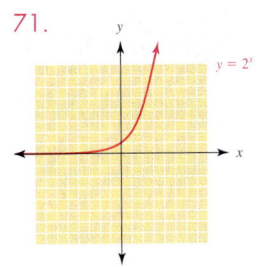

65. 2.7048
66. 2.7169
67. 2.71815
68. 2.71827
69. 2.71828

71.

72.
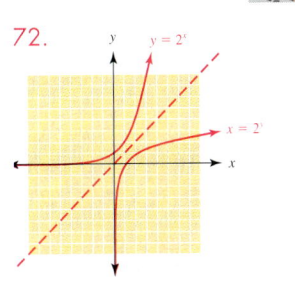

73. (a) $h = (0.003)(2^n)$
    (b) 12
    (c) 8.192 ft

t weeks of training, with the equation $N = 80(1 - e^{-0.06t})$. Use a calculator to find the following.

61. (a) N after 10 weeks, (b) N after 20 weeks, (c) N after 30 weeks.

62. Graph the relationship between the number of words per minute N and the number of weeks of training t.

63. Find two different calculators that have $\boxed{e^x}$ keys. Describe how to use the function on each of the calculators.

64. Are there any values of x for which e^x produces an exact answer on the calculator? Why are other answers not exact?

A possible calculator sequence for evaluating the expression

$$\left(1 + \frac{1}{n}\right)^n$$

where $n = 10$ is

$\boxed{(}\ \boxed{1}\ \boxed{+}\ \boxed{1}\ \boxed{\div}\ \boxed{10}\ \boxed{)}\ \boxed{\wedge}\ \boxed{10}\ \boxed{=}$

In Exercises 65 to 69, use that sequence to find $\left(1 + \frac{1}{n}\right)^n$ for the following values of n.

65. $n = 100$ **66.** $n = 1000$ **67.** $n = 10,000$

68. $n = 100,000$ **69.** $n = 1,000,000$

70. What did you observe from the experiment above?

71. Graph the exponential function defined by $y = 2^x$.

72. Graph the function defined by $x = 2^y$ on the same set of axes as the previous graph. What do you observe? (*Hint*: To graph $x = 2^y$, choose convenient values for y and then the corresponding values for x.)

73. Suppose you have a large piece of paper whose thickness is 0.003 in. If you tear the paper in half and stack the pieces, the height of the stack is $(0.003)(2)$ in., or 0.006 in. If you now tear the stack in half again, and then stack the pieces, the stack is $(0.003)(2)(2) = (0.003)(2^2)$ in., or 0.012 in. high.

(a) Define a function that gives the height h of the stack (in inches) after n tears.

(b) After which tear will the stack's height exceed 8 in.?

(c) Compute the height of the stack after the fifteenth tear. You will need to convert your answer to the appropriate units.

74. Use your graphing calculator to find all three points of intersection of the graphs of $f(x) = x^2$ and $g(x) = 2^x$. Give coordinates accurate to two decimal places.

74. $(-0.77, 0.59)$; $(2, 4)$; $(4, 16)$

SECTION 13.3 Logarithmic Functions

13.3 OBJECTIVES

1. Graph a logarithmic function
2. Convert between logarithmic and exponential equations
3. Evaluate a logarithmic expression
4. Solve an elementary logarithmic equation

Napier also coined the word "logarithm" from the Greek words "logos"—a ratio—and "arithmos"—a number.

Recall that f is a one-to-one function, so its inverse is also a function.

Given our experience with the exponential function in Section 13.2 and our earlier work with the inverse of a function in 13.1, we now can introduce the logarithmic function.

John Napier (1550–1617), a Scotsman, is credited with the invention of logarithms. The development of the logarithm grew out of a desire to ease the work involved in numerical computations, particularly in the field of astronomy. Today the availability of inexpensive scientific calculators has made the use of logarithms as a computational tool unnecessary.

However, the concept of the logarithm and the properties of the logarithmic function that we describe in a later section still are very important in the solutions of particular equations, in calculus, and in the applied sciences.

Again, the applications for this new function are numerous. The Richter scale for measuring the intensity of an earthquake and the decibel scale for measuring the intensity of sound both make use of logarithms.

To develop the idea of a logarithmic function, we must return to the definition of an exponential function

$$f = \{(x, y) | y = b^x, b > 0, b \neq 1\} \quad (1)$$

Interchanging the roles of x and y, we have the inverse function

$$f^{-1} = \{(x, y) | x = b^y\} \quad (2)$$

Presently, we have no way to solve the equation $x = b^y$ for y. So, to write the inverse, equation (2), in a more useful form, we offer the following definition.

The *logarithm of x to base b* is denoted

$$\log_b x$$

and

$$y = \log_b x \quad \text{if and only if} \quad x = b^y$$

Note that the restrictions on the base are the same as those used for the exponential function.

We can now write an inverse function, using this new notation, as

$$f^{-1} = \{(x, y) | y = \log_b x, b > 0, b \neq 1\} \quad (3)$$

In general, any function defined in this form is called a **logarithmic function.**

At this point we should stress the meaning of this new relationship. Consider the equivalent forms illustrated here.

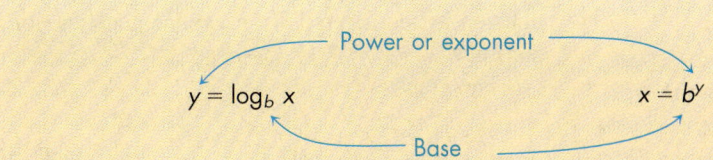

The logarithm y is the power to which we must raise b to get x. In other words, a logarithm is simply a power or an exponent. We return to this thought later when using the exponential and logarithmic forms of equivalent equations.

We begin our work by graphing a typical logarithmic function.

Example 1 Graphing a Logarithmic Function

Graph the logarithmic function

$$y = \log_2 x$$

Since $y = \log_2 x$ is equivalent to the exponential form

$$x = 2^y$$

The base is 2, and the logarithm or power is y.

we can find ordered pairs satisfying this equation by choosing convenient values for y and calculating the corresponding values for x.

Letting y take on values from -3 to 3 yields the table of values shown below. As before, we plot points from the ordered pairs and connect them with a smooth curve to form the graph of the function.

What do the vertical- and horizontal-line tests tell you about this graph?

x	y
$\frac{1}{8}$	-3
$\frac{1}{4}$	-2
$\frac{1}{2}$	-1
1	0
2	1
4	2
8	3

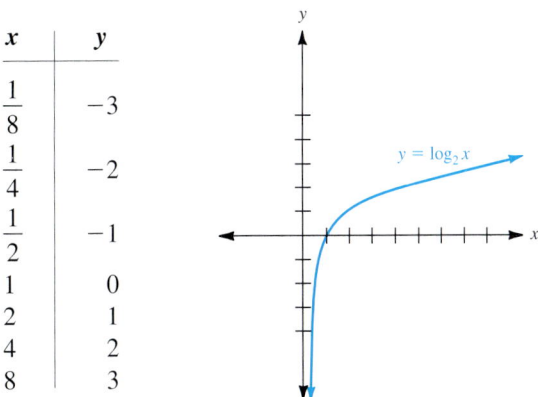

We observe that the graph represents a one-to-one function whose domain is $\{x \mid x > 0\}$ and whose range is the set of all real numbers.

For base 2 (or for any base greater than !) the function will always be increasing over its domain.

Recall from Section 13.1 that the graphs of a function and its inverse are always reflections of each other about the line $y = x$. Since we have defined the logarithmic function as the inverse of an exponential function, we can anticipate the same relationship.

The graphs of

$$f(x) = 2^x \quad \text{and} \quad f^{-1}(x) = \log_2 x$$

are shown in the figure.

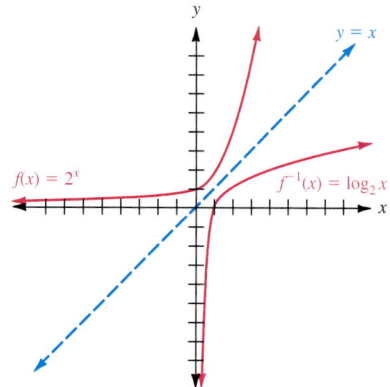

We see that the graphs of f and f^{-1} are indeed reflections of each other about the line $y = x$. In fact, this relationship provides an alternate method of sketching $y = \log_b x$. We can sketch the graph of $y = b^x$ and then reflect that graph about line $y = x$ to form the graph of the logarithmic function.

✓ CHECK YOURSELF 1

Graph the logarithmic function defined by

$$y = \log_3 x$$

(*Hint*: Consider the equivalent form $x = 3^y$.)

For our later work in this chapter, it will be necessary for us to be able to convert back and forth between exponential and logarithmic forms. The conversion is straightforward. You need only keep in mind the basic relationship

Again, this tells us that a logarithm is an exponent or a power.

$$y = \log_b x \quad \text{means the same as} \quad x = b^y$$

Look at the following example.

Example 2 — Writing Equations in Logarithmic Form

Convert to logarithmic form.

The base is 3, the exponent or power is 4.

(a) $3^4 = 81$ is equivalent to $\log_3 81 = 4$.

(b) $10^3 = 1000$ is equivalent to $\log_{10} 1000 = 3$.

(c) $2^{-3} = \dfrac{1}{8}$ is equivalent to $\log_2 \dfrac{1}{8} = -3$.

(d) $9^{1/2} = 3$ is equivalent to $\log_9 3 = \dfrac{1}{2}$.

✓ CHECK YOURSELF 2

Convert each statement to logarithmic form.

(a) $4^3 = 64$ (b) $10^{-2} = 0.01$ (c) $3^{-3} = \dfrac{1}{27}$ (d) $27^{1/3} = 3$

Example 3 shows how to write a logarithmic expression in exponential form.

Example 3 — Writing Equations in Exponential Form

Convert to exponential form.

Here, the base is 2; the logarithm, which is the power, is 3.

(a) $\log_2 8 = 3$ is equivalent to $2^3 = 8$.

(b) $\log_{10} 100 = 2$ is equivalent to $10^2 = 100$.

(c) $\log_3 \dfrac{1}{9} = -2$ is equivalent to $3^{-2} = \dfrac{1}{9}$.

(d) $\log_{25} 5 = \dfrac{1}{2}$ is equivalent to $25^{1/2} = 5$.

✓ **CHECK YOURSELF 3**

Convert to exponential form.

(a) $\log_2 32 = 5$ (b) $\log_{10} 1000 = 3$

(c) $\log_4 \dfrac{1}{16} = -2$ (d) $\log_{27} 3 = \dfrac{1}{3}$

Certain logarithms can be directly calculated by changing an expression to the equivalent exponential form, as Example 4 illustrates.

Example 4 Evaluating Logarithmic Expressions

(a) Evaluate $\log_3 27$.

If $x = \log_3 27$, in exponential form we have

Recall that $b^m = b^n$ if and only if $m = n$.

$$3^x = 27$$
$$3^x = 3^3$$
$$x = 3$$

We then have $\log_3 27 = 3$.

(b) Evaluate $\log_{10} \dfrac{1}{10}$.

If $x = \log_{10} \dfrac{1}{10}$, we can write

Rewrite each side as a power of the same base.

$$10^x = \dfrac{1}{10}$$
$$= 10^{-1}$$

We then have $x = -1$ and

$$\log_{10} \dfrac{1}{10} = -1$$

✓ **CHECK YOURSELF 4**

Evaluate each logarithm.

(a) $\log_2 64$ (b) $\log_3 \dfrac{1}{27}$

The relationship between exponents and logarithms also allows us to solve certain equations involving logarithms where two of the quantities in the equation $y = \log_b x$ are known, as Example 5 illustrates.

Example 5 **Solving Logarithmic Equations**

(a) Solve $\log_5 x = 3$ for x.

Since $\log_5 x = 3$, in exponential form we have

$$x = 5^3$$
$$= 125$$

(b) Solve $y = \log_4 \dfrac{1}{16}$ for y.

The original equation is equivalent to

$$4^y = \dfrac{1}{16}$$
$$= 4^{-2}$$

We then have $y = -2$ as the solution.

(c) Solve $\log_b 81 = 4$ for b.

In exponential form the equation becomes

$$b^4 = 81$$
$$b = 3$$

Keep in mind that the base must be *positive*, so we do not consider the possible solution $b = -3$.

✓ **CHECK YOURSELF 5**

Solve each of the following equations for the variable.

(a) $\log_4 x = 4$ (b) $\log_b \dfrac{1}{8} = -3$ (c) $y = \log_9 3$

Loudness can be measured in **bels (B)**, a unit named for Alexander Graham Bell. This unit is rather large, so a more practical unit is the **decibel (dB)**, a unit that is one-tenth as large.

Variable I_0 is the intensity of the minimum sound level detectable by the human ear.

The **decibel scale** is used in measuring the loudness of various sounds.

If I represents the intensity of a given sound and I_0 represents the intensity of a "threshold sound," then the decibel (dB) rating of the given sound is given by

$$L = 10 \log_{10} \frac{I}{I_0}$$

where $I_0 = 10^{-16}$ watt per square centimeter (W/cm^2). Consider Example 6.

Example 6 A Decibel Application

(a) A whisper has intensity $I = 10^{-14}$. Its decibel rating is

$$L = 10 \log_{10} \frac{10^{-14}}{10^{-16}}$$

$$= 10 \log_{10} 10^2$$

$$= 10 \cdot 2$$

$$= 20$$

(b) A rock concert has intensity $I = 10^{-4}$. Its decibel rating is

$$L = 10 \log_{10} \frac{10^{-4}}{10^{-16}}$$

$$= 10 \log_{10} 10^{12}$$

$$= 10 \cdot 12$$

$$= 120$$

✓ CHECK YOURSELF 6

Ordinary conversation has intensity $I = 10^{-12}$. Find its rating on the decibel scale.

The scale was named after Charles Richter, a U.S. geologist.

Geologists use the **Richter scale** to convert seismographic readings, which give the intensity of the shock waves of an earthquake, to a measure of the magnitude of that earthquake.

The magnitude M of an earthquake is given by

$$M = \log_{10} \frac{a}{a_0}$$

A "zero-level" earthquake is the quake of least intensity that is measurable by a seismograph.

where a is the intensity of its shock waves and a_0 is the intensity of the shock wave of a zero-level earthquake.

Example 7 A Richter Scale Application

How many times stronger is an earthquake measuring 5 on the Richter scale than one measuring 4 on the Richter scale?

Suppose a_1 is the intensity of the earthquake with magnitude 5 and a_2 is the intensity of the earthquake with magnitude 4. Then

$$5 = \log_{10} \frac{a_1}{a_0} \quad \text{and} \quad 4 = \log_{10} \frac{a_2}{a_0}$$

We convert these logarithmic expressions to exponential form.

$$10^5 = \frac{a_1}{a_0} \quad \text{and} \quad 10^4 = \frac{a_2}{a_0}$$

or

$$a_1 = a_0 \cdot 10^5 \quad \text{and} \quad a_2 = a_0 \cdot 10^4$$

On your calculator, the log key is actually $\log_{10} x$.

The ratio of a_1 to a_2 is
$$\frac{a_1}{a_2}$$

We want the ratio of the intensities of the two earthquakes, so

$$\frac{a_1}{a_2} = \frac{a_0 \cdot 10^5}{a_0 \cdot 10^4} = 10^1 = 10$$

The earthquake of magnitude 5 is *10 times stronger* than the earthquake of magnitude 4.

✓ CHECK YOURSELF 7

How many times stronger is an earthquake of magnitude 6 than one of magnitude 4?

✓ CHECK YOURSELF ANSWERS

1. $y = \log_3 x.$

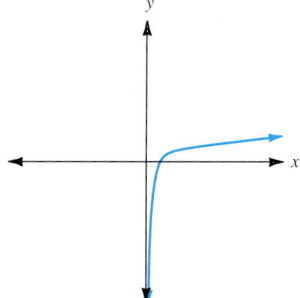

2. (a) $\log_4 64 = 3$; (b) $\log_{10} 0.01 = -2$; (c) $\log_3 \frac{1}{27} = -3$; (d) $\log_{27} 3 = \frac{1}{3}$.
3. (a) $2^5 = 32$; (b) $10^3 = 1000$; (c) $4^{-2} = \frac{1}{16}$; (d) $27^{1/3} = 3$.
4. (a) $\log_2 64 = 6$; (b) $\log_3 \frac{1}{27} = -3$. **5.** (a) $x = 256$; (b) $b = 2$; (c) $y = \frac{1}{2}$.
6. 40 dB.   **7.** 100 times.

Exercises • 13.3

In Exercises 1 to 6, sketch the graph of the function defined by each equation.

1. $y = \log_4 x$

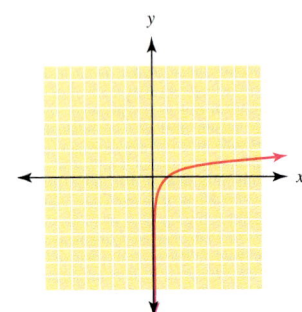

2. $y = \log_{10} x$

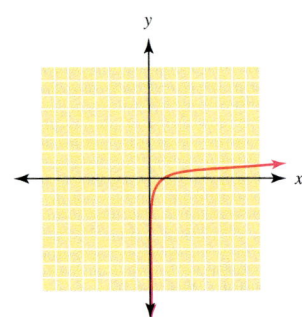

3. $y = \log_2 (x - 1)$

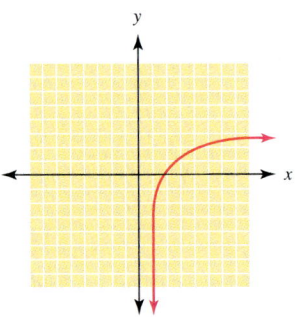

4. $y = \log_3 (x + 1)$

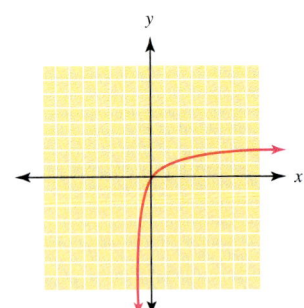

5. $y = \log_8 x$

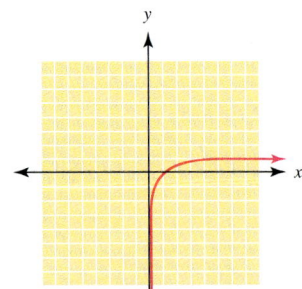

6. $y = \log_3 x + 1$

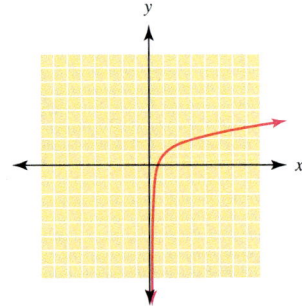

7. $\log_2 16 = 4$
8. $\log_3 243 = 5$
9. $\log_{10} 100 = 2$
10. $\log_4 64 = 3$
11. $\log_3 1 = 0$
12. $\log_{10} 1 = 0$
13. $\log_4 \left(\dfrac{1}{16}\right) = -2$
14. $\log_3 \left(\dfrac{1}{81}\right) = -4$

In Exercises 7 to 24, convert each statement to logarithmic form.

7. $2^4 = 16$

8. $3^5 = 243$

9. $10^2 = 100$

10. $4^3 = 64$

11. $3^0 = 1$

12. $10^0 = 1$

13. $4^{-2} = \dfrac{1}{16}$

14. $3^{-4} = \dfrac{1}{81}$

Section 13.3 ■ Logarithmic Functions **841**

15. $\log_{10}\left(\dfrac{1}{1000}\right) = -3$

16. $\log_2\left(\dfrac{1}{32}\right) = -5$

17. $\log_{16} 4 = \dfrac{1}{2}$

18. $\log_{125} 5 = \dfrac{1}{3}$

19. $\log_{64}\left(\dfrac{1}{4}\right) = -\dfrac{1}{3}$

20. $\log_{36}\left(\dfrac{1}{6}\right) = -\dfrac{1}{2}$

21. $\log_8 4 = \dfrac{2}{3}$

22. $\log_9 27 = \dfrac{3}{2}$

23. $\log_{27}\left(\dfrac{1}{9}\right) = -\dfrac{2}{3}$

24. $\log_{16}\left(\dfrac{1}{64}\right) = -\dfrac{3}{2}$

25. $2^4 = 16$ 26. $3^1 = 3$
27. $5^0 = 1$ 28. $3^3 = 27$
29. $10^1 = 10$ 30. $2^5 = 32$
31. $5^3 = 125$ 32. $10^0 = 1$
33. $3^{-3} = \dfrac{1}{27}$ 34. $5^{-2} = \dfrac{1}{25}$

35. $10^{-2} = 0.01$

36. $10^{-3} = \dfrac{1}{1000}$

37. $16^{1/2} = 4$

38. $125^{1/3} = 5$

39. $8^{2/3} = 4$

40. $9^{3/2} = 27$

41. $25^{-1/2} = \dfrac{1}{5}$

15. $10^{-3} = \dfrac{1}{1000}$

16. $2^{-5} = \dfrac{1}{32}$

17. $16^{1/2} = 4$

18. $125^{1/3} = 5$

19. $64^{-1/3} = \dfrac{1}{4}$

20. $36^{-1/2} = \dfrac{1}{6}$

21. $8^{2/3} = 4$

22. $9^{3/2} = 27$

23. $27^{-2/3} = \dfrac{1}{9}$

24. $16^{-3/2} = \dfrac{1}{64}$

In Exercises 25 to 42, convert each statement to exponential form.

25. $\log_2 16 = 4$ 26. $\log_3 3 = 1$ 27. $\log_5 1 = 0$

28. $\log_3 27 = 3$ 29. $\log_{10} 10 = 1$ 30. $\log_2 32 = 5$

31. $\log_5 125 = 3$ 32. $\log_{10} 1 = 0$ 33. $\log_3 \dfrac{1}{27} = -3$

34. $\log_5 \dfrac{1}{25} = -2$ 35. $\log_{10} 0.01 = -2$ 36. $\log_{10} \dfrac{1}{1000} = -3$

37. $\log_{16} 4 = \dfrac{1}{2}$ 38. $\log_{125} 5 = \dfrac{1}{3}$ 39. $\log_8 4 = \dfrac{2}{3}$

40. $\log_9 27 = \dfrac{3}{2}$ 41. $\log_{25} \dfrac{1}{5} = -\dfrac{1}{2}$ 42. $\log_{64} \dfrac{1}{16} = -\dfrac{2}{3}$

In Exercises 43 to 52, evaluate each logarithm.

43. $\log_2 32$ 44. $\log_3 81$ 45. $\log_4 64$ 46. $\log_{10} 1000$

47. $\log_3 \dfrac{1}{81}$ 48. $\log_4 \dfrac{1}{64}$ 49. $\log_{10} \dfrac{1}{100}$ 50. $\log_5 \dfrac{1}{25}$

51. $\log_{25} 5$ 52. $\log_{27} 3$

In Exercises 53 to 74, solve each equation for the unknown variable.

53. $y = \log_5 25$ 54. $\log_2 x = 4$ 55. $\log_b 64 = 3$

56. $y = \log_3 1$ 57. $\log_{10} x = 2$ 58. $\log_b 125 = 3$

59. $y = \log_5 5$ 60. $y = \log_3 81$ 61. $\log_{3/2} x = 3$

62. $\log_b \dfrac{4}{9} = 2$ 63. $\log_b \dfrac{1}{25} = -2$ 64. $\log_3 x = -3$

65. $\log_{10} x = -3$ 66. $y = \log_2 \dfrac{1}{16}$ 67. $y = \log_8 \dfrac{1}{64}$

42. $64^{-2/3} = \dfrac{1}{16}$ 51. $\dfrac{1}{2}$ 52. $\dfrac{1}{3}$ 61. $\left\{\dfrac{27}{8}\right\}$ 62. $\left\{\dfrac{2}{3}\right\}$

43. 5   44. 4   53. {2}   54. {16}   63. {5}   64. $\left\{\dfrac{1}{27}\right\}$

45. 3   46. 3   55. {4}   56. {0}   65. $\left\{\dfrac{1}{1000}\right\}$ 66. {−4}

47. −4   48. −3   57. {100}   58. {5}   67. {−2}

49. −2   50. −2   59. {1}   60. {4}

68. {10}

69. {3}

70. $\left\{\dfrac{1}{2}\right\}$

71. {25}

72. {16}

73. $\left\{-\dfrac{2}{3}\right\}$

74. {16}

75. 50 dB

76. 140 dB

77. 70 dB

78. 80 dB

79. 10^{-8}

80. 10^{-9} W/cm^2

81. 10

82. 100

83. 1000

84. $I = I_0 \, 10^{L/10}$

85. 6

86. 8.3

---

68. $\log_b \dfrac{1}{100} = -2$ 69. $\log_{27} x = \dfrac{1}{3}$ 70. $y = \log_{100} 10$

71. $\log_b 5 = \dfrac{1}{2}$ 72. $\log_{64} x = \dfrac{2}{3}$ 73. $y = \log_{27} \dfrac{1}{9}$

74. $\log_b \dfrac{1}{8} = -\dfrac{3}{4}$

Use the decibel formula

$$L = 10 \log_{10} \dfrac{I}{I_0}$$

to solve Exercises 75 to 78.

75. Sound. A television commercial has a volume with intensity $I = 10^{-11}$ W/cm^2. Find its rating in decibels.

76. Sound. The sound of a jet plane on takeoff has an intensity $I = 10^{-2}$ W/cm^2. Find its rating in decibels.

77. Sound. The sound of a computer printer has an intensity of $I = 10^{-9}$ W/cm^2. Find its rating in decibels.

78. Sound. The sound of a busy street has an intensity of $I = 10^{-8}$ w/cm^2. Find its rating in decibels.

The formula for the decibel rating L can be solved for the intensity of the sound as $I = I_0 \cdot 10^{L/10}$. Use this formula in Exercises 79 to 83.

79. Sound. Find the intensity of the sound in an airport waiting area if the decibel rating is 80.

80. Sound. Find the intensity of the sound of conversation in a crowded room if the decibel rating is 70.

81. Sound. What is the ratio of intensity of a sound of 80 dB to that of 70 dB?

82. Sound. What is the ratio of intensity of a sound of 60 dB to one measuring 40 dB?

83. Sound. What is the ratio of intensity of a sound of 70 dB to one measuring 40 dB?

84. Derive the formula for intensity provided above. (*Hint*: First divide both sides of the decibel formula by 10. Then write the equation in exponential form.)

Use the earthquake formula

$$M = \log_{10} \dfrac{a}{a_0}$$

to solve Exercises 85 to 88.

85. Earthquakes. An earthquake has an intensity a of $10^6 \cdot a_0$ where a_0 is the intensity of the zero-level earthquake. What was its magnitude?

86. Earthquakes. The great San Francisco earthquake of 1906 had an intensity of $10^{8.3} \cdot a_0$. What was its magnitude?

87. $10^5 \cdot a_0$

88. $10^6 \cdot a_0$

91. 24,000 yr

92. 28 yr

93. 77,000 yr

94. 2,000,000 yr

95. Pu-239: 240,000 yr; SR-90: 280 yr; Th-230: 770,000 yr; Cs-135: 20,000,000 yr

96. (a) 5
    (b) 2; 3
    (c) $\log_2 (4 \cdot 8) = \log_2 4 + \log_2 8$

97. (a) 6
    (b) 2; 4
    (c) $\log_3 (9 \cdot 81) = \log_3 9 + \log_3 81$

98. $\log_a(MN) = \log_a M + \log_a N$

---

87. **Earthquakes.** An earthquake can begin causing damage to buildings with a magnitude of 5 on the Richter scale. Find its intensity in terms of a_0.

88. **Earthquakes.** An earthquake may cause moderate building damage with a magnitude of 6 on the Richter scale. Find its intensity in terms of a_0.

89. The **learning curve** describes the relationship between learning and time. Its graph is a logarithmic curve in the first quadrant. Describe that curve as it relates to learning.

90. In which scientific fields would you expect to again encounter a discussion of logarithms?

The *half-life* of a radioactive substance is the time it takes for half the original amount of the substance to decay to a nonradioactive element. The half-life of radioactive waste is very important in figuring how long the waste must be kept isolated from the environment in some sort of storage facility. Half-lives of various radioactive waste products vary from a few seconds to millions of years. It usually takes at least 10 half-lives for a radioactive waste product to be considered safe.

The half-life of a radioactive substance can be determined by the following formula.

$$\ln \frac{1}{2} = -\lambda x$$

where

λ = radioactive decay constant

x = half-life

In Exercises 91 to 95, find the half-lives of the following important radioactive waste products given the radioactive decay constant (RDC).

91. Plutonium 239. RDC = 0.000029

92. Strontium 90. RDC = 0.024755

93. Thorium 230. RDC = 0.000009

94. Cesium 135. RDC = 0.00000035

95. How many years will it be before each waste product will be considered safe?

96. (a) Evaluate $\log_2 (4 \cdot 8)$. (b) Evaluate $\log_2 4$ and $\log_2 8$. (c) Write an equation that connects the result of part a with the results of part b.

97. (a) Evaluate $\log_3 (9 \cdot 81)$. (b) Evaluate $\log_3 9$ and $\log_3 81$. (c) Write an equation that connects the result of part a with the results of part b.

98. Based on Exercises 96 and 97, propose a statement that connects $\log_a (mn)$ with $\log_a m$ and $\log_a n$.

SECTION 13.4 Properties of Logarithms

13.4 OBJECTIVES

1. Apply the properties of logarithms
2. Evaluate logarithmic expressions with any base.
3. Solve applications involving logarithms
4. Estimate the value of an antilogarithm

As mentioned earlier, logarithms were developed as aids to numerical computations. The early utility of the logarithm was due to the properties that we will discuss in this section. Even with the advent of the scientific calculator, that utility remains important today. We can apply these same properties to applications in a variety of areas that lead to exponential or logarithmic equations.

Since a logarithm is, by definition, an exponent, it seems reasonable that our knowledge of the properties of exponents should lead to useful properties for logarithms. That is, in fact, the case.

Logarithmic Properties

We start with two basic facts that follow immediately from the definition of the logarithm.

The properties follow from the facts that

$b^1 = b$ and $b^0 = 1$

For $b > 0$ and $b \neq 1$,

1. $\log_b b = 1$

2. $\log_b 1 = 0$

We know that the logarithmic function $y = \log_b x$ and the exponential function $y = b^x$ are inverses of each other. So, for $f(x) = b^x$, we have $f^{-1}(x) = \log_b x$.

For any one-to-one function f,

The inverse has "undone" whatever f did to x.

$$f^{-1}(f(x)) = x \qquad \text{for any } x \text{ in the domain of } f$$

and

$$f(f^{-1}(x)) = x \qquad \text{for any } x \text{ in the domain of } f^{-1}$$

Since $f(x) = b^x$ is a one-to-one function, we can apply the above to the case where

$$f(x) = b^x \qquad \text{and} \qquad f^{-1}(x) = \log_b x$$

to derive the following.

Section 13.4 ■ Properties of Logarithms   845

For Property 3,
$$f^{-1}(f(x)) = f^{-1}(b^x) = \log_b b^x$$
But in general, for any one-to-one function f,
$$f^{-1}(f(x)) = x$$

3. $\log_b b^x = x$

4. $b^{\log_b x} = x$ for $x > 0$

Since logarithms are exponents, we can again turn to the familiar exponent rules to derive some further properties of logarithms. Consider the following.

We know that

$$\log_b M = x \quad \text{if and only if} \quad M = b^x$$

and

$$\log_b N = y \quad \text{if and only if} \quad N = b^y$$

Then

$$M \cdot N = b^x \cdot b^y = b^{x+y} \tag{1}$$

From equation (1) we see that $x + y$ is the power to which we must raise b to get the product MN. In logarithmic form, that becomes

$$\log_b MN = x + y \tag{2}$$

Now, since $x = \log_b M$ and $y = \log_b N$, we can substitute in equation (2) to write

$$\log_b MN = \log_b M + \log_b N \tag{3}$$

This is the first of the basic logarithmic properties presented here. The remaining properties may all be proved by arguments similar to those presented in equations (1) to (3).

In all cases, $M, N > 0$, $b > 0$, $b \neq 1$, and p is any real number.

Properties of Logarithms

Product Property

$$\log_b MN = \log_b M + \log_b N$$

Quotient Property

$$\log_b \frac{M}{N} = \log_b M - \log_b N$$

Power Property

$$\log_b M^p = p \log_b M$$

846 Chapter 13 ■ Exponential and Logarithmic Functions

Many applications of logarithms require using these properties to write a single logarithmic expression as the sum or difference of simpler expressions, as Example 1 illustrates.

Example 1 — Using the Properties of Logarithms

Expand, using the properties of logarithms.

(a) $\log_b xy = \log_b x + \log_b y$ Product property

(b) $\log_b \dfrac{xy}{z} = \log_b xy - \log_b z$ Quotient property

$\quad\quad = \log_b x + \log_b y - \log_b z$ Product property

(c) $\log_{10} x^2 y^3 = \log_{10} x^2 + \log_{10} y^3$ Product property

$\quad\quad = 2 \log_{10} x + 3 \log_{10} y$ Power property

Recall $\sqrt{a} = a^{1/2}$

(d) $\log_b \sqrt{\dfrac{x}{y}} = \log_b \left(\dfrac{x}{y}\right)^{1/2}$ Definition of exponent

$\quad\quad = \dfrac{1}{2} \log_b \dfrac{x}{y}$ Power property

$\quad\quad = \dfrac{1}{2} (\log_b x - \log_b y)$ Quotient property

✓ **CHECK YOURSELF 1**

Expand each expression, using the properties of logarithms.

(a) $\log_b x^2 y^3 z$ (b) $\log_{10} \sqrt{\dfrac{xy}{z}}$

In some cases, we will reverse the process and use the properties to write a single logarithm, given a sum or difference of logarithmic expressions.

Example 2 — Rewriting Logarithmic Expressions

Write each expression as a single logarithm with coefficient 1.

(a) $2 \log_b x + 3 \log_b y$

$\quad = \log_b x^2 + \log_b y^3$ Power property

$\quad = \log_b x^2 y^3$ Product property

(b) $5 \log_{10} x + 2 \log_{10} y - \log_{10} z$

$= \log_{10} x^5 y^2 - \log_{10} z$

$= \log_{10} \dfrac{x^5 y^2}{z}$ Quotient property

(c) $\dfrac{1}{2}(\log_2 x - \log_2 y)$

$= \dfrac{1}{2}\left(\log_2 \dfrac{x}{y}\right)$

$= \log_2 \left(\dfrac{x}{y}\right)^{1/2}$ Power property

$= \log_2 \sqrt{\dfrac{x}{y}}$

✓ CHECK YOURSELF 2

Write each expression as a single logarithm with coefficient 1.

(a) $3 \log_b x + 2 \log_b y - 2 \log_b z$ (b) $\dfrac{1}{3}(2 \log_2 x - \log_2 y)$

Example 3 illustrates the basic concept of the use of logarithms as a computational aid.

Example 3

Evaluating Logarithmic Expressions

We have written the logarithms correct to three decimal places and will follow this practice throughout the remainder of this chapter.

Suppose $\log_{10} 2 = 0.301$ and $\log_{10} 3 = 0.477$. Given these values, find the following:

(a) $\log_{10} 6$

Since $6 = 2 \cdot 3$,

$$\begin{aligned}\log_{10} 6 &= \log_{10} (2 \cdot 3) \\ &= \log_{10} 2 + \log_{10} 3 \\ &= 0.301 + 0.477 \\ &= 0.778\end{aligned}$$

Keep in mind, however, that this is an approximation and that $10^{0.301}$ will only approximate 2. Verify this with your calculator.

(b) $\log_{10} 18$

Since $18 = 2 \cdot 3 \cdot 3$,

We have extended the product rule for logarithms.

$$\log_{10} 18 = \log_{10} (2 \cdot 3 \cdot 3)$$
$$= \log_{10} 2 + \log_{10} 3 + \log_{10} 3$$
$$= 1.255$$

(c) $\log_{10} \dfrac{1}{9}$

Since $\dfrac{1}{9} = \dfrac{1}{3^2}$,

$$\log_{10} \dfrac{1}{9} = \log_{10} \dfrac{1}{3^2}$$
$$= \log_{10} 1 - \log_{10} 3^2$$
$$= 0 - 2 \log_{10} 3$$
$$= -0.954$$

Note that $\log_b 1 = 0$ for any base b.

(d) $\log_{10} 16$

Since $16 = 2^4$

$$\log_{10} 16 = \log_{10} 2^4 = 4 \log_{10} 2$$
$$= 1.204$$

(e) $\log_{10} \sqrt{3}$

Verify each answer with your calculator.

Since $\sqrt{3} = 3^{1/2}$

$$\log_{10} \sqrt{3} = \log_{10} 3^{1/2} = \dfrac{1}{2} \log_{10} 3$$
$$= 0.239$$

✓ CHECK YOURSELF 3

Given the values above for $\log_{10} 2$ and $\log_{10} 3$, find each of the following.

(a) $\log_{10} 12$ (b) $\log_{10} 27$ (c) $\log_{10} \sqrt[3]{2}$

Logarithms to Particular Bases

You can easily check the results in Example 3 by using the $\boxed{\log}$ key on a scientific calculator. For instance, in Example 3, part d, to find $\log_{10} 16$, enter

$$16 \;\boxed{\log}$$

> On a graphing calculator, enter the $\boxed{\log}$ first, then the 16.

and the result (to three decimal places) will be 1.204. As you can see, the $\boxed{\log}$ key on your calculator provides logarithms to base 10, which is one of two types of logarithms used most frequently in mathematics:

Logarithms to base 10

Logarithms to base e

Of course, the use of logarithms to base 10 is convenient because our number system has base 10. We call logarithms to base 10 **common logarithms,** and it is customary to omit the base in writing a common (or base 10) logarithm. So

> Note: When no base for "log" is written, it is assumed to be 10.

$$\log N \quad \text{means} \quad \log_{10} N$$

The following table shows the common logarithms for various powers of 10.

Exponential Form	Logarithmic Form
$10^3 = 1000$	$\log 1000 = 3$
$10^2 = 100$	$\log 100 = 2$
$10^1 = 10$	$\log 10 = 1$
$10 = 1$	$\log 1 = 0$
$10^{-1} = 0.1$	$\log 0.1 = -1$
$10^{-2} = 0.01$	$\log 0.01 = -2$
$10^{-3} = 0.001$	$\log 0.001 = -3$

Example 4 — Approximating Logarithms with a Calculator

Verify each of the following with a calculator.

(a) $\log 4.8 = 0.681$

> The number 4.8 lies between 1 and 10, so log 4.8 lies between 0 and 1.

(b) $\log 48 = 1.681$

(c) $\log 480 = 2.681$

(d) $\log 4800 = 3.681$

(e) $\log 0.48 = -0.319$

Note that

$$480 = 4.8 \times 10^2$$

and

$$\begin{aligned}&\log(4.8 \times 10^2)\\&= \log 4.8 + \log 10^2\\&= \log 4.8 + 2\\&= 2 + \log 4.8\end{aligned}$$

The value of log 0.48 is really $-1 + 0.681$. Your calculator will combine the signed numbers.

Note: A solution is **neutral** with pH = 7, **acidic** if the pH is less than 7, and **basic** if the pH is greater than 7.

✓ CHECK YOURSELF 4

Use your calculator to find each of the following logarithms, correct to three decimal places.

(a) log 2.3   (b) log 23   (c) log 230
(d) log 2300 (e) log 0.23 (f) log 0.023

Let's look at an application of common logarithms from chemistry. Common logarithms are used to define the pH of a solution. This is a scale that measures whether the solution is acidic or basic.

The pH of a solution is defined as

$$pH = -\log[H^+]$$

where $[H^+]$ is the hydrogen ion concentration, in moles per liter (mol/L), in the solution.

Example 5

Note the use of the product rule here.

Also, in general, $\log_b b^x = x$, so $\log 10^{-7} = -7$.

A pH Application

Find the pH of each of the following. Determine whether each is a base or an acid.

(a) Rainwater: $[H^+] = 1.6 \times 10^{-7}$

From the definition,

$$\begin{aligned}pH &= -\log[H^+]\\&= -\log(1.6 \times 10^{-7})\\&= -(\log 1.6 + \log 10^{-7})\\&= -[0.204 + (-7)]\\&= -(-6.796) = 6.796\end{aligned}$$

The rain is just slightly acidic.

(b) Household ammonia: $[H^+] = 2.3 \times 10^{-8}$

$$\begin{aligned}pH &= -\log(2.3 \times 10^{-8})\\&= -(\log 2.3 + \log 10^{-8})\\&= -[0.362 + (-8)]\\&= 7.638\end{aligned}$$

The ammonia is slightly basic.

(c) Vinegar: $[H^+] = 2.9 \times 10^{-3}$

$$\begin{aligned} pH &= -\log(2.9 \times 10^{-3}) \\ &= -(\log 2.9 + \log 10^{-3}) \\ &= 2.538 \end{aligned}$$

The vinegar is very acidic.

✓ CHECK YOURSELF 5

Find the pH for the following solutions. Are they acidic or basic?

(a) Orange juice: $[H^+] = 6.8 \times 10^{-5}$

(b) Drain cleaner: $[H^+] = 5.2 \times 10^{-13}$

Example 6 — Using a Calculator to Estimate Antilogarithms

Suppose that $\log x = 2.1567$. We want to find a number x whose logarithm is 2.1567. Using a scientific calculator requires one of the following sequences:

$$2.1567 \boxed{10^x} \quad \text{or} \quad 2.1567 \boxed{\text{INV}} \boxed{\log}$$

Because it is a one-to-one function, the logarithmic function has an inverse.

Both give the result 143.45, often called the **antilogarithm** of 2.1567.

✓ CHECK YOURSELF 6

Find the value of the antilogarithm of x.

(a) $\log x = 0.828$ (b) $\log x = 1.828$

(c) $\log x = 2.828$ (d) $\log x = -0.172$

Let's return to the application from chemistry for an example requiring the use of the antilogarithm.

Example 7   A pH Application

Suppose that the pH for tomato juice is 6.2. Find the hydrogen ion concentration [H$^+$].

Recall from our earlier formula that

$$\text{pH} = -\log [\text{H}^+]$$

In this case, we have

$$6.2 = -\log [\text{H}^+]$$

or

$$\log [\text{H}^+] = -6.2$$

The desired value for [H$^+$] is then the antilogarithm of -6.2, and we use the following scientific calculator sequence:

$$6.2 \; \boxed{+/-} \; \boxed{\text{INV}} \; \boxed{\log}$$

On a graphing calculator, enter

$\boxed{\text{2nd}} \; \boxed{\log} \; \boxed{(-)} \; 6.2$

The result is 0.00000063, and we can write

$$[\text{H}^+] = 6.3 \times 10^{-7}$$

 CHECK YOURSELF 7

The pH for eggs is -7.8. Find [H$^+$] for eggs.

As we mentioned, there are two systems of logarithms in common use. The second type of logarithm uses the number e as a base, and we call logarithms to base e the **natural logarithms.** As with common logarithms, a convenient notation has developed, as the following definition shows.

Natural logarithms are also called **napierian logarithms** after Napier. The importance of this system of logarithms was not fully understood until later developments in the calculus.

> The *natural logarithm* is a logarithm to base e, and it is denoted ln x, where
>
> $$\ln x = \log_e x$$

The restrictions on the domain of the natural logarithmic function are the same as before. The function is defined only if $x > 0$.

By the general definition of a logarithm,

$$y = \ln x \qquad \text{means the same as} \qquad x = e^y$$

and this leads us directly to the following rules.

$\ln 1 = 0$ since $e^0 = 1$

In general,
$\log_b b^x = x$ $b \neq 1$

$\ln e = 1$ since $e^1 = e$

$\ln e^2 = 2$ and $\ln e^{-3} = -3$

Example 8

Estimating Natural Logarithms

To find other natural logarithms, we can again turn to a calculator. To find the value of ln 2, use the sequence

$$2 \;\boxed{\ln}$$

or, with a graphing calculator,

$$\boxed{\ln} \; 2$$

The result is 0.693 (to three decimal places).

 CHECK YOURSELF 8

Use a calculator to find each of the following.

(a) ln 3   (b) ln 6   (c) ln 4   (d) $\ln \sqrt{3}$

Of course, the properties of logarithms are applied in an identical fashion, no matter what the base.

Example 9

Evaluating Logarithms

If $\ln 2 = 0.693$ and $\ln 3 = 1.099$, find the following.

(a) $\ln 6 = \ln (2 \cdot 3) = \ln 2 + \ln 3 = 1.792$

Recall that
$\log_b MN = \log_b M + \log_b N$
$\log_b M^p = p \log_b M$

(b) $\ln 4 = \ln 2^2 = 2 \ln 2 = 1.386$

(c) $\ln \sqrt{3} = \ln 3^{1/2} = \dfrac{1}{2} \ln 3 = 0.550$

Again, verify these results with your calculator.

✓ CHECK YOURSELF 9

Use $\ln 2 = 0.693$ and $\ln 3 = 1.099$ to find the following.

(a) $\ln 12$ (b) $\ln 27$

The natural logarithm function plays an important role in both theoretical and applied mathematics. Example 10 illustrates just one of the many applications of this function.

Example 10 — A Learning Curve Application

A class of students took a final mathematics examination and received an average score of 76. In a psychological experiment, the students are retested at weekly intervals over the same material. If t is measured in weeks, then the new average score after t weeks is given by

$$S(t) = 76 - 5 \ln (t + 1)$$

Recall that we read $S(t)$ as "S of t" which means that S is a function of t.

Complete the following.

(a) Find the score after 10 weeks.

$$S(t) = 76 - 5 \ln (10 + 1)$$
$$= 76 - 5 \ln 11 \approx 64$$

(b) Find the score after 20 weeks.

$$S(t) = 76 - 5 \ln (20 + 1) \approx 61$$

(c) Find the score after 30 weeks.

$$S(t) = 76 - 5 \ln (30 + 1) \approx 59$$

This is an example of a **forgetting curve.** Note how it drops more rapidly at first. Compare this curve to the learning curve drawn in Section 13.2, exercise 62.

✓ CHECK YOURSELF 10

The average score for a group of biology students, retested after time t (in months), is given by

$$S(t) = 83 - 9 \ln (t + 1)$$

Find the average score after

(a) 3 months. (b) 6 months.

We conclude this section with one final property of logarithms. This property will allow us to quickly find the logarithm of a number to any base. Although work with logarithms with bases other than 10 or e is relatively infrequent, the relationship between logarithms of different bases is interesting in itself. Consider the following argument.

Suppose that

$$x = \log_2 5$$

or

$$2^x = 5 \quad (4)$$

Taking the logarithm to base 10 of both sides of equation (4) yields

$$\log 2^x = \log 5$$

or

$$x \log 2 = \log 5 \quad \text{Use the power property of logarithms.} \quad (5)$$

(Note that we omit the 10 for the base and write log 2, for example.) Now, dividing both sides of equation (5) by log 2, we have

$$x = \frac{\log 5}{\log 2}$$

We can now find a value for x with the calculator. Dividing with the calculator log 5 by log 2, we get an approximate answer of 2.3219. Since $x = \log_2 5$ and $x = \dfrac{\log 5}{\log 2}$, then

Caution
Do not cancel the logs.

$$\log_2 5 = \frac{\log 5}{\log 2}$$

856 Chapter 13 ■ Exponential and Logarithmic Functions

Generalizing our result, we find the following.

Change-of-Base Formula

For the positive real numbers a and x,

$$\log_a x = \frac{\log x}{\log a}$$

Note that the logarithm on the left side has base a while the logarithms on the right side have base 10. This allows us to calculate the logarithm to base a of any positive number, given the corresponding logarithms to base 10 (or any other base), as Example 11 illustrates.

Example 11 Evaluating Logarithms

Find $\log_5 15$.

From the change-of-base formula with $a = 5$ and $b = 10$,

We have written the $\log_{10} 15$ rather than $\log 15$ to emphasize the change-of-base formula.

$$\log_5 15 = \frac{\log 15}{\log 5}$$

$$= 1.683$$

Note: $\log_5 5 = 1$ and $\log_5 25 = 2$, so the result for $\log_5 15$ must be between 1 and 2.

The graphing calculator sequence for the above computation is

On a scientific calculator, we type

$15 \boxed{\log} \boxed{\div} 5 \boxed{\log} \boxed{=}$

✓ **CHECK YOURSELF 11**

Use the change-of-base formula to find $\log_8 32$.

Caution

A *common error* is to write

$$\frac{\log 15}{\log 5} = \log 15 - \log 5$$

This is *not* a logarithmic property. A true statement would be

$$\log \frac{15}{5} = \log 15 - \log 5$$

but

$$\log \frac{15}{5} \quad \text{and} \quad \frac{\log 15}{\log 5}$$

are *not* the same.

Note: The $\log_e x$ is called the **natural log** of x. We use "ln x" to designate the natural log of x. A special case of the change-of-base formula allows us to find natural logarithms in terms of common logarithms:

$$\ln x = \frac{\log x}{\log e}$$

so

$$\ln x \approx \frac{\log x}{0.434} \quad \text{or, since} \quad \frac{1}{0.434} \approx 2.304, \text{ then } \ln x \approx 2.304 \log x$$

Of course, since all modern calculators have both the log function key and the ln function key, this conversion formula is now rarely used.

✓ CHECK YOURSELF ANSWERS

1. (a) $2 \log_b x + 3 \log_b y + \log_b z$; (b) $\frac{1}{2}(\log_{10} x + \log_{10} y - \log_{10} z)$.

2. (a) $\log_b \frac{x^3 y^2}{z^2}$; (b) $\log_2 \sqrt[3]{\frac{x^2}{y}}$.   **3.** (a) 1.079; (b) 1.431; (c) 0.100.

4. (a) 0.362; (b) 1.362; (c) 2.362; (d) 3.362; (e) -0.638; (f) -1.638.

5. (a) 4.17, acidic; (b) 12.28, basic.   **6.** (a) 6.73; (b) 67.3; (c) 673; (d) 0.673.

7. $[H^+] = 1.6 \times 10^{-8}$.   **8.** (a) 1.099; (b) 1.792; (c) 1.386; (d) 0.549.

9. (a) 2.485; (b) 3.297.   **10.** (a) 70.5; (b) 65.5.

11. $\log_8 32 = \frac{\log 32}{\log 8} \approx 1.667$.

Exercises · 13.4

1. $\log_b 5 + \log_b x$
2. $\log_3 7 + \log_3 x$
3. $\log_4 x - \log_4 3$
4. $\log_b 2 - \log_b y$
5. $2 \log_3 a$ 6. $4 \log_5 y$
7. $\dfrac{1}{2} \log_5 x$ 8. $\dfrac{1}{3} \log z$
9. $3 \log_b x + 2 \log_b y$
10. $2 \log_5 x + 4 \log_5 z$
11. $2 \log_4 y + \dfrac{1}{2} \log_4 x$
12. $3 \log_b x + \dfrac{1}{3} \log_b z$
13. $2 \log_b x + \log_b y - \log_b z$
14. $\log_5 3 - \log_5 x - \log_5 y$
15. $\log x + 2 \log y - \dfrac{1}{2} \log z$
16. $3 \log_4 x + \dfrac{1}{2} \log_4 y - 2 \log_4 z$
17. $\dfrac{1}{3} (\log_5 x + \log_5 y - 2 \log_5 z)$
18. $\dfrac{1}{2} \log_b x + \dfrac{1}{4} \log_b y - \dfrac{3}{4} \log_b z$
19. $\log_b xy$ 20. $\log_5 \dfrac{x}{y}$
21. $\log_2 \dfrac{x^2}{y}$ 22. $\log_b x^3 z$
23. $\log_b x \sqrt{y}$
24. $\log_b \dfrac{\sqrt[3]{x}}{z^2}$

In Exercises 1 to 18, use the properties of logarithms to expand each expression.

1. $\log_b 5x$
2. $\log_3 7x$
3. $\log_4 \dfrac{x}{3}$
4. $\log_b \dfrac{2}{y}$
5. $\log_3 a^2$
6. $\log_5 y^4$
7. $\log_5 \sqrt{x}$
8. $\log \sqrt[3]{z}$
9. $\log_b x^3 y^2$
10. $\log_5 x^2 z^4$
11. $\log_4 y^2 \sqrt{x}$
12. $\log_b x^3 \sqrt[3]{z}$
13. $\log_b \dfrac{x^2 y}{z}$
14. $\log_5 \dfrac{3}{xy}$
15. $\log \dfrac{xy^2}{\sqrt{z}}$
16. $\log_4 \dfrac{x^3 \sqrt{y}}{z^2}$
17. $\log_5 \sqrt[3]{\dfrac{xy}{z^2}}$
18. $\log_b \sqrt[4]{\dfrac{x^2 y}{z^3}}$

In Exercises 19 to 30, write each expression as a single logarithm.

19. $\log_b x + \log_b y$
20. $\log_5 x - \log_5 y$
21. $2 \log_2 x - \log_2 y$
22. $3 \log_b x + \log_b z$
23. $\log_b x + \dfrac{1}{2} \log_b y$
24. $\dfrac{1}{3} \log_b x - 2 \log_b z$

25. $\log_b \dfrac{xy^2}{z}$

26. $\log_5 \dfrac{x^2}{y^3 z}$

27. $\log_6 \dfrac{\sqrt{y}}{z^3}$

28. $\log_b \dfrac{x}{z^4 \sqrt[3]{y}}$

29. $\log_b \sqrt[3]{\dfrac{x^2 y}{z}}$

30. $\log_4 \sqrt[5]{\dfrac{x^2 z^3}{y}}$

31. 1.380   **32.** 1.556

33. 0.903   **34.** 1.908

35. 0.151   **36.** 0.159

37. −0.602

38. −1.431

39. 0.833

40. 1.833

41. 2.833

42. 3.833

43. −0.167

44. −1.167

45. 7.42, basic

46. 2.19, acidic

47. 5.61

48. 56.1

49. 5610

50. 0.561

51. 2×10^{-5}

52. 1.6×10^{-8}

53. 0.693

54. 1.099

55. 2.303

56. 3.401

25. $\log_b x + 2 \log_b y - \log_b z$

26. $2 \log_5 x - (3 \log_5 y + \log_5 z)$

27. $\dfrac{1}{2} \log_6 y - 3 \log_6 z$

28. $\log_b x - \dfrac{1}{3} \log_b y - 4 \log_b z$

29. $\dfrac{1}{3}(2 \log_b x + \log_b y - \log_b z)$

30. $\dfrac{1}{5}(2 \log_4 x - \log_4 y + 3 \log_4 z)$

In Exercises 31 to 38, given that log 2 = 0.301 and log 3 = 0.477, find each logarithm.

31. log 24   **32.** log 36   **33.** log 8

34. log 81   **35.** $\log \sqrt{2}$ **36.** $\log \sqrt[3]{3}$

37. $\log \dfrac{1}{4}$ **38.** $\log \dfrac{1}{27}$

In Exercises 39 to 44, use your calculator to find each logarithm.

39. log 6.8   **40.** log 68   **41.** log 680

42. log 6800   **43.** log 0.68   **44.** log 0.068

In Exercises 45 and 46, find the pH, given the hydrogen ion concentration [H$^+$] for each solution. Use the formula pH = −log [H$^+$]

Are the solutions acidic or basic?

45. Blood: [H$^+$] = 3.8×10^{-8} **46.** Lemon juice: [H$^+$] = 6.4×10^{-3}

In Exercises 47 to 50, use your calculator to find the antilogarithm for each logarithm.

47. 0.749   **48.** 1.749

49. 3.749   **50.** −0.251

In Exercises 51 and 52, given the pH of the solutions, find the hydrogen ion concentration [H$^+$].

51. Wine: pH = 4.7   **52.** Household ammonia: pH = 7.8

In Exercises 53 to 56, use your calculator to find each logarithm.

53. ln 2   **54.** ln 3

55. ln 10   **56.** ln 30

57. 74

58. 64

59. Between 2 and 3

60. Between 2 and 3

61. Between 6 and 7

62. Between 4 and 5

63. Between 2 and 3

64. Between 3 and 4

65. 1

66. 2

67. 2

68. 2.930

69. 2.113

70. 5.6 kg

71. 0.25 kg

72. 60 kg

73. 160 kg

76. (a) $\dfrac{\log 5}{\log 2}$; 2.322

(b) $\dfrac{\ln 5}{\ln 2}$; 2.322

(c) Same

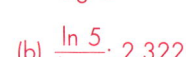

The average score on a final examination for a group of psychology students, retested after time t (in weeks), is given by

$$S = 85 - 8 \ln (t + 1)$$

In Exercises 57 and 58, find the average score on the retests:

57. After 3 weeks   **58.** After 12 weeks

Estimate each logarithm by "trapping" it between consecutive integers. To estimate $\log_4 52$, we would note that $4^2 = 16$ and $4^3 = 64$, so $\log_4 52$ must lie between 2 and 3.

59. $\log_3 25$ **60.** $\log_5 30$ **61.** $\log_2 70$

62. $\log_2 19$ **63.** $\log 680$ **64.** $\log 6800$

Without a calculator, use the properties of logarithms to evaluate each expression.

65. $\log 5 + \log 2$ **66.** $\log 25 + \log 4$ **67.** $\log_3 45 - \log_3 5$

In Exercises 68 and 69, use the change-of-base formula to find each logarithm.

68. $\log_3 25$ **69.** $\log_5 30$

The amount of a radioactive substance remaining after time t is given by the following

$$A = e^{\lambda t + \ln A_0}$$

where A is the amount remaining after time t, A_0 is the original amount of the substance, and λ is the radioactive decay constant.

70. How much plutonium 239 will remain after 50,000 years if 24 kg was originally stored? Plutonium 239 has a radioactive decay constant of -0.000029.

71. How much plutonium 241 will remain after 100 years if 52 kg was originally stored? Plutonium 241 has a radioactive decay constant of -0.053319.

72. How much strontium 90 was originally stored if after 56 years it is discovered that 15 kg still remains? Strontium 90 has a radioactive decay constant of -0.024755.

73. How much cesium 137 was originally stored if after 90 years it is discovered that 20 kg still remains? Cesium 137 has a radioactive decay constant of -0.023105.

74. Which keys on your calculator are function keys and which are operation keys? What is the difference?

75. How is the pH factor relevant to your selection of a hair care product?

76. (a) Use the change-of-base formula to write $\log_2 5$ in terms of base 10 logarithms. Then use your calculator to find $\log_2 5$ accurate to three decimal places.

(b) Use the change-of-base formula to write $\log_2 5$ in terms of base e logarithms. Then use your calculator to find $\log_2 5$ accurate to three decimal places.

(c) Compare your answers to parts a and b.

SECTION 13.5 Logarithmic and Exponential Equations

13.5 OBJECTIVES

1. Solve a logarithmic equation
2. Solve an exponential equation
3. Solve an application involving an exponential equation

Much of the importance of the properties of logarithms developed in the previous section lies in the application of those properties to the solution of equations involving logarithms and exponents. Our work in this section will consider solution techniques for both types of equations. Let's start with a definition.

> A **logarithmic equation** is an equation that contains a logarithmic expression.

We solved some simple examples in Section 13.3. Let's review for a moment. To solve $\log_3 x = 4$ for x, recall that we simply convert the logarithmic equation to exponential form. Here,

$$x = 3^4$$

so

$$x = 81$$

{81} is the solution set for the given equation.

Now, what if the logarithmic equation involves more than one logarithmic term? Example 1 illustrates how the properties of logarithms must then be applied.

Example 1

Solving a Logarithmic Equation

Solve each logarithmic equation.

(a) $\log_5 x + \log_5 3 = 2$

The original equation can be written as

$$\log_5 3x = 2$$

We apply the product rule for logarithms:
$\log_b M + \log_b N = \log_b MN$

861

Now, since only a single logarithm is involved, we can write the equation in the equivalent exponential form:

$$3x = 5^2$$
$$3x = 25$$
$$x = \frac{25}{3}$$

$\left\{\dfrac{25}{3}\right\}$ is the solution set.

(b) $\log x + \log (x - 3) = 1$

Write the equation as

Since no base is written, it is assumed to be 10.

$$\log x(x - 3) = 1$$

or

Given the base of 10, this is the equivalent exponential form.

$$x(x - 3) = 10^1$$

We now have

$$x^2 - 3x = 10$$
$$x^2 - 3x - 10 = 0$$
$$(x - 5)(x + 2) = 0$$

Possible solutions are $x = 5$ or $x = -2$.
Note that substitution of -2 into the original equation gives

Checking possible solutions is particularly important here.

$$\log (-2) + \log (-5) = 1$$

Since logarithms of negative numbers are *not* defined, -2 is an extraneous solution and we must reject it. The only solution for the original equation is 5. The solution set is {5}.

✓ CHECK YOURSELF 1

Solve $\log_2 x + \log_2 (x + 2) = 3$ for x.

The quotient property is used in a similar fashion for solving logarithmic equations. Consider Example 2.

Example 2 Solving a Logarithmic Equation

Solve each equation for x.

(a) $\log_5 x - \log_5 2 = 2$

Rewrite the original equation as

We apply the quotient rule for logarithms:

$\log_b M - \log_b N = \log_b \dfrac{M}{N}$

Now,

$$\log_5 \frac{x}{2} = 2$$

$$\frac{x}{2} = 5^2$$

$$\frac{x}{2} = 25$$

$$x = 50$$

The solution set is $\{50\}$.

(b) $\log_3 (x + 1) - \log_3 x = 3$

$$\log_3 \frac{x+1}{x} = 3$$

$$\frac{x+1}{x} = 27$$

$$x + 1 = 27x$$

$$1 = 26x$$

$$x = \frac{1}{26}$$

Again, you should verify that substituting $\dfrac{1}{26}$ for x leads to a positive value in each of the original logarithms.

$\left\{\dfrac{1}{26}\right\}$ is the solution set.

✓ CHECK YOURSELF 2

Solve $\log_5 (x + 3) - \log_5 x = 2$ for x.

The solution of certain types of logarithmic equations calls for the one-to-one property of the logarithmic function.

864 Chapter 13 ■ Exponential and Logarithmic Functions

> If $\quad \log_b M = \log_b N$
> then $\quad M = N$

Example 3 Solving a Logarithmic Equation

Solve the following equation for x.

$$\log(x + 2) - \log 2 = \log x$$

Again, we rewrite the left-hand side of the equation. So

$$\log \frac{x+2}{2} = \log x$$

Since the logarithmic function is one-to-one, this is equivalent to

$$\frac{x+2}{2} = x$$

or

$$x = 2$$

So, $\{2\}$ is the solution set.

 CHECK YOURSELF 3

Solve for x.

$$\log(x + 3) - \log 3 = \log x$$

The following algorithm summarizes our work in solving logarithmic equations.

Solving Logarithmic Equations

Step 1 Use the properties of logarithms to combine terms containing logarithmic expressions into a single term.
Step 2 Write the equation formed in step 1 in exponential form.
Step 3 Solve for the indicated variable.
Step 4 Check your solutions to make sure that possible solutions do not result in the logarithms of negative numbers.

Let's look now at **exponential equations,** which are equations in which the variable appears as an exponent.

We solved some particular exponential equations in Section 13.2. In solving an equation such as

$$3^x = 81$$

we wrote the right-hand member as a power of 3, so that

$$3^x = 3^4$$

Again, we want to write both sides as a power of the same base, here 3.

or

$$x = 4$$

The technique here will work only when both sides of the equation can be conveniently expressed as powers of the same base. If that is not the case, we must use logarithms to solve the equation, as illustrated in Example 4.

Example 4

Solving an Exponential Equation

Solve $3^x = 5$ for x.

We begin by taking the common logarithm of both sides of the original equation.

$$\log 3^x = \log 5$$

Again:

if $M = N$, then

$\log_b M = \log_b N$

Now we apply the power property so that the variable becomes a coefficient on the left.

$$x \log 3 = \log 5$$

Dividing both sides of the equation by $\log 3$ will isolate x, and we have

Caution
This is *not* log 5 − log 3, a common error.

$$x = \frac{\log 5}{\log 3}$$

$$\approx 1.465 \quad \text{(to three decimal places)}$$

Note: You can verify the approximate solution by using the $\boxed{y^x}$ key on a scientific calculator. Raise 3 to power 1.465.

 CHECK YOURSELF 4

Solve $2^x = 10$ for x.

Example 5 Solving an Exponential Equation

On the left, we apply
$\log_b M^p = p \log_b M$

Solve $5^{2x+1} = 8$ for x.

The solution begins as in Example 4.

$$\log 5^{2x+1} = \log 8$$
$$(2x + 1) \log 5 = \log 8$$
$$2x + 1 = \frac{\log 8}{\log 5}$$
$$2x = \frac{\log 8}{\log 5} - 1$$
$$x = \frac{1}{2}\left(\frac{\log 8}{\log 5} - 1\right)$$
$$x \approx 0.146$$

On a graphing calculator, the sequence would be

[(] [log] [8] [)] [÷] [log] [5] [)] [−] [1] [)] [÷] [2] [=]

A scientific calculator sequence to find x would be

8 [log] [÷] 5 [log] [−] 1 [=] [÷] 2 [=]

✓ **CHECK YOURSELF 5**

Solve $3^{2x-1} = 7$ for x.

The procedure is similar if the variable appears as an exponent in more than one term of the equation.

Example 6 Solving an Exponential Equation

Solve $3^x = 2^{x+1}$ for x.

Use the power property to write the variables as coefficients.

We now isolate x on the left.

To check the reasonableness of this result, use your calculator to verify that
$3^{1.710} \approx 2^{2.710}$

$$\log 3^x = \log 2^{x+1}$$
$$x \log 3 = (x + 1) \log 2$$
$$x \log 3 = x \log 2 + \log 2$$
$$x \log 3 - x \log 2 = \log 2$$
$$x(\log 3 - \log 2) = \log 2$$
$$x = \frac{\log 2}{\log 3 - \log 2}$$
$$\approx 1.710$$

✓ CHECK YOURSELF 6

Solve $5^{x+1} = 3^{x+2}$ for x.

The following algorithm summarizes our work with solving exponential equations.

> **Solving Exponential Equations**
>
> Step 1 Try to write each side of the equation as a power of the same base. Then equate the exponents to form an equation.
> Step 2 If the above procedure is not applicable, take the common logarithm of both sides of the original equation.
> Step 3 Use the power rule for logarithms to write an equivalent equation with the variables as coefficients.
> Step 4 Solve the resulting equation.
> Step 5 Check for extraneous solutions.

There are many applications of our work with exponential equations. Consider the following.

Example 7 — An Interest Application

If an investment of P dollars earns interest at an annual interest rate r and the interest is compounded n times per year, then the amount in the account after t years is given by

$$A = P\left(1 + \frac{r}{n}\right)^{nt} \tag{1}$$

If $1000 is placed in an account with an annual interest rate of 6%, find out how long it will take the money to double when interest is compounded annually and how long when compounded quarterly.

(a) Compounding interest annually.

Since the interest is compounded once *a year, $n = 1$.*

Using equation (1) with $A = 2000$ (we want the original 1000 to double), $P = 1000$, $r = 0.06$, and $n = 1$, we have

$$2000 = 1000(1 + 0.06)^t$$

Dividing both sides by 1000 yields

$$2 = (1.06)^t$$

We now have an exponential equation that can be solved by our earlier techniques.

$$\log 2 = \log (1.06)^t$$
$$= t \log 1.06$$

From accounting, we have the **rule of 72,** which states that the doubling time is approximately 72 divided by the interest rate as a percentage. Here $\dfrac{72}{6} = 12$ years.

or

$$t = \frac{\log 2}{\log 1.06}$$
$$\approx 11.9 \text{ years}$$

It takes just a little less than 12 years for the money to double.

(b) Compounding interest quarterly.

Since the interest is compounded four times per year, $n = 4$.

Now $n = 4$ in equation (1), so

$$2000 = 1000\left(1 + \frac{0.06}{4}\right)^{4t}$$
$$2 = (1.015)^{4t}$$
$$\log 2 = \log (1.015)^{4t}$$
$$\log 2 = 4t \log 1.015$$
$$\frac{\log 2}{4 \log 1.015} = t$$
$$t \approx 11.6 \text{ years}$$

Note that the doubling time is reduced by approximately 3 months by the more frequent compounding.

✓ CHECK YOURSELF 7

Find the doubling time in Example 7 if the interest is compounded monthly.

Problems involving rates of growth or decay can also be solved by using exponential equations.

Section 13.5 ■ Logarithmic and Exponential Equations 869

Example 8 A Population Application

A town's population is presently 10,000. Given a projected growth rate of 7% per year, t years from now the population P will be given by

$$P = 10{,}000e^{0.07t}$$

In how many years will the town's population double?

We want the time t when P will be 20,000 (doubled in size). So

$$20{,}000 = 10{,}000e^{0.07t}$$

or

Divide both sides by 10,000.

$$2 = e^{0.07t}$$

In this case, we take the *natural* log*arithm* of both sides of the equation. This is because e is involved in the equation.

$$\ln 2 = \ln e^{0.07t}$$

Apply the power property.

$$\ln 2 = 0.07t \ln e$$

Note: $\ln e = 1$

$$\ln 2 = 0.07t$$

$$\frac{\ln 2}{0.07} = t$$

$$t \approx 9.9 \text{ years}$$

The population will double in approximately 9.9 years.

✓ CHECK YOURSELF 8

If $1000 is invested in an account with an annual interest rate of 6%, compounded continuously, the amount A in the account after t years is given by

$$A = 1000e^{0.06t}$$

Find the time t that it will take for the amount to double ($A = 2000$). Compare this time with the result of the Check Yourself 7 exercise. Which is shorter? Why?

✓ CHECK YOURSELF ANSWERS

1. $\{2\}$. **2.** $\left\{\dfrac{1}{8}\right\}$. **3.** $\left\{\dfrac{3}{2}\right\}$. **4.** $\{3.322\}$. **5.** $\{1.386\}$. **6.** $\{1.151\}$.

7. 11.58 years. **8.** 11.55 years. The doubling time is shorter, because interest is compounded more frequently.

Exercises • 13.5

Answers (left column):

1. {64} 2. $\left\{\dfrac{1}{9}\right\}$
3. {99} 4. {13}
5. {8} 6. {20}
7. {162} 8. {512}
9. {2} 10. $\left\{\dfrac{3}{2}\right\}$
11. {6} 12. {6}
13. $\left\{\dfrac{20}{9}\right\}$ 14. $\left\{\dfrac{5}{24}\right\}$
15. $\left\{\dfrac{19}{8}\right\}$ 16. $\left\{\dfrac{12}{19}\right\}$
17. $\left\{\dfrac{15}{4}\right\}$ 18. {6}
19. {5, 3} 20. {3, 7}
21. {4} 22. {3}
23. {−4} 24. $\left\{\dfrac{1}{2}\right\}$
25. $\left\{\dfrac{1}{3}\right\}$ 26. {2}
27. {1.771} 28. {2.113}
29. {0.792} 30. {0.732}
31. {0.670} 32. {4.699}
33. {0.894} 34. {0.220}
35. {3.819} 36. {1.513}
37. {4.419} 38. {0.869}
39. 8.04 yr 40. 7.87 yr
41. 7.79 yr 42. 7.73 yr

In Exercises 1 to 20, solve each logarithmic equation for x.

1. $\log_4 x = 3$
2. $\log_3 x = -2$
3. $\log(x+1) = 2$
4. $\log_5(2x-1) = 2$
5. $\log_2 x + \log_2 8 = 6$
6. $\log 5 + \log x = 2$
7. $\log_3 x - \log_3 6 = 3$
8. $\log_4 x - \log_4 8 = 3$
9. $\log_2 x + \log_2(x+2) = 3$
10. $\log_3 x + \log_3(2x+3) = 2$
11. $\log_7(x+1) + \log_7(x-5) = 1$
12. $\log_2(x+2) + \log_2(x-5) = 3$
13. $\log x - \log(x-2) = 1$
14. $\log_5(x+5) - \log_5 x = 2$
15. $\log_3(x+1) - \log_3(x-2) = 2$
16. $\log(x+2) - \log(2x-1) = 1$
17. $\log(x+5) - \log(x-2) = \log 5$
18. $\log_3(x+12) - \log_3(x-3) = \log_3 6$
19. $\log_2(x^2-1) - \log_2(x-2) = 3$
20. $\log(x^2+1) - \log(x-2) = 1$

In Exercises 21 to 38, solve each exponential equation for x. If your solution is an approximation, give your solutions in decimal form, correct to three decimal places.

21. $5^x = 625$
22. $4^x = 64$
23. $2^{x+1} = \dfrac{1}{8}$
24. $9^x = 3$
25. $8^x = 2$
26. $3^{2x-1} = 27$
27. $3^x = 7$
28. $5^x = 30$
29. $4^{x+1} = 12$
30. $3^{2x} = 5$
31. $7^{3x} = 50$
32. $6^{x-3} = 21$
33. $5^{3x-1} = 15$
34. $8^{2x+1} = 20$
35. $4^x = 3^{x+1}$
36. $5^x = 2^{x+2}$
37. $2^{x+1} = 3^{x-1}$
38. $3^{2x+1} = 5^{x+1}$

Use the formula

$$A = P\left(1 + \dfrac{r}{n}\right)^{nt}$$

to solve Exercises 39 to 42.

39. **Interest.** If $5000 is placed in an account with an annual interest rate of 9%, how long will it take the amount to double if the interest is compounded annually?

40. Repeat Exercise 39 if the interest is compounded semiannually.

41. Repeat Exercise 39 if the interest is compounded quarterly.

42. Repeat Exercise 39 if the interest is compounded monthly.

43. 1.47 h
44. 4.12 h
45. 3.17 h
46. 4.64 h
47. 20.6 yr
48. 65 yr
49. 11.6 yr
50. 93 yr
51. 7.5 yr
52. 15.4 yr
53. 23.1 h
54. 46.2 h
55. 4558 ft
56. 10,137 ft

Suppose the number of bacteria present in a culture after t hours is given by $N(t) = N_0 \cdot 2^{t/2}$, where N_0 is the initial number of bacteria. Use the formula to solve Exercises 43 to 46.

43. How long will it take the bacteria to increase from 12,000 to 20,000?

44. How long will it take the bacteria to increase from 12,000 to 50,000?

45. How long will it take the bacteria to triple? (*Hint*: Let $N = 3N_0$.)

46. How long will it take the culture to increase to 5 times its original size? (*Hint*: Let $N = 5N_0$.)

The radioactive element strontium 90 has a half-life of approximately 28 years. That is, in a 28-year period, one-half of the initial amount will have decayed into another substance. If A_0 is the initial amount of the element, then the amount A remaining after t years is given by

$$A(t) = A_0 \left(\frac{1}{2}\right)^{t/28}$$

Use the formula to solve Exercises 47 to 50.

47. If the initial amount of the element is 100 g, in how many years will 60 g remain?

48. If the initial amount of the element is 100 g, in how many years will 20 g remain?

49. In how many years will 75% of the original amount remain? (*Hint*: Let $A = 0.75A_0$.)

50. In how many years will 10% of the original amount remain? (*Hint*: Let $A = 0.1A_0$.)

Given projected growth, t years from now a city's population P can be approximated by $P(t) = 25{,}000e^{0.045t}$. Use the formula to solve Exercises 51 and 52.

51. How long will it take the city's population to reach 35,000?

52. How long will it take the population to double?

The number of bacteria in a culture after t hours can be given by $N(t) = N_0 e^{0.03t}$, where N_0 is the initial number of bacteria in the culture. Use the formula to solve Exercises 53 and 54.

53. In how many hours will the size of the culture double?

54. In how many hours will the culture grow to four times its original population?

The atmospheric pressure P, in inches of mercury (in. Hg), at an altitude h feet above sea level is approximated by $P(t) = 30e^{-0.00004h}$. Use the formula to solve Exercises 55 and 56.

55. Find the altitude if the pressure at that altitude is 25 in. Hg.

56. Find the altitude if the pressure at that altitude is 20 in. Hg.

57. 2876 yr

58. 12,979 yr

61. The graph is that of $y = x$. The two operations cancel each other.

62. The graph is that of $y = x$. The two operations cancel each other.

63. The graph is that of $y = x$. The two operations cancel each other.

64. The graph is that of $y = x$. The two operations cancel each other.

Carbon 14 dating is used to measure the age of specimens and is based on the radioactive decay of the element carbon 14. This decay begins once a plant or animal dies. If A_0 is the initial amount of carbon 14, then the amount remaining after t years is $A(t) = A_0 e^{-0.000124t}$. Use the formula to solve Exercises 57 and 58.

57. Estimate the age of a specimen if 70% of the original amount of carbon 14 remains.

58. Estimate the age of a specimen if 20% of the original amount of carbon 14 remains.

59. In some of the earlier exercises, we talked about bacteria cultures that double in size every few minutes. Can this go on forever? Explain.

60. The population of the United States has been doubling every 45 years. Is it reasonable to assume that this rate will continue? What factors will start to limit that growth?

In Exercises 61 to 64, use your calculator to find the graph for each equation, then explain the result.

61. $y = \log 10^x$

62. $y = 10^{\log x}$

63. $y = \ln e^x$

64. $y = e^{\ln x}$

65. In this section, we solved the equation $3^x = 2^{x+1}$ by first applying the logarithmic function, base 10, to each side of the equation. Try this again, but this time apply the natural logarithm function to each side of the equation. Compare the solutions that result from the two approaches.

Summary Exercises • 13

This summary exercise set is provided to give you practice with each of the objectives in the chapter. Each chapter is keyed to the appropriate chapter section.

[13.1] Write the inverse relation for each of the following functions. Which inverses are also functions?

1. $\{(1, 5), (2, 7), (3, 9)\}$ **2.** $\{(3, 1), (5, 1), (7, 1)\}$

3. $\{(2, 4), (4, 3), (6, 4)\}$

[13.1] Write an equation for the inverse of the relation defined by each of the following equations.

4. $y = 3x - 6$ **5.** $y = \dfrac{x + 1}{2}$ **6.** $y = x^2 - 2$

[13.1] Write an equation for the inverse of the relation defined by each of the following equations. Which inverses are also functions?

7. $y = 3x + 6$ **8.** $y = -x^2 + 3$ **9.** $4x^2 + 9y^2 = 36$

[13.2] Graph the exponential functions defined by each of the following equations.

10. $y = 3^x$ **11.** $y = \left(\dfrac{3}{4}\right)^x$

[13.3] Solve each of the following exponential equations for x.

12. $5^x = 125$ **13.** $2^{2x+1} = 32$ **14.** $3^{x-1} = \dfrac{1}{9}$

[13.3] If it takes 2 h for the population of a certain bacteria culture to double (by dividing in half), then the number N of bacteria in the culture after t hours is given by $N = 1000 \cdot 2^{t/2}$, where the initial population of the culture was 1000. Using this formula, find the number in the culture:

15. After 4 h **16.** After 12 h **17.** After 15 h

---

1. $\{(5, 1), (7, 2), (9, 3)\}$; function

2. $\{(1, 3), (1, 5), (1, 7)\}$; not a function

3. $\{(4, 2), (3, 4), (4, 6)\}$; not a function

4. $y = \dfrac{1}{3}x + 2$

5. $y = 2x - 1$

6. $y = \pm\sqrt{x + 2}$

7. $y = \dfrac{1}{3}x - 2$; function

8. $y = \pm\sqrt{-x + 3}$; not a function

9. $\dfrac{y^2}{9} + \dfrac{x^2}{4} = 1$; not a function

10.–11. See Instructor's Solutions Manual

12. $\{3\}$

13. $\{2\}$

14. $\{-1\}$

15. 4000

16. 64,000

17. 181,019

20. $\log_3 81 = 4$

21. $\log 1000 = 3$

22. $\log_5 1 = 0$

23. $\log_5 \left(\dfrac{1}{25}\right) = -2$

24. $\log_{25} 5 = \dfrac{1}{2}$

25. $\log_{16} 8 = \dfrac{3}{4}$

26. $3^4 = 81$

27. $10^0 = 1$

28. $81^{1/2} = 9$

29. $5^2 = 25$

30. $10^{-3} = 0.001$

31. $32^{-1/5} = \dfrac{1}{2}$

32. $\{3\}$

33. $\{3\}$

34. $\{64\}$

35. $\{0\}$

36. $\{9\}$

37. $\left\{\dfrac{1}{4}\right\}$

38. $\left\{\dfrac{1}{3}\right\}$

[13.3] Graph the logarithmic functions defined by each of the following equations.

18. $y = \log_3 x$

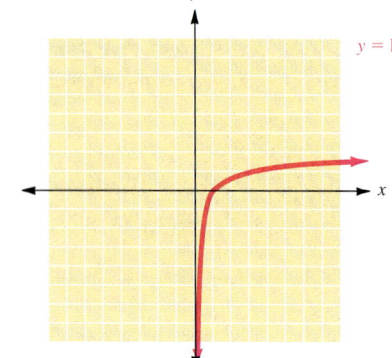

19. $y = \log_2 (x - 1)$

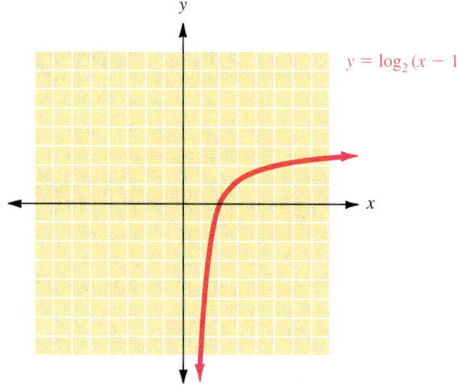

[13.3] Convert each of the following statements to logarithmic form.

20. $3^4 = 81$
21. $10^3 = 1000$
22. $5^0 = 1$
23. $5^{-2} = \dfrac{1}{25}$
24. $25^{1/2} = 5$
25. $16^{3/4} = 8$

[13.3] Convert each of the following statements to exponential form.

26. $\log_3 81 = 4$
27. $\log 1 = 0$
28. $\log_{81} 9 = \dfrac{1}{2}$
29. $\log_5 25 = 2$
30. $\log 0.001 = -3$
31. $\log_{32} \dfrac{1}{2} = -\dfrac{1}{5}$

[13.3] Solve each of the following equations for the unknown variable.

32. $y = \log_5 125$
33. $\log_b \dfrac{1}{9} = -2$
34. $\log_8 x = 2$
35. $y = \log_5 1$
36. $\log_b 3 = \dfrac{1}{2}$
37. $y = \log_{16} 2$
38. $y = \log_8 2$

[13.3] The decibel (dB) rating for the loudness of a sound is given by

$$L = 10 \log \dfrac{I}{I_0}$$

where I is the intensity of that sound in watts per square centimeter and I_0 is the intensity of the "threshold" sound $I_0 = 10^{-16}$ W/cm². Find the decibel rating of each of the given sounds.

39. 100 dB
40. 80 dB
41. 10
42. 100
43. 8.4
44. 10
45. $2 \log_b x + \log_b y$
46. $3 \log_4 y - \log_4 5$
47. $\log_3 x + 2 \log_3 y - \log_3 z$
48. $3 \log_5 x + \log_5 y + 2 \log_5 z$
49. $\log x + \log y - \frac{1}{2} \log z$
50. $\frac{2}{3} \log_b x + \frac{1}{3} \log_b y - \frac{1}{3} \log_b z$
51. $\log xy^2$
52. $\log_b \frac{x^3}{z^2}$
53. $\log_b \frac{xy}{z}$
54. $\log_5 \frac{x^2}{y^3 z}$
55. $\log \frac{x}{\sqrt{y}}$
56. $\log_b \sqrt[3]{\frac{x}{y^2}}$
57. 1.255
58. 1.204
59. −0.903
60. 0.239

---

39. A table saw in operation with intensity $I = 10^{-6}$ W/cm².

40. The sound of a passing car horn with intensity $I = 10^{-8}$ W/cm².

[13.3] The formula for the decibel rating of a sound can be solved for the intensity of the sound as

$$I = I_0 \cdot 10^{L/10}$$

where L is the decibel rating of the given sound.

41. What is the ratio of intensity of a 60-dB sound to one of 50 dB?

42. What is the ratio of intensity of a 60-dB sound to one of 40 dB?

[13.3] The magnitude of an earthquake on the Richter scale is given by

$$M = \log \frac{a}{a_0}$$

where a is the intensity of the shock wave of the given earthquake and a_0 is the intensity of the shock wave of a zero-level earthquake. Use that formula to solve the following.

43. The Alaskan earthquake of 1964 had an intensity of $10^{8.4} a_0$. What was its magnitude on the Richter scale?

44. Find the ratio of intensity of an earthquake of magnitude 7 to an earthquake of magnitude 6.

[13.4] Use the properties of logarithms to expand each of the following expressions.

45. $\log_b x^2 y$
46. $\log_4 \frac{y^3}{5}$
47. $\log_3 \frac{xy^2}{z}$
48. $\log_5 x^3 y z^2$
49. $\log \frac{xy}{\sqrt{z}}$
50. $\log_b \sqrt[3]{\frac{x^2 y}{z}}$

[13.4] Use the properties of logarithms to write each of the following expressions as a single logarithm.

51. $\log x + 2 \log y$
52. $3 \log_b x - 2 \log_b z$
53. $\log_b x + \log_b y - \log_b z$
54. $2 \log_5 x - 3 \log_5 y - \log_5 z$
55. $\log x - \frac{1}{2} \log y$
56. $\frac{1}{3}(\log_b x - 2 \log_b y)$

[13.4] Given that $\log 2 = 0.301$ and $\log 3 = 0.477$, find each of the following logarithms. Verify your results with a calculator.

57. $\log 18$
58. $\log 16$
59. $\log \frac{1}{8}$
60. $\log \sqrt{3}$

876 Chapter 13 ■ Exponential and Logarithmic Functions

61. 5.301, acidic
62. 9.495, basic
63. 3.2×10^{-4}
64. 6.3×10^{-11}
65. 70.2
66. 66.6
67. 64.4
68.

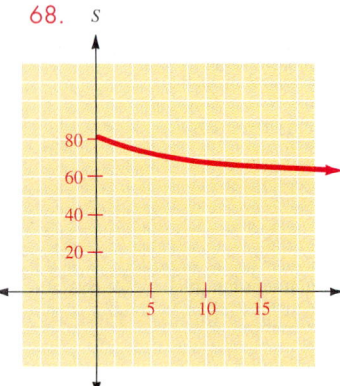

69. 2.161
70. 1.969
71. $\left\{\dfrac{27}{5}\right\}$
72. {250}
73. {3}
74. {2}
75. $\left\{\dfrac{10}{9}\right\}$
76. {3}
77. {5}
78. {−2}
79. {1.431}
80. {2.500}
81. {1.710}
82. {4.419}

[13.4] Use your calculator to find the pH of each of the following solutions, given the hydrogen ion concentration [H⁺] for each solution, where

$$pH = -\log [H^+]$$

Are the solutions acidic or basic?

61. Coffee: $[H^+] = 5 \times 10^{-6}$ **62.** Household detergent: $[H^+] = 3.2 \times 10^{-10}$

[13.4] Given the pH of the following solutions, find the hydrogen ion concentration [H⁺].

63. Lemonade: pH = 3.5   **64.** Ammonia: pH = 10.2

[13.4] The average score on a final examination for a group of chemistry students, retested after time t (in weeks), is given by

$$S(t) = 81 - 6 \ln (t + 1)$$

Find the average score on the retests after the given times.

65. After 5 weeks   **66.** After 10 weeks   **67.** After 15 weeks

68. Graph these results.

[13.4] The formula for converting from a logarithm with base a to logarithms with base b is

$$\log_a x = \frac{\log_b x}{\log_b a}$$

Use that formula to find each of the following logarithms.

69. $\log_4 20$ **70.** $\log_8 60$

[13.5] Solve each of the following logarithmic equations for x.

71. $\log_3 x + \log_3 5 = 3$ **72.** $\log_5 x - \log_5 10 = 2$
73. $\log_3 x + \log_3 (x + 6) = 3$ **74.** $\log_5 (x + 3) + \log_5 (x - 1) = 1$
75. $\log x - \log (x - 1) = 1$ **76.** $\log_2 (x + 3) - \log_2 (x - 1) = \log_2 3$

[13.5] Solve each of the following exponential equations for x. Give your results correct to three decimal places.

77. $3^x = 243$ **78.** $5^x = \dfrac{1}{25}$ **79.** $5^x = 10$
80. $4^{x-1} = 8$ **81.** $6^x = 2^{2x+1}$ **82.** $2^{x+1} = 3^{x-1}$

83. 5.86 yr
84. 8.21 yr
85. 66 yr
86. 166 yr
87. 4.2 yr
88. 8.7 yr
89. 3.8 mi
90. 6.6 mi

[13.5] If an investment of P dollars earns interest at an annual rate of 12% and the interest is compounded n times per year, then the amount A in the account after t years is

$$A(t) = P\left(1 + \frac{0.12}{n}\right)^{nt}$$

Use that formula to solve each of the following.

83. If $1000 is invested and the interest is compounded quarterly, how long will it take the amount in the account to double?

84. If $3000 is invested and the interest is compounded monthly, how long will it take the amount in the account to reach $8000?

[13.5] A certain radioactive element has a half-life of 50 years. The amount A of the substance remaining after t years is given by

$$A(t) = A_0 \cdot 2^{-t/50}$$

where A_0 is the initial amount of the substance. Use this formula to solve each of the following.

85. If the initial amount of the substance is 100 milligrams (mg), after how long will 40 mg remain?

86. After how long will only 10% of the original amount of the substance remain?

[13.5] A city's population is presently 50,000. Given the projected growth, t years from now the population P will be given by $P(t) = 50{,}000 e^{0.08t}$. Use this formula to solve each of the following.

87. How long will it take the population to reach 70,000?

88. How long will it take the population to double?

[13.5] The atmospheric pressure, in inches of mercury, at an altitude h miles above the surface of the earth, is approximated by $P(h) = 30 e^{-0.021h}$. Use this formula to solve the following exercises.

89. Find the altitude at the top of Mt. McKinley in Alaska if the pressure is 27.7 in. Hg.

90. Find the altitude outside an airliner in flight if the pressure is 26.1 in. Hg.

Self-Test ■ 13

The purpose of this self-test is to help you check your progress and to review for a chapter test in class. Allow yourself about 1 hour to take the test. When you are done, check your answers in the back of the book. If you missed any answers, be sure to go back and review the appropriate sections in the chapter and the exercises that are provided.

1. Use $f(x) = 4x - 2$ and $g(x) = x^2 + 1$ in each of the following.

   (a) Find the inverse of f. Is the inverse also a function?

   (b) Find the inverse of g. Is the inverse also a function?

   (c) Graph f and its inverse on the same set of axes.

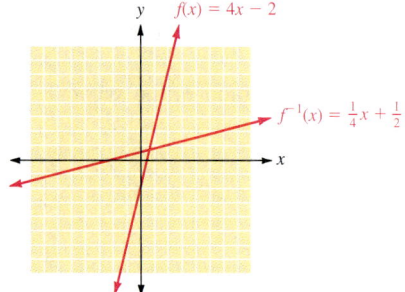

Graph the exponential functions defined by each of the following equations.

2. $y = 4^x$

3. $y = \left(\dfrac{2}{3}\right)^x$

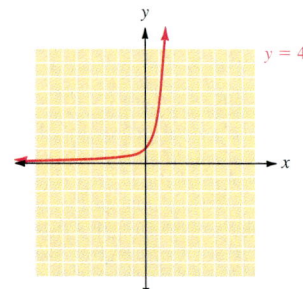

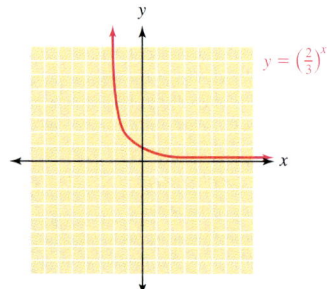

4. Solve each of the following exponential equations for x.

   (a) $5^x = \dfrac{1}{25}$ (b) $3^{2x-1} = 81$

Answers in margin:

1. (a) $f^{-1} = \{(x, y) | y = \dfrac{1}{4}x + \dfrac{1}{2}\}$, function;
   (b) $g^{-1} = \{(x, y) | y = \pm\sqrt{x - 1}$; not a function

4. (a) $\{-2\}$
   (b) $\left\{\dfrac{5}{2}\right\}$

878

6. $\log 10{,}000 = 4$

7. $\log_{27} 9 = \dfrac{2}{3}$

8. $5^3 = 125$

9. $10^{-2} = 0.01$

10. $\{6\}$

11. $\{4\}$

12. $\{5\}$

13. $2 \log_b x + \log_b y + 3 \log_b z$

14. $\dfrac{1}{2}(\log_5 x + 2 \log_5 y - \log_5 z)$

15. $\log(xy^3)$

16. $\log_b \sqrt[3]{\dfrac{x}{z^2}}$

17. $\{8\}$

18. $\left\{\dfrac{11}{8}\right\}$

19. $\{0.262\}$

20. $\{2.151\}$

5. Graph the logarithmic function defined by the following equation.

$$y = \log_4 x$$

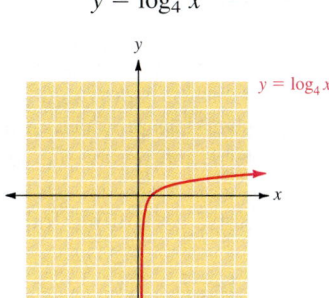

Convert each of the following statements to logarithmic form.

6. $10^4 = 10{,}000$ **7.** $27^{2/3} = 9$

Convert each of the following statements to exponential form.

8. $\log_5 125 = 3$ **9.** $\log 0.01 = -2$

Solve each of the following equations for the unknown variable.

10. $y = \log_2 64$ **11.** $\log_b \dfrac{1}{16} = -2$ **12.** $\log_{25} x = \dfrac{1}{2}$

Use the properties of logarithms to expand each of the following expressions.

13. $\log_b x^2 y z^3$ **14.** $\log_5 \sqrt{\dfrac{xy^2}{z}}$

Use the properties of logarithms to write each of the following expressions as a single logarithm.

15. $\log x + 3 \log y$ **16.** $\dfrac{1}{3}(\log_b x - 2 \log_b z)$

Solve each of the following logarithmic equations for x.

17. $\log_6 (x + 1) + \log_6 (x - 4) = 2$ **18.** $\log (2x + 1) - \log(x - 1) = 1$

Solve each of the following exponential equations for x. Give your results correct to three decimal places.

19. $3^{x+1} = 4$ **20.** $5^x = 3^{x+1}$

Cumulative Test ▪ 1–13

1. {11}
2. {x|x < $\frac{2}{3}$ or x > 4}
3. {$\frac{10}{9}$}
4.–5. See Answers
6. 6x + 5y = 7
7. {x|x ≥ 10}
8. 2x² − 6x + 1
9. 6x² − 13x − 5
10. (2x − 5)(x + 2)
11. x(5x + 4y)(5x − 4y)
12. $\frac{-x+2}{(x-4)(x-5)}$
13. $\frac{x+2}{x-2}$
14. −4√2
15. 22² + 12√2
16. $\frac{5}{3}$(√5 + √2)
17. 77, 79, 81
18. {−2, 1}
19. {$\frac{3 \pm \sqrt{19}}{2}$}
20. {x|−$\frac{3}{2}$ ≤ x ≤ 1}

This test is provided to help you in the process of reviewing the previous chapters. Answers are provided in the back of the book. If you missed any answers, be sure to go back and review the appropriate chapter section.

Solve each of the following.

1. $2x - 3(x + 2) = 4(5 - x) + 7$
2. $|3x - 7| > 5$
3. $\log x - \log(x - 1) = 1$

Graph each of the following.

4. $5x - 3y = 15$
5. $-8(2 - x) \geq y$

6. Find the equation of the line that passes through the points $(2, -1)$ and $(-3, 5)$.

7. Solve the linear inequality

$$3x - 2(x - 5) \geq 20$$

Simplify each of the following expressions.

8. $4x^2 - 3x + 8 - 2(x^2 + 5) - 3(x - 1)$
9. $(3x + 1)(2x - 5)$

Factor each of the following completely.

10. $2x^2 - x - 10$
11. $25x^3 - 16xy^2$

Perform the indicated operations.

12. $\frac{2}{x-4} - \frac{3}{x-5}$
13. $\frac{x^2 - x - 6}{x^2 + 2x - 15} \div \frac{x-2}{x+5}$

Simplify each of the following radical expressions.

14. $\sqrt{18} + \sqrt{50} - 3\sqrt{32}$
15. $(3\sqrt{2} + 2)(3\sqrt{2} + 2)$
16. $\frac{5}{\sqrt{5} - \sqrt{2}}$

17. Find three consecutive odd integers whose sum is 237.

Solve each of the following equations.

18. $x^2 + x - 2 = 0$
19. $2x^2 - 6x - 5 = 0$

20. Solve the following inequality:

$$2x^2 + x - 3 \leq 0$$

Answers to Odd Exercises, Self-Tests, and Cumulative Tests

Chapter 0

Section 0.1
1. $\frac{6}{14}, \frac{9}{21}, \frac{12}{28}$ 3. $\frac{8}{18}, \frac{16}{36}, \frac{40}{90}$ 5. $\frac{10}{12}, \frac{15}{18}, \frac{50}{60}$
7. $\frac{20}{34}, \frac{30}{51}, \frac{100}{170}$ 9. $\frac{18}{32}, \frac{27}{48}, \frac{90}{160}$ 11. $\frac{14}{18}, \frac{35}{45}, \frac{140}{180}$
13. $\frac{2}{3}$ 15. $\frac{5}{7}$ 17. $\frac{2}{3}$ 19. $\frac{7}{8}$ 21. $\frac{1}{4}$
23. $\frac{1}{3}$ 25. $\frac{8}{9}$ 27. $\frac{4}{5}$ 29. $\frac{5}{7}$ 31. $\frac{4}{5}$
33. $\frac{7}{9}$ 35. $\frac{15}{44}$ 37. $\frac{21}{20}$ 39. $\frac{3}{7}$ 41. $\frac{8}{39}$
43. $\frac{7}{33}$ 45. $\frac{1}{6}$ 47. $\frac{4}{15}$ 49. $\frac{8}{15}$ 51. $\frac{2}{3}$
53. $\frac{63}{50}$ 55. $\frac{4}{3}$ 57. $\frac{4}{15}$ 59. $\frac{13}{20}$ 61. $\frac{13}{15}$
63. $\frac{19}{24}$ 65. $\frac{7}{12}$ 67. $\frac{107}{90}$ 69. $\frac{7}{8}$ 71. $\frac{5}{9}$
73. $\frac{1}{2}$ 75. $\frac{5}{24}$ 77. $\frac{7}{18}$ 79. $\frac{11}{24}$ 81. $\frac{13}{42}$

Section 0.2
1. True 3. True 5. True 7. False
9. False 11. False 13. True 15. True
17. False 19. True 21. True 23. False
25. 10 27. 20 29. 7 31. -30 33. 6
35. 50 37. 3 39. -7 41. $>$ 43. $<$
45. $=$ 47. $<$ 49. $-3, 2, 0$ 51. $0, 2$
53. $-2, 0, 1$ 55. $0, 1$ 57.

59. (a) 3 (b) -3 (c) 3 (d) Answers will vary (e) -7

Section 0.3
1. -11 3. 4 5. -2 7. 16 9. -6
11. -15 13. -8 15. 1 17. 0 19. -6
21. -3 23. 0 25. $\frac{9}{4}$ 27. $\frac{3}{2}$ 29. $\frac{23}{8}$
31. -8 33. -12 35. -7 37. -2 39. 6
41. -11 43. -20 45. 1 47. 0 49. 50
51. 10 53. -10 55. 35 57. 63 59. $\frac{17}{2}$
61. $\frac{3}{2}$ 63. -8 65. -11 67. -24.883
69. -11.494 71. -5.702 73. 6.494 75. 33°F
77. 28th floor 79. 925 ft 81. 98.6°
83. 85. 87.

Section 0.4
1. 56 3. -12 5. -72 7. 42 9. 0
11. 64 13. -80 15. 108 17. -20
19. -200 21. 150 23. $\frac{1}{4}$ 25. 22 27. $\frac{1}{6}$
29. 120 31. -80 33. -150 35. 144

37. 15 39. -48 41. 70 43. 36 45. -5
47. 6 49. -10 51. 8 53. -4 55. -18
57. 0 59. 12 61. Undefined 63. -25
65. 5 67. $\frac{1}{6}$ 69. $-\frac{5}{4}$ 71. -5
73. -2.349 75. 3.341 77. -0.668
79. 14 weeks 81. $1.98 83. 2.4°F 85. -2
87. -2 89. 4 91. 93.

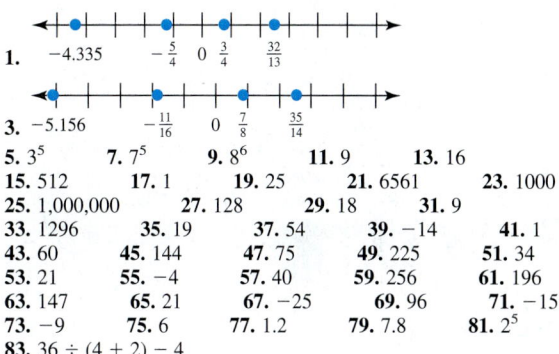

Section 0.5
1. -4.335, $-\frac{5}{4}$, 0, $\frac{3}{4}$, $\frac{32}{13}$

3. -5.156, $-\frac{11}{16}$, 0, $\frac{7}{8}$, $\frac{35}{14}$

5. 3^5 7. 7^5 9. 8^6 11. 9 13. 16
15. 512 17. 1 19. 25 21. 6561 23. 1000
25. 1,000,000 27. 128 29. 18 31. 9
33. 1296 35. 19 37. 54 39. -14 41. 1
43. 60 45. 144 47. 75 49. 225 51. 34
53. 21 55. -4 57. 40 59. 256 61. 196
63. 147 65. 21 67. -25 69. 96 71. -15
73. -9 75. 6 77. 1.2 79. 7.8 81. 2^5
83. $36 \div (4 + 2) - 4$

Chapter 1

Section 1.1
1. $c + d$ 3. $w + z$ 5. $x + 2$ 7. $y + 10$
9. $a - b$ 11. $b - 7$ 13. $r - 6$ 15. wz
17. $5t$ 19. $8mn$ 21. $3(p + q)$ 23. $2(x + y)$
25. $2x + y$ 27. $2(x - y)$ 29. $(a + b)(a - b)$
31. $m(m - 3)$ 33. $\frac{x}{5}$ 35. $\frac{a + b}{7}$ 37. $\frac{p - q}{4}$
39. $\frac{a + 3}{a - 3}$ 41. $x + 5$ 43. $x - 7$ 45. $9x$
47. $3x + 6$ 49. $2(x + 5)$ 51. $(x + 2)(x - 2)$
53. $\frac{x}{7}$ 55. $\frac{x + 5}{8}$ 57. $\frac{x + 6}{x - 6}$ 59. $4s$
61. $\pi r^2 h$ 63. $\frac{1}{2} h (b_1 + b_2)$ 65. Expression
67. Not an expression 69. Not an expression
71. Expression 73. $2x$ 75. $I = prt$
77.

Section 1.2
1. -22 3. 32 5. -20 7. 12 9. 4
11. 83 13. 6 15. -40 17. 14 19. -9
21. 2 23. 2 25. 11 27. 1 29. 11
31. 91 33. 1 35. 91 37. 29 39. 9
41. 16 43. 2 45. -7 47. -72 49. -15.3
51. -11.5 53. 1.1 55. 14.0 57. -90
59. -77922 61. True 63. False 65. 3.75Ω

A-1

67. 30 in. **69.** $1875 **71.** 14°F **73.**

75. (a) 0, 80, 96, 72, 32, 0 (b) 1.8 (c) 94.5, 95.744, 96.492, 96.768, 96.596, 96, 95.004, 93.632, 91.908, 89.856, 87.5
77.

Section 1.3
1. $5a, 2$ **3.** $4x^3$ **5.** $3x^2, 3x, -7$ **7.** $5ab, 4ab$
9. $2x^2y, -3x^2y, 6x^2y$ **11.** $10m$ **13.** $17b^3$
15. $28xyz$ **17.** $6z^2$ **19.** 0 **21.** $-n^2$
23. $-15p^2q$ **25.** $6x^2$ **27.** $2a + 4b$ **29.** $-3x - y$
31. $4a - 10b - 1$ **33.** $-2a - 3b$ **35.** $5a - 2b + 3c$
37. $6r - 5s$ **39.** $8p - 2q$ **41.** $9a + 4$
43. $13b^2 - 18b$ **45.** $-2x^2$ **47.** $5x^2 - 2x + 1$
49. $2b^2 + 5b + 16$ **51.** $8y^3 - 2y$ **53.** $-a^3 + 4a^2$
55. $-2x^2 - x + 3$ **57.** $x - 7$ **59.** $m^2 - 3m$
61. $-2y^2$ **63.** $2x^2 - x + 1$ **65.** $8a^2 - 12a - 7$
67. $-6b^2 + 8b$ **69.** $2x^2 + 12$ **71.** $6b - 1$
73. $10x - 9$ **75.** $2x^2 + 5x - 12$ **77.** $-6y^2 - 8y$
79. $5x^2 - 3x + 9$ **81.** 206.8 **83.** 260.6
85. $28x + 4$ **87.** $-x^2 + 65x - 150$ **89.**
91. **93.**

Section 1.4
1. {Monday, Tuesday, Wednesday, Thursday, Friday, Saturday, Sunday} **3.** {1, 2, 4, 8, 16}
5. {1, 2, 3, 5, 7, 11, 13, 17, 19, 23, 29}
7. {-5, -4, -3, -2, -1} **9.** {2, 4, 6, 8, 10, 12}
11. {3, 4, 5, 6} **13.** {-3, -2}
15. {-5, -4, -3, -2, -1, 0, 1, 2} **17.** {1, 3, 5, 7, ...}
19. {2, 4, 6, ..., 96, 98} **21.** {5, 10, 15, 20, ...}
23. $\{x \mid x > 8\}$ **25.** $\{x \mid x \geq -5\}$
27. $\{x \mid 2 < x < 7\}$ **29.** $\{x \mid -4 \leq x \leq 4\}$
31.
33.
35.
37.
39.
41.
43.
45.
47. $\{x \mid x \leq 1\}$ **49.** $\{x \mid x > -2\}$
51. $\{x \mid -2 \leq x \leq 2\}$ **53.** $\{-3 < x < 2\}$
55. $\{x \mid -2 \leq x < 4\}$

Chapter 2

Section 2.1
1. Yes **3.** No **5.** No **7.** Yes **9.** No
11. No **13.** Yes **15.** Yes **17.** Yes **19.** No
21. Yes **23.** Linear equation **25.** Expression
27. Linear equation **29.** {2} **31.** {11} **33.** {-2}
35. {6} **37.** {4} **39.** {-10} **41.** {-3}
43. {2} **45.** {4} **47.** {6} **49.** {6}
51. {6} **53.** {-18} **55.** {16} **57.** {8}
59. {2} **61.** $x + 3 = 7$ **63.** $3x - 7 = 2x$
65. $2(x + 5) = x + 18$ **67.** c **69.** a **71.** True
73. $x + 7 = 33; 26$ **75.** $x - 15 = 7; 22$
77. $1840 + x = 3260; 1420$ **79.** $x + 360 = 650; \$290$
81. $x + 225 = 965; \$740$ **83.**
85. **87.**

Section 2.2
1. {4} **3.** {6} **5.** {7} **7.** {-4} **9.** {-8}
11. {-9} **13.** {3} **15.** {-7} **17.** {9}
19. {8} **21.** {15} **23.** {42} **25.** {-20}
27. {-24} **29.** {9} **31.** {-20} **33.** {-25}
35. {4} **37.** {-6} **39.** {4} **41.** {4}
43. {-3} **45.** {3} **47.** {4} **49.** $5x = 40$
51. $\dfrac{x}{7} = 6$ **53.** $\dfrac{1}{3}x = 8$ **55.** $\dfrac{3}{4}x = 18$
57. $\dfrac{2x}{5} = 12$ **59.** 116 **61.** 26 min **63.** 3
65. 35 in.

Section 2.3
1. {4} **3.** {3} **5.** {-2} **7.** {3} **9.** {8}
11. {18} **13.** {3} **15.** {6} **17.** {-2}
19. {-4} **21.** {6} **23.** {5} **25.** {-4}
27. {5} **29.** {4} **31.** {5} **33.** $\left\{\dfrac{5}{2}\right\}$
35. {5} **37.** $\left\{-\dfrac{4}{3}\right\}$ **39.** {4} **41.** {7}
43. {30} **45.** {6} **47.** {15} **49.** {3}
51. $\left\{\dfrac{3}{2}\right\}$ **53.** Conditional **55.** Contradiction
57. Identity **59.** Contradiction **61.** Identity
63. $2x + 3 = 7$ **65.** $4x - 7 = 41$ **67.** $\dfrac{2}{3}x + 5 = 21$
69. $3x = x + 12$ **71.** 13 **73.** 18 **75.** 35,36
77. 20,21,22 **79.** 32,34 **81.** 25,27 **83.** 12,13
85. 1550,1710 **87.** $360, $290 **89.** 18,9
91.

93. A value for which the original equation is true.
95. Multiplying by 0 would always give $0 = 0$.
97. (a) $-\dfrac{3}{2}$ (b) $-\dfrac{7}{4}$ (c) $\dfrac{1}{6}$ (d) $\dfrac{2}{5}$ (e) $\dfrac{8}{3}$ (f) $-\dfrac{9}{5}$

Section 2.4
1. $\dfrac{p}{4}$ **3.** $\dfrac{E}{I}$ **5.** $\dfrac{V}{LW}$ **7.** $180 - A - C$
9. $-\dfrac{b}{a}$ **11.** $\dfrac{2s}{t^2}$ **13.** $\dfrac{15 - x}{5}$ or $-\dfrac{1}{5}x + 3$

Answers **A-3**

15. $\dfrac{P-2W}{2}$ or $\dfrac{P}{2} - W$ **17.** $\dfrac{PV}{K}$ **19.** $2x - a$
21. $\dfrac{5}{9}(F-32)$ or $\dfrac{5(F-32)}{9}$ **23.** $\dfrac{S-2\pi r^2}{2\pi r}$ or $\dfrac{S}{2\pi r} - r$
25. 3 cm   **27.** 5%   **29.** 25°C   **31.** $2(x+4) = 20$
33. $3(x-5) = 21$ **35.** $2x + 3(x+1) = 48$ **37.** 10, 18
39. 8, 15   **41.** 5, 6   **43.** 12 in., 25 in.
45. 6 m, 22 m   **47.** Legs, 13 cm; base, 10 cm
49. 200 $6 tickets, 300 $8 tickets
51. 30 20¢ stamps, 50 35¢ stamps
53. 60 coach, 40 berth, and 20 sleeping room
55. 40 mi/h, 30 mi/h
57. 6 PM   **59.** 2 PM   **61.** 3 PM
63. 360 Douglas fir, 140 hemlock
65.   **67.**

Section 2.5

1. $5 < 10$ **3.** $7 > -2$ **5.** $0 < 4$ **7.** $-2 > -5$
9. x is less than 3   **11.** x is greater than or equal to -4
13. -5 is less than or equal to x
15.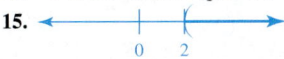
17.
19.
21.
23.
25.
27.
29. $x < 13$
31. $x \geq 2$
33. $x < 7$
35. $x \leq 8$
37. $x \geq 8$
39. $x < -9$
41. $x \leq 3$
43. $x > -7$

45. $x \leq -3$
47. $x > 6$
49. $x > 20$
51. $x \leq 6$
53. $x < 9$
55. $x > 4$
57. $x < 1$
59. $x < -1$
61. $x \leq -6$
63. $x \leq 9$
65. $x > -5$
67. $x < \dfrac{7}{4}$
69. $x < 8$
71. $x \geq -\dfrac{5}{2}$
73. $x \leq \dfrac{3}{2}$
75. $x < -\dfrac{2}{3}$
77. $x + 5 > 3$ **79.** $2x - 4 \leq 7$ **81.** $4x - 15 > x$
83. a   **85.** c   **87.** b   **89.** $P < 1000$
91. $x \geq 88$ **93.** $10,000   **95.**
97.

Section 2.6

1. 15   **3.** 15   **5.** 5   **7.** 4   **9.** 31   **11.** 8
13. 25   **15.** $\{5, -5\}$ **17.** $\{5, -3\}$
19. $\{3.5, -2.5\}$ **21.** $\{1.5, 4.5\}$ **23.** No solution
25. $\{0, 4\}$ **27.** $\{7, 1\}$ **29.** $\left\{-\dfrac{2}{3}, 4\right\}$
31. $\left\{-\dfrac{2}{3}, \dfrac{6}{7}\right\}$ **33.** $\left\{2, -\dfrac{4}{9}\right\}$
35. $2 \leq x \leq 4$

37. $-4 < x < 2$

39. $2 \le x \le \dfrac{9}{2}$

41. $-2 < x < 1$

43. $x < -2$ or $x > 4$

45. $x < -3$ or $x > 4$

47. $x < -2$ or $x > \dfrac{8}{3}$

49. $-5 < x < 5$

51. $x \le -7$ or $x \ge 7$

53. $x < 2$ or $x > 6$

55. $-10 \le x \le -2$

57. $x < -2$ or $x > 8$

59. No solution

61. $1 < x < 4$

63. $x < -3$ or $x > \dfrac{1}{3}$

65. $x < -\dfrac{4}{5}$ or $x > 2$

67. $3 \le x \le \dfrac{13}{3}$

69. $|x| < 3$ **71.** $|x| \ge 5$ **73.** $|x-(-2)| < 7$
75. $|x-(-4)| \ge 3$

Cumulative Test—Chapters 0–2
1. -23 **2.** 11 **3.** 2 **4.** 80 **5.** 7 **6.** 7
7. -16 **8.** 4 **9.** 63 **10.** -16 **11.** 0
12. -5 **13.** 9 **14.** -9 **15.** 13 **16.** 3
17. 7 **18.** $15x - 9y$ **19.** $10x^2 - 13x + 2$
20. $\{4\}$ **21.** $\{-2, 8\}$ **22.** $x \le -4$ **23.** $-3 \le x < 3$
24. $-3 \le x \le 5$ **25.** $x < -9$ or $x > 7$
26. $\dfrac{I}{Pt}$ **27.** $\dfrac{2A}{b}$ **28.** $\dfrac{c-ax}{b}$ **29.** $4x - 7 = 45;\ 13$
30. $x + (x+1) = 85;\ 42, 43$ **31.** $3x - 12 = x + 2;\ 7$
32. $x + (x + 120) = 720;\ \$420$
33. $2x + 2(3x+2) = 44;\ 5$ cm, 17 cm
34. $x + (x+5) + 2x = 37;\ 8$ in., 13 in., 16 in.

Chapter 3

Section 3.1
1. $(4, 2), (0, 6), (-3, 9)$ **3.** $(5, 2), (4, 0), (6, 4)$
5. $(2, 0), (1, 3)$ **7.** $(3, 0), (6, 2), (0, -2)$
9. $(4, 0), \left(\dfrac{2}{3}, -5\right), \left(5, \dfrac{3}{2}\right)$ **11.** $(0, 0), (2, 8)$
13. $(3, 5), (3, 0), (3, 7)$ **15.** $8, 7, 12, 12$ **17.** $0, 0, 4, 9$
19. $3, -5, 5, 2$ **21.** $4, 0, -3, 6$ **23.** $-3, 11, 9, 7$
25. $-4, 3, \dfrac{4}{3}, 1$ **27.** $(0, -7), (2, -5), (4, -3), (6, -1)$
29. $(0, -6), (3, 0), (6, 6), (9, 12)$
31. $(8, 0), (-4, 3), (0, 2), (4, 1)$
33. $(-5, -4), (0, -2), (5, 0), (10, 2)$
35. $(0, 3), (1, 5), (2, 7), (3, 9)$
37. $(-5, 0), (-5, 1), (-5, 2), (-5, 3)$
39. $(2, -3, 1)$ **41.** $(1, -6, 5)$ **43.** $(-2, 5, 1)$
45. $\$9.50, \$11.75, \$15.50, \$19.25, \$23$
47. 25 cm², 100 cm², 144 cm², 225 cm²   **49.**

Section 3.2
1. $(5, 6)$ **3.** $(2, 0)$ **5.** $(-4, -5)$ **7.** $(-5, -3)$
9. $(-3, 5)$
11–21.

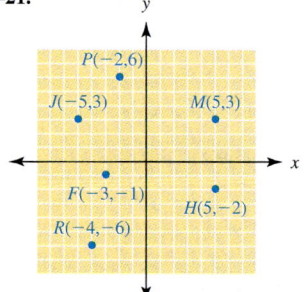

23. Quadrant I   **25.** Quadrant III   **27.** x axis
29. Quadrant II   **31.** y axis   **33.** Quadrant IV
35. Points are $(1, 30), (2, 45), (3, 60), (4, 60), (5, 75), (6, 90),$
$(7, 95)$ **37.** Points are $(7, 100), (15, 70), (20, 80), (30, 70),$
$(40, 50), (50, 40), (60, 30), (70, 40), (80, 25)$
39. The points lie on a line; $(1,2)$
41. The points lie on a line; $(2,-6)$
43.

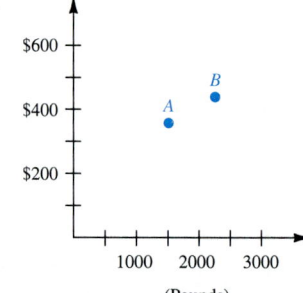

■ Answers **A-5**

45.

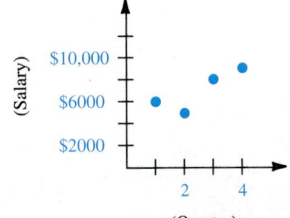

47.   **49.**

Section 3.3

1. $x + y = 6$

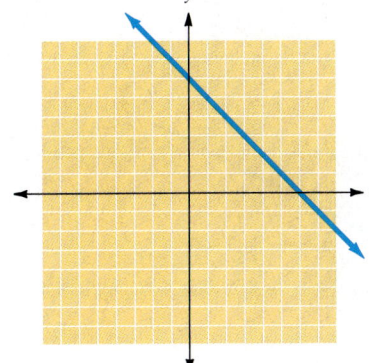

3. $x - y = -3$

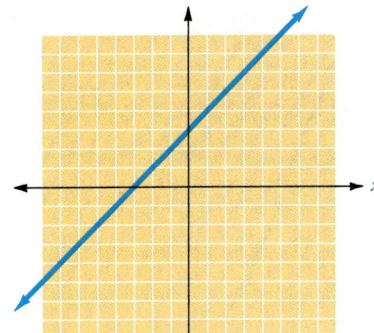

5. $2x + y = 2$

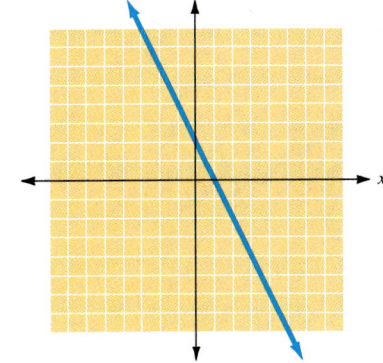

7. $3x + y = 0$

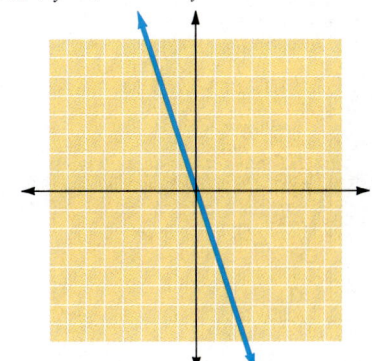

9. $x + 4y = 8$

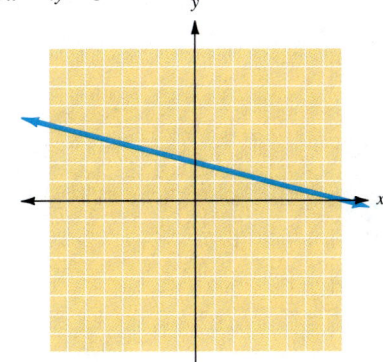

11. $y = 5x$

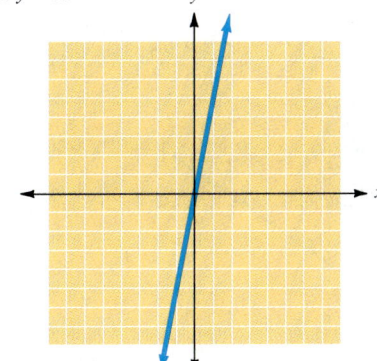

13. $y = 2x - 1$

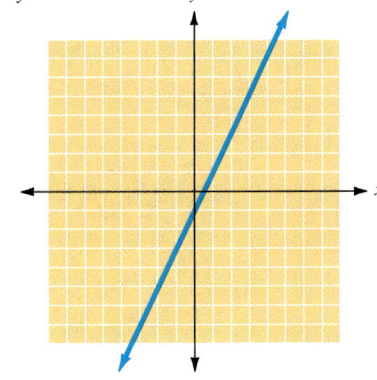

15. $y = -3x + 1$

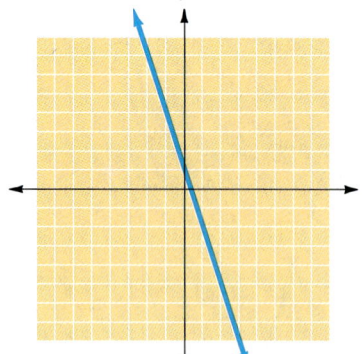

17. $y = \frac{1}{3}x$

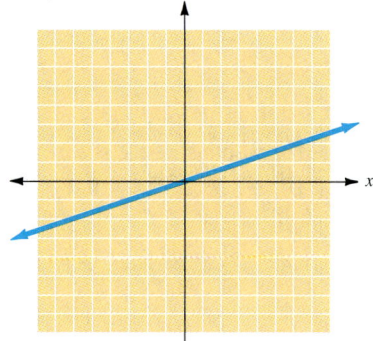

19. $y = \frac{2}{3}x - 3$

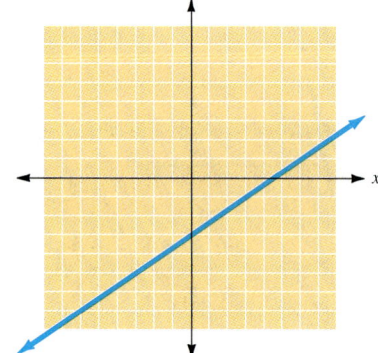

21. $x = 5$

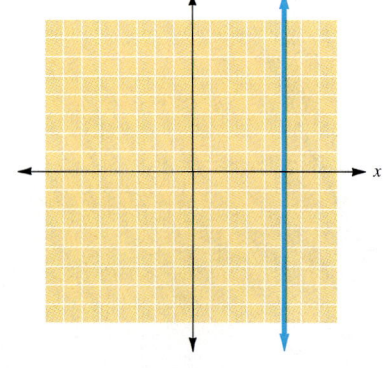

23. $y = 1$

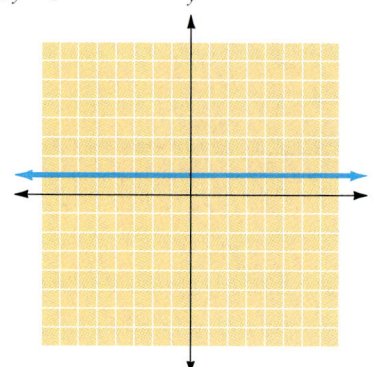

25. $x - 2y = 4$

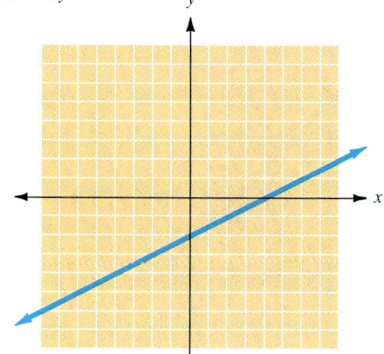

27. $5x + 2y = 10$

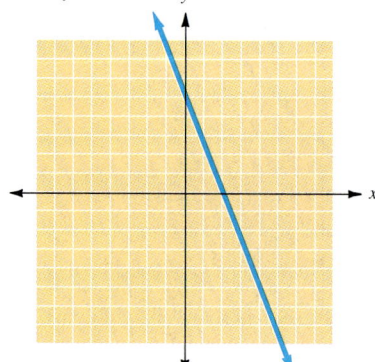

29. $3x + 5y = 15$

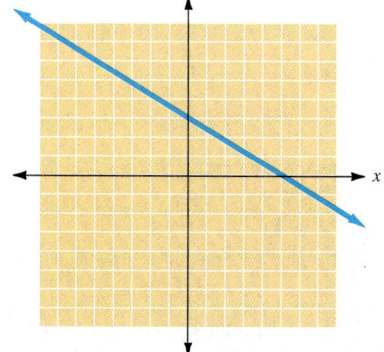

31. $y = 2 - \dfrac{x}{3}$

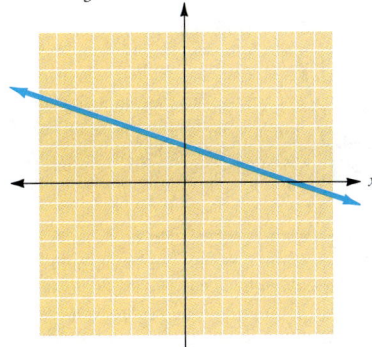

33. $y = 3 - \dfrac{3}{4}x$

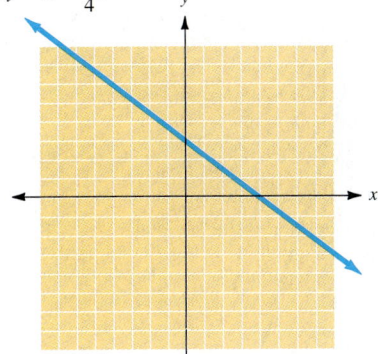

35. $y = -5 + \dfrac{5}{4}x$

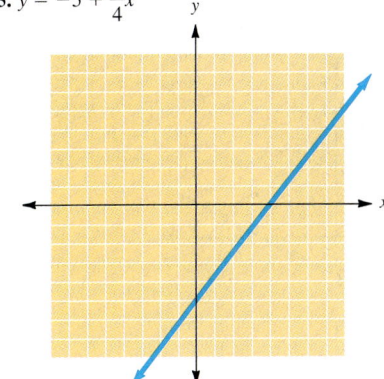

37. $y = 2x$ **39.** $y = x + 3$ **41.** $y = 3x - 3$
43. $x - 4y = 12$
45. (3, 1)

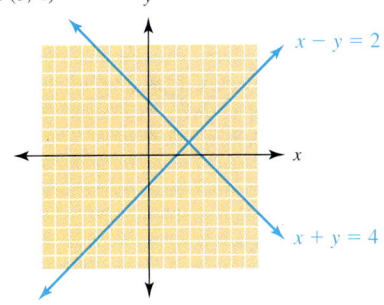

47. Parallel lines

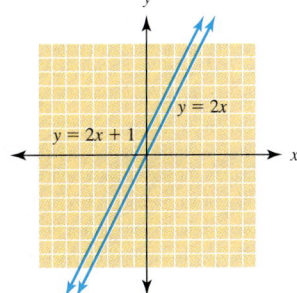

49. Perpendicular lines.

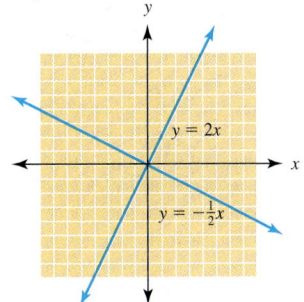

51. Graph

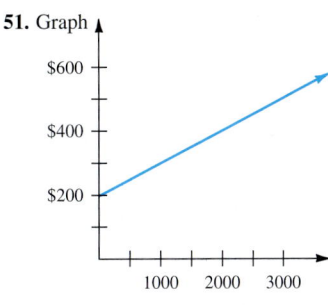

53. (a)

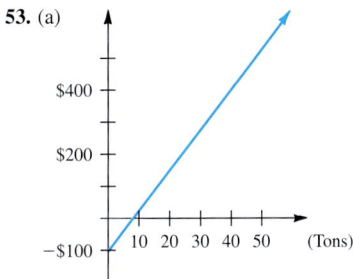

(b) 17 tons (c) $140 (d) $y = 17x - 125$

55. $C = 0.08s + 12$

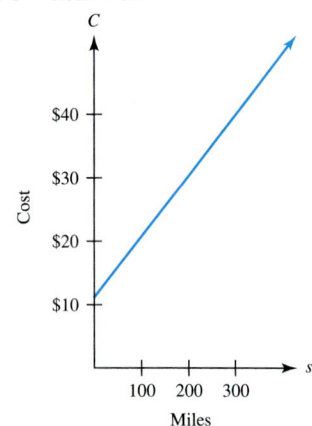

57. (a) $T = 35h + 75$ and (b) see graph

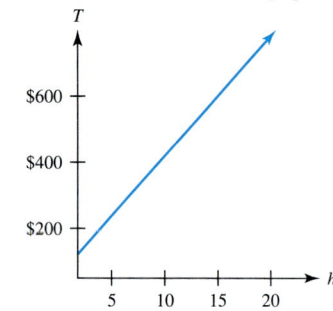

59. (a)

POINT	x	y
A	5	13
B	6	15
C	7	17
D	8	19
E	9	21

(b) increases by 2
(c) yes
(d) grows by 2 units

61. (a)

POINT	x	y
A	5	13
B	6	16
C	7	19
D	8	22
E	9	25

(b) increases by three
(c) yes
(d) grows by 3 units

63. (a)

POINT	x	y
A	5	30
B	6	26
C	7	22
D	8	18
E	9	14

(b) decreases by 4
(c) yes
(d) decreases by 4 units

Section 3.4

1. 1 **3.** 5 **5.** $\frac{4}{5}$ **7.** -2 **9.** $-\frac{3}{2}$
11. Undefined **13.** $\frac{5}{7}$ **15.** 0 **17.** $-\frac{4}{3}$
19. Slope 3, y intercept 5   **21.** Slope -2, y intercept -5
23. Slope $\frac{3}{4}$, y intercept 1   **25.** Slope $\frac{2}{3}$, y intercept 0
27. Slope $-\frac{4}{3}$, y intercept 4   **29.** Slope 0, y intercept 9

31. Slope $\frac{3}{2}$, y intercept -4

33. $y = 3x + 5$

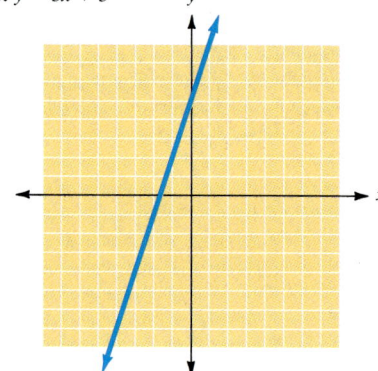

35. $y = -3x + 4$

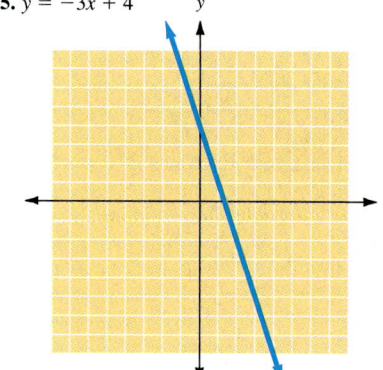

37. $y = \frac{1}{2}x - 2$

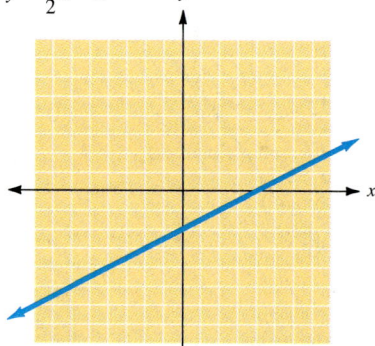

39. $y = \frac{2}{3}x$

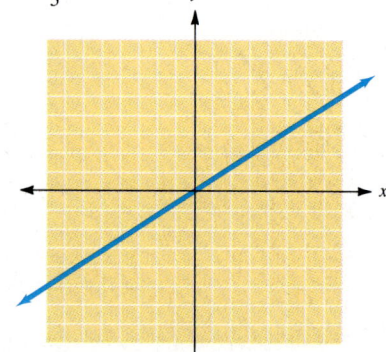

41. $y = \frac{3}{4}x + 3$

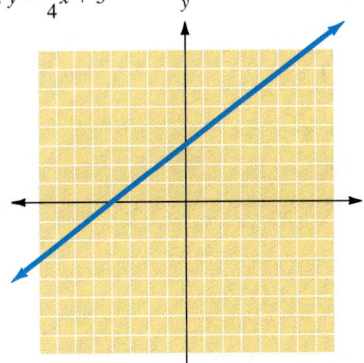

33. $y = 4x - 2$ **35.** $y = -\frac{1}{2}x + 2$ **37.** $y = 4$
39. $y = 5x - 13$ **41.** $y = 3x + 3$ **43.** $y = \frac{1}{2}x + 4$
45. $y = 3$ **47.** $y = 2x + 8$ **49.** $y = \frac{4}{3}x - 2$
51. $y = \frac{1}{3}x - \frac{11}{3}$ **53.** $y = -\frac{1}{2}x$ **55.** Yes, yes
57. Yes, no   **59.** $F = \frac{9}{5}C + 32$ **61.** (a) $C = 35x + 50$,
(b) $172.50, (c) 3.15 hours   **63.** Slope = 1; y-int = 3
65. Slope = 2; y-int = 1   **67.** Slope = -3; y-int = 1
69. Slope = -2; y-int = -3

Cumulative Test—Chapters 0–3
1. 17   **2.** 8   **3.** $6x + 3y$ **4.** $\{-11\}$
5. $\{100\}$ **6.** $\{3, 4\}$ **7.** $C = \frac{5}{9}(F - 32)$
8. $\{x \mid x < 4\}$
9. $\{x \mid x > -4\}$
10. $\{x \mid 5 \leq x \leq 9\}$
11. $\{x \mid x < 2 \text{ or } x > 7\}$
12. $\{x \mid -\frac{13}{2} \leq x \leq \frac{3}{2}\}$
13. $\{x \mid x < 1 \text{ or } x > 11\}$
14. -2 **15.** $y = 5x - 6$ **16.** $10y + x = 86$
17. $y = -\frac{2}{3}x + 6$ **18.** $4y + 5x = 26$ **19.** 4   **20.** 7

43. IV   **45.** III   **47.** I   **49.** III and IV   **51.** g
53. e   **55.** h   **57.** c
59.

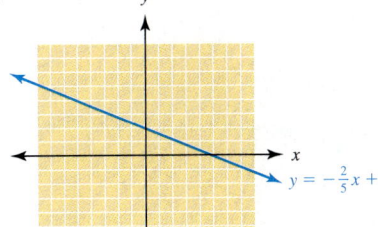

61.

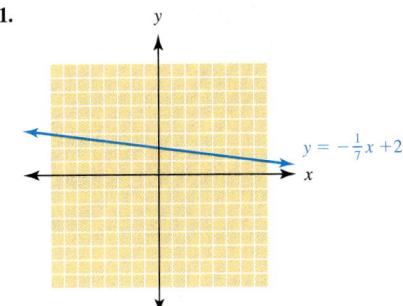

63. 2   **65.** -3 **67.** 3   **69.** $-\frac{5}{2}$
71. $m = 0.10$, market price of jugs; y intercept = 200, the 200 award.   **73.** Slope represents the price of newsprint; the y intercept the cost of the truck.   **75.** -0.30
77. 2.14¢/yr   **79.**   **81.**
83. Parallel lines; no   **85.** Perpendicular lines; -1

Section 3.5
1. Parallel   **3.** Neither   **5.** Perpendicular   **7.** $\frac{1}{3}$
9. 12   **11.** $y = \frac{5}{4}x - 5$ **13.** $y = 3x - 1$
15. $y = -3x - 9$ **17.** $y = \frac{2}{5}x - 5$ **19.** $x = 2$
21. $y = -\frac{4}{5}x + 4$ **23.** $y = x + 1$ **25.** $y = \frac{3}{4}x - \frac{3}{2}$
27. $y = 2$ **29.** $y = \frac{3}{2}x - 3$ **31.** $y = \frac{5}{2}x + 4$

Chapter 4

Section 4.1
1. (a) and (c)   **3.** (b)   **5.** D: $\{1, 3, 5, 7, 9\}$;
R: $\{2, 4, 6, 8, 10\}$ **7.** D: $\{1, 3, 4, 5, 6\}$; R: $\{1, 2, 3, 4, 6\}$
9. D: $\{1\}$; R: $\{2, 3, 4, 5, 6\}$ **11.** D: $\{1, 2, 3, 4\}$; R: $\{4, 5, 6\}$
13. D: $\{-3, -2, -1, 4, 5\}$; R: $\{3, 4, 5, 6\}$
15. D: reals; R: reals   **17.** D: reals; R: $\{5\}$
19. D: $\{23\}$; R: reals   **21.** $\{(1, 9\frac{1}{8}), (2, 8), (3, 8\frac{7}{8}),$
$(4, 9\frac{1}{4}), (5, 9)\}$ **23.** In ordered pairs, the order of elements is important.

Section 4.2
1. (a) -2, (b) 4, (c) -2 **3.** (a) 9, (b) -1, (c) 3
5. (a) -62, (b) -2, (c) 2   **7.** (a) 45, (b) 3, (c) -75
9. (a) -1, (b) 2, (c) 13   **11.** $f(x) = -3x + 2$
13. $f(x) = 4x - 8$ **15.** $f(x) = -\frac{3}{25}x + 3$
17. $f(x) = \frac{1}{3}x + \frac{3}{2}$ **19.** $f(x) = -\frac{5}{8}x + \frac{9}{8}$

21. $f(x) = 3x + 7$

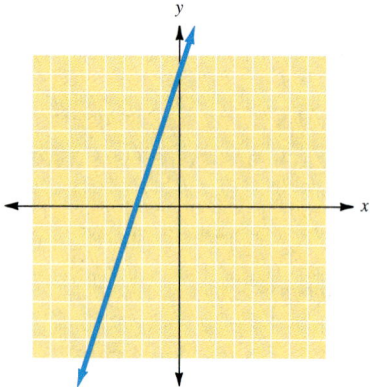

23. $f(x) = -2x + 7$

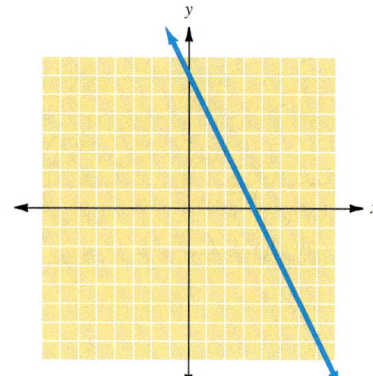

25. $f(x) = -x - 1$

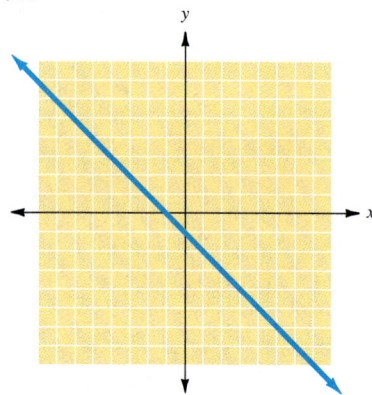

27. 17   **29.** 13   **31.** -19
33. $5a - 1$ **35.** $5x + 4$ **37.** $5x + 5h - 1$
39. $-3m + 2$ **41.** $-3x - 4$ **43.** 5
45. (1, 5), (3, 9)   **47.** 107, 75, 49, 29, 15, 7, 5, 9, 19, 35, 57
49. $-221, -114, -51, -20, -9, -6, 1, 24$

Section 4.3
1. Function   **3.** Function   **5.** Not a function
7. Not a function   **9.** Function   **11.** Not a function
13. Function

15. Function

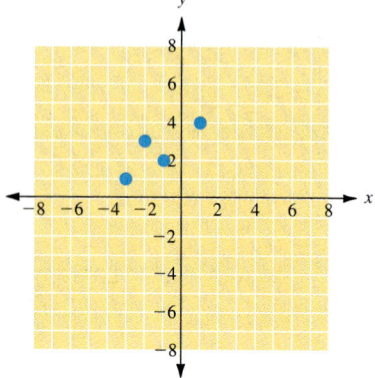

17. Function

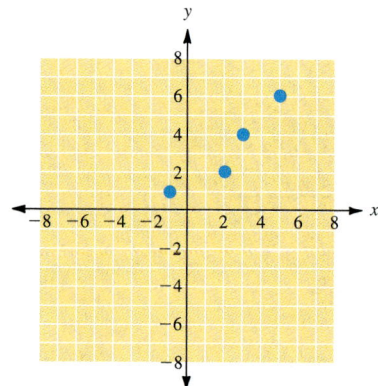

19. Not a function

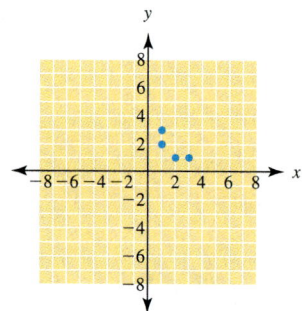

21. Function   **23.** Not a function   **25.** Function
27. Function   **29.** (a) D: $-2 < x \leq 2$; R: $1 \leq y \leq 2$;
(b) Yes; (c) Answers will vary.   **31.** Independent: size of tank; dependent: cost   **33.** Independent: length of time; dependent: amount of penalty   **35.** Independent: length of winter; dependent: amount of snowfall   **37.**

Section 4.4
1. A (3, 3), B (2, -4)   **3.** A (2, 5), B ($-2, -4$)
5. A (0, 5), B (3, 0)   **7.** A (2, 0), B (6, 4)

9. $A(3, 3)$, $B(-3, -3)$ **11.** $A(3, 6)$, $B(3, 0)$
13. (a) 3, (b) 1, (c) 2, (d) 5, (e) 0 **15.** (a) 0.5, (b) 0.5, (c) 0, (d) 4, (e) 1.5 **17.** (a) 1, (b) 3, (c) 2, (d) 1, (e) 4
19. (a) 3, (b) 3, (c) 3, (d) 3, (e) 3 **21.** (a) 1, (b) 2, (c) 4
23. (a) 2, -2, (b) 3, -3, (c) 4.5, -4.5 **25.** (a) -1.5, 1.5, (b) 1, -1, (c) 0 **27.** (a) 1.5 (b) 2.5 (c) 5.5
29.

Section 4.5
1. (a) $3x + 1$, (b) $-11x + 9$, (c) 10, (d) -13 **3.** (a) $3x + 4$, (b) $13x - 8$, (c) 13, (d) 18 **5.** (a) $6x + 4$, (b) all reals
7. (a) $3x + 2 + \dfrac{1}{x-2}$, (b) $\{x | x \neq 2\}$
9. (a) $(2x - 1)(x - 3)$, (b) $\dfrac{2x-1}{x-3}$, (c) $\{x | x \neq 3\}$
11. (a) $(3x + 2)(2x - 1)$, (b) $\dfrac{3x+2}{2x-1}$, (c) $\{x | x \neq \dfrac{1}{2}\}$
21. (b) $V = 10 - 4.9t^2$

Cumulative Test—Chapters 0–4
1. $\{15\}$ **2.** $\{-108\}$ **3.** $\{5, -\dfrac{5}{3}\}$ **4.** $\{-3, 1\}$
5. $R = \dfrac{R_1 R_2}{R_1 + R_2}$
6. $x \leq 1$
7. $x < -2$
8. $-3 < x < 6$
9. $x > 13$ or $x < -3$
10. $x < \dfrac{1}{3}$ or $x > 4$
11. $y = 2x - 3$
12. $y = \dfrac{2}{3}x + \dfrac{7}{3}$ **13.** $y = -\dfrac{5}{4}x - 2$ **14.** 22; 10
15. $f(x) = -3x + 5$

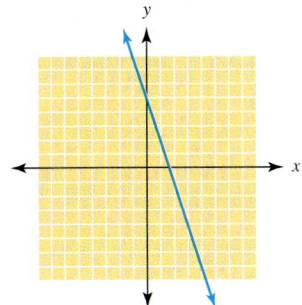

16. Function **17.** Not a function

18. (a) Not a function (b) Function **19.** (a) -3 (b) 0 (c) 3
20. 5 cm by 17 cm

Chapter 5

Section 5.1
1. x^9 **3.** x^{10} **5.** 3^7 **7.** $(-2)^8$ **9.** $4x^{13}$
11. $\left(\dfrac{1}{2}\right)^6$ **13.** $(-2)^5 x^9$ **15.** $(2x)^9$ **17.** $x^6 y^5$
19. $x^9 y^7$ **21.** $-24x^{10}$ **23.** $-30x^9$ **25.** $30x^4 y^5$
27. $x^9 y^7 z^3$ **29.** x^3 **31.** $x^3 y^8$ **33.** $x^4 y^2 z$
35. $3x^3 y^3$ **37.** 64 **39.** 81 **41.** 256 **43.** 384
45. 512 **47.** 1536 **49.** 51; 7 **51.** 250; 46
53. 1; 2; -56 **55.** 3; 372; 15 **57.** $-15x^6$
59. $8x^3$ **61.** x^{21} **63.** $-24x^4$ **65.** $32x^{15}$
67. $-216x^{12}$ **69.** $50x^9$ **71.** $144x^{20}$ **73.** $\dfrac{4}{9}$
75. $\dfrac{a^4}{16}$ **77.** $\dfrac{a^{16}}{b^{12}}$ **79.** $\dfrac{x^{15} y^6}{z^{12}}$ **81.** $\dfrac{8x^6}{27}$
83. $-675 x^{19} y^{14}$ **85.** $\dfrac{6x^{10} y^9}{5}$ **87.** \$5909.82
89. 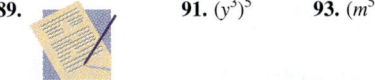 **91.** $(y^3)^5$ **93.** $(m^5)^4$

95. 9^4; 9^7; 9^{20}; 9^{14} **97.** $3x^3 y^2 z^4$

Section 5.2
1. Polynomial **3.** Polynomial **5.** Polynomial
7. Not a polynomial **9.** $2x^2$, $-3x$; 2, -3
11. $4x^3$, $-3x$, 2; 4, -3, 2 **13.** Binomial
15. Trinomial **17.** Not classified **19.** Monomial
21. Not a polynomial **23.** $4x^5 - 3x^2$; 5
25. $-5x^9 + 7x^7 + 4x^3$; 9 **27.** $4x$; 1
29. $x^6 - 3x^5 + 5x^2 - 7$; 6 **31.** 7, -5 **33.** 4, -4
35. 62, 30 **37.** 0, 0 **39.** Always **41.** Sometimes
43. Sometimes **45.** Sometimes **47.** 4 **49.** 11
51. 14 **53.** 5 **55.** 10 **57.** 7 **59.** -2
61. $3x + 20$, \$170 **63.** \$337

Section 5.3
1. $9a + 4$ **3.** $13b^2 - 18b$ **5.** $-2x^2$
7. $5x^2 - 2x + 1$ **9.** $2b^2 + 5b + 16$ **11.** $8y^3 - 2y$
13. $-a^3 + 4a^2$ **15.** $-2x^2 - x + 3$ **17.** $-2a - 3b$
19. $5a - 2b + 3c$ **21.** $6r - 5s$ **23.** $8p - 2q$
25. $x - 7$ **27.** $m^2 - 3m$ **29.** $-2y^2$
31. $2x^2 - x + 1$ **33.** $8a^2 - 12a - 7$ **35.** $-6b^2 + 8b$
37. $2x^2 + 12$ **39.** $6b - 1$ **41.** $10x - 9$
43. $2x^2 + 5x - 12$ **45.** $-6y^2 - 8y$ **47.** $6w^2 - 2w + 2$
49. $9x^2 - x$ **51.** $2a^2 + 5a$ **53.** $3x^2 + x$
55. $3x^2 + 3x - 9$ **57.** $5x^2 - 3x + 9$
59. (a) $9x + 4$, (b) 13, (c) 13 **61.** (a) $7x^2 + 7x$, (b) 14, (c) 14 **63.** (a) $-x^2 - 2x - 2$, (b) -5, (c) -5
65. (a) $5x^3 - 5x^2 - 11x - 25$, (b) -36, (c) -36
67. (a) $2x + 13$, (b) 15, (c) 15 **69.** (a) $12x^2 - x$, (b) 11, (c) 11 **71.** (a) $3x^2 - 2x - 7$, (b) -6, (c) -6
73. (a) $-3x^2 + 7x - 5$, (b) -1, (c) -1 **75.** $a = 3$, $b = 5$, $c = 0$, $d = -1$ **77.** $28x + 4$ **79.** $-x^2 + 65x - 150$

Section 5.4
1. $15x^5$ **3.** $-28b^{10}$ **5.** $40p^{13}$ **7.** $-12m^6$
9. $32x^5 y^3$ **11.** $-6m^9 n^3$ **13.** $10x + 30$
15. $12a^2 + 15a$ **17.** $12s^4 - 21s^3$ **19.** $8x^3 - 4x^2 + 2x$
21. $6x^3 y^2 + 3x^2 y^3 + 15x^2 y^2$ **23.** $18m^4 n^2 - 12m^3 n^2 + 6m^3 n^3$
25. $x^2 + 5x + 6$ **27.** $m^2 - 14m + 45$ **29.** $p^2 - p - 56$

31. $w^2 + 30w + 200$ **33.** $3x^2 - 29x + 40$
35. $6x^2 - x - 12$ **37.** $12a^2 - 31ab + 9b^2$
39. $21p^2 - 13p - 20q^2$ **41.** $6x^2 + 23xy + 20y^2$
43. $x^2 + 10x + 25$ **45.** $w^2 - 12w + 36$
47. $z^2 + 24z + 144$ **49.** $4a^2 - 4a + 1$
51. $36m^2 + 12m + 1$ **53.** $9x^2 - 6xy + y^2$
55. $4r^2 + 20rs + 25s^2$ **57.** $64a^2 - 144ab + 81b^2$
59. $x^2 + x + \frac{1}{4}$ **61.** $x^2 - 36$ **63.** $m^2 - 144$
65. $x^2 - \frac{1}{4}$ **67.** $p^2 - 0.16$ **69.** $a^2 - 9b^2$
71. $16r^2 - s^2$ **73.** $64w^2 - 25z^2$ **75.** $25x^2 - 81y^2$
77. $24x^3 - 10x^2 - 4x$ **79.** $80a^3 - 45a$
81. $60s^3 - 39s^2 + 6s$ **83.** $x^3 - 4x^2 + x + 6$
85. $a^3 - 3a^2 + 3a - 1$ **87.** $\frac{x^2}{5} + \frac{11x}{45} - \frac{4}{15}$
89. $x^2 - y^2 + 4y - 4$ **91.** False   **93.** True
95. $6x^2 - 11x - 35$ cm^2 **97.** $10x - 3x^2$
99. $25x^2 - 40x + 16$ **101.** $x(x + 2)$, or $x^2 + 2x$
103.   **105.**   **107.**

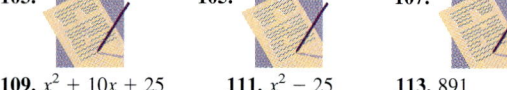

109. $x^2 + 10x + 25$ **111.** $x^2 - 25$ **113.** 891
115. 9996

Cumulative Test—Chapters 0–5
1. 7   **2.** -36 **3.** 4   **4.** 23   **5.** 9, -1
6. 3, $-\frac{19}{3}$ **7.** $x < -15$ **8.** $x < 5$ **9.** $3 < x < 13$
10. $x \geq 6$ or $x \leq \frac{2}{3}$ **11.** $4x - 5y = 2$
12.

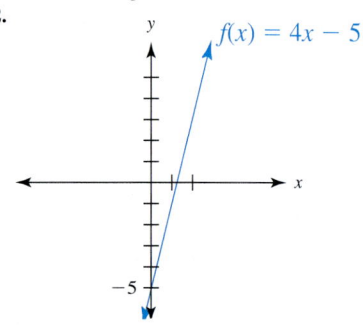

13. $3x^2 + 2x - 4$
14. $x^2 - 3x - 18$ **15.** $12x^2 - 20x$ **16.** $6x^2 + x - 40$
17. $x^3 - x^2 - x + 10$ **18.** $4x^2 - 49$
19. $9x^2 - 30x + 25$ **20.** $20x^3 - 100x^2 + 125x$
21. 4, 0, -6 **22.** 9   **23.** 8   **24.** 65, 67
25. 4 cm by 24 cm

Chapter 6

Section 6.1
1. 2   **3.** 8   **5.** x **7.** a^3 **9.** $5x^4$
11. $2a^4$ **13.** $3xy$ **15.** $5b$ **17.** $3abc^2$
19. $(x + y)^2$ **21.** $4(2a + 1)$ **23.** $8(3m - 4n)$
25. $4m(3m + 2)$ **27.** $5s(2s + 1)$ **29.** $12x(x + 2)$
31. $5a^2(3a - 5)$ **33.** $6pq(1 + 3p)$ **35.** $7mn(m^2 - 3n^2)$
37. $6(x^2 - 3x + 5)$ **39.** $3a(a^2 + 2a - 4)$
41. $3m(2 + 3n + 5n^2)$ **43.** $5xy(2x + 3 - y)$
45. $5r^2s^2(2r + 5 - 3s)$ **47.** $3a(3a^4 - 5a^3 + 7a^2 - 9)$

49. $5mn(3m^2n - 4m + 7n^2 - 2)$ **51.** $(x - 2)(x + 3)$
53. $(p - 2q)(p - q)$ **55.** $(y - z)(x + 3)$
57. $(b - c)(a + b)$ **59.** $(r + 2s)(6r - 1)$
61. $(a - 2)(b^2 + 3)$ **63.** $(x + 2)(x - 5y)$
65. $(m - 3n)(m + 2n^2)$ **67.** Correct   **69.** Incorrect
71. Correct   **73.** 10   **75.** $6x$ **77.**
79. $33 - t$ **81.**
83.

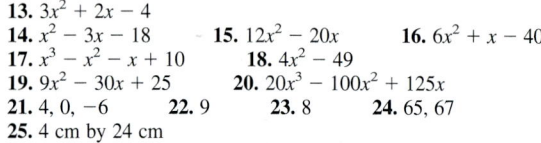

Section 6.2
1. No   **3.** Yes   **5.** No   **7.** No   **9.** Yes
11. $(m + n)(m - n)$ **13.** $(x + 7)(x - 7)$
15. $(7 + y)(7 - y)$ **17.** $(3b + 4)(3b - 4)$
19. $(4w + 7)(4w - 7)$ **21.** $(2s + 3r)(2s - 3r)$
23. $(3w + 7z)(3w - 7z)$ **25.** $(4a + 7b)(4a - 7b)$
27. $(x^2 + 6)(x^2 - 6)$ **29.** $(xy + 4)(xy - 4)$
31. $(5 + ab)(5 - ab)$ **33.** $(r^2 + 2s)(r^2 - 2s)$
35. $(9a + 10b^3)(9a - 10b^3)$ **37.** $2x(3x + y)(3x - y)$
39. $3mn(2m + 5n)(2m - 5n)$ **41.** $(a^2 + 4b^2)(a + 2b)(a - 2b)$
43. $(x + 4)(x^2 - 4x + 16)$ **45.** $(m - 5)(m^2 + 5m + 25)$
47. $(ab - 3)(a^2b^2 + 3ab + 9)$ **49.** $(2w + z)(4w^2 - 2wz + z^2)$
51. $(r - 4s)(r^2 + 4rs + 16s^2)$
53. $(2x - 3y)(4x^2 + 6xy + 9y^2)$
55. $3(a + 3b)(a^2 - 3ab + 9b^2)$
57. $4(x - 2y)(x^2 + 2xy + 4y^2)$
59. (a) -16; (b) $-2x(3x^2 + 5)$; (c) -16
61. (a) -30; (b) $5x^3(x^2 - 7)$; (c) -30
63. (a) -10; (b) $2x^5(3x - 8)$; (c) -10
65. $(b - c)(a + 4b)(a - 4b)$
67. $3a(2a + b)(a + 3b)(a - 3b)$
69. 49   **71.** 45   **73.**   **75.** $a^3 - b^3$
77.

Section 6.3
1. True   **3.** False   **5.** True   **7.** False   **9.** True
11. $a = 1, b = 4, c = -9$ **13.** $a = 1, b = -3, c = 8$
15. $a = 3, b = 5, c = -8$ **17.** $a = 4, b = 8, c = 11$
19. $a = -3, b = 5, c = -10$ **21.** Factorable; 3, -2
23. Not factorable   **25.** Factorable; $-3, -2$
27. Factorable; 6, -1 **29.** Factorable; $-15, -4$
31. $x^2 + 2x + 4x + 8$; $(x + 2)(x + 4)$ **33.** $x^2 - 5x - 4x + 20$; $(x - 5)(x - 4)$ **35.** $x^2 - 9x + 7x - 63$; $(x - 9)(x + 7)$
37. $(x + 3)(x + 5)$ **39.** $(x - 4)(x - 7)$
41. $(s + 10)(s + 3)$ **43.** $(a - 8)(a + 6)$
45. $(x - 1)(x - 7)$ **47.** $(x - 10)(x + 4)$
49. $(x - 7)(x - 7)$ **51.** $(p - 12)(p + 2)$
53. $(x + 11)(x - 6)$ **55.** $(c + 4)(c + 15)$
57. $(n + 10)(n - 5)$ **59.** $(x + 2y)(x + 5y)$
61. $(a - 7b)(a + 6b)$ **63.** $(x - 5y)(x - 8y)$
65. $(3x + 2)(2x + 5)$ **67.** $(5x - 3)(3x + 2)$
69. $(6m - 5)(m + 5)$ **71.** $(3x - 2)(3x - 2)$
73. $(6x + 5)(2x - 3)$ **75.** $(3y - 2)(y + 3)$
77. $(8x + 5)(x - 4)$ **79.** $(2x + y)(x + y)$
81. $(5a + 2b)(a - 2b)$ **83.** $(9x - 5y)(x + y)$
85. $(3m - 4)(2m - 3n)$ **87.** $(12a - 5b)(3a + b)$
89. $(x + 2y)^2$ **91.** $5(2x - 3)(2x + 1)$
93. $4(2m + 1)(m + 1)$ **95.** $3(5r - 2s)(r - s)$
97. $2x(x - 2)(x + 1)$ **99.** $y^2(2y + 3)(y + 1)$
101. $6a(3a - 1)(2a - 3)$ **103.** $3(p + q)(3p + 7q)$
105. 6 or 9   **107.** 8 or 10 or 17   **109.** 4   **111.** 2

113. 3, 8, 15, 24, ... **115.** (a) $(x-4)(x+2)$ (b) 4, -2
117. (a) $(2x-3)(x+1)$ (b) $\frac{3}{2}$, -1

Section 6.3*
1. True **3.** False **5.** True **7.** False **9.** True
11. $a=1, b=4, c=-9$ **13.** $a=1, b=-3, c=8$
15. $a=3, b=5, c=-8$ **17.** $a=4, b=8, c=11$
19. $a=-3, b=5, c=-10$ **21.** $(x+2)(x+4)$
23. $(x-5)(x-4)$ **25.** $(x-9)(x+7)$
27. $(x+3)(x+5)$ **29.** $(x-4)(x-7)$
31. $(s+10)(s+3)$ **33.** $(a-8)(a+6)$
35. $(x-1)(x-7)$ **37.** $(x-10)(x+4)$
39. $(x-7)(x-7)$ **41.** $(p-12)(p+2)$
43. $(x+11)(x-6)$ **45.** $(c+4)(c+15)$
47. $(n+10)(n-5)$ **49.** $(x+2y)(x+5y)$
51. $(a-7b)(a+6b)$ **53.** $(x-5y)(x-8y)$
55. $(3x+2)(2x+5)$ **57.** $(5x-3)(3x+2)$
59. $(6m-5)(m+5)$ **61.** $(3x-2)(3x-2)$
63. $(6x+5)(2x-3)$ **65.** $(3y-2)(y+3)$
67. $(8x+5)(x-4)$ **69.** $(2x+y)(x+y)$
71. $(5a+2b)(a-2b)$ **73.** $(9x-5y)(x+y)$
75. $(3m-4)(2m-3n)$ **77.** $(12a-5b)(3a+b)$
79. $(x+2y)^2$ **81.** $5(2x-3)(2x+1)$
83. $4(2m+1)(m+1)$ **85.** $3(5r-2s)(r-s)$
87. $2x(x-2)(x+1)$ **89.** $y^2(2y+3)(y+1)$
91. $6a(3a-1)(2a-3)$ **93.** $3(p+q)(3p+7q)$
95. 6 or 9 **97.** 8 or 10 or 17 **99.** 4 **101.** 2
103. 3, 8, 15, 24, ... **105.** (a) $(x-4)(x+2)$ (b) 4, -2
107. (a) $(2x-3)(x+1)$ (b) $\frac{3}{2}$, -1

Section 6.4
1. $2x^4$ **3.** $5m^2$ **5.** $a+2$ **7.** $3b^2-4$
9. $4a^2-6a$ **11.** $-4m-2$ **13.** $3a^3+2a^2-a$
15. $4x^2y-3y^2+2x$ **17.** $x+3$ **19.** $x-5$
21. $x+3$ **23.** $2x+3+\dfrac{4}{x-3}$ **25.** $4x+2+\dfrac{-5}{x-5}$
27. $2x+3+\dfrac{5}{3x-5}$ **29.** x^2-x-2
31. $x^2+2x+3+\dfrac{8}{4x-1}$ **33.** $x^2+x+2+\dfrac{9}{x-2}$
35. $5x^2+2x+1+\dfrac{2}{5x-2}$ **37.** $x^2+4x+5+\dfrac{2}{x-2}$
39. x^3+x^2+x+1 **41.** $x-3$ **43.** $x^2-1+\dfrac{1}{x^2+3}$
45. y^2-y+1 **47.** x^2+1 **49.** $c=-2$
51. 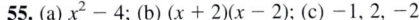 **53.** (a) (x^2+3x-4); (b) $(x+4)(x-1)$; (c) $-2, -4, 1$
55. (a) x^2-4; (b) $(x+2)(x-2)$; (c) $-1, 2, -2$

Section 6.5
1. $\{-3, -1\}$ **3.** $\{-3, 5\}$ **5.** $\{5, 6\}$ **7.** $\{-3, 7\}$
9. $\{-5, 10\}$ **11.** $\{-5, 7\}$ **13.** $\{0, 8\}$
15. $\{0, -10\}$ **17.** $\{0, 5\}$ **19.** $\{-5, 5\}$
21. $\{-8, 8\}$ **23.** $\left\{-\dfrac{3}{2}\right\}$ **25.** $\left\{4, \dfrac{9}{2}\right\}$ **27.** $\left\{-1, \dfrac{4}{3}\right\}$
29. $\left\{\dfrac{1}{2}, \dfrac{2}{3}\right\}$ **31.** $\{-3, 9\}$ **33.** $\{0, 6\}$ **35.** $\{0, 3\}$
37. $\{-3, 5\}$ **39.** $\left\{-\dfrac{3}{2}, 3\right\}$ **41.** $\left\{-\dfrac{7}{3}, 2\right\}$
43. $\{-2, 6\}$ **45.** $\left\{-\dfrac{3}{2}, 5\right\}$ **47.** $\{2, 6\}$

49. $\{-5, 1\}$ **51.** 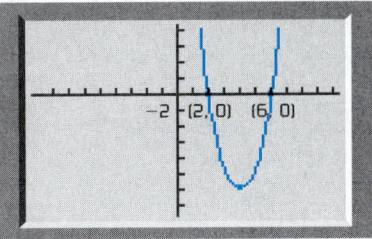 **53.** $x^2+x-6=0$
55. $x^2-8x+12=0$ **57.** $\{-2, 0, 5\}$ **59.** $\{-3, 0, 3\}$
61. $\{-2, -1, 2\}$ **63.** $\{-3, -1, 1, 3\}$
65. $\{0\text{ cm}, 100\text{ cm}\}$ **67.** $\{0\text{ cm}, 132\text{ cm}\}$ **69.** $\{30\}$
71.

2, 6 are the zeros of Exercise 55.

73.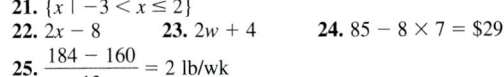

Cumulative Test—Chapters 0–6
1. $2x^2y+5xy$ **2.** $3a^4b^6$ **3.** $4x^2+x+3$
4. x^2+9x-2 **5.** 27 **6.** -1 **7.** a^2-9b^2
8. $x^2-4xy+4y^2$ **9.** $x^2+3x-10$ **10.** a^2+a-12
11. $3x+2$ **12.** $(3x+3)+\dfrac{1}{x-1}$
13. 11 **14.** -33 **15.** $\{15, -3\}$ **16.** $\{1, -10\}$
17. $x>-6$ **18.** $\dfrac{1}{2}<x<\dfrac{13}{2}$ **19.** $x>-2$ or $x<8$
20. $4(3x+5)$ **21.** $(5x+7y)(5x-7y)$
22. $(4x-5)(3x+2)$ **23.** $(x-5)(2x-3)$
24. $3x-5=46; 17$ **25.** $x+(x-5)=81; \$43$

Self-Test: Chapter 0
1. $\dfrac{1}{3}$ **2.** $\dfrac{3}{11}$ **3.** $\dfrac{9}{11}$ **4.** $\dfrac{25}{16}$ **5.** $\dfrac{29}{30}$
6. $\dfrac{21}{80}$ **7.** $\dfrac{9}{44}$ **8.** $\dfrac{14}{5}$ **9.** 2 **10.** -12
11. 0 **12.** -3 **13.** -12 **14.** -14 **15.** 11
16. 15 **17.** -35 **18.** 54 **19.** 300 **20.** 36
21. 0 **22.** 7 **23.** -4 **24.** Undefined
25. 15 **26.** 6 **27.** $<$ **28.** $>$ **29.** 7
30. 86

Self-Test: Chapter 1
1. $x+y$ **2.** $m-n$ **3.** $a \cdot b$ **4.** $\dfrac{p}{q}$ **5.** $a+7$
6. $c-5$ **7.** $3(a+b)$ **8.** $3(m-n)$ **9.** 0
10. 17 **11.** -4 **12.** 144 **13.** 5 **14.** 16
15. $3a-b$ **16.** $11x+3y$ **17.** $2x^2+8$
18. $\{-4, -3, -2, -1\}$ **19.** $\{x \mid -2 \leq x \leq 4\}$
20.
$\begin{array}{c}\xleftarrow{}\!\!\!\!\vert\!\!-\!\!\vert\!\!-\!\!\vert\!\!-\!\!\vert\!\!-\!\!\vert\!\!-\!\!\vert\!\!-\!\!\vert\!\!\xrightarrow{}\\ -503\end{array}$
21. $\{x \mid -3 < x \leq 2\}$
22. $2x-8$ **23.** $2w+4$ **24.** $85-8\times 7=\$29$
25. $\dfrac{184-160}{12}=2$ lb/wk

Self-Test: Chapter 2

1. No 2. Yes 3. {11} 4. {12} 5. {7}
6. {7} 7. {−12} 8. {25} 9. {4}
10. $\left\{-\dfrac{2}{3}\right\}$ 11. {−5} 12. $\left\{\dfrac{4}{5}\right\}$ 13. {2}
14. $\left\{\dfrac{5}{2}\right\}$ 15. $\left\{\dfrac{14}{5}\right\}$ 16. $\left\{\dfrac{C}{2\pi}\right\}$ 17. $\left\{\dfrac{3V}{B}\right\}$

18. $\{x \mid x \le 14\}$

19. $\{x \mid x < -4\}$

20. $\{x \mid -10 < x < -2\}$

21. $\{x \mid x < 2 \text{ or } x > 6\}$

22. 7 23. 21, 22, 23 24. 6, 12, 17
25. 10 in. by 21 in. 26. 3:30 p.m.
27. $\{x \mid 1 \le x \le 4\}$ 28. $\{x \mid x < -5 \text{ or } x > 4\}$
29. $\{x \mid -3 < x < 5\}$ 30. $\{x \mid x \le -4 \text{ or } x \ge 3\}$

Self-Test: Chapter 3

1. (4, 0), (5, 4) 2. (3, 0), (0, 4), $\left(\dfrac{3}{4}, 3\right)$
3. (4, 2) 4. (−4, 6) 5. (0, −7)

6–8.

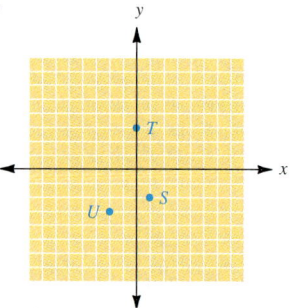

9. $x + y = 4$

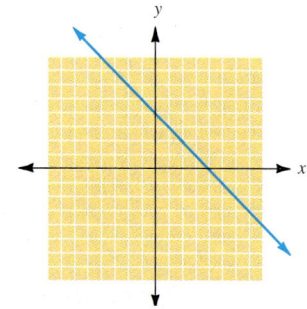

10. $y = 3x$

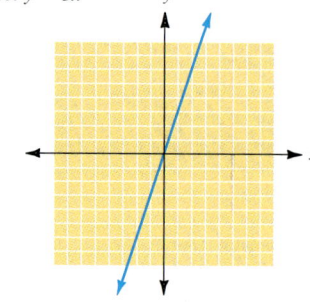

11. $y = \dfrac{3}{4}x - 4$

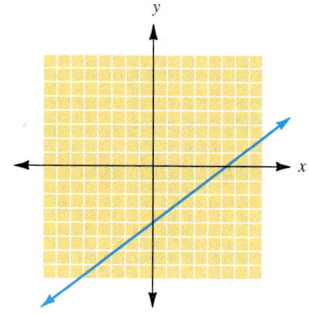

12. $x + 3y = 6$

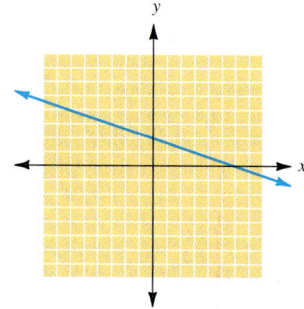

13. $2x + 5y = 10$

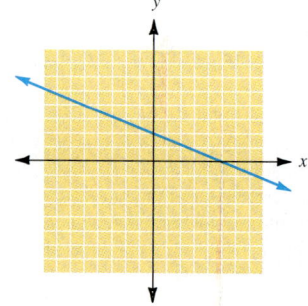

14. $y = -4$

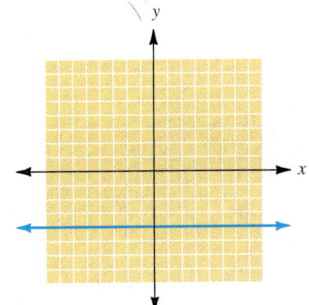

15. 1   **16.** $\frac{3}{4}$
17. $y = -3x + 6$

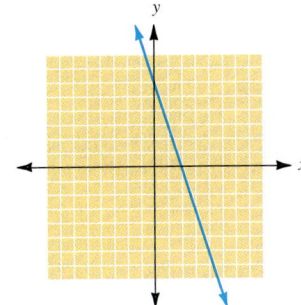

18. $y = \frac{2}{5}x - 3$

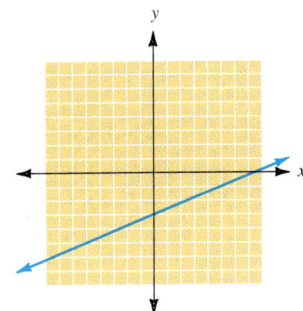

19. Slope: 5; y-int: -9 **20.** Slope: $-\frac{6}{5}$; y-int: 6
21. Slope: 0; y-int: 5   **22.** $y = 5x - 2$
23. $y = -4x - 16$ **24.** $y = 4x + 3$
25. $y = -\frac{5}{2}x - 17$

Self-Test: Chapter 4
1. b   **2.** (a) D: $\{-3, 1, 2, 3, 4\}$; R: $\{-2, 0, 1, 5, 6\}$;
(b) D: All reals; R: all reals   **3.** $-5, 7,$ **4.** $-15, -7$
5. $-15, 6$ **6.** $3a - 25, 3(x-1) - 25 = 3x - 28$
7. Function   **8.** Not a function   **9.** Function
10. Function   **11.** Function   **12.** Not a function
13. (a) $f(x) = -6x + 9$; (b) $f(x) = \frac{4x - 35}{7}$

14. $f(x) = 9x - 2$

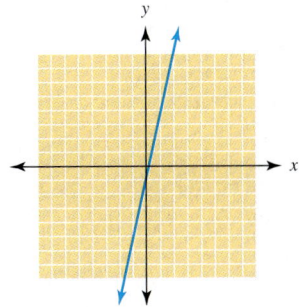

15. $f(x) = -7x + 3$

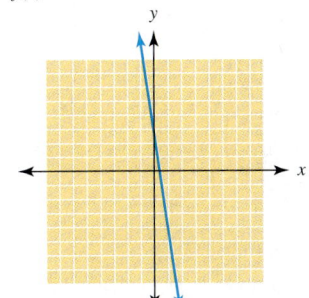

16. $f(x) = -3x - 2$

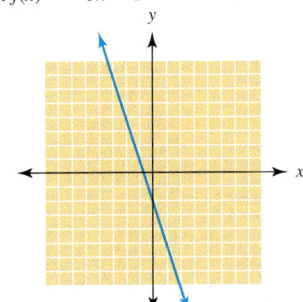

17. A: $(1, 0)$; B: $(-3, -4)$ **18.** A: $(-4, -2)$; B: $(1, 2)$
19. (a) 3, (b) 0, (c) -3 **20.** (a) -3, (b) -2, (c) 0
21. 17   **22.** $5x - 3$ **23.** -7 **24.** $x - 5$

Self-Test: Chapter 5
1. $-6x^3y^4$ **2.** $\frac{16m^4n^{10}}{p^6}$ **3.** x^8y^{10} **4.** $\frac{c^3d}{2}$
5. $108x^8y^7$ **6.** $-128x^7y^5$ **7.** 25   **8.** Binomial
9. Trinomial   **10.** $8x^4 - 3x^2 - 7$; 8, -3, -7; 4
11. $10x^2 - 12x - 7$ **12.** $7a^3 + 11a^2 - 3a$
13. $3x^2 + 11x - 12$ **14.** $b^2 - 7b - 5$ **15.** $7a^2 - 10a$
16. $4x^2 + 5x - 6$ **17.** $2x^2 - 7x + 5$
18. $15a^3b^2 - 10a^2b^2 + 20a^2b^3$ **19.** $3x^2 + x - 14$
20. $a^2 - 49b^2$ **21.** $4x^2 + 7xy - 15y^2$
22. $9m^2 + 12mn + 4n^2$ **23.** $2x^3 + 7x^2y - xy^2 - 2y^3$
24. $4x^2 - 25y^2$ **25.** $x^2 - 10xy + 25y^2$

Self-Test: Chapter 6
1. $6(2b + 3)$ **2.** $3p^2(3p - 4)$ **3.** $5(x^2 - 2x + 4)$
4. $6ab(a - 3 + 2b)$ **5.** $(a + b)(a - 5)$
6. $(8m + n)(8m - n)$ **7.** $(7x + 4y)(7x - 4y)$
8. $2b(4a + 5b)(4a - 5b)$ **9.** $(2x - 3y)(4x^2 + 6xy + 9y^2)$

10. $2xy(4x + 5y)(16x^2 + 20xy + 25y^2)$ **11.** $(a − 7)(a + 2)$
12. $(b + 3)(b + 5)$ **13.** $(x − 4)(x − 7)$
14. $(y + 10z)(y + 2z)$ **15.** $(x + 2)(x − 5)$
16. $(2x − 3)(3x + 1)$ **17.** $(2x − 1)(x + 8)$
18. $(3w + 7)(w + 1)$ **19.** $(4x − 3)(2x + y)$
20. $3x(2x + 5)(x − 2)$ **21.** $(3x + 2)(2x − 5)$
22. $(x + 1)(4x + 5)$ **23.** $(x − 1) + \dfrac{-3}{3x + 1}$
24. $4x^2 + 3x + 13 + \dfrac{17}{x - 2}$ **25.** $3x^2 − 5$ **26.** $\{-1, 3\}$
27. $\{-2, 7\}$ **28.** $\{5, 6\}$ **29.** $\{-5, -3\}$
30. $\left\{-\dfrac{1}{3}, \dfrac{3}{2}\right\}$

Final Exam—Chapters 0–6
1. 27 **2.** -1 **3.** 6 **4.** $a + 6b$
5. $2x^2y + 3x + 2xy$ **6.** $x^2 + 9x$ **7.** $a^2 − 9b^2$
8. $x^2 − 4xy + 4y^2$ **9.** $x^2 + 3x − 10$ **10.** $a^2 + a − 12$
11. $3x + 2$ **12.** $(3x + 3) + \dfrac{1}{x - 1}$ **13.** 180
14. 118 **15.** -16 **16.** $\{11\}$ **17.** $\{-33\}$
18. $\left\{\dfrac{8}{5}\right\}$ **19.** $\{1\}$ **20.** $\{-3, 15\}$ **21.** $\{-10, 1\}$
22. $x > -6$ **23.** $\dfrac{1}{2} < x < \dfrac{13}{2}$ **24.** $x < -8$ or $x > -2$
25. $x \geq \dfrac{14}{3}$ or $x \leq -2$ **26.** $4(3x + 5)$
27. $(5x + 6y)(5x − 6y)$ **28.** $(4x − 5)(3x + 2)$
29. $(2x − 3)(x − 5)$ **30.** Function **31.** Not a function
32. Not a function **33.** Function
34. $f(x) = 3x$

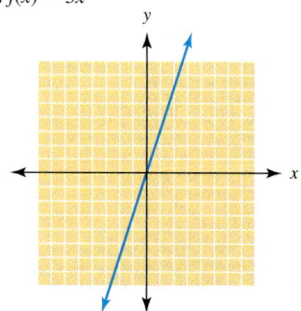

35. $f(x) = x + 3$

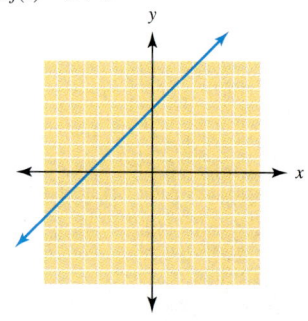

36. $f(x) = -3x − 2$

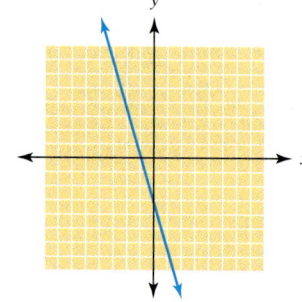

37. $f(x) = 7x + 3$

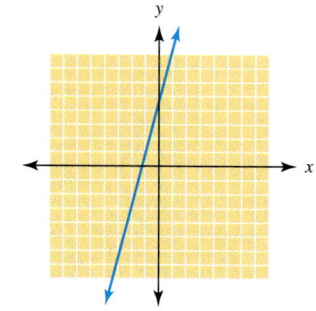

38. $124 = 2(x + 1)$; 30.5 in. By 31.5 in.
39. $x + (x + 5) = 81$; Biology: $43, Math: $38
40. $\{-4, 7\}$ **41.** $\{-2, 5\}$ **42.** $\left\{-\dfrac{3}{7}, \dfrac{3}{7}\right\}$
43. $\left\{-8, \dfrac{1}{2}\right\}$

Chapter 7

Section 7.1
1. 5 **3.** Never undefined **5.** $\dfrac{1}{2}$ **7.** 0 **9.** -2
11. 0 **13.** 2 **15.** 3 **17.** $\dfrac{2}{3}$ **19.** $\dfrac{2x^3}{3}$
21. $\dfrac{2xy^3}{5}$ **23.** $\dfrac{-3x^2}{y^2}$ **25.** $\dfrac{a^3b^2}{3c^2}$ **27.** $\dfrac{6}{x + 4}$
29. $\dfrac{x + 1}{6}$ **31.** $\dfrac{x + 2}{x + 7}$ **33.** $3b + 1$ **35.** $\dfrac{y - z}{y + 3z}$
37. $\dfrac{x^2 + 4x + 16}{x + 4}$ **39.** $\dfrac{(a^2 + 9)(a - 3)}{a + 2}$ **41.** $\dfrac{y - 2}{x + 5}$
43. $\dfrac{x + 6}{x^2 - 2}$ **45.** $\dfrac{-2}{m + 5}$ **47.** $-\dfrac{x + 7}{2x + 1}$
49. Rational **51.** Rational **53.** Not rational
55. (a) $f(x) = x − 2$, (b) $(-1, -3)$
57. (a) $f(x) = 3x − 1$, (b) $(-2, -7)$
59. (a) $f(x) = \dfrac{x + 2}{5}$, (b) $(-2, 0)$

61.
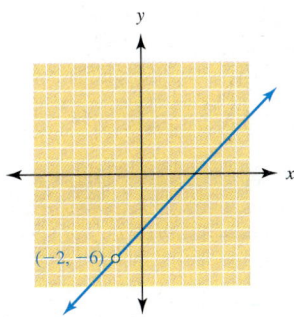

63.
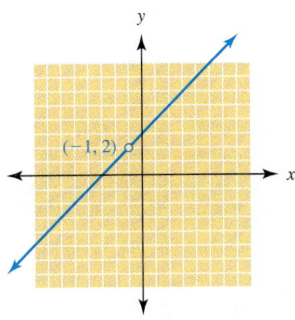

65.
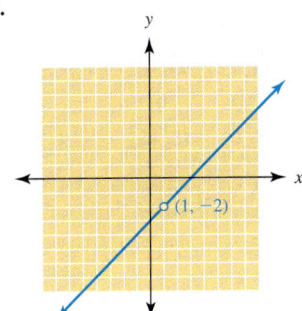

67.  **69.**  **71.** 2 **73.** 3

75. $2x + h$ **77.** $x + 5$ **79.** (a) $3000, (b) $9231, (c) $15,000, (d) $623 **81.**

Section 7.2

1. $\dfrac{2}{x}$ **3.** $\dfrac{3}{a^4}$ **5.** $\dfrac{5}{12x}$ **7.** $\dfrac{16b^3}{3a}$ **9.** $5mn$
11. $\dfrac{15x}{2}$ **13.** $\dfrac{9b}{8}$ **15.** $x^2 + 2x$ **17.** $\dfrac{3(c-2)}{5}$

19. $\dfrac{5x}{2(x-2)}$ **21.** $\dfrac{5d}{4(d-3)}$ **23.** $\dfrac{x-5}{2x+3}$
25. $\dfrac{2a+1}{2a}$ **27.** $\dfrac{-6}{w+2}$ **29.** $-\dfrac{a}{6}$ **31.** $\dfrac{2}{x}$
33. $\dfrac{3}{m}$ **35.** $\dfrac{5}{x}$ **37.** (a) -4, (b) undefined, (c) $(x+1)(x-4)$, $x \neq -2$, $x \neq 4$, (d) -4, (e) undefined
39. (a) -6, (b) undefined, (c) $(2x-5)(3x+1)$, $x \neq -2$, $x \neq -1$, (d) -6, (e) undefined **41.** (a) $-\dfrac{4}{5}$, (b) $\dfrac{5}{4}$, (c) $\dfrac{(3x-2)(x+4)}{(x-2)(x-5)}$, (d) $2, -4, -1, 5$ **43.** $\dfrac{x^2+2}{4}$
45. $\dfrac{x+2}{x(x+3)}$

Section 7.3

1. $\dfrac{6}{x^2}$ **3.** $\dfrac{7}{3a+7}$ **5.** 2 **7.** $\dfrac{y-1}{2}$ **9.** 2
11. $\dfrac{3}{x+2}$ **13.** $\dfrac{19}{6x}$ **15.** $\dfrac{3(2a+1)}{a^2}$ **17.** $\dfrac{2(n-m)}{mn}$
19. $\dfrac{9b-20}{12b^3}$ **21.** $\dfrac{a-4}{a(a-2)}$ **23.** $\dfrac{5x+7}{(x+1)(x+2)}$
25. $\dfrac{4(y+2)}{(y-3)(y+1)}$ **27.** $\dfrac{w(3w-11)}{(w-7)(w-2)}$
29. $\dfrac{7x}{(3x-2)(2x+1)}$ **31.** $\dfrac{4}{m-7}$ **33.** $\dfrac{2x+11}{(x+4)(x-4)}$
35. $\dfrac{3m+1}{(m-1)(m-2)}$ **37.** $\dfrac{15}{y-5}$ **39.** $\dfrac{-12}{x-4}$
41. $\dfrac{5z+14}{(z+2)(z-2)(z+4)}$ **43.** (a) $\dfrac{1}{2}$, (b) $\dfrac{5x^2-7x}{(x+1)(x-3)}$, $\dfrac{+1}{1^2}$, (c) $\left(1, \dfrac{3}{4}\right)$ (c) $\left(1, \dfrac{1}{2}\right)$ **45.** (a) $\dfrac{3}{4}$, (b) $\dfrac{x^2+x}{(x+1)^2}$
47. (a) $-\dfrac{5}{6}$, (b) $\dfrac{20x}{(x-5)(x+5)}$, (c) $\left(1, \dfrac{5}{6}\right)$
49. (a) $-\dfrac{3}{16}$, (b) $\dfrac{x+5}{4(x-9)}$, (c) $\left(1, -\dfrac{3}{16}\right)$ **51.** 1
53. $\dfrac{49}{25}$ **55.** 2 **57.** 8 **59.** Undefined

Section 7.4

1. $\dfrac{8}{9}$ **3.** $\dfrac{14}{5}$ **5.** $\dfrac{5}{6}$ **7.** $\dfrac{1}{2x}$ **9.** $\dfrac{m}{2}$
11. $\dfrac{2(y+1)}{y-1}$ **13.** $\dfrac{3b}{a^2}$ **15.** $\dfrac{x}{(x+2)(x-3)}$
17. $\dfrac{2x-1}{2x+1}$ **19.** $y - x$ **21.** $\dfrac{x-y}{y}$ **23.** $\dfrac{a+4}{a+3}$
25. $\dfrac{x^2y(x+y)}{x-y}$ **27.** $\dfrac{x}{x-2}$ **29.** $\dfrac{y+2}{(y-1)(y+4)}$
31. $\dfrac{x}{3}$ **33.** 1 **35.** $\dfrac{2a}{a^2+1}$ **37.** $\dfrac{2x+1}{x+1}$
39. $\dfrac{3x+2}{2x+1}$ **41.** $\dfrac{5x+3}{3x+2}$
43.  **45.** $\dfrac{-1}{3+h}$ **47.**

49. $44\dfrac{4}{9}$ mi/h **51.** $54\dfrac{6}{9}$ mi/h **53.**

55.

Section 7.5

1. Equation, 36 **3.** Expression, $\dfrac{3x}{10}$ **5.** Equation, 5
7. Equation, 3 **9.** {5} **11.** {8} **13.** $\left\{\dfrac{3}{2}\right\}$
15. {−1} **17.** $\left\{-\dfrac{9}{5}\right\}$ **19.** $\left\{-\dfrac{2}{3}\right\}$ **21.** No solution
23. {−23} **25.** {6} **27.** {4} **29.** {8}
31. {4} **33.** $\left\{\dfrac{3}{2}\right\}$ **35.** No solution **37.** {7}
39. {5} **41.** $\left\{-\dfrac{5}{2}\right\}$ **43.** $\left\{-\dfrac{1}{2}, 6\right\}$ **45.** $\left\{-\dfrac{1}{2}\right\}$
47. $\left\{-\dfrac{1}{3}, 7\right\}$
49. {−8, 9} **51.** $\dfrac{ab}{b-a}$ **53.** $\dfrac{RR_2}{R_2 - R}$ **55.** $\dfrac{y+1}{y-1}$
57. $\dfrac{A}{1+rt}$
59. $-1 < x < 2$
61. $x < 2$ or $x < 4$
63. $-3 < x \le 5$
65. $x < -3$ or $x \ge \dfrac{1}{2}$
67. $-2 \le x < 3$
69. $-3 < x \le 4$
71. $x < 2$ or $x > 3$
73. 24 **75.** $\dfrac{25}{7}$ **77.** $\dfrac{3}{4}$ **79.** 3, 13
81. 6, 24 **83.** 9 **85.** 4 mi/h **87.** 150 mi/h
89. 36 min **91.** **93.**

Self-Test: Chapter 7

1. $-\dfrac{3x^4}{4y^2}$ **2.** $\dfrac{w+1}{w-2}$ **3.** $\dfrac{x+3}{x+2}$
4.

5. $\dfrac{4a^2}{7b}$ **6.** $\dfrac{m-4}{4}$ **7.** $\dfrac{2}{x-1}$ **8.** $\dfrac{2x+y}{x(x+3y)}$
9. $\dfrac{-5(3x+1)}{3x-1}$ **10.** $\dfrac{2(2x+1)}{x(x-2)}$ **11.** $\dfrac{2}{x-3}$
12. $\dfrac{4}{x-2}$ **13.** $\dfrac{8x+17}{(x-4)(x+1)(x+4)}$ **14.** $\dfrac{y}{3y+x}$
15. $\dfrac{z-1}{2(z+3)}$ **16.** $\dfrac{1}{xy(x-y)}$ **17.** {2, 6}
18. $-4 \le x < 3$
19. $-3 < x \le -1$ **20.** 3

Cumulative Test—Chapters 0–7

1. {2} **2.** −77 **3.** x-int.: (6, 0); y-int.: (0, 4)

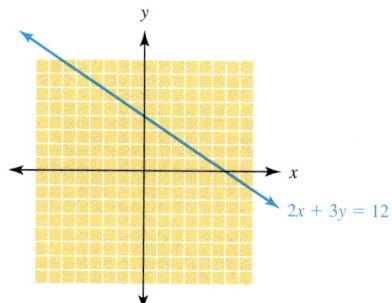

4. $y = \dfrac{5}{4}x - \dfrac{3}{4}$ **5.** $256x^{13}y^{12}$ **6.** 6
7. Domain: all reals; range: all reals
8. {−6, 1} **9.** $-x^2 - 5x + 9$ **10.** $10x^2 + 7x - 12$
11. $x(3x+2)(x-1)$ **12.** $(4x+5y)(4x-5y)$
13. $(x-y)(3x+1)$ **14.** (a) $(3x+7)$, (b) $-13x - 5$,
(c) $-40x^2 - 22x + 6$, (d) $\dfrac{5x+1}{8x+6}$ **15.** {5}
16. {−18, 28} **17.** {2} **18.** $x \ge -1$
19. $-13 < x < -5$ **20.** $-6 < x < -4$

Chapter 8

Section 8.1

1. (5, 1)

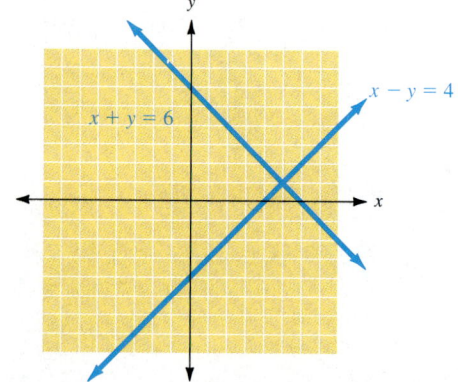

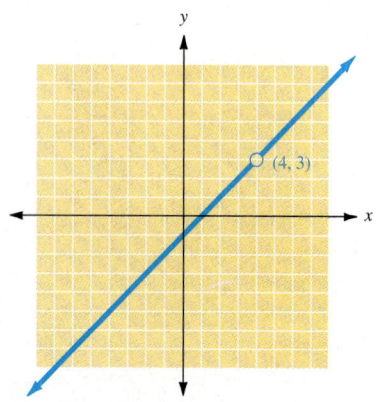

3. (1, 4)

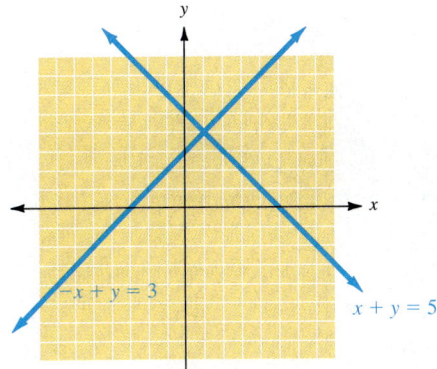

5. (2, 1)

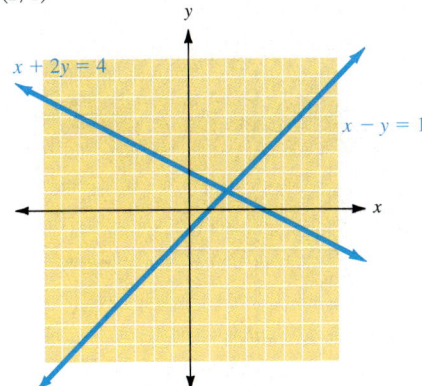

7. (2, 4)

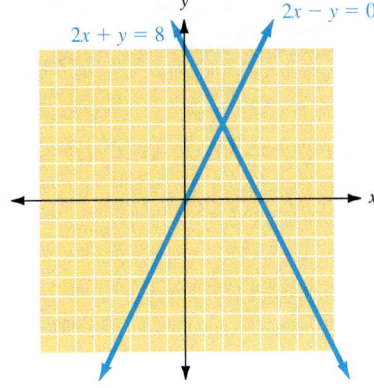

9. (6, 2)

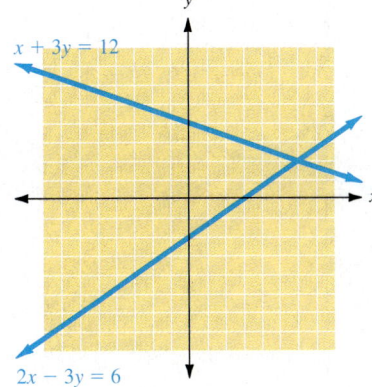

11. (2, 3)

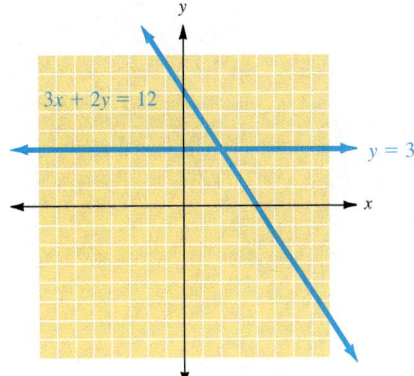

13. Dependent

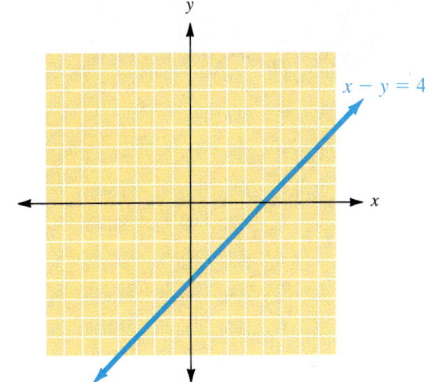

15. (4, 2)

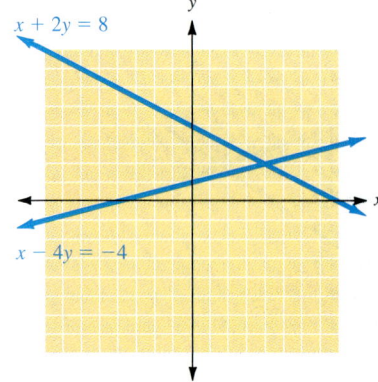

17. (4, 3)

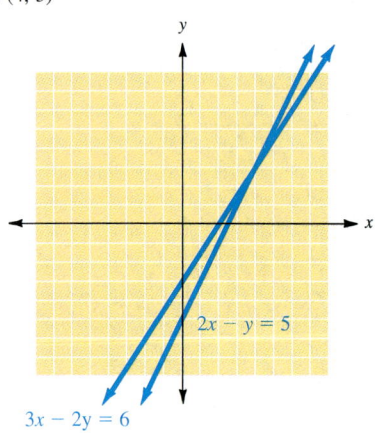

19. Inconsistent

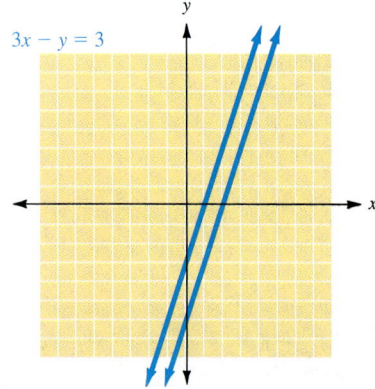

21. $\left(0, \dfrac{3}{2}\right)$

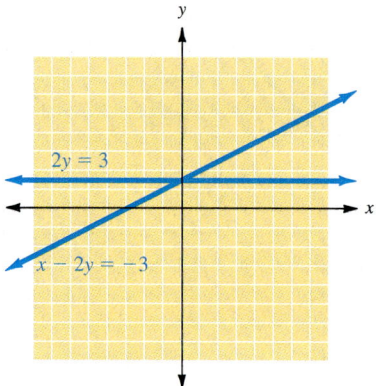

23. (4, −6)

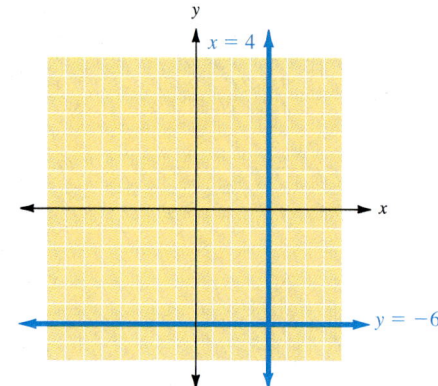

25. (3.18, 28.60) **27.** (−445.35, −253.01)
29. Consistent **31.** Inconsistent **33.** Consistent
35. Dependent **37.** $m = 2, b = 5$ **39.**

41.   **43.** (a) $-\dfrac{A}{B}$ (b) $-\dfrac{D}{E}$ (c) $AE - BD \ne 0$

Section 8.2

1. (2, 3) **3.** $\left(-5, \dfrac{3}{2}\right)$ **5.** (2, 1) **7.** Dependent
9. (5, −3) **11.** Inconsistent **13.** (−8, −2)
15. (5, −2) **17.** (−4, −3) **19.** Dependent
21. $\left(\dfrac{1}{3}, 2\right)$ **23.** (10, 1) **25.** Inconsistent
27. $\left(-2, -\dfrac{8}{3}\right)$ **29.** (3, 4) **31.** (4, −5)
33. (12, −6) **35.** (9, −15) **37.** (d) **39.** (g)
41. (h) **43.** (c) **45.** 550 adult tickets, 200 student tickets
47. 27 in. by 15 in. **49.** Mulch: $1.80; fertilizer: $3.20
51. 105 lb of $4 beans, 45 lb of $6.50 beans

53. $7000 time deposit, $5000 bond
55. 100 mL of 10%, 300 mL of 50%
57. 15 mi/h boat, 3 mi/h current **59.** 26
61. 15 battery powered, 20 solar models **63.** 15
65. 100 mi **67.** $\left(\dfrac{2}{3}, \dfrac{2}{5}\right)$ **69.** $\left(\dfrac{1}{3}, -\dfrac{3}{2}\right)$
71. $y = \dfrac{3}{2}x - 2$
73. $(1.3, -0.5)$

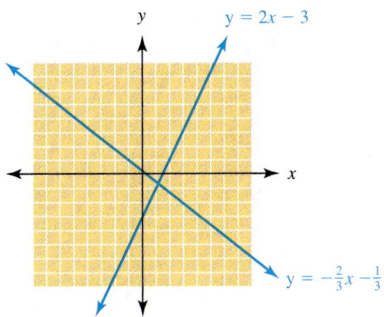

75. $(6, 2)$

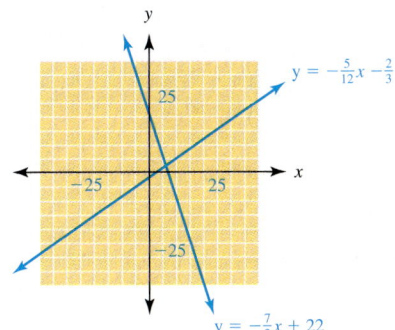

77. **79.** (a) $y = \dfrac{AF - CD}{AE - BD}$; $AE - BD \neq 0$

(b) $x = \dfrac{CE - BF}{AE - BD}$; $AE - BD \neq 0$

Section 8.3
1. $(1, 2, 4)$ **3.** $(-2, 1, 2)$ **5.** $(-4, 3, 2)$
7. Infinite number of solutions **9.** $\left(3, \dfrac{1}{2}, -\dfrac{7}{2}\right)$
11. $(3, 2, -5)$ **13.** $(2, 0, -3)$ **15.** $\left(4, -\dfrac{1}{2}, \dfrac{3}{2}\right)$
17. Inconsistent **19.** $\left(2, \dfrac{5}{2}, -\dfrac{3}{2}\right)$ **21.** 3, 5, 8
23. 3 nickels, 5 dimes, 17 quarters **25.** 4 cm, 7 cm, 8 cm
27. $3000 savings, $9000 bond, $5000 money market
29. 243 **31.** 30 **33.** 110 single, 230 carpool
35. Roy 8 mi, Sally 16 mi, Jeff 26 mi
37. $(1, 2, -2)$ **39.** $(7, 2)$
41. $T = 20$, $C = 25$, $B = 40$
43. (a) $y = 2x^2 - x + 4$ (b) $y = 3x^2 - 2x + 1$

Section 8.4
1. $x + y < 4$

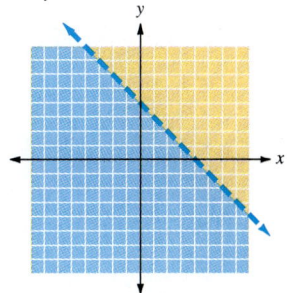

3. $x - y \geq 3$

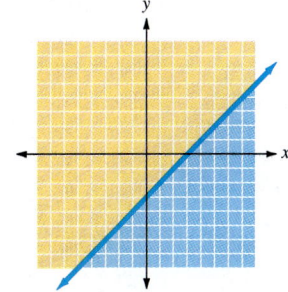

5. $y \geq 2x + 1$

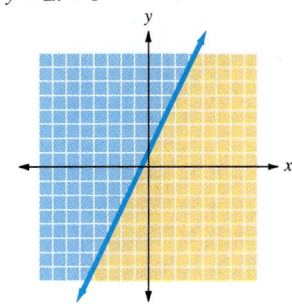

7. $2x + 3y < 6$

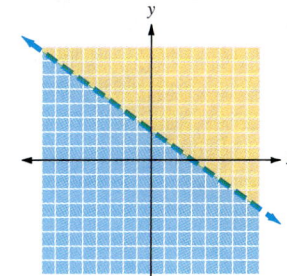

9. $x - 4y > 8$

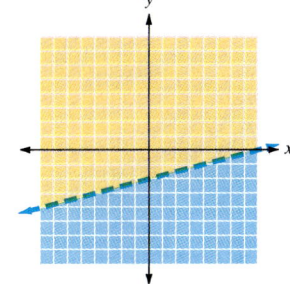

11. $y \geq 3x$

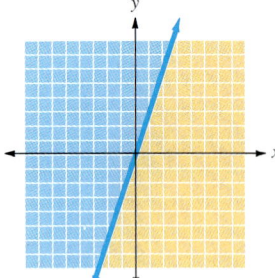

13. $x - 2y > 0$

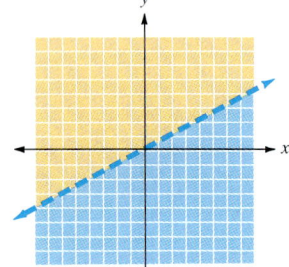

15. $x < 3$

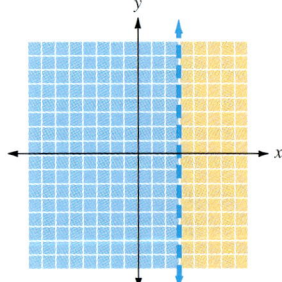

17. $y > 3$

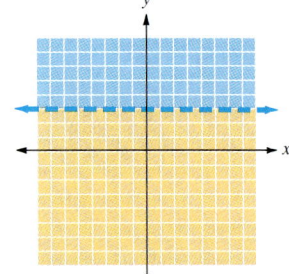

19. $3x - 6 \leq 0$

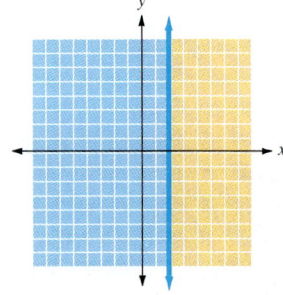

21. $0 < x < 1$

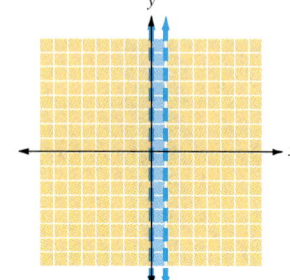

23. $1 \leq x \leq 3$

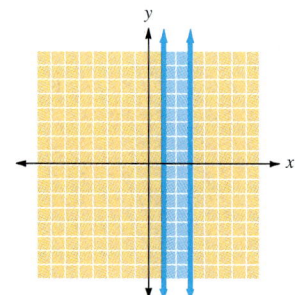

25. $0 \leq x \leq 3$
    $2 \leq y \leq 4$

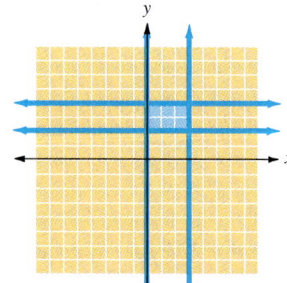

27. $x + 2y \leq 4$
    $x \geq 0$
    $y \geq 0$

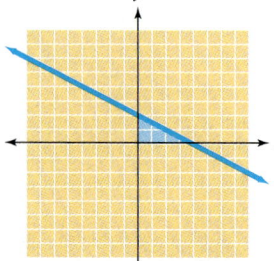

29.

31.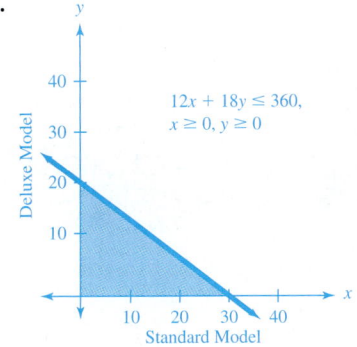

33.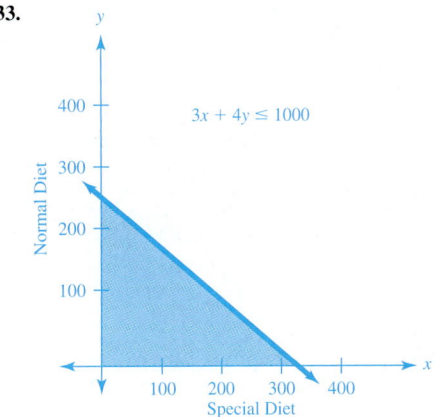

35. $y \geq -x + 4$ **37.** $y > \frac{1}{2}x - 3$

Section 8.5

1.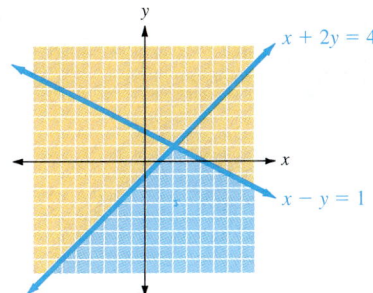

3.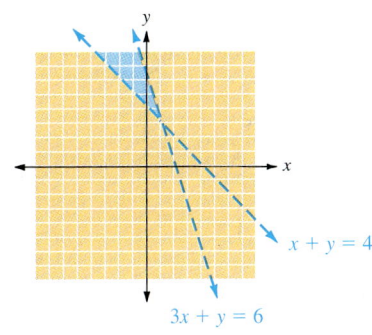

5.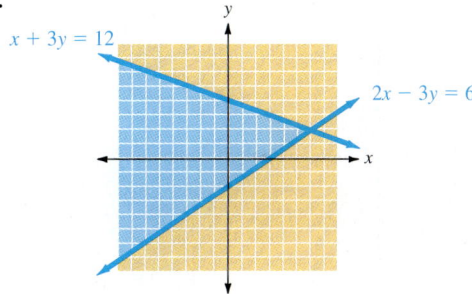

7.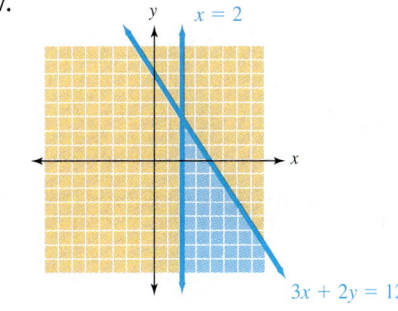

9.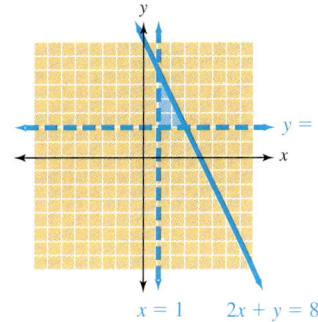

11.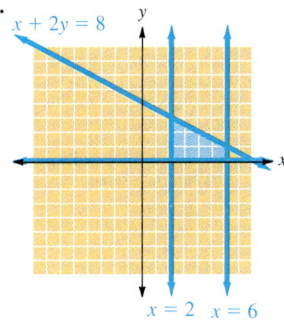

13.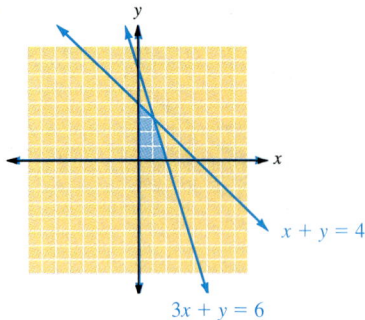

15.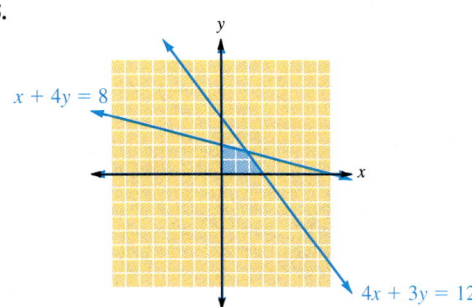

17.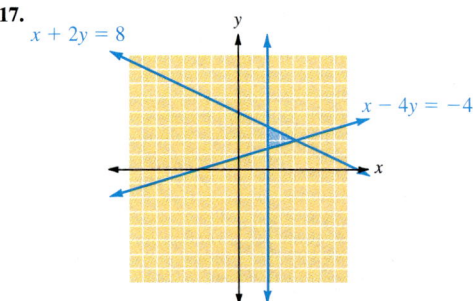

19.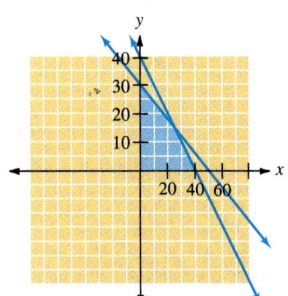

21.

23.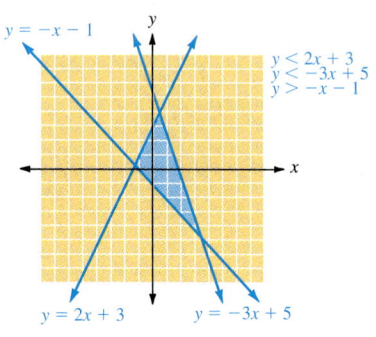

Self-Test: Chapter 8

1. $(-3, 4)$ **2.** Dependent **3.** Inconsistent
4. $(-2, -5)$ **5.** $(5, 0)$ **6.** $\left\{3, -\dfrac{5}{3}\right\}$
7. $(-1, 2, 4)$ **8.** $\left(2, 3, -\dfrac{1}{2}\right)$
9. Disks $2.50, ribbons $6 **10.** 60-lb jawbreakers, 40-lb licorice **11.** Four 5-in. sets, six 12-in. sets
12. $8000 savings, $4000 bond, $2000 mutual fund
13. 50 ft by 80 ft

14.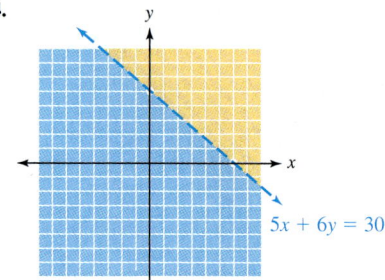

15.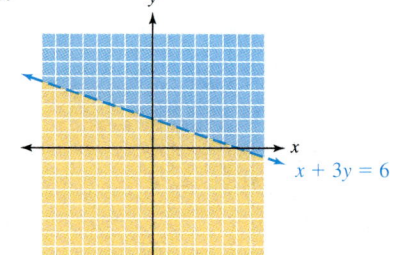

16.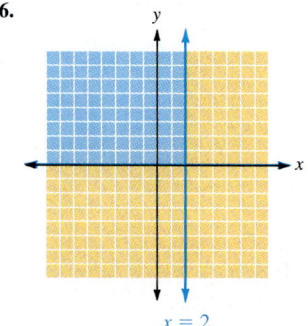

17.

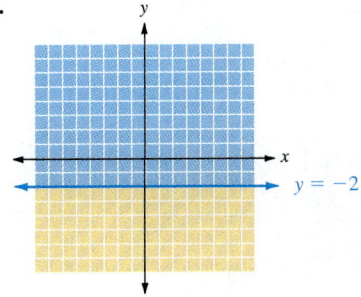

18.

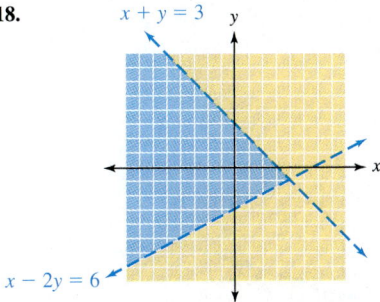

19.

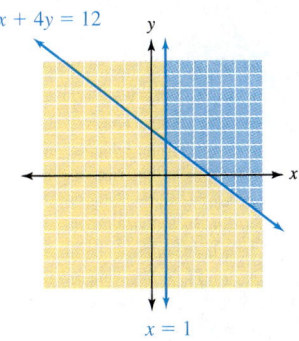

20.

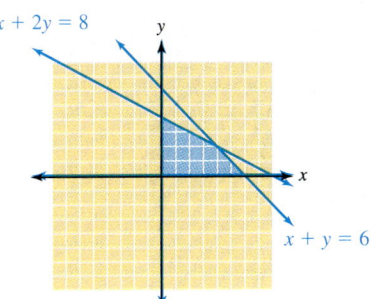

Cumulative Test—Chapters 0–8

1. $\left\{\dfrac{22}{4}\right\}$ **2.** $x > -2$ **3.** $\{-1, 4\}$ **4.** $-4 \leq x \leq \dfrac{2}{3}$

5. $x < -\dfrac{17}{5}$ or $x > 5$ **4.** $-4 \leq x \leq \dfrac{2}{3}$

5. $x < -\dfrac{17}{5}$ or $x > 5$

6. 2

7.

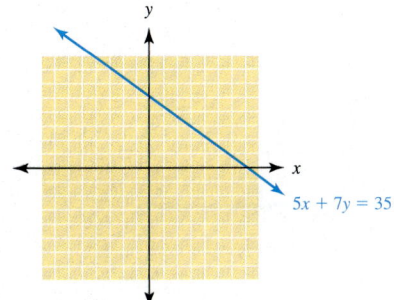

8.

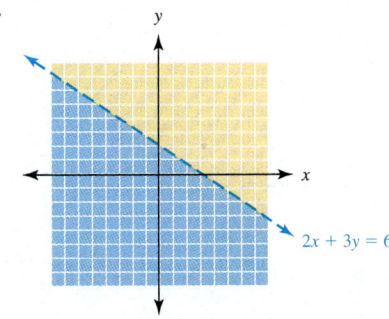

9. $\dfrac{P - P_0}{It}$ **10.** 7   **11.** $y = -x + 3$
12. $y = 2x^2 - 5x - 3$ **13.** $9x^2 - 12x + 4$
14. $(x - 3)(x^2 - 5)$ **15.** $y = \dfrac{4}{5}x - \dfrac{2}{5}$ **16.** 90
17. $\left(-10, \dfrac{26}{3}\right)$ **18.** $(2, 2, -1)$ **19.** 8 cm by 19 cm
20. 73

Chapter 9

Section 9.1

1.

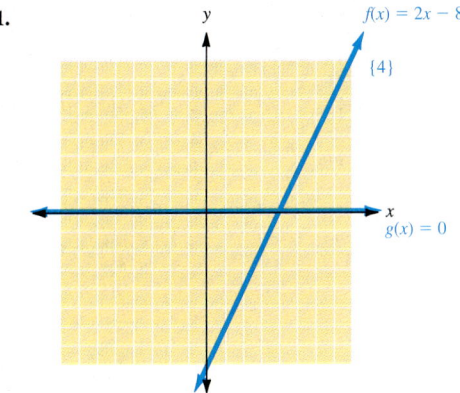

3. 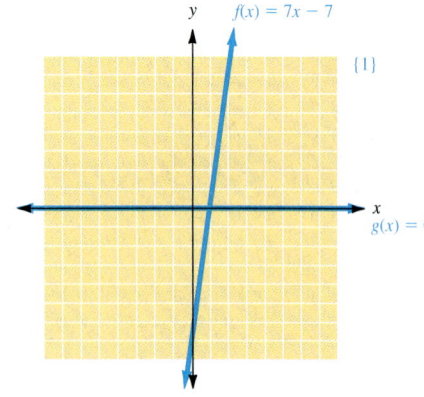 {1}

5. 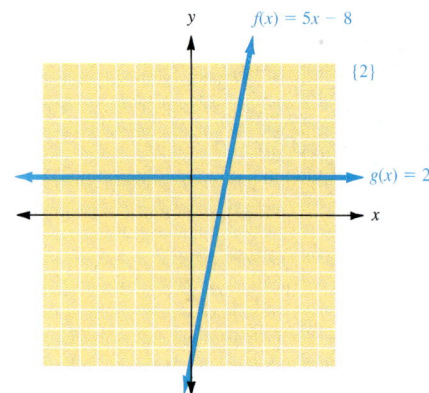 {2}

7. 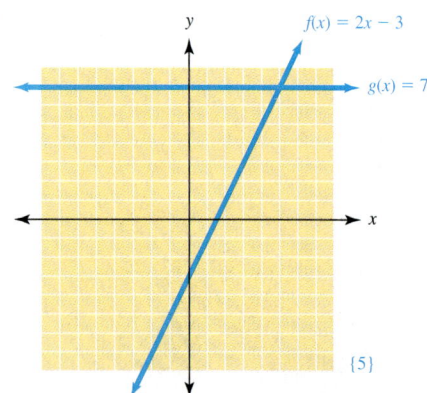 {5}

9. 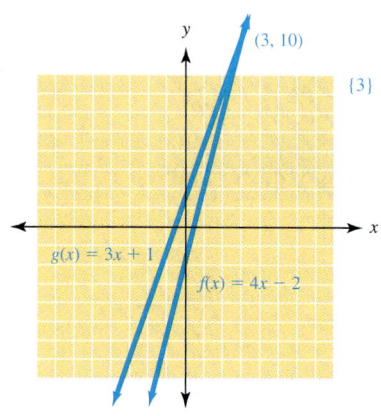 {3}

11. 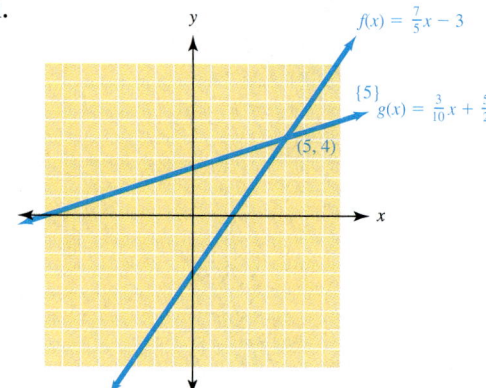 {5}

13. 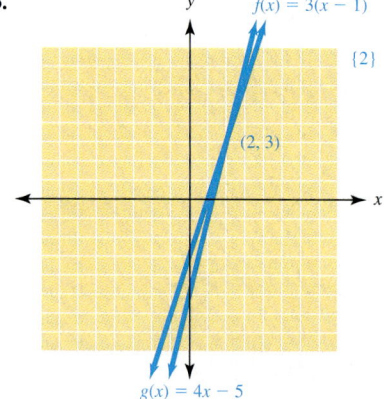 {2}

15. 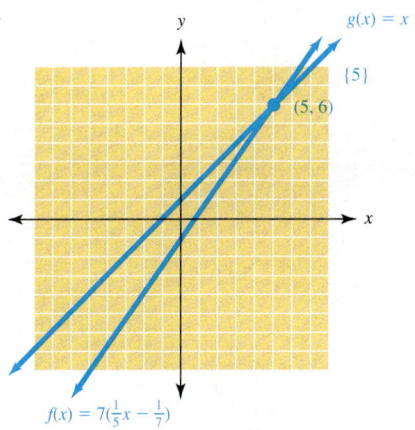 {5}

17. Always use the lower graph to determine the cheaper cost.
19. 250 sets   **21.**

Section 9.2

1. 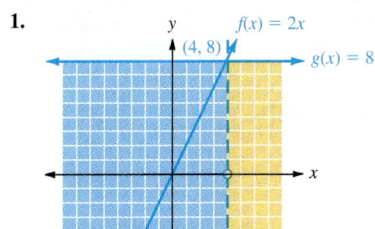 $\{x | x < 4\}$

3. 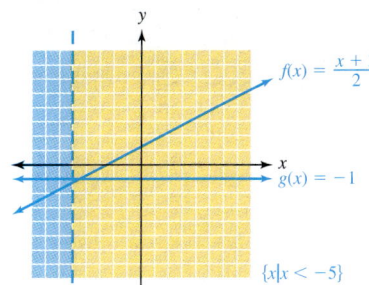 $\{x | x < -5\}$

5. 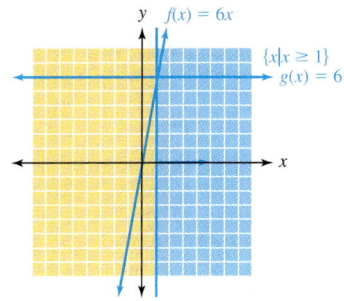 $\{x | x \geq 1\}$

7. 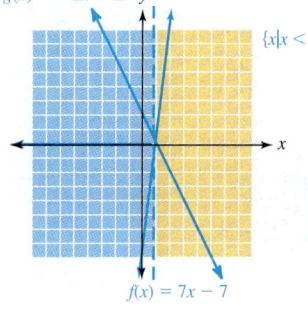 $\{x | x < 1\}$

9. 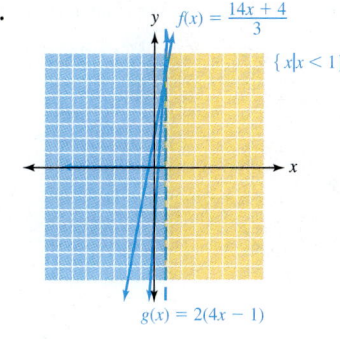 $\{x | x < 1\}$

11. 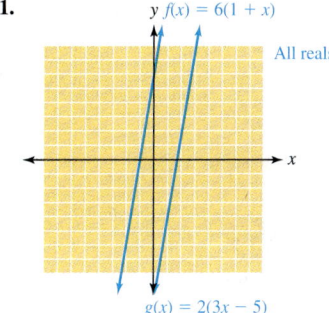 All reals

13. 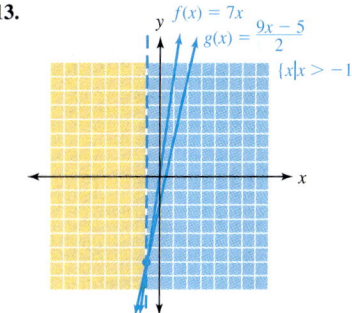 $\{x | x > -1\}$

A-28 Answers

15.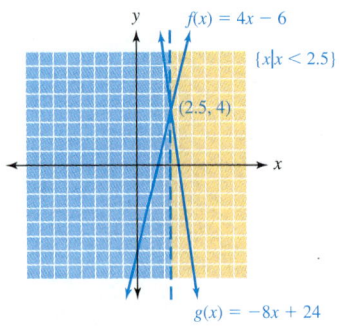

17. $x \geq 500$
19. If miles are under 700, Downtown Edsel; if over 700, Wheels, Inc.; $W = \$28 \times 7 = \$196$; $DE = 98 + 0.14x$ (x is number of miles)   **21.** 110 people   **23.**

Section 9.3

1.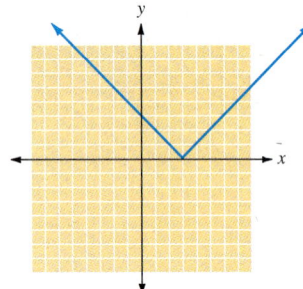

3. $f(x) = |x + 3|$
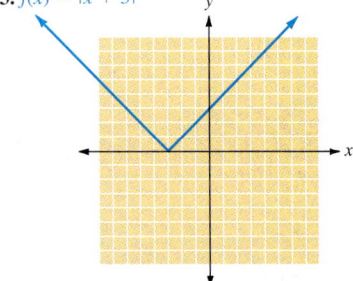

5.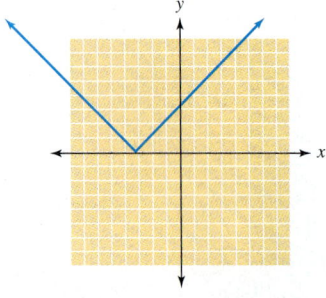

7.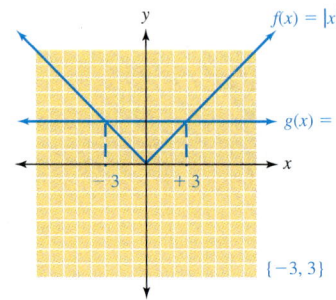

9. $f(x) = |x - 2|$
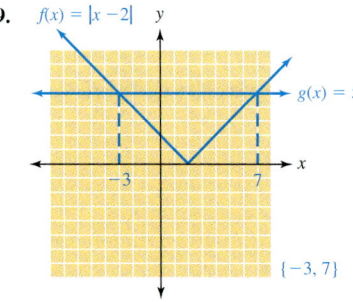

11.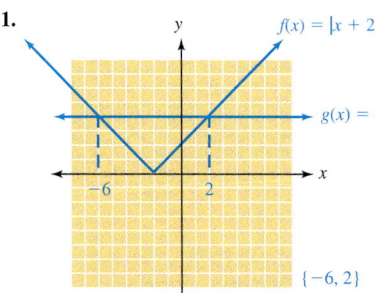

13. $f(x) = |x - 2|$ **15.** $f(x) = |x + 2|$

17.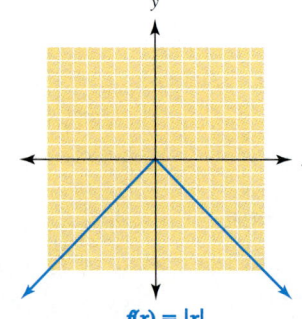

19.
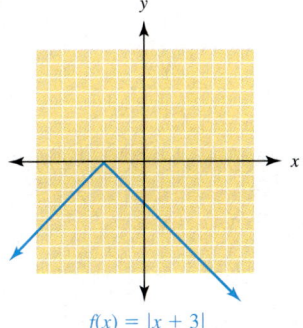
$f(x) = |x + 3|$

21.
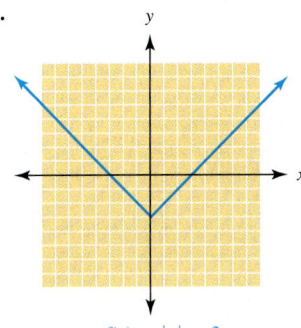
$f(x) = |x| - 3$

23.
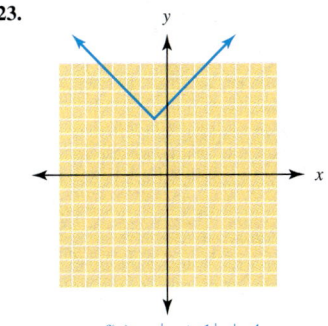
$f(x) = |x + 1| + 4$

25.

27.
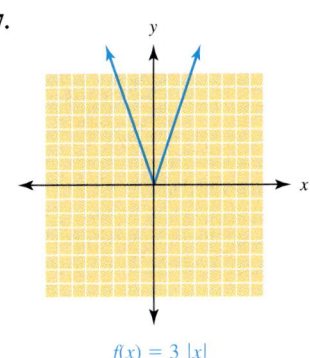
$f(x) = 3|x|$

29.
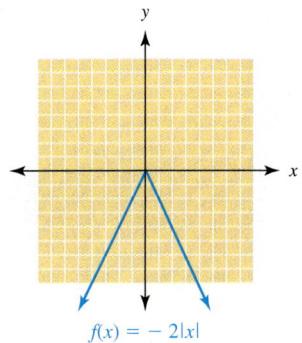
$f(x) = -2|x|$

31.   **33.** -1 **35.** 1

Section 9.4

1.
$\{x \mid -4 < x < 4\}$

3.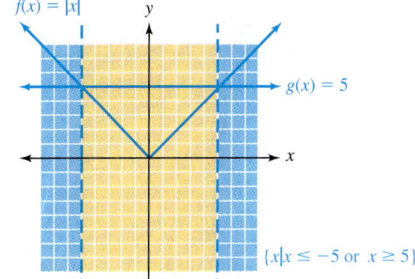
$\{x \mid x \leq -5 \text{ or } x \geq 5\}$

5.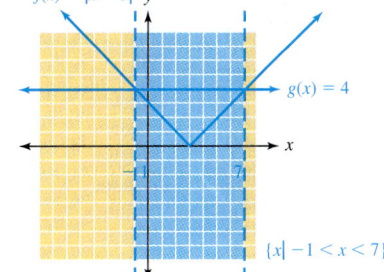
$\{x \mid -1 < x < 7\}$

7.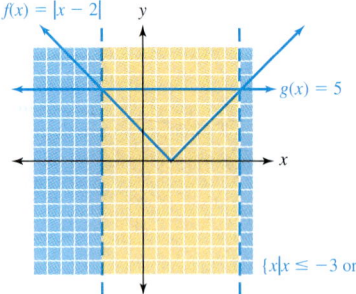
{x | x ≤ −3 or x ≥ 7}

9.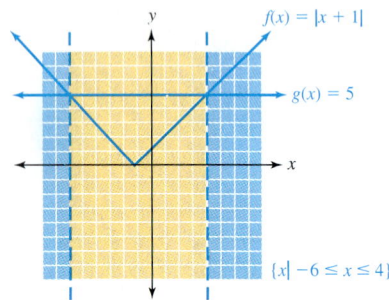
{x | −6 ≤ x ≤ 4}

11.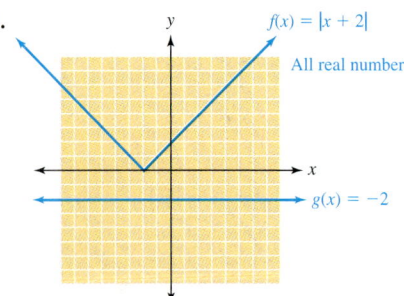
All real numbers

13. $-5 < x < 5$

15. $x \leq -7$ or $x \geq 7$

17. $x < 2$ or $x > 6$

19. $-10 \leq x \leq -2$

21. $x < -2$ or $x > 8$

23. No solution

25. $1 < x < 4$

27. $x \leq -3$ or $x \geq \frac{1}{3}$

29. $x < -\frac{4}{5}$ or $x > 2$

31. $-3 \leq x \leq \frac{13}{3}$

33. $|x| < 3$ 35. $|x| \geq 5$
37. $|x - (-2)| < 7$ 39. $|x - (-4)| \geq 3$

Self-Test: Chapter 9

1.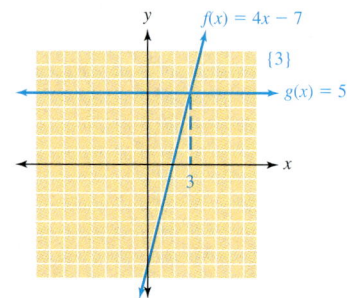
{3}

2.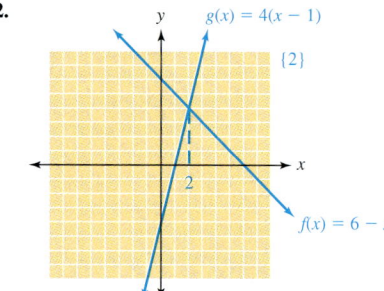
{2}

3.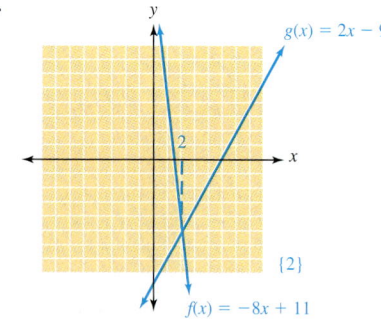
{2}

4.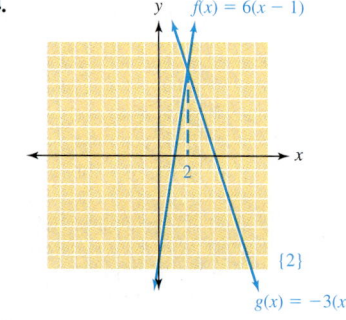
{2}

5. 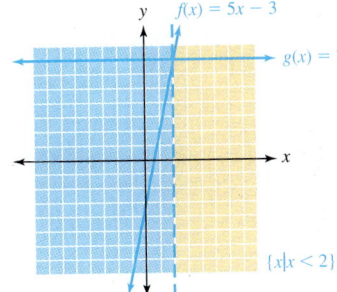 $\{x | x < 2\}$

6. 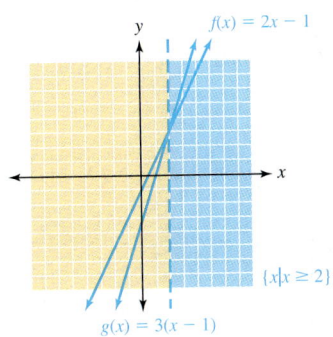 $\{x | x \geq 2\}$

7. $\{-10, 4\}$ 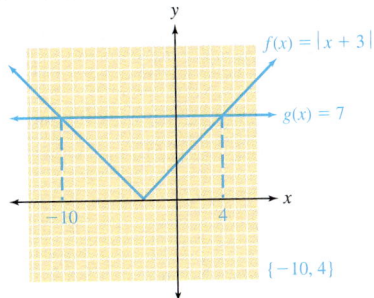 $\{-10, 4\}$

8. $\{-4, 14\}$ 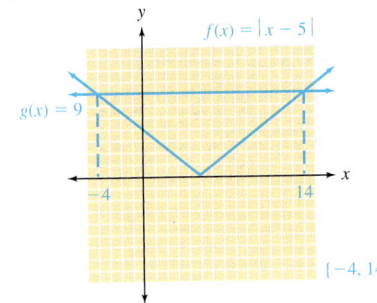 $\{-4, 14\}$

9. 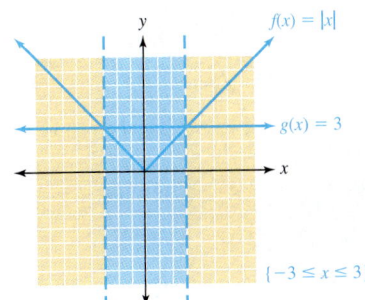 $\{-3 \leq x \leq 3\}$

10. 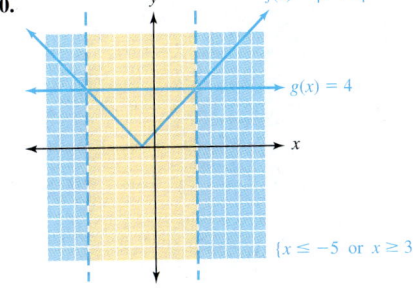 $\{x \leq -5 \text{ or } x \geq 3\}$

11. 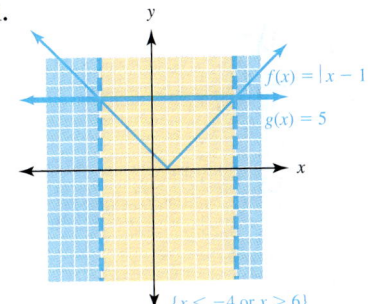 $\{x < -4 \text{ or } x > 6\}$

12. 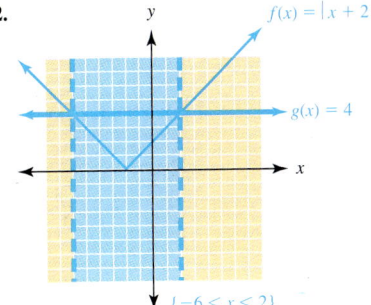 $\{-6 < x < 2\}$

13.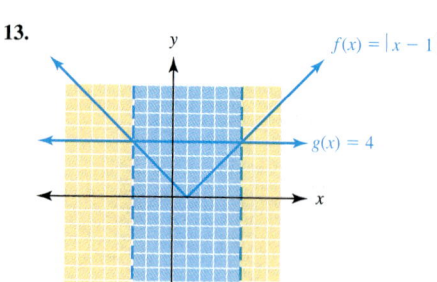

14.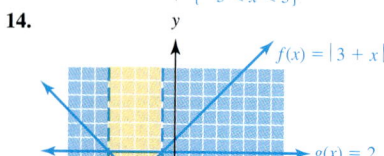

15.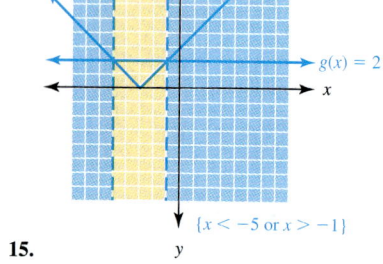

16.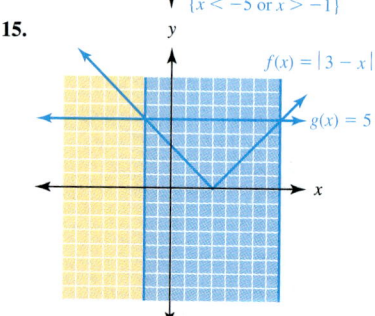

17. 3:30 p.m.   **18.** 20,000 boxes
19. $\{x | -\frac{9}{2} < x < \frac{7}{2}\}$
20. $\{x | x \leq -\frac{7}{3} \text{ or } x \geq \frac{11}{3}\}$

Cumulative Test — Chapters 0–9

1. $\{-\frac{4}{3}\}$ **2.** $\{8, -2\}$ **3.** -21 **4.** $y = \frac{2}{3}x - 2$
5. $4x^2 - 5x - 3$ **6.** $x(x+3)(x-3)$
7. $(2x+1)(x-3)$ **8.** $(x-2)(x+5)$

9. $\frac{1}{x+1}$ **10.** $\frac{-x^2 + x + 5}{(x+3)(x-2)}$
11. $\frac{1}{(x-4)(x+1)}$ **12.** $(2, -2)$ **13.** $x = $ int: 3; y-int: 6
14. $\{x | x \geq -8\}$ **15.** $\{x | -3 \leq x \leq 9\}$
16. $\{x | x < -3 \text{ or } x > 1\}$
17. $\frac{1}{x-5} + x + 6$ **18.** $\frac{1}{(x-5)(x+6)}$
19. $y = -\frac{2}{3}x + 5$

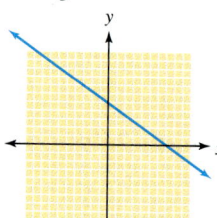

20. $f(x) = |x+4|$

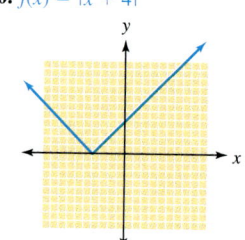

Chapter 10

Section 10.1

1. $\frac{1}{x^5}$ **3.** $\frac{1}{25}$ **5.** $\frac{1}{25}$ **7.** $-\frac{1}{8}$ **9.** $\frac{27}{8}$
11. $\frac{3}{x^2}$ **13.** $-\frac{5}{x^4}$ **15.** $\frac{1}{9x^2}$ **17.** x^3
19. $\frac{2x^3}{5}$ **21.** $\frac{y^4}{x^3}$ **23.** x^2 **25.** $\frac{1}{a^3}$ **27.** $\frac{1}{z^{10}}$
29. 1 **31.** $\frac{1}{x^3}$ **33.** x^{15} **35.** $2x^5$ **37.** $3a$
39. $x^{10}y$ **41.** $a^{17}b^8c^{13}$ **43.** $\frac{1}{x^{15}}$ **45.** b^8
47. $\frac{x^{10}}{y^6}$ **49.** $x^{12}y^6$ **51.** $\frac{x^{15}}{32}$ **53.** $\frac{y^4}{x^2}$ **55.** $\frac{y^2}{x^4}$
57. $\frac{48}{x^8}$ **59.** $16x^{26}$ **61.** $\frac{72}{x^3}$ **63.** $x^{42}y^{33}z^{25}$
65. $75x^2$ **67.** $144w^2$ **69.** $\frac{3x^3}{2y^4}$ **71.** $-567x^{22}y^{25}$
73. $\frac{6}{x^2y^5}$ **75.** $\frac{y^5}{x^3}$ **77.** $\frac{3xy^5}{4z^6}$ **79.** $\frac{1}{x^5y^3}$
81. $\frac{y^8}{x^7}$ **83.** x^{5n} **85.** x^2 **87.** y^{3n^2} **89.** x^2
91. 9.3×10^7 **93.** 1.3×10^{11} **95.** 28 **97.** 0.008
99. 0.000028 **101.** 5×10^{-4} **103.** 3.7×10^{-4}
105. 8×10^{-8} **107.** 3×10^5 **109.** 8×10^9
111. 2×10^2 **113.** 6×10^{16} **115.** 66 years

117. 1.55×10^{23}, 2.9×10^{13}
119. 6.5×10^{14} **121.**    **123.**

Section 10.2
1. 7   **3.** -6 **5.** Not a real number   **7.** 4
9. -6 **11.** 3   **13.** 2   **15.** -2 **17.** $\frac{2}{3}$
19. $\frac{2}{3}$ **21.** 6   **23.** 9   **25.** -5 **27.** $2\sqrt{3}$
29. $-6\sqrt{3}$ **31.** $4\sqrt{2}$ **33.** $-2\sqrt[3]{6}$ **35.** $2\sqrt[4]{6}$
37. $3x^2\sqrt{7}$ **39.** $5a^2\sqrt{3a}$ **41.** $4xy\sqrt{5y}$
43. $5x^2\sqrt[3]{2x^2}$ **45.** $2x^2yz\sqrt[3]{7y^2z}$ **47.** $3y^3\sqrt[4]{2}$
49. $2\sqrt[5]{2w}$ **51.** $\frac{\sqrt{5}}{4}$ **53.** $\frac{x^2}{5}$ **55.** $\frac{\sqrt{5}}{3y^2}$
57. $\frac{\sqrt[3]{3}}{4}$ **59.** $\sqrt[3]{42}$ **61.** $\sqrt[3]{42}$ **63.** $\sqrt[3]{36}$
65. 6   **67.** $4x\sqrt{2x}$ **69.** $5x\sqrt[3]{2x}$ **71.** 3.9
73. 14.6   **75.** Not a real number   **77.** 9.1
79. -3.9 **81.** 2.1   **83.** 8.7   **85.** 2.2   **87.** 1.9
89. 1.9   **91.** False   **93.** True   **95.** False
97. 1   **99.** 0, 1   **101.**   **103.**

Section 10.3
1. 6   **3.** -5 **5.** Not a real number   **7.** 3
9. 3   **11.** $\frac{2}{3}$ **13.** 9   **15.** 16   **17.** 4
19. 729   **21.** $\frac{4}{9}$ **23.** $\frac{1}{5}$ **25.** $\frac{1}{3}$ **27.** $\frac{1}{27}$
29. $\frac{1}{32}$ **31.** $\frac{5}{2}$ **33.** x **35.** $y^{4/5}$ **37.** $b^{13/6}$
39. $x^{1/3}$ **41.** s **43.** $w^{3/4}$ **45.** x **47.** $a^{3/5}$
49. $\frac{1}{y^6}$ **51.** a^4b^9 **53.** $32xy^3$ **55.** $st^{1/3}$
57. $4pq^{5/3}$ **59.** $x^{2/5}y^{1/2}z$ **61.** $a^{1/2}b^{1/4}$ **63.** $\frac{s^2}{r^4}$
65. $\frac{x^3}{y^2}$ **67.** $\frac{1}{mn^2}$ **69.** $\frac{s^3}{r^2t}$ **71.** $\frac{2xz^3}{y^2}$
73. $\frac{2n}{m^{1/5}}$ **75.** $xy^{3/4}$ **77.** $\sqrt[4]{a^3}$ **79.** $2\sqrt[3]{x^2}$
81. $3\sqrt[5]{x^2}$ **83.** $\sqrt[5]{9x^2}$ **85.** $(7a)^{1/2}$ **87.** $2m^2n^3$
89. 9.946   **91.** 0.370   **93.**

95. $a^2 + a^{5/4}$ **97.** $a - 4$ **99.** $m - n$
101. $x + 4x^{1/2} + 4$ **103.** $r + 2r^{1/2}s^{1/2} + s$
105. $(x^{1/3} + 1)(x^{1/3} + 3)$ **107.** $(a^{2/5} - 3)(a^{2/5} - 4)$
109. $(x^{2/3} - 2)(x^{2/3} + 2)$
111. x^{5n} **113.** y^{4n} **115.** r^2 **117.** $a^{6n}b^{4n}$
119. x **121.** $\sqrt[4]{x}$ **123.** $\sqrt[8]{y}$ **125.** 2×10^4
127. 2×10^{-3} **129.** 4×10^{-4} **131.** 40
133.   **135.** (a) 81 (b) Not defined

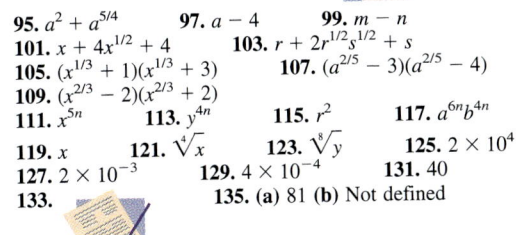

Section 10.4
1. $4i$ **3.** $-8i$ **5.** $i\sqrt{21}$ **7.** $2i\sqrt{3}$
9. $-6i\sqrt{3}$ **11.** $8 + 3i$ **13.** $1 + 5i$ **15.** $2 + 2i$
17. $5 - 7i$ **19.** $7 + 11i$ **21.** $3 + 11i$ **23.** $6 - 3i$

25. $0 + 0i$ **27.** $-15 + 9i$ **29.** $28 + 12i$
31. $-6 - 8i$ **33.** $-5 + 4i$ **35.** $13i$ **37.** $23 + 14i$
39. $18 + i$ **41.** $21 - 20i$ **43.** $3 + 2i$, 13
45. $2 - 3i$, 13   **47.** $-3 + 2i$, 13   **49.** $-5i$, 25
51. $2 - 3i$ **53.** $-2 - 3i$ **55.** $\frac{6}{29} - \frac{15}{29}i$
57. $2 - 3i$ **59.** $\frac{17}{25} + \frac{6}{25}i$ **61.** $-\frac{7}{25} - \frac{24}{25}i$
63.   **65.** $-\sqrt{35}$ **67.** -6 **69.** $-3\sqrt{10}$
71. -10 **73.** -1 **75.** 1   **77.** -1 **79.** $-i$

Self-Test: Chapter 10
1. x^8y^{-10} **2.** $\frac{c^2d^7}{2}$ **3.** 4.23×10^9
4. 2.5×10^{-5} **5.** $7a^2$ **6.** $-3w^2z^3$ **7.** $\frac{7x}{8y}$
8. 2   **9.** $x^3\sqrt{14x}$ **10.** $2x^3\sqrt{4x^2}$ **11.** 6.6
12. 1.4   **13.** $\frac{1}{64x^6}$ **14.** $\frac{9m}{n^4}$ **15.** $\frac{8s^3}{r}$
16. $a^2b\sqrt{a^4b}$ **17.** $5p^3q^2$ **18.** $3 + 10i$
19. $-14 + 22i$ **20.** $5 - 5i$

Cumulative Test — Chapters 0–10
1. $\{-15\}$ **2.** 5   **3.** $3x - 2y = 12$
4. $\left\{3, \frac{1}{3}\right\}$ **5.** $10x^2 - 2x$ **6.** $10x^2 - 39x - 27$
7. $x(x - 1)(2x + 3)$ **8.** $9(x^2 + 2y^2)(x^2 - 2y^2)$
9. $(x + 2y)(4x - 5)$ **10.** $\frac{x + 5}{3x - 1}$ **11.** $\frac{x + 7}{(x - 5)(x - 1)}$
12. $\frac{a + 2}{2a}$ **13.** 0   **14.** $2x^4y^3\sqrt{3y}$
15. $\sqrt{6} - 5\sqrt{2} + 3\sqrt{3} - 15$
16.

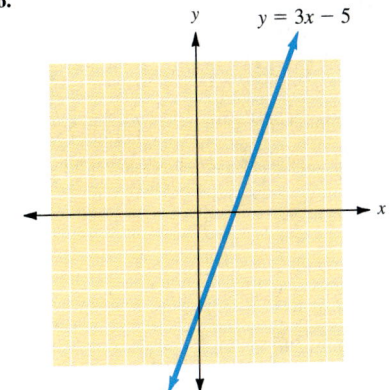

17.

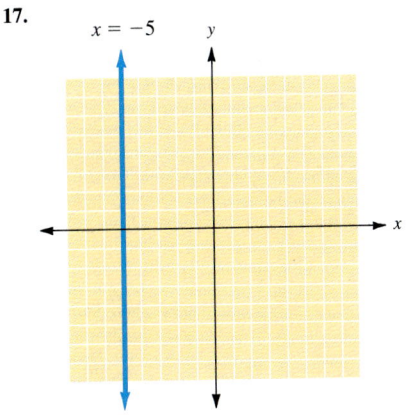

18.

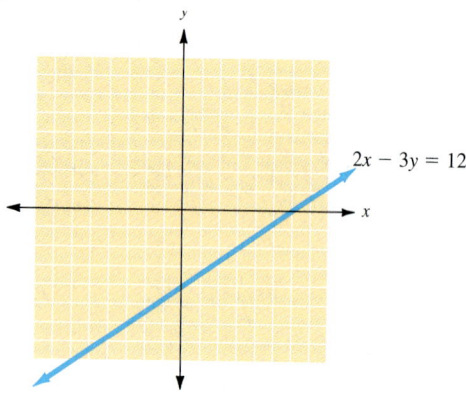

19. {3, −1}   **20.** (a) {x|x ≤ −4}, (b){x|−6 < x < 10}

Chapter 11

Section 11.1
1. {−5, −1}   **3.** {−5, 7}   **5.** $\left\{-\frac{1}{2}, 3\right\}$
7. $\left\{-\frac{1}{2}, \frac{2}{3}\right\}$ **9.** {±6}   **11.** {−√7, √7}
13. {−√6, √6}   **15.** {±2i}   **17.** {−1 ± 2√3}
19. $\left\{-\frac{1 \pm \sqrt{3}}{2}\right\}$ **21.** 36   **23.** 16   **25.** $\frac{9}{4}$
27. $\frac{1}{4}$ **29.** $\frac{1}{16}$ **31.** $\frac{1}{16}$ **33.** {−6 ± √38}
35. {−2, 4}   **37.** {1 ± √6}   **39.** {−5 ± 2√3}
41. $\left\{\frac{5 \pm \sqrt{53}}{2}\right\}$ **43.** $\left\{\frac{1 \pm \sqrt{13}}{2}\right\}$ **45.** $\left\{\frac{-1 \pm \sqrt{17}}{4}\right\}$
47. $\left\{\frac{-1 \pm \sqrt{3}}{2}\right\}$ **49.** $\left\{\frac{4 \pm \sqrt{22}}{3}\right\}$ **51.** $\left\{\frac{1 \pm i\sqrt{35}}{3}\right\}$
53. {−4 ± 2i}   **55.**   **57.** a^2 **59.** $\frac{9}{4}a^2$

61. 1   **63.** {−a + √(4 + a²)}
65. −12.2, 0.2, the same as Exercise 33

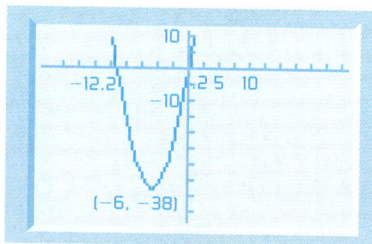

67. −2, 4, the same as Exercise 35

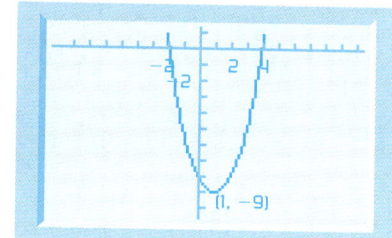

69. (a) no x-int, (b) ±i
71. (a) $x^2 + 8x + 8x$, (b) 64, (c) $(x + 8)^2$

Section 11.2
1. {−2, 7}   **3.** {−13, 5}   **5.** $\left\{-1, \frac{1}{5}\right\}$ **7.** $\left\{\frac{3}{4}\right\}$
9. {1 ± √6}   **11.** $\left\{\frac{-3 \pm 3\sqrt{13}}{2}\right\}$ **13.** $\left\{\frac{3 \pm \sqrt{15}}{2}\right\}$
15. $\left\{\frac{1 \pm \sqrt{3}}{2}\right\}$ **17.** $\left\{-\frac{2}{3}, 1\right\}$ **19.** $\left\{\frac{1 \pm \sqrt{41}}{4}\right\}$
21. {1, 3}   **23.** {4}   **25.** $\left\{\frac{1 \pm \sqrt{7}}{2}\right\}$
27. $\left\{\frac{1 \pm i\sqrt{2}}{3}\right\}$ **29.** $\left\{\frac{1 \pm \sqrt{161}}{8}\right\}$ **31.** $\left\{\frac{3 \pm \sqrt{39}}{15}\right\}$
33. $\left\{\frac{-1 \pm \sqrt{41}}{2}\right\}$ **35.** $\left\{\frac{1 \pm \sqrt{23}}{2}\right\}$ **37.** $\left\{\frac{5 \pm \sqrt{37}}{6}\right\}$
39. $\left\{\frac{3 \pm \sqrt{33}}{4}\right\}$ **41.** {1 ± √6}   **43.** 25, two
45. 0, one   **47.** 37, two   **49.** −63, none   **51.** {4}
53. $\left\{\frac{7 \pm \sqrt{37}}{6}\right\}$ **55.** $\left\{0, \frac{2}{5}\right\}$ **57.** $\left\{\frac{-5 \pm \sqrt{33}}{4}\right\}$
59. {−3i, 3i}   **61.** {−2, 5}   **63.** {1.2 or 5.8 s}
65. 1.4 s   **67.** −9, −8, or 8, 9   **69.** 7 by 10 ft
71. 5 by 17 cm   **73.** 2.7 cm, 5.4 cm
75. 5.5, 6.5, 8.5 in.   **77.** 3.2 m, 21.8 m
(b) 1 s   **81.** 50 chairs   **83.** 5.5 s   **79.** (a) 4 s,   **85.** 5 s
87. $h(t) = 100 + 20t − 5t^2$ **89.** 5 s
91. $h(t) = 100 + 20t − 16t^2$ **93.** 1.9 s   **95.** 63 or 27
97. $0.94 or $17.06   **99.**

101. {±√(z² − y²)}   **103.** {−6a, 6a}   **105.** $\left\{-3a, \frac{a}{2}\right\}$
107. $\left\{\frac{-a \pm a\sqrt{17}}{4}\right\}$ **109.** {−1 ± √6}
111. $\left\{-1, \frac{1 \pm i\sqrt{3}}{2}\right\}$ **113.** $\left\{-1, 1, \frac{1 \pm i\sqrt{3}}{2}, \frac{-1 \pm i\sqrt{3}}{2}\right\}$

115. (a) {4.2, −1.2}, (b) $y = x^2 − 3x − 5$,

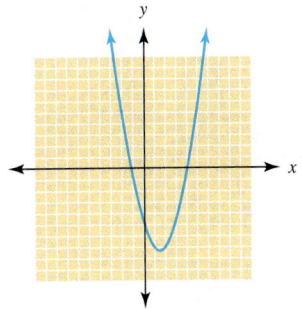

(c) answers are the same

Section 11.3
1. (f)   **3.** (g)   **5.** (h)   **7.** (c)   **9.** (a), (c)
11. (a), (c)   **13.** (b), (c)
15. $x = 2$; vertex: (2, −4); (0, 0) and (4, 0)

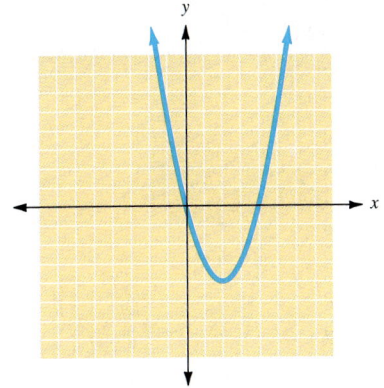

17. $x = 0$; vertex: (0, 4); (−2, 0) and (2, 0)

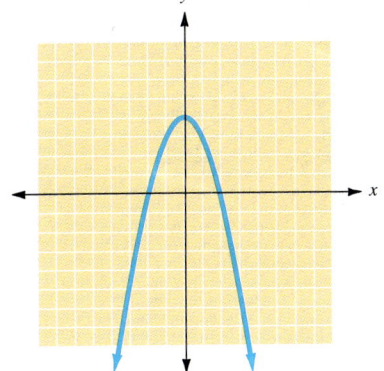

19. $x = −1$; vertex: (−1, 1); (−2, 0) and (0, 0)

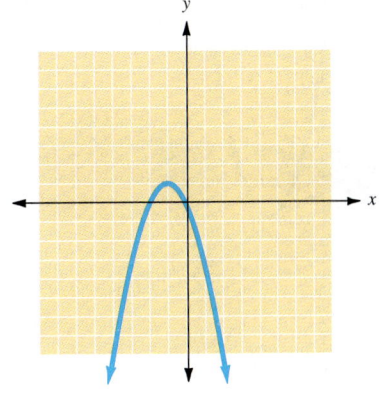

21. $x = 3$; vertex: (3, −4); (1, 0) and (3, 0)

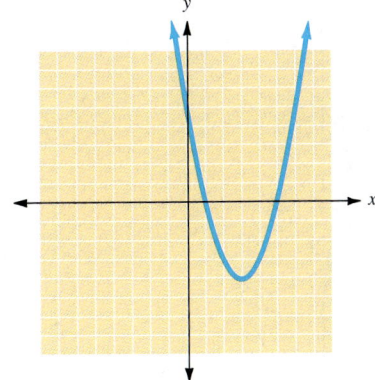

23. $x = \dfrac{5}{2}$; vertex: $\left(\dfrac{5}{2}, -\dfrac{1}{4}\right)$; (3, 0) and (2, 0)

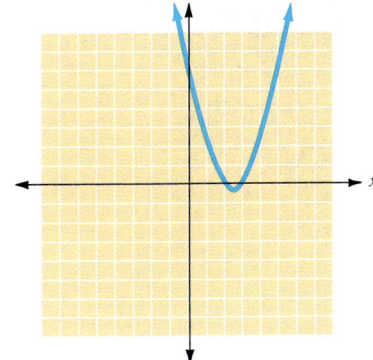

A-36 Answers

25. $x = 3$; vertex: $(3, -1)$; $(2, 0)$ and $(4, 0)$

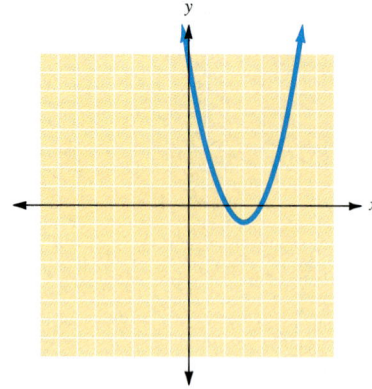

27. $x = -3$; vertex: $(-3, 4)$; $(-5, 0)$ and $(-1, 0)$

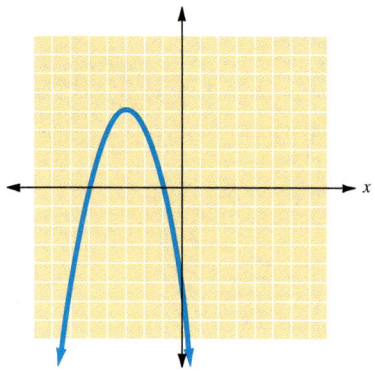

29. $x = 1$; vertex: $(1, -2)$

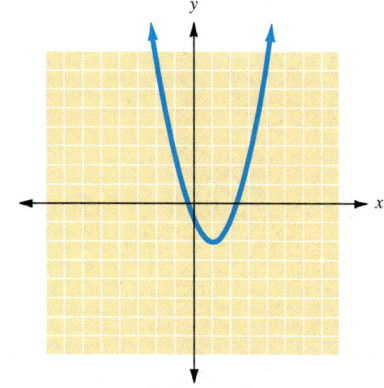

31. $x = 2$; vertex: $(2, -5)$

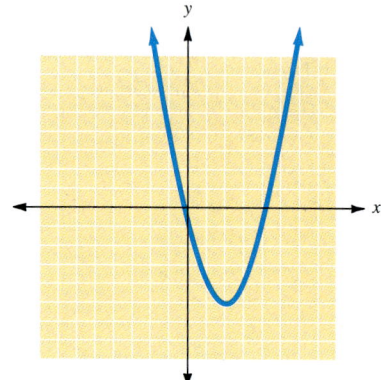

33. $x = \dfrac{3}{2}$; vertex: $\left(\dfrac{3}{2}, -\dfrac{3}{4}\right)$

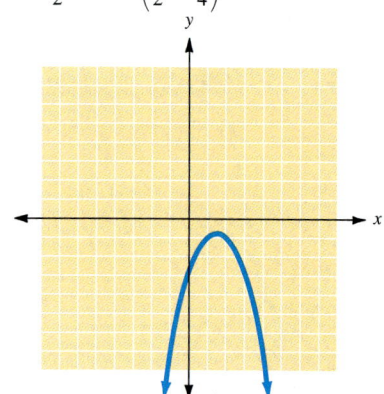

35. $x = -1$; vertex: $(-1, -3)$

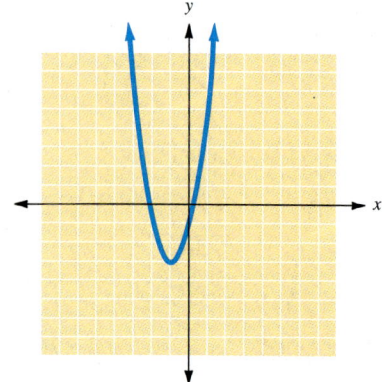

37. $x = \frac{3}{2}$; vertex: $\left(\frac{3}{2}, -\frac{9}{4}\right)$

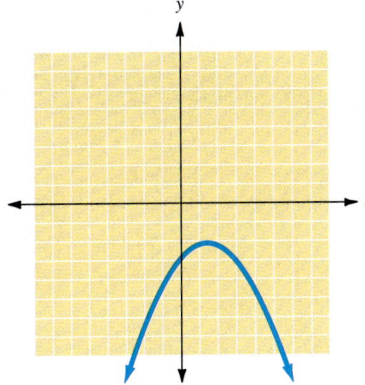

39. $x = -2$; vertex: $(-2, -7)$

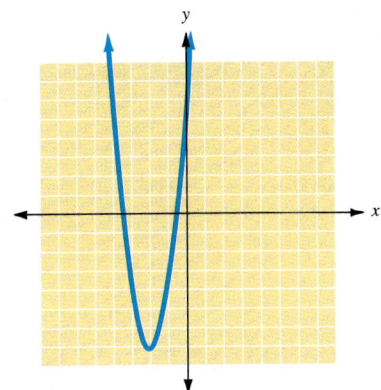

41. 2; 1 positive, 1 negative **43.** 2; 1 positive
45. None **47.** 2; 1 positive, 1 negative
49. 2; 1 positive, 1 negative **51.** $\{-4, 3\}$ **53.** $\{0, 3.2\}$
55. $\{0.4, -1, 8\}$ **57.** $\{-3.8, 1.8\}$ **59.** 100 items; $2600
61. 500 ft by 500 ft; 250,000 ft^2 **63.** 144 ft
65. 88 ft/sec **67.** $(-4, 2)$; $y \geq 2$ **69.** $(-2, -1)$; $y \leq -1$
71. $(4, 0)$; $y \leq 0$ **73.** $-3 \leq x \leq 3$; $-25 \leq y \leq 0$
75. $-2 \leq x \leq 4$; $-10 \leq y \leq 0$

Section 11.4

1. $-4 < x < 3$
3. $x < -4$ or $x > 3$
5. $-4 \leq x \leq 3$
7. $x \leq -4$ or $x \geq 3$
9. $x < -1$ or $x > 4$
11. $-4 \leq x \leq 3$
13. $x \leq 2$ or $x \geq 3$
15. $-6 \leq x \leq 4$
17. $x < -9$ or $x > 3$
19. $-2 \leq x \leq \frac{3}{2}$
21. $-1 < x < \frac{3}{4}$
23. $-4 \leq x \leq 4$
25. $x \leq -5$ or $x \geq 5$
27. $x < -2$ or $x > 2$
29. $0 \leq x \leq 4$
31. $x \leq 0$ or $x \geq 6$
33. $0 < x < 4$
35. $x = 2$
37. $-4 \leq x \leq 7$
39.
41. $x < -1$ or $0 < x < 2$
43. $-2 \leq x \leq 1$ or $x \geq 3$
45. $x \leq 3$ or $0 \leq x \leq 5$
47. $50 \leq x \leq 60$ **49.** $2 \leq t \leq 3$
51.

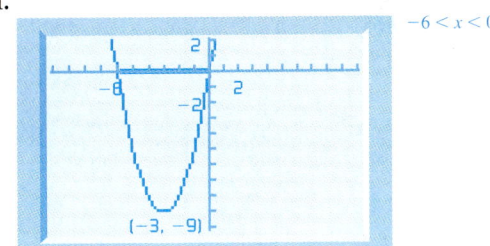

53.

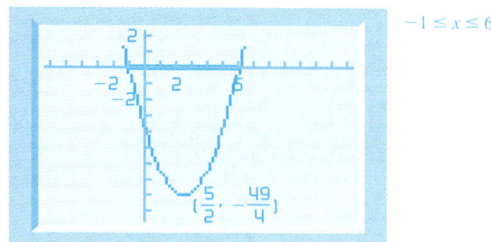

Self-Test: Chapter 11

1. $\left\{-3, -\dfrac{1}{2}\right\}$ 2. $\left\{-\dfrac{5}{2}, \dfrac{2}{3}\right\}$ 3. $\left\{0, \dfrac{3}{2}, -\dfrac{3}{2}\right\}$
4. $\{\pm\sqrt{5}\}$ 5. $\{1 \pm \sqrt{10}\}$ 6. $\left\{\dfrac{2 \pm \sqrt{23}}{2}\right\}$
7. $\left\{\dfrac{-3 \pm \sqrt{13}}{2}\right\}$ 8. $\left\{\dfrac{5 \pm \sqrt{19}}{2}\right\}$ 9. $\left\{\dfrac{5 \pm \sqrt{37}}{2}\right\}$
10. $\{-2 \pm \sqrt{11}\}$ 11. $\left\{-\dfrac{2}{3}, 4\right\}$ 12. 7, 9
13. 2.7 s 14. $x = -2, (-2, 1)$ 15. $x = 2, (2, 9)$
16. $x = \dfrac{3}{2}, \left(\dfrac{3}{2}, \dfrac{3}{2}\right)$ 17. $x = 3, (3, -2)$
18. $x = 3, (3, -7)$
19. $f(x) = (x - 5)^2$

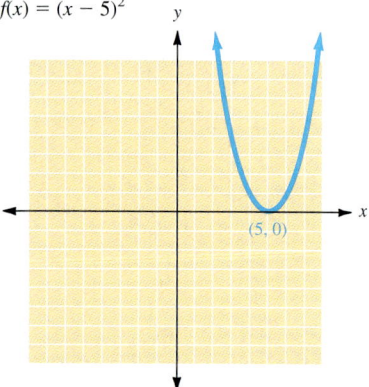

20. $f(x) = (x + 2)^2 - 3$

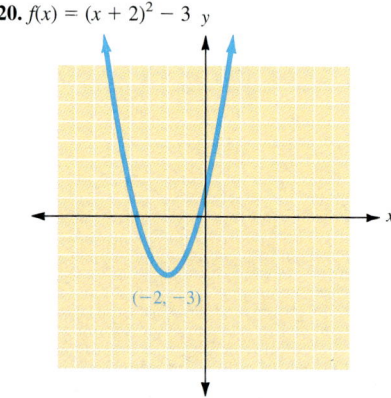

21. $f(x) = -2(x - 3)^2 - 1$

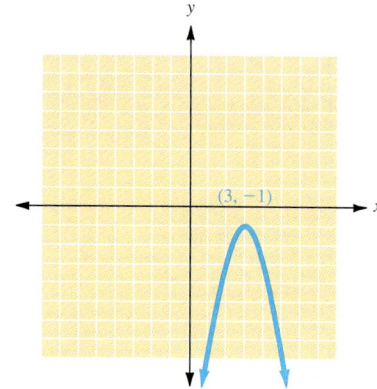

22. $f(x) = 3x^2 + 9x + 2$

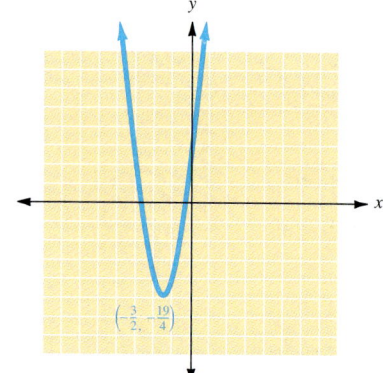

23. $\{-4.5, 1.5\}$ 24. $\{-1.4, 0.9\}$
25. $\{x \mid x < -3 \text{ or } x > 1\}$
26. $\{x \mid -7 < x < 2\}$
27. $\{x \mid x \leq -3 \text{ or } x \geq 6\}$
28. $\{x \mid -2 \leq x < \dfrac{5}{3}\}$
29. $\{x \mid 0 \leq x \leq 3\}$
30. $\{x \mid x < -5 \text{ or } x > 0\}$

Cumulative Test—Chapters 0–11

1. $2x - 3y = 6$

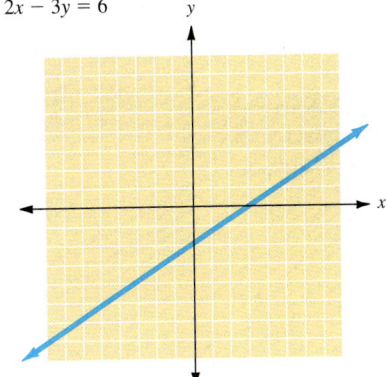

2. $y = -\frac{1}{3}x - 2$

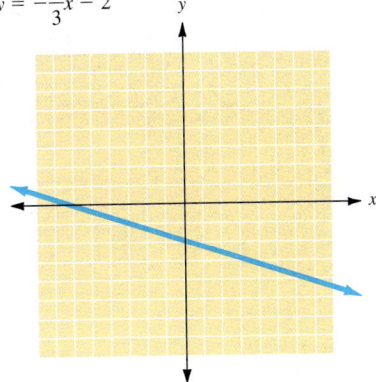

3. $y = 4$

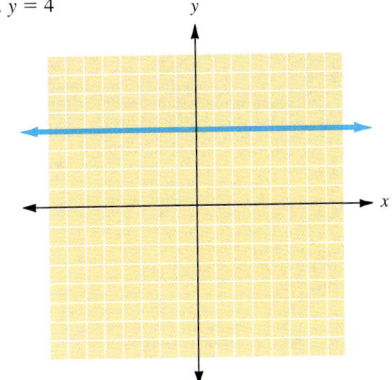

4. -3 **5.** $\frac{4}{3}$ **6.** 35 **7.** $f(x) = x^3 + 3x^2 - x - 3$
8. $x(x + 3)(x - 2)$ **9.** $\dfrac{-x - 4}{(x - 3)(x + 2)}$
10. $\sqrt{21} - \sqrt{6} + \sqrt{42} - \sqrt{3}$ **11.** $\left\{\dfrac{7}{2}\right\}$
12. $\{-4\}$ **13.** $\{-5, 3\}$ **14.** $\{1, 2\}$ **15.** $\{-10, 3\}$
16. $\left\{\dfrac{3 \pm \sqrt{21}}{2}\right\}$ **17.** $\{3 \pm \sqrt{5}\}$ **18.** $\{-2, 6\}$
19. $\left\{\dfrac{13}{6}\right\}$ **20.** $\{x | x \leq 9\}$ **21.** $\{x | x \leq -7 \text{ or } x \geq 9\}$
22. $\{x | -2 \leq x \leq 3\}$ **23.** -13
24. $-2 + 2\sqrt{97}, 2 + \sqrt{97}$ or 17.7 ft, and 21.7 ft
25. Between 21 and 59 units

Chapter 12

Section 12.1
1. $y = 3x^2 + 2x - 1$ **3.** $y = \frac{1}{2}x^2 - \frac{3}{2}x - \frac{5}{2}$
5. $y = -x^2 + 2x + \frac{3}{2}$
7. $y = -2$; $V(-4, -2)$

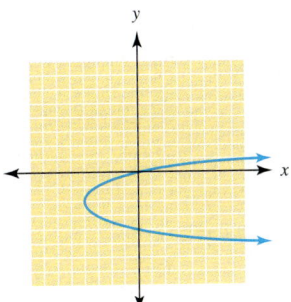

9. $y = -4$; $V(-4, -4)$

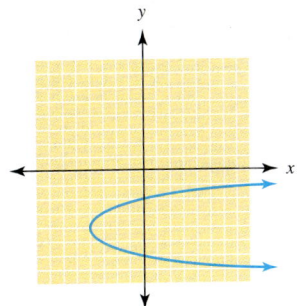

11. $y = 3$; $V(4, 3)$

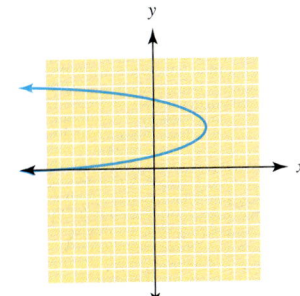

13. $y = 1$; $V(-3,-1)$

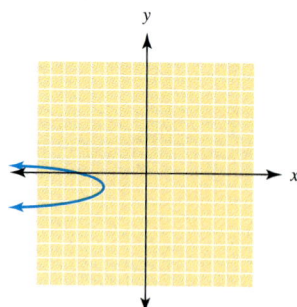

15. $y = 1$; $V(-4,1)$

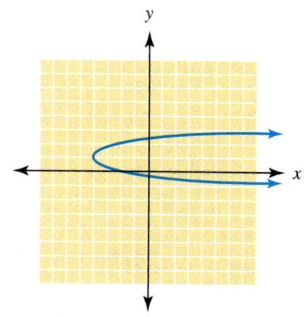

17. $a > 0$ and $h > 0$ or $a < 0$ and $h < 0$

19.

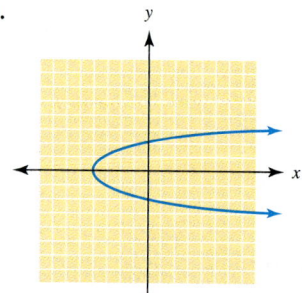

21.

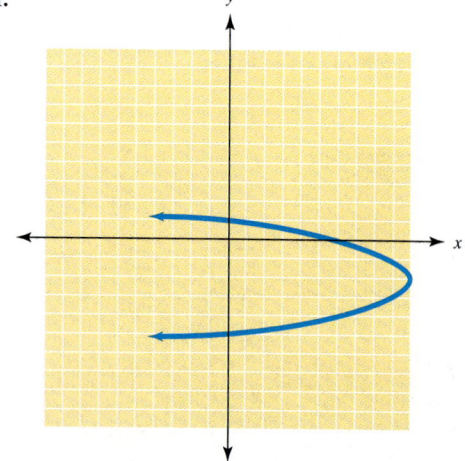

23.

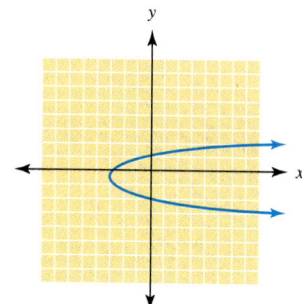

25. Domain: $\{x | x \geq 9\}$
27. Domain: $\{x | x \leq 2\}$

Section 12.2

1. Parabola **3.** Line **5.** Circle **7.** Circle
9. None of these **11.** Parabola **13.** Center: $(0, 0)$; radius: 5 **15.** Center: $(3, -1)$; radius: 4 **17.** Center: $(-1, 0)$; radius: 4 **19.** Center: $(3, -4)$; radius: $\sqrt{41}$
21. Center: $(0,0)$; radius: 2

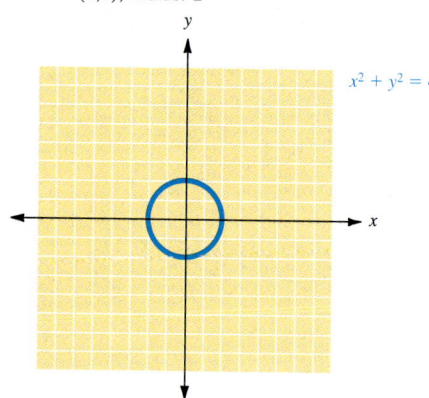

$x^2 + y^2 = 4$

23. Center: $(0,0)$; radius: 3

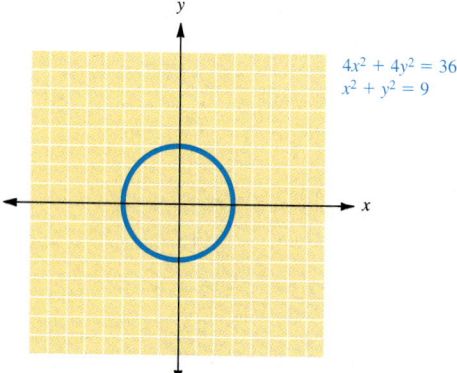

$4x^2 + 4y^2 = 36$
$x^2 + y^2 = 9$

25. Center: (1,0); radius: 3

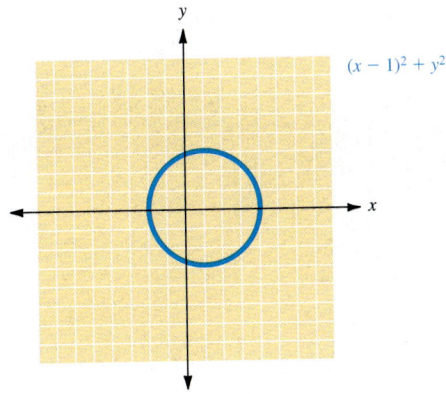

27. Center: (4,−1); radius: 4

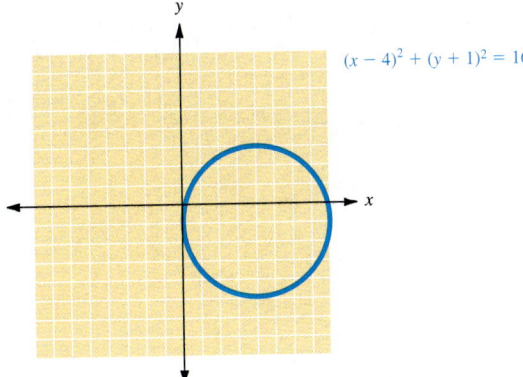

29. Center: (0,2); radius: 4

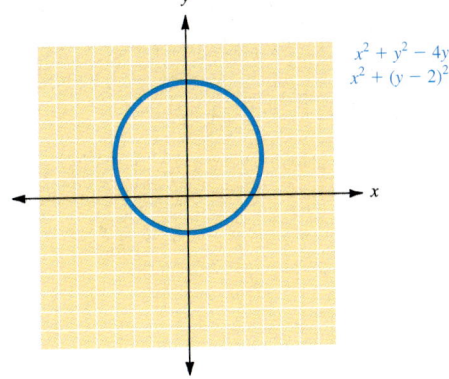

31. Center: (2,−1); radius: 2

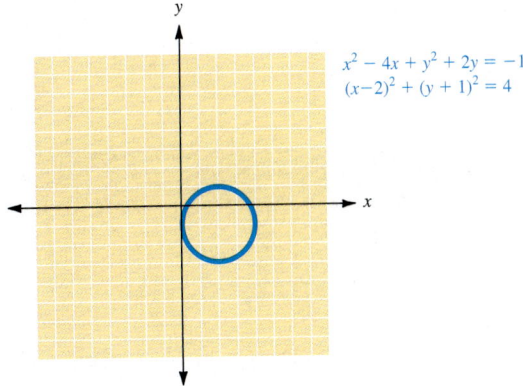

33.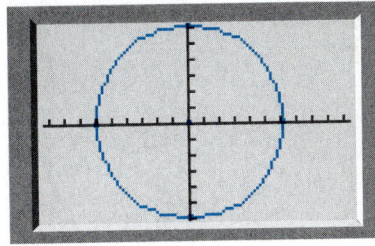

35. $\sqrt{500} = 10\sqrt{5}$ cm $\cong 22.4$ cm

37. $x^2 + y^2 = \dfrac{4}{9}$

39. $x^2 + y^2 = 36$

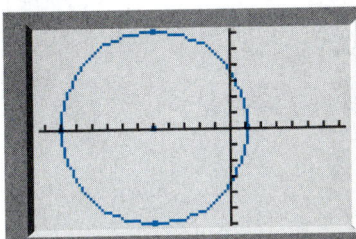

41. $(x+5)^2 + y^2 = 36$

43. D: $\{x|-1 \leq x \leq 6\}$; R: $\{y|-2 \leq y \leq 6\}$

45. D: $\{x|-5 \leq x \leq 5\}$; R: $\{y|-2 \leq y < 8\}$

Section 12.3

1. $\dfrac{x^2}{4} + \dfrac{y^2}{9} = 1$
x-intercepts: ± 2
y-intercepts: ± 3

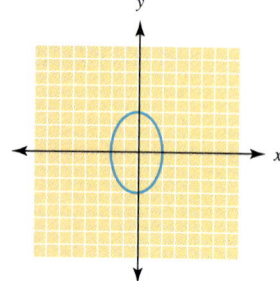

3. $\dfrac{x^2}{9} + \dfrac{y^2}{25} = 1$
x-intercepts: ± 3
y-intercepts: ± 5

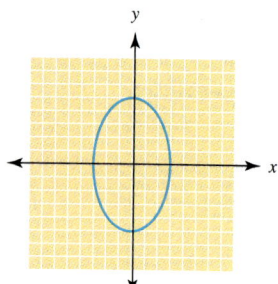

5. $x^2 - 9y^2 = 36$
$\dfrac{x^2}{36} + \dfrac{y^2}{4} = 1$
x-intercepts: ± 6
y-intercepts: ± 2

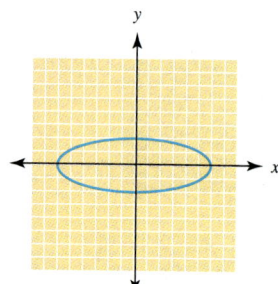

7. $4x^2 + 9y^2 = 36$
$\dfrac{x^2}{9} + \dfrac{y^2}{4} = 1$
x-intercepts: ± 3
y-intercepts: ± 2

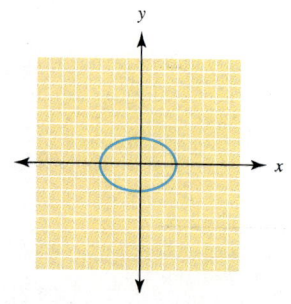

9. $4x^2 + 25y^2 = 100$

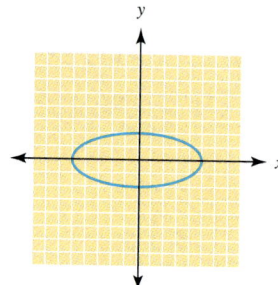

11. $25x^2 + 9y^2 = 225$

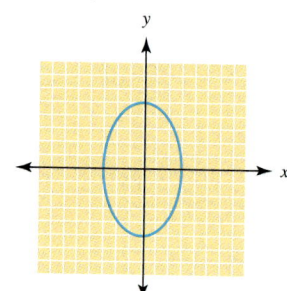

13. Yes

15. $4x^2 + 16y^2 = 64$

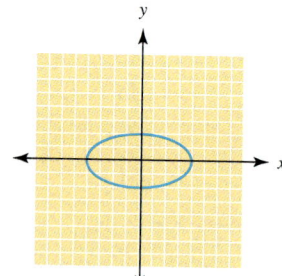

17. $25x^2 + 9y^2 = 225$

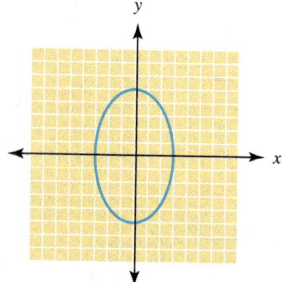

Section 12.4

1. Circle **3.** Parabola **5.** Hyperbola
7. Hyperbola **9.** Circle **11.** Hyperbola **13.** d
15. h **17.** b **19.** c

21. $\dfrac{x^2}{9} - \dfrac{y^2}{9} = 1$

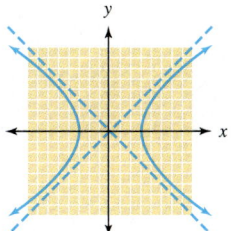

23. $\dfrac{y^2}{16} - \dfrac{x^2}{9} = 1$

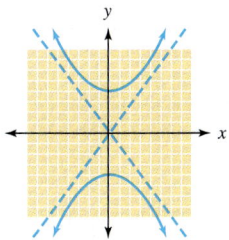

25. $\dfrac{x^2}{36} - \dfrac{y^2}{9} = 1$

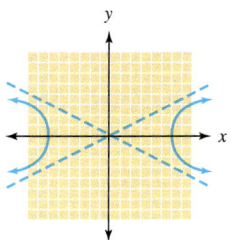

27. $x^2 - 9y^2 = 36$

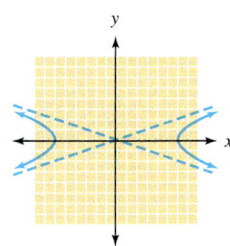

29. $9x^2 - 4y^2 = 36$

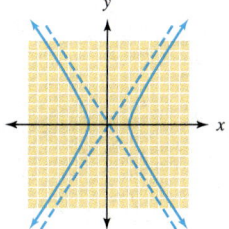

31. $16y^2 - 9x^2 = 144$

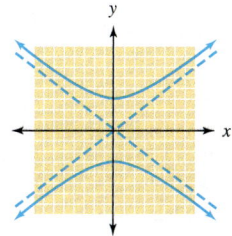

33. $25y^2 - 4x^2 = 100$

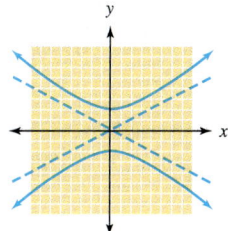

35. $y^2 - 16x^2 = 16$

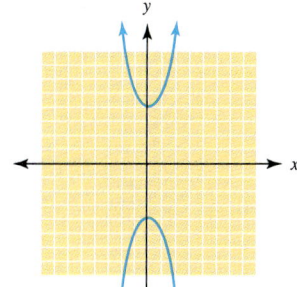

37. $16x^2 - 9y^2 = 144$

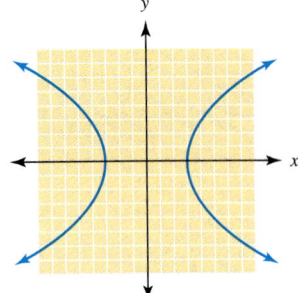

Self-Test: Chapter 12

1. $y = -x^2 + 3x - \dfrac{3}{2}$

2. $y = -2; (-4, -2)$

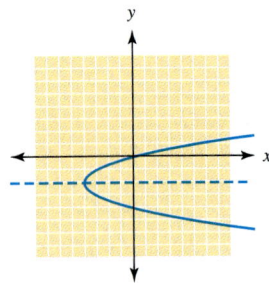

3. $y = 3; (-1, 3)$

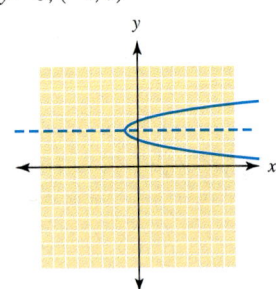

4. $y = 2; (6, 2)$

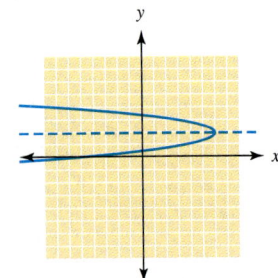

5. C: $(3, -2)$; r = 6

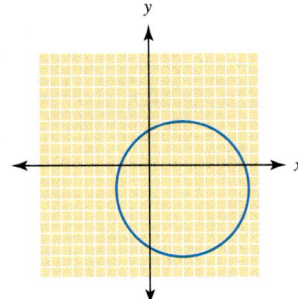

6. C: $(-1, 2)$; r = 5

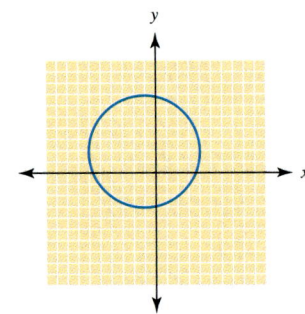

7. C: $(-3, 0)$; r = 2

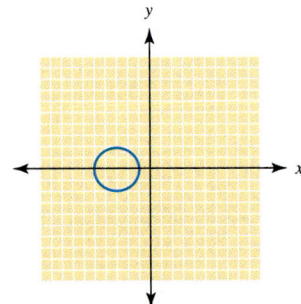

8. C$(2, -3)$; r = 3

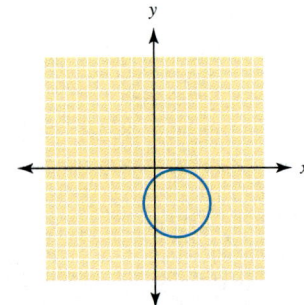

9. $\frac{x^2}{25} + \frac{y^2}{9} = 1$

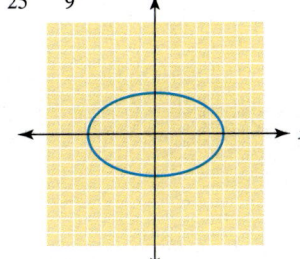

10. $16x^2 + 4y^2 = 64$

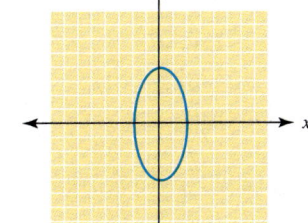

11. $\frac{x^2}{9} - \frac{y^2}{16} = 1$

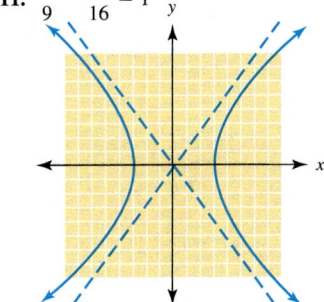

12. $4y^2 + 25y^2 = 1y$

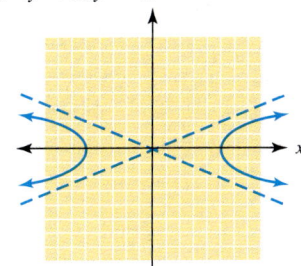

13. Line   **14.** Parabola   **15.** Circle   **16.** Circle
17. Parabola   **18.** Ellipse   **19.** Hyperbola
20. Circle

Cumulative Test — Chapters 1–12

1. $\left\{\frac{11}{2}\right\}$ **2.** $\{x|x > -2\}$ **3.** $\{-1, 4\}$
4. $\left\{x|-4 \leq x \leq \frac{2}{3}\right\}$ **5.** $\left\{x|x < \frac{17}{5} \text{ or } x > 5\right\}$
6. $\{8, -3\}$

7.

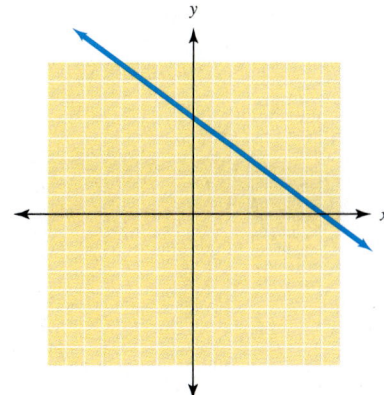

8.

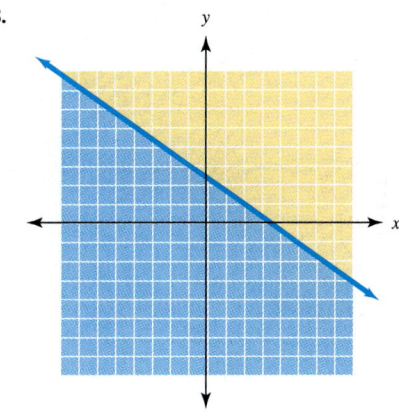

9. 24.5   **10.** 7   **11.** $f(x) = -x + 3$

12. $2x^2 - 5x - 3$ **13.** $9x^2 - 12x + 4$

14. $(x - 3)(x^2 - 5)$

15. $y = x^2 - 6x + 5$

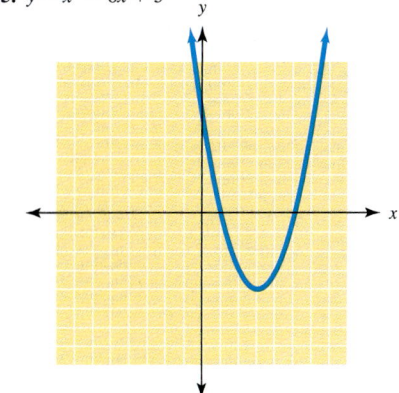

16.

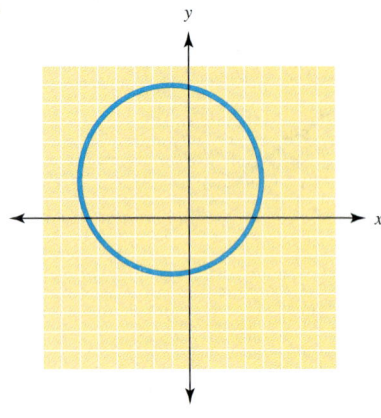

17.

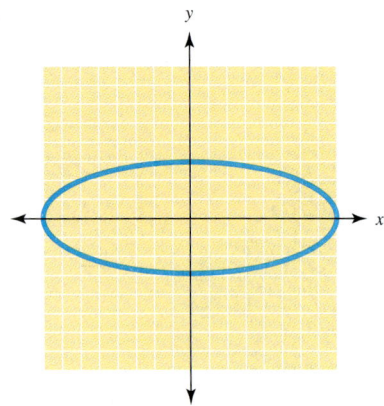

18. $\left(-10, \dfrac{26}{3}\right)$ **19.** 8 cm by 19 cm **20.** 37

Chapter 13

Section 13.1

1. {(3, 2), (4, 3), (5, 4)}; function **3.** {(2, 1), (2, 2), (2, 3)}; not a function **5.** {(4, 2), (9, 3), (16, 4)}; function
7. $y = \dfrac{1}{2}x - 4$ **9.** $y = 2x + 1$
11. $y = \pm\sqrt{x+1}$ or $x = y^2 - 1$ **13.** $4x^2 + y^2 = 36$
15. $y^2 - x^2 = 9$
17. Inverse is a function.

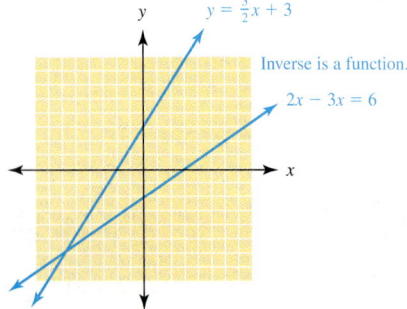

19. Inverse is a function.

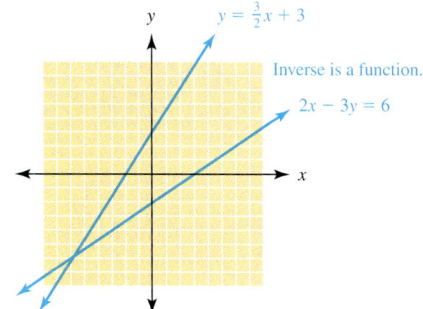

21. Inverse is not a function.

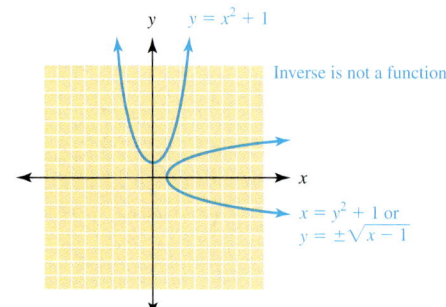

23. **25.** 12 **27.** 6 **29.** x **31.** 2
33. 3 **35.** x **37.** 16 **39.** 4 **41.** x
43. 5 **45.** $\dfrac{x-b}{m}$ **47.** $\dfrac{5}{2}$ **49.** 7 **51.** x

Section 13.2

1. (c) **3.** (b) **5.** (h) **7.** (f) **9.** 1
11. 16 **13.** 4 **15.** 64 **17.** 2 **19.** 17
21. $\dfrac{1}{4}$ **23.** 16
25. $y = 4^x$

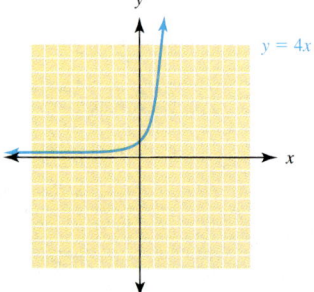

27. $y = \left(\dfrac{2}{3}\right)^x$

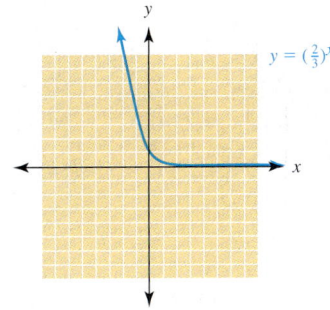

29. $y = 3 \cdot 2^x$

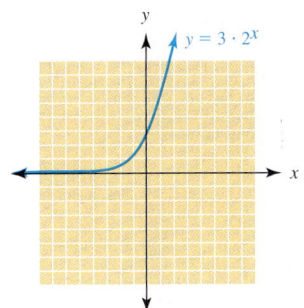

31. $y = 3^x$

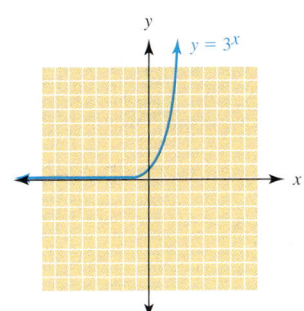

33. $y = 2^{2x}$

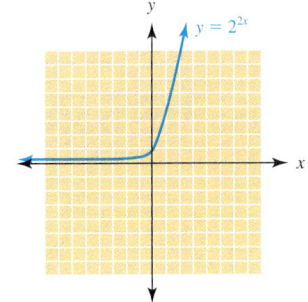

35. $y = e^{-x}$

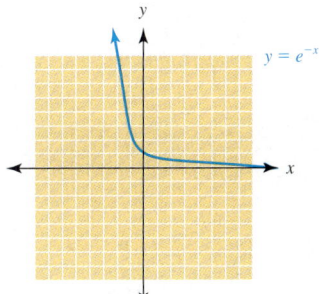

37. $\{5\}$ **39.** $\{4\}$ **41.** $\{-2\}$ **43.** $\{3\}$
45. $\{5\}$ **47.** $\{-2\}$ **49.** 400   **51.** 3200
53. 32 g   **55.** 8 g   **57.** $1166.40   **59.** $1999
61. (a) 36, (b) 56, (c) 67   **63.**

65. 2.7048   **67.** 2.71815   **69.** 2.71828
71.

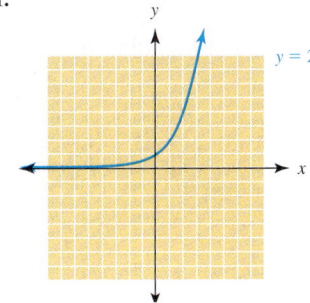

73. (a) h = (0.003) (2^n), (b) 12, (c) 8.192 ft

Section 13.3

1. $y = \log_4 x$

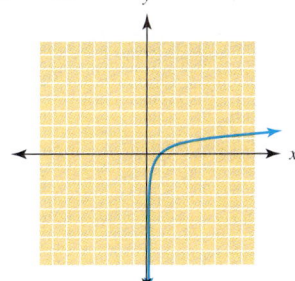

3. $y = \log_2(x - 1)$

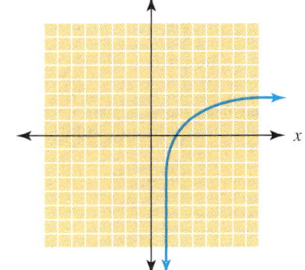

5. $y = \log_8 x$

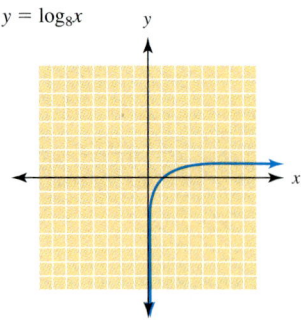

7. $\log_2 16 = 4$ **9.** $\log_{10} 100 = 2$ **11.** $\log_3 1 = 0$
13. $\log_4 \frac{1}{16} = -2$ **15.** $\log_{10} \frac{1}{1000} = -3$
17. $\log_{16} 4 = \frac{1}{2}$ **19.** $\log_{64} \frac{1}{4} = -\frac{1}{3}$ **21.** $\log_8 4 = \frac{2}{3}$
23. $\log_{27} \frac{1}{9} = -\frac{2}{3}$ **25.** $2^4 = 16$ **27.** $5^0 = 1$
29. $10^1 = 10$ **31.** $5^3 = 125$ **33.** $3^{-3} = \frac{1}{27}$
35. $10^{-2} = 0.01$ **37.** $16^{1/2} = 4$ **39.** $8^{2/3} = 4$
41. $25^{-1/2} = \frac{1}{5}$ **43.** 5 **45.** 3 **47.** -4
49. -2 **51.** $\frac{1}{2}$ **53.** $\{2\}$ **55.** $\{4\}$ **57.** $\{100\}$
59. $\{1\}$ **61.** $\left\{\frac{27}{8}\right\}$ **63.** $\{5\}$ **65.** $\left\{\frac{1}{1000}\right\}$
67. $\{-2\}$ **69.** $\{3\}$ **71.** $\{25\}$ **73.** $\left\{-\frac{2}{3}\right\}$
75. 50 dB **77.** 70 dB **79.** 10^{-8} **81.** 10
83. 1000 **85.** 6 **87.** $10^5 \cdot a_0$ **89.**

91. 24,000 yr **93.** 77,000 yr **95.** Pu239: 240,000 yr, Sr90: 280 yr, Th230: 770,000 yr, Cs135: 20,000,000 yr
97. (a) 6, (b) 2, 4, (c) $\log_3 (9 \cdot 81) = \log_3 9 + \log_3 81$

Section 13.4
1. $\log_b 5 + \log_b x$ **3.** $\log_4 x - \log_4 3$ **5.** $2 \log_3 a$
7. $\frac{1}{2} \log_5 x$ **9.** $3 \log_b x + 2 \log_b y$
11. $2 \log_4 y + \frac{1}{2} \log_4 x$ **13.** $2 \log_b x + \log_b y - \log_b z$
15. $\log x + 2 \log y - \frac{1}{2} \log z$
17. $\frac{1}{3}(\log_5 x + \log_5 y - 2 \log_5 z)$
19. $\log_b xy$ **21.** $\log_2 \frac{x^2}{y}$ **23.** $\log_b x \sqrt{y}$
25. $\log_b \frac{xy^2}{z}$ **27.** $\log_6 \frac{\sqrt{y}}{z^3}$ **29.** $\log_b \sqrt[3]{\frac{x^2 y}{z}}$
31. 1.380 **33.** 0.903 **35.** 0.151 **37.** -0.602
39. 0.833 **41.** 2.833 **43.** -0.167
45. 7.42, basic **47.** 5.61 **49.** 5610
51. 2×10^{-5} **53.** 0.693 **55.** 2.303 **57.** 74
59. between 2 and 3 **61.** between 6 and 7
63. between 2 and 3 **65.** 1 **67.** 2 **69.** 2.113
71. 0.25 kg **73.** 160 kg **75.**

Section 13.5
1. $\{64\}$ **3.** $\{99\}$ **5.** $\{8\}$ **7.** $\{162\}$ **9.** $\{2\}$
11. $\{6\}$ **13.** $\left\{\frac{20}{9}\right\}$ **15.** $\left\{\frac{19}{8}\right\}$ **17.** $\left\{\frac{15}{4}\right\}$
19. $\{5, 3\}$ **21.** $\{4\}$ **23.** $\{-4\}$ **25.** $\left\{\frac{1}{3}\right\}$
27. $\{1.771\}$ **29.** $\{0.792\}$ **31.** $\{0.670\}$
33. $\{0.894\}$ **35.** $\{3.819\}$ **37.** $\{4.419\}$ **39.** 8.04 yr
41. 7.79 yr **43.** 1.47 h **45.** 3.17 h **47.** 20.6 yr
49. 11.6 yr **51.** 7.5 yr **53.** 23.1 h **55.** 4558 ft
57. 2876 yr **59.**

61.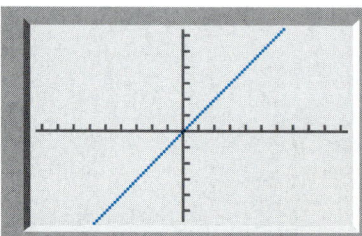

The graph is that of $y = x$. The two operations cancel each other.

63.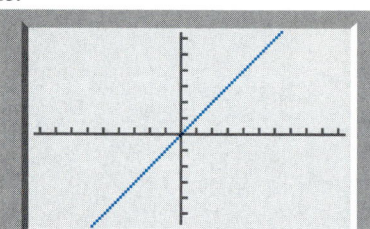

The graph is that of $y = x$. The two operations cancel each other.

65.

Self-Test: Chapter 13
1. (a) $f^{-1} = \left\{(x,y) \mid y = \frac{1}{4}x + \frac{1}{2}\right\}$; a function,
(b) $g^{-1} = \left\{(x,y) \mid y = \pm \sqrt{x - 1}\right\}$; not a function,
(c) $f(x) = 4x - 2$

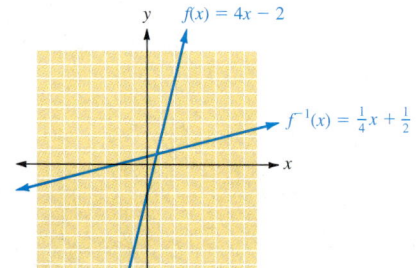

2. $y = 4x$

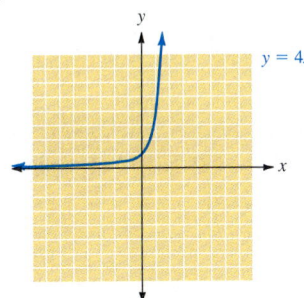

3. $y = \left(\dfrac{2}{3}\right)^x$

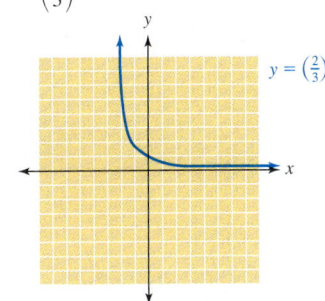

4. (a) $\{-2\}$ (b) $\left\{\dfrac{5}{2}\right\}$

5. $y = \log_4 x$

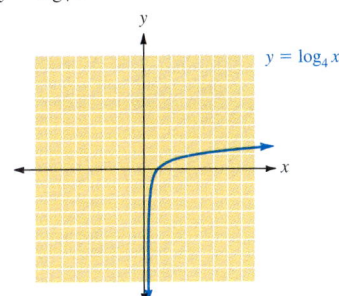

6. $\log 10{,}000 = 4$ **7.** $\log_{27} 9 = \dfrac{2}{3}$ **8.** $5^3 = 125$

9. $10^{-2} = 0.01$ **10.** $\{6\}$ **11.** $\{4\}$ **12.** $\{5\}$

13. $2 \log_b x + \log_b y + 3 \log_b z$

14. $\dfrac{1}{2}(\log_b x + 2 \log_5 y - \log_5 z)$ **15.** $\log xy^3$

16. $\log_b \sqrt[3]{\dfrac{x}{z^2}}$ **17.** 8 **18.** $\dfrac{11}{8}$ **19.** $\{0.262\}$

20. $\{2.151\}$

Cumulative Test: Chapters 0–13.

1. $\{11\}$ **2.** $\left\{x \mid x < \dfrac{2}{3} \text{ or } x > 4\right\}$ **3.** $\left\{\dfrac{10}{9}\right\}$

4. $5x - 3y = 15$

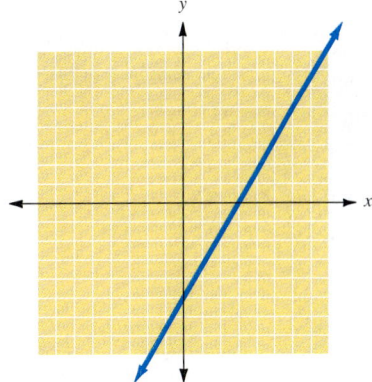

5.

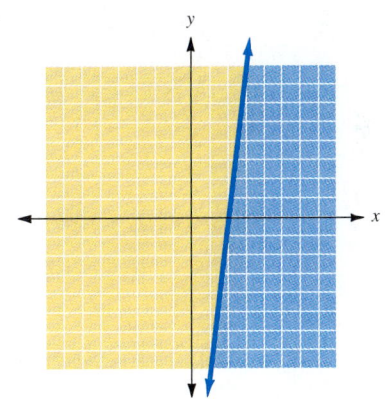

6. $6x + 5y = 7$ **7.** $\{x \mid x \geq 10\}$ **8.** $2x^2 - 6x + 1$

9. $6x^2 - 13x - 5$ **10.** $(2x - 5)(x + 2)$

11. $x(5x + 4y)(5x - 4y)$ **12.** $\dfrac{-x + 2}{(x - 4)(x - 5)}$

13. $\dfrac{x + 2}{x - 2}$ **14.** $-4\sqrt{2}$ **15.** $22 + 12\sqrt{2}$

16. $\dfrac{5}{3}(\sqrt{5} + \sqrt{2})$ **17.** $77, 79, 81$ **18.** $\{-2, 1\}$

19. $\left\{\dfrac{3 \pm \sqrt{19}}{2}\right\}$ **20.** $\left\{x \mid -\dfrac{3}{2} \leq x \leq 1\right\}$

Index

A

Abscissa, 195
Absolute value
   equations, 173–177, 458
      properties of, 174, 176, 458
   functions, 637–641
   inequalities, 177–181, 459, 645–654
      properties of, 647, 650
   of numbers, 12–13, 173–181, 449
Acidic solution, 850
ac test, 402–405
Addition
   of algebraic expressions, 75–82, 452
   associative property of, 19–20, 450
   commutative property of, 19, 450
   of complex numbers, 701
   definition of, 56, 452
   distributing multiplication over, 33–35, 450
   of fractions, 5–6
   of functions, 321, 467
   of polynomials, 353–354, 357, 471
   properties of, 19–20, 33–35, 450
   of rational expressions, 503–514
   of rational functions, 511–512
   of signed numbers, 16–22, 449
   solving
      equations by, 99–111
      system of linear equations by, 561–564, 579–582
   symbol of, 56–57
   vertical method of, 356–357
Addition property
   of equality, 102–106
   of inequality, 158

Additive identity property, 20
Additive inverse property, 20–21
Age word problems, 94, 139, 186
Algebra
   compared to arithmetics, 56
   history of, 55
   transition to, 56–61
   translating words to, 109–110
Algebraic expressions
   addition of, 75–82, 452–453
   definition of, 75
   evaluation of, 65–70, 452
   removing grouping symbols in, 79–81
   subtraction of, 75–82, 452–453
Algorithm, 132, 404
Antilogarithm, 851
Appolonius, 778
Area word problems
   circle, 72
   rectangle, 33, 375, 392, 758
   square, 199, 375
   triangle, 72, 374
Arithmetic mean, 684
Arithmetics, compared to algebra, 56
Associative property
   of addition, 19–20, 450
   of multiplication, 31–32, 450
Asymptotes
   definition of, 797
   horizontal, 822
Average of set of numbers, 684
Axes, scaling, 205–207
Axis
   of symmetry of parabolas, 742
   x, 201
   y, 201

B

Banking word problems, 26, 27. *See also* Interest rate word problems; Investment word problems; Money word problems
Base, 335
Basic solution, 850
Bel (B), 838
Bell, Alexander Graham, 838
Binomials
    definition of, 347
    division of polynomial by, 424–428
    finding greatest common factor of, 385
    multiplication of, 363–364, 471
    square of, 368–369, 472
Boundary line, 593
Bounded regions, 603
Brackets, 31, 45, 59
Break-even point, 589, 613, 625–626, 628
Business word problems, solving. *See also* Costs word problems
    by algebraic expressions, 85
    by graphing, 625–627, 628, 635
    by linear equations, 212, 270
    by polynomials, 345, 361
    by systems of linear equations, 570–572, 576, 589, 604–605

C

Calculator, graphing, 24, 205, 291–292, 440–441
Calculator exercises
    approximating logarithms, 849–850
    approximating value of radical expressions, 683
    division of signed numbers, 38
    estimating antilogarithms, 851
    estimating powers, 691
    evaluation of algebraic expressions, 67–70
    evaluation of expressions, 46–47
    evaluation of expressions with exponents, 44–45
    evaluation of rational expressions, 481–482
    scientific notation, 673
    solving linear equation, 132–133
    subtraction of signed numbers, 24
Cartesian coordinate system, 201–207
Catenary curve, 713
Change-of-base formula, 856
Circle
    area of, 72
    center of, 785, 786–788
    definition of, 778, 785
    diameter of, 123
    equation of, 786
    radius of, 785, 786–788
Coefficients
    definition of, 141, 470
    identifying in factoring trinomials, 401–402
    numerical, 76, 346, 470
Common monomial factor, 383, 474
Commutative property
    of addition, 19, 450
    of multiplication, 31, 450
Complex conjugates, 702
Complex fractions, 518–522
Complex numbers
    addition of, 701
    definition of, 700
    division of, 703–704
    multiplication of, 701–703
    subtraction of, 701
Compound inequality, 88, 177–179
Conditional equation, 99, 130–131
Conic sections, 777–808
    definition of, 778
Conjecture, 27

Consecutive integers, 133–134, 138–139, 152–153
Consistent system, 555, 557–558
Contradictions, 130–131
Coordinates
   negative, 201–202
   positive, 201–202
   x, 193
   y, 193
   zero, 201–202
Coordinate system
   Cartesian, 201–207
   rectangular, 201
Corresponding values, 193
Costs word problems, solving. *See also* Business word problems
   by equations in one variable, 114, 123, 139, 171
   by equations in two variables, 238, 266–267
   by polynomials, 351
   by rational expressions, 492
Cubes
   difference of, factoring, 395–396
   sum of, factoring, 395–396
   volume of, 50

D
Decay function, 823
Decibel (dB), 838
Decimals, 42
Degrees of polynomials, 347, 471
Denominator(s)
   addition of two rational expressions with same, 503
   of complex fraction, 518
   in exponential form, 338, 340–341, 470, 666, 670, 692
   finding prime factors of, 3–4
   least common, 5–6, 504–506
   multiplying by same number, 2–3, 483
   in multiplying rational expressions, 495
   in simplifying rational expressions, 486
   subtraction of two rational expressions with same, 503
   zero as, 480–481
Dependent system, 557
Dependent variable, 302, 303
Descartes, René, 201, 700
Descending-exponent (power) form, 348, 471
Dieting word problem, 40
Difference. *See also* Subtraction
   definition of, 6
Difference of squares, factoring, 393–395, 474
Discriminant, analyzing, 730
Distance formula, 535, 569–570, 785
Distributive property, 33–35, 450
   in combining like terms, 78
   in factoring, 383–384
   and solving equations, 107
Division
   of complex numbers, 703–704
   definition of, 60, 452
   of fractions, 5, 448
   of functions, 323, 467
   of monomials, 422–423
   of polynomials, 422–428, 475
   of rational expressions, 495–496, 498–499
   of rational functions, 498–499
   of signed numbers, 36–38, 449
   solving equations by, 116–121
   symbol of, 60
   by zero, 37, 480–481
   of zero, 37
Domain, 281–283, 315, 322–323, 466
Double inequalities, 177
Double solution, of equations, 435
Doubling, 334

E

Earnings word problems, solving
  by equations in one variable, 114, 139, 171
  by equations in two variables, 199, 239
  by graphing, 308
Earthquake, measuring intensity of, 838–839
Election word problems, 114, 134–135, 139
Electrical resistance word problem, 72
Electric power costs, 754
Elements
  definition of, 86, 453
  listing, 86–87
  plotting on number line, 88–89
Elevator stops word problem, 26
Ellipse
  definition of, 778, 791
  graphing, 792–793
Ellipsis, 87
Empty set, 86, 131
Equality
  addition property of, 102–106
  applying properties of, 125–130
  multiplication property of, 116–119
Equations, 97–190. *See also* System of linear equations
  absolute value, 173–177, 458
    property of, 174, 176, 458
  conditional, 99, 130–131
  definition of, 99, 456
  equivalent, 102, 456
  exponential, 825–826, 835–836, 865–867
  first-degree, 101
  fractions in, 129–130, 528–541
  as functions, 289
  identifying, 101–102
  linear. *See* Linear equations
  literal, 141–150, 457, 534–535
  logarithmic, 835, 837, 861–865
  quadratic. *See* Quadratic equations
  rational, 528–538
  repeated solution of, 435
  solving
    by addition and subtraction, 99–111
    by combining like terms, 106–107, 120–121, 127
    by combining rules, 124–136
    by multiplying and dividing, 116–121
    word problems with, 108–110, 458
  in two variables, 192–197, 462
Equivalent equations, 102, 456
Euler, Leonhard, 700, 824
Euler's number, 824
Evaluating
  exponential terms, 44
  expressions, 46–48, 132–133, 285–287
    algebraic, 65–70, 452
    logarithmic, 836–837, 847–848, 853–854, 856
    radical, 680–684
    rational, 481–482
  functions, 291–292, 342
  polynomials, 348–349
Exponential equation, 825–826, 835–836, 865–867
  application of, 867–868
Exponential functions, 821
  applications of, 824–825
  graphing, 821–824
Exponential notation (form), 43–44, 335–336, 450, 694
Exponents
  definition of, 335, 450
  negative, 667–673
  in order of operations, 42–48
  positive integer, 334–342
  power rule for, 340, 470, 692
  product-power rule for, 339, 470, 692
  product rule for, 336–337, 470, 692
  properties of, 337–339, 470, 667–673, 692

quotient-power rule for, 340–341, 470, 670, 692
quotient rule for, 338, 470, 692
rational, 688–694
zero, 666–667

Expressions
algebraic. *See* Algebraic expressions
definition of, 45, 58, 75
evaluating. *See* Evaluating, expressions
geometric, 61
identifying, 59, 101–102
logarithmic, 836–837, 847–848, 853–854, 856
with more than one operation, 59–60
radical, 680–684, 689
rational. *See* Rational expressions

F

Factor(s)
common monomial, 383, 474
definition of, 43, 58
greatest common, 384
of binomial, 385
of polynomial, 386
removing, 394–395
removing common, 407–408

Factoring
definition of, 383
difference of cubes, 395–396
difference of squares, 393–395, 474

by grouping, 387–389, 474
monomial from polynomial, 384, 474
polynomials, 381–446
solving quadratic equations by, 432–437
sum of cubes, 395–396
by trial and error method, 412–418
trinomials, 400–408, 412–418, 474
ac test, 402–405
of form $ax^2 + bx + c$, 401–408
identifying coefficients, 401–402
removing common factors, 407–408
rewriting middle terms, 406

Feasible region, 605
Fireworks, 777
First-degree equations, 101
FOIL method, 365–367
Forgetting curve, 854
Form
exponential, 43–44, 335–336, 450, 694
roster, 86, 453

Formulas
change-of-base, 856
definition of, 141
distance, 535, 569–570, 785
quadratic, 725–734
solving of, 143

Fractions
addition of, 5–6, 448
complex, 518–522
division of, 5, 448
equations involving, 129–130, 528–541
equivalent, 448
multiplication of, 4, 448
rewriting, 2–3
simplifying, 3–4, 448, 518–522
subtraction of, 6, 448

Function notation, 286–287
Functions
absolute value, 637–641
addition of, 321, 466
decay, 823

definition of, 296, 466
division of, 323, 467
exponential, 821–826
graphing. See Graphing, functions
growth, 823
identifying, 296–303
inverse of, 813–818
linear. See Linear functions
logarithmic, 832–839
multiplication of, 323, 467
one-to-one, 813, 814, 816–817
polynomial, factoring, 397
quadratic, 436–437, 713–775
   applications of, 751–753
   graphing, 815
   inverse of, 815
rational. See Rational function
reading values of, 309–315
subtraction of, 321, 467
writing equations as, 289

G

Gauss, Carl Friedrich, 700
GCF (greatest common factor)
   of binomial, 385
   definition of, 384
   of polynomial, 386
   removing, 394–395
Geometric applications, solving. See also Area word problems; Perimeter word problems
   by algebraic expressions, 72, 85
   by equations in one variable, 123
   by literal equations, 145–146, 153
   by systems of linear equations, 575, 589

Geometric expressions, 61
Geometric mean, 665, 684
Graphing, 215, 466
   ellipse, 792–793
   functions
      absolute value, 638–640
      exponential, 821–824
      linear, 289–290, 814–815
      logarithmic, 833–834
      quadratic, 815
      rational, 488
   hyperbola, 797–800
   intercept method of, 224–227
   linear equations, 463
      in one variable, 621–627
      in two variables, 214–230, 260–261
      using slope of line, 240–251
   linear inequalities
      absolute value, 645–654
      in one variable, 157–158, 161–165, 630–633
      in two variables, 592–597, 602–604
   parabolas, 744–747, 780–781
   reading values from, 309–315, 467
   solving
      quadratic equations by, 741–754
      quadratic inequalities by, 760
      system of linear equations by, 554–558
      using slope-intercept form, 246
Graphing calculator, 24, 205, 291–292, 440–441
Greater than (>) symbol, 156, 453
Greatest common factor (GCF)
   of binomial, 385
   definition of, 384
   of polynomial, 386
   removing, 394–395
Grouping, factoring by, 387–389, 474
Grouping symbols
   brackets, 31, 45, 59
   fraction bar, 45, 69
   parentheses. See Parentheses
   removing

in addition and subtraction of algebraic expressions, 79–81
in addition and subtraction of polynomials, 352, 354, 471
Growth function, 823

H
Half-life of drugs, 810
Half plane, 592
Heating costs, 653–654
Height, of thrown ball, 761–765
Horizontal asymptote, 822
Horizontal change, 241
Horizontal line, 243, 262–263
as result of equation, 223–225
Hyperbola
definition of, 778, 796
graphing, 797–800

I
Identities, 130–131
Imaginary numbers, 699
Imaginary part of complex numbers, 700

Imaginary unit, 699
Inconsistent system, 556
Independent variable, 302
Indeterminate form of zero exponent, 666
Index, 681
Inequalities
absolute value, 177–181, 459, 645–654
addition property of, 158
compound, 88, 177–179
double, 177
linear. *See* Linear inequalities
multiplication property of, 160
properties of, 158, 160, 647, 650
quadratic, 759–765
rational, 538–541
solving, 156–167, 457–458
Inspection, finding least common denominator by, 504
Integers, 10–13, 449
consecutive, 133–134, 138–139, 152–153
identifying, 13
and monomials, 334–342
set of, 13
symbol of, 282
Intercept method of graphing, 224–227
Intercepts
x, 224–225, 314
zero, 314
y, 224–225, 246–247, 314
zero, 249, 314
Interest rate word problems, solving. *See also* Banking word problems; Investment word problems; Money word problems
by equations in one variable, 144–145
formula for, 63, 72
by logarithmic equations, 867–868
by systems of linear equations, 586–587
Inverse functions, 813–818
Inverse relations, 811–812

Investment word problems, solving.
See also Banking word problems;
Interest rate word problems;
Money word problems
   by signed numbers, 40
   by systems of linear equations,
     576, 586–587, 589, 611,
     612
Irrational numbers, 704

L

LCD (least common denominator),
   5–6, 504–506
LCD (lowest common denominator),
   128
LCM (least common multiple),
   128
Learning curve, 854–856
Least common denominator (LCD),
   5–6, 504–506
Least common multiple (LCM),
   128
Less than ($<$) symbol, 156, 453
Light-year, 675
Like terms
   combining, 78–80, 353
     and solving equations, 106–107,
       120–121, 127
   definition of, 76, 453
   identifying, 77–78
Linear equations. *See also* System of
   linear equations
   definition of, 101
   forms of, 259–267, 463
   graphing. *See* Graphing, linear
     equations
   solving, 132, 456, 462,
     621–625
   in two variables, 216
Linear functions
   application of, 266–267
   definition of, 288
   graphing, 289–290, 814–815
   inverse of, 814–815
   writing equations as, 289
Linear inequalities
   graphing. *See* Graphing, linear
     inequalities
   in one variable, 156–167
   system of, 602–606
   in two variables, 592–597
Lines
   domain of, 282–283
   horizontal, 223–225, 243
   number, 10, 42, 88–89
   parallel, 264–265, 463
   perpendicular, 259, 265–266,
     463
   range of, 282–283
   slope-intercept form for,
     246
   slope of, 240–251, 463
   vertical, 222
     test for, 297–298, 466
Literal equations, 141–150, 457,
   534–535
Logarithm
   definition of, 832
   napierian, 852
   natural, 852–853, 869
   to particular bases, 849
   properties of, 844–857
Logarithmic equations, 835, 837,
   861–865
Logarithmic functions
   applications of, 838–839, 850–851,
     852, 854–856
   definition of, 832
   graphing, 833–834
Loudness, measuring, 838
Lowest common denominator (LCD),
   128

M

Maximum point of parabola, 742
Mean
   arithmetic, 684
   geometric, 665, 684
Minimum point of parabola, 742
Mixture problems, solving
   by equations in one variable, 146–148
   by systems of linear equations, 567–569, 575
Money word problems, solving. *See also* Banking word problems; Interest rate word problems; Investment word problems
   by algebraic expressions, 94
   by equations in one variable, 144–145, 153
   by logarithmic equations, 867–868
   by systems of linear equations, 586–587
Monomials
   common factor, 383, 474
   definition of, 347
   division of, 422–424
   division of polynomial by, 423–424, 475
   factoring from polynomial, 384, 474
   multiplication of, 362–364
   positive integer exponents and, 334–342
Motion word problems, solving
   by equations in one variable, 147–150, 187
   by quadratic functions, 758
   by rational expressions, 535–536, 545, 549
   by systems of linear equations, 575, 590, 611
Multiplication
   associative property of, 31–32, 450
   commutative property of, 31, 450
   of complex numbers, 701–703
   definition of, 58, 452
   distributive property of, 33–35, 450
   of fractions, 4
   of functions, 323, 467
   of negative numbers, 29–30, 35
   of polynomials, 362–371, 471, 472
   properties of, 31–34, 450
   of radicals, 682
   of rational expressions, 493–495, 496–498
   of rational functions, 496–498
   of signed numbers, 28–35, 449
   solving equations by, 116–121
   symbol of, 58
   vertical method of, 367–368
Multiplication property
   of equality, 116–119
   of inequality, 160
Multiplicative identity property, 32
Multiplicative inverse property, 667

N

Napier, John, 832
Napierian logarithms, 852
Natural logarithm, 852–853
Natural numbers, 2, 10
   symbol of, 282
Negative coordinates, 201–202
Negative exponents, 667–673
Negative numbers, 10, 448
   addition of, 17–18
   multiplication of, 29–30, 35
   opposite of, 11–12
   squared value of, 69
Negative reciprocals, 259, 463
Negative slope, 243
Neutral solution, 850
Notation
   exponential, 43–44, 335–336, 450, 694

function, 286–287
ordered-pair, 193, 280
scientific, 674–675
set-builder, 87–88, 99, 453
Number line, 10, 42, 88–89, 453
Number problems, solving
by algebraic expression, 94
by equations in one variable, 113, 123, 152
by quadratic functions, 737, 770
by rational expressions, 545, 549
by systems of linear equations, 576, 585–586, 589, 611, 612, 613
Numbers. *See also* Signed numbers
absolute value of, 173–181
complex. *See* Complex numbers
with different signs
addition of, 18
division of, 36
multiplication of, 28–29
Euler's, 824
imaginary, 699
irrational, 704
natural, 2, 10
negative. *See* Negative numbers
opposite of, 11–12
of ordinary arithmetic, 2
positive, 10
addition of, 17–18
opposite of, 11–12
pure imaginary, 700, 704
rational, 42, 704
as exponents, 688–694
real, 704
with same signs
addition of, 17–18
division of, 36
multiplication of, 29–30, 35
whole, 2, 10
Numerator
of complex fraction, 518
finding prime factors of, 3–4
multiplying by same number, 2–3, 483

in multiplying rational expressions, 495
in simplifying rational expressions, 486
Numerical coefficient, 76, 346, 470

O
One-to-one function, 813, 814, 816–817
Operations, order of, 42–48, 450
Opposite of number, 11–12, 20–21, 22–23, 449
Ordered-pair notation, 193
Ordered pairs, 280–283, 288, 462, 466
Ordered triple, 579
Order of operations, 42–48, 450
Ordinate, 195
Origin, 201

P
Pairs, ordered, 280–283, 288, 462, 466
Parabolas
axis of symmetry of, 742
and conic sections, 778–781

definition of, 713
graphing, 744–747, 780–781
maximum point of, 742
minimum point of, 742
opening rightward or leftward, 780
opening upward or downward, 741
vertex of, 741, 742, 743
Parallel lines, 264–265, 463
Parentheses, 28, 45, 59
removing
in adding and subtracting algebraic expressions, 79–81
in adding and subtracting polynomials, 79–81
in distributing multiplication over addition, 34–35
solving equations containing, 128
Pari-mutuel betting, 514
Perfect-square, 393
Perimeter word problems
rectangle, 72, 75, 85, 145, 361, 611, 613
triangle, 85, 361
Perpendicular lines, 259, 265–266, 463
pH word problems, 850–851, 852
Point plotting, 203
Point-slope form for equation of line, 261
Polynomials
addition of, 353–354, 471
definition of, 346
degrees of, 347, 471
descending-exponent (power) form of, 348
division of, 422–428, 475
factoring, 381–446
finding greatest common factor, 386
multiplication of, 362–371, 471, 472
subtraction of, 355–358, 471
types of, 347, 470

Population word problems, 50, 63, 824–825, 869
Positive coordinates, 201–202
Positive integer exponents, 334–342
Positive numbers, 10, 448
addition of, 17–18
opposite of, 11–12
Positive slope, 243, 248
Power, 75, 76, 335
Power property of logarithms, 845, 846, 847
Power rule for exponents, 340, 470, 692
Probability, 500, 514
Product. *See also* Multiplication
definition of, 4
Product-power rule for exponents, 339, 470, 692
Product property of logarithms, 845, 846
Product rule for exponents, 336–337, 470, 692
Product theorem, 681
Property
of absolute value equations, 174, 176
of absolute value inequalities, 647, 650
of addition, 19–20, 33–35, 450
additive identity, 20
additive inverse, 20–21
definition of, 18
of equality, 102–106, 116–119, 125–130
of exponents, 337–339, 470, 667–673, 692
of inequality, 158, 160, 647, 650
logarithmic, 844–857
of multiplication, 31–34, 450
multiplicative identity, 32
multiplicative inverse, 667
square root, 715
trichotomy, 730
Pure imaginary numbers, 700, 704
Pythagorean theorem, 733
applications of, 733–734

Q

Quadrants, 201
Quadratic equations
   definition of, 101, 432
   number of solutions to, 748–751
   rewriting, 779–780
   solving
      by completing square, 717–721
      by factoring, 432–437, 475, 714
      by graphing, 741–754
      by quadratic formula, 729
      by square root method, 715–717
Quadratic functions, 713–775
   applications of, 751–753
   definition of, 436
   graphing, 815
   inverse of, 815
Quadratic inequalities
   definition of, 759
   solving
      algebraically, 761–765
      by graphing, 760
Quotient. *See also* Division
   definition of, 5
   raised to negative power, 671
   raised to positive power, 670
Quotient-power rule for exponents, 340–341, 470, 670, 692
Quotient property of logarithms, 845, 846, 847
Quotient rule for exponents, 338, 470, 692
Quotient theorem, 682

R

Radical expressions, 680–684, 689
Radicals, 665–712
Radicand, 681
Radius of circle, 785, 786–788
Range, 281–283, 315, 466
Rational equations, 528–538
Rational exponents, 688–694
Rational expressions
   addition of, 503–514
   definition of, 480
   division of, 495–496, 498–499
   equations involving, 530–534
   fundamental principle of, 483
   multiplication of, 493–495, 496–498
   simplifying, 482–488
   subtraction of, 503–514
Rational function, 486, 538
   addition of, 511–512
   division of, 498–499
   graphing, 488
   identifying, 486–487
   multiplication of, 496–498
   simplifying, 487
   subtraction of, 512–513
Rational inequalities, 538–541
Rational numbers, 42, 704
   as exponents, 688–694
Real numbers, 704
   symbol of, 282
Real part of complex numbers, 700
Reciprocals
   negative, 259, 463
   solving equations by, 119–120
Rectangle
   area of, 33, 375, 392, 758
   dimensions of, 145, 153, 575, 737, 770
   perimeter of, 72, 75, 85, 145, 361, 611, 613
Rectangular coordinate system, 201, 462
Relation
   definition of, 280, 466
   inverse of, 811–812
Repeated solution, of equation, 435
Richter, Charles, 838
Richter scale, 838–839
Rise, 241
Root, of equations, 99, 433
Roster form, 86, 453
Rule of 72, 868
Run, 241

S

Salary word problems. *See* Earnings word problems
Scientific notation, 674–675
Score word problems, 165–166, 171
Set-builder notation, 87–88, 99, 453
Set of integers, 13
Sets
   definition of, 86, 453
   listing elements of, 86–87
   plotting elements of, on number line, 88–89
   union of, 179
Signed numbers
   absolute values of, 12–13
   addition of, 16–24, 449
   arithmetic of, 1–54
   definition of, 10, 448
   division of, 36–38, 449
   identification of, 10–11
   multiplication of, 28–35, 449
   opposite of, 11–12
   subtraction of, 16–24, 449
Signs
   of absolute numbers, 12
   of negative numbers, 10
   of positive numbers, 10
Simplifying
   compound inequality, 179
   expressions with rational exponents, 690, 692–693
   fractions, 3–4, 448, 518–522
   polynomials, 337–341
   rational expressions, 482–488
   rational function, 487
Slope-intercept form for a line, 246, 463
Slope of line, 240–251, 463
   negative, 241, 243
   positive, 241, 243, 248
   undefined, 244, 262–263
   zero, 243–244, 262–263
Solution
   checking, 124
   definition of, 99, 554
   double, 435
   repeated, 435
   verifying, 100–101
Solution by elimination, 561–564
Solution set, 99, 100
Speed word problems, 154, 186
Square, area of, 199, 375
Square root method, 715–717
Square roots
   of binomials, 368–369, 472
   difference of, 393–395, 474
   of negative numbers, 69
   solving quadratic equations by, 714–721
Standard form of complex numbers, 700
Steinmetz, Charles, 700
Submarine word problem, 26
Substitution
   definition of, 107
   of nonnumeric values for x, 290–291
   solving
      system of linear equations by, 565–567
      word problems by, 108
Subtraction
   of algebraic expressions, 75–82, 452
   of complex numbers, 701
   definition of, 57, 452
   distributing multiplication over, 35
   of fractions, 6
   of functions, 321, 467
   of polynomials, 355–358, 471
   of rational expressions, 503–513
   of rational functions, 512–514
   of signed numbers, 22–24, 449
   solving equations by, 99–111
   symbol of, 57
   vertical method of, 357–358
Sum. *See also* Addition
   definition of, 5
Symbol, of set continuing in pattern, 87
Symbol(s)
   of addition, 56–57
   in arithmetic, 56
   of division, 60
   of empty set, 86
   grouping. *See* Grouping symbols
   inequality, 156

integers, 282
of multiplication, 31, 58
of natural numbers, 282
of real numbers, 282
of square root, 680
of subtraction, 57
Symmetry, axis of, of parabolas, 742
System of linear equations. *See also* Linear equations
applications of, 561–572, 585–587
consistent, 555, 557–558
definition of, 554
dependent, 557, 583–584
inconsistent, 556, 584–585
solving
by addition, 561–564, 579–582
by graphing, 554–558
by substitution, 565–567
in three variables, 579–587
in two variables, 561–572
System of linear inequalities, 602–606

T

Temperature word problem, 26, 40, 72
Terms
definition of, 75, 346, 470
identifying, 76
like. *See* Like terms
Test point, 593
Thrown-ball word problems, 731–732, 738, 739, 764–765, 770, 771
Ticket word problems, solving
by equations in one variable, 146–147, 153
by systems of linear equations, 611, 612

Time word problems, 154, 187, 601
Tolerance, 641
Trajectory, of thrown ball, 761–765
Trial and error method of factoring, 412–418
Triangles
area of, 72, 374
in literal equations, 141–142
perimeter of, 85, 361
Trichotomy property, 730
Trinomials
definition of, 347
factoring, 400–408, 412–418, 474
ac test, 402–405
of form $ax^2 + bx + c$, 401–408
identifying coefficients, 401–402
removing common factors, 407–408
rewriting middle terms, 406

U

Undefined slope, 244
Union of two sets, 179

V

Variables
definition of, 56, 285

dependent, 302, 303
equations in two, 192–197
independent, 302
Vertex of parabola, 741, 742, 743
Vertical change, 241
Vertical line, 244, 262–263
as result of equation, 222
Vertical line test, 297–298, 466
Vertical method
of addition, 357
of multiplication, 367–368
of subtraction, 357–358
Vertices, 797

W

Wage word problems. *See* Earnings word problems
Wallis, John, 667
Whole numbers, 2, 10
Word problems, 107
age, 94, 139, 186
area. *See* Area
banking, 26, 27
business. *See* Business word problems
costs. *See* Costs word problems
dieting, 40
distance, 535, 569–570, 785
earnings. *See* Earnings word problems
election, 114, 134–135, 139
electrical resistance, 72
elevator stops, 26
geometry. *See* Geometric applications
interest rate. *See* Interest rate word problems

investment. *See* Investment word problems
mixture. *See* Mixture problems
money. *See* Money word problems
motion. *See* Motion word problems
number. *See* Number problems
perimeter. *See* Perimeter word problems
pH, 850–851, 852
population, 50, 63, 824–825
scores, 165–166, 171
solving
by equations, 108–110, 458
steps of, 108, 145
by substitution, 108
speed, 154, 186
submarines, 26
temperature, 26, 40, 72
thrown-ball, 731–732, 738, 739, 764–765, 770, 771
ticket. *See* Ticket word problems
time, 154, 187, 601
volume of cube, 50
Work problems, solving
by equations in one variable, 135–136
by rational expressions, 536–537, 545, 549
by systems of linear equations, 611

X

x
as independent variable, 302
substituting nonnumeric values for, 290–291

x axis, 201
x coordinate, 193
x intercept, 224–225, 314

Y
y
    as dependent variable, 302
    solving equations for, 227–228
y axis, 201
y coordinate, 193
y intercept, 224–225, 246–247, 314
    zero, 249

Z
Zero
    as additive identity, 20
    coordinate, 201–202
    division by, 37, 480–481
    division of, 37
    as exponent, 666–667
    as integer, 10
    line with slope of, 243–244, 262–263
    multiplication by, 32–33
    x intercept, 314
    y intercept, 249, 314
Zero-product principle, 432–437